U.S. CUSTOMARY UNITS AND SI EQUIVALENTS

Quantity	U.S. customary unit	SI equivalent
Area	mi^2	[illegible]0 km^2
	acre	[illegible]7 m^2
	ft^2	[illegible]29 m^2
	in^2	[illegible]2 mm^2
Concentration	lb/milli[illegible]	[illegible]200 mg/l
Energy	ft · lb	1.356 J
Force	lb	4.448 N
Flow	ft^3/s	0.0283 m^3/s
	gal/min	0.003785 m^3/min
Length	ft	0.3048 m
	in	25.40 mm
	mi	1.609 km
Mass	grain	64.80 mg
	oz	28.35 g
	lb	0.4536 kg
	ton	907.2 kg
Power	ft · lb/s	1.356 W
	hp	745.7 W
Pressure	lb/ft^2	47.88 Pa
	lb/in^2	6.895 kPa
	ft (of water)	2.988 kPa
Velocity	ft/s	0.3048 m/s
	in/s	0.0254 m/s
	gal/ft^2 per min	0.0407 m/min
		58.678 m/day
	gal/ft^2 per day	0.0407 m/day
	ft^3	0.02832 m^3
	yd^3	0.7646 m^3
Volume	gal	0.003785 m^3
	acre · ft	1233.5 m^3

WATER SUPPLY AND SEWERAGE

FIFTH EDITION

E. W. Steel
Formerly University of Texas

Terence J. McGhee
Tulane University

McGraw-Hill Book Company

New York St. Louis San Francisco Auckland Bogotá Düsseldorf
Johannesburg London Madrid Mexico Montreal New Delhi
Panama Paris São Paulo Singapore Sydney Tokyo Toronto

WATER SUPPLY AND SEWERAGE

1 2 3 4 5 6 7 8 9 0 D O D O 7 8 3 2 1 0 9 8

This book was set in Times Roman. The editors were B. J. Clark and Julienne V. Brown; the production supervisor was Gayle Angelson.
The jacket was designed by Tom Hoffman.
R. R. Donnelley & Sons Company was printer and binder.

Library of Congress Cataloging in Publication Data

Steel, Ernest William, date
Water supply and sewerage.

(McGraw-Hill series in water resources and environmental engineering)
Includes index.
1. Water-supply engineering. 2. Water-supply.
3. Sewerage. 4. Sewage. I. McGhee, Terence J., joint author. II. Title.
TD145.S8 1979 628 78-4528
ISBN 0-07-060929-2

CONTENTS

Preface **xi**

List of Abbreviations **xiii**

1 Introduction **1**

Work of the Sanitary Engineer. Water Supply. Sewerage. Effects of Sanitary Engineering Upon City Life. Magnitude of the Problem.

2 Quantity of Water and Sewage **7**

Water. Forecasting Population. Consumption for Various Purposes. Factors Affecting Consumption. Size of City. Characteristics of Population. Industries and Commerce. Climatic Conditions. Metering. Variations in Rate of Consumption. Fire Demand. Density of Population. Zoning. Periods of Design and Water Consumption Data Required. Sources of Sewage. Relation to Water Consumption. Infiltration and Inflow. Fluctuations in Sewage Flow. Design Periods and Use of Sewage Flow Data.

3 Rainfall and Runoff **26**

Hydrology. Sources of Information. Precipitation. Measurement of Precipitation. Types of Storms. Precipitation Data and Their Adjustment. Rainfall Intensity-Duration-Frequency Relationships. Depth-Area Relationship. Precipitation Deficiency. Artificial Rainfall Control. Runoff. Measurement of Runoff. The Hydrograph. Factors Which Affect Runoff. Infiltration. Determination of Infiltration Rates. Evaporation from Water Surfaces. Evaporation from Land Surfaces. Transpiration. Evapotranspiration. Multiple Correlation Method of

Forecasting Runoff. Yield. Reservoir Storage by Mass Diagram. Flood Flows. The Rational Method. The Overland Flow Hydrograph. The Unit Hydrograph. Flood Frequencies. Silting of Reservoirs.

4 Groundwater **55**

Groundwater. Occurrence of Aquifers. Types of Wells. Groundwater Flow. Groundwater Hydraulics. The Equilibrium Equations for Flow of Water into Wells. Specific Capacity. The Nonequilibrium Analysis. The Nonequilibrium Formula. Modified Nonequilibrium Formula. Interferences in Aquifers. Appraisal of Groundwater Resources. Recharge of Aquifers. Shallow Well Construction. Deep Well Construction. Well Screens. Development of Wells. The Gravel-packed Well. The Ranney Method. Cementing Wells. Groundwater Projects. Well Specifications. Well Troubles. Springs.

5 Aqueducts and Water Pipes **84**

Water Carriage. Open Channels. Aqueducts. Stresses in Pipes. Pipe Lines. Cast-iron Pipe. Lined Cast-iron Pipe. Flanged Cast-iron Pipe. Joints in Cast-iron Pipe. Cast-iron Fittings. Steel Pipe. Protection and Maintenance of Steel Pipe. Concrete Pipe. Asbestos Cement Pipe. Plastic Pipe. Corrosion of Metal. Protection Against Corrosion. Grounding of Electrical Wiring to Water Pipes. Use of Inert Materials.

6 Collection and Distribution of Water **104**

Intakes. Intakes from Impounding Reservoirs. Lake Intakes. River Intakes. The Intake Conduit. Methods of Distribution. Storage Necessary. Fire Flow. Pressures. Pressure Zones. The Motor Pumper. The Pipe System. Design of Water Distribution Systems. Flow in Pipes. Equivalent-pipe Method. The Circle Method. Fire-flow Tests. Hardy Cross Method of Analysis of Flow in a Pipe Network. Computers and Electric Analyzers for Pipe Networks. Two-main System. The Service Pipe. Meters. Valves. Fire Hydrants. Construction Excavation and Backfilling. Pipe Handling and Laying. Submerged Pipes. Hydrant Placement. Valve Placement. Maintenance of Valves. Disinfection of New and Old Water Mains. Testing. Water Waste Surveys. Leak and Pipe Location. Cleaning of Water Mains. Thawing Pipes.

7 Pumps and Pumping Stations **153**

Need for Pumping. Classification of Pumps. Work and Efficiency of Pumps. Action of the Centrifugal Pump. Effect of Varying Speed. Characteristics of Centrifugal Pumps. Suction Lift. Cavitation. Pump Selection. Variation in Capacity. Deep-well Pumps. Miscellaneous Pumps. Choice of Prime Mover. Steam Power. Steam Turbines. The Diesel Engine. The Gasoline Engine. Electric Motors. Pumping Station Location. Architecture. Capacity and Operation. Auxiliary Pumping Stations. Valves and Piping Details.

8 Quality of Water Supplies **179**

Impurities of Water. Waterborne Diseases. Methemoglobinemia. Lead Poisoning. Fluoride. Radioactivity in Water. Water Bacteria. Relationship Between Pathogens and Coliforms. Determination of Bacterial Numbers. Fecal Coliform/Fecal Strep Ratio. Multiple Tube Fermentation. Viruses in Water. Microscopic Organisms. Worms. Turbidity. Color. Alkalinity and Acidity. pH. Soluble Mineral Impurities. Iron and Manganese. Chlorides. Gaseous Impurities. Sampling Methods. EPA Standards. Governmental Control of Water Supplies. Liability for Unsafe Water. Characteristics of Waterborne Epidemics. The Sanitary Survey. Watershed and Reservoir Sanitation. Lake Overturns. The Water Treatment Plant. Groundwater Supplies. Pumping Plant, Pipes, and Reservoirs. Cross Connections. Plumbing Defects. The Piping System. Drinking Fountains. Tracing Pollution.

9 Treatment of Water—Clarification **206**

Clarification. Storage. Screens. Principles of Sedimentation—Discrete Particles. Flocculent Suspensions. Hindered Settling. Scour. Sedimentation Tank Details. Presedimentation. Purposes and Action of Coagulants. Simplified Coagulation Reactions. Polymeric Coagulants. Coagulant Aids. Mixing. Flocculation. Suspended Solids Contact Clarifiers. Design Criteria. Chemical Feeding Methods.

10 Treatment of Water—Filtration **240**

Filter Types. The Rapid Filter. Theory of Filtration Through Coarse Media. Filter Media. Mixed Media. Gravel. Filtration Rates. The Underdrain System. The Filter Unit and Washwater Troughs. The Washing Process. The Control System. Pipe Galleries and Piping. Negative Head. Operating Difficulties. Laboratory Control and Records. The Clear Well and Plant Capacity. Filter Galleries. Pressure Filters. Diatomaceous Earth Filter. Upflow Filters. Other Filtration Processes.

11 Miscellaneous Water Treatment Methods **274**

Quality Considerations. Chlorine in Water. Chlorination. Chloramines. Use of Chlorine Gas. Hypochlorination. Ozonation. Other Disinfection Techniques. Algae Control. Aeration. Prevention of Odors. Activated Carbon. Removal of Iron and Manganese. Water Softening. The Lime-soda Method. Suspended Solids Contact Units. The Cation-exchange Method. Hydrogen Exchange and Demineralizing. Desalination. Color Removal. Treatment of Boiler Waters. Stabilization of Water. Fluoridation. Defluoridation of Water. Radioisotope Removal by Water Treatment. Water Treatment Wastes.

12 Sewerage—General Considerations **316**

Introduction. Definitions. General Considerations. Combined vs. Separate Sewers. Liability for Damages Caused by Sewage.

13 Amount of Storm Sewage **320**

Estimation of Storm Flow. The Rational Method. Runoff Coefficients. Time of Concentration. Rainfall Intensity. Intensity Curves and Formulas. Use of Intensity-Duration-Frequency Data. Other Techniques.

14 Sewer Pipes **331**

Pipe Materials. Clay Sewer Pipe. Strength and Loading of Vitrified Clay Pipe. Plain Concrete Sewer Pipe. Reinforced Concrete Sewer Pipe. Asbestos Cement

Pipe. Plastic Truss Pipe. Other Sewer Materials. Infiltration and Sewer Joints. Sewers Built In Place. Design of Concrete Sewers. Corrosion of Sewers.

15 Flow in Sewers **354**

Conditions of Flow. Flow Formulas. The Hydraulic Grade Line, or Piezometric Head Line. Required Velocities. Flow Diagrams. Partial-flow Diagrams. Sewer Shapes.

16 Sewer Appurtenances **365**

Operational Requirements. Manholes. Inlets. Catch Basins. Flushing Devices. Sand, Grease, and Oil Traps. Regulators. Junctions. Sewer Outlets. Inverted Siphons. Sewer Crossings. Need for Pumping. Pumps for Sewage. Pumping Stations. Sump Pumps. Sewage, Ejectors. Vacuum Collection Systems. Grinder Pumps.

17 Design of Sewer Systems **392**

Design Stages. Preliminary Investigations. The Underground Survey. The Survey and Map. Layout of The System. The Profile. Design of a Sanitary Sewer System. Design of a Storm Sewer System.

18 Sewer Construction **404**

Responsibility. Lines and Grades. Classification of Excavation. Hand Excavation. Machine Excavation. Rock Excavation. Sheeting and Bracing. Removal of Sheeting and Bracing. Dewatering of Trenches. Pipe Laying. Jointing. Jacking and Boring. Backfilling. Concrete Sewers. Tunneling.

19 Maintenance of Sewers **418**

Scope and Cost. Protective Ordinances. Equipment. Stoppage Clearing. Sewer Cleaning. Inspection Practice. Making Repairs and Connections. Cleaning Catch Basins. Gases in Sewers.

20 Characteristics of Sewage **426**

Analyses Required. Physical Characteristics. Solids Determinations. Chemical Characteristics. Microbiology of Sewage and Sewage Treatment. Anaerobic Processes. Aerobic Processes. Other Microorganisms. Biochemical Oxygen Demand (BOD). Chemical Oxygen Demand (COD). Total Organic Carbon (TOC). Sampling. Population Equivalents.

21 Sewage Disposal **438**

Disposal Techniques. Effects of Stream Discharge. Factors in Self-purification. Self-purification of Lakes. The Oxygen Sag Curve. The Critical Deficit. Evaluation of Rate Constants. Application of Formulas. The Multiple Correlation Technique. Ocean Discharges. Submarine Outfalls. Land Disposal and Treatment. Disposal Site Selection. Disposal Site Preparation. Spray Irrigation.

Rapid Infiltration. Overland Runoff. Soil Response to Wastewater Disposal. Evaporation (Total Retention) Systems. Selection of a Disposal System.

22 Preliminary Treatment Systems **456**

Purpose. The Parshall Flume. Racks and Screens. Comminutors. Grit Removal. Grease Removal. Flotation. Preaeration.

23 Primary Treatment Systems **477**

Primary Treatment. Plain Sedimentation. Rectangular Tanks. Circular and Square Tanks. Solids Quantity. Chemical Addition. Other Primary Processes. Secondary Clarifiers.

24 Secondary Treatment Systems **520**

Purpose and Classification. Attached-Growth Processes. Intermittent Sand Filters. Trickling Filters. Filter Classification. Recirculation. Efficiency of Trickling Filters. Design of Trickling-Filter Systems. Trickling Filter Final Clarifier Design. Operational Problems of Trickling Filters. Rotating Biological Contacters. Suspended-Growth Processes. Principles of Suspended-Growth Systems. Completely Mixed Process with Solids Recycle. Activated-Sludge Process Details. Aeration and Mixing Techniques. Suspended-Growth Process Clarifier Design. Operational Problems of Activated-Sludge Systems. Completely Mixed Processes without Solids Recycle. Oxidation Ponds. Upgrading Stabilization Pond Effluent.

25 Sludge Treatment and Disposal **524**

Importance. Amount and Characteristics of Sludge. Sludge Conditioning. Anaerobic Digestion. Aerobic Digestion. Chemical Conditioning. Elutriation. Heat Treatment. Other Conditioning Processes. Thickening. Dewatering. Drying and Combustion. Ultimate Disposal.

26 Advanced Waste Treatment **552**

Scope. Suspended Solids Removal. Nitrogen Removal. Biological Nitrogen Removal. Chemical Nitrogen Removal. Phosphorus Removal. Biological Phosphorus Removal. Chemical Phosphorus Removal. Refractory Organics. Dissolved Solids.

27 Miscellaneous Sewage Treatment Problems **569**

Disinfection. Odor Control. Garbage Disposal with Sewage. Industrial Wastes. Excreta Disposal in Unsewered Sections. The Pit Privy. Septic Tanks. Aerobic Units. Subsurface Disposal Fields. Sand Filtration.

28 Financial Considerations **580**

Viewpoint in Economic Analyses. Cost Estimates. Cost Comparisons. Optimization of Process Selection. Financing of Waterworks. Charges for Service. Accounting. Financing of Sewage Works. Sewer Charges.

Appendixes

I National Drinking Water Regulations 595

II Saturation Values of Dissolved Oxygen in Fresh Water and Seawater 611

III Cost Estimation of Sewage Treatment Facilities 613

IV Physical Properties of Water 642

Index 643

PREFACE

It is both a privilege and a formidable challenge to follow in the footsteps of Professor Steel, whose text has been a standard in the field of sanitary engineering for over forty years.

In writing the book the original format has been followed. The emphasis is upon design of water and sewerage works, and the intended audience is the undergraduate student in Civil Engineering. The text has been thoroughly revised to incorporate the latest techniques of water and wastewater conveyance and treatment, and outdated materials and methods have been dropped.

The present author trusts that the revised work will continue to serve the needs of the profession in the same fashion as did the earlier editions, and will welcome comments and suggestions from users of the text.

Terence J. McGhee

LIST OF ABBREVIATIONS

A	ampere
AASHTO	American Association of State Highway and Transportation Officials
ABS	Acrylonitrile-Butadiene-Styrene
ac	alternating current
ASA	American Standards Association
ASTM	American Society for Testing Materials
AWWA	American Water Works Association
BOD	Biochemical Oxygen Demand
°C	degrees Celsius
Ci	curies
cm	centimeters
COD	Chemical Oxygen Demand
dc	direct current
DO	Dissolved Oxygen
D_{10}	effective size
D_{60}	60 percent size
E. coli	*Escherichia Coli*
EPA	Environmental Protection Agency
ft	feet
ft^3	cubic feet
ft/s	feet per second
ft^3/s	cubic feet per second
FWPCA	Federal Water Pollution Control Administration
g	grams
gal	gallons
gpm	gallons per minute
h	hours
hp	horsepower
HTH	High Test Hypochlorite
in	inches
J	joules
kc	kilocycles
kg	kilograms
km	kilometers
kW	kilowatts
L	ultimate BOD
l	liters

lb	pounds
ln	natural logarithm
m	meters
MCL	Maximum Contaminant Level
meq	milliequivalents
mgd	million gallons per day
mg/l	milligrams per liter
mi	miles
min	minutes
ml	milliliters
MLSS	Mixed Liquor Suspended Solids
MLVSS	Mixed Liquor Volatile Suspended Solids
mm	millimeters
MPN	Most Probable Number
N	newtons
NEMA	National Electrical Manufacturers Association
NPDES	National Pollutional Discharge Elimination System
NPSH	Net Positive Suction Head
N_R	Reynolds Number
OR	Overland Runoff
Pa	pascals
pH	negative logarithm (base 10) of the hydrogen ion concentration
psi	pounds per square inch
psig	pounds per square inch (gauge)
PVC	polyvinylchloride
rem	roentgen equivalent—man
RI	Rapid Infiltration
r/min	revolutions per minute
RPP	Reduced Pressure Principle
s	seconds
SF	Safety Factor
sg	specific gravity
SI	Spray Irrigation
SOR	Surface Overflow Rate
SS	Suspended Solids
STP	Standard Temperature and Pressure
TDH	Total Dynamic Head
TDS	Total Dissolved Solids
TKN	Total Kjeldahl Nitrogen
TOC	Total Organic Carbon
TSS	Total Suspended Solids
tu	turbidity units
TVS	Total Volatile Solids
USGS	United States Geological Survey
USPHS	United States Public Health Service
uv	ultraviolet
W	watts
ν	kinematic viscosity
μ	absolute viscosity
ρ	mass density
θ	liquid retention time
θ_c	sludge age
θ_c^m	minimum sludge age
θ_c^d	design sludge age

WATER SUPPLY AND SEWERAGE

CHAPTER

ONE

INTRODUCTION

1-1 Work of the Sanitary Engineer

The development of sanitary engineering has paralleled and contributed to the growth of cities. Without an adequate supply of safe water, the great city could not exist, and life in it would be both unpleasant and dangerous unless human and other wastes were promptly removed. The concentration of population in relatively small areas has made the task of the sanitary engineer more complex. Groundwater supplies are frequently inadequate to the huge demand and surface waters, polluted by the cities, towns, and villages on watersheds, must be treated more and more elaborately as the population density increases. Industry also demands more and better water from all available sources. The rivers receive ever-increasing amounts of sewage and industrial wastes, thus requiring more attention to sewage treatment, stream pollution, and the complicated phenomena of self-purification.

The design, construction, and operation of water and sewage works are treated in this book, but the field of sanitary engineering extends beyond these limits. The public looks to the sanitary engineer for assistance in such matters as the control of malaria by mosquito control, the eradication of other dangerous insects, rodent control, collection and disposal of municipal refuse, industrial hygiene, and sanitation of housing and swimming pools. The activities just given, which are likely to be controlled by local or state health departments, are sometimes known as public health or environmental engineering, terms which, while descriptive, are not accepted by all engineers. The terms, however, are indicative of the important place the engineer holds in the field of public health and in the prevention of diseases.

1-2 Water Supply

Throughout recorded history large cities have been concerned with their water supplies. Even ancient cities found that local sources of supply—shallow wells, springs, and brooks—were inadequate to meet the very modest sanitary demands of the day, and the inhabitants were constrained to build aqueducts[1] which could bring water from distant sources. Such supply systems could not compare with modern types, for only a few of the wealthier people had private taps in their homes or gardens, and most citizens carried water in vessels to their homes from fountains or public outlets. Medieval cities were smaller than the ancient cities, and public water supplies were practically nonexistent. The existing aqueducts of ancient Athens, Rome, and the Roman provincial cities fell into disuse, and their purposes were even forgotten.

The waterworks engineer of ancient times labored under the severe handicap of having no type of pipe that could withstand even moderate pressures. He used pipe of clay, lead, and bored wood in small sizes, but even with these, as with masonry aqueducts and tunnels, he followed the hydraulic grade line and rarely placed conduits under pressure.

In the seventeenth century the first experiments were made with cast-iron pipe but it was not until the middle of the eighteenth century that these pipes were cheap enough for wide use. The durability of cast iron and its freedom from breaks and leakages soon made its use almost universal, although steel and other materials were also used. This advance, together with improved pumping methods, made it economically possible for all but the smallest villages to obtain water supplies and to deliver the water into the homes of the citizens.

Although some cities were able to collect safe water from uninhabited regions and thereby reduce waterborne disease to a low level, many others found that their supplies were dangerously polluted and that the danger was increasing as population increased upon watersheds. Accordingly treatment methods were developed that, when properly applied, reduced the hazard.

Coagulants have been used in water treatment since at least 2000 B.C., as has filtration, however their use in municipal treatment in the United States was not common until about 1900. The application of various treatment techniques in the early part of the twentieth century resulted in the marked decrease in waterborne disease illustrated in Fig. 1-1.

Philadelphia's water supply came, without treatment of any kind, from increasingly polluted rivers until 1906, when slow sand filters were completed. An immediate reduction in typhoid fever followed over a period of 7 years. A tendency to increase, possibly caused by further increases in the pollution of the untreated water, was checked by disinfection of the filtered water with chlorine. A still greater decrease was accomplished after 1920 by careful control over infected persons who had become carriers.

Outbreaks of waterborne disease still occur in the United States and other countries with generally modern treatment systems. The average number of such incidents in the United States in the period 1971 to 1974 was 25 per year.[2] Most of

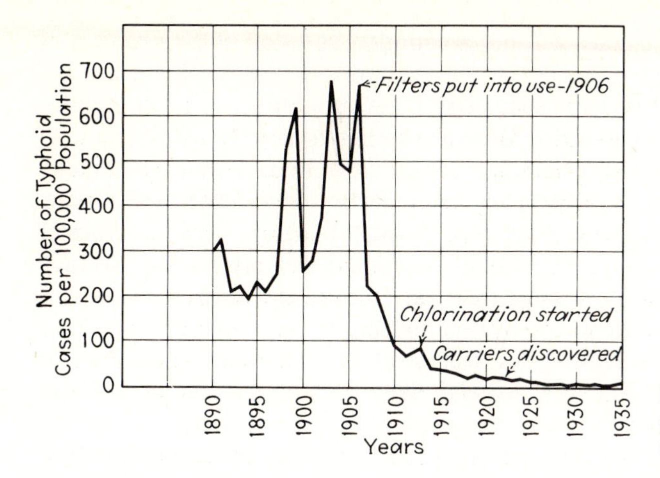

Figure 1-1 Typhoid fever cases per 100,000 population from 1890 to 1935, Philadelphia, Pa.

these outbreaks were associated with obvious deficiencies in treatment or distribution systems.

1-3 Sewerage

Remains of sanitary sewers are to be found in the ruins of the ancient cities of Crete and Assyria. Rome also had sewers, but they were primarily drains to carry away storm water. It was the practice to deposit all sorts of refuse in the streets, and accordingly the storm sewers also carried much organic matter at times. Sewerage was practically unknown during the Middle Ages, and construction of sewers was not resumed until modern times. At first, these were storm sewers not intended to carry domestic sewage. As late as 1850, the discharge of household wastes into the sewers of London was forbidden. The water courses in or near towns apparently were used as convenient places of refuse disposal, for many writers comment upon the offensive condition of the London brooks, with their burden of dead dogs and filth† of all sorts. In the course of time it was recognized that sanitation would best be served by permitting the use of sewers to convey human excreta away from dwellings as promptly as possible, and the original storm drains became combined sewers which carried both storm-water runoff and the liquid wastes from occupied buildings. The development of water supplies, of course, played a large part in the greater use of plumbing systems with water-flush toilets. The commonly used vault toilets, which frequently overflowed and always produced odors, were soon legislated out of existence in the larger cities in favor of the water-carried system. This improvement together with safer water supplies caused a sharp decline in the urban death rate.

† Queen Elizabeth I once decreed that the rushes which were used as floor coverings in many homes were not to be disposed of by dumping into streams, as the practice was polluting the drinking water of Her Majesty's subjects. History does not record how well the decree was enforced.

Providing sewers for the cities was not a complete solution of the problem of excreta disposal. The offensive and dangerous materials were discharged into the streams where they decomposed to cause discomfort and danger to rural populations or to cities located downstream. Most cities, therefore, soon found it necessary to treat the sewage before releasing it. Even cities located along the ocean were in many cases obliged to protect bathing beaches or shellfish beds. Some, however, were able to discharge their sewage untreated into very large bodies of water or into streams that traversed relatively uninhabited regions. Still others were indifferent to the need for sewage treatment and in the absence of laws or proper enforcement spoiled the beauty of streams, made them unusable for recreational purposes, and endangered lives. A later development of this sanitary problem was the pollution of streams by industrial plants located not only in cities but in previously unspoiled rural sections. Streams have been spoiled for fishing, camping, and swimming by the putrescible and toxic wastes of industrial plants.

When the problem of sewage treatment first attracted attention, a difference of opinion existed among engineers as to the completeness of treatment that should be given to sewage before discharge into a body of water. Some engineers maintained that the public interest required the most complete treatment possible. Others held the opinion that treatment should be based upon local conditions and that no more treatment need be provided than would give reasonable assurance, with a factor of safety, that danger and nuisance would not exist. So far as safety of water supplies is concerned, this viewpoint placed upon the waterworks authorities some of the burden of safeguarding and treating their raw water. When it is considered that water of streams and lakes may often be polluted or made unsuitable for use otherwise than by city sewage, it is obviously inequitable to require all cities to produce a sewage treatment plant effluent comparable to drinking water. Therefore, sewage treatment has been based upon local conditions rather than idealistic standards.

Present regulations in the United States establish which bodies of water are quality, and which effluent limited. Those waters which are of a quality suitable for their highest intended use are defined as *effluent limited*. Wastes discharged into such waters must be treated to the degree obtained in secondary systems. Waters which are not suitable for their highest intended use under such effluent limitations are governed by *water quality* and are analyzed to establish the allowable total pollutional load which can be assimilated without degradation. This allowable waste load is then allocated to present and future discharges. Treatment at each discharge point is then tailored to meet this *waste load allocation*. Treatment to a level less than that provided by secondary systems is never permitted under either system.

Waste discharges are regulated by the National Pollutional Discharge Elimination System (NPDES) under which permits are issued either by the U.S. Environmental Protection Agency (EPA) or state agencies approved by EPA. Each discharge must have an NPDES permit specifying the required waste quality, and regular reports must be submitted to the regulatory agency by the permittee.

1-4 Effects of Sanitary Engineering upon City Life

The bills of mortality of London in the seventeenth century, which were the vital statistics of the day, indicate that the death rate in large cities at that time was greater than the birth rate. Cities, therefore, grew slowly and only by migration from country to city. This condition can be ascribed to prevailing insanitary conditions combined with crowding of people into a small area and the resulting prevalence of communicable disease. The first municipal sanitary improvement both in England and elsewhere was the construction of water supplies which, in large cities, were soon followed by sewerage. Small cities and towns have also installed waterworks and sewerage systems during the last few decades until, at the present time, there are few communities that do not have a public water supply and in most cases a sewer system.

The construction grants program of EPA and, earlier, the Federal Water Pollution Control Administration (FWPCA), under which the construction of sewage works is largely financed by federal funds, has hastened the construction of collection and treatment systems in communities which had lacked them.

EPA is also charged with ensuring the provision of suitable drinking water and, as successor to the U.S. Public Health Service (USPHS), is conducting research into the health effects of minute concentrations of various contaminants and has established water quality standards for public water supplies.

The important effects of waterworks and sewerage upon cities are not confined to safeguarding of health. Safety of life and property against fire has been obtained. Street cleaning and flushing are possible. Swimming pools, fountains, and other ornamental and recreational uses of water are now commonplace. Industries will locate in cities where they are assured of an ample supply of water and where there are sewers to remove their liquid wastes. Some industries, and this should be recognized by municipal authorities, may make unreasonably high demands for water or may produce wastes which are unsuitable for joint treatment and disposal.

The unthinking citizen, accustomed to the comforts of civilization, has little conception of the significance of the stream of water that he obtains when he turns on a tap and even less of the vast network of underground conduits available to receive that water as it escapes into the drainpipe. To the student, this book, brief as it is, will convey some idea of the investigation, planning, accumulated experience, and plain hard work behind the waterworks and sewerage works that give our citizen water in the amounts that he wants when he wants it and remove it when it has served his purpose.

1-5 Magnitude of the Problem

Supply of water to the cities of the United States is a huge engineering problem. According to the U.S. Geological Survey[3] the cities of the United States, with a total population of 165 million in 1970, produced and distributed 104 billion liters (27.4 billion gallons) of water daily to their domestic, commercial, and industrial consumers. Of this, 68 billion liters (18 billion gallons) were from surface water sources which usually require elaborate treatment. Of the 36 billion liters (9.4 billion gallons) from ground sources a portion would require treatment. Large as these figures are, they are increasing yearly. It is estimated that in the year 2000 public supply requirements will be about 193 billion liters (51 billion gallons) daily.[4]

Water use is not confined to cities. In 1965 industrial users and such miscellaneous users as motels, resorts, army installations, and mines, all self-supplied, withdrew 174 billion liters (46 billion gallons) daily, and this use is estimated to reach 481 billion liters (127 billion gallons) in the year 2000.

Water supply is only a part of the task. Nearly all the water will become sewage that must be collected and disposed of after such treatment as the local situation may require. A third problem is the collection and disposal of the storm sewage resulting from rainfall upon cities growing in area as well as population.

A large amount of capital is required to construct adequate works for a city. Such works will include a source (which may be wells, impounding reservoirs, lakes, or rivers), pumping equipment, treatment plant, and the distribution system (which will include the storage reservoirs, system of pipes or mains, and such appurtenances as valves and fire hydrants). The total per capita investment will vary according to local conditions as to source, need for elaborate water treatment, and topography.

No general rule for the cost of water and sewerage works can be given. The first cost of water and sewer lines, in place, in moderate sizes was about $1.30/cm diameter per m of length ($1.00/in per ft) in 1976. This cost does not include valves or fittings, household connections, nor provision for unusual construction conditions.

The cost of treatment facilities is dependent upon the degree of treatment required, the process selected, and the equipment used. Cost per unit treated varies widely with total plant capacity. Estimation of costs is dealt with in Chap. 28.

REFERENCES

1. Abrahams, Harold J., "The Water Supply of Rome," *Journal American Water Works Association* **67**:633, December, 1975.
2. Craun, G. F., L. J. McCabe, and J. M. Hughes, "Waterborne Disease Outbreaks in the U.S. 1971–1974," *Journal American Water Works Association* **68**:420, August, 1976.
3. Murray, C. R., and E. B. Reeves, "Estimated Use of Water in the U.S., 1970," *U.S. Geological Survey Circular 676*, 1972.
4. "The Nation's Water Resources", United States Water Resources Council, 1968.

CHAPTER

TWO

QUANTITY OF WATER AND SEWAGE

WATER

2-1

In the design of any waterworks project it is necessary to estimate the amount of water that is required. This involves determining the number of people who will be served and their per capita water consumption, together with an analysis of the factors that may operate to affect consumption.

It is usual to express water consumption in liters or gallons per capita per day, obtaining this figure by dividing the total number of people in the city into the total daily water consumption. For many purposes the average daily consumption is convenient. It is obtained by dividing the population into the total daily consumption averaged over one year. It must be realized, however, that using the total population may, in some cases, result in serious inaccuracy, since a large proportion of the population may be served by privately owned wells. A more accurate figure would be the daily consumption per person served.

2-2 Forecasting Population

Prior to design of a waterworks one must establish the length of time the improvement will serve the community before it is abandoned or enlarged.

For example, an impounding reservoir may be constructed of such capacity that it will furnish a sufficient amount of water for 30 years, or the capacity of a water purification plant may be adequate for 10 years. These periods are known as periods of design, and they have an important bearing upon the amount of funds that may be invested in construction of both waterworks and sewerage works. Since most American cities are growing in population, the period of design depends mainly upon the rate of population growth; i.e., the water purification plant mentioned above will just serve the population expected 10 years hence. The problem, accordingly, is to forecast as accurately as possible the population 10, 20, or 30 years in the future.

One source of population figures is the U.S. Bureau of the Census, which makes decennial counts and publishes reports covering its enumerations. While the Census Reports give the data upon which to base estimates of population, it is frequently necessary to estimate present population in a year subsequent to the last decennial census. Finding present population is sometimes done by plotting the line of population increase as shown by the last two preceding enumerations and continuing the line to the year in question. Population figures shown by city directories may also be used. Possibly the best method is to obtain the ratio of population to the number of children in the schools or to the number of telephone services in the census year and apply the same ratio to the number of school children or services of the present year.

It is more difficult to estimate the population in some future year. Several methods are used, but it should be pointed out that judgment must be exercised by the engineer as to which method is most applicable. A knowledge of the city and its environs, its trade territory, whether or not its industries are expanding, the state of development in the surrounding country, location with regard to rail or water shipment of raw materials and manufactured goods will all enter into the estimation of future population. Of course, extraordinary events, such as discovery of a nearby oil field or sudden development of a new industry, upset all calculations of future growth and necessitate hasty extension of existing water and sewage facilities.

Arithmetic method This method is based upon the hypothesis that the rate of growth is constant. The hypothesis may be tested by examining the growth of the community to determine if approximately equal incremental increases have occurred between recent censuses. Mathematically this hypothesis may be expressed as:

$$\frac{dP}{dt} = K \qquad (2\text{-}1)$$

in which dP/dt is the rate of change of population with time and K is a constant. K is determined graphically or from consideration of actual populations in successive censuses as

$$K = \frac{\Delta P}{\Delta t} \qquad (2\text{-}2)$$

The population in the future is then estimated from

$$P_t = P_0 + Kt \qquad (2\text{-}3)$$

where P_t is the population at some time in the future, P_0 is the present population, and t is the period of the projection.

Uniform percentage method The hypothesis of geometric or uniform percentage growth assumes a rate of increase which is proportional to population:

$$\frac{dP}{dt} = K'P \tag{2-4}$$

Integration of this equation yields

$$\ln P = \ln P_0 + K' \, \Delta t \tag{2-5}$$

This hypothesis is best tested by plotting recorded population growth on semilog paper. If a straight line can be fitted to the data, the value of K' can be determined from its slope. Alternately K' may be estimated from recorded population using

$$K' = \frac{\ln P - \ln P_0}{\Delta t} \tag{2-6}$$

in which P and P_0 are recorded populations separated by a time interval Δt.

Curvilinear method This technique involves the graphical projection of the past population growth curve, following whatever tendencies the graph indicates. The commonly used variant of this method includes comparison of the projected growth to that of other cities of larger size. The cities chosen for the comparison should be as similar as possible to the city being studied. Geographical proximity, likeness of economic base, access to similar transportation systems, and other such factors should be considered. As an example, in Fig. 2-1, city A, the city being

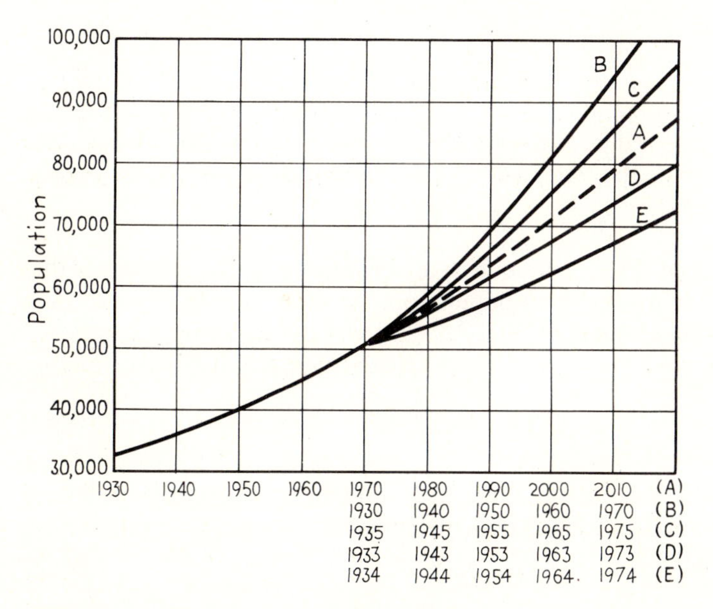

Figure 2-1 Curvilinear method of predicting population. The dashed line is the forecast for city A. Scales A, B, C, D, and E apply to the corresponding cities.

studied, is plotted up to 1970, the year in which its population was 51,000. City B reached 51,000 in 1930, and its curve is plotted from 1930 on; similarly curves are drawn for cities C, D, and E from the years in which they reached A's 1970 population. A's curve can then be continued, allowing it to be influenced by the rates of growth of the larger cities. So far as possible the larger cities chosen should reflect conditions as they are in the city being studied.

Logistic method The logistic curve used in modeling population trends has an S shape. The hypothesis of logistic growth may be tested by plotting recorded population data on logistic paper—on which it will appear as a straight line if the hypothesis is valid. In the short term, a logistic projection can be made based upon the equation:

$$P = \frac{P_{sat}}{1 + e^{a + b\Delta t}} \tag{2-7}$$

in which P_{sat} is the saturation population of the community and a and b are constants. P_{sat}, a and b may be determined from three successive census populations and the equations:

$$P_{sat} = \frac{2P_0P_1P_2 - P_1^2(P_0 + P_2)}{P_0P_2 - P_1^2} \tag{2-8}$$

$$a = \ln \frac{P_{sat} - P_0}{P_0} \tag{2-9}$$

$$b = \frac{1}{n} \ln \frac{P_0(P_{sat} - P_1)}{P_1(P_{sat} - P_0)} \tag{2-10}$$

where n is the time interval between successive censuses. Substitution of these values in Eq. (2-7) permits the estimation of population for any period Δt beyond the base year corresponding to P_0.

Declining growth method This technique, like the logistic method, assumes that the city has some limiting saturation population, and that its rate of growth is a function of its population deficit:

$$\frac{dP}{dt} = K''(P_{sat} - P) \tag{2-11}$$

Following estimation of the saturation population upon some rational basis such as land available and existing population density, K'' may be determined from successive censuses and

$$K'' = -\frac{1}{n} \ln \frac{P_{sat} - P}{P_{sat} - P_0} \tag{2-12}$$

where P and P_0 are populations recorded n years apart. Future population can then be estimated using this value and

$$P = P_0 + (P_{sat} - P_0)(1 - e^{K'' \Delta t}) \tag{2-13}$$

Ratio method The ratio method of forecasting relies upon the population projections of state or federal demographers and the presumption that the city in question will maintain the same trend in the change of the ratio of its population to that of the larger entity. Application of the method requires calculation of the ratio of the local to regional population in a series of census years, projection of the trend line using any of the techniques above, and application of the projected ratio to the estimated regional population in the year of interest.

Accuracy in population estimation is important since if the estimate is too low the engineering works will quickly be inadequate and redesign, reconstruction, and refinancing will be necessary. Overestimation of population, on the other hand, results in excess capacity which must be financed by a smaller population at a considerably higher unit cost.

The selection of an appropriate technique is not always easy and many engineers will test all methods against the recorded growth, consider whatever local factors may be known to have influenced the rate, and eliminate those techniques which are clearly inapplicable. The growth of a community with limited land area for future expansion might be modeled using the declining growth or logistic technique, while another, with large resources of land, power, water, and good transportation might be best predicted by the geometric or uniform percentage growth model. In nearly all cases comparison to similar cities is used, and the results obtained by this technique are favored by most engineers.

2-3 Consumption for Various Purposes

The water furnished to a city can be classified according to its ultimate use or end. The uses are:

Domestic. This includes water furnished to houses, hotels, etc., for sanitary, culinary, drinking, washing, bathing, and other purposes. It varies according to living conditions of consumers, the range usually being considered as 75 to 380 l (20 to 100 gal) per capita per day, averaging 190 to 340 l (50 to 90 gal) per capita. These figures include air conditioning of residences and irrigation or sprinkling of privately owned gardens and lawns, a practice that may have a considerable effect upon total consumption in some parts of the country. The domestic consumption may be expected to be about 50 percent of the total in the average city; but where the total consumption is small, the proportion will be much greater.

Commercial and Industrial. Water so classified is that furnished to industrial and commercial plants. Its importance will depend upon local conditions, such as the existence of large industries, and whether or not the industries patronize the public waterworks. Self-supplied industrial water requirements are estimated to be more than 200 percent of municipal water supply demand.[1]

The quantity of water required for commercial and industrial use has been related to the floor area of buildings served. Symons[2] proposes an average of 12.2 $m^3/1000\ m^2$ of floor area per day (0.3 gal/ft^2 per day). In cities of over 25,000 population commercial consumption may be expected to amount to about 15 percent of the total consumption.

Public Use. Public buildings, such as city halls, jails, and schools, as well as public service—flushing streets and fire protection—require much water for which, usually, the city is not paid. Such water amounts to 50 to 75 l per capita. The actual amount of water used for extinguishing fires does not figure greatly in the average consumption, but very large fires will cause the rate of use to be high for short periods.

Loss and Waste. This water is sometimes classified as "unaccounted for," although some of the loss and waste may be accounted for in the sense that its cause and amount are approximately known. Unaccounted-for water is due to meter and pump slippage, unauthorized water connections and leaks in mains. It is apparent that the unaccounted-for water, and also waste by customers, can be reduced by careful maintenance of the water system and by universal metering of all water services. In a system 100 percent metered and moderately well maintained, the unaccounted-for water, exclusive of pump slippage, will be about 10 percent.

The total consumption will be the sum of the foregoing uses and the loss and waste. The probable division of this consumption is shown in Table 2-1. Table 2-2 shows some total consumptions as reported in cities in various parts of the country. The average daily per capita consumption may be taken to be 670 l (175 gal). This means little, however, as individual figures vary widely. Each city has to be studied, particularly with regard to industrial and commercial uses and actual or probable loss and waste. Care must also be taken in considering per capita figures since the figure may be based upon persons actually served or upon census population of the city.

Table 2-1 Projected consumption of water for various purposes in the year 2000[1]

Use	Liters per capita/day†	Percentage of total
Domestic	300	44
Industrial	160	24
Commercial	100	15
Public	60	9
Loss and waste	50	8
Total	670	100

† Gallons = liters × 0.264

Table 2-2 Recorded rates of water consumption in some American cities

City	Average daily per capita consumption, l†	Maximum one-day consumption in a 3-year period, l	Maximum in proportion to average, %
Rochester, N.Y.	451	637	141
Syracuse, N.Y.	728	917	126
Hartford, Conn.	671	887	132
Albany, N.Y.	671	860	128
El Paso, Tex.	447	739	165
Portland, Me.	572	773	135
Camden, N.J.	641	963	150
Albuquerque, N.M.	402	766	190
Winston-Salem, N.C.	447	580	130
Waterloo, Iowa	383	625	163
Passaic, N.J.	807	1016	126
South Gate, Calif.	550	891	162
Fort Smith, Ark.	474	652	138
Poughkeepsie, N.Y.	569	728	128
Tyler, Tex.	371	743	200
Monroe, La.	584	875	150
Spartanburg, S.C.	754	955	126
Pomona, Calif.	629	1092	173
St. Cloud, Minn.	277	701	254
Salina, Kan.	603	1357	225
Alliance, Ohio	796	1114	140
Ashtabula, Ohio	766	985	129

† Gallons = liters × 0.264

2-4 Factors Affecting Consumption

The average daily per capita water consumption varies from 130 to 2000 l (35 to 530 gal) in American cities. Such variations depend upon a number of important factors, including size of city, presence of industries, quality of the water, its cost, its pressure, the climate, characteristics of the population, whether supplies are metered, and efficiency of the waterworks administration. The more important of these factors will be separately treated below, but some can be briefly discussed here.

The efficiency of the waterworks management will affect consumption by decreasing loss and waste. Leaks in the water mains and services and unauthorized use of water can be kept to a minimum by surveys (see Art. 6-33). A water supply that is both safe and attractive in quality will be used to a greater extent than one of poor quality. In this connection it should be recognized that improvement of the quality of water supply will probably be followed by an increase in consumption. Increasing the pressure will have a similar effect. Changing the rates charged for water has little effect upon consumption, at least in prosperous periods.

2-5 Size of City

The effect of size of the city is probably indirect. It is true that a small per capita water consumption is to be expected in a small city, but this is usually due to the fact that there are only limited uses for water in small towns. On the other hand, the presence of an important water-using industry may result in high consumption. A small city is likely to have a relatively larger area that is inadequately served by both the water and sewer systems than a large city. Sewerage or its absence will have considerable effect. In the unsewered home, water consumption will rarely exceed 40 l (10 gal) per capita/day; while in the average sewered home, it will equal or exceed 300 l (80 gal). The extension of sewers may, therefore, necessitate additional water supply facilities.

2-6 Characteristics of the Population

Although the average domestic use of water may be expected to be about 300 l (80 gal) per capita daily, wide variations are found. These are largely dependent upon the economic status of the consumers and will differ greatly in various sections of a city. In the high-value residential districts of a city or in a suburban community with a similar population the water consumption per capita will be high. In apartment houses, which may be considered as representing the maximum domestic demand to be expected, the average consumption should be about 380 l (100 gal) per capita. In areas of moderate- or high-value single residences even higher consumption may be expected, since to the ordinary domestic demand there will be added an amount for watering lawns. The slum districts of large cities will have low per capita consumptions, perhaps 100 l (25 gal), but consumptions as low as 50 l (13 gal) per capita have been reported. The lowest figures of all will be found in low-value districts where sewerage is not available and where perhaps a single faucet serves one or several homes.

2-7 Industries and Commerce

The presence of industries in a city has a great effect upon total consumption. As already indicated, domestic use is comparatively small, and the difference between it and the total water consumption is largely caused by industrial use and loss and waste. Since industrial use has no direct relation to population, great care must be taken when estimating present or future water consumption in any restricted portion of a city. It is necessary to study the existing industries of the section, their actual use of water, and the probability of establishment of more industrial plants in the neighborhood. Zoning of the city (see Art. 2-13) will make the prediction of future consumption in various districts far more accurate. The use of auxiliary water supplies, frequently untreated, for certain industrial processes is quite common. This will, of course, decrease the consumption of the municipal supply.

Commercial consumption is that of the retail and wholesale mercantile houses and office buildings. Although it is largely dependent upon the number of people working in the business districts, who may be very numerous, it cannot be estimated on the basis of the residents of such districts, who may be very few in numbers. Figures are few and widely divergent as to the commercial consumption of water; and if the consumption is desired for any district, a special investigation should be made. Use of floor area as a unit has already been mentioned. Another convenient unit for this consumption is ground area. Some investigations indicate that in the highly developed business sections of large cities the water consumption may reach 94,000 m^3/km^2 per day (100,000 gal/acre per day).

Use of water for air conditioning of buildings, including residences, is common, and its importance is enhanced by the fact that it coincides with the normally high hot-weather use for other purposes. If the water is under 16°C, it may be used directly for cooling. This provides cheap cooling from an energy standpoint but the water cost and use are high. In some areas water so used must be returned to the ground (Art. 4-13).

2-8 Climatic Conditions

Where summers are hot and dry, much water will be used for watering lawns. Domestic use will be further increased by more bathing, while public use will be affected by use in parks and recreation fields for watering grass and for ornamental fountains. On the other hand, in cold weather water may be wasted at the faucets to prevent freezing of pipes, thereby greatly increasing consumption. High temperatures may also lead to high water use for air conditioning.

2-9 Metering

Every waterworks should have some means at the pumping plant of accurately measuring all the water that is delivered to the city. If the meters are of the recording type, valuable information regarding hourly rates of consumption will be available. If all services are metered, the difference between the total amount pumped and the sum of service meter readings and any unmetered publicly used water will be the unaccounted-for water.

Metering of services consists of placing a recording meter in the line leading from the water main to the building served. Consumers are then billed for the water that they use. The alternative to this method is charging by some form of flat rate which has no relation to the actual amount of water used or wasted. The advantages of metering are apparent. Pumping and treatment of water cost money, and wasting of water means a greater cost to be distributed among customers. If services are unmetered, the careful consumers bear some of the burden imposed by the careless and wasteful. It is almost impossible to construct a good system of water charges unless they are based upon actual consumption of water.

Lack of service meters has a definite effect upon water consumption. In fact, the installation of meters may so reduce consumption that provision of more water may be indefinitely postponed. Comparison of figures[3] in 22 cities, 90 to 100 percent metered, with 13 cities, 20 percent metered, showed that the former group had an average consumption of 366 l (96 gal); and the latter, 824 l (217 gal). These were all cities having over 100,000 population. Metering all services of a city should reduce consumption to about 50 percent of the consumption without meters. Although metering reduces water consumption there is a tendency for consumption to increase gradually after all services are metered.

2-10 Variations in Rate of Consumption

The per capita daily water consumption figures discussed above have been based upon annual consumption. The annual average daily consumption, while useful, does not tell the full story. Climatic conditions, the working day, etc., tend to cause wide variations in water use. Through the week, Monday will usually have the highest consumption, and Sunday the lowest. Some months will have an average daily consumption higher than the annual average. In most cities the peak month will be July or August. Especially hot, dry weather will produce a week of maximum consumption, and certain days will place still greater demand upon the water system. Peak demands also occur during the day, the hours of occurrence depending upon the characteristics of the city. There will usually be a peak in the morning as the day's activities start and a minimum about 4 A.M. A curve showing hourly variations in consumption for a limited area of a city may show a characteristic shape. If the district is commercial or industrial, no pronounced peak will be seen, but there will be a fairly high consumption through the working

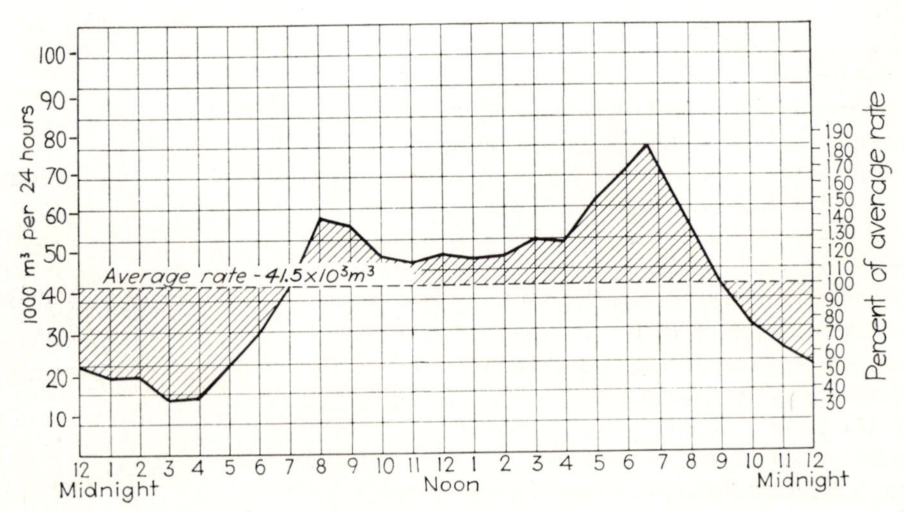

Figure 2-2 Variation in water consumption throughout the day at Sheboygan, Wis. ($m^3 \times 264.2 =$ gal.)

day. The night flow, excluding industries using much water at night, is a good indication of the magnitude of the loss and waste.

The importance of keeping complete records of water pumpage of a city for each day and fluctuations of demand throughout the day cannot be overemphasized. So far as possible the information should be obtained for specific areas. These are the basic data required for planning of waterworks improvements. If obtained and analyzed, they will also indicate trends in per capita consumptions and hourly demands for which future provision must be made.

In the absence of data it is sometimes necessary to estimate the maximum water consumption during a month, week, day, or hour. The maximum daily consumption is likely to be 180 percent of the annual average and may reach 200 percent. The formula suggested by R. O. Goodrich is convenient for estimating consumption and is:

$$p = 180t^{-0.10} \tag{2-14}$$

in which p is the percentage of the annual average consumption for the time t in days from 2/24 to 360. The formula gives the consumption for the maximum day as 180 percent of the average, the weekly consumption as 148 percent, and the monthly as 128 percent. These figures apply particularly to smaller residential cities. Other cities will generally have smaller peaks.

The maximum hourly consumption is likely to be about 150 percent of the average for that day. Therefore, the maximum hourly consumption for a city having an annual average consumption of 670 l/day per capita would occur on the maximum day and would be 670 × 1.80 × 1.50, or 1800 l/day. The fire demand must also be added, according to the method indicated in the following article. Minimum rate of consumption is of less importance than maximum flow but is required in connection with design of pumping plants. It will depend largely upon loss and waste, night industrial uses of water, and the proportion of peak demand provided from storage. Usually it will vary from 25 to 50 percent of the daily average.

Table 2-2 gives consumption statistics of cities. It will be noted that there are wide variations in the rates, suggesting that each city is a problem in itself. The very high rates of some Western cities are probably caused by much use of water for irrigating gardens and lawns.

Peaks of water consumption in certain areas of a city will affect design of the distribution system. High peaks of hourly consumption can be expected in residential or predominantly residential sections because of heavy uses of water for lawn watering, especially where underground systems are used, air conditioning, or in other water-using appliances. Since use of such appliances is increasing, peak hourly consumptions are also increasing. A study[4] indicated hourly peaks as high as 1000 percent of the annual average for a suburban-type city, with hourly peaks of 300 to 400 percent not uncommon in residential areas of large cities. The lower the population density of a fully developed residential area, the higher will be the hourly peak in terms of the average. Commercial and industrial users will reduce the peak, although the total use will be affected (see also Art. 6-13).

2-11 Fire Demand

Although the actual amount of water used in a year for fire fighting is small, the rate of use is large. The Insurance Services Office[5] uses the formula

$$F = 18C(A)^{0.5} \tag{2-15}$$

in which F is the required fire flow in gpm (l/min/3.78), C is a coefficient related to the type of construction, and A is the total floor area in ft^2 (m^2 × 10.76) excluding the basement of the building.

A variety of special factors which can affect the required fire flow are presented in reference 5. The general considerations are as follows:

C ranges from a maximum of 1.5 for wood frame to a minimum of 0.6 for fire resistive construction. The fire flow calculated from the formula is not to exceed 8000 gpm (30,240 l/min) in general, nor 6000 gpm (22,680 l/min) for one story construction. The minimum fire flow is not to be less than 500 gpm (1890 l/min). Additional flow may be required to protect nearby buildings. The total for all purposes for a single fire is not to exceed 12,000 gpm (45,360 l/min) nor be less than 500 gpm (1890 l/min).

For groups of single and two-family residences the following table may be used to determine the required flow.

The fire flow must be maintained for a minimum of 4 hours as shown in Table 2-4. Most communities will require a duration of 10 hours.

In order to determine the maximum water demand during a fire, the fire flow must be added to the maximum daily consumption. If it is assumed that a community with a population of 22,000 has an average consumption of 600 l per capita/day and a fire flow dictated by a building of ordinary construction with a floor area of 1000 m^2 and a height of 6 stories, the calculation is as follows:

Average domestic demand = 22,000 × 600 = 13.2 × 10^6 l/day

Maximum daily demand = 1.8 × avg = 23.76 × 10^6 l/day

Table 2-3 Residential fire flows

Distance between adjacent units		Required fire flow	
ft	m	gpm	l/min
> 100	> 30.5	500	1890
31–100	9.5–30.5	750–1000	2835–3780
11–30	3.4–9.2	1000–1500	3780–5670
≤ 10	≤ 3.0	1500–2000†	5670–7560†

† For continuous construction use 2500 gpm (9450 l/min).

Table 2-4 Fire flow duration

Required fire flow		Duration, h
gpm	l/min	
< 1000	< 3780	4
1000–1250	3780–4725	5
1250–1500	4725–5670	6
1500–1750	5670–6615	7
1750–2000	6615–7560	8
2000–2250	7560–8505	9
> 2250	> 8505	10

$$F = 18(1)(1000 \times 10.76 \times 6)^{0.5} = 4574 \text{ gpm}$$
$$= 17{,}288 \text{ l/min} = 24.89 \times 10^6 \text{ l/day}$$

$$\text{Maximum rate} = 23.76 \times 10^6 + 24.89 \times 10^6$$
$$= 48.65 \times 10^6 \text{ l/day}$$
$$= 2211 \text{ l per capita/day for 10 hours}$$

The total flow required during this day would be

$$23.76 + 24.89 \times 10/24 = 34.13 \times 10^6 \text{ l}$$
$$= 1551 \text{ l per capita/day}$$

The difference between the maximum domestic rate and the values above is frequently provided from elevated storage tanks. Article 6-7 discusses this matter further.

2-12 Density of Population

Population density, considering a whole city, rarely exceeds an average of 7500 to 10,000 per km^2 (30 to 40 per acre). More important to the engineer, in solving water and sewerage problems, are the densities in particular areas, since he must design sewers and water mains so that each section of the city will be adequately served. Densities vary widely within a city, the general range being from 3800 per km^2 (15 per acre) in the sparsely built-up residential sections to 8800 to 10,000 per km^2 (35 to 40 per acre) in closely built-up single-family residential areas with small lots. In apartment and tenement districts the populations will be 25,000 to 250,000 per km^2 (100 to 1000 per acre). In commercial districts the day population will be highly variable according to development.

2-13 Zoning

Zoning is that feature of city planning which regulates the height and bulk of buildings and the uses to which they may be put. A city plan, therefore, controls the character of districts and prevents, directs, or foresees changes in them. The advantages of this degree of certainty in the solution of water distribution and sewerage problems are important. In residential sections the density of population at maximum development will be known. A residential district of high or medium class will not become a slum or apartment house district. Industrial districts will be set aside on the plan and not allowed to encroach upon residential areas. Commercial districts will be largely decentralized, and the main business district will grow in a planned direction. Water mains and sewer systems can then be planned only for actual needs and with some certainty that future changes in the character of districts will not overtax them.

2-14 Periods of Design and Water Consumption Data Required

The economic design period of a structure depends upon its life, first cost, ease of expansion, and likelihood of obsolescence. In connection with design, the water consumption at the end of the period must be estimated. Overdesign is not conservative since it may burden a relatively small community with the cost of extravagant works designed for a far larger population. Different segments of the water treatment and distribution systems may be appropriately designed for differing periods of time using differing capacity criteria.

1. Development of source The design period will depend upon the source. For groundwater, if it is easy to drill additional wells, the design period will be short, perhaps 5 years. For surface waters requiring impoundments, the design period will be longer, perhaps as much as 50 years. The design capacity of the source should be adequate to provide the maximum daily demand anticipated during the design period, but not necessarily upon a continuous basis.

2. Pipe lines from source The design period is generally long since the life of pipe is long and the cost of material is only a portion of the cost of construction. Twenty-five years or more would not be unusual. The design capacity of the pipe line should be based upon average consumption at the end of the design period with consideration being given to provision of suitable velocities under all anticipated flow conditions.

3. Water treatment plant The design period is commonly 10 to 15 years since expansion is generally simple if it is considered in the initial design. Most treatment units will be designed for average daily flow at the end of the design period since overloads do not result in major losses of efficiency. Hydraulic design should be based upon maximum anticipated flow.

4. Pumping plant The design period is generally 10 years since modification and expansion are easy if initially considered. Pump selection requires knowledge of maximum flow including fire demand, average flow, and minimum flow during the design period.

5. Amount of storage The design period may be influenced by cost factors peculiar to the construction of storage vessels, which dictate a minimum unit cost for a tank of specific size.[6] Design requires knowledge of average consumption, fire demand, maximum hour, maximum week, and maximum month, as well as the capacity of the source and pipe lines from the source.

6. Distribution system The design period is indefinite and the capacity of the system should be sized to accommodate the maximum anticipated development of the area served. Anticipated population densities, zoning regulations, and other factors affecting per capita flow should be considered. Maximum hourly flow including fire demand is the basis for design.

SEWAGE

2-15 Sources of Sewage

Sewage is defined as a combination of (*a*) the liquid wastes conducted away from residences, business buildings, and institutions; and (*b*) the liquid wastes from industrial establishments; with (*c*) such ground, surface, and storm water as may be admitted to or find its way into the sewers. Sewage *a* is frequently known as *sanitary sewage* or *domestic sewage*. Sewage *b* is usually called *industrial waste*. Sewers are classified according to the type of sewage that they are designed to carry. *Sanitary sewers* carry sanitary sewage and the industrial wastes produced by the community and only such ground, surface, and storm water as may enter through poor joints, around manhole covers, and through other deficiencies. *Storm sewers* are designed to carry the surface and storm water which runs off the area that they serve. *Combined sewers* carry all types of sewage in the same conduits.

The following discussion has for its purpose the development of methods of estimating the quantity of sewage that is or would be carried by sanitary sewers, i.e., wastes from residences, business buildings, institutions, industrial plants, and such water as may enter incidentally. The surface water runoff that must be carried by combined and storm sewers will be discussed in Chap. 13.

2-16 Relation to Water Consumption

Sanitary sewage and industrial waste will obviously be derived largely from the water supply. Accordingly, an estimate of the amount of such wastes to be expected must be prefaced by a study of water consumption, either under present conditions or at some time in the future. The proportion of the water consumed which will reach the sewer must be decided upon after careful consideration of local conditions. Water used for steam boilers in industries, air conditioning, and that used to water lawns and gardens may or may not reach the sewers. On the other hand, many industrial plants may have their own supplies but discharge their wastes into the sewers. Although the sewage may vary in individual cities from 70 to 130 percent of the water consumed, designers frequently assume that the average rate of sewage flow, including a moderate allowance for infiltration, equals the average rate of water consumption.

2-17 Infiltration and Inflow

Infiltration is the water that enters sewers through poor joints, cracked pipes, and the walls of manholes. Inflow enters through perforated manhole covers, roof drains, and drains from flooded cellars during runoff events. Because infiltration may be nonexistent during dry weather, the dry-weather flow may be considered as the sanitary sewage plus the industrial wastes. In wet weather, infiltration will be greatly increased as groundwater levels rise, and may be augmented by the inflow from roofs which reaches the sewers by rain leaders from the roof gutters. Most cities prohibit such connections, but they are sometimes made illegally, and some may remain from a period when they were not forbidden. Some sewers may be located below the groundwater table and therefore have some infiltration at all times. Sewers that are constructed in or close to stream beds are especially likely to have high infiltration.

The amount of infiltration to be expected will depend upon the care with which the sewer system is constructed, the height of the groundwater table, and the character of the soil. Special types of joints tend to reduce the infiltration. A soil that heaves with varying water content will pull joints apart and so permit water to enter. A pervious soil permits easy travel of percolating water to the sewers where it will travel along them until it reaches a crack or open joint. Since conditions of construction and soil differ widely, the infiltration found in sewer systems varies considerably. Sewer size apparently has little effect. The large sewers present more joint length for leakage, but the joints are more likely to be of better workmanship. Infiltration rates are likely to vary from 35 to 115 m^3/km of sewer per day (15,000 to 50,000 gal/mi per day) in old systems, but even higher rates have been noted where sewers are below the water table and are poorly constructed. Specifications for sewer projects now limit infiltration to 45 l/km per day per mm of diameter (500 gal/mi per in/day).

Since sewers deteriorate, however, engineers are liberal in estimating the infiltration for design purposes. The figures given are based upon length of the public sewers and do not include the house sewers which extend to the buildings. It should be recognized that they will also permit infiltration, and their construction should be carefully controlled. In order to obtain federal funds for the construction of sewage treatment plants it is necessary to demonstrate that the sewer system does not permit excessive infiltration or inflow.

2-18 Fluctuations in Sewage Flow

Wherever possible, gaugings of flow in existing sewers should be made in order to determine actual variations. Recording gauges are available or can be devised that will give depths of sewage in the outfall sewer or in the main leading from a district. In order to design a system for a previously unsewered town or section of a city, an estimate must be made of the fluctuations to be expected in the flow. This is of importance, as the sewers must be large enough to accommodate the maximum rate, or there may be a backing up of sewage into the lower plumbing fixtures of buildings.

As in water consumption, the rate of sewage production will vary according to the season of the year, weather conditions, day of the week, and time of day. The variations do not depart so far from the average as for water because of the storage space in the sewers and because of the time required for the sewage to run to the point of gauging. That is, the peaks are flattened because it requires considerable sewage to fill the sewers to the high flow point, and the high flows from various sections will reach the gauging point after various times of flow. When the peak occurs will depend upon the flow time in the sewers and the type of district served. In a residential district the greatest use of water is in the early morning. Hence if the sewage is gauged near its origin, the peak flow will be quite pronounced and occur about 9 A.M., whereas if the sewage must travel a long distance, the peak will be deferred. In commercial and industrial districts the water is used throughout the working day, and accordingly the peak is less pronounced. Observation of fluctuations in various cities indicates that the peak for a small residential area is likely to be 225 percent of the average for that day. For commercial areas the peak may reach 150 percent of the average, and for industrial areas somewhat less. The flow in the outfall line of a sewer system serving a city having a normal population and commercial and industrial activities will have a peak flow of about 150 percent of the daily average. Figure 20-8 illustrates the greater peak which may be expected in sewage flow from a small residential district.

Some designers use the following formula[7] to estimate the maximum rate of domestic sewage flow from small areas:

$$M = 1 + \frac{14}{4 + P^{1/2}} \tag{2-16}$$

in which M is the ratio of the maximum sewage flow to the average, and P is the population served in thousands. Some engineers use 22 as the numerator of the fraction.

The maximum sewage flow will be the hourly maximum, or the peak rate of the maximum day plus the maximum infiltration. In relation to water supply this will mean the peak rate of the maximum day, multiplied by the proportion of the water supply reaching the sewers, plus the infiltration. Note that the fire demand does not enter into sewage calculations. Minimum rates of sewage flow are useful in the design of sewage pumping plants and occasionally to investigate the velocities in sewers during low flow periods. In the absence of gauging, minimum flow may be taken as 50 percent of the average.

For design purposes a number of state health departments specify the following as minimum requirements: normal infiltration to be cared for but not from rain leaders or unpolluted cooling water (which should not be discharged to sanitary sewers); laterals and submains to be designed on the basis of 1500 l (400 gal) per capita/day, including normal infiltration; main, trunk, and outfall sewers on the basis of 950 l (250 gal) per capita/day, to include normal infiltration but with additions for industrial wastes if known to be in large amounts.

2-19 Design Periods and Use of Sewage Flow Data

As in waterworks design the engineer must adopt periods of design for sewage works and make proper use of flow data.

1. Design of a sewer system Period of design is indefinite as the system is designed to care for the maximum development of the area which it serves. It is necessary to estimate maximum population densities expected in various districts and locations of commercial and industrial districts together with maximum rates of sewage flow per second and maximum infiltration per day.

2. Sewage pumping plant Design period is usually 10 years. Rates of flow required are average daily, peak, and minimum flow rates, including infiltration.

3. Sewage treatment plant Design period is 15 to 20 years. Flow rates required are average and peak rates, both including infiltration.

PROBLEMS

2-1 Using the methods described in Art. 2-2, estimate the population of a nearby city 5, 10, 15, and 20 years in the future. Explain why certain of the answers are less likely to be correct for this particular community.

2-2 For the community of Prob. 2-1 determine the design flow which should be used for:

(*a*) A groundwater source development
(*b*) Pipe lines to the community from the source
(*c*) Water treatment plant
(*d*) Pumping plant

Present the answer in liters per day, gallons per day, cubic feet per second, and cubic meters per second.

2-3 A community has an estimated population 20 years hence which is equal to 35,000. The present population is 28,000, and present average water consumption is 16×10^6 l/day. The existing water treatment plant has a design capacity of 5 mgd. Assuming an arithmetic rate of population growth determine in what year the existing plant will reach its design capacity.

2-4 Determine the fire flow required for a residential area consisting of homes of ordinary construction, 2500 ft^2 in area, 10 ft apart. What total volume of water must be provided to satisfy the fire demand of this area?

2-5 Determine the required fire flow for a 3-story wood frame building covering 700 m^2 which connects with a 5-story building of fire resistive construction covering 900 m^2.

2-6 From the following table of recorded average monthly flows at a community's water and sewage treatment plants estimate the percentage of sewage flow contributed by infiltration and inflow.

	Flow in m^3/month × 10^{-5}	
Month	Total water pumped	Total sewage treated
Jan.	6.3	6.6
Feb.	6.0	6.3
Mar.	6.4	7.2
Apr.	6.2	9.2
May	6.8	9.3
Jun.	7.0	8.7
Jul.	8.6	7.2
Aug.	8.9	7.5
Sep.	7.3	7.3
Oct.	6.3	7.6
Nov.	6.0	7.7
Dec.	6.2	6.5

2-7 A residential area of a city has a population density of 15,000 per km^2 and an area of 120,000 m^2. If the average sewage flow is 300 l per capita/day estimate the maximum rate to be expected in m^3/s.

REFERENCES

1. "The Nation's Water Resources," United States Water Resources Council, 1968.
2. Symons, G. E., "Water Works Practice–Design Criteria, Part I," *Water and Sewage Works* **101,** May, 1954.
3. Wolpert, W. N., "Dollars for Your Water," *Water Works Engineering,* **85:** 23, November, 1932.
4. Wolff, J. B., "Forecasting Residential Requirements," *Journal American Water Works Association,* **49,** March, 1957.
5. Carl, K. J., R. A. Young, and G. C. Anderson, "Guidelines for Determining Fire-Flow Requirements," *Journal American Water Works Association,* **65:** 335, May, 1973.
6. Babbitt, H. E., J. J. Doland, and J. L. Cleasby, *Water Supply Engineering* 6th ed., McGraw-Hill, New York, 1962.
7. Stanley, W. E., and W. J. Kaufman, "Sewer Design Capacity Practice," *Journal Boston Society of Civil Engineers,* **40:** 4, October, 1953.

CHAPTER

THREE

RAINFALL AND RUNOFF

3-1

Hydrology is the science which treats of the water of the earth and its occurrence as rain, snow, and other forms of precipitation. It includes study of the movement of water on the ground surface and underground to the sea, transpiration from vegetation, and evaporation from land and water surfaces, back to the atmosphere from which it embarks upon another hydrologic cycle. In recent years systems analysis has led to improved understanding of hydrologic processes both from a physical and theoretical standpoint, and to extensive use of mathematical or computer simulation models which can predict hydrologic events.

The sanitary engineer is concerned with hydrology since water supplies are taken from streams, impounding reservoirs, and wells which are fed directly or indirectly by precipitation. He is also concerned with the maximum and minimum rates of runoff, total volume of flows for various flood, drought, and normal periods, as these data are necessary to design reservoirs, spillways, storm sewers, and other hydraulic structures. Hydrology is a broad science and is increasing in scope. Only a few principles can be given in Chaps. 3, 4, and 13 of this book. The treatment presented herein is intended to be largely descriptive. For more thorough study the reader is referred to the publications listed in the References.

3-2 Sources of Information

The engineer, in using hydrology, is dependent upon data collected from many sources and, when confronted by a problem, should ascertain whether pertinent information is available. Information dealing with surface and groundwater quality and quantity has been compiled by the U.S.G.S., and values of various parameters together with statistical analyses of the data used are available from the STORET system.[1] The U.S. Environmental Data Service has long-term records of precipitation and other climatological information,[2,3,4] while the U.S.G.S. and the U.S. Army Corps of Engineers serve as a source of storm runoff and flood flow data. Individual states may also compile data, either selectively from federal sources, or from a combination of these and their own agencies.[5]

PRECIPITATION

3-3

Precipitation, which includes rainfall, snow, hail, and sleet, is the primary source of water in streams, lakes, springs, and wells, and the engineer is concerned principally with precipitation data in the absence of stream-flow records. The U.S. Geological Survey operates an extensive network of stream-flow gauging stations throughout the country and use of their records[6, 7] may make it unnecessary to study precipitation records and estimate runoff therefrom. The smaller streams of the country usually are not gauged, although short-term records for specific streams may be available from state agencies, the U.S. Army Corps of Engineers, the Bureau of Reclamation, or the Soil Conservation Service. Where no stream-flow records are available precipitation data must be obtained and the watershed studied in order to determine the stream-flow characteristics. This is less desirable than the use of actual runoff data and even a few years of stream-flow records are of great value in relating precipitation and runoff.

3-4 Measurement of Precipitation

The U.S. National Weather Service maintains observation stations throughout the country, and publications of the Environmental Data Service are available which give daily, monthly, and annual precipitation at these stations. The records are expressed in millimeters or inches of rainfall per hour, day, month, or year. Time averages reported are arithmetic means for 30 years, while space averages are usually statewide averages of little use to engineers.

The standard rain gauge used by the Weather Service consists of a cylinder 200 mm (8 in) in diameter and 610 mm (2 ft) high (see Fig. 3-1). The upper open end contains a funnel which discharges into a receiving tube which has a cross-sectional area one-tenth that of the cylinder. A measuring stick is used to determine the height of the water in the tube. The reading, divided by 10, is the rainfall. The volume of water displaced by the stick increases the reading by about 1 percent. For large rainfalls there is an overflow opening from the tube into the cylinder.

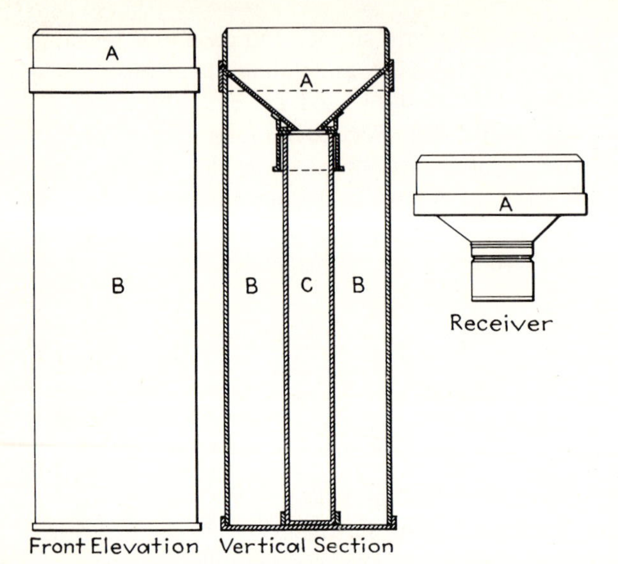

Figure 3-1 Standard rain gauge: *A*, receiver; *B*, overflow attachment; *C*, measuring tube.

The standard rain gauge gives the rainfall only between the daily readings. At some stations hourly readings are made. For determining the precipitation during shorter periods of time, recording gauges are used. Figure 3-2 shows one type of recording gauge. The rain is caught in a funnel-shaped pan. This discharges into a double-tipping bucket, which on being filled tips and empties into a cylinder below. The tipping of the bucket moves a pen which traces a line upon a moving chart. Each tip, which corresponds to 0.25 mm (0.01 in) of rain, causes a jog in the line.

Remote recorders located in the office are sometimes used with this type of gauge, although the weighing gauge is much more common in the United States. The latter catches rain or snow in a bucket which is set on a weighing mechanism. The increase of the weight of the bucket and its contents is recorded on a chart attached to a moving drum, thus showing the rate of precipitation. A variety of float recording gauges are also in use in which the motion of a float records either direct rainfall or displacement of another fluid by accumulated precipitation. Determining maximum intensities for short periods is important in design of storm sewers while maximum average intensities over longer periods are desirable for prediction of flood flows in streams.

Rain gauge measurements are subject to a variety of errors and generally yield results that are too low.[8] The most important factor is error introduced by wind effects, and protective devices such as the Alter shield are now in common use on gauges installed in exposed locations, particularly where snowfall must be measured. Natural protection, such as that afforded by trees and shrubs is considered best, provided it is not so close as to intercept the rain.

Radar can be used to measure the extent and relative intensity of thunderstorms. Estimations of rainfall are made by photographic or electronic integration of echoes which are then correlated with measured rainfall.[8]

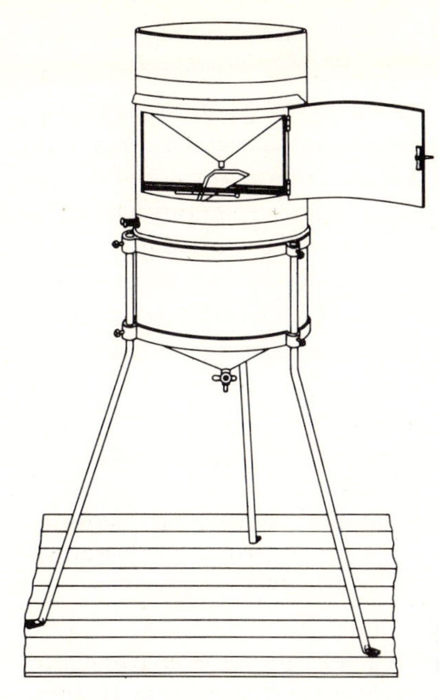

Figure 3-2 Tipping-bucket gauge. *(Courtesy U.S. Weather Service.)*

Snow is an important part of the precipitation in the northern part of the country. It is sometimes a source of flood flows, while in some mountainous areas the winter accumulations melt during the warmer months to give a dependable source of supply. Its separate measurement, therefore, may be of importance. In Weather Service reports snowfall is included in the precipitation. The standard rain gauge can be used to measure it with the volume of snow being changed to depth of water, or a " cut " sample may be taken from the snowfall by removing a core of snow from the snow on the ground, which is then melted or weighed and expressed as depth. Special snow gauges employing pressure transducers or radioactive attenuation are also used. Stick measurements of the snow accumulation on the ground have been largely superseded by snow surveys, which consist of sampling the snow along courses in the area.[9,10] Applications of aerial and satellite photography and electromagnetic analysis[11] have also been evaluated.

3-5 Types of Storms

There are three recognized types of rainstorms. They are (*a*) thermal convection, (*b*) orographic, and (*c*) frontal or cyclonic.[12]

The thermal convection type of storm results from the heating of air near the ground surface during warm weather, its rise, expansion, cooling, and the resulting condensation and precipitation of its moisture. This is the cause of thunderstorms or cloudbursts, which frequently cause intense precipitation over relatively small areas. When a thunderstorm first develops it may cover only 8 to 10 km^2 (3 to 4 mi^2), but as it travels, usually from west to east at about 50 km/h (30 mi/h), it may extend at a frontal length of 80 to 160 km (50 to 100 mi) and a width one-half as great. Such storms are most likely to occur in mountainous areas, but they are also common in the prairie sections of the Middle West.

Orographic storms are induced by mountain barriers. As air masses are forced over them the same conditions result as in the convectional storms with accompanying precipitation. This is the cause of the large annual rainfall on the Northwest coast of the United States. It is present but less conspicuous along the Southern coast and may occur inland wherever high lands project above flat country. In this situation, however, the storms may be of combined types.

The frontal, or cyclonic, type of storm results from contact of masses of air of varying characteristics. Several theories are advanced as to their generation, but the storm consists of a mass of air whirling about a center having a low barometric pressure. In the Northern Hemisphere the air approaches the center spirally in a counterclockwise direction with a vertical component. At the center there is a pronounced rise, the air expands and cools, and condensation and precipitation occur. These storms move in a general easterly direction at a speed of about 50 km/h (30 mi/h).

Both orographic and cyclonic storms cover large areas, usually with low to moderate intensities of precipitation, and may continue for several days. Large engineering works, concerned with runoffs from large drainage areas, place emphasis upon study of storms of these types. For smaller catchment areas, say less than 2500 km^2 (1000 mi^2) the convection type of storm may be most important.

3-6 Precipitation Data and Their Adjustment

The commonest expression of rainfall data is annual precipitation expressed in depth. Figure 3-3 shows mean annual precipitation for the various parts of the United States. Lines of equal precipitation are known as isohyets, and Fig. 3-3 is an isohyetal map.

While figures for annual precipitation are useful, the mean rainfall on a particular catchment area may be more important. It should also be recognized that figures given for observing stations are "point" figures and apply only to the point of location of the rain gauge. The mean rainfall over the surrounding territory may be considerably different. This applies especially to rainfall for a particular storm. Several methods are used to obtain the mean rainfall over an area when there are a number of observing stations located on it. One is to compute the arithmetical mean of the observations. The second, or Thiessen method,[13] weights the observations as follows. Lines are drawn connecting all neighboring stations. Perpendicular bisectors are drawn to these lines, and the areas of the polygons thus formed around each station are measured. The observed precipitation at the observation point located in each polygon is multiplied by its area, and the sum is divided by the total area of the basin. Figure 3-4 shows a Thiessen network applied to a drainage area with rainfalls for a storm shown for each observing station. The mean precipitation over the whole area is found to be 2.54 mm. Note that one station outside the basin exerts its influence on the result. The third method consists of drawing an isohyetal map of the area and computing the mean

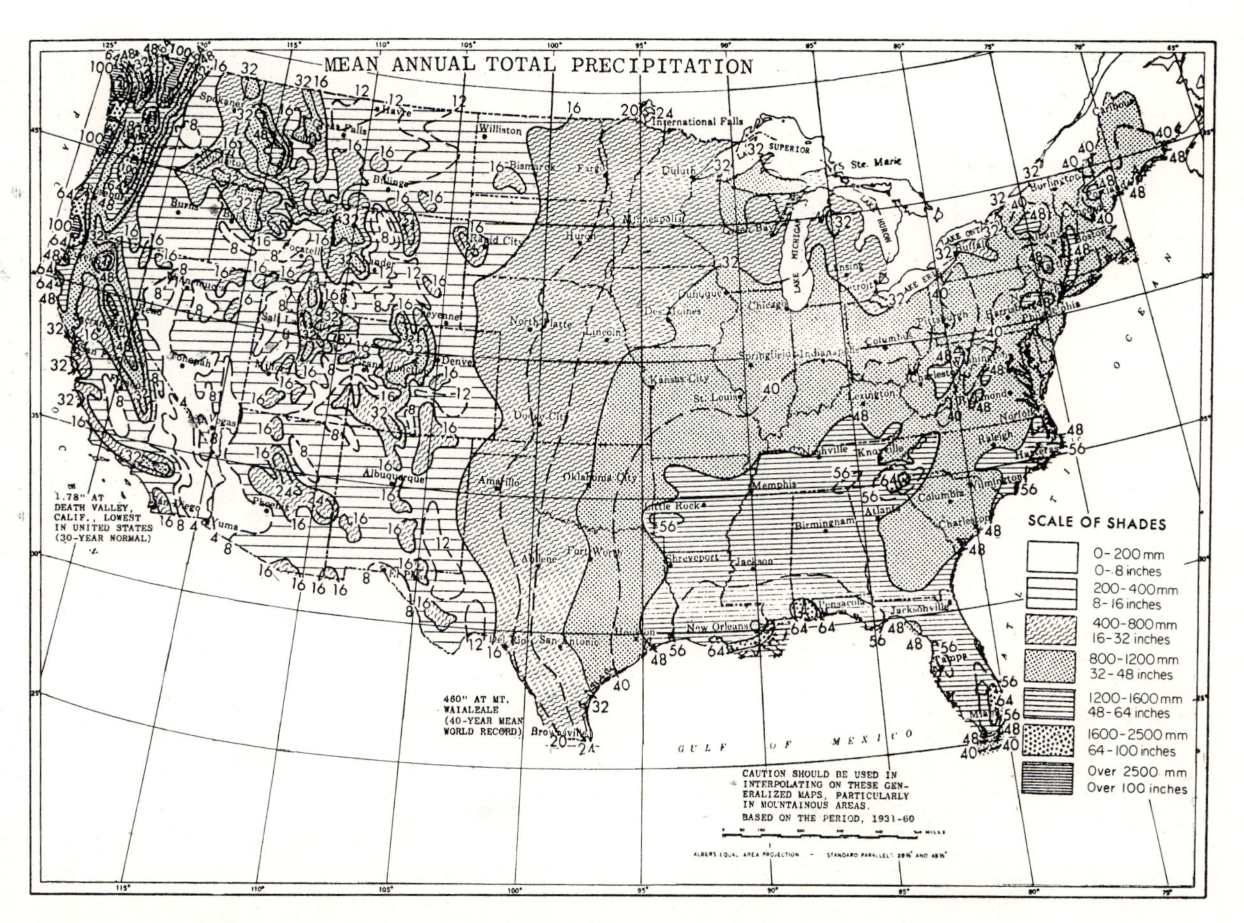

Figure 3-3 Normal annual precipitation 1931–1960. *(From Climates of the United States.[3])*

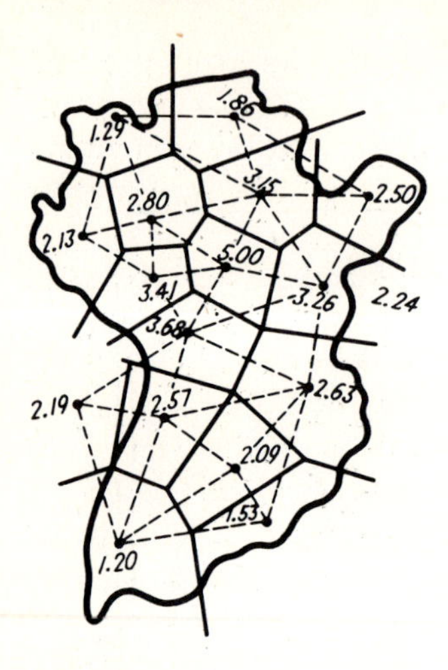

Figure 3-4 Thiessen method of determining average precipitation over an area. *("Hydrology Handbook," American Society of Civil Engineers.)*

rainfall by measuring the areas between isohyets. This may be more accurate since the isohyets can be drawn with regard to topographic and other features, as well as rainfall observations. It is also recommended by some authorities for areas of over 5000 km^2 (2000 mi^2) where differences in topography are likely to be more pronounced. Mathematical models avoid averaging by using the point data independently in computer simulation of rainfall-runoff events. In cyclonic storms there is typically little variation in rainfall intensity over radii of 15 km (10 mi) or more. Convectional storms, on the other hand, may vary widely over short distances—as much as 30 mm/km (2 in/mi).[14]

It is sometimes necessary to supplement existing records that are missing at one or more stations. With regard to mean annual records extending over a long period, missing annual figures from one or more stations may be obtained by using the annual figure for the nearest station multiplied by the ratio of the means of the two stations. To obtain missing figures for precipitation during a storm the isohyets of the storm are drawn from the known figures, and this will allow an estimation of the rainfall at the station whose figure is missing.

Relocation of gauges, even over apparently small distances, may produce a break in the data since the new site may not yield equivalent results from similar storms due to changes in exposure and local turbulence. Such problems may be handled by use of double mass curves which plot measured precipitation at one station versus that at others located nearby. A linear relationship should exist which will permit adjustment of data. Variations will be shown by a change in the slope of the double mass curve after the time of gauge relocation. It should be noted that only marked changes in slope or changes in slope of considerable duration should be accepted as valid.

3-7 Rainfall Intensity-Duration-Frequency Relationships

The rate at which rain falls over an area and its duration are of importance for several reasons. As developed more fully in Art. 13-4, a duration equal to the time of concentration, or the time required for the water to run from the farthest part of the catchment basin to the point in question, is critical, since the shorter the duration of a rainfall the greater may be its average intensity. The greatest intensity to be expected for the critical duration therefore will produce the greatest runoff.

For drainage districts and storm sewers it may not be economical to design for the greatest rainfall, and therefore it is usual to consider that such structures may be overtaxed at intervals of 2, 5, 25, 50 years, or other periods depending upon the cost of the structure and the damage that will result from overtaxing.

The formulas given in Table 13-3, Chap. 13, used in conjunction with Fig. 13-5 may be useful in approximation of rainfall intensities of various durations and frequencies. If sufficient local data are available, frequencies can be computed by the engineer using a formula of the form

$$T = \frac{n+1}{m} \tag{3-1}$$

in which T is the recurrence interval, or the average period in years between rainfalls of a given intensity that will be reached or exceeded, n is the length of the record in years, and m is the rank of each event or intensity when arranged in descending order of magnitude.

3-8 Depth-Area Relationship

If sufficient rainfall data are available, the isohyet method given in Art. 3-6 can be used to determine the average depth over the area covered for any duration by a particular storm. Lack of areal rainfall data, particularly from recording gauges for durations less than 24 hours, contributes to the engineer's difficulty in estimating runoff. Extensive frontal type storms will produce uniform average depths over several square kilometers, but thunderstorms are known to precipitate heavy

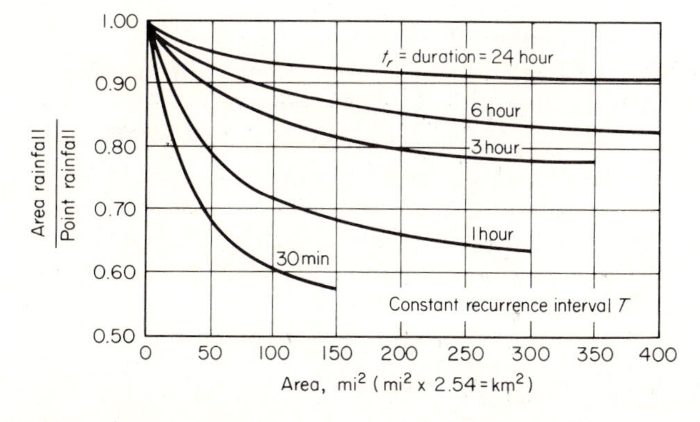

Figure 3-5 Geographically fixed areal variability of rainfall. *(From U.S. Weather Bureau, Rainfall Intensity-Frequency Regime, Tech. Paper 29, pt. 1–5, 1957–1960.)*

rains over areas as small as a few city blocks. If data are available for a storm, the point rainfall figures may be applied to a surrounding area of 2 km^2 (0.75 mi^2). For larger areas the methods described in Art. 3-6 can be used. Depth-area curves are constructed for various rainfall durations, and the larger the area, the smaller will be the average depth of rainfall over it. Figure 3-5 gives depth-area curves for several storm durations.

3-9 Precipitation Deficiency

The two preceding articles have been concerned with rainfalls during storms. Deficiency studies, however, cover seasonal or annual precipitations for a series of years which may be below the mean. It is considered that the mean derived from a 30-year record will not be far from the true mean. Attempts have been made to establish cycles or fluctuations of more or less regular occurrence of annual precipitation, but no general relationships of widespread applicability have been obtained.

It appears, however, that trends of annual precipitation above or below the mean annual may occur for any station and should be investigated by plotting the figures. Several statistical methods are available to make trends more apparent.

Deficiencies in annual precipitation are of importance to the engineer, since they may prevent an impounding reservoir from filling or refilling and restrict the amount of water available for water supplies, water power, or irrigation. The latter may be of particular importance where the rainfall is lower than 500 mm (20 in) per year. He may also be concerned with rainfall deficiencies in certain months, such as the crop-growing season from April to August.

The problem of determining frequencies of rainfalls that will reach or be less than those represented in the period of record can be solved by the method described in Art. 3-7. The frequencies can be obtained for seasonal rainfalls, i.e., part of the year, or for annual rainfalls. A record of at least 30 years is considered necessary for such a statistical analysis, and such records are now available for most parts of the United States.

Records should also be examined for periods of persistently low rainfall. This may be done by making a mass curve of cumulative departures from the mean rainfall of the record. Two or three successive years of rainfall below the mean may seriously affect surface storage in impounding reservoirs. This may be important in arid regions where a very large proportion of the runoff may be required for a project.

3-10 Artificial Rainfall Control

Clouds which will produce rainfall are made up of very small water droplets which have resulted from cooling of expanded rising air. It appears that if solid nuclei are present in the cloud the water vapor will condense and freeze about them to fall out as snow or rain. This, of course, requires temperatures below freezing. Natural nuclei, which may be ice crystals or dust particles, require temperatures of $-15°C$ or less, whereas other particles, like dry ice or silver iodide, will cause condensation at higher temperatures. Silver iodide is effective at $-4°C$.

The technique has possible applications in hail suppression and modest (10–20 percent) increases in precipitation under favorable conditions. The possible coincidence of damaging storms with cloud seeding activity, like that which occurred at Rapid City, S. Dakota in 1972, may lead to restrictions upon its use.

RUNOFF

3-11

Of the precipitation upon a catchment area, some runs off immediately to appear in streams as storm or flood flow; some evaporates from land and water surfaces; some, the snow, remains where it falls, with some evaporation, until it melts; some, known as interception, is caught on leaves of vegetation and evaporates; some, called depression storage, is held in low-lying areas; and some, termed infiltration, seeps into the ground. Of the infiltration a part is taken up by vegetation and transpired through the leaves; some percolates through the soil to emerge again to form springs and streams which make up the dry-weather flow; a part is held by capillarity of the soil; another part is held in the soil particles by molecular action; while a small portion may penetrate into deep porous underground strata and be lost so far as the catchment area is concerned. The last three factors usually are of little or no consequence so far as contribution to runoff is concerned. The difference between the total flow of a stream, as indicated by gaugings, and the rainfall of its watershed will be the water lost by evaporation and the other mechanisms above.

3-12 Measurement of Runoff

The units which have been used in stream-flow data are: (*a*) cubic feet per second (also called second-feet); (*b*) second-feet per square mile of area drained; (*c*) runoff in inches per year or some other period of time, which is useful for comparison with the precipitation on the same drainage area; and (*d*) the acre-foot, which is the quantity of water required to cover an acre to the depth of one foot. Appropriate metric units are m^3/s, $m^3/s/km^2$, mm or cm of runoff per unit time, and m^3.

Since runoff appears as stream flow, the method of determining it directly is by measuring the mean velocity by current meter or otherwise, and calculating the discharge. This should be done, if possible, at high water, low water, and intermediate stages. From these data a rating curve is then constructed showing discharge against the stage, or stream depth. The depth or stage may then be measured continuously by a float in a stilling well or a pressure sensing device. In either case the flow is usually continuously recorded. As stream beds change in shape additional observations are made and new rating curves established. Obviously the longer the period of observation at the station the greater will be the value of the records. Investigators of the U.S. Geological Survey conclude that a 1-year record is likely to vary so far from the mean annual flow that it may involve a very serious error. A 5-year period, while in many cases giving results within 10 percent of the actual mean, may be as far as 30 percent or more in error. A 10-year period can be depended upon to give better results, for in a 10-year group there will probably occur a year of what may be called *average high water* and a year of *average low water*. The 10-year period will not, however, allow for the abnormal year which can be expected once in many years. Since runoff events are assumed to be random occurrences, for adequate results a minimum of 30 years data should be available. This is generally the case in the United States.

Long-continued records not only give a more accurate mean flow but also give information regarding what may be expected in the way of successive dry years and outstanding flood flows. Two or more successive dry years may cause serious depletion of an impounding reservoir unless suitable provision has been made in the design.

Where only a few gaugings are available, some relation between rainfall and runoff may be recognizable. The runoff may then be calculated over a series of years on the basis of observed rainfall. This, of course, is more hazardous than actual runoff observations, and, if it is used, areal distribution of the rainfall should be given careful consideration.

Where no gaugings are available, runoff records of drainage areas of similar characteristics may be used to predict the discharge of the watershed in question. The characteristics which are important in making such comparisons are climate, vegetal cover, topography, soil, and surface geology.

3-13 The Hydrograph

The hydrograph is a graphical, chronological representation of the flow of a stream. Usually the ordinates are expressed in daily discharge in one of the units mentioned in Art. 3-12. Annual hydrographs, like Fig. 3-6, are particularly informative. A base flow or runoff will be noted. This is the dry-weather flow of the stream and is sustained by outflow from the groundwater and from large lakes or swamps. The lower direct flows and flood flows will result from the causative rainfalls. In some areas there will be flows caused by melting snows with or without rainfall, in which case the stream flow is related in part to temperature. By observation of the hydrograph, and rainfall, and study of the stream basin, rainfall-runoff relations can be established, infiltration can be arrived at, and groundwater recharge and basin storage determined.

3-14 Factors Which Affect Runoff

The total runoff from a drainage area is made up of surface runoff and flows from groundwater sources. Surface runoff is the water that has not passed through the

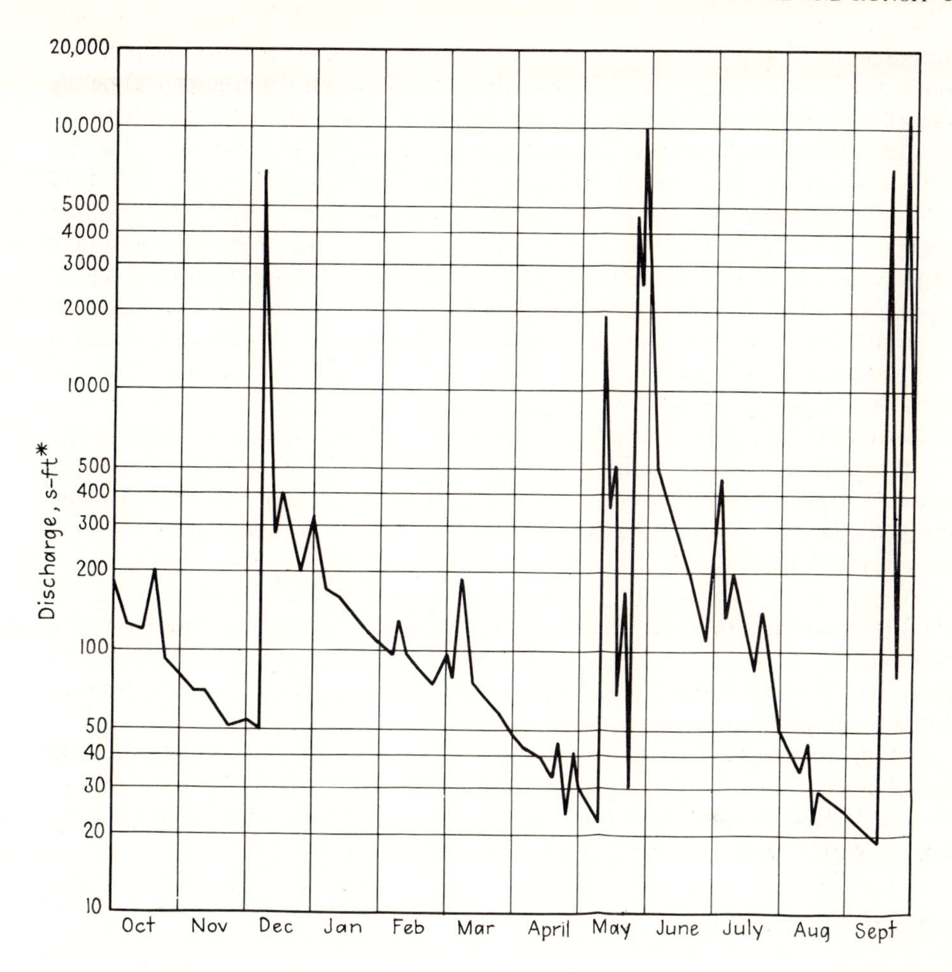

Figure 3-6 Hydrograph, San Gabriel River, Tex. *s-ft × 0.0283 = m^3/s.

soil since precipitation. The groundwater runoff is that which has been a part of the groundwater. The flood flow is almost entirely derived directly from the rainfall and possibly melting ice and snow. The dry-weather flow is produced by the groundwater runoff and drainage from surface ponds. The drainage area, or watershed, is the area that contributes its runoff to the stream above the point where runoff figures are being obtained. The total runoff is, of course, the rainfall less all losses, although this, owing to the varying amounts of water held in the ground, holds good exactly only over a period of years. There may be a permanent loss due to percolation to groundwater which discharges below the gauging point.

The factors that control the amount of runoff and modify the stream-flow rates are numerous and very largely interrelated. They must be carefully studied in connection with projects where extended records of stream flow are not available.

Of these, the rainfall is most important. The total precipitation should be learned and its character, i.e., the proportion of rain and snow. The distribution is important since rains occurring during the growing season for vegetation may contribute little to runoff. The intensity, duration, extent, and usual path of storms will have an effect upon the amount of precipitation and flood flows.

Solar radiation and its variation on the watershed will affect evaporation. Low radiation causes low temperature and accumulations of ice and snow with increased runoff during periods of higher temperatures. Low temperatures, causing freezing of the ground and thereby reducing infiltration, are likely to cause excessive runoffs from later heavy rains. The amount and type of vegetal cover affect infiltration and the loss of water by transpiration.

The topography of the drainage area, its degree of roughness and slope, affects the time of concentration of the direct runoff and thereby causes high or low runoff rates. Evaporation is not very important to floods because they generally occur in a time scale so short that the amount of evaporation is small. Peak rates of runoff are greater when concentration times are short because all the runoff is delivered in a short period.

The geology of the area, including the perviousness or imperviousness of the subterranean formations, is important. If pervious, the slopes of the strata and their extent and points of discharge require study. The condition of the channel of the stream, whether it is pervious and whether there is an extensive underflow, will greatly affect the surface flow. Such underflow may contribute materially to an impounding reservoir. Some streams lose water to pervious formations under their beds, while others receive groundwaters from their banks.

The shape and character of the drainage area will affect concentration and runoff. Shapes providing quickest concentration will tend to produce greatest peak flow. The relation to mountains, glaciers, oceans, large lakes, and wooded areas may also have important effects.

The character of the stream and its tributaries, the slopes and cross section, whether deep or shallow; artificial uses of the stream by irrigation or otherwise; ice formation with accompanying ice jams; and floods—all modify the runoff.

3-15 Infiltration

The entrance of rain water or melted snow into the ground is known as infiltration. Percolation is movement through the soil after entrance. It is by infiltration that groundwater in its various forms is replenished. Some of the groundwater percolates through the soil to appear again at the surface to form dry-weather flow of streams. A small amount percolating through shallow soil may enter streams or surface channels shortly after precipitation. This is known as subsurface flow and is included in direct runoff. The amount of infiltration depends upon a number of conditions on the watershed, the more important of which are discussed below.

1. *Rainfall characteristics.* A small rainfall may all be absorbed and produce no runoff. Heavy rains compact the soil surface by impact of the raindrops and reduce entrance. This is especially noted on bare soils even where cultivated. The increased porosity resulting from plowing or cultivation is soon lost by compaction. Storms of low intensity are likely to have high infiltration rates. Antecedent rains, by increasing soil moisture, decrease infiltration.
2. *Soil characteristics.* The smaller the pore sizes in the surface soil, the smaller will be infiltration. Small soil particles, as in clay, mean small pores, while sand or gravel is at the other extreme. Mixtures of large and small particles tend to pack and reduce pore size. It should be recognized that porosity in percent is not of as much importance as pore size. The fact that many fine soils cement together into clusters or large particles of sand size, 0.2 mm or larger, and thus increase their permeability is also important. Some soils contain colloidal clay, which swells on wetting and reduces infiltration.
3. *Soil cover.* The type of surface cover is important in several ways. It protects the soil from compaction by rain and also provides detention on the surface and thereby increases infiltration opportunity, the degree of action depending upon the density of the cover. The root systems make the soil more pervious. A vigorous sod or dense forest is best,[15] while a thin sod or forest may provide little protection. Cultivated crops vary in effect. Corn, cotton, and tobacco may give considerable protection during their most luxuriant development but little or none during much of the time. Some vegetation, especially grass, encourages the soil crumbs or clusters mentioned above, while others, like soybeans, cause dispersion. Residues of dead vegetation or mulches increase detention time. Snow cover also increases detention and infiltration, provided that the soil is not frozen.

3-16 Determination of Infiltration Rates

Infiltration capacity is defined as the maximum rate at which a given soil in a given condition can absorb rain as it falls. Thus the capacity will be equal to an observed rate of infiltration if the rainfall intensity equals or exceeds it. Runoff occurs if the rainfall intensity exceeds the infiltration capacity. The relation between the two is shown in Fig. 3-7, which is based upon observations of the soils indicated. In general such curves have high infiltration capacities at first, but decrease exponentially to a minimum after a short period of time. According to Linsley, Kohler, and Paulhus,[16] initial moisture greatly affects the shape of the infiltration-capacity curve, particularly in the first 10 min but with some modification for a considerable period. The curve when there is initial moisture in the soil is not the last portion of the curve for a soil initially dry, although this may not apply to all soils.

A number of methods of determining infiltration rates are used. Laboratory experiments are made on soils with artificial rainfall, or field experiments are made on small soil plots using artificial rain. Natural rainfall on isolated runoff plots is observed in the field, or rainfall and runoff records are noted for small

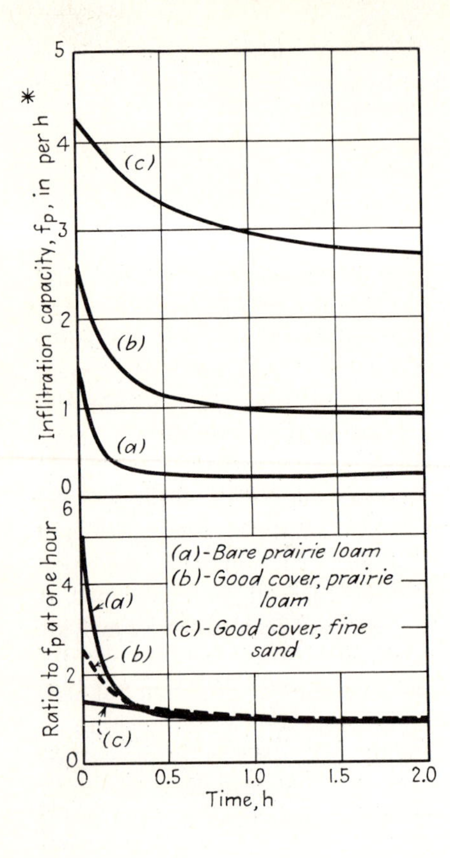

Figure 3-7 Variations in shape of some standard infiltration capacity curves. (*"Hydrology Handbook," American Society of Civil Engineers.*)* in × 25.4 = mm.

watersheds with homogeneous soils. Rainfall and runoff records for larger heterogeneous basins are observed, and an average equivalent infiltration capacity is determined.[17,18]

The rational method of arriving at drainage from urban districts is discussed in Art. 13-2. It makes use of runoff coefficients which are related to the surfaces in the drainage area. The determination of losses by infiltration is more logical and would eliminate the coefficients. Several studies of this method have been made. It is recognized that antecedent moisture in the soil is especially important through its effect upon the infiltration curve, since the time of concentration is short in urban areas. The infiltration-capacity values are deducted from the related rainfall rates, a further correction is made for depression storage out of the first excess rainfall, and the effect of the hydraulics of the flow paths upon the hydrograph at the design point is analyzed.

The infiltration approach to the calculation of surface runoff from a storm is basic, and it is considered as the most reasonable method of attacking the problem by many eminent hydrologists although it requires detailed computations best handled by computers. It also has value in determination of recharge of underground water-bearing formations. There are difficulties in applying the method because of variations in intensity and duration of rainfall over a given watershed, variations in infiltration characteristics and the surface-storage characteristics. These objections are less applicable to small watersheds, especially if infiltration studies can be correlated with observed rainfall and runoff.

3-17 Evaporation from Water Surfaces

The engineer must estimate the loss of water by evaporation from lakes but more especially from impounding reservoirs, where it may reduce the yield from a catchment area by a considerable amount. The amount lost depends upon temperature of the air and water, wind velocity, and atmospheric humidity.

The most commonly used method of estimating evaporation from water surfaces is the Weather Service class A pan, which is made of galvanized steel, 1.22 m (4 ft) in diameter, 250 mm (10 in) deep, and is set on a 150-mm (6-in) wood grill so that the water surface is a little more than 300 mm (1 ft) above the ground. The water is kept within 50 to 75 mm (2 to 3 in) of the rim of the pan. Results are expressed in depth and are multiplied by 0.7 to reduce them to equivalent reservoir evaporation. This is an average annual factor which varies from month to month and region to region. Monthly evaporations are most useful as they can be applied to the drier months of the year when reservoir drawdown by use of water may be expected to be greatest. Figure 3-8 gives evaporation from standard pans in the United States.

The high evaporation loss from reservoirs in arid regions has stimulated experiments in methods of reducing it by application of thin chemical films, floating covers, or floating granular materials. None of these techniques have proven to be practical in large-scale applications, but are useful on small reservoirs.

3-18 Evaporation from Land Surfaces

The amount of water evaporated from land surfaces depends upon the amount of moisture available, which depends upon the rainfall and those characteristics of the soil which affect infiltration, absorption, and percolation. Climatic conditions, particularly radiation and humidity also play an important part. Such evaporation also includes interception, or the precipitation which lodges upon leaves of trees, blades of grass, etc., and is quickly evaporated. Shading by vegetation can greatly reduce evaporation from underlying soil. Evaporation, in soils not in contact with a free water surface, is limited to that available in the pores of the upper layer of soil. Upon evaporation this moisture is not rapidly replaced from the lower layers and evaporation ceases. Soils in contact with a free water surface will replace surface moisture by capillary action, and evaporation will continue as long as a difference in vapor pressure exists if the water surface is reasonably close to the soil surface. Even under these conditions evaporation is less significant than transpiration.

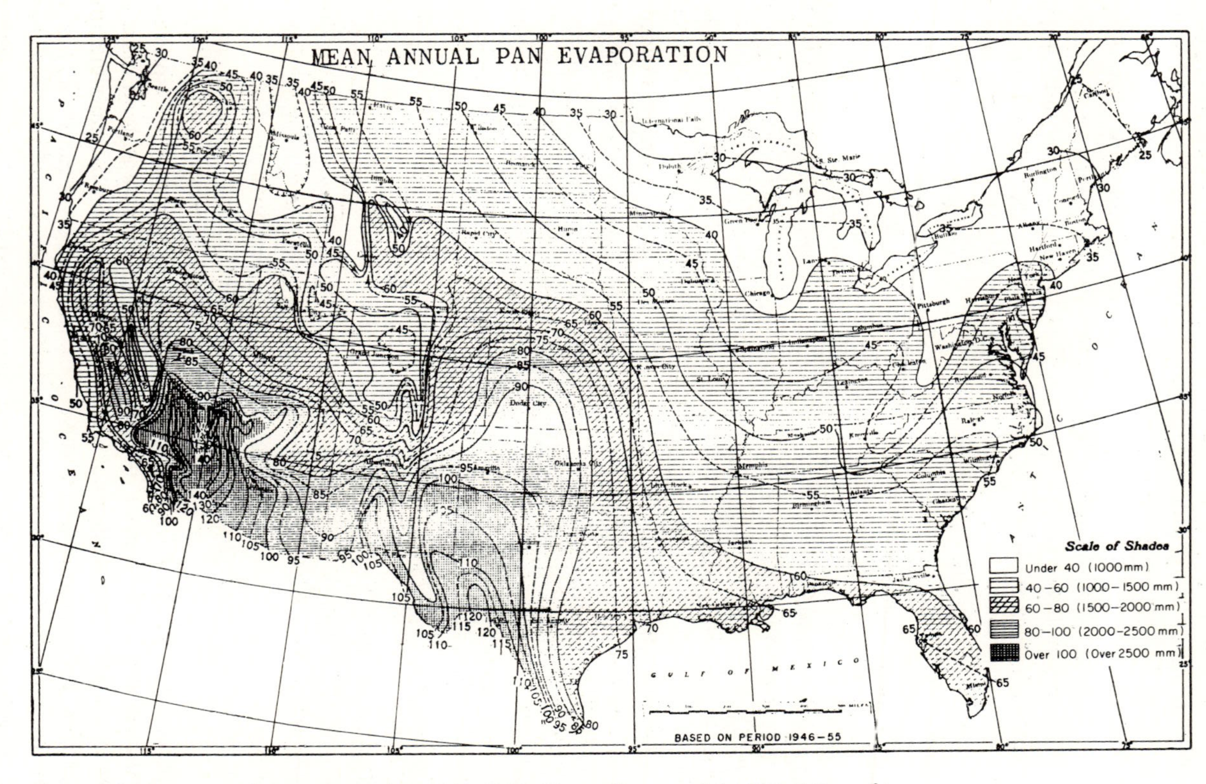

Figure 3-8 Mean annual pan evaporation, 1946–1955. *(From Climates of the United States.*[3]*)*

3-19 Transpiration

Water taken from the soil by roots of vegetation and discharged into the atmosphere as vapor is lost by transpiration. This causes a considerable reduction of runoff. In fact, it can be said that vegetation always reduces the water yield of a catchment area, and it is particularly true where the groundwater is close to the surface. The *rate* of transpiration is independent of plant species, but *total* transpiration depends upon root depth and percent cover, and thus upon the plant type. Attempts have been made to relate transpiration to the dry weight of crops and other vegetation, however there are few usable data concerning the relationship between the weights of crops, their dry weights, and their transpiration potential.

3-20 Evapotranspiration

Evapotranspiration is the total amount of water taken up by vegetation for transpiration and building of plant tissue and the evaporation of soil moisture, during the growing of some particular crop. Valley evapotranspiration is the difference between the annual inflow into the valley (consisting principally of precipitation but possibly including surface and subsurface movement of water into the valley) and the total outflow therefrom in the same period (consisting of surface and subsurface movement out of the valley) with correction for changes in surface and subsurface storage.

The difficulty of separating evaporation, interception, and transpiration has led many engineers to consider evapotranspiration only in arriving at the total water loss from a catchment area. Such losses will, of course, be affected by those conditions which govern evaporation and transpiration, and the rate of loss will be great during the growing season. Table 3-1[19] gives evapotranspiration in millimeters per year with adequate amounts of water available.

Lowry and Johnson[20] made a study of 22 areas of the United States and concluded that evapotranspiration could be related to the effective heat during the growing season. Their findings are closely approximated by the equation

$$V_c = 0.082H + 274 \tag{3-2}$$

in which V_c is the average annual consumptive use of water per year in millimeters and H is the accumulated degree days during the growing season computed from the maximum temperature in degrees Celsius. H is the sum of the daily maxima in degrees Celsius during the growing season. The technique neglects the effects of wind, humidity, and cloud cover. Linsley, Kohler, and Paulhus[16] recommend that potential evapotranspiration or consumptive use be taken as equal to that of an equivalent free water surface with negligible heat storage capacity.

Table 3-1 Annual evapotranspiration on areas with different crops and vegetal covers

mm × 0.039 = in

Vegetation	Location	Annual evapotranspiration, mm	Remarks
Alfalfa	New Mexico	1220	
Alfalfa and clover	Various	760–1220	
Field crops	New Mexico	760	
Field crops and native grasses	Colorado	533	During crop season
Field crops and native grasses	Oregon	305–460	During crop season
Garden truck and small grains	Idaho	305–460	During crop season
Meadows	Wyoming	380	
Citrus trees	California	660	
Peach trees	California	860	
Brush and grass	Texas and New Mexico	690–760	
Brush and grass	California	255–510	Located in a dry section
Alder, maple, and sycamore trees	California	1190	Ample water available

3-21 Multiple Correlation Method of Forecasting Runoff

Kohler and Linsley[21] have presented a multiple correlation technique for prediction of runoff events. Storms which have occurred over a period of record are examined for effects upon runoff of the amount of precipitation, the duration of the rainfall, the antecedent precipitation, the time of year in which the storms occurred, and the basin demand. Basin demand is sometimes called basin recharge and is the difference between the total precipitation of a storm and the total runoff it produces. This factor is more closely related to the parameters mentioned than is the runoff itself. Coaxial curves are constructed which permit forecasting the recharge from a storm of any total precipitation and duration at any time of year. A final result in terms of runoff is directly calculable from the recharge and the rainfall.

3-22 Yield

The portion of the precipitation on a watershed that can be collected for use is the *yield*. This includes direct runoff and that water which passes underground before appearing as stream flow. *Safe yield* is the minimum yield recorded for a given past period. *Draft* is the intended or actual quantity of water drawn for use. Unless the minimum daily flow of a stream is well above the maximum daily draft which must be satisfied in a water supply project, the minimum flow must be supplemented by water from the higher flows which has been impounded in a reservoir.

Impounding reservoirs have two functions: (*a*) to impound water for beneficial use and (*b*) to retard flood flows. The two functions may be combined to some extent by careful operation.

An impounding reservoir presents a water surface for evaporation, and this loss must be considered. Estimates of yield must allow for this loss. In addition, the possibility of large seepage losses must be considered. If it is economically impossible to prevent them at the proposed reservoir site, the project may have to be abandoned or a more favorable site found. There will be some loss by seepage through and under the dam itself.

There are other considerations. Riparian owners downstream, by common law or prior appropriation, may be entitled to have the water of the stream come to them undiminished in quantity so that they may make their accustomed use of it. Impounding of flood waters will not, of course, cause these owners any injury, but interference with the dry-weather flow may result in damage suits against the municipality or water company. This trouble may be avoided by buying the water rights or by allowing water in agreed amounts to pass by. Thereby the water users are enabled to continue their legitimate use of water for irrigation, power, or domestic purposes. Where such compensation in kind is made, the water passed must be added to the draft or subtracted from the stream flow in calculating reservoir storage capacities.

3-23 Reservoir Storage by Mass Diagram

The ordinary hydrograph has been described in Art. 3-13. An integrated, or summation, hydrograph, also known as a mass diagram, may be constructed to show the accumulated yield from month to month and from year to year in the most convenient units.

Mass diagrams are of little use, however, unless the records of the stream are available for substantial periods of time, generally more than 30 years. With so many years of record, periods of low stream flow can be easily recognized, and the necessary storage obtained graphically with much less effort than required by arithmetical means.

Figure 3-9 is an integrated hydrograph extending over a period of 3 years. The curved line *OA* represents the accumulated stream flow during the period. The decreasing runoff of the summer months is very apparent and is shown by the flat slopes of the curve during that period of the year. Strictly speaking, the consumption varies, but great accuracy in this respect is not justified if the record is of short duration. Losses by evaporation from the reservoir and other water surfaces, very important in arid and semiarid areas, and compensation water to satisfy prior water users may be subtracted from the mean flow (which may result in a negative flow during the summer months) or may be added to the consumption. This may be called the regulated flow.

The usual problem is to determine the storage necessary to satisfy a draft equal to *OC*. To find this, tangents parallel to *OC* are drawn at low points and summits in *OA*. The lines *DE* and *MN* have been so drawn. The vertical distance *FN* will be the storage required to maintain this particular draft. At *D* the reservoir will be full; thereafter the runoff is less than the draft, and the depletion of the reservoir begins, reaching a maximum at *N* in October. During the balance of the

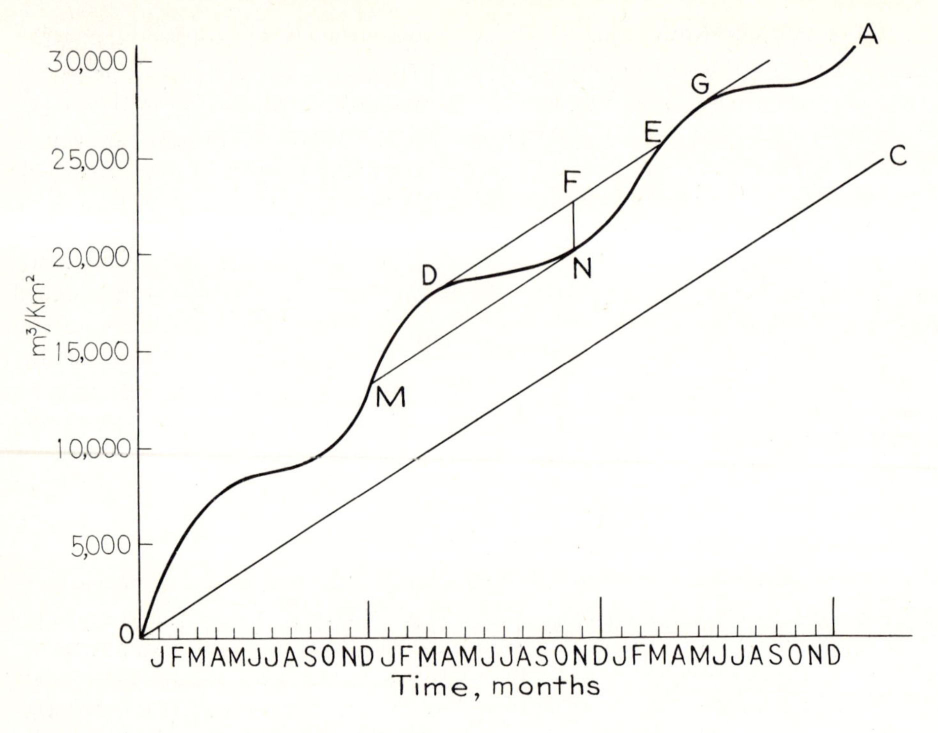

Figure 3-9 Integrated hydrograph. ($m^3/km^2 \times 6.84 \times 10^2 = gal/mi^2$.)

year the flow exceeds the draft, and the reservoir refills, becoming full at *E*. Water will then be wasted over the spillway until the draft again exceeds the runoff, which will occur at *G*. If the tangent *DE*, when extended upward, does not intersect the curve at any point, it indicates that the runoff is insufficient to supply the proposed draft. The extension of the tangent may not intersect the curve for several years, in which case a prolonged period of low water in the reservoir is indicated. Similarly the point *M* is the latest date in this particular year at which the reservoir could start filling and collect enough water to supply the demand of the summer months. By investigation of all sections of the plotted record the maximum storage required to sustain any draft may be obtained. Naturally, the driest year or group of years of the record will be the governing period. The selection of the design year or group of years requires the analysis of runoff duration data for an extensive period so that the probability of given events may be assessed. It may be desirable to consider flows over a given period of consecutive days rather than considering each recorded flow as a discrete event.[22]

With a fixed size storage reservoir it is sometimes necessary to determine the maximum draft that can be maintained. This is done by drawing parallel lines at summits and low points of *OA* which will give ordinates between them equal to the storage available. The pair of parallel lines with the minimum slope will give the draft that can be maintained.

An alternative to the integrated hydrograph or mass diagram is the residual mass diagram or residual mass tabulation.[19] The method consists of investigating the storage required for a number of regulated stream flows, usually in 10 percent increments, up to the annual average regulated flow. Regulated flows are actual flows adjusted or reduced to satisfy downstream requirements, reservoir evaporation, etc. Then, by inspection of the stream-flow record for consecutive periods of days or months of low runoff, the required storage can be obtained for the required outflow for that period. If properly done, the results will be the same as for the mass diagram, and the curve obtained showing storage for any regulated flow is more convenient.

FLOOD FLOWS

3-24

The problems of flood control for large rivers and forecasting of floods and prevention or mitigation of damages from them is largely a federal matter. Here floods and high runoff rates in general are considered in relation to impounding reservoirs for relatively small watersheds and for urban areas (Chap. 13).

Long-time gauge records of a stream are valuable in indicating flood flows, and they are used as much as possible. For smaller streams they may be non-existent or of such short duration that they are useful only as a basis for further estimates by means of unit hydrographs or frequency studies.

Reservoir designs make provision for wasting water at the dam during or after periods of heavy precipitation. Should the capacity of the spillway be exceeded, serious consequences may result, particularly if the dam is of earth construction. Many failures of earth dams have occurred owing to flood flows overtopping the dam and cutting it away. The quick release of the impounded water in such cases has caused heavy economic damage and sometimes loss of life. Consequently, spillways must be designed with extreme flood conditions in mind. Discharges from catchments of various sizes may be estimated by the techniques below:

Catchment area, km^2*	Present practice
Less than 3	Overland flow hydrograph; rational method
3–250	Unit hydrograph; flood frequencies
250–5000	Unit hydrograph; flood frequencies
Over 5000	Flood routing[16,23]; flood frequencies

* $km^2 \times 0.386 = mi^2$

Of the four area classifications only the three smallest will be discussed here.

3-25 The Rational Method

This consists of considering all factors which contribute to maximum runoffs and combining them to obtain an expected amount. The rainfall intensity to be designed for is that which has a duration equal to the time of concentration. The time of concentration is the time required for water to flow from the farthest point of the watershed to the point in question; in other words, when all the area is contributing water at the average rainfall intensity assumed for the duration. Intensity and watershed area are multiplied and modified by a coefficient which depends upon the perviousness of the watershed to obtain expected actual runoff. Application of this method to urban areas is described in Chap. 13.

Characteristics of the drainage area in question must be carefully studied in so far as they affect direct runoff and time of concentration. Time of concentration will include that required for overland flow and in the channels. If there is extensive storage in the flood plain or lakes and ponds, they will not only retard the flow but also reduce the flood peak by furnishing surface storage, and the rational method should not be used in such cases. The previousness or expected infiltration, which will usually be variable over the area, must be estimated. While it is of little importance on urban areas, the rainfall distribution over larger watersheds may be of importance. For example, the heaviest precipitation may occur on the least pervious portion of the area. Presence of ice and snow and effect of antecedent rainfall may have to be considered, but cannot be considered effectively with the rational method. After arriving at time of concentration the expected maximum rainfall can be obtained from the formulas of Chap. 13, and the runoff coefficient for the area is applied. In general this technique is applicable only to very small areas. The procedure discussed in Art. 3-26 is superior.

3-26 The Overland Flow Hydrograph

Izzard's overland flow hydrograph[24] permits the determination of the time to equilibrium flow (equivalent to the time of concentration) and the flow at times other than t_e, both before and after cessation of the rain. The time to equilibrium, or the time of concentration, in minutes may be calculated from

$$t_e = 526.76 k L^{1/3} i_e^{-2/3} \tag{3-3}$$

in which k is defined by

$$k = \frac{2.76 \times 10^{-5} i_e + c}{s^{1/3}} \tag{3-4}$$

where L = the distance of flow in meters
s = the slope of the land
c = a retardence coefficient, typical values for which are presented in Table 3-2, and
i_e = the excess of rainfall over infiltration in mm/h. This is equivalent, in a sense, to the product $C(i)$ in the rational formula.

Table 3-2 Izzard's retardence coefficient

Surface	c
Very smooth asphalt pavement	0.0070
Tar and sand pavement	0.0075
Concrete pavement	0.012
Tar and gravel pavement	0.017
Closely clipped sod	0.046
Dense bluegrass sod	0.060

The experimental verification of Izzard's formulation is limited to $i_e(L) < 3880$.

3-27 The Unit Hydrograph

The unit hydrograph, first suggested by Sherman,[25] has been found very useful in prediction of flood flows. It has been defined as follows:

> The unit hydrograph is the hydrograph of surface runoff (not including groundwater runoff) on a given basin, due to an effective rain falling for a unit of time. The term "effective rain" means rain producing surface runoff. The unit of time may be one day or preferably a fraction of a day. It must be less than the time of concentration.

The effective rain is sometimes expressed as "rainfall excess," which is the volume of rainfall available for direct runoff or the total rainfall minus infiltration and detention losses. This excess should be close to, or greater than, 25.4 mm (1 in).

The term "lag time" is also used. It is variously defined as the time from the center of mass (or beginning) of the rainfall to the peak (or center of mass) of runoff.

The usefulness of the unit hydrograph is based upon the fact that all single storms of some chosen duration will produce runoff during equal lengths of time. For example, all storms occurring over a given watershed with a duration of, say, 12 hours may result in runoff extending over 5 days. Furthermore, the ordinates of the hydrograph will be proportional to the amount of rainfall excess, and the volumes of runoff in corresponding increments of time are always in the same proportion to total runoff regardless of the actual depth of rainfall and total volume of runoff.

The procedure used in constructing a unit hydrograph is as follows. Existing rainfall and stream gauge records are studied and some duration for the unit storm is chosen. This must be less than the lag time, and the shorter the period, the better. It is common practice to use 1-hour unit hydrographs for large areas. For small areas a unit time of one-third to one-fourth of the lag time of the drainage area should be used.

If available in the records, isolated unit storms are studied since these are uncomplicated by the effects of antecedent rainfall. The hydrographs of the storms are analyzed, and the base flow is eliminated. The base flow is the normal dry-weather flow of the stream which is derived from groundwater or from other delayed sources. If there has been an antecedent rainfall its effects can be noted from recession curves of other rainfalls, and ordinates of the base curve and the recession curve can be subtracted from ordinates of the storm under study. The base curve tends to rise after the peak of surface runoff from a storm, but this is so small a proportion of the flood flow that often the base flow is considered a straight line.

Hydrographs of direct runoff plotted for various storms of like duration will have approximately the same durations, but the ordinates will depend upon the volume of runoff. If the ordinates are divided by the total volume of runoff in millimeters or inches, there will be a number of graphs, each representing 1 mm or 1 in of runoff—"unit hydrographs."

In some cases the lag times vary because of variations in rainfall intensity over the watershed. In these cases an average duration time to the peak is obtained, the average peak flow is taken as the mean of the actual peaks, and the shape of the curve is drawn in to conform to the shapes of the observed curves.

Figure 3-10(*a*) from *A* to *C* is the hydrograph of a storm of unit duration, assumed as 12 hours in this case, and the time unit intervals in the figure are also

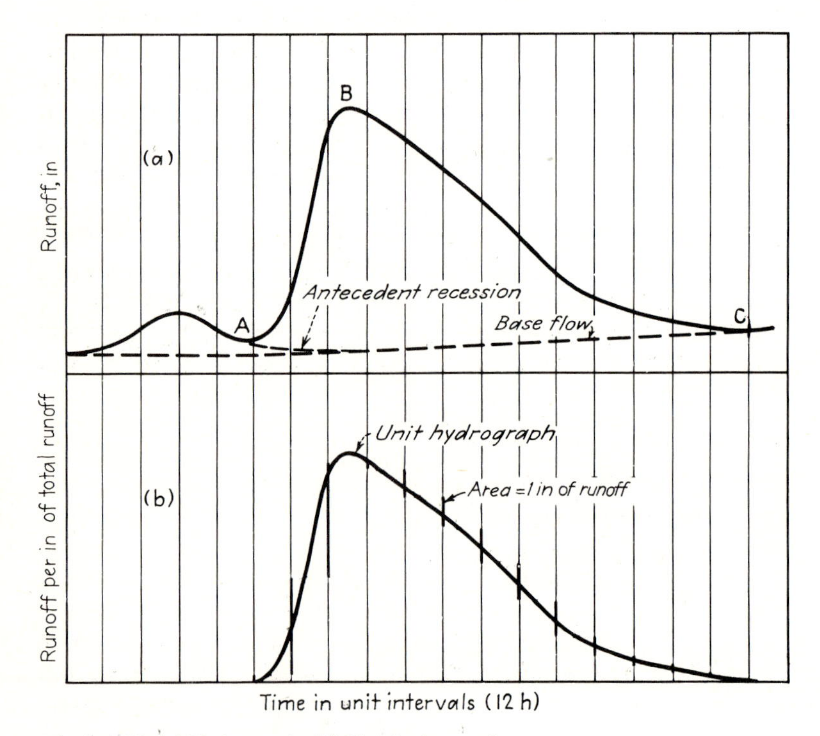

Figure 3-10 (*a*) Hydrograph. (*b*) Unit hydrograph.

12 hours. The runoff curve preceding *A* is caused by an antecedent rainfall, and the antecedent recession curve is a continuation to the base flow. The base flow will have been located as a result of study of the long-time hydrograph of the stream. The base flow can be expected to rise to point *C* where the direct runoff from the storm ceases and base flow is the only flow in the stream. There will then be a slow recession of base flow until there is another rain. Figure 3-10(*b*) shows the unit hydrograph derived from the runoff observations. This indicates the rate of runoff resulting from an effective rainfall of 1 in occurring during the unit time in question. The values for the lower curve are obtained by subtracting from the total flow the flow due to antecedent rainfall, if any, and the base flow, and dividing each resulting ordinate by the average runoff due to the rainfall event.

Before the unit hydrograph can be used to predict the runoff from any assumed or expected rainstorm of unit duration, an estimate must be made of infiltration and other losses. These are discussed in Arts. 3-15 to 3-19. The difference between these losses and the volume of rainfall available for direct runoff is known as excess rainfall, or net rainfall.

3-28 Flood Frequencies

For many purposes it may be useful to know the frequency of occurrence of floods of known magnitude, and the method of Art. 3-7 as used for rainfalls will be useful within the limitations of the record. Another problem is to determine the maximum flood which may be expected in 100 years or some other period. This has led to application of statistical methods based upon probability relationships or formulas. Various methods[16] of attack have been used. The two probability distributions most frequently used in the analysis of flood frequency data are the Gumbel distribution and the log-Pearson Type 3. A special case of the log-Pearson Type 3 is the log normal distribution in which skewness is zero. This distribution will plot as a straight line on log-normal probability paper. Data which fit the Gumbel distribution will plot as a straight line on extreme value paper.

Figure 3-11 illustrates a curve drawn on probability paper. It is assumed that the stream-flow records cover a period of 50 years. The flood flow equaled or exceeded 10 times in the period is plotted; in this case it is 0.2 m^3/s per km^2, and the percent of years exceeded is 20. In like manner the flood flows equaled or

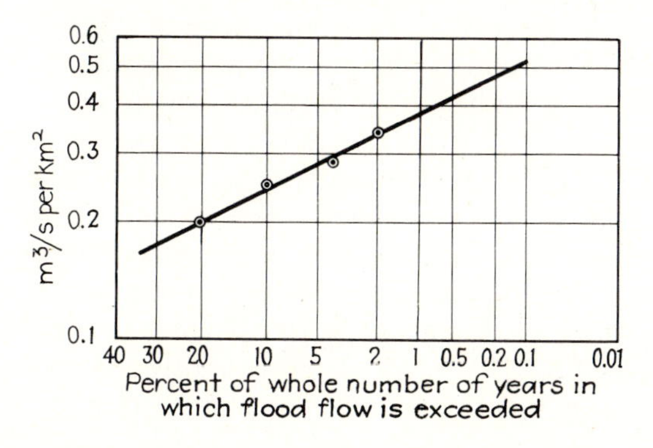

Figure 3-11 Flood flows plotted on probability paper.

exceeded five times, two times, and one time are plotted, with their percents exceeding as 10, 4, and 2, respectively. The data may also be plotted in total flow rather than flow per unit area. If the data conform to the normal probability curve, they will plot in an approximately straight line. This may be extended to allow prediction of extreme flood flows to be expected, say, once in 100 years. In this case the 100-year flood would be 0.38 m^3/s per km^2, while the 200-year flood would be 0.42 m^3/s per km^2. It should be recognized that this does not mean that such a flood will occur only once in 200 years but that in any one year there is only one chance in 200 that it will occur.

When loss of life is not threatened by possible failure of the dam, it may be more economical to design for floods which may be exceeded in shorter periods since the cost of repairs will be less than the compound interest on the excess cost due to greater flood provision. In some populous districts, however, where a dam failure would jeopardize many lives, provision for the probable maximum flood may be made. The probable maximum flood is the runoff resulting from the probable maximum rainfall.

The probable maximum rainfall is estimated based upon maximum water content of the atmosphere and maximum rates of inflow to the study area. Rainfalls exceeding previously estimated "maxima" have occurred. Extrapolation to long recurrence intervals of data based upon 30 to 60 years of observations may be misleading. Hydrologic data do not always follow a known probability distribution, and floods have occurred which indicate this.

3-29 Silting of Reservoirs

A number of impounding reservoirs have had their usefulness decreased or even terminated by accumulations of silt in a few years. This is especially likely where the soil on the watershed is easily erodible, vegetal cover is sparse, and the reservoir is small in relation to the amount of silt-bearing water which passes through it. Therefore, consideration should be given to the amount and character of the material which will be carried by the stream in suspension and in traction or contact with the stream bed. Theoretical approaches to calculating silt loads and sampling methods[13] have been developed but will not be considered here. Silt records for many areas are also available from the U.S. Geological Survey, the Soil Conservation Service, and the U.S. Army Corps of Engineers.

Silt-bearing water is heavier than the clear water of a reservoir and tends to form a density current near the bottom, although this may be modified at the entrance by a delta of larger materials dropped from suspension or from traction. The density current moves along the bottom until it reaches the dam where it is forced upward. If the reservoir is shallow, some of the sediment may escape over the spillway but much will remain and slowly settle out. Silting may be reduced by placing outlet gates in the dam at such points that the density currents may be wasted and thus allow perhaps 20 percent of the silt to escape. Another preventive measure is the control of erosion by all the soil-saving measures which are practical and economically possible. Removal of silt already deposited is usually too expensive, and construction of a new reservoir is more economical.

PROBLEMS

3-1 A 40-year record of rainfalls of 5-hour duration shows that a rainfall of 45 mm was third in order of magnitude. What is the recurrence interval to be expected for this rainfall?

3-2 What is the time of concentration for a concrete street, 150 m long with a slope of 0.01, for rainfall intensities of 10 and 25 mm/h? Assume that C in the rational formula would be 1.0 for this surface.

3-3 Using the values of Prob. 3-2, plot the time of concentration versus rainfall intensity for values of i_e from 2 to 50 mm/h.

3-4 Calculate the time of concentration for a rainfall of 25 mm/h which runs 100 m over closely clipped sod on a slope of 0.005, and then 30 m down a paved asphalt street with a slope of 0.025. Assume that C in the rational formula is equal to 0.5 for the sod, and 0.9 for the asphalt.

3-5 From the following record of average monthly stream flows determine the required reservoir size to provide a uniform flow of 10,000 m^3/day.

Month	Jan.	Feb.	Mar.	April	May	June	July	
Monthly flow (10^6 m^3)	0.18	1.02	1.32	0.51	0.87	0.67	0.19	
Month	Aug.	Sep.	Oct.	Nov.	Dec.	Jan.	Feb.	Mar.
Monthly flow (10^6 m^3)	0.08	0.07	0.04	0.10	0.26	0.20	1.10	1.01

REFERENCES

1. U.S. Environmental Protection Agency, "Water Quality Control Information System (STORET)."
2. U.S. Environmental Data Service, "Climatological Data, National Summary," (Monthly).
3. Baldwin, John L., *Climates of the United States*, U.S. Environmental Data Service, 1973.
4. U.S. Environmental Data Service, "Guide to Standard Weather Summaries and Climatic Services," January, 1973.
5. Sauer, Vernon B., *Floods in Louisiana—Magnitude and Frequency*, 2d ed., United States Department of the Interior, Geological Survey, Baton Rouge, La., 1964.
6. U.S. Geological Survey, "Water Resources Data."
7. U.S. National Weather Service, "Daily River Stages."
8. Gray, Donald M., *Principles of Hydrology*, Water Information Center, Port Washington, New York, 1973.
9. U.S. Soil Conservation Service, "Snow Survey and Water Supply Forecasting," *SCS National Engineering Handbook, Sec* 22, 1972.
10. Work, R. A., H. J. Stockwell, and R. T. Beaumont, "Accuracy of Field Snow Surveys," U.S. Army Corps of Engineers, Cold Regions Research Laboratory, *Tech. Rep.* **163,** 1965.
11. Linlor, William I., "Remote Sensing and Snowpack Measurement," *Journal American Water Works Association*, **66:**553, September, 1974.
12. Petterssen, S., *Introduction to Meteorology*, 2d ed., McGraw-Hill, New York, 1958.
13. Thiessen, A. H., "Precipitation Averages for Large Areas," *Monthly Weather Rev.*, **39:**1082, 1911.
14. Hershfield, D. M., "Some Meteorological Requirements in Watershed Engineering," *Proceedings 3rd Annual Conference, American Water Resources*, 1967.
15. Ward, R. C., *Principles of Hydrology* 2d ed., McGraw-Hill, London, 1975.
16. Linsley, R. K., Jr., M. A. Kohler, and J. L. H. Paulhus, *Hydrology for Engineers* 2d ed., McGraw-Hill, New York, 1975.

17. Cook, H. L., "The Infiltration Approach to the Calculation of Surface Runoff," *Transactions American Geophysical Union* **27**:726, 1946.
18. Gardner, W. R., "Development of Modern Infiltration Theory and Application in Hydrology," *Transactions American Society of Agricultural Engineers*, **10**:379, 1967.
19. *Hydrology Handbook*, American Society of Civil Engineers, Manual of Engineering Practice 28, New York, 1949.
20. Lowry, R. L., and A. F. Johnson, "Consumptive Use of Water in Agriculture," *Transactions American Society of Civil Engineers*, **107**:1252, 1942.
21. Kohler, M. A., and R. K. Linsley, Jr., "Predicting the Runoff from Storm Rainfalls," *U.S. Weather Bureau Research Paper*, **34,** 1951.
22. Hudson, H. E., and R. Hazen, "Droughts and Low Streamflow," *Handbook of Applied Hydrology*, Ven Te Chow, Ed., McGraw-Hill, New York, 1964.
23. Lawler, E. A., "Hydrology of Flow Control—Part II. Flood Routing," *Handbook of Applied Hydrology*, Ven Te Chow, Ed., McGraw-Hill, New York, 1964.
24. Izzard, C. F., "Hydraulics of Runoff from Developed Surfaces," *Proceedings Highway Research Board*, December, 1946.
25. Sherman, L. K., "Streamflow from Rainfall by the Unitgraph Method," *Engineering News Record*, **108:**501, April, 1932.
26. Eagleson, Peter S., *Dynamic Hydrology*, McGraw-Hill, New York, 1970.

CHAPTER
FOUR
GROUNDWATER

4-1

Groundwaters are important sources of water supply which have a number of advantages. They may require no treatment, have uniform temperature throughout the year, are cheaper than impounding reservoirs, and amounts of water available are more certain. Practically speaking they are not affected by drought in the short run. Lowering of water levels in wells sometimes causes alarm and their abandonment, but this may be unnecessary, since modern methods of groundwater investigation will permit a close approximation of groundwater resources for long-time production.

Of the water that falls upon the earth as rain a part percolates into the soil to become subsurface water. Some of the subsurface water is taken up by plants to be transpired through their leaves, a portion is evaporated directly, and some, the hygroscopic water, resists evaporation and is held by the soil. The remainder of the percolating water passes downward under the influence of gravity until it reaches an impervious stratum or *aquiclude*. It then begins to move in a lateral direction toward some outlet. The portion of the earth through which the lateral movement takes place is known as the *zone of saturation*, and its water is the *groundwater*. The water-bearing stratum or formation is an *aquifer*. A *water table* or phreatic surface is the upper surface of the zone of saturation, except when the aquifer is overlaid by an aquiclude. The level of the water table is likely to fluctuate considerably. A long period of dry weather will result in lowering the level, whereas a rain will cause it to rise. During dry periods wells, springs, and streams draw upon the stored groundwater, which is again replenished by percolation during rains.

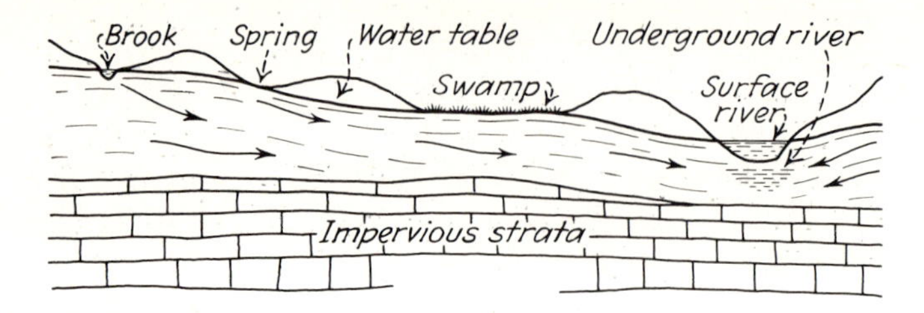

Figure 4-1 Profile of groundwater table.

Groundwater outlets occur whenever the water table intersects the ground surface to form a pond, spring, swamp, or dry-weather surface stream. Figure 4-1 shows a profile of a water table with several outcrops. The most important is the river at the lowest point of the valley.

Above the main water table, supported by aquicludes, small bodies of water are sometimes found. The upper limit of the saturated zone in this case is known as a *perched water table*. Such water accumulations sometimes serve as sources of supply for shallow wells.

Under some conditions groundwater may be fed by streams. A surface stream having a pervious bottom may lose some of its water by percolation into the groundwater if the water table is below the surface of the river. A river may at some points lose water, thus recharging the aquifer, and at other points, where the water table is higher, regain groundwater.

A permeable rock may outcrop somewhere on the earth's surface and dip between aquicludes. Infiltration from rainfall or bodies of surface water will percolate through the aquifer toward some point of discharge between the aquicludes as *confined water*, i.e., under pressure. If the upper aquiclude is tapped by a well or other hole the water will rise into the hole to some point known as the *piezometric* or *static level*. If there is movement of water through the aquifer, the piezometric levels will lower in the direction of flow. The slope of these levels is the *hydraulic gradient* of the confined water. The same term is applied to the slope of the water table in unconfined flow.

4-2 Occurrence of Aquifers

The value of soils or rocks as water bearers depends upon their porosity and the size of their particles. The porosity is a measure of the absorptive power of the material, but if the pores are small the resistance to water movement is so great that it is difficult to collect the water in a well. Accordingly, the compactness of the material and gradation of the sizes of the grains, both of which greatly affect the size of the pores, will have considerable effect. The porosities of common soils and rocks are as follows: sands and gravels of fairly uniform size and moderately compacted, 35 to 40 percent; well-graded and compacted sands and gravels, 24 to 30 percent; sandstone, 4 to 30 percent; chalk, 14 to 45 percent; granite, schist, and gneiss, 0.02 to 2 percent; slate and shale, 0.5 to 8 percent; limestone, 0.5 to 17 percent; clay, 44 to 47 percent; topsoils, 37 to 65 percent.

So far as public water supplies are concerned, sand, gravel, and sandstone are the most important aquifers, in that order. Clay, although highly porous, is so fine grained as to be practically impervious. Limestone and shale are not sufficiently permeable to furnish much water unless they have caverns, fissures, or faults.

Permeable chalk deposits constitute an important source in England. Igneous rocks in general are not water producers, although lava is an exception as it is permeable, and upper portions of other rocks that are decomposed or fissured frequently are sources of springs.

Aquifers may be divided into three classes depending upon their origin and extent.

1. The most important aquifers are extensive and thick formations of porous material laid down in past geological epochs by water and wind. They are fairly uniform, and reliable information regarding them is available or can be easily obtained by a few borings or records of existing wells. Important examples of this class are (*a*) the Tertiary deposits of sand and gravels which underlie the Western plains; (*b*) the sand and gravel deposits of the Eastern coastal plain, a strip bordering on the Atlantic and Gulf coasts 160 to 320 km (100 to 300 mi) wide and extending from Long Island into Texas and up the Mississippi Valley to the Ohio River; (*c*) the sandstones of the eastern part of the Dakotas and parts of Nebraska and Kansas; (*d*) the sandstones of southern Wisconsin, northern Illinois, and eastern Iowa.
2. Old lakes and river beds often contain deposits of sand and gravel. These deposits collect water from surface runoff and groundwater from higher land and conduct it underground in the general direction of surface stream flow. The old stream bed may be approximately parallel to an existing surface stream, or there may be no surface evidence of the aquifer. The sand and gravel may be a uniform deposit, or there may be alternating layers of clay and more permeable materials. Such deposits are far less extensive than aquifers of the first class. Many cities situated beside rivers tap such aquifers by means of comparatively shallow wells and in some cases by filtration galleries.
3. In Northern states deposits of glacial drift or till which were left by the glaciers at the edge of the ice cap are numerous. These sand and gravel deposits are very irregular and occur in old river beds, in thin strata in valleys, and in or along moraines. They are likely to be interspersed with and covered by layers of clay.

4-3 Types of Wells

In the ordinary or *water-table well* the water rises to the height of the saturated material surrounding it, and there is no pressure other than atmospheric upon the water in the surrounding aquifer. An *artesian well* is one in which the water rises above the level at which it is encountered in the aquifer because of pressure in the confined water of the aquifer. A flowing well is an artesian well where the pressure

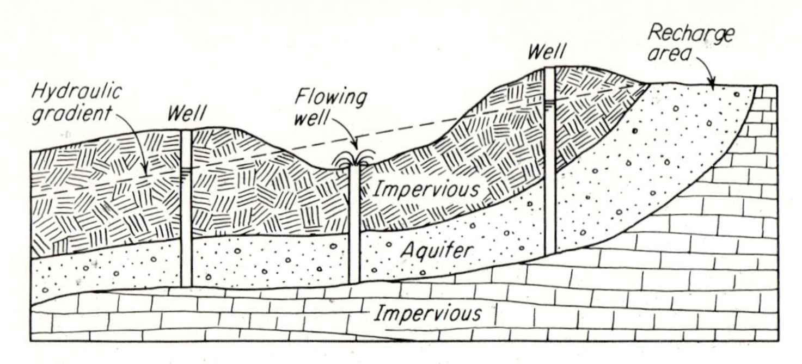

Figure 4-2 Diagram illustrating flowing and nonflowing artesian wells.

raises the water above the casing head. Heavy draft upon the aquifer may so lower the hydraulic gradient that a flowing well will cease to flow. Figure 4-2 illustrates artesian conditions.

4-4 Groundwater Flow

The water table corresponds to the water level of a surface stream. Therefore, as in a surface stream, the water table assumes the slope or hydraulic gradient necessary to cause flow. It can be no lower than the surface of the water into which the aquifer is discharging. In the case of confined flow the piezometric levels will indicate the hydraulic gradient. The greater the hydraulic gradient, the greater will be the velocity and the greater the amount of water carried by the aquifer. Obviously, also, the finer the material, the greater will be the resistance to flow, and the gradient must be steeper if the aquifer is to carry much water. Aquifers are not uniform in fineness, with the result that the hydraulic gradient is not likely to be a straight line. In general, the hydraulic gradient of the shallower groundwaters roughly follows the slope of the ground surface. Under natural conditions hydraulic gradients over 0.2 to 0.4 percent are rarely found. Wells of good yield are often developed in aquifers where the velocity is only 1.5 m/day (5 ft/day). In gravels, velocities of 9 to 18 m/day (30 to 60 ft/day)[1] have been noted in the laboratory with hydraulic gradients of 0.1 to 0.2 percent. Field tests in coarse gravels have shown velocities as high as 122 m/day (400 ft/day) while velocities in water-bearing sandstone may be as low as 15 m/year (50 ft/year).

4-5 Groundwater Hydraulics

The velocity of groundwater flow can be estimated by the use of hydraulic principles or by direct measurements. Direct measurement is limited in value because it can give the velocity only at the point of measurement, whereas aquifers vary considerably in permeability within short distances. Darcy's investigations indicated that flow in water-bearing sands varied directly with the slope of the hydraulic gradient. His conclusions can be expressed as the equation

$$V = k\frac{h}{l} \qquad \text{or} \qquad V = ks \tag{4-1}$$

in which V is the velocity of the moving water, h is the difference in head between two points separated a distance l, s is the slope of the hydraulic gradient, and k is a constant which depends upon the character of the aquifer and is determined experimentally for each type of material.

There has been considerable research with respect to values of k. Hazen related it to effective size of the material, arriving at a formula for velocity through sands which included effective size and a constant depending upon porosity. Permeability is clearly a function of both fluid and particle properties. For water, the permeability may be related to the mechanical properties of the solid medium.[2] At normal temperatures k varies from 3×10^{-6} m/s (10^{-5} ft/s) for coarse sand to 3×10^{-8} m/s (10^{-7} ft/s) for dense clay,[3] yielding apparent velocities of 1.5 m/year (5 ft/year) to 1.5 m/day (5 ft/day).[4] It should be recognized, however, that generalized formulas for the permeability of soils apply only to relatively uniform sands and not to the nonuniform materials normally encountered in aquifers.

Permeability, expressed as the coefficient P, has been determined for various materials in the laboratory. Difficulties of both sampling and reproduction of field conditions in the laboratory have led to large errors when this method is used.

Thiem[5] proposed a field determination of permeability which was tested in the United States by Wenzel.[6] The method consists of drilling observation wells in the cone of depression of a well and noting the drawdowns in the observation wells. The Thiem formula is

$$P = \frac{Q \ln (r_2/r_1)}{2\pi m(d_1 - d_2)} \tag{4-2}$$

in which P is the coefficient of permeability, Q is the pumping rate from the well, r_1 and r_2 are the distances of the two observation wells from the pumped well, d_1 and d_2 are the respective drawdowns† at the observation wells, m is the average thickness of the bed at r_1 and r_2 for water-table conditions, and for artesian conditions the average vertical thickness of the aquifer in any commensurate units.

A value of P so obtained will apply to a large portion of the aquifer. Specifically it is the rate of flow through a unit area of the cross section of the material under a hydraulic gradient of 100 percent.

Darcy's law may then be expressed as

$$Q = Ps \tag{4-3}$$

† Before it is pumped, water stands in a well at what is called its *static level*. When it is pumped the water level falls, and the vertical distance between the new level and the static level is the *drawdown*.

where Q is flow per unit area, and s is the slope of the hydraulic gradient, also

$$V = \frac{Q}{p} = \frac{Ps}{p} \tag{4-4}$$

where V is the average velocity in the voids of the water-bearing material, and p is the porosity of the material expressed as a decimal.

This formula will permit calculation of the amount of water flowing through an aquifer, and its velocity, if P, p, and the slope of the hydraulic gradient are accurately known.

4-6 The Equilibrium Equations for Flow of Water into Wells

When a water-table well is pumped, there is a drawdown of water in its vicinity and the hydraulic gradient assumes a slope toward the well, thus forming an inverted cone of depression with the well at the apex. The base of the cone is called the *circle of influence.* In the equilibrium analysis it is assumed that no further drawdown will occur, and all water passing through the cylinder extending downward from the circle of influence will be pumped out of the well. The formula for flow into the well is obtained by assuming that all the water pumped from the well passes through a succession of cylinders having diameters varying from r, the radius of the well, to R, the radius of the circle of influence, and heights varying from h to H (see Fig. 4-3). The formula is

$$Q = \pi P \frac{H^2 - h^2}{\ln (R/r)} \tag{4-5}$$

in which Q is flow per day, other quantities as shown in Figure 4-3, except P, which is the permeability coefficient as defined in Art. 4-5.

The formula for flow into an artesian well is derived in similar fashion, modified by the fact that the water, in this case, flows through the confined aquifer having a thickness m as indicated in Fig. 4-4. The formula is

$$Q = 2\pi Pm \frac{H - h}{\ln (R/r)} \tag{4-6}$$

The two equations are seldom used since field tests have indicated that they do not always conform to observed conditions. There are several reasons. The value of P is highly uncertain unless the material in the aquifer is tested by actual pumping, and the material may vary widely in closely adjacent portions of the aquifer. The value of R is also uncertain since the circle of influence is never a circle and a long pumping period is necessary, usually years, before the cone of influence reaches an approximate equilibrium far from the well. In the analysis it is further assumed that the well completely penetrates the aquifer and that the drawdown is small compared to the aquifer thickness. Neither may be true in a real well. A number of modifications of the basic Dupuit Technique exist, but these are not so generally applicable as the nonequilibrium analysis below.

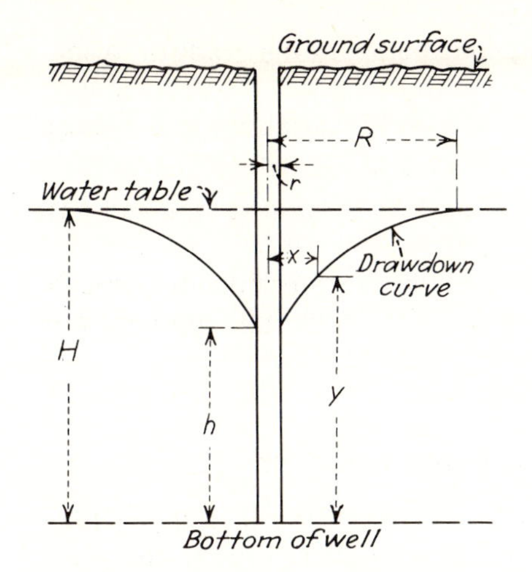

Figure 4-3 Diagram illustrating the hydraulics of a water-table well.

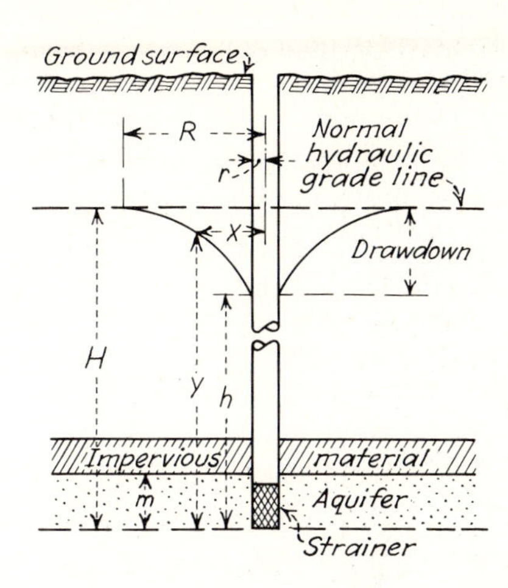

Figure 4-4 Diagram illustrating the hydraulics of an artesian well.

4-7 Specific Capacity

A relationship suggested by the equilibrium formulas is that the output varies directly as the drawdown. This has been found to be approximately true in practice, and it is therefore possible to express the output of a well in terms of flow per unit drawdown, and this is known as its specific capacity. For example, if a well after prolonged pumping produces 600 m^3/h and the drawdown from the static level is 65 m, the specific capacity is 9.2 m^3.

4-8 The Nonequilibrium Analysis

The discussion in Art. 4-6 assumes that equilibrium will occur when water flowing into a fixed and measurable cone of depression is in equilibrium with the water discharged by the well. Theis[7] has developed a method which is now used by the U.S. Geological Survey and others in making groundwater investigations leading toward prediction of long-time yields from aquifers. Certain assumptions are made in the application of the method. These are:

1. In an aquifer before wells are constructed there is equilibrium between the discharge to natural bodies of water from the aquifer and the recharge.
2. Any withdrawal by a well or wells will upset this equilibrium by reducing the natural discharge or adding previously rejected recharge and must necessarily change the hydraulics of the aquifer. This will result in changes of the piezometric surface, which will not become stabilized until a new equilibrium is established.

3. The cone of depression will enlarge during a long period and will finally extend to the boundary of the aquifer. At long distances from the well the effect of such extension may be slight. Also the effects of other wells, variations in pumpage from them, and changes in recharge rates will tend to obscure the depression due to any one well.
4. Much of the water discharged by a well, over an indefinite period, will be from storage in the aquifer. In the case of water-table wells it will be from the dewatered cone of depression. Elasticity of the overlying aquiclude and the compression and compaction of the aquifer accompanying reduction of water pressure will provide water from storage. This retards enlargement of the cone of depression, in addition to furnishing much water in large aquifers. Meinzer[8] held that after several decades of pumping the wells in Dakota sandstone were still drawing on storage and accordingly cones of depression were still enlarging.

4-9 The Nonequilibrium Formula

The formula for drawdown is based upon several assumptions:

1. The aquifer is homogeneous in all directions. This implies that its transmissibility, or ability to convey water, is constant at all points. This quality is expressed quantitatively by its *coefficient of transmissibility*, which is the coefficient of permeability (Art. 4-5), multiplied by the thickness of the aquifer ($T = mP$). The units are volume per day which will move through a strip of the aquifer of unit width with a hydraulic gradient of unity.
2. The water taken from storage in any vertical column of the aquifer having a unit base and height equal to that of the aquifer will be assumed to be constant for each unit lowering of the piezometric surface. This is the *coefficient of storage*, S. In water-table aquifers it is approximately the specific yield, which is the volume of the formation. In artesian aquifers the coefficient of storage represents water released from storage by compression of the aquifer, and it is presumed to be directly proportional to its thickness and also to be constant. It is also assumed that water is released from storage instantaneously upon drawdown. This is not so for water-table conditions, but except for short pumping periods it does not cause serious errors.

The nonequilibrium formula as expressed by Theis for the ideal aquifer is

$$d = \frac{Q}{4\pi T} \int_{u}^{\infty} \frac{e^{-u}}{u}\, du = \frac{Q}{4\pi T} W(u) \tag{4-7}$$

in which d is the drawdown at any point in the vicinity of a well pumped at a uniform rate, Q is the discharge of the well, T is the coefficient of transmissibility, u is $0.25r^2S/Tt$, r is the distance from the pumped well to the point of observation, S is the coefficient of storage of the aquifer expressed as a decimal, and t is the time the well has been pumped, in any commensurate units.

This is an exponential integral which in this usage is written as $W(u)$ and called the "well function of u." Values of the integral may be calculated from

$$W(u) = -0.5772 - \ln u + u - u^2/4 + u^3/18 - u^4/96 + \cdots$$
$$+ (-1)^{n-1}\frac{u^n}{(n)(n!)} + \cdots \tag{4-8}$$

If the coefficient of transmissibility and the coefficient of storage are known, the drawdown can be computed for any time and any point on the cone of depression including the pumped well. If one or more drawdowns are known, the two coefficients can be computed.

Solution of the nonequilibrium formula to obtain the coefficients would be extremely laborious by trial, and a graphical method devised by Theis is used. In applying the method it should be remembered that

$$d = \frac{Q}{4\pi T} W(u) \tag{4-9}$$

and

$$u = \frac{r^2 S}{4Tt} \tag{4-10}$$

First a type curve is plotted on log-log paper with the values of the integral $W(u)$, from Eq. (4-9) plotted against the quantity u.

This general curve may be used in the solution of multiple problems. Values of T and S may be obtained from a series of drawdown observations on one well with times known, or from a line of observation wells. In the first case values of d are plotted against $1/t$ and in the latter case against r^2/t, both on transparent log-log paper and to the same scale. The Q, or discharge from the pumped well, must be known and be constant during the test. Since d is related to r^2/t in the same manner as $W(u)$ is related to u, the curve of the observed data will be similar to the type curve, and solution of the exponential equation is possible by superimposing the curve of the observed data over the type curve and determining coordinates of a matching point. These coordinates can then be used in Eqs. (4-9) and (4-10).

The following example will indicate how Fig. 4-5 is constructed and used. An artesian well was pumped at a uniform rate of 1.5 m^3/min for 30 days. Observations of drawdown were made in an unused test hole 300 m away. The following data were obtained.

r, m	t, days	r^2/t	d, m
300	1	9×10^4	0.13
	2	4.5×10^4	0.53
	3	3×10^4	0.95
	5	1.8×10^4	1.60
	10	9×10^3	2.75
	20	4.5×10^3	4.10
	30	3×10^3	4.80

A type curve is plotted on log-log paper using the values of $W(u)$ and u from Eq. (4-8). Values of d and r^2/t are also plotted on log-log paper. The curves are plotted on separate sheets, to the same scale, and superposed so that a match point can be found in the region of which the curves nearly coincide when their coordinate axes are parallel. The coordinates of the match point are marked on both curves yielding corresponding values of $W(u)$, u, d, and r^2/t. In this case the values obtained (from Fig. 4-5) are

$$u = 6 \times 10^{-1} \qquad r^2/t = 3 \times 10^4$$

$$W(u) = 0.45 \qquad d = 0.95$$

These values are substituted in Eqs. (4-9) and (4-10) to yield

$$0.45 \frac{1.5 \times 1440}{4\pi T} = 0.95, \qquad T = 81.36 \text{ m}^3\text{/m per day}$$

$$\frac{S}{4 \times 81.36}(3 \times 10^4) = 6 \times 10^{-1}, \qquad S = 0.0065$$

The curve of the field observation shown in the figure may be extended in the same shape as the type curve. This will give the drawdowns for further values of r^2/t, so long as Q is 1.5 m^3/min.

Other values of T and S can be obtained in the same manner from records of time and drawdown at the pumped well and at other observation wells, and these can be averaged. The averages will be more representative of the aquifer.

When T and S for an aquifer are known, it is possible to determine the drawdown after any period of pumping and any pumping rate. For example: It is desired to determine the drawdown in a well, located 1200 m from the pumped well after 2 years of pumping at the rate of 2 m^3/min. From Eq. (4-10)

$$u = \frac{r^2 S}{4Tt} = \frac{1200^2(0.0065)}{4(81.4)(365)(2)} = 3.94 \times 10^{-2}$$

$$W(u) = -0.5772 - \ln 3.94 \times 10^{-2} + 3.94 \times 10^{-2} \cdots = 2.69$$

$$d = \frac{Q}{4\pi T} W(u) = \frac{2 \times 1440}{4\pi(81.4)} \times 2.69 = 7.58 \text{ m}$$

If the drawdown in the pumped well is required after pumping 1 year at 2 m^3/min, in this case r is the radius of the well, assumed to be 0.15 m. Substituting in Eq. (4-10),

$$u = \frac{0.15^2(0.0065)}{4(81.4)(365)(2)} = 6.16 \times 10^{-10}$$

$$W(u) = 20.63$$

$$d = \frac{Q}{4\pi T} W(u) = \frac{2 \times 1440}{4\pi(81.4)} \times 20.63 = 58.1 \text{ m}$$

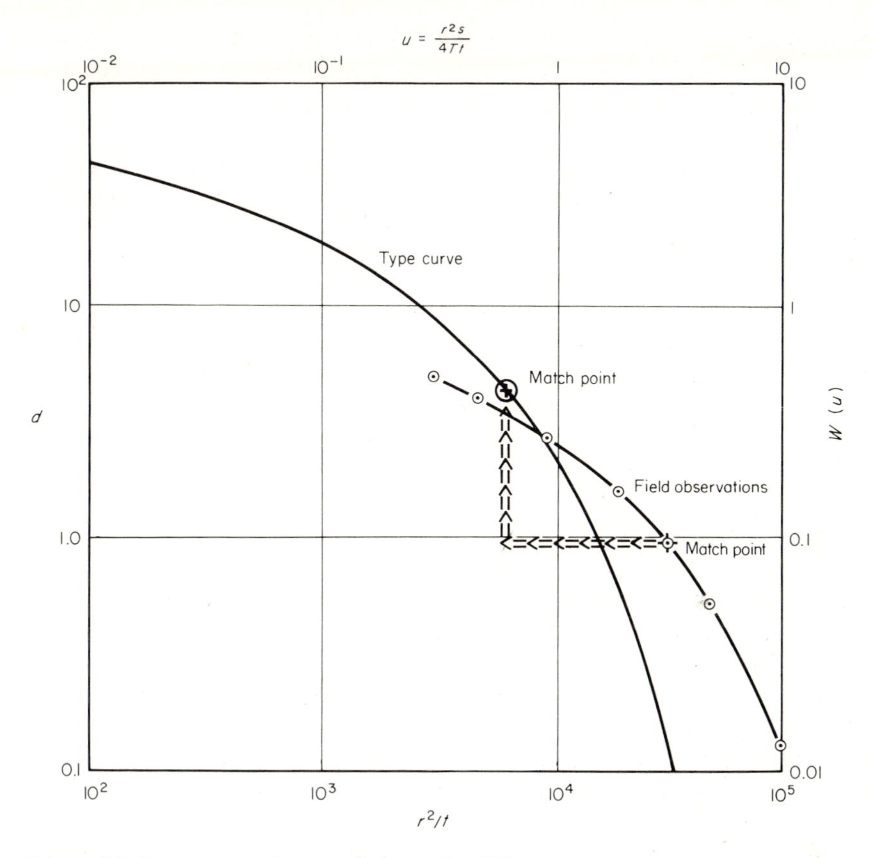

Figure 4-5 Type curve and curve of observed well data.

This is a large drawdown and might indicate that such a pumping rate should not be attempted.

4-10 Modified Nonequilibrium Formula

The nonequilibrium formula may be modified as shown by Jacob.[9] For small values of u, that is, for large values of t, $W(u)$ may be approximated by

$$W(u) = -0.5772 - \ln u = -0.5772 - \ln \frac{r^2 S}{4Tt} \tag{4-11}$$

Therefore

$$d = \frac{Q}{4\pi T} W(u) = \frac{Q}{4\pi T}\left[-0.5772 - \ln \frac{r^2 S}{4Tt}\right]$$

The drawdown is thus seen to be a linear function of ln t, hence a plot of drawdown versus ln time will yield a straight line with slope $Q/4\pi T$ and an intercept at $d = 0$, t_0, defined by

$$-0.5772 - \ln \frac{r^2 S}{4Tt_0} = 0 \qquad (4\text{-}12)$$

If the slope is measured over one log cycle of time (that is, $t_2/t_1 = e$), the slope will equal the change in drawdown, Δh. Thus,

$$\Delta h = Q/4\pi T, \text{ and } T = Q/4\pi \, \Delta h \qquad (4\text{-}13)$$

Similarly, from Eq. (4-12)

$$S = 2.25 T t_0 / r^2 \qquad (4\text{-}14)$$

in which S is the coefficient of storage as defined in Art. 4-9, T is the coefficient of transmissibility, r is the distance to the observation well, and t_0 is the intercept of the straight line of time against drawdown on the line of zero drawdown.

Jacob's formulas do not apply to periods immediately after pumping starts, and this will be indicated by the fact that the graph of the drawdown against time will not be a straight line at first. As indicated above, the straight line part of the curve must be extended to obtain t_0. The time required before the data are applicable will be relatively short for artesian conditions, an hour or less, while for water-table aquifers, because of the time required for voids to drain, 12 hours or more may be needed.

4-11 Interferences in Aquifers

Obviously if several wells are drilled into the same aquifer there will be some interference in output and drawdown, the extent depending primarily upon the distances between the wells and secondarily upon their relative location. There will be less interference if the wells are located far apart or, when they must be close together, if they are located in a line which is perpendicular to the direction of the underground flow. The effect of interference upon drawdowns can be obtained by using the appropriate equations for the individual wells, and the resulting piezometric levels will be sums of the calculated drawdowns determined at any location.

Flow in aquifers may be interfered with or modified by faults, surface rivers which produce recharge in certain areas, geological unconformities, etc. Their effects are determined by the method of images, or imaginary recharge wells or producing wells. The effects of these images are then subtracted or added to the theoretical effects of the actual producing well to obtain the expected piezometric levels. The details of this method are discussed by Ferris and Todd.[10,4]

4-12 Appraisal of Groundwater Resources

Consideration of the principles above makes it apparent that any investigation of the groundwater resources of an area must be based upon observation of actual yield from one or more producing wells or test wells. The records of existing wells will be valuable, particularly if variations of drawdown and static level are available over a long period. If a controlled test can be made with drawdowns related to constant yield over a period of time, the formulas of Thiem or Theis can be applied, and from these, one can determine the long-term yields and whether the groundwater resources of the district are diminishing or existing withdrawals can be maintained. If records are kept continuously, they can be scrutinized periodically and estimates modified accordingly.

An important factor which may affect the groundwater flow where two or more aquifers are tapped by a single well is the possibility that water from the aquifer under higher pressure will discharge into that with lower pressure when the well is not in use. The static heads of each aquifer should be determined to check this and its possible effect upon pumpage or yield.

An appraisal should also cover seasonal changes of static level which can sometimes be noted in relation to rainfall and variations of recharge. The possibilities of artificial recharge methods can also be noted, particularly in water-table aquifers.

4-13 Recharge of Aquifers

Return of water to aquifers has been done for several reasons:

1. Aquifers have been recharged to conserve water. Some states have legislation requiring that well water employed for cooling be returned to the ground, and this is done by pumping the water into recharge wells. In California, floodwater or water drawn from reservoirs has been spread on sandy or gravelly natural or artificial basin areas to refill depleted aquifers.[11] In some cases turbid floodwaters are excluded. Such projects should be preceded by studies to assure that the recharge water will go where it is required.
2. Saltwater intrusion from the ocean into an aquifer[12] has been prevented by recharging it with freshwater to set up a barrier with a high piezometric level. Treated sewage can also be used for the same purpose.[13]
3. Groundwater recharge through land disposal of treated sewage is an alternative which is routinely considered in wastewater facility planning.[14] A variety of industrial wastes and oil field brines may be similarly handled. In such cases the recharge may be incidental to wastewater disposal.

Recharge wells used in connection with air cooling not only conserve water but allow the warmed water to be returned and replaced with cooler water from the ground. The recharge well must have its outlet strainer in permeable material so that a cone of influence is built up above the water table. Should this cone

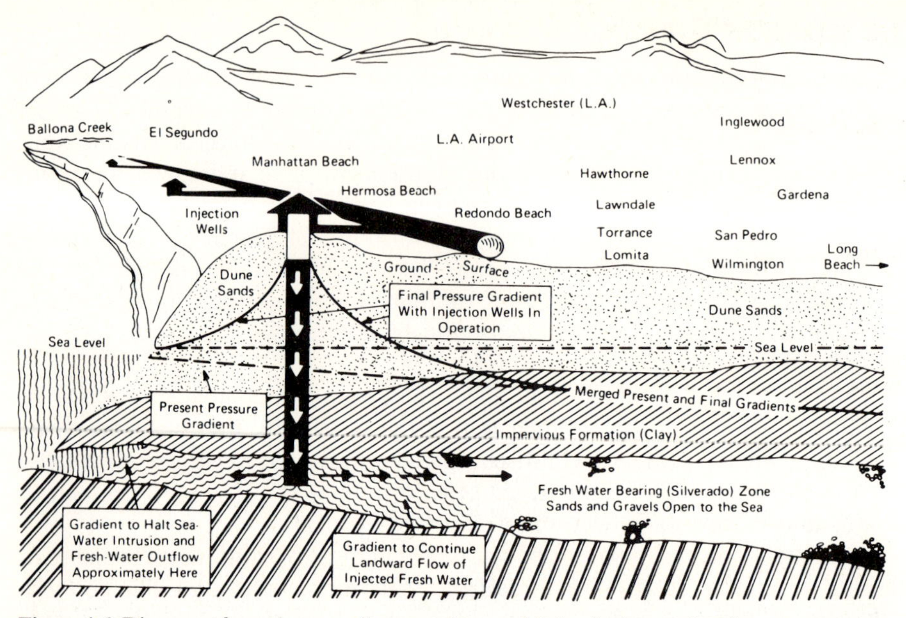

Figure 4-6 Diagram of a recharge well operated to establish a freshwater barrier to encroachment by seawater. *(Reprinted from Journal American Water Works Association* **66** *by permission of the Association. Copyright 1974, American Water Works Association, Inc. 6666 W. Quincy Avenue, Denver, Colorado 80235).*

overlay in part the cone of depression of the supply well, some warm water will enter the latter. Recharge wells should be constructed and developed as carefully as supply wells, and all possible precautions should be taken to exclude sand and silt, or the screen will clog and require cleaning. Chlorination of the water or other treatments may be necessary to prevent slimes or other organic growths from obstructing the screens. If the water contains iron or other precipitating constituents, there may be clogging of the aquifer or screen as a result of oxidation, or pressure and temperature changes. If chemical wastes are disposed of underground, there should be no chemicals present that will precipitate or react with the aquifer to reduce its porosity.

Recharge wells must set up a pressure cone of influence above the water table or piezometric level of a pressure aquifer by natural head, or more likely by pumping. This will force the injected water out of the well and into the aquifer. Figure 4-6 is a diagram which indicates recharging near Los Angeles, Calif. As a result of heavy pumping of inland wells the piezometric level of the aquifer had dropped, and saltwater moved into and through the aquifer, necessitating the abandoning of some wells. Operation of a recharge well resulted in a pressure cone, and a barrier of injected freshwater was established as shown.[12]

Recharging to water in aquifers not under pressure is also done from seepage pits. In this case flow of saltwater from the sea is prevented by formation of a mound of freshwater.

WELL CONSTRUCTION

4-14 Shallow Wells

Wells under 30 m (100 ft) in depth are classed as shallow wells. They include a wide variety of types, ranging from the dug wells used on farms to types used for municipal supplies. Shallow wells are open to the following objections to use for public supplies: (*a*) The water table of shallow groundwater which they tap is likely to fluctuate considerably, thereby making the yield uncertain. (*b*) The municipal wells themselves will lower the water table and may affect private wells, resulting in damage suits. (*c*) Possible infiltration of seawater may occur, if the wells are near the ocean, unless they are very carefully operated. (*d*) Sanitary quality of the water is likely to be poor.

Dug wells are those excavated by picks and shovels or excavating machinery. They are generally more than 0.5 m (1.5 ft) in diameter and not over 15 m (50 ft) in depth. Lining or casing is usually of concrete or brick. Brick with cement mortar is sometimes used for a distance of 3 m (10 ft) below the ground surface with dry joints below. The wells should also be covered to prevent contamination from entering. The pumps should be provided with a net positive suction head so that priming will be unnecessary. Dug wells are ocasionally used by small towns for public supply. In this case they are likely to be of large diameter and extend into shallow strata of sand or gravel. Wells of this type have the advantage of furnishing considerable storage and will therefore allow a heavy draft for a short period.

Shallow wells of small diameter are also driven and bored. A driven well consists of a pipe casing having on its end a driving point, slightly greater in diameter than the casing. Above the driving point are perforations or a screen through which water enters the casing. The driving is done by means of mauls or a dropping weight. Since the screen may become clogged with clay or damaged in the driving, a precaution sometimes taken is to arrange the screen, casing, and drive point so that the casing can be raised until it clears the screen. Driven wells are usually 25 to 75 mm (1 to 3 in) in diameter and are used only in unconsolidated materials. For protection against frost and contamination an outer casing is sometimes placed for 3 m (10 ft) below the surface.

Bored wells are 25 to 900 mm (1 to 36 in) in diameter. They are also constructed in unconsolidated materials, the boring being accomplished by augers which are raised to the surface for clearing when filled. Casing for the well is placed by driving after the auger has reached the water-bearing sand. In sandy material, clay may be dumped into the hole to make the sand sticky, or the auger may be discarded for the sand bucket.

4-15 Deep Wells

Cities using groundwater usually depend upon deep wells. These have the advantage of tapping deep and extensive aquifers, a condition that prevents quick fluctuations in the water-table or piezometric level and results in a large and uniform yield. Such deep water is likely to be of good sanitary quality unless it is contaminated by seepage into the aquifer from caverns or fissures in overlying rock. The disadvantages are the cost of the wells and the fact that long underground travel of the water may have given it a heavy load of dissolved solids which may make it hard, corrosive, or otherwise undesirable.

Deep wells are constructed by several methods of which the following are most important: the standard; California, or stovepipe; rotary; jetting; and core drill.

Standard Method. This method is suitable for any material. It requires that a derrick be set up over the well so that casing and strings of tools can be raised and lowered. The drilling is accomplished by a bit suspended by a rope, which is attached to a walking beam or lever. The walking beam is raised and lowered, thus causing the drill bit to strike a blow at the bottom of the hole. The spring of the rope causes a rebound of the bit and thereby prevents

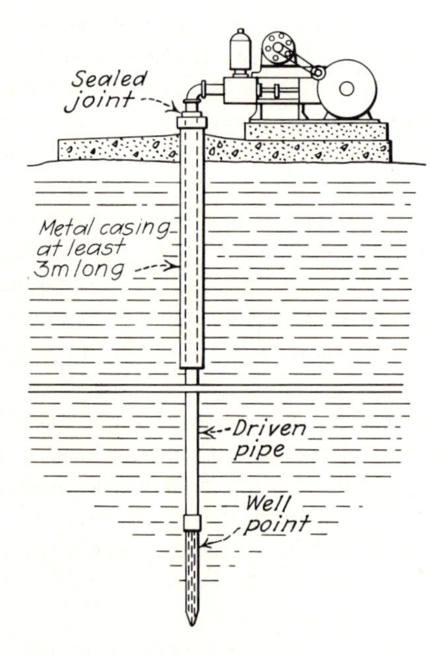

Figure 4-7 A driven well.

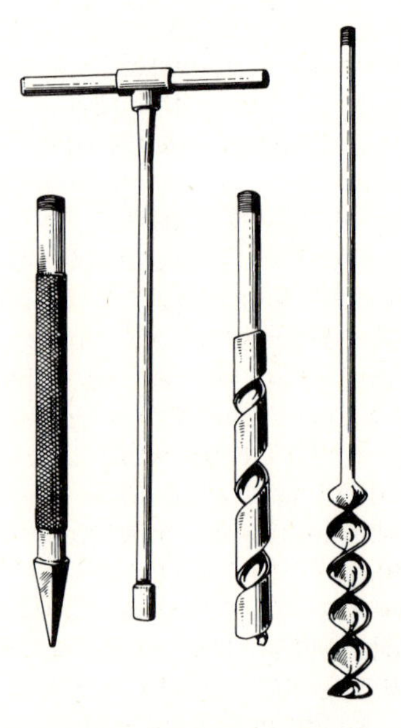

Figure 4-8 Drive point and screen. Auger bits for bored wells. *(U.S. Geol. Survey Water Supply paper 257.)*

jamming in the hole. When required, weight is added to the string of tools by means of a steel sinker bar. As the drilling proceeds, the cable is lengthened by running out the temper screw a few turns at a time. Wells are sunk first to 30 m (100 ft) by spudding, a process that is necessary before the rope is long enough to allow use of the walking beam. While spudding, the blows of the drill bit are delivered by suspending the tools from the top of the derrick and alternately raising them and allowing them to drop.

After the temper screw has been run out its full length, the tools are withdrawn, and the loose material in the hole is bailed out by means of the sand bucket. This is a long bucket having a flap valve at the bottom. When dropped in the well it fills with cuttings, but the valve closes as it is raised for emptying. After bailing, a sharpened bit may be substituted, the rope is lengthened, and drilling is resumed.

Where caving material is encountered, it is usual to sink the casing as fast as the drilling proceeds. A steel shoe is screwed or shrunk to the lower end of the casing so that driving can be accomplished without injury. A driving head is placed on the upper end of the casing to prevent damage at that point. A modification, the *hollow-rod method* employs hollow steel rods through which the broken material, mixed with water, can rise to the surface. As the rods fall, the water and broken material rise in the hollows, check valves preventing their sinking when the tools are raised.

California, or Stovepipe, Method. This method, first used in California, is useful in unconsolidated alluvial deposits. It consists of pushing short lengths of steel casing into the earth by means of hydraulic jacks. Two diameters are used, one size just slipping within the other, the joints of the outer pipe falling midway between those of the inner casings. The inner and outer casings are united by indenting them with a pick, and successive lengths of pipe are added and sunk.

As sinking progresses, the casing is kept even or ahead of the excavation, the material inside the casing being removed by a sand bucket or clamshell. Boulders are broken by a drill or worked to one side. When the required depth has been reached, a tool is lowered which cuts vertical slots in the casing at the selected aquifers. Wells up to 1 m (40 in) in diameter have been constructed by this method.

Rotary Method. This method is used for unconsolidated materials of fine texture. It also requires a derrick, and drilling is accomplished by rotating the string of casing, by means of a rotary table on the drilling floor. The casing is equipped with a cutting shoe at the lower end. Water is pumped into the well to rise between the side of the hole and the casing carrying the loosened material with it. The drill water is used over and over again, the sand and heavier particles settling out in a slush pit.

In firmer materials either a fishtail bit or a diamond-shaped bit is rotated in the well, to cut and loosen the material to a hole of the required diameter. Water or a mixture of water and mud, known as drilling mud, is pumped through the hollow drill stem, and it, together with the loosened material, is washed up through the casing. The mud runs to a sump in which the cuttings settle out while the liquid is pumped into the well for re-use. The casing is sunk as excavation proceeds, or the drilling mud may give enough stability to the sides of the hole so that the hole may be completed before the casing is placed.

Jetting Method. Jetting is accomplished by means of a drill pipe with a nozzle or drill bit on the lower end. Water is pumped down through the drill pipe and escapes through the drill bit, which is raised, lowered, and turned slowly. The stream of water loosens materials and carries the finer portion out of the hole. The casing is lowered by its own weight or driven, as necessary. It may accompany the excavation or, in clay and other fine-textured materials, may be inserted after the well is jetted down to full depth.

Core Drilling. This method is sometimes used for drilling in hard materials. A core drill consists of a ring with diamond or steel teeth. The ring is attached to a drill rod and rotated, water being used to remove cuttings. As the cutter advances, a core rises inside the ring which must be broken off from time to time and removed. The core is of value in that it provides a representative sample of the material encountered.

4-16 Well Screens

In unconsolidated materials, that portion of the casing which is in the one or more aquifers to be tapped must be replaced with some type of screen or strainer which will allow water to enter but exclude sand or gravel. The size of the openings will depend upon the character of the material encountered, the outer widths in practice varying from 0.01 mm (0.004 in) upward. The openings should always be flared toward the inside to prevent packing of fine particles in them. In coarse gravel, large perforations in the casing may suffice. Usually, however, more elaborate strainers are needed. They are obtainable from various commercial firms and vary in design. Some are of brass tubing with vertical or horizontal slots. Others consist of a frame of metal rods or perforated pipe wound around with wire, the openings between each round of wire providing the slots. Since wells sometimes fail because of corrosion of screens, a corrosion-resistant metal should be specified. Monel metal (70 percent nickel, 30 percent copper), Everdur metal (copper 96 percent, silicon 3 percent, manganese 1 percent), silicon red brass, and Anaconda red brass, in the order named, are to be recommended. Concrete strainers, which are used in connection with concrete casings, are also obtainable. Figure 4-10 illustrates several types of commercial metal strainers.

The net total area of the strainer openings should be such that the entering velocity of the water will not exceed 50 to 100 mm/s (2 to 4 in/s). This precaution will tend to prevent clogging of the strainer by sand which would be carried by greater velocities. The openings should be of such size that 20 to 30 percent of the material in the aquifer will be larger unless it is very uniform in size, in which case

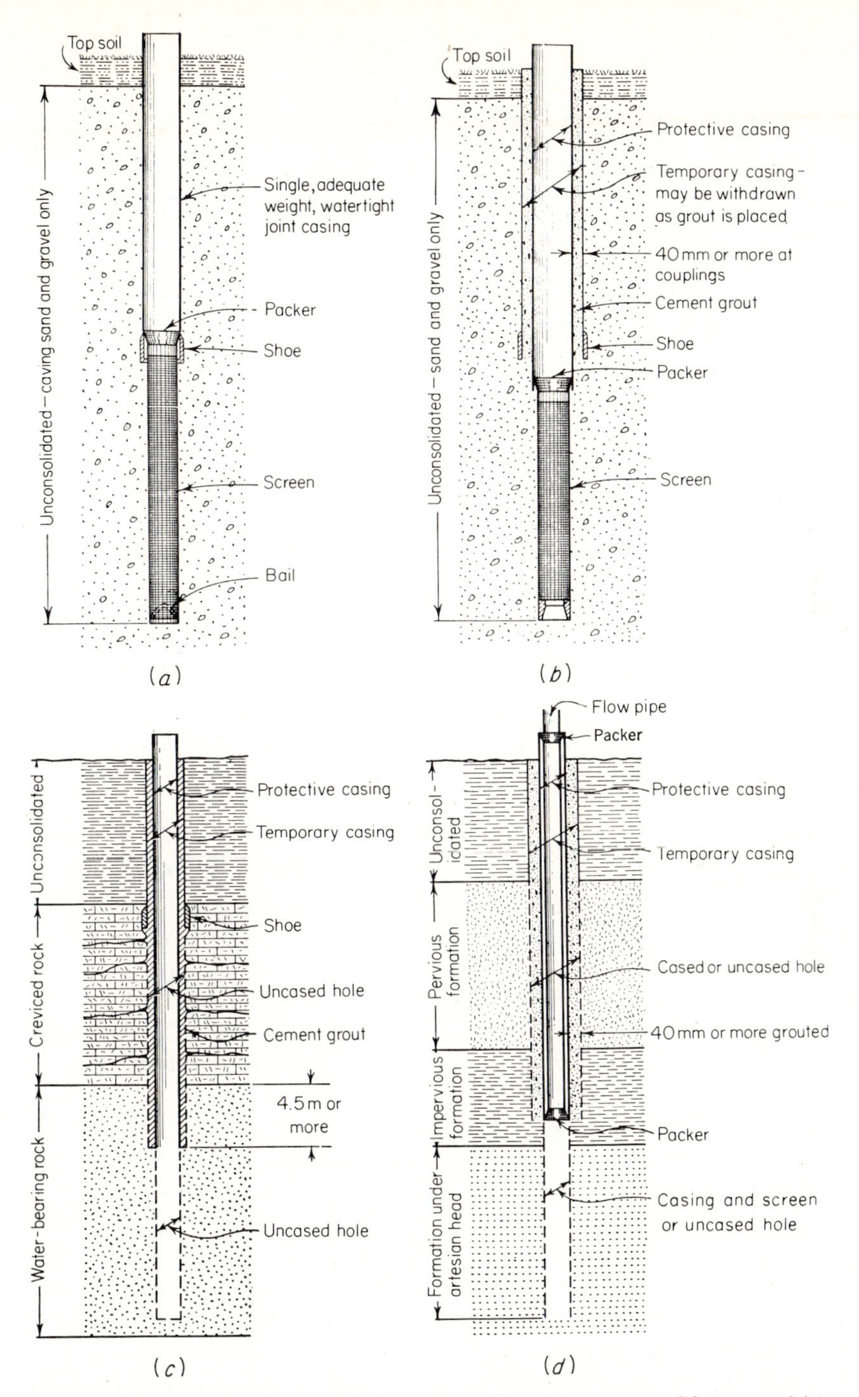

Figure 4-9 Some types of wells. (*a*) For unconsolidated formations. (*b*) As type (*a*) but grouted for protection against pollution. (*c*) Well in consolidated formation overlain by creviced rock. (*d*) Well, flowing or artesian type. Prevents loss of water to previous formation. Casing and screen in aquifer required if erosion is possible. The inside or eduction pipe may be replaced. *(From AWWA A100, Standard Specification for Deep Wells, by permission of the American Water Works Association.)*

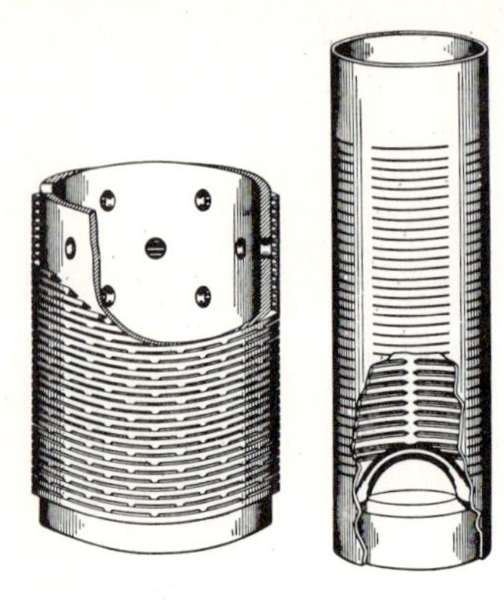

Figure 4-10 Two types of metal well screens.

Figure 4-11 Well screen in aquifer showing removal of fine material by development. *(Courtesy Edward E. Johnson Inc.)*

a finer screen is used. Screens should not be placed so that the level of drawdown will be below the top of the screen as this permits contact with air and encourages corrosion.

4-17 Development of Wells

As a new well is pumped the fine sand in the vicinity of the screen will move toward it and pass through, sometimes in such amounts that it must be removed by bailing. Overpumping, i.e., at a rate greater than planned for normal operation, is sometimes practiced to facilitate this process, which is characterized by increased yield per unit drawdown. Skillful drillers, however, realize that more positive measures are required to remove fine sand from the formation and reduce friction near the screen. It is recognized that "bridging" of fine particles occurs over the screen openings and between large particles and that a reversed motion of the water is needed to break up the bridges and promote flow of fine particles to and through the screen. Surging is done by raising and lowering a plunger which on the downstroke forces water outward through the screen. Air pressure is used for the same effect. A turbine pump without a foot valve may be operated and then stopped, and the water in the pump column will cause the backflow. Well specifications usually require development of new wells. The work must be skillfully done, however, for water may be forced up between the well casing and the hole and perhaps ruin the well.

4-18 The Gravel-packed Well

In aquifers that contain some coarse material and whose screen openings are of proper size, natural gravel-packing results during development. The fine material is pumped out leaving an envelope of highly permeable coarse material around the screen. Such a well should not pump sand after it is in normal operation. If the aquifer is of very fine, uniform material, very fine screen openings will be needed and partial clogging may occur together with sand pumping. Greater yield and elimination of sand pumping can be accomplished by artificial gravel-packing. This is done by drilling the well somewhat larger than the screen and filling the annular space around it with fine gravel placed preferably through a string of tubing starting at the bottom of the casing screen, or casing, if the lining is to be withdrawn to the top of the screen. In some cases the gravel is fed into the annular space between hole and casing at the top of the well.

The thickness of the gravel and its size are important. Too great thickness prevents proper development of the well, since surging will not effectively reach the sand of the aquifer. A 75-mm (3-in) thickness is as small as can be properly placed. Recommended thicknesses are 75 to 225 mm (3 to 9 in). Coarse gravels may permit sand pumping and may clog. A fine gravel with a low uniformity coefficient—say, 1.5 to 1.7—is desirable.

Gravel-packed wells are not considered necessary by some authorities,[15] where the natural water-bearing formation has an effective size of more than 0.25 mm and uniformity coefficient of more than 2.0.

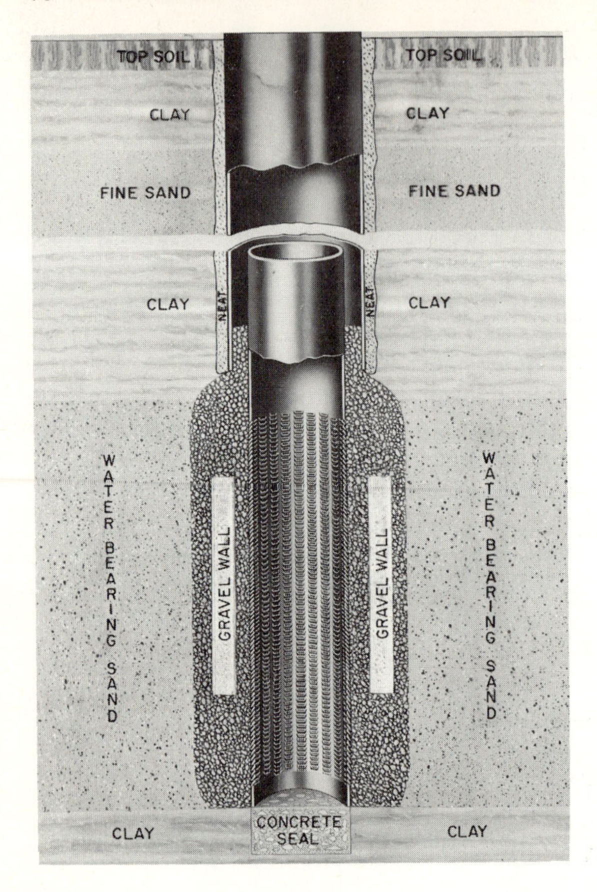

Figure 4-12 Under-reamed type gravel-wall well, showing shutter-type screen and cement grout between casing and hole. *(Courtesy Layne and Bowler, Inc.)*

4-19 The Ranney Method

Figure 4-13 illustrates this method of obtaining water from aquifers which have a direct connection with a surface water source. A caisson is sunk into a water-bearing sand or gravel adjacent to a stream or lake. The bottom of the caisson is then plugged, and horizontal screens are projected into the aquifer through ports. The screens are placed by combined pumping and backwashing, a large part of the fine material being removed in the process to develop a natural gravel wall around the screen. This results in a number of horizontal collectors which approximate a filter gallery of considerable length and induce infiltration from the surface water source. The collectors may be arranged as shown in the figure, or all or some may be extended under the stream bed. This arrangement will also intercept water which has entered the aquifer by infiltration on the landward side of the stream. The great length of screen allows a high production of water with small loss of head. The amount of water available will be related principally to the stream rather than to fluctuations in the piezometric levels of the ordinary aquifer.

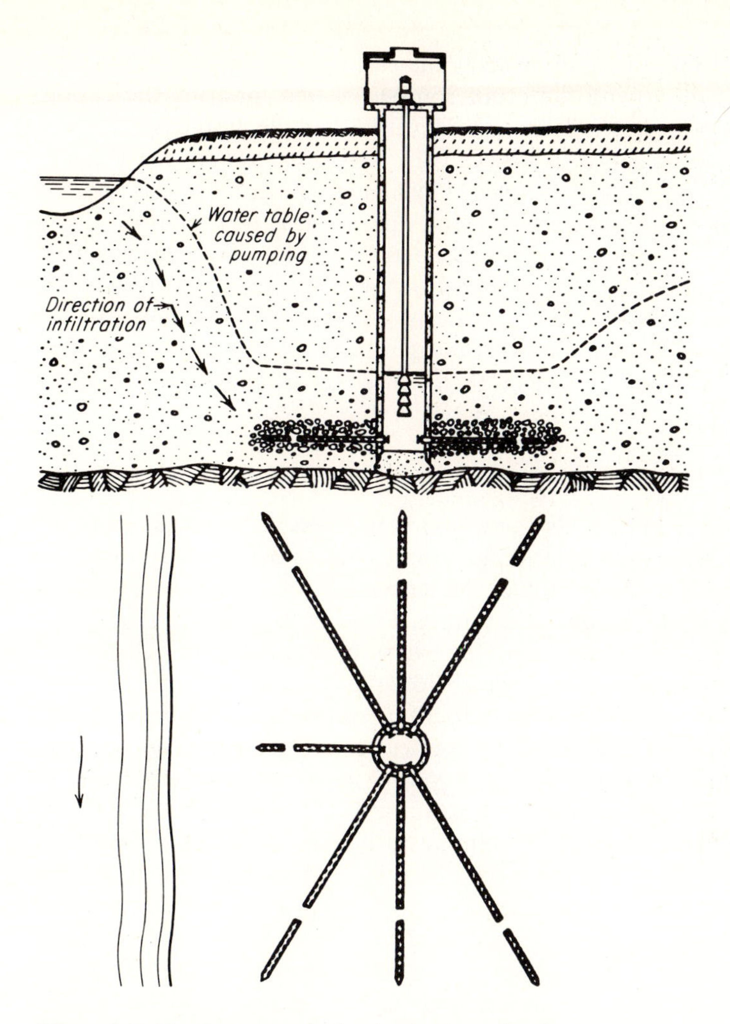

Figure 4-13 The Ranney method, using horizontal wells.

This method is used by a number of cities and many industrial plants which are favorably located. River movement tends to prevent silt from clogging the intake area on the river bottom. Output will depend upon the permeability of the aquifer and water temperature. Filtration through the sand may make treatment of the water unnecessary.

4-20 Cementing Wells

This consists of placing cement grout between the well casing and the hole from the surface to the level where the casing enters the first impervious stratum. It has the advantage of preventing the seepage of surface or shallow groundwater down to the water-bearing formation and protects the casing from corrosion from the outside. One method of placing the grout consists of first lowering the casing, which has guides welded to the outside to center it in the hole, to the point where grouting is to start. It has a cap with a check valve in it at its lower end. The lower portion of the hole, if already drilled, is closed by a plug. The grout pipe is lowered in the casing to connect with the check valve, and grout is then pumped down until it rises to the surface in the annular space between casing and hole, the check valve preventing any backflow of grout. The grout pipe is immediately unscrewed and raised a few millimeters, and the casing is lowered or forced down to the bottom of the hole. Water is then flushed through the grout pipe and up through the casing to clear both pipes. The well is then cleared by drilling out the plug and check valve and any grout remaining in the well.[15]

4-21 Groundwater Projects

A city or industry sometimes finds it necessary to make a thorough study of groundwater possibilities before drilling supply wells. Before the supply is developed investigations should be made to determine the extent and nature of the aquifers. This may include geophysical testing. Existing geological information and maps may indicate in a general way the existence and depth of well-defined aquifers. Test holes should always be drilled before plans for well construction and purchase of pumps are completed. These should be sufficient in number to establish the depth, thickness, and nature of the aquifers at and near the proposed area which is to be developed. Prospect holes 50 mm (2 in) in diameter are drilled to get general information and samples of materials. Test holes are made 100 mm (4 in) in diameter and are pumped to determine output. Representative samples of the natural formations are obtained from the holes and indexed for depth and location. Water-table levels encountered during drilling are also recorded and water samples are taken from each aquifer. The record of formations passed through during drilling and whether they are water-bearing or not is known as the "log" of the well. The driller's records are likely to be inaccurate and they should be checked by an electric log. This charts the resistance of the various formations, and when interpreted by qualified persons determines the exact location of various formations, and whether they contain water. Electric logging is done by firms which specialize in this work.

Analyses of the water-bearing formations and of the water are made by qualified persons. A groundwater map may be made showing all pertinent information. Thereafter the number, location, and type of wells can be decided. In connection with such a project the information already available from the U.S. Geological Survey and the advice of its engineers and geologists should be obtained.

4-22 Well Specifications

Since wells are usually constructed by contract, it is advisable that careful specifications be written to insure a satisfactory well.

Before specifications are written there should be considerable investigation made by the engineer or water superintendent. Some estimate should be made of the well size; and by means of methods outlined in this chapter or by consulting records of wells already tapping the same aquifer, the yield to be expected may be estimated. If test holes have been made it will be possible to specify the type of well, the casing diameters and lengths, and length of screen. There should also be investigation of proximity of sewers and other structures which may affect the safety of the well and its water. The contract should cover the following:

1. The contractor is often required to guarantee an output from the well. To this it is advisable to add a clause giving a stated bonus for each unit output above the guarantee and a penalty for each unit under the guarantee, with a limit of bonus and penalty.
2. The well should not vary from the vertical in excess of the smallest diameter of that part of the well being tested per 30 m (100 ft) of length.
3. Proper alignment of the well should be guaranteed, and a test of alignment required. A 10 m (30 ft) dummy having an outside diameter 12 mm (0.5 in) less than the casing shall move freely through the well to the lowest anticipated pump setting.
4. The contractor should be required to state the trade name of the screen that he proposes to use. The character of the metal used and the proper beveling of slots should be specified. A separate price per unit length of screen placed should be made by the contractor.
5. The type and weight of the casing should be specified. Where casings change in size, an overlap should be specified, and a packer required to prevent leakage of undesirable water into the well or loss of water into the ground. Such packers are usually collars expanded to fill the annular opening. They are of importance not only to prevent leakage but also to prevent shallow ground or surface water, which may be contaminated, from entering the well. Seepage downward from the surface outside the casing should also be prevented by requiring a seal between the casing and the first rock stratum encountered. If no rock strata are passed, puddled clay should be placed between the casing and hole. A seal or packer will usually be needed where the strainer joins the casing. The requirement of cementing should be considered.
6. Sanitary requirements should be recognized by closing the top of the well so that no surface water can enter (Art. 8-32). The upper terminal of the well should extend at least 150 mm (6 in) and preferably 300 mm (12 in) above the pump house floor or any similar cover. The well is cleaned of all debris, grease, and mud, using a brush if necessary. Disinfection of the well is also required, the following procedure being recommended. A chlorine solution of at least 50 mg/l strength is made, and enough is poured into the well to fill it to the static level. It is thoroughly agitated and allowed to stand for at least 2 hours. Flushing water is then introduced to the bottom of the well. As the chlorine solution rises it will come in contact with the upper casing and disinfect it, particularly if the water is allowed to rise and fall several times. After flushing,

the well is pumped until there is no chlorine odor in the water. Before the well is accepted, a water sample should show freedom from coliform bacteria (Art. 8-7).

7. The contractor should be required to keep a record of the strata encountered and the elevation of each. This record, together with properly labeled samples of excavated materials, taken at 3 m (10 ft) intervals, should be turned over to the waterworks authorities.
8. The method of developing and testing the well for yield and drawdown which the contractor intends to use should be described by him, including pump details and methods of measuring flow. Minimum duration of the final test may be specified. This portion of the well's cost may be a separate item, in which case prices quoted should include equipment, materials, and labor to be used.

4-23 Well Troubles

Failure of wells or great reduction in their yield is not uncommon. This may be due to a draft greater than the percolation of water at the recharge area of the aquifer. More often it is caused by pumpage greater than P (Art. 4-5) in the vicinity of the well. Where this is the case there is an excessive drawdown in the neighborhood of the well in order to move water toward it. In water-table wells the water table will, of course, be lowered. Lowering the pumps may increase the yield for a time, but the permanent remedy is to reduce pumpage.

The casing or screen may cave in or collapse, partially or completely blocking the well. This trouble will necessitate replacement of the casing and screen. If the material in which the well is drilled is unconsolidated, it may have to be abandoned.

The casing may corrode and leak, allowing water to escape into the ground or contaminated water to enter. The remedy here is to withdraw and replace the casing. If the well is of large diameter, a smaller casing may be placed inside the old one. Grouting the well between the casing and the hole has been used to prevent corrosion. The life of casing will depend upon the character of the water handled. If it is highly corrosive, cast-iron or concrete casing may be used. Under ordinary conditions steel casing should last 15 to 20 years.

Screens cause trouble by corrosion or incrustation. There is little that can be done about corrosion in wells, but some remedies are available against incrustation.[18] This occurs not only on the screen but on the sand near it. It results from the reduction of pressure at or near the screen, which lowers the ability of the water to hold certain compounds in solution, particularly calcium carbonate, although a particularly hard scale may form in waters high in sulfates. A third form of deposit, occasionally noted in iron-bearing water, is caused by crenothrix, a bacterium. All forms of incrustation may be remedied by acid treatment. This consists of placing commercial hydrochloric acid, inhibited with gelatin[18] to minimize attack upon iron fittings, in the well to fill the screen and a little above. Every 2 hours the acid is agitated. After 8 to 12 hours the well is surged with a solid block for a few minutes to loosen incrustation. After 2 to 4 hours more the well is surged heavily for a brief time and then cleaned out with a bailer. Thereafter the well is pumped until there are no traces of acid in the water. In a few localities, where the aquifer is high in organic matter and clayey materials, double acid treatment is used, the first as described, and the second using commercial sulfuric acid. As an alternative to acid treatment dry ice has been used to generate back pressure to dislodge deposits. Not more than 1 kg of dry ice should be used per 11 kg of water in the well or freezing may occur, and suitable pressure-relief devices should be used to prevent damage. Where air is available a back pressure may be built up with it. Deposits of crenothrix can be destroyed by use of chlorine, using the method of disinfection described in Art. 4-22. Polyphosphates have also been successful in removing incrustation at dosages of 30 to 40 g/l of water in the well.[18]

4-24 Springs

A few cities obtain their water supplies from springs. In all cases springs must be carefully safeguarded from surface water. This will necessitate a tight collecting basin, preferably covered. Shallow groundwater or seepage insufficiently filtered may enter the springwater stream just before it emerges at the surface. A spring whose water is muddy after rains is receiving such water. This condition may require a special underground gallery or pipe to intercept deeper water. Several types of springs are recognizable.

1. The water in a shallow aquifer may be dammed by a less pervious or an impervious formation, thereby causing the water table to intersect the ground surface (see Fig. 4-14). A somewhat similar type called a depression spring is pictured in Fig. 4-1. Springs of these types are likely to have fluctuating yields owing to the variations in height of the water table resulting from rainfall irregularities. The yield of such a spring may be increased by excavating horizontal galleries across the direction of underground flow and at the line of seepage, thereby intercepting more water.
2. A deep aquifer may carry water under pressure, i.e., artesian water. A fault or fissure in the overlying impervious stratum, if it extends to the ground surface

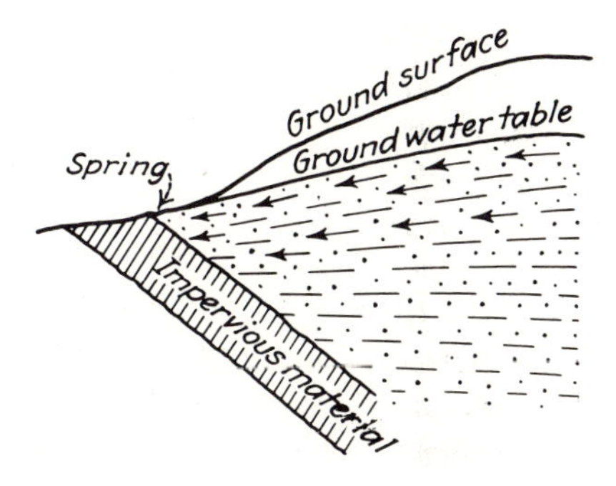

Figure 4-14 Overflow spring.

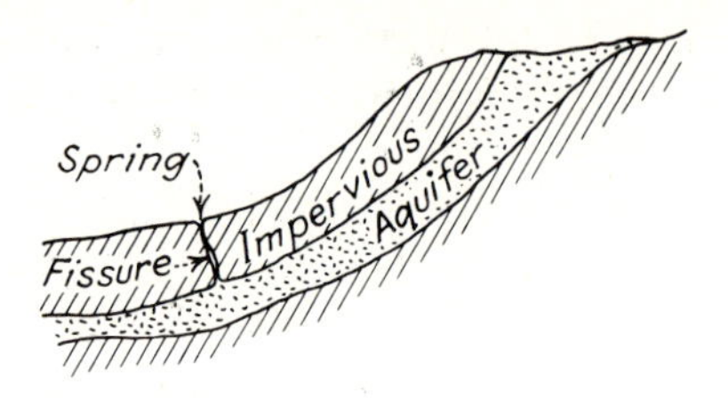

Figure 4-15 Fissure spring.

and the hydraulic gradient is high enough, will form, in effect, a natural artesian well (see Fig. 4-15). A spring of this type is likely to be very uniform in yield, showing little or no seasonal variation. The yield of such a spring may be increased by cleaning or otherwise smoothing the fissure so that the friction head lost in rising will be minimized. This may be difficult if not impossible, and increase of flow may be more easily obtained by sealing the spring and drilling a well to the aquifer.

3. Some springs of large flow issue from caverns in limestone. The original source of the water may be an outcrop of porous rock on higher ground, or the outcrop itself may be the limestone formation, the water seeping downward and horizontally through a fissure that it has dissolved. In the latter case the water may be clear and sparkling but insufficiently filtered to be safe.

PROBLEMS

4-1 Well A is 750 m due north of well B. Well C is 635 m northeast of B. The static levels are 260, 265, and 258 m in wells A, B, and C, respectively. Determine the direction of flow and slope of the groundwater table.

4-2 An artesian well is pumped at the rate of 1.6 m^3/min. At observation wells 150 m and 300 m away the drawdowns noted are 0.75 and 0.60 m, respectively. The average thickness of the aquifer at the observation wells is 6 m. Compute the coefficient of permeability of the aquifer.

4-3 If the aquifer of Prob. 4-2 has a porosity of 0.3 and a hydraulic gradient of 0.002, how many liters of water will pass per day through a strip 1 m wide with a depth equal to the thickness of the aquifer? What is the velocity in m/day?

4-4 A well is pumped at the rate of 0.75 m^3/min. At an observation well 30 m away the following drawdowns were noted at times after pumping began:

Days	m	Days	m	Days	m
1	0.75	6	3.45	30	7.47
2	1.30	8	4.02	40	8.24
3	1.90	10	4.57	60	9.34
4	2.45	15	5.60	80	10.10
5	3.00	20	6.37	100	10.66

Determine the values of T and S using Jacob's formulas. Use these values to predict the drawdown in a well 300 m from the pumped well after pumping for 2 years at a rate of 1.5 m^3/min. Also determine the drawdown in the well itself assuming its diameter is 0.2 m.

4-5 The match point in a nonequilibrium analysis is found to be at $u = 10^{-2}$, $W(u) = 4.04$, $d = 5$ m, and $r^2/t = 4.5 \times 10^6$ m^2/day. Determine T and S for this aquifer and the drawdown in a 0.2 m well pumped at a rate of 1 m^3/min after 1, 3, and 10 years.

REFERENCES

1. *Hydrology Handbook*, American Society of Civil Engineers, Manual of Engineering Practice 28, New York, 1949.
2. Brooks, R. H., and A. T. Corey, "Properties of Porous Media Affecting Fluid Flow," *Proceedings American Society of Civil Engineers Journal Irrigation and Drainage Division*, **92: IR2:**61, June, 1966.
3. Eagleson, Peter S., *Dynamic Hydrology*, McGraw-Hill, New York, 1970.
4. Todd, David K., *Groundwater Hydrology*, John Wiley & Sons, Inc., New York, 1959.
5. Thiem, G., *Hydrologische Methoden*, J. M. Gebhardt, Leipzig, 1906.
6. Wenzel, L. K., "Methods for Determining Permeability of Water-bearing Materials," U.S. Geological Survey *Water Supply Paper 887*, 1942.
7. Theis, C. V., "The Significance of the Cone of Depression in Ground-water Bodies," *Economic Geology*, **33:**889, December, 1938.
8. Meinzer, O. E., "Compressibility and Elasticity of Artesian Aquifers," *Economic Geology*, **28:**280, 1928.
9. Jacob, C. E., "Drawdown Test to Determine Effective Radius of Artesian Well," *Proceedings American Society of Civil Engineers*, **72:**5:629, 1940.
10. Wisler, C. O., and E. F. Brater, *Hydrology* 2d ed., John Wiley & Sons, Inc., New York, 1959.
11. Toups, John M., "Water Quality and Other Aspects of Ground-Water Recharge in Southern California," *Journal American Water Works Association*, **66:**149, March, 1974.
12. Todd, David K., "Salt Water Intrusion and Its Control," *Journal American Water Works Association*, **66:**180, March, 1974.
13. Reed, S., Coordinator, "Wastewater Management by Disposal on the Land," U.S. Army Corps of Engineers Cold Regions Research and Engineering Laboratory *Special Report No. 171*, 1972.
14. Bouwer, Herman, "Renovating Municipal Wastewater by High Rate Infiltration for Ground Water Recharge," *Journal American Water Works Association*, **66:**159, March, 1974.
15. Bennison, E. W., *Ground Water, Its Development, Uses, and Conservation*, Edward E. Johnson, Inc., St. Paul, Minn., 1947.
16. AWWA A-100, "Standard Specifications for Deep Wells," American Water Works Association, Denver, Colo., 1958.
17. "Deep Well Specifications," American Water Well Drillers Association, South Bend, Ind.
18. Arceneaux, W., "Operation and Maintenance of Wells," *Journal American Water Works Association*, **66:**199, March, 1974.

CHAPTER
FIVE
AQUEDUCTS AND WATER PIPES

5-1

A complete waterworks system includes the *source*, which may be a natural or artificial lake or impounding reservoir, a river, or groundwater. An *intake* will be needed. Wells and springs may be considered as intakes, but special intake structures are required for lakes, reservoirs, and rivers. If the source is a long distance from the city, an *aqueduct*, *pipe line*, or *open channel* will be needed to convey the water. If the water is from surface sources, a *treatment plant* will usually be required. In some cases groundwater is also treated. A *pumping station* will be needed in most cases to generate sufficient head to force the water through the network of street mains. A few cities are able to depend entirely upon gravity for distribution and need no pumps; others have pumps at the source and again pump the water after treatment. The pump may discharge all or a part of the water into *elevated storage tanks* or *reservoirs*. These furnish water for emergencies and also equalize demand with the pumping. Finally there is the *water distribution system*, which includes the mains, valves, service pipes, and fire hydrants to be found in the city streets.

This chapter is devoted to a discussion of the conduits and materials used to convey water in the various portions of the system.

5-2 Open Channels

Open channels are occasionally used to convey water from a source to the pumping station or treatment plant. They have the advantages of using low-cost materials and saving roof cost. They have the following disadvantages: The hydraulic grade line must be followed. There is a loss of water by seepage and evaporation. There is danger of contamination of the water, especially in a populous region. Ice troubles can be expected in a cold climate. Cattle, tree roots, and holes of burrowing animals may cause serious damage.

The sections used are the semicircle and the half hexagon or some other form of trapezoid. The half hexagon, being the most hydraulically efficient trapezoid, is most used. The channels, if unlined, almost always deviate from the half hexagon owing to the need of flatter side slopes in loose materials. Where water is valuable and seepage must be kept to a minimum, the channels are lined, usually with concrete. Lining 50 mm (2 in) thick has been successfully used where the side slopes are 1 vertically to 1.5 horizontally or flatter. For steep side slopes, thicker linings, 100 or 150 mm (4 to 6 in), with reinforcement on the fills, will be needed.

5-3 Aqueducts

The term aqueduct is usually restricted to closed conduits of masonry built in place. The types used may be classified as (*a*) cut-and-cover at the hydraulic gradient; (*b*) cut-and-cover beneath the hydraulic gradient and therefore under pressure; (*c*) tunnel at the hydraulic grade line; (*d*) tunnel beneath the hydraulic grade line and under pressure.

The alternative to the aqueduct is the pipe line, which employs conduit manufactured elsewhere than on the job. The advantages of the aqueduct over the pipe line are (*a*) the possibility of using local materials, such as sand and gravel for concrete; (*b*) longer life of masonry than metal conduits; (*c*) lower loss of carrying

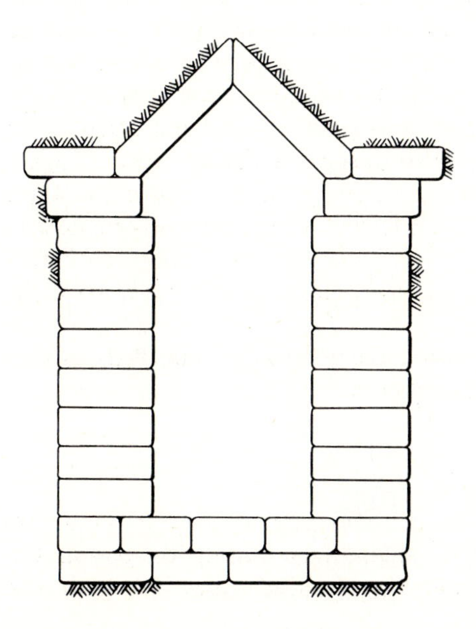

Cross section of the ancient conduit at Athens. Construction was started by Hadrian in A.D. 134. It is 1.1 m high and 0.4 m wide, lined with brick and rock, and plastered on the interior. *(Trans. Am. Soc. Civil Engrs., vol. 91, 1927.)*

capacity with age. These advantages may result in lower first cost which, combined with lower maintenance cost, may result in a more economical installation than a pipe line. There are some disadvantages: (*a*) A conduit must be built full size at first, whereas relatively small pipe lines can be used and parallel units added as the requirements increase; (*b*) balancing of cuts and fills often places aqueducts at the ground surface, causing interference with natural drainage and unsightly embankments.

Masonry aqueducts, unless reinforced with steel, are usually constructed in horseshoe cross section, a shape that has fair hydraulic properties, is stable, resists earth pressure well, is economical of materials, and is easy to build. The bottom is frequently dished to facilitate cleaning and to take upward water pressure and side thrust of earth at the bottom of the side walls. Earlier aqueducts were made of stone or brick; modern construction utilizes concrete.

Tunnels not under pressure are also usually constructed in horseshoe shape. If they convey water under pressure, circular cross sections are used. In pressure tunnels, the depth of cover is generally such that the weight of overlying material will overcome the bursting pressure. Tunnels are now used by some large cities to convey water into the cities themselves, connections to the system of mains being accomplished by vertical shafts. Such tunnels must be watertight and are often constructed with a riveted steel lining imbedded in the concrete of the tunnel lining.

5-4 Stresses in Pipes

Stresses in pipe are caused by the static pressure of the water, by centrifugal forces produced by changes in direction of flow, by temperature changes, by external loads, and by overpressures produced by water hammer when velocity of flow is quickly changed. The magnitude of the stress resulting from these varied causes may be calculated by the methods of applied mechanics.

Internal pressure of any kind produces a hoop stress given by

$$\sigma_h = rP/t \tag{5-1}$$

and a longitudinal stress given by

$$\sigma_l = rP/2t \tag{5-2}$$

in which r is the radius, P the internal pressure, and t the pipe wall thickness.

Water hammer produces a maximum pressure, additive to static pressure, given by

$$P_h = \rho V \sqrt{\frac{1}{\rho(1/K + d/Et)}} \tag{5-3}$$

in which V is the velocity of flow, d and t are the diameter and thickness, respectively, of the pipe, K is the bulk modulus of elasticity of water, and E is the modulus of elasticity of the pipe. The actual pressure produced depends upon the pipe length, the rate of valve closure, and the degree of relief provided by relief or air valves, surge tanks, air chambers, or similar devices.[1]

Changes in direction develop additional longitudinal tension in pipes equal to

$$\sigma_l' = \frac{d}{2t} \rho V^2 \sin \alpha/2 \tag{5-4}$$

in which ρ is the density of water and α is the angle of deflection of the pipe.

Deformations may be produced at changes in pipe direction due to reaction of the force producing this stress in the pipe. Such deformation may be prevented by suitable buttressing. The design force for such a buttress is given by

$$F = (\pi d^2/2)(V^2 \rho - P) \sin \alpha/2 \tag{5-5}$$

Thermal stresses may be calculated in the usual fashion from

$$\sigma_T = C\theta E \tag{5-6}$$

in which C is the coefficient of thermal expansion, θ is the change in temperature, and E is the modulus of elasticity. The force produced in restrained pipe subjected to temperature changes will be reacted in buttresses and by soil friction.

The vertical load upon a pipe may be calculated from the equations developed by Marston and Schlick[1] in which the load is related to the rigidity of the pipe, the dimensions of the trench, the properties of the fill material, the care of bedding (Fig. 6-29) and the imposed surface loads. Cast-iron pipe is manufactured in a variety of thicknesses for specified loading conditions.[2] For the bedding conditions shown in Fig. 6-29 and known working pressures the pipe class (which specifies its thickness) may be selected from prepared tables in AWWA or ASA Manuals.[2] The thickness includes an allowance for water hammer and impact loads from vehicles.

5-5 Pipe Lines

Pipe lines may be used for conveying large amounts of water, in which case they serve the same purpose as the open channels and aqueducts mentioned previously. When so used, they may be constructed of reinforced concrete, cast iron, or steel. Pipe is also used for the distribution system in a city. In this case the pipes are of varying sizes, having many connections and branches. For the supply network, cast iron is mostly used, with some steel, plastic, and asbestos cement pipe.

The pipe line must generally follow the profile of the ground, and a location is chosen that will be most favorable with regard to construction cost and resulting pressures. A profile of the pipe location is drawn, and the pipe line is located, with particular attention to the hydraulic grade line. The closer to the hydraulic grade line, the lower will be the pressure in the pipe, a condition that may result in lower pipe costs. High pressures can be avoided at times by breaking the hydraulic grade line with overflows or auxiliary reservoirs. Such overflows will also prevent full static pressure caused by closing the lower end of the pipe line. This difficulty is overcome in some instances by using valves at the outlet end which automatically bypass water to a waste channel when they are closed. Velocities should be high enough to prevent deposits of silt in the pipe; 0.6 to 0.8 m/s (2 to 2.5 ft/s) is satisfactory for this purpose.

At low points in the pipe, blowoff branches with valves are placed to drain the line or remove accumulated sediment. Summits in the pipe above the hydraulic grade line should be avoided, as siphoning will occur. In general, all high points in long pipe lines accumulate air which has come from solution in the water. Air and vacuum valves and air relief valves (Art. 6-23) are placed at such points. The former are available to allow air to escape when filling the pipe and automatically allow entrance of air when emptying, a matter of particular importance with steel pipes, since a partial vacuum due to unbalanced outside air pressure may cause collapse. The latter allow accumulated air to escape. Water hammer is guarded against by use of a surge tank or standpipe at the end of the line.

In deciding upon the type of pipe to be used in a long gravity pipe line the carrying capacity, durability, maintenance cost, and first cost must be considered. The type of water to be conveyed and its possible corrosive effect upon the pipe material with, perhaps, a serious reduction in length of life, must be taken into account. The pipe material giving the smallest annual cost or capitalized cost will be most economical. In pipe lines carrying water being pumped, the energy required is not fixed entirely by the topography but depends largely upon the velocity of the water in the pipe. The problem here is to adopt both a material and a size of pipe that will give minimum annual or capitalized cost. Larger pipe with smaller velocity and friction losses will give lower power costs but higher first costs and carrying charges. Probable length of life of various materials will be found in this chapter.

5-6 Cast-iron Pipe

Cast iron is the material most used for city water pipes. It is resistant to corrosion and accordingly is long-lived. Cast-iron pipe at Versailles, France, was dug up after being in use for 250 years and found in good condition. More of this pipe, in use since 1664, is still in service. The average life of cast-iron pipe is difficult to state owing to the varying conditions to which it may be exposed, but it is safe to assume it as 100 years. It is, however, subject to tuberculation in certain waters, which may reduce its carrying capacity as much as 70 percent. Nearly all cast-iron pipe is manufactured by one of three methods: vertical or pit casting, centrifugal casting in metal molds, or centrifugal casting in sand-lined molds.

Pit casting of pipe consists of pouring the molten metal into flasks, which are boxes in which molds have been shaped by ramming sand around patterns, each mold containing a core which forms the inner surface of the pipe. The core consists of a pipe or barrel which is wrapped with paper, straw, or excelsior, then daubed with wet clay and sand and brought to cylindrical shape. It has holes through the barrel to allow air and gases to escape as the molten metal is poured. The flask is placed vertically, and the space between the core and mold is filled

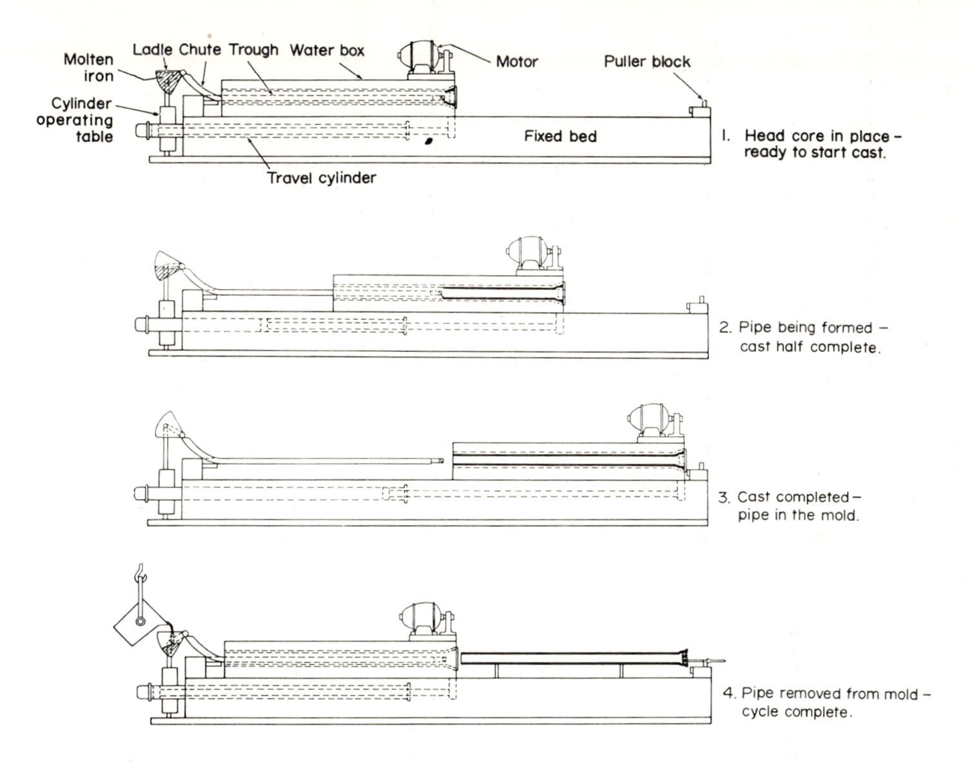
Molten iron
Ladle Chute Trough Water box
Motor
Puller block
Cylinder operating table
Fixed bed
Travel cylinder
1. Head core in place - ready to start cast.
2. Pipe being formed - cast half complete.
3. Cast completed - pipe in the mold.
4. Pipe removed from mold - cycle complete.

Figure 5-1 Cast-iron pipe manufacture. *(Courtesy Clow Corporation.)*

with hot metal poured from a ladle. After cooling, the pipe is removed from the flask, cleaned of sand and dust, and heated to about 150°C. At this temperature it is dipped into a vat of coal tar and oil, and then placed in an inclined position to drain. The coating is for protection against corrosion. The tar coating is standard for inside and outside, but the pipe buyer may specify cement mortar coating for the inside (Art. 5-7) or bituminous enamel for inside and outside. Each length is weighed, and before or after coating it is subjected to a hydrostatic test for one-half minute. Pit-cast pipe is made in 3.7 and 4.9 m (12 and 16 ft) lengths. Because fewer joints are required, the longer lengths are preferred.

Centrifugal casting in metal molds is sometimes called the Delavaud process. The molten metal is applied to the interior of a water-cooled cylindrical metal mold, which is rotated rapidly. Centrifugal force holds the metal against the mold and forms a homogeneous pipe with a cylindrical bore. After the pipe has cooled to 800°C it is taken from the machine to travel through a furnace where it is raised to 925°C and then slowly cooled to 650°C. This is to anneal it. After hydrostatic testing, it is coated like pit-cast pipe. This pipe is made in 5.5 m (18 ft) lengths.

Centrifugal casting in sand-lined molds, also called the Mono-Cast process, is done in a metal flask which contains a pattern. Foundry sand is rammed between the pattern and the flask, and when this is completed the pattern is withdrawn. The completed mold assembly is placed horizontally in the casting machine, which spins it about its horizontal axis. Molten metal is applied from a pouring trough to the interior of the mold. When the metal has solidified, the pipe is cooled in the mold to 650°C; thereafter it is weighed, tested, and coated. This pipe is made in 4.9-m (16-ft) lengths.

Fittings are made with solid patterns and core boxes placed in sand molds in flasks. After the fittings have been poured and allowed to cool in the mold the sand is shaken out, and the casting is taken out to be cleaned, either in a tumbler where it rubs against other castings, or by brushing or sandblasting. It is then weighed, coated, and given a final inspection.

Pipes and fittings are inspected by the manufacturers, but large purchasers may send their own inspectors to the factory. Engineering firms with offices at the factories may be retained by purchasers to make tests and inspections. The inspection consists of checking sizes and thicknesses, seeking defects, obtaining specimens for tensile and transverse tests and witnessing tensile, transverse, and hydrostatic tests.

5-7 Lined Cast-iron Pipe

Although cast-iron pipes resist corrosion sufficiently to prevent failure, they are sometimes seriously affected in carrying power by formation of tubercles of rust. These may become so numerous and large that interference with the water flow and loss of pressure result, making expensive cleaning operations necessary. The tar coating gives considerable protection against tuberculation; but at small cracks or other imperfections, trouble may be expected, especially under severe conditions. Hence with corrosive waters it is good practice to obtain fuller protection. This may be obtained by using bituminous, plastic,[3] or cement-lined pipes. Treating of water to prevent corrosion is discussed in Art. 11-22.

The cement lining, which is actually a 1 : 2 Portland cement mortar, is applied centrifugally. The thickness is tapered slightly at each end of a pipe length. Unless otherwise specified, curing of the lining is effected by applying a bituminous seal coat which retains moisture. An alternative is to store the pipe in a moist atmosphere, but the bituminous seal is especially desirable when the water may affect the cement lining. Cement-lined pipe retains good hydraulic properties, where ordinary cast-iron pipe would have a high coefficient of friction, and retains the carrying qualities of new, clean pipe. The value of C in the Hazen-Williams formula may be assumed as 145 for both new and old cement-lined pipe. No injury is done to the coating by cutting the pipe or drilling holes in it. Steel and wrought-iron pipes are also protected with cement coatings and have been used successfully for many years.

Cast-iron and steel pipes of large diameter have been given cement linings in place after years of use by means of a centrifugal machine. Small pipes have also been lined by using mandrels which press the mortar against the pipe. Removal of existing tubercles is necessary first. A 1 : 2 mortar is used in both processes. Cracks in cement linings have been shown to be self-healing when placed in contact with water.[4]

5-8 Flanged Cast-iron Pipe

This pipe is used in pipe galleries of water treatment plants, pumping plants, and wherever rigidity, strength, and joint tightness are required.

Both of the flanges are screwed to the pipe. It is made in 3.7 and 5.5 m (12 and 18 ft) lengths with the joints formed by flanges (Fig. 5-2*d*). The flanges have smooth, machined faces, and a gasket is placed between to insure a close fit when the bolts have been tightened. For waterworks use, the flanges are generally the American Standard for 860 kPa (125 psi) pressure. For higher pressures, 1720 kPa (250 psi) and stronger flanges are obtainable.

5-9 Joints in Cast-iron Pipe

The bell-and-spigot joint, using lead as the filling material, is still commonly used. The first step in making the joint is wrapping yarn around the spigot of one length before it is placed in the bell of the previously laid length. After insertion of the spigot, the yarn is straightened and adjusted. Sufficient yarn should be used to fill the joint within 50 mm (2 in) of the face of the bell. A gasket or joint runner is then clamped in place around the joint so that it fits tightly against the outer edge of the bell. Wet clay may be used to make a tight contact between the runner and the pipe so that the hot lead will not run out of the joint space. The lead is then poured into the V-shaped opening left in the top by the clamped joint runner and fills the space between the yarn and the runner. After the lead has hardened, the runner is removed, and the lead, which shrinks while cooling, is expanded by calking until it makes a tight fit. By means of calking tools, using the smaller size first, and a hammer, operated manually or mechanically, the lead is driven into the joint space. Calking requires skill, or the joints may not be tight. On the other hand, too heavy blows when calking may split the bell. Sterilized yarning material is available. Processed paper is also used.

Lead wool, or lead in shredded form, is sometimes used when water is encountered in the trench. In this process the joint is filled with yarn to within 30 mm (1.25 in) of the face. Joints of this type are more expensive than poured joints because material cost is higher and more labor is needed for calking.

Lead substitutes have been developed which are supplanting lead to a considerable degree. Leadite is the oldest and best known of these compounds, but Metalium and Hydrotite have been successfully used. They are mixtures of sulfur, sand, and other substances and are obtained as a black powder or cake. They are melted on the job and poured into a joint, prepared as for a cast-lead joint. It is recommended that hemp yarn be used rather than tarry or oily materials. The advantages of Leadite, etc., are that they require no calking, are light in weight,

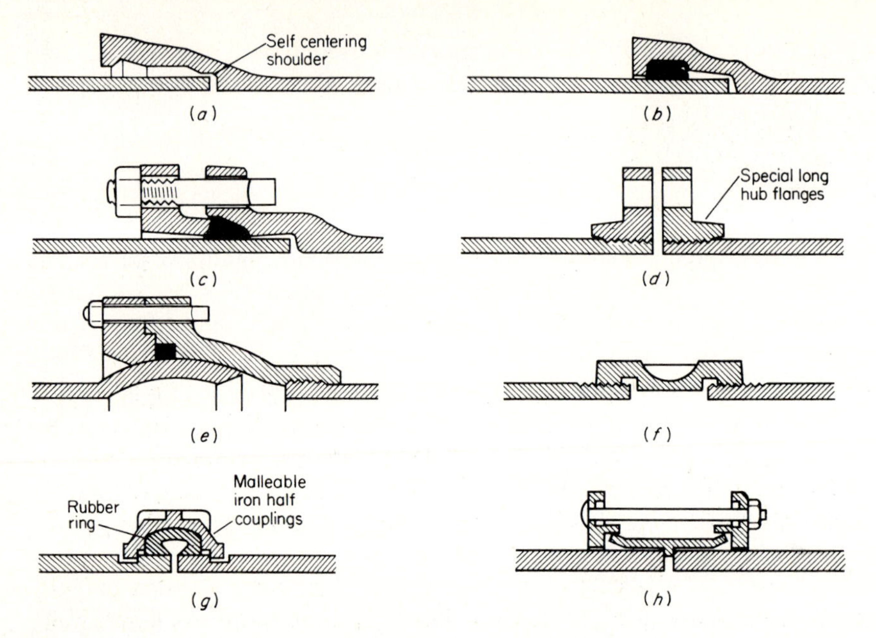

Figure 5-2 Joints in cast-iron pipe. (*a*) Standard lead joint in bell-and-spigot pipe. (*b*) Push-on joint. (*c*) Mechanical joint. (*d*) Flanged joint. (*e*) Ball joint. (*f*) Threaded I.P.S. joint. (*g*) Victaulic coupling. (*h*) Dresser coupling.

therefore easy to transport, and give strong joints that are not likely to blow out. Their lighter weight and avoidance of calking may make them cheaper than lead. Joints of these materials leak or "sweat" slightly at first but tighten up in a short time. Since the joints are rigid, they should not be used to connect a newly laid line to an old one, as the settlement of the new line may cause a broken pipe. A lead joint should be used at such connections. In a few instances lead substitutes have deteriorated, apparently as a result of bacterial action.

Cement joints for cast-iron water pipes have been used by various California cities for years. In making cement joints the joint is first yarned. The cement is prepared by placing enough for one joint in a pan and sprinkling it with water until it is wet enough to mold but still dry enough to crumble when dropped 300 mm (1 ft). The cement mortar is rammed into the joint with a calking tool. This is done all around the joint until it is filled to half its depth. The cement is then calked as hard as possible. The filling is completed, and the cement calked again. The joint is protected with wet cloth until the cement hardens. A man can make 16 joints in 200-mm (8-in) pipe in a day. After 24 hours, water can be turned into the main; and, though the cement joints are likely to sweat for a time, they soon close up.

Cast-iron pipe is obtainable with a special type of slip-on bell-and-spigot joint. A rubber gasket is inserted in a specially grooved bell and a lubricant is

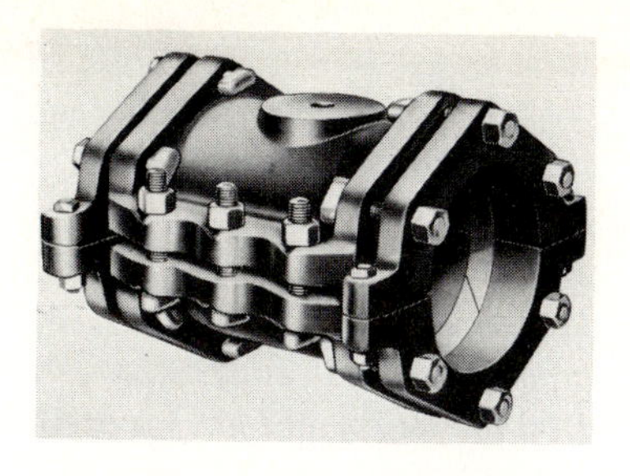

Figure 5-3 Split sleeve. *(Courtesy Clow corporation.)*

applied to it. The spigot is then forced into the bell, and the joint is complete. This type of joint, known as Tyton, All-Tite, Bell-Tite, or Fastite, is immediately tight and requires no lead, yarning, calking, or bell holes. Deflection is possible to form curves, and jointing proceeds rapidly. The joint is not strong longitudinally, however, and blocking or anchoring is especially necessary at all changes of direction.

Flexible joints are often used for pipe lines on river beds, where settlement will occur after the pipe is laid. Such joints also allow the pipe to be joined together on rafts or barges and sunk, the raft being moved along as a new length is added, while the pipe line slopes from the rear of the raft to the river bottom. Figure 5-2(*e*) shows a commonly used type of flexible joint.

Dresser couplings consist of a middle ring and two followers which are connected by bolts and force gaskets tightly beneath the ends of the middle ring. They are made for both steel and cast-iron pipe and are used where flexibility and tightness against high pressures are required. One type of Dresser coupling is shown in Fig. 5-2(*h*).

Victaulic joints consist of two half housings, which are bolted together around the pipe. They engage grooves near the pipe ends and enclose a leakproof ring or gasket. When the joint is installed, there is sufficient clearance between the keys of the housing and the grooves to allow for expansion, contraction, and deflection. This type of joint is used for cast-iron, steel, or ductile iron pipe. Such joints are frequently used on exposed pipe, especially where considerable vibration is expected.

The mechanical joint has a ring which, when the bolts are tightened, expands the rubber gasket to make a tight fit.

Precalked joints can be obtained in smaller sizes of cast-iron pipe. The joints are yarned and poured in the factory. The spigots are inserted on the job, and the joint is tightened by calking the upper half only. Several lengths can be inserted and calked before being placed in the trench. This pipe is quickly laid and can be jointed in wet trenches.

Split sleeves (Fig. 5-3) are somewhat similar to couplings. They are used to repair breaks in cast-iron pipes without making it necessary to install a new length and with minimum interruption to service.

5-10 Cast-iron Fittings

Changes in direction of flow are made by means of fittings, the most commonly used types being shown in Fig. 5-4. They are obtainable as bell-and-spigot or all bell or with flanged ends or, as specials, with bell and flange or flange and spigot. Tees, crosses, and 90° bends can also be obtained with side outlets and also with reducing branches.

Fittings are of two types, short-body and long-body. The former are favored for water distribution because of their light weight and compactness. Under the ASA specifications bell-and-spigot fittings are obtainable only in one thickness in sizes up to 305 mm (12 in). In larger sizes two class thicknesses are available. Dimensions of fittings are obtainable from the Cast-Iron Pipe Research Association and from publications of manufacturers. Dimensions involving lengths are complied with by manufacturers with a plus or minus tolerance of 1.6 mm ($\frac{1}{16}$ in).

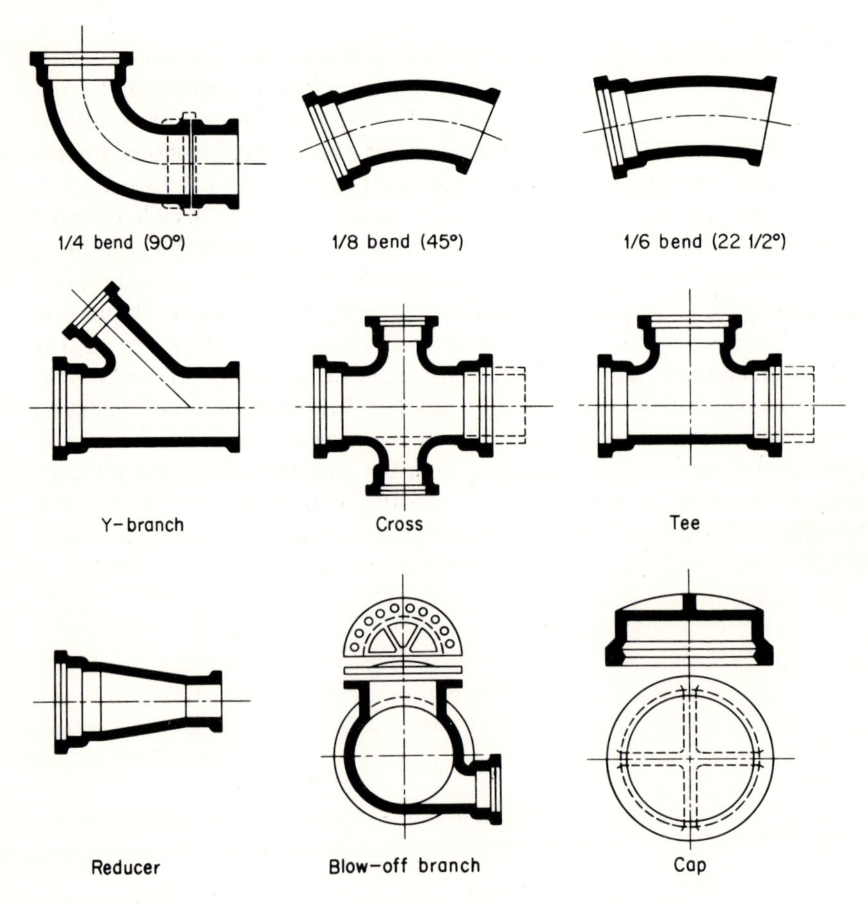

Figure 5-4 Some standard bell-and-spigot fittings.

5-11 Steel Pipe

Steel is frequently used for water pipes, especially for pipe lines, trunk mains, and inverted siphons where pressures are high and sizes large. Owing to difficulty in making connections, steel is seldom used for distribution mains, although there have been exceptions, notably in California. Steel being much stronger than cast iron, large steel pipes constructed for high pressures are much lighter in weight. Also steel pipes are cheaper, easier to construct, and more easily transported than cast iron. There are, however, important disadvantages. The pipes are not adapted to withstanding external loads, while a partial vacuum caused by emptying a pipe rapidly may cause collapse or distortion. Riveted pipes suffer loss of capacity owing to the effect of joints and rivets in reducing the net diameter of the pipe. The thinner walls and greater susceptibility to corrosion are likely to cause high maintenance charges and shortened life. Steel pipe under average conditions should have a life of 25 to 50 years. Unfavorable conditions are highly corrosive water and corrosive soil. Sand, clay, shale, and ashes are especially likely to cause pitting from the outside.

Riveted-steel pipes are generally made with longitudinal lap seams, single riveted for pipes under 1220 mm (48 in) in diameter and double riveted above that size. If the steel plates are over 15 mm ($\frac{5}{8}$ in) thick, butt joints are used. Large pipes may have two longitudinal seams. The latter have better hydraulic properties but are more expensive. The joints of riveted pipes must be calked to prevent leakage. Shop joints are calked in the shop, but field joints must, of course, be calked on the job. This is accomplished by pneumatic tools which spread or expand the metal edges at the joint.

Spiral riveted pipe is made in diameters from 75 to 1020 mm (3 to 40 in) and of various thicknesses. It has the advantages of withstanding greater collapsing pressures than other steel pipes of like thickness. It is also stiffer and can be used to span ravines and withstand heavy earth loads. It has been used for high pressures in waterworks and hydroelectric plants. The pipes are laid with the laps in the direction of the flow. The transverse joints may be made by screwing flanges on the steel plates, using Dresser couplings or similar joints.

Steel pipes of all sizes up to 2440 mm (96 in) in diameter can be obtained with welded longitudinal joints. The circular joints may be welded in the field, or

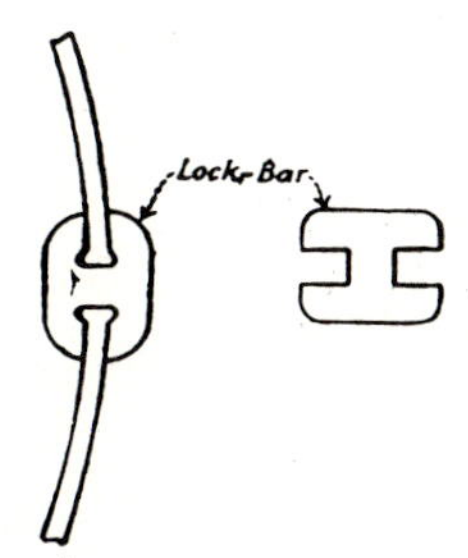

Figure 5-5 Lock-bar pipe, longitudinal seams.

connection may be made by means of flanges or couplings of the Dresser type or with standard screwed couplings. Bell-and-spigot ends have been used for sizes up to 760 mm (30 in). Standard specifications are available for welded steel pipes.[5]

Lock-bar joint pipe is made of steel sheets which are bent into cylinders and locked by means of a bar along the longitudinal seams. The edges of the plate are upset or slightly thickened, and the locking bar is squeezed shut around the edges by means of jacks. The circular joints are made as in riveted pipes.

5-12 Protection and Maintenance of Steel Pipe

It is even more important to furnish protection for steel pipe than for cast iron. Steel plates have more or less mill scale on their surfaces after rolling and must be cleaned. The pipe is then coated with tar or a bituminous enamel.[6] Scars or other defects in the coating caused by handling and laying of the pipe must be repaired before the pipe is put into operation. If the defect is small, use of a blowtorch will cause the coating to flow over the scratch. Large defects are remedied by applying more material and then fusing old and new with the blowtorch.

Bituminous enamel, a coal tar with 20 percent or more inert filler, has been found of value when applied by experienced workmen. Methods have been developed for coating large cast-iron and steel water mains in place after they have been in use. The pipe must first be thoroughly cleaned and dried, a priming coat placed on it, and then hot enamel put on by hand brushing to a thickness of 1.6 mm ($\frac{1}{16}$ in). Ventilation is necessary during cleaning and coating. Graphite paint, water-gas and coal-gas tar paint, and asphaltum have been used. All coatings tend in time to lose elasticity and adhesive properties. Some are pervious to water at high pressures and allow corrosion of the metal beneath. Hence recoating may be necessary after a varying number of years.

Greater protection against soil corrosion than is afforded by ordinary coating has been obtained by winding strips of burlap soaked in asphalt spirally around the pipe. Roofing paper held with wire has been used for pipe of smaller sizes.

Steel pipes have been protected both inside and outside with Portland cement mortar.[7] Not only has good protection been obtained, but compared with pipes of equal clear waterway, the carrying capacity is better.

5-13 Concrete Pipe

For water pipes not under pressure where leakage inward or outward is of no consequence, plain concrete pipe with bell-and-spigot joints, similar to sewer pipe, has been used. Above 610 mm (24 in) diameter such pipe is reinforced.

Concrete pipe has the advantage of not being subject to tuberculation; hence its carrying capacity remains high. With good joints, it presents a smoother surface than new steel pipe. Tests made on reinforced concrete pipes showed C, the friction coefficient in the Hazen-Williams formula to be 138 to 152. The life of concrete pipe, under normal conditions, should be at least 75 years.

For precast pipes reinforced concrete is used for sizes 400 to 2740 mm (16 to 108 in) in lengths of 3.7, 4.3, or 4.9 m (12, 14, or 16 ft). It can be obtained for static heads from 300 to 1800 kPa (43 to 260 psi). The reinforcement consists of a welded steel cylinder to insure watertightness and about this is wound high-tensile wire,[8] which may be wound tightly enough to cause prestressing of the core. The core consists of the reinforcing and the concrete pipe lining, which may be placed centrifugally. An outer coating of concrete is placed over the reinforcing. Cylinder pipe is also made without prestressing, and pipes without cylinders are also made for heads up to 360 kPa (52 psi). The Lock Joint Pipe Company asserts that leakages will not exceed 0.018 m^3/day per km/mm dia (75 gal/day per mi/in) for cylinder pipe and 0.036 m^3/day per km/mm (150 gal/day per mi/in) for noncylinder pipe. Joints are made by welding circular steel rings to the cylinders which are so shaped and placed that they will be self-centering. A rubber gasket or a fiber-filled lead gasket is placed between the joint rings. Figure 5-6 shows types of joints and pipe sections that are used by the Lock Joint Pipe Company. The joints can be deflected slightly to allow curves of long radius. Fittings required for bends, branches, blowoffs, air valves, etc., are made in the same fashion as the pipe or may be of welded steel plates of necessary thickness with exterior and interior mortar coatings. Concrete pipe is frequently manufactured at or near the project where it will be used.

5-14 Asbestos Cement Pipe

Asbestos pipe is composed of a mixture of Portland cement and asbestos fiber, built upon a rotating steel mandrel to proper thickness and then compacted into a dense, homogeneous structure by means of steel pressure rolls. Pipe of this type has been used with success for many years in Europe. In Great Britain it has been manufactured since 1928, and has an accepted life in excess of 30 years. It is made in 4 m (13 ft) lengths, in sizes 100 to 410 mm (4 to 16 in), in three classes for working pressures of 690, 1034, and 1380 kPa (100, 150, and 200 psi). It has the advantage of being immune to the action of ordinary soil and also to acids and salts; and, being a non-conductor of electricity, it is not affected by electrolysis. Tuberculation will not occur, and accordingly the pipe will remain smooth even when conveying corrosive water, tests indicating that C in the Hazen-Williams formula can safely be taken as 140.

The joints used consist of a sleeve which fits over the plain ends of the lengths, watertightness being obtained by two rubber rings compressed between the sleeve and pipe barrels. Such joints are highly flexible, capable of being deflected to an angle of 12°, and are claimed to allow high pressure without leakage of calked joints. Standard cast-iron fittings are used with joints of Leadite or similar material. The pipe can be tapped for service lines.

Crushing strengths are specified by the American Society for Testing Materials. Installation methods are given by the manufacturers. The economies of using this type of pipe are based upon its possibilities of smaller cost for laying and jointing, in use of smaller sizes for equal carrying power, smaller pumping costs

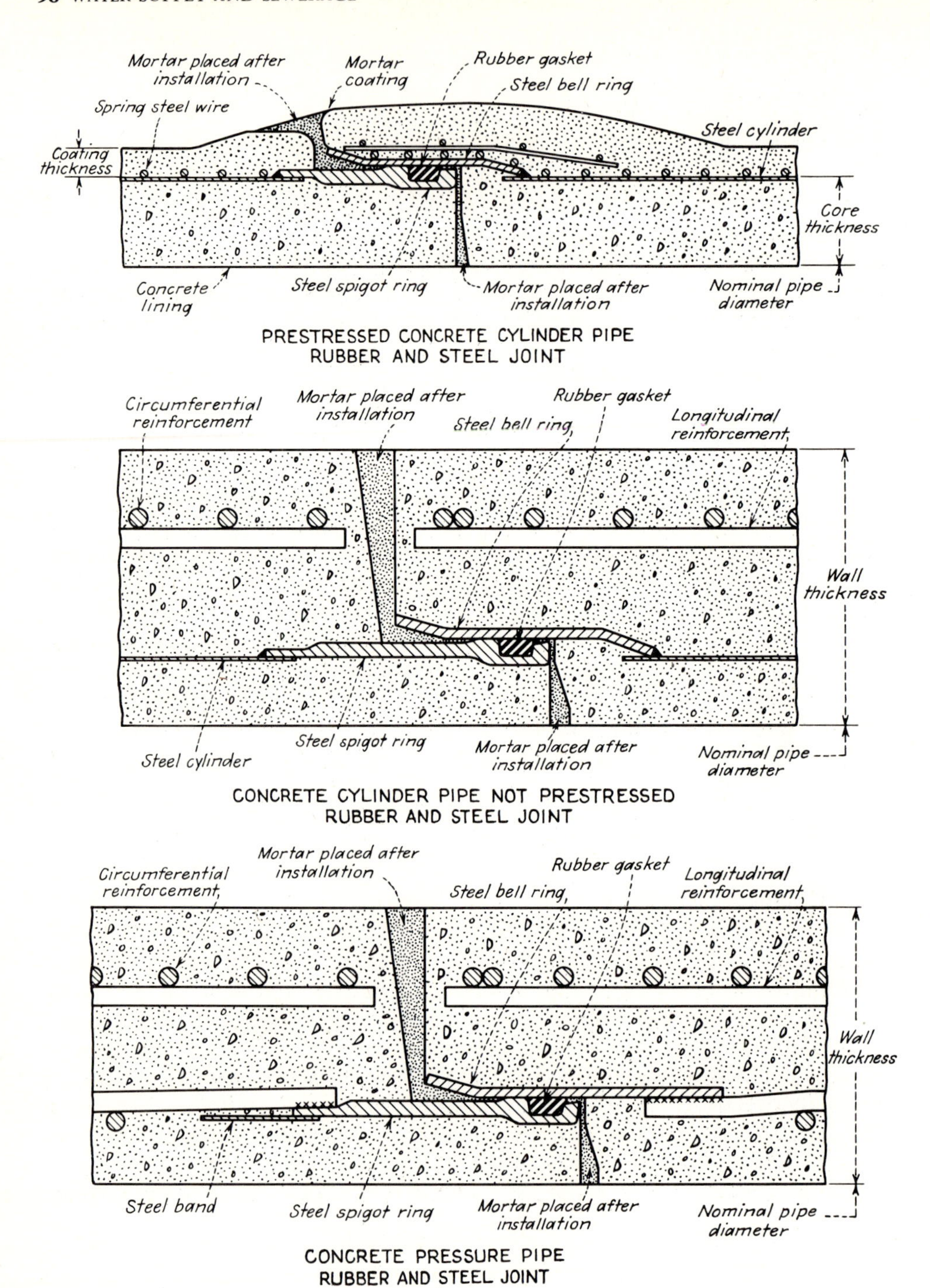

Figure 5-6 Wall sections and joints of Lock Joint pressure pipe.

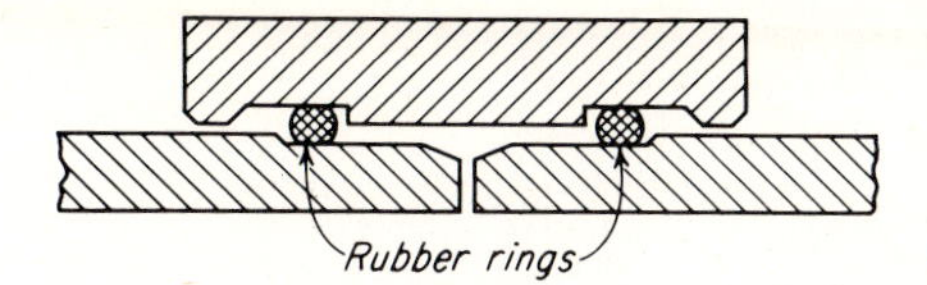

Figure 5-7 Joint in asbestos-cement pipe.

because of less friction, no reduction of carrying power with age, and elimination of leakage. Despite recent concern with respect to asbestos in public water supplies, there is no evidence that asbestos pipe contributes significant or potentially harmful levels of this material to the water it conveys.[9] Over 2.4 million km (1.5 million mi) of such pipe is in service worldwide.

5-15 Plastic Pipe

Plastic pipe manufactured of both uniform and fiber reinforced materials is now widely used both in domestic plumbing and water distribution systems.[10,11] Such pipes are far easier to install, more easily handled, and generally cheaper in material cost than traditional materials. The long-term performance of these lines can only be established by the passage of time. Cold flow, age embrittlement, and affect of installation stresses are suggested difficulties which might occur. Some manufacturers may offer a 25-year pro-rated warranty on both materials and labor. Standard specifications for polyvinyl chloride pipe have been published by the American Water Works Association.[12]

CORROSION IN WATER AND SEWAGE WORKS

5-16 Corrosion of Metal

Corrosion may be defined as the conversion of a metal to a salt or oxide with a loss of desirable properties such as mechanical strength.[13] Corrosion may occur over an entire exposed surface or may be localized at micro- or macroscopic discontinuities in the metal. In all types of corrosion an electron transfer must occur, and in most reactions of interest in sanitary engineering this transfer occurs either between dissimilar metals or between different areas upon a single material. The zone which releases electrons is called anodic, while that which accepts them is called cathodic, as in other electrical circuits.

At the anode an oxidation reaction takes place which can be represented as:

$$\underset{\text{(metal)}}{\mathrm{Fe}} \rightarrow \underset{\text{(ionized in solution)}}{\mathrm{Fe}^{++} + 2\mathrm{e}^-} \tag{5-7}$$

in which it is seen that the iron enters into solution. For this reaction to proceed a simultaneous reduction reaction must occur, which commonly is the reduction of

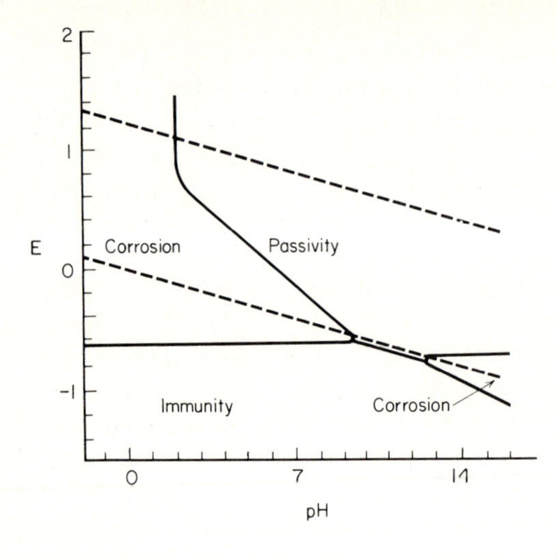

Figure 5-8 Simplified Pourbaix diagram for iron *(From "The Fundamentals of Corrosion", 2d ed. by J. C. Scully, Copyright 1975, J. C. Scully. Used with permission of the author and publisher.)*

the dissolved oxygen present in the water or the reduction of hydrogen ion to hydrogen gas.

$$\underset{\text{(in solution)}}{O_2} + 4e^- + 2H_2O \rightarrow \underset{\text{(in solution)}}{4OH^-} \tag{5-8}$$

$$\underset{\text{(in solution)}}{2H^+} + 2e^- \rightarrow \underset{\text{(gaseous)}}{H_2} \tag{5-9}$$

Other reactions may occasionally occur with chlorine or permanganate serving as the electron acceptor if they are present.

Subsequent reactions between the oxidized iron and the hydroxide ion, produced by oxygen reduction or liberated by hydrogen reduction, may occur leading to the precipitation of insoluble products such as Fe_2O_3, $Fe(OH)_2$, $Fe(OH)_3$, etc. Accumulation of these products may slow the rate of the reaction by interfering with oxygen diffusion to the metal surface.

A variety of complex reactions can occur in the oxidation of iron. These reactions are dependent upon either pH, electrode potential, or both. A simplified Pourbaix diagram[13] is presented in Fig. 5-8 which indicates the zones in which corrosion will or will not occur in systems involving iron and pure water. Corrosion can occur in the region labeled passivity, but ordinarily does not. In real systems the diagram is complicated by reactions with other dissolved salts. The most important observation which can be made from this diagram is that corrosion of iron is possible under all conditions occurring in water and waste treatment save, perhaps, softening and phosphorus precipitation with lime.

5-17 Protection Against Corrosion

Since two reactions, an oxidation and a reduction, must occur for aqueous corrosion to proceed, the process can be halted or slowed by interfering with either. The techniques employed include cathodic protection, anodic protection, inhibition, and application of metallic or chemical coatings.

Cathodic protection forces the entire metallic surface to act as a cathode. Since corrosion occurs at anodic areas, such a procedure prevents loss of metal. Cathodic protection can be provided by either the impressed current or the galvanic technique. *Impressed current* protection involves applying a dc voltage to the metal so that electrons will flow into it at a rate at least equal to that at which they left under conditions of corrosion. The anode may be any conducting material. If it is corrosion prone it may be replaced at intervals. A number of commercial firms design and manufacture impressed current cathodic protection systems. Design details are presented elsewhere.[14]

Galvanic protection utilizes a sacrificial anode—a material of higher corrosion potential than that which it is desired to protect. Since such a material will be anodic with respect to the protected metal, the latter will be entirely cathodic and will not corrode. Materials used to protect iron and steel include magnesium, zinc, and aluminum. Magnesium is most commonly used in waterworks practice since it does not form dense oxide films nor exhibit passivation with respect to iron. The anode will corrode and must eventually be replaced.

Anodic protection is a passivation technique in which impression of an external voltage greatly reduces the rate of corrosion of a metal. A relatively high current density may be necessary to achieve passivation, however, far less is required for maintenance of protection. A breakdown of the polarized film on the metallic surface may produce extremely rapid corrosion since a small area may become anodic to the entire surface. Variations in applied voltage once passivation has been achieved may also produce rapid corrosion.[15]

Inhibition is effected by the deposition or adsorption of ions on metallic surfaces. Chemicals used for this purpose include chromates, nitrates, phosphates, molybdates, tungstates, silicates, benzoates, etc.[13] Such coatings are effective in reducing corrosion in neutral or alkaline (but not acid) solutions.

Metallic coatings can be applied by hot dipping, metal spraying, cladding, vapor deposition, electroplating, metalliding, and mechanical plating. The film formed ranges from 2×10^{-4} to 5 mm and may serve as a final cover or as a base for another protective coating. Coatings may be either anodic or cathodic with respect to the base metal. If the coating is anodic it will provide galvanic protection to the base metal in the event the coating is broken. Noble (cathodic) coatings may produce rapid pitting in the base metal when discontinuities occur.

Chemical coatings include paints, coal tar preparations,[16] asphalt, epoxy materials,[17] and, as mentioned earlier, cement.[4,7] These materials generally serve to isolate the metal from the aqueous environment. Paints, in addition to isolating the surface, may provide other protection. Zinc chromate paints, for example, include an element of cathodic (galvanic) protection, an element of inhibition provided by the chromate, and are mildly alkaline as well.

5-18 Grounding of Electrical Wiring to Water Pipes

The National Electrical Code requires that, where it is available, a metallic underground water pipe be used as the grounding electrode. If the electrode so provided is less than 3 m (10 ft) in length it is supplemented by an additional electrode.[18]

Corrosion, while an electrochemical phenomenon, is associated with direct, not alternating current and it is the latter which is used nearly exclusively in the United States. Corrosion may, however, result from the bonding provided between dissimilar metals in a grounding installation.

Copper, which is widely used as an electrical conductor, is cathodic with respect to iron and connection between the two metals can produce rapid corrosion of the iron in the presence of an electrolyte. It should be observed that an identical problem can arise when household copper plumbing is directly connected to iron distribution lines.

5-19 Use of Inert Materials

There has been a marked increase in recent years in the use of materials which are not subject to corrosion. Asbestos cement and plastic pipe are common examples of such replacement, although the corrosion properties are not the sole factor involved. Plastic materials reinforced with fiberglass and compounded to have superior resistance to ultraviolet light (an earlier drawback to the use of such materials) are now used in many applications in which sheet metal was earlier required.[18] Weirs, launders, and other light structures are typical examples in which reinforced plastics are of use.

REFERENCES

1. Babbitt, Harold E., James J. Doland, and John L. Cleasby, *Water Supply Engineering*, 6th ed., McGraw-Hill, New York, 1962.
2. AWWA C101, "American National Standard for Thickness Design of Cast-Iron Pipe," American Water Works Association, Denver, Colo., 1972.
3. AWWA C105, "American National Standard for Polyethylene Encasement for Gray and Ductile Cast-Iron Piping for Water Mains and Other Liquids," American Water Works Association, Denver, Colo., 1973.
4. Wagner, Ernest F., "Autogenous Healing of Cracks in Cement Mortar Linings for Gray-Iron and Ductile-Iron Water Pipe," *Journal American Water Works Association*, **66:**358, June, 1974.
5. AWWA C200, "American National Standard for Steel Water Pipe 6 Inches and Larger," American Water Works Association, Denver, Colo., 1975.
6. Kinsey, William R., "Steel Water Pipe Design, Lining, Coating, Joints, and Installation," *Journal American Water Works Association*, **65:** 786, December, 1973.
7. AWWA C205, "AWWA Standard for Cement-Mortar Protective Coating for Steel Pipe," American Water Works Association, Denver, Colo., 1970.
8. AWWA C303, "AWWA Standard for Reinforced Concrete Water Pipe—Steel Cylinder Type, Pretensioned," American Water Works Association, Denver, Colo., 1970.
9. Olson, Harold L., "Asbestos in Potable Water Supplies," *Journal American Water Works Association*, **66:**515, September, 1974.
10. Lincoln, David A., "Experience With Plastic Mains and Services," *Journal American Water Works Association*, **68:**234, May, 1976.
11. Nesbitt, William D., "PVC Pipe in Water Distribution: Reliability and Durability," *Journal American Water Works Association*, **67:**576, October, 1975.
12. AWWA C900, "AWWA Standard for Polyvinyl Chloride (PVC) Pressure Pipe, 4 in through 12 in for Water," American Water Works Association, Denver, Colo., 1975.

13. Scully, J. C., *The Fundamentals of Corrosion*, 2d ed., Pergamon Press, Oxford, 1975.
14. Baeckmann, Walter G. v., "Cathodic Protection of Underground Pipelines with Special Reference to Urban Areas," *Journal American Water Works Association*, **66:**466, August, 1974.
15. Henthorne, Michael, "Cathodic and Anodic Protection for Corrosion Control," *Chemical Engineering*, **78:**27:73, December, 1971.
16. AWWA C203, "AWWA Standard for Coal Tar Enamel Protective Coatings for Steel Pipe," American Water Works Association, Denver, Colo., 1970.
17. Frye, S. C., "Epoxy Lining for Steel Water Pipe," *Journal American Water Works Association*, **66:**498, August, 1974.
18. Szymanski, Walter A., and Robert C. Taylor, "Fiber Glass Reinforced Polyesters and Furan Resins for Water and Waste Treatment Systems," *Journal American Water Works Association*, **68:**228, May, 1976.

CHAPTER

SIX

COLLECTION AND DISTRIBUTION OF WATER

INTAKES

6-1

Intakes consist of the opening, strainer, or grating through which the water enters, and the conduit conveying the water, usually by gravity, to a well or sump. From the well the water is pumped to the mains or treatment plant. Intakes should be so located and designed that possibility of interference with the supply is minimized and where uncertainty of continuous serviceability exists, intakes should be duplicated. The following must be considered in designing and locating intakes: (*a*) the source of supply, whether impounding reservoir, lake, or river (including the possibility of wide fluctuation in water level); (*b*) the character of the intake surroundings, depth of water, character of bottom, navigation requirements, the effects of currents, floods, and storms upon the structure and in scouring the bottom; (*c*) the location with respect to sources of pollution; and (*d*) the prevalence of floating material such as ice, logs, and vegetation.

6-2 Intakes from Impounding Reservoirs

The water of impounding reservoirs is likely to vary in quality at different levels, making it usually desirable to take water from about a meter below the surface. This, with the fluctuations of water level which may be expected in reservoirs, makes it advisable to have ports at various heights. Where the dam is of earth, the

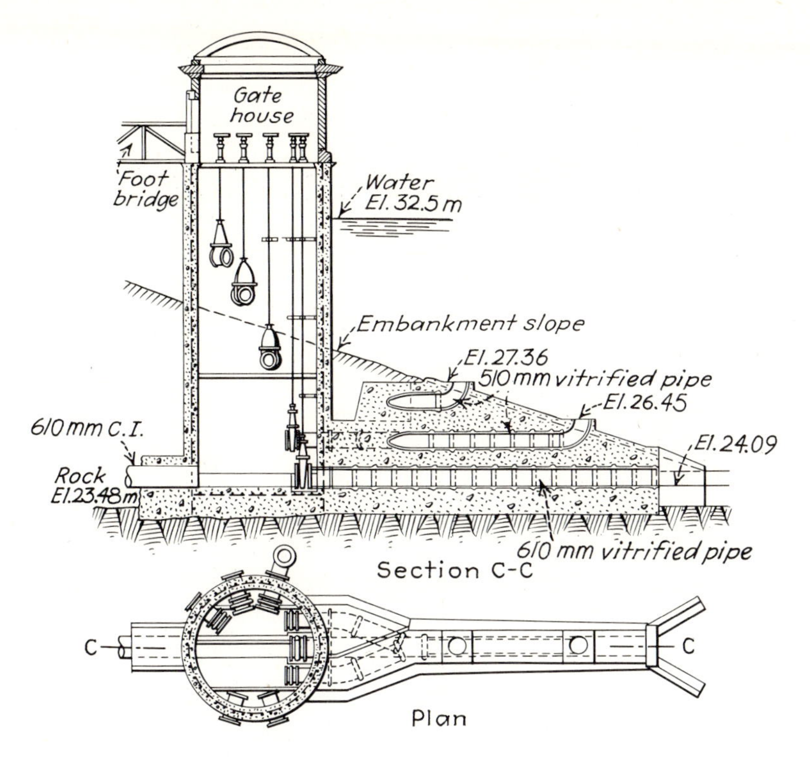

Figure 6-1 Intake in impounding reservoir, Lexington, Ky.

intake is usually a concrete tower located in deep water near the upstream toe of the dam. Access to the tower so that the gates of the various openings may be manipulated is obtained by means of a footbridge. The ports may be closed by sluice gates or by gate valves on short lengths of pipe. Where the dam is of masonry, the intake may be a well in the dam structure itself, also with openings at various heights.

6-3 Lake Intakes

If the lake shore is inhabited, the intake should be so located that danger of pollution will be minimized. This may require study of currents and effects of winds with particular attention to movement of sewage or industrial wastes, if these are discharged into the lake. It is also advisable to have the intake opening 2.5 m (8 ft) or more above the bottom so that large amounts of silt will not be carried in with the water. Entering velocities must be low, or excessive amounts of floating matter, sediment, fish, and frazil ice may be carried in. Entering velocities less than 0.15 m/s (0.5 ft/s) have been used successfully. Offshore winds tend to stir up sediment which will be carried out for long distances. On this account Great Lakes intakes must be located at a distance not less than 600 to 900 m (2000 to

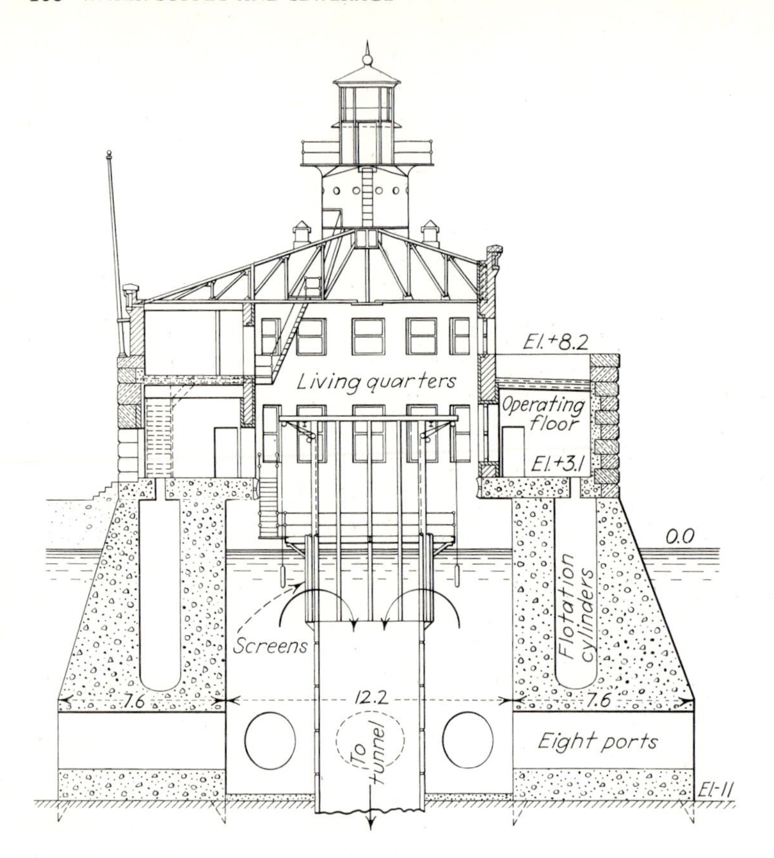

Figure 6-2 Wilson Avenue intake, Chicago.

3000 ft) from shore, while Chicago, because of the great amount of pollution entering the lake nearby, has found it necessary to go 3 to 6 km (2 to 4 mi) out. A depth of 6 to 9 m (20 to 30 ft) is required to avoid trouble caused by ice jams which may solidly fill the water at shallow depths. Ice may also form upon the screens, gates, and valves. This is known as anchor ice and may be expected during cold, clear nights and cloudy days. Frazil is formed as long ice crystals in the water. When drawn into the intakes, it may adhere to the surfaces and add to the anchor ice accumulations. Compressed air has been used to remove ice accumulations, and preventive measures include steam piping and space heaters within the intake structure so arranged that the surfaces exposed are kept at a temperature slightly above the freezing point of water.

Figure 6-2 shows the Wilson Avenue intake of Chicago. It is located 3.2 km (2 mi) from shore in 11 m (35 ft) of water. It has eight ports and movable bar screens. Living quarters are furnished for an attendant who needs additional assistance during periods when ice accumulates. Continuous attendance is made necessary by the high entrance velocity, frequently 0.45 m/s (1.5 ft/s).

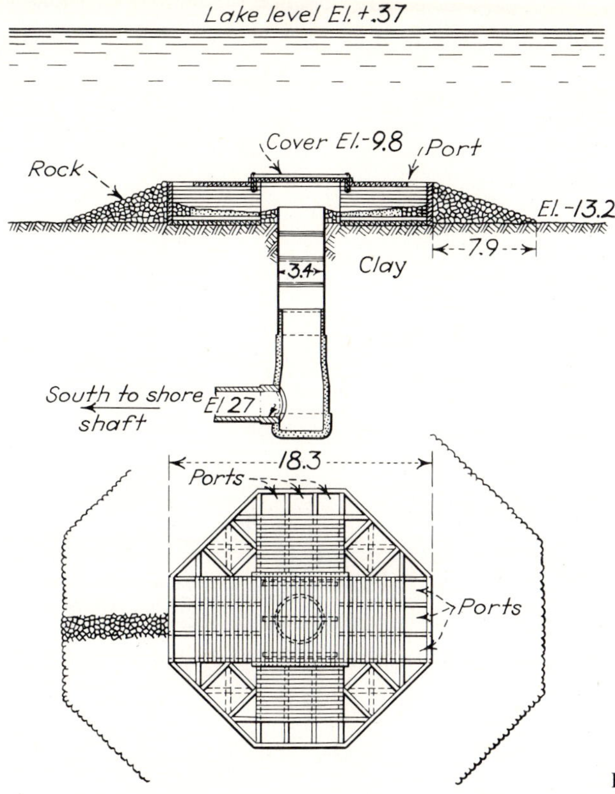

Figure 6-3 Submerged crib intake.

Submerged cribs are used by smaller cities. They must be placed at depths dictated by navigation requirements. With low entrance velocities they require no attendance. Usually they are square or octagonal wooden cribs protected by riprap. The ports may be screened by wood bars.

Pipe intakes have also been used by small cities. The simplest type consists of a pipe projecting into the water and having a special flaring fitting on the end with the opening protected by a cast-iron strainer. More elaborate intakes consist of horizontal pipes supported above the bottom on piers. If the bottom is unstable, the piers may require pile foundations. The water may enter through perforations in a strainer or through tees in the line having upturned open ends protected by screens.

6-4 River Intakes

Where rock foundations are available, some large cities, notably St. Louis and Cincinnati, have built elaborate river intakes, resembling bridge piers with ports at various depths, to allow for great fluctuations in river stage. Small cities may use pipe intakes similar to those described under lake intakes. The bottom must be sufficiently stable, and the water deep enough to allow a submergence of at

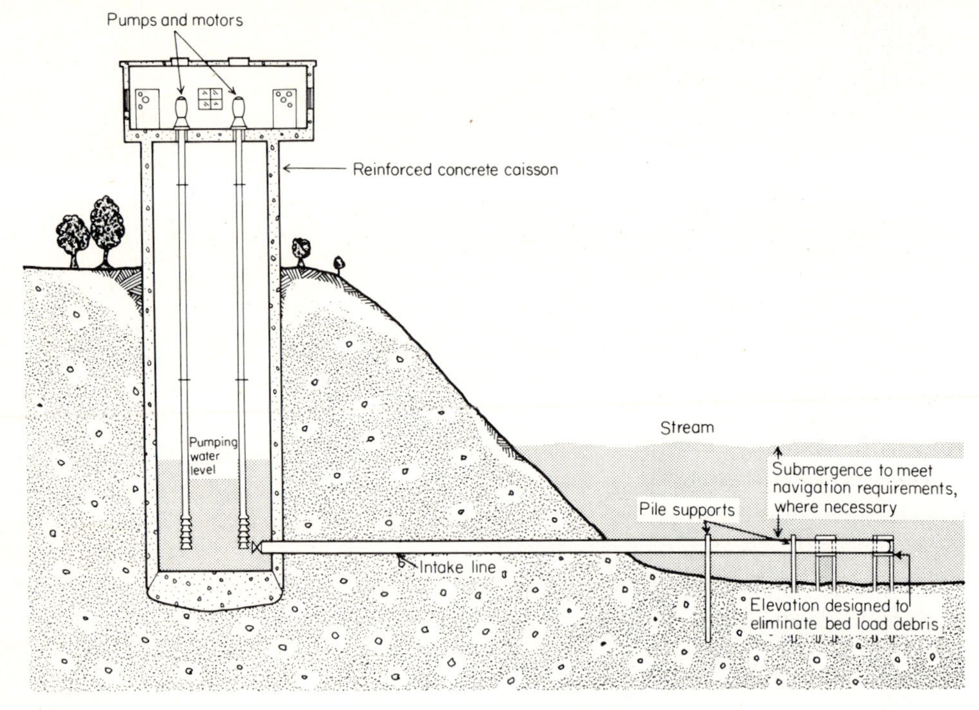

Figure 6-4 Screened pipe intake. *(Courtesy The Ranney Company.)*

least 1 m (40 in) at all times with a clear opening beneath the pipe so that any tendency to form a bar is overcome. Intakes of this type should have provision for reversing flow to clear the strainer openings or screens. On rock bottoms where it is inadvisable, on account of navigation requirements, to have the intake project above the river bottom, an intake box of steel plates has been placed in a trench. On soft bottoms intake boxes have been supported on piles.

Where the river bottom changes considerably, a shore intake is used. It is located on the low-water bank and may consist of a trench or tunnel paralleling the stream with one or more ports leading the river water in. The ports may be protected by bar gratings. It may be advisable to provide facilities for quick removal of silt from the vicinity of the ports. To prevent such silting, low diversion dams have been built to deflect the main current past the intake during low-river stages.

River intakes are especially likely to need screens to exclude large floating matter which might injure pumps. Submerged crib intakes may use wood bars for this purpose; other intakes employ vertical gratings of steel bars. Automatically cleaned bar screens can be obtained from various manufacturers and are sometimes used (see Chap. 9). It may not be necessary to operate the automatic cleaning feature at all times. Movable fine screens, if installed, are usually placed in the more accessible pump well at the entrance to the suction pipes.

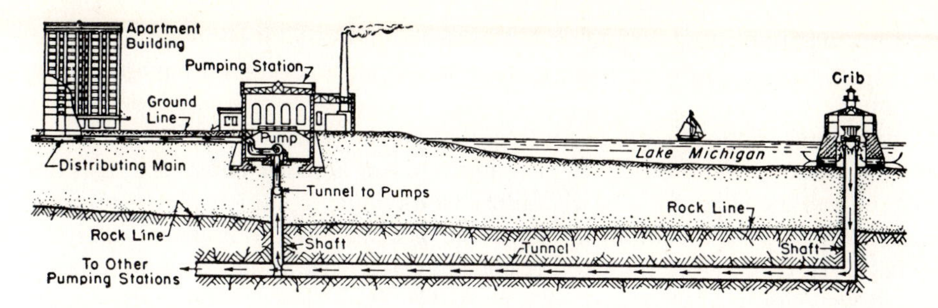

Figure 6-5 Intake, tunnel, and pumping station, Chicago. *(Civil Engrng vol. 20, no. 11, Nov. 1950.)*

6-5 The Intake Conduit

A shore intake may also be the supply well for the suction pipe leading to the pumps. Intakes located long distances from the pumps usually deliver their water to the pump well at the shore end by gravity. This necessitates a large pipe or conduit so that velocities will be low but not low enough to allow sedimentation. The conduit may be a submerged pipe or a tunnel. Tunnels are expensive but less likely to be damaged than are pipes. A submerged pipe should be protected by burying it in a trench or by surrounding it with rock or holding it in place with piling.

THE DISTRIBUTION SYSTEM

6-6 Methods of Distribution

Water is distributed to consumers in several different ways, as local conditions or other considerations may dictate. The methods are:

1. Gravity distribution. This is possible when the source of supply is a lake or impounding reservoir at some elevation above the city so that sufficient pressure can be maintained in the mains for domestic and fire service. This is the most reliable method if the conduit leading from source to city is adequate in size and well safeguarded against accidental breaks. High pressure for fire fighting, however, may be obtainable only by using the motor pumpers of the fire department.
2. Distribution by means of pumps with more or less storage. In this method the excess water pumped during periods of low consumption is stored in elevated tanks or reservoirs. During periods of high consumption the stored water is drawn upon to augment that pumped. This method allows fairly uniform rates of pumping and hence is economical, for the pumps may be operated at their rated capacity. Since the water stored furnishes a reserve to care for fires and pump breakdowns, this method of operation is fairly reliable. Motor pumpers

must ordinarily be used for higher fire pressure, although it is possible to close the valves leading to the elevated storage tanks and operate a fire pump at the pumping plant.

3. Use of pumps without storage. In this method the pumps force water directly into the mains with no other outlet than the water actually consumed. It is the least desirable system, for a power failure would mean complete interruption in water supply. As consumption varies, the pressure in the mains is likely to fluctuate. To conform to the varying consumption several pumps are available to add water output when needed, a procedure requiring constant attendance.

   The peak power consumption of the water plant is likely to occur during periods of otherwise high current consumption and thus increase power costs. An advantage of direct pumping is that a large fire service pump may be used which can run up the pressure to any desired amount permitted by the construction of the mains.

6-7 Storage Necessary

Water is stored to equalize pumping rates over the day, to equalize supply and demand over a long period of high consumption, and to furnish water for such emergencies as fire fighting or accidental breakdowns. Some of the storage will be elevated, except where direct pumping is used.

Elevated storage is furnished in earth or masonry reservoirs situated on high ground or in elevated tanks or standpipes. Where standpipes are used the amount of water available for fire use is that above the level which gives a residual pressure of 140 kPa (20 psi) at the fire pumps. Elevated tanks, usually of steel and in capacities up to 15,000 m^3 (4×10^6 gal) can be purchased from firms specializing in their design and erection. Several elevated tanks may be used at suitable points. The best and most economical arrangement is as shown in Fig. 6-6, the tank being situated on the opposite side of the high-consumption district from the pumping station. During periods of high water use, the district will be fed from both sides, a condition that will reduce the loss of head in the water mains to about one-quarter of what it would be without elevated storage. If the elevated tank is placed at the

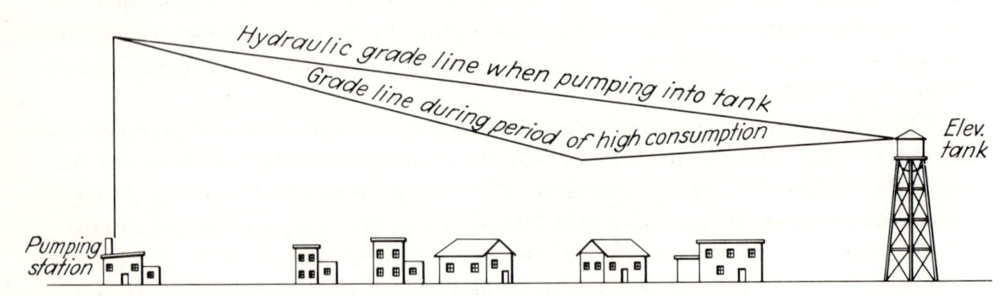

Figure 6-6 Hydraulic grade line with use of an elevated tank.

pumping plant, which is not uncommon, the results are a poor hydraulic gradient and low pressures in outlying areas.

The capacity of the elevated tank or tanks will depend upon the load characteristics of the system, which should be carefully studied before any decision is made. To equalize the pumping rate, i.e., allow a uniform rate throughout the day, will ordinarily require storage of 15 to 30 percent of the maximum daily use. Future increases in demand must, of course, be considered. Figure 2-2 shows the variation in water consumption on the maximum day in Sheboygan, Wis. The average rate of use is 41,466 m^3, or, with 43,000 persons served, 964 l/day per capita. The maximum hourly rate is 1525 l per capita, occurring at 6:30 P.M. Obviously if the pumps can operate at the average rate, rather than at the maximum rate, economies will result through allowing smaller pumps and operation at better efficiencies. To equalize the flow the excess water pumped during the periods from 9 P.M. to 7 A.M., when consumption is small, should be stored for draft during the period of more than average consumption. The cross-hatched area below the average line will equal that above the line and will give the elevated storage needed. In this case the area represents 60,480 m^3—147 l per capita, or 28 percent of the annual average daily pumpage (the Sheboygan annual average is 529 l per capita daily), and 15.8 percent of the maximum. This amount should, of course, be increased to allow for future growth.

Additional storage may have to be provided for fire fighting. The Insurance Services Office, which grades cities upon their fire defense facilities, considers adequacy, so far as water supply is concerned, to be the ability to furnish the required fire flow, in addition to the average consumption for the maximum day.

The required flow may be entirely pumped or may be provided partially from both pumps and storage. Deficiencies in pumping capacity can be offset by storage, elevated and other, which is available for fire flow and which is in addition to that required to equalize normal demand.

Storage may also be needed to equalize demand over a long-continued period of high use, such as a cold period in winter or a dry period in summer. Such storage is particularly needed when wells of limited capacity are the sources of supply or when water must be filtered. The storage required can be arrived at only through a study of demand records covering at least several years, and again future increases in demand must be considered. Periods of high use must be tabulated, and by constructing mass-diagram curves (Art. 3-23) the storage necessary can be computed. When flow data is not available the Goodrich formula (Art. 2-10) is useful. This matter is also discussed under clear-well capacities for filters (Art. 10-16).

Elevated storage tanks in areas of high consumption and low pressure will increase pressure during peak loads without an increase in sizes of the mains. The tanks have a riser or supply pipe connected to the mains and fill during the night when consumption is small and pressure is high, or they may be filled by a booster pump operating either from the main during low consumption or direct from the source of supply. When consumption is high, the stored water is discharged into the mains and raises the hydraulic grade line in the vicinity. The tanks can be

Figure 6-7 Horton radial-cone bottom elevated tank. *(Courtesy Chicago Bridge and Iron Company.)*

equipped with valves which close the entrance pipe when the tank is full and reopen when the pressure drops during peak demand. Figure 6-7 shows a modern elevated tank. It provides a large amount of storage with little fluctuation in water level.

6-8 Fire Flow[1]

The amount of water required for fire control is dependent upon the character of construction in the area being considered (see Chap. 2). The maximum flow required for an individual fire is 45.4 m^3/min (12,000 gpm) for all purposes. In large communities the possibility of concurrent fires should also be considered. In residential districts the required flow ranges from a minimum of 1.9 m^3/min (500 gpm) to a maximum of 9.5 m^3/min (2500 gpm). Failure to provide the specified flow results in unfavorable insurance rates.

Hydrant spacing is dictated by the required fire flow since the capacity of a single hydrant is limited. A single hose stream is considered to be 0.95 m^3/min

(250 gpm), hence maximum fire flow requires 48 hose connections. The minimum area served by a single hydrant is commonly taken as 3720 m^2 (40,000 ft^2), thus minimum spacing is approximately 60 m (200 ft). In no case should hydrants be more than 150 m (500 ft) apart. Ordinarily hydrants are located at street intersections where streams can be taken in any direction. In high value districts additional hydrants may be necessary in the middle of long blocks.

In some cities tall buildings in high value districts may provide their own elevated storage for maintenance of useful pressure on upper floors as well as for fire protection. Such systems may be connected to external wall hydrants which provide a limited high pressure source of water for fire flow.

6-9 Pressures

There are wide differences in the pressures maintained in distribution systems in various cities. For ordinary service they range from 150 to 300 kPa (20 to 40 psi) in residential districts having houses not over four stories in height; about 400 kPa (60 psi) in residential districts where direct hose streams are used for fire fighting to 500 kPa (75 psi) for commercial districts. Pressures under 350 kPa (50 psi) will not supply a 150 kPa (20 psi) pressure at the faucets of the top floor of a six-story building, while pressures under 200 kPa (30 psi) are unable properly to supply the upper floors of four-story buildings. During heavy fire demands when a pumper is used, a drop in pressure to not less than 150 kPa (20 psi) is permissible in the vicinity of the fire.

The American Water Works Association recommends a normal static pressure of 400 to 500 kPa (60 to 75 psi) as presenting the following advantages:

1. It will supply ordinary consumption for buildings up to 10 stories in height.
2. Effective automatic sprinkler service is possible in four- and five-story buildings.
3. It permits direct hydrant service for a few fire-hose streams, thus insuring quick action by the fire department.
4. A larger margin is allowed in fluctuations of local pressures to meet sudden drafts and to offset losses due to partial clogging or excessive length of service pipes.

While low pressures are adequate for normal service and are used for fire fighting in small towns, the heavy drafts entailed by fires must be met in various ways. Three methods are used: (*a*) The main pumping station has extra pumps which are operated only during fires to raise pressure and furnish more water. This method complicates the pumping station and introduces an element of unreliability, as pumps may fail, check valves may not open completely, and pipes have been known to burst under pressures over 700 kPa (100 psi). In addition, the flow necessary to fight a great conflagration will not be developed. (*b*) Dual systems are used, one of very high pressure for fire service only. This method is expensive and is economically possible only in districts of highest value. A few large cities use this method, developing pressures up to 2100 kPa (300 psi) in the high-pressure system. (*c*) Motor pumpers of the fire department are used to force water from the fire hydrants through the fire-hose lines. This is the most economical and most widely used method of obtaining fire streams.

6-10 Pressure Zones

The topography of a city may require pressure zoning. Most of the city may have normal pressures for all purposes but a low area, if directly connected, may have pressures that are too high, with danger of pipe leaks and breakage, both in the public system and house plumbing. This is remedied by supplying the low area through one or several feeder mains and installing automatic pressure-regulating valves that will maintain any desired pressure on the downstream side. (See Art. 6-26).

6-11 The Motor Pumper

Motor pumpers with pumps capable of supplying 1.9 to 5.7 m^3/min (500 to 1500 gpm) are used. Class A pumpers are rated at 1000 kPa (150 psi) and class B at 800 kPa (120 psi). For cities over 10,000 population, pumps of 2.8 m^3/min (750 gpm) or over are best adapted to requirements. The pumper, when in use, is connected to the fire hydrant, preferably through a 100-mm (4-in) opening by means of a hard suction hose.

6-12 The Pipe System[2]

The system serving the consumers consists of:

The *primary feeders*, sometimes called the arterial mains, form the skeleton of the distribution system. They are so located that they will carry large quantities of water from the pumping plant, to and from the storage tanks and to the various parts of the area to be served. In small cities they should form a loop about 1000 m (3000 ft) or two-thirds of the distance from the center of the town to the outskirts. They should have valves not over 1.5 km (1 mi) apart, and mains connecting to them should also be valved so that interruptions of service in them will not require shutting down the feeder main. In large cities the primary feeders should be constructed as several interlocking loops with the mains not over 1000 m (3000 ft) apart. Looping allows continuous service through the rest of the primary mains even though one portion is shut down temporarily for repairs. Under normal conditions looping also allows supply from two directions for large fire flows. Large and long feeders should be equipped with blowoffs at low points and air relief valves at high points.

The *secondary feeders* carry large quantities of water from the primary feeders to the various areas to care for normal supply and fire fighting. They form smaller loops within the loops of the primary mains by running from one primary feeder to another. They should be only a few blocks apart and thus serve to allow concentration of large amounts of water for fire fighting without excessive headloss and resulting low pressure.

The *small distribution mains* form a grid over the area to be served and supply water to the fire hydrants and service pipes of the residences and other buildings. Their sizes will usually be determined by fire flow requirements. In residential areas, however, particularly where there are heavy water uses for lawn sprinkling, it may be necessary to determine the maximum customer demand.

Pipe sizes. Velocities at maximum flows, including fire flow, in feeders and distribution mains usually do not exceed 1 m/s (3 ft/s), with 2 m/s (6 ft/s) as the upper limit, which may be reached near large fires. Ordinarily the sizes of distribution mains will be not less than 150 mm (6 in), with the cross pipes also 150 mm (6 in) in size at intervals of not more than 180 m (600 ft). Where initial pressures are high, a liberal percentage of larger cross mains may be permitted at longer intervals. In high-value districts 200 mm (8 in) is the minimum size, with cross mains as given above, with 305 mm (12 in) and larger to be used on principal streets and for all long lines not cross-connected. Lines furnishing domestic supply only may be 100 mm (4 in); in this case they should not exceed 400 m (1300 ft) in length if dead-ended, or 600 m (2000 ft) if connected at each end. Lines 50 and 75 mm (2 and 3 in) in size may also be used; in these cases allowable lengths are 100 m (300 ft) and 200 m (600 ft) respectively for lines dead-ended or connected at both ends. Since dead ends frequently cause complaints of odors and red water, they should be avoided if possible.

The importance of the *valves* is frequently overlooked, both in design of new systems and extensions and in maintenance. A sufficient number of properly located valves is a necessity for proper operation and control of the pipe system. (See Art. 6-26.)

6-13 Design of Water Distribution Systems

For small communities which do not have a water distribution system, the design requirements are indicated in Art. 6-12, however, other matters must be considered. Topography will affect pressures, while existing and expected population densities, and commercial and industrial needs, will affect both pipe size and the location and capacity of storage tanks. Expected per capita consumptions plus fire fighting requirements and industrial and commercial needs will permit calculation of pumping capacity, required storage and pipe sizes. Pipe sizes can be computed using methods of Arts. 6-14 to 6-18. The hydraulic computations involved in designing a system can only be approximate, as it is impossible to consider all factors affecting loss of head in a complicated network of pipes. Most communities have a distribution system, but inadequacies frequently develop and future growth is to be expected. With either of these conditions the following procedure is recommended.

1. *Preparation of a master plan.* The city plan, if it exists, can be applied to the water distribution system. If not, a plan must be prepared. The plan will cover a period of about 20 years. It should include expected additions of territory to be served and populations of those additions, expected per capita consumption annually, daily, and hourly, following any trends noted in the existing populations, and maximum consumptions to be expected over a series of days, depending upon distribution and storage conditions, changes in character of present neighborhoods and their probable effects upon water consumption, industrial and commercial needs in various areas, proposed locations of schools and parks and their effects upon water demand, and the possibility of provision of an additional source or sources of supply.
2. *Hydraulic study of the existing system.* This will cover (*a*) analysis of present inadequacies, which will be apparent from records of pressure in the system, pumpage, storage reservoir levels, and consumer complaints; (*b*) prediction of future inadequacies, which will become apparent from step 1 above; (*c*) design of improvements including additional main station pumping capacity, additional primary and secondary feeders, and additional elevated tanks and other storage and booster pumping stations, if necessary.
3. *The improvement program.* The required improvements indicated in step 2 will be related to the master plan but this, for practical use, should be translated into an improvement program which sets forth chronologically when each improvement will be needed and its cost. Good planning will also relate this to the financial picture of the water utility and suggest how the improvements are to be financed and their effects upon waterworks revenues.

6-14 Flow in Pipes

Since flow will be turbulent in pipes used for water supply the friction factors depend upon the roughness of the pipe and also upon the Reynolds number, which, in turn, depends in part upon the velocity in the pipe and its diameter.

Various pipe-flow formulas are available to predict headlosses as a function of velocity in pipes, and of these the Hazen-Williams formula is most used in the design of water distribution systems:

$$v = k(C)r^{0.63}s^{0.54} \tag{6-1}$$

in which v is the velocity in the pipe in distance per second, r is the hydraulic radius of the pipe in the distance unit common to velocity, s is the hydraulic gradient, C is a constant depending upon the relative roughness of the pipe, and k is an experimental coefficient and unit conversion equal to 1.318 (ft/s, ft) or 0.849 (m/s, m) for example. The exponents of the formula were selected based upon records of experiments with the pipes and channels ordinarily used in waterworks practice.

The following table contains values of C.

Table 6-1 Hazen-Williams coefficients for various pipe materials

Description of the Pipe	Values of C
Extremely smooth and straight	140
Cast iron:	
New	130
5 years old	120
10 years old	110
20 years old	90–100
30 years old	75–90
Concrete or cement lined	120–140
Welded steel, as for cast-iron pipe, 5 years older	
Riveted steel, as for cast-iron pipe, 10 years older	
Plastic	150
Asbestos cement	120–140

Some engineers use a value of 100 for cast-iron pipe for designing water distribution systems—considering this to represent its relative roughness at some future time. This may be too large where very corrosive or incrusting water is encountered. Where such waters are encountered the use of cement-lined, plastic, or asbestos cement pipe or treatment of the water to reduce its corrosiveness will be advantageous.

Although the commonly available hand calculator permits ready manipulation of the Hazen-Williams formula, nomograms such as that shown in Fig. 6-8 remain a useful adjunct in the solution of pipe flow problems. The errors introduced in reading the nomogram are no greater than those involved in estimating C. The figure shown is based upon a value of $C = 100$, but may be used for any pipe when the results are modified as indicated below.

The diagram can be used to solve the Hazen-Williams formula for velocity (or flow) when pipe size and slope of the hydraulic grade line are known, to determine the slope of the hydraulic grade line given the flow (or velocity) and pipe size, or to find the required pipe size given the flow and headloss. A straightedge is used to connect the known quantities and the unknown are read at the intersection of the straightedge with the other axes. Consider a 200-mm pipe, 1500 m in length which carries a flow of 2 m^3/min. A straightedge is laid intersecting the flow on the left and the diameter on the second line from the left. The loss of head per unit length or slope of the hydraulic grade line is read from the third line and the velocity, if desired, from the fourth. The value of s is found to be 10×10^{-3}, from which the headloss is 15 m of water ($\sim$ 150 kPa). In like manner pipe size can be determined from a prescribed flow and headloss. The pipe selected must, of course, be a commercially available size.

To apply Fig. 6-8 to pipe for which C is other than 100, the values obtained should be multiplied by the factors calculated from Eqs. (6-2) through (6-4).

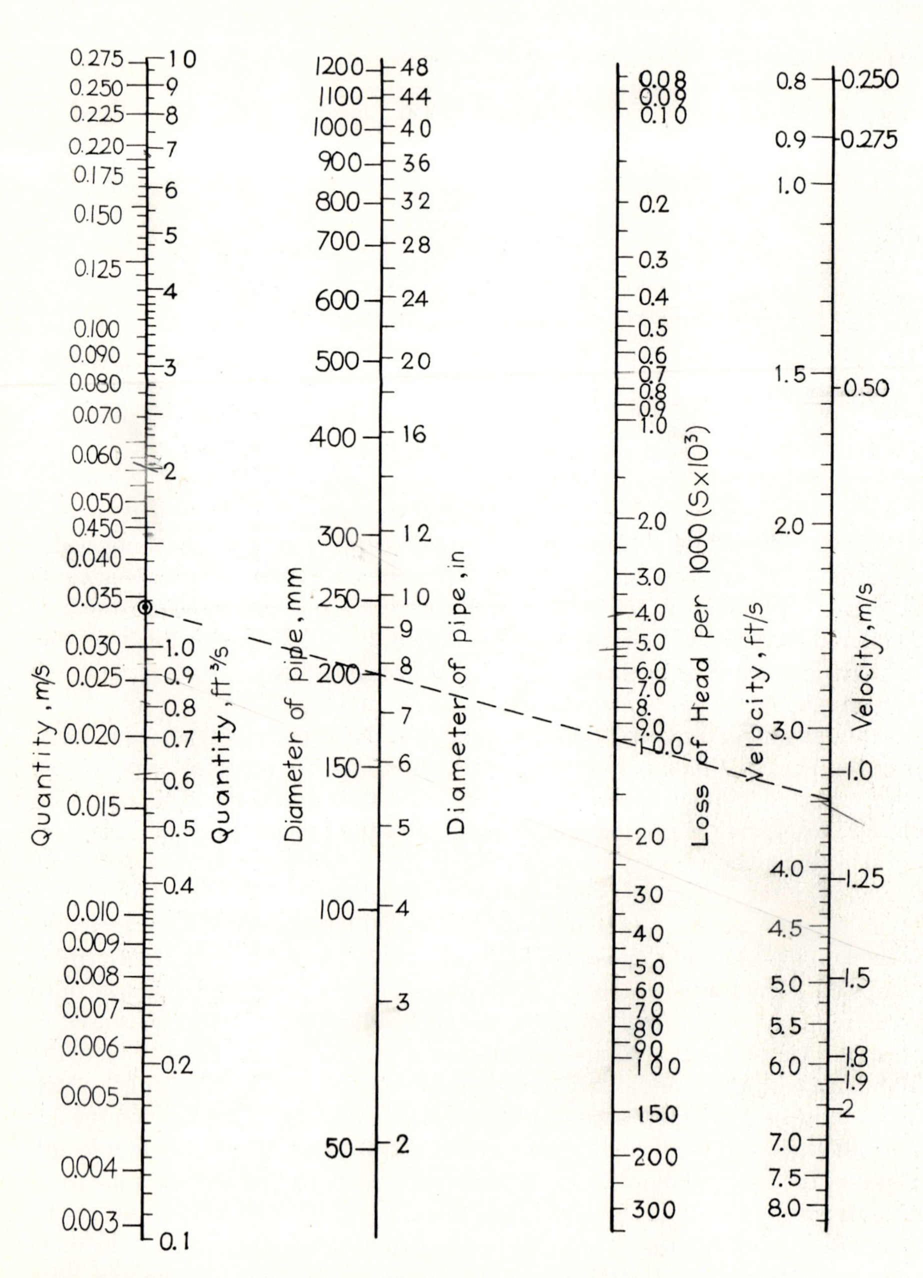

Figure 6-8 Flow in old cast-iron pipes. (Hazen-Williams $C = 100$.)

Given flow and diameter, find s from nomogram.

$$s_c = s_{100}(100/C)^{1.85} \tag{6-2}$$

Given flow and s find diameter from nomogram.

$$d_c = d_{100}(100/C)^{0.38} \tag{6-3}$$

Given diameter and s, find flow from nomogram.

$$Q_c = Q_{100}(C/100) \tag{6-4}$$

The Manning formula (Art. 15-2) is also applicable to water pipes if the proper value of n, the roughness coefficient, is used. For cast-iron pipe that will be used over a period of years it should be not less than 0.015.

6-15 Equivalent-pipe Method

Computations are easier if small loops within the grid are replaced with hydraulically equivalent pipes. For example, to determine the diameter and length of pipe that will replace, in later computations, the system shown in Fig. 6-9, the following procedure can be followed:

First equivalents to internal lines in series are found. It is necessary, for example, to find a pipe which will produce the same headloss as lines *BD* and *DE* when carrying the same flow. Assuming a flow of 2 m^3/min exists from *B* to *E*, the headloss, from the nomogram, is:

$$\begin{aligned} \text{Loss of head in } BD &= 0.01 \times 500 = 5 \text{ m} \\ \text{Loss of head in } DE &= 0.04 \times 250 = 10 \text{ m} \\ \hline \text{Total loss} &= 15 \text{ m} \\ \text{Average loss of head} &= 15/750 = 0.02 \end{aligned}$$

From Fig. 6-8 a flow of 2 m^3/min can be carried at a headloss of 0.02 m/m in a pipe with a diameter of 170 mm. Since this is a hypothetical pipe the noncommercial diameter is unimportant. Using the same procedure lines *BC* and *CE* are found to be equivalent to a 160-mm pipe 500 m long. It is now necessary to find a single pipe which is equivalent to the parallel lines *BDE* and *BCE*. Since the loss of head from *B* to *E* is independent of the path, an assumption of headloss will permit determination of the flow in each of the parallel lines. A single line which will carry the total flow at the same headloss can then be selected. Assuming the

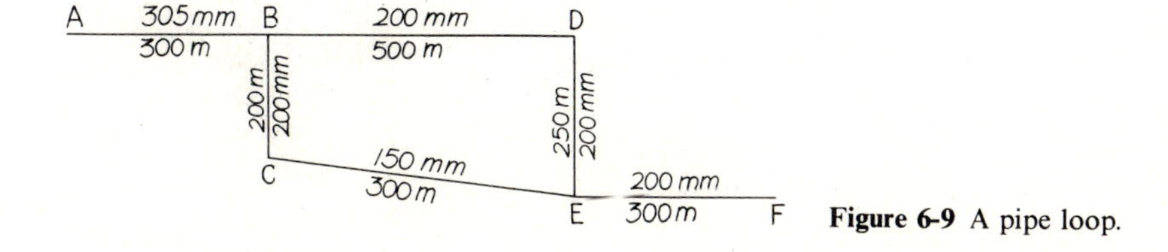

Figure 6-9 A pipe loop.

headloss from B to E to be 7.5 m the slope of the hydraulic grade line in BDE is 0.01 and in BCE, 0.015. From the nomogram a 170-mm line will carry 1.3 m^3/min with $s = 0.01$, and a 160-mm line will carry 1.4 m^3/min with $s = 0.015$. The total flow carried by the equivalent pipe is thus 2.7 m^3/min at a total headloss of 7.5 m. Selecting a length of 600 m for the equivalent pipe, the value of s is $7.5/600 = 0.0125$, and from the nomogram the diameter of the pipe is 213 mm. It is now possible to combine AB, BE_{eq}, and EF. Assuming a flow of 3 m^3/min:

$$\begin{aligned} \text{Headloss in } AB &= 0.003 \times 300 = 0.90 \text{ m} \\ \text{Headloss in } BE &= 0.013 \times 600 = 7.80 \text{ m} \\ \text{Headloss in } EF &= 0.020 \times 300 = 6.00 \text{ m} \\ \hline \text{Total} &= 14.7 \text{ m} \end{aligned}$$

Taking a length for the final equivalent pipe of 1200 m, the value of s is 0.01225, from which $d = 226$ mm. Thus it has been determined that a single line 226 mm in diameter and 1200 m in length is hydraulically equivalent to the system shown in Fig. 6-9.

This method can be applied to many pipe systems, especially long and narrow ones, by eliminating the smaller cross pipes. These play only a small part in carriage of water, and their elimination from the computations has the effect of adding a factor of safety.

6-16 The Circle Method

This method allows design or investigation of the minor pipes of a gridiron. Figure 6-10(a) shows an arrangement of pipes, and the following assumptions are made. The ordinary domestic flow of water is neglected, and only fire demand is considered. The blocks are 75 by 150 m. The large feeder pipes (not shown) of the system comply with the recommendations of the Insurance Services Office, are 1000 m apart, and are themselves gridironed. The pipes cut by a circle of 150-m radius are those furnishing the water needed at the fire. This is based upon the facts that hose lines over 150 m long will furnish little water and that pressure at the hydrants should not be less than 150 kPa (20 psi) if motor pumpers are to be used.

Figure 6-10(a) shows 14 pipes cut by the circle, counting points of tangency as two pipes. If the district is a densely built-up residential section, a maximum of 9.5 m^3/min (2500 gpm) will be needed for fire flow. Each of the 14 pipes cut will carry 0.68 m^3/min which, if they are 150 mm, from Fig. 6-8, gives a loss of head of 5.5×10^{-3}. Each pipe, if the fire is at the mid-point between the large feeders, will be about 500 m long to the feeder supplying it; hence the loss of head in each pipe will be 28 kPa. With a normal pressure in the feeder of 280 kPa, it is evident that pressure at the hydrants in the vicinity of the fire will be ample. An unknown amount of water will reach the area after flowing a large part of the way through pipes paralleling the lines running direct to the feeders. With a 150-mm network, however, this would be small, and in addition the normal domestic demand would tend to prevent any appreciable reduction of the pressure loss from this cause.

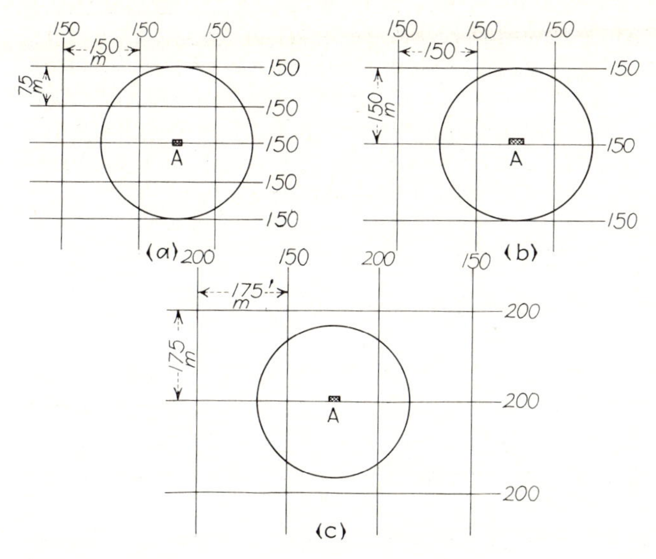

Figure 6-10 The circle method of investigating flow in water mains.

Figure 6-10(*b*) shows blocks 150-m square. The circle in this case cuts only 10 pipes, and each pipe will carry 0.95 m^3/min giving, for 150-mm size, a loss of head of 10 m per 1000 m. This amounts to 50 kPa from the circle to the feeders, which would bring the pressure at the hydrant to 230 kPa. If blocks were 175 m square, only six pipes would be cut, as shown in Fig. 6-10(*c*), and it might be desirable in such a circumstance to replace some of the 150-mm pipe with larger lines to insure usable pressures in the event of fire.

The foregoing discussion applies to a residential area of highest requirements. Usually smaller fire demands are encountered, except in the high-value or business districts, where 200-mm pipes are the minimum required by the Insurance Services Office. The large feeder system is of great importance. With feeders on two sides only, it would be reasonable to consider that all the water needed would reach the area in only half to three-fourths of the pipes cut, thereby increasing the pressure loss and necessitating larger pipes leading to the feeders.

6-17 Fire-flow Tests[3]

Fire-flow tests can be used to determine headlosses in the distribution system during fire flow and to search out points of weakness in the distribution system. The test consists of determining the pressure drop at a fire hydrant for a measured flow from a nearby hydrant or hydrants and using a ratio to determine the headloss for other flows. The equipment required includes a cap to cover a 63.5-mm ($2\frac{1}{2}$-in) outlet of the hydrant at which the pressure is to be read. Into the cap is tapped a 6.5-mm ($\frac{1}{4}$-in) copper pipe which has a tee connection for a finely graduated pressure gauge. At the end of the pipe is a cock to allow air to escape. In some cases a valve is placed between the hydrant cap and the gauge. At the hydrant or hydrants where the flow is measured pitot tubes are used, with gauges attached to measure the velocity pressures. The gauges should read to 350 kPa (50 psi). The assembly is obtainable from manufacturers of pitot equipment. To measure the discharge the pitot orifice is placed at the center of the stream about 30 mm from the outlet. A 63.5-mm ($2\frac{1}{2}$-in) outlet should be used as its cross section will always be full when a hydrant is opened. The formula used to compute the discharge from gauge readings is

$$Q = K\,d^2\sqrt{p} \tag{6-5}$$

in which Q is the flow, d the diameter of the hydrant opening, p the pressure at the gauge, and K a coefficient dependant upon units and equal to 59.01 (m^3/min, m, kPa) or 26.44 (gpm, in, psi). All gauges should be tested at various readings so that correction factors can be applied if necessary.

When a test is made, one or more hydrants are chosen for measuring discharge and one is chosen for reading pressure. If several hydrants are opened for flow, they are grouped about the one observed for pressure. If a dead-end line is being tested, the end hydrant should be the residual pressure hydrant, and the next nearest will be the opened one. The test is begun by opening the air-escape cock of the pressure gauge at the residual hydrant and then opening the hydrant valve. After the air has escaped the cock is closed and the pressure is read. If the needle fluctuates, the center point is read. Then at a signal one hydrant is opened, and both the residual and pitot pressures are read. At a signal the next hydrant is opened and readings are made, and so on until all hydrants to be opened are flowing. Then the hydrants are closed, slowly to prevent water hammer, and the final pressure at the residual should return to the pressure before the test was started. There should be sufficient flow to insure a pressure drop of at least 70 kPa (10 psi) and also not so much that the residual will be reduced to less than 150 kPa (20 psi).

The data used are the corrected residual pressure when all the fire hydrants are open and the total flow. The problem usually is to determine the fire flow which can be furnished with a residual pressure of, say, 150 kPa. Inspection of the Hazen-Williams formula indicates that it may be written as $Q = Kh^{0.54}$, in which K includes the constants, the area of a given pipe and $r^{0.63}$, so long as the pipe size is not changed. It can also be said that the distribution system will be equivalent to a single pipe conveying water to the point where the test was made. If Q_F is the discharge at the pressure loss h_F noted during the test, and Q_R is the unknown discharge at h_R, the desired drop in pressure, then

$$\frac{Q_F}{Q_R} = \frac{Kh_F^{0.54}}{Kh_R^{0.54}} = \frac{h_F^{0.54}}{h_R^{0.54}} \tag{6-6}$$

Example The normal pressure at a hydrant is 350 kPa. At a test, with a flow of 4.5 m^3/min the pressure drop was 100 kPa. What is the allowable fire flow with a residual pressure of 150 kPa?

$$Q_R = 4.5 \frac{(350 - 150)^{0.54}}{100^{0.54}} = 6.54 \text{ m}^3/\text{min}$$

When tests are made on a system supplied by pumps, a record of the discharge pressure at the pumping station is kept during the time of the test. If the pressure at the pumps falls during the test it means that the pipe system is capable of delivering more water than the pumps can deliver at their normal operating pressure. If the capacity of the system is desired in such a case, the h_F used is that noted at the test hydrant minus the drop at the pumping station.

It sometimes happens that an elevated storage tank is located so that it will furnish a considerable quantity of the water discharged during a test. If the storage tank is in service during the test, the fire-flow results will indicate the quantity available only while the tank is in service. If the tank will deliver its contents into the system in a time shorter than the fire-flow duration, the quantity available after the tank is empty should be investigated.

6-18 Hardy Cross Method of Analysis of Flow in a Pipe Network

This method, which was developed by Hardy Cross,[4] consists of assuming a distribution of flow in the network and balancing the resulting headlosses. Pipe-flow formulas are used to determine the headlosses and successive corrections are made in the flows until the heads are essentially balanced.

The Manning, Chezy, and Hazen-Williams formulas are used. They can be expressed in general terms as

$$h = kQ^x \tag{6-7}$$

in which h is the headloss in the pipe, Q is the quantity flowing, and k is a constant, depending upon the size of the pipe, its internal condition, and the units used. The Hazen-Williams formula written in this form will be

$$h = kQ^{1.85} \tag{6-8}$$

It can be said of any pipe in a circuit that

$$Q = Q_1 + \Delta \tag{6-9}$$

in which Q is the actual amount of water flowing, Q_1 is the assumed amount, and Δ is the required flow correction. Then substituting (6-9) in (6-7),

$$kQ^x = k(Q_1 + \Delta)^x = k[Q_1^x + xQ_1^{(x-1)}\Delta + \cdots] \tag{6-10}$$

The remaining terms in the expansion may be neglected if Δ is small compared with Q. For a pipe circuit the sum of the headlosses must be zero:

$$\Sigma kQ^x = 0 \tag{6-11}$$

and from (6-10) above,

$$\Sigma kQ^x = \Sigma kQ_1^x + \Sigma xkQ_1^{(x-1)}\Delta = 0 \tag{6-12}$$

whence

$$\Delta = -\frac{\Sigma k Q_1^x}{\Sigma x k Q_1^{(x-1)}} = -\frac{\Sigma h}{k\Sigma h/Q} \tag{6-13}$$

The procedure can be expressed as follows:

1. Assume any internally consistent distribution of flow. The sum of the flows entering any junction must equal the sum of the flows leaving.
2. Compute the headloss in each pipe by means of an equation or diagram. Conventionally, clockwise flows are positive and produce positive headloss.
3. With due attention to sign, compute the total headloss around each circuit: $\Sigma h = \Sigma k Q_1^x$.
4. Compute, without regard to sign, for the same circuit, the sum: $\Sigma k x Q_1^{(x-1)}$.
5. Apply the corrections obtained from Eq. (6-13) to the flow in each line. Lines common to two loops receive both corrections with due attention to sign.

It will be noted that the heads which make up $\Sigma k Q_1^x$, are obtainable directly from Fig. 6-8, the nomograph, if it is assumed that $C = 100$. It may also be shown that the solution is not dependent upon C, that is, the flow distribution will be the same provided C is the same for all pipes in the network. The method is illustrated below.

Example Figure 6-11 is a network of pipes which has been simplified by elimination of the smaller pipes which form the grid. At point G, the most

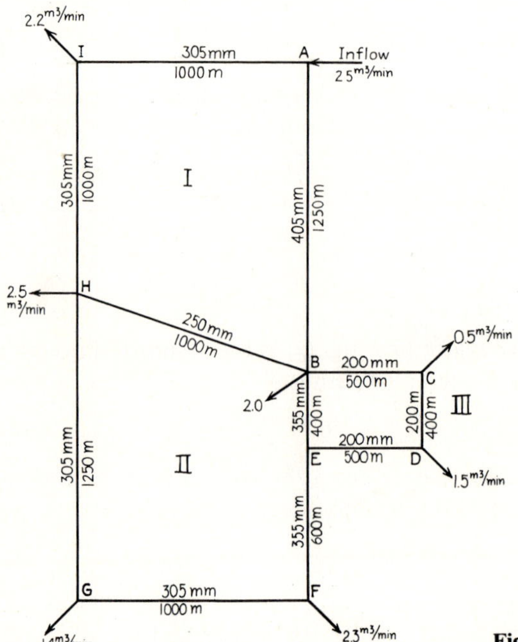

Figure 6-11 Simplified distribution system.

unfavorable as far as loss of head is concerned, it is assumed that a fire plus the domestic flow requires 14 m^3/min. At the other lettered points it is estimated that the amounts shown will be required to satisfy the normal domestic demand. These points are chosen at intersections of the larger cross mains, if any, or at intermediate points between such intersections. In either case the amount of water required is obtained by scaling from a map the area served and using the population density of the area. The demands so obtained can only be approximate, and their accuracy will depend upon the experience and judgment of the engineer. The water enters the system at *A* and in amount equals the total of the domestic demand and fire flow. It is assumed that *C* for the Hazen-Williams formula will be 100.

Figure 6-11, the simplified pipe system, also shows the estimated demands, pipe lines, and their diameters. Figure 6-12 indicates the steps taken in the investigation of the system. The first procedure is to assume directions and amounts of flow in each pipe, remembering that the flow into and away from each intersection must be equal. A number of techniques for correcting the flows are used. The most straightforward involves determining corrections to each loop, applying all the corrections, recalculating corrections for each loop, and continuing until the largest correction is less than some arbitrary amount such as 0.2 m^3/min (50 gpm) which is the approximate limit to which the nomogram may be read. When digital computers are used (Art. 6-19) the limit of correction is specified in the program.

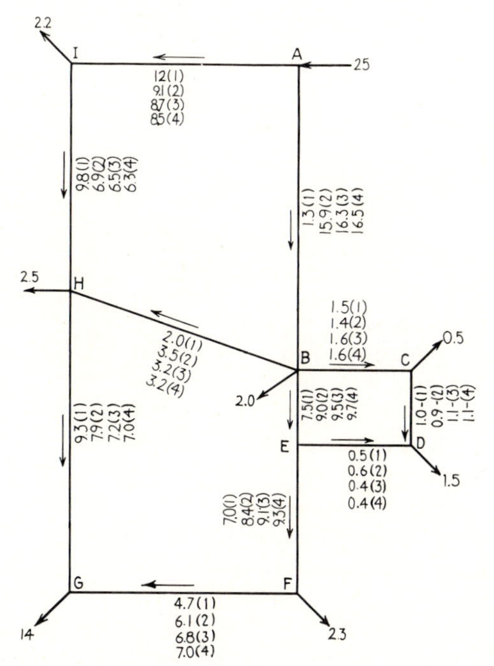

Figure 6-12 Steps taken in investigation of distribution system.

Table 6-2 Hardy Cross analysis
First Correction

Loop I

Line	Flow, m³/min	Dia, m	Length, m	s	h, m	h/Q, m/m³/min
AB	13	0.40	1250	0.0110	13.75	1.058
BH	2	0.25	1100	0.0033	3.63	1.815
HI	−9.8	0.30	1000	−0.0260	−26.00	2.653
IA	−12	0.30	1000	−0.0380	−37.80	3.150
					−46.42	8.676

$$\Delta_{\mathrm{I}} = -\frac{-46.42}{1.85(8.676)} = 2.9$$

Loop II

Line	Flow, m³/min	Dia, m	Length, m	s	h, m	h/Q, m/m³/min
BE	7.5	0.35	400	0.0075	3.00	0.400
EF	7.0	0.35	600	0.0066	3.96	0.566
FG	4.7	0.30	1000	0.0067	6.68	1.423
GH	−9.3	0.30	1250	−0.0236	−29.54	3.177
HB	−2.0	0.25	1100	−0.0033	−3.63	1.815
					−19.53	7.381

$$\Delta_{\mathrm{II}} = -\frac{-19.53}{1.85(7.381)} = 1.4$$

Loop III

Line	Flow, m³/min	Dia, m	Length, m	s	h, m	h/Q, m/m³/min
BC	1.5	0.20	500	0.0058	2.91	1.937
CD	1.0	0.20	400	0.0028	1.10	1.110
DE	−0.5	0.20	500	−0.0008	−0.38	0.762
EB	−7.5	0.35	400	−0.0075	−3.00	0.400
					0.63	4.209

$$\Delta_{\mathrm{III}} = -\frac{0.63}{1.85(4.209)} = -0.1$$

Table 6-3 Hardy Cross analysis
Second Correction

Loop I

Line	Flow, m^3/min	Dia, m	Length, m	s	h, m	h/Q, m/m^3/min
AB	15.9	0.40	1250	0.0157	19.65	1.236
BH	3.5	0.25	1100	0.0094	10.34	2.954
HI	−6.9	0.30	1000	−0.0136	−13.60	1.971
IA	−9.1	0.30	1000	−0.0227	−22.70	2.495
					−6.31	8.656

$\Delta_I = 0.4$

Loop II

Line	Flow, m^3/min	Dia, m	Length, m	s	h, m	h/Q, m/m^3/min
BE	9.0	0.35	400	0.0105	4.20	0.467
EF	8.4	0.35	600	0.0093	5.58	0.664
FG	6.1	0.30	1000	0.0108	10.80	1.770
GH	−7.9	0.30	1250	−0.0175	−21.88	2.769
HB	−3.5	0.25	1100	−0.0094	−10.34	2.954
					−11.64	8.624

$\Delta_{II} = 0.7$

Loop III

Line	Flow, m^3/min	Dia, m	Length, m	s	h, m	h/Q, m/m^3/min
BC	1.4	0.20	500	0.0051	2.55	1.821
CD	0.9	0.20	400	0.0023	0.92	1.022
DE	−0.6	0.20	500	−0.0011	−0.55	0.917
EB	−9.0	0.35	400	−0.0105	−4.20	0.467
					−1.28	4.227

$\Delta_{III} = 0.2$

In Fig. 6-12 the network is subdivided into three loops (*ABHI*, *BEFGH*, and *BCDE*). Any other system might be used, such as *ABCDEFGHI*, *ABHI*, and *BCDE*, provided all lines are included in at least one loop. It is frequently simpler to use internal loops as in Fig. 6-12, and this procedure is followed here. The assumed directions and magnitudes of flows are shown for each line with the first assumption labeled (1). Table 6-2 illustrates a tabular computation useful in such problems. Assumed flows, pipe diameters and lengths are entered, the slope of the hydraulic grade line for each pipe under the assumed flow condition is determined from the nomogram, headloss is calculated with

Table 6-4 Hardy Cross analysis

Third Correction

Loop I

Line	Flow, m^3/min	Dia, m	Length, m	s	h, m	h/Q, m/m^3/min
AB	16.3	0.40	1250	0.0165	20.63	1.265
BH	3.2	0.25	1100	0.0080	8.80	2.750
HI	−6.5	0.30	1000	−0.0122	−12.20	1.877
IA	−8.7	0.30	1000	−0.0209	−20.90	2.402
					−3.67	8.294

$\Delta_I = 0.2$

Loop II

Line	Flow, m^3/min	Dia, m	Length, m	s	h, m	h/Q, m/m^3/min
BE	9.5	0.35	400	0.0116	4.64	0.488
EF	9.1	0.35	600	0.0107	6.42	0.705
FG	6.8	0.30	1000	0.0132	13.20	1.941
GH	−7.2	0.30	1250	−0.0147	−18.38	2.552
HB	−3.2	0.25	1100	−0.0080	−8.80	2.750
					−2.92	8.436

$\Delta_{II} = 0.2$

Loop III

Line	Flow, m^3/min	Dia, m	Length, m	s	h, m	h/Q, m/m^3/min
BC	1.6	0.20	500	0.0066	3.30	2.063
CD	1.1	0.20	400	0.0033	1.32	1.200
DE	−0.4	0.20	500	−0.0005	−0.25	0.625
EB	−9.5	0.35	400	−0.0116	−4.64	0.488
					−0.27	4.376

$\Delta_{III} = 0.03$

attention to sign, and the headloss in each line is divided by the assumed flow. The headlosses in the lines in each loop are then totaled, as are the headlosses divided by flow. The corrections are found as shown, to be $\Delta_I = +2.9$ m^3/min, $\Delta_{II} = +1.4$ m^3/min, and $\Delta_{III} = -0.1$ m^3/min. The corrections are applied to each pipe in each loop. Thus 2.9 m^3/min in a clockwise direction is added to each pipe in loop I, 1.4 m^3/min in a clockwise direction is added to each pipe in loop II, and 0.1 m^3/min in a counterclockwise direction is added to each pipe in loop III. Pipes common to two loops receive two corrections. To the assumed flow in line HB one adds 2.9 m^3/min to the left and 1.4 m^3/min to the right for a net correction of 1.5 m^3/min in the direction initially assumed. Line *BE* is handled similarly, the net correction being 1.5 m^3/min in the direction initially assumed.

The corrected flows are then reanalyzed in precisely the same fashion, and new corrections are obtained as shown in Table 6-3. The procedure is repeated (Table 6-4) until the corrections obtained become sufficiently small, in this case 0.2 m^3/min. The final corrections are applied and the distribution of flow in the balanced network is known. Headlosses from point to point can then be calculated from the nomogram. The pressure distribution in the network is then determined from the headlosses, the pressure at the pumps, and a knowledge of the pipe elevation.

Occasionally the assumed direction of flow will be incorrect. In such circumstances the method will produce corrections larger than the original flow and in subsequent calculations the direction will be reversed. Even when the initial flow assumptions are poor, the convergence will usually be rapid. Only in unusual cases will more than three iterations be necessary. The method is applicable to the design of a new system or to evaluation of proposed changes in an existing system. It may be desired, for example, with a system tentatively designed as in Fig. 6-11 to determine the residual pressure at *G* if the static pressure at *A* is 650 kPa. The headloss may be found for any flow path and is, variously, 49 m of water or 480 kPa. The residual pressure at *G* is thus 170 kPa which is adequate for fire supply assuming *G* and *A* are at equal elevation. If this is not the case the calculation must be modified to include static pressure differences. If the pressure at *G* were found to be inadequate it would be necessary to enlarge some of the pipes and repeat the calculation.

6-19 Computers and Electric Analyzers for Pipe Networks

Both analog and digital computers are useful in water distribution system analysis. The McIlroy pipe network analyzer is an analog type and has been used by a number[5,6] of cities. It requires the use of special vacuum tubes, called fluistors, in which the voltage drop is proportional to the current to the 1.85 power. Such a voltage drop is analogous to headloss caused by pipe friction. Current inputs and take-offs equivalent to water inputs and take-offs proposed for a water system can be made, and voltage changes noted. Effects of proposed new water mains can also be investigated by replacement of tubes with others equivalent to larger lines. Such an electric network is expensive, but large cities have found it convenient to set up one permanently. Some are also available on a rental basis. Other more general types of analog computers are more elaborate and depend upon more complicated electronic circuits. These computers may allow a wider range of variables to be introduced instantaneously, and the machine can solve a wider range of problems.

The application of digital computation techniques to the Hardy Cross method is direct, and a wide variety of programs are available ranging from relatively simple ones[7] to models capable of handling the largest and most complex systems, including booster pumps, storage towers, pressure reducing valves, and other such features.[8] Other analytical techniques which may be advantageously applied through the use of digital computers include the finite element method[9] in which convergence is more rapidly and certainly achieved than with the basic Hardy Cross procedure, and minimum route matrix analysis[10] which will select the least expensive combination of standard pipe sizes to serve selected nodes in the distribution system.

6-20 Two-main System

The two-main system of water distribution is illustrated by the practice of Chicago, where two lines are laid when placing new mains in wide boulevards, newly widened traffic arteries, and all streets 25 m (80 ft) or more in width. Destruction of paving when repairs are needed is the main consideration in adopting this system. The pipes are located near the curbs, either on the street or on the sidewalk side, usually the latter, especially if there is a grass plot between sidewalk and curb. One pipe is the feeder main and supplies the fire hydrants and domestic service on its side of the street, while the other serves only the domestic supplies on the other side. Figure 6-13 illustrates a method of laying out a two-main system.

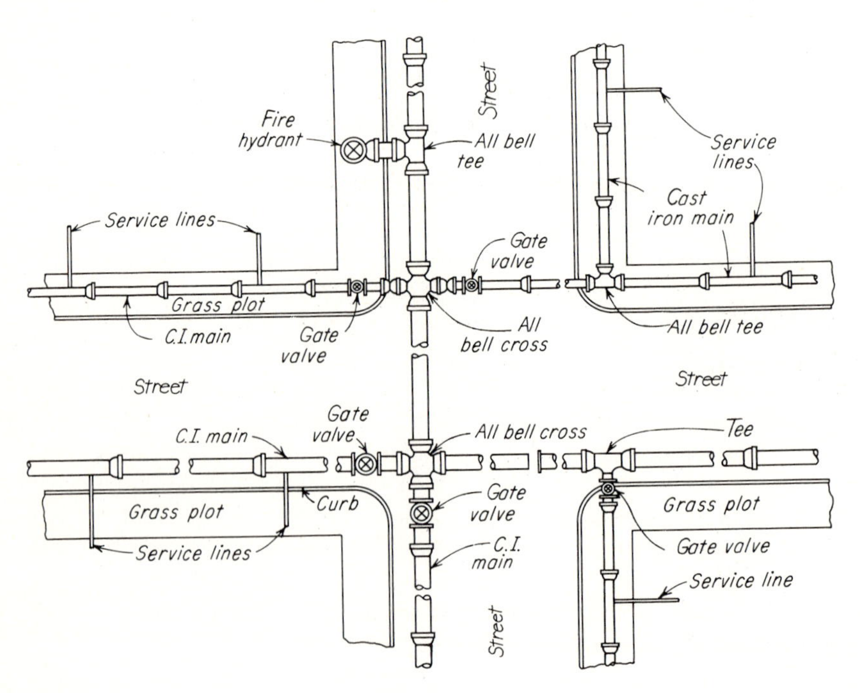

Figure 6-13 Diagram of two-main system. *(Courtesy W. T. Miller, Cast-Iron Pipe Research Assoc.)*

The two-main system frequently will compare favorably in cost with single mains. The advantages are that repairs can be made and new services laid without interfering with traffic or damaging the pavement, repairs can be made at smaller cost, leakage is probably reduced, since service lines are reduced in length, and these are especially liable to have heavy leakage, and service pipes need not be laid where the main passes vacant lots, thereby reducing idle investment. Such unused services not uncommonly show serious leaks also.

6-21 The Service Pipe

The pipe extending from the main to the consumer's meter, or curb stop if no meter is used, is known as the service pipe. With the high-grade paving now in use in cities, service pipes must be durable, or repairs will necessitate expensive and unsightly breaks in the paving. Cities have learned by costly experience that it is advisable to replace old services of short-lived material with copper, or cast iron. Cast iron for services is obtainable, however, only in 32-mm ($1\frac{1}{4}$-in) or larger sizes, with screw joints, bell-and-spigot joints, or precast bell-and-spigot joints. Plastic pipe is now widely used for services (Art. 5-15).

Figure 6-14 shows the usual arrangement of the service connection. A corporation cock is tapped into the main, and a short length of flexible pipe, the "gooseneck," is used to make a connection between the service and the cock. Unless flexibility is provided, unequal settlement will break the service from the main. Soldered or cemented joints unite the gooseneck and the cock and service. A cutoff or curb cock is placed just inside the curb.

When copper pipe is used, the gooseneck can be dispensed with, as the pipe itself can be bent to a curve that will give flexibility. The corporation cock is located in an upper section of the pipe to make a flexible connection more easily obtained and also to reduce possibility of entrance of sediment from the main to the service.

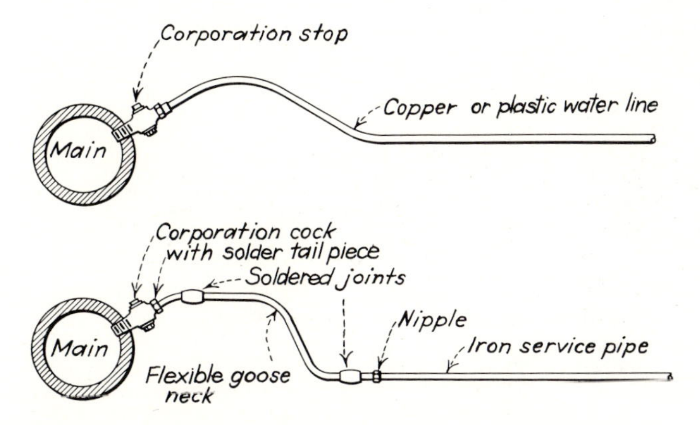

Figure 6-14 Arrangement of service pipes.

Service pipes should be liberal in size to insure good pressure in the building served. Pipes smaller than 20 mm ($\frac{3}{4}$ in) should never be used; and if the service pipe is long, over 15 m (50 ft), 32-mm ($1\frac{1}{4}$-in) or larger pipe is needed for the single-family house, up to 40-mm ($1\frac{1}{2}$-in) for the two-family house, and 50-mm (2-in) for an apartment house having not more than 25 families. Corporation cocks requiring over a 50-mm hole are not ordinarily tapped into the mains. To avoid weakening the mains, large services are supplied by a number of cocks with flexible goosenecks uniting at the beginning of the service pipe.

6-22 Meters

While not all cities measure the water furnished to each consumer, the practice of metering is well established. In large cities where universal metering is not practiced, large consumers, such as factories and hotels, are usually metered. Meters are generally furnished and maintained by the water department, although there are exceptions to this rule, the consumer in a few cities being required to pay for the meter or to pay a rental charge. City ownership is a far simpler arrangement.

Meters may be actuated by a nutating (wobbling) disc, an oscillating piston, or a rotor. The first two have been widely used in American waterworks practice, but are likely to be replaced in future applications by the latter, which has been developed to the point at which it is sufficiently accurate even under low flow conditions.[11]

In *nutating disc meters* a hard rubber or polymeric disc oscillates in guides, rotating once with each filling and emptying of the meter chamber. *Piston meters*, similarly, are positive displacement devices which measure filling and emptying of a chamber of known volume. *Rotor meters* are, in effect, current meters which measure the velocity of flow, with volume being inferred from the velocity and meter dimensions.

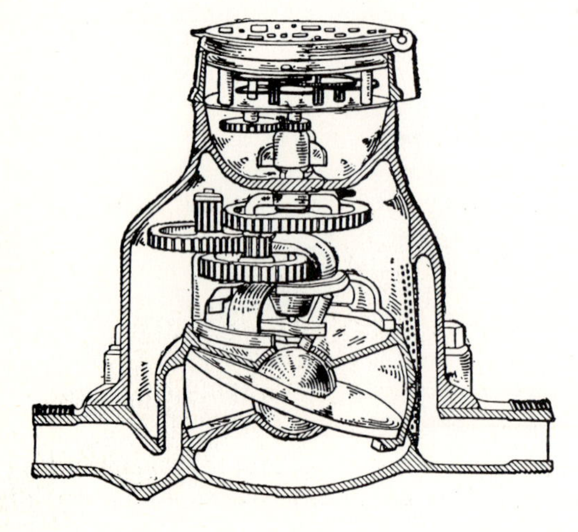

Figure 6-15 Worthington disk meter.

Water meters have traditionally been constructed of bronze and other corrosion resistant alloys. Such materials are expensive and the recent development of plastic meters is likely to result in their gradual replacement. Plastic meters, in addition to being lower in cost, are noncorrosive, smooth surfaced, light, strong, and easily serviced.[12]

Meters are still generally read by monthly or bimonthly visual examination. Systems have been developed, however, which permit readings to be taken by direct transmission,[13] or direct recording on magnetic tape via a receptacle at or near the meter installation.[14] As replacement of existing hardware occurs it is likely that such equipment will become more widely used. Meters are fairly accurate in the ranges for which they are designed. Large- and medium-sized cities usually test the meters they buy for over- and underregistration according to the standards of the American Water Works Association.

A compound meter includes a meter of large capacity placed in the main service line to register large flows and a small bypass meter to measure small flows, together with automatic valves for diverting the small flows to the bypass. In this manner fair accuracy is obtained under all conditions.

Meter upkeep is usually paid for by cities, except for damage caused through negligence of the consumer, such as violence, freezing, or allowing hot water to back up from a heater, an occurrence that may cause permanent damage. Frost damage is overcome in some meters by allowing the expansive force of the ice to be relieved through breakage of the bottom of the meter. The bottom can then be replaced at nominal cost. Placing meters in cellars rather than at the curb reduces freezing troubles but increases reading costs.

Practice varies in testing of meters, with some waterworks superintendents holding that the saving does not justify the expense of periodical testing. The character of the water will have much to do with the matter, a corrosive water or one bearing much sediment being most likely to cause trouble. Many cities test regularly at intervals of 5, 3, or 2 years. Some cities test meters 50 mm (2 in) in size and over each year. Other cities test when the meter has registered 3000 m^3 (100,000 ft^3) or when fluctuations in readings indicate inaccuracy. The inaccuracies are usually underregistration.

6-23 Valves

Various types of valves are used in the conveyance of water. Descriptions and fields of usefulness of the more commonly used valves are given here.

Gate valves. When repairs are needed in a distribution system, it is necessary that the water be shut off. Consequently valves must be numerous so that only a small portion of the system need be shut off at one time. Valves are generally placed at street corners where lines intersect. In the usual gridiron system a valve placed on every line at the intersection would mean that only one block would be cut off when a break occurred. This would be costly, however, and

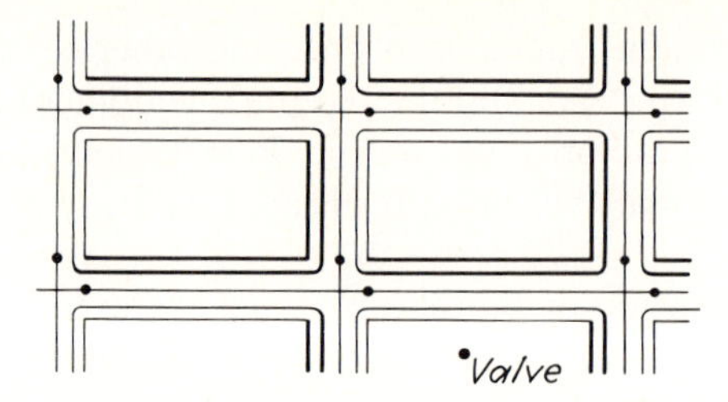

Figure 6-16 Location of valves in distribution system.

commonly the arrangement is as shown in Fig. 6-16, with two valves at each intersection. The Insurance Services Office requires valves to be so located that a shutdown will not necessitate closing a primary feeder (except in the case of a break in the feeder) or of a length of pipe longer than 150 m (500 ft) in a high-value district or 250 m (750 ft) in other districts. It is good practice to standardize valve locations, as shown, with the valve on the extension of the property line. It will be seen that this arrangement will affect only two blocks in case of breakage.

Gate valves are used for the foregoing purpose. Most cities use the double-disk or gate type which is shown in Fig. 6-17. It consists of a corrosion resistant stem and disk in which the disks are hinged so that they will wedge themselves between the tapering seats. The valves may have rising or nonrising stems. The standard type opens when turned to the left, and it therefore has a "right-hand" thread. Left-handed threaded valves may also be obtained. It is highly desirable that all valves in the system work the same way, or confusion and damage will result. The smaller valves generally have their stems set in boxes (see Fig. 6-18) of plastic or cast iron with no portion of the

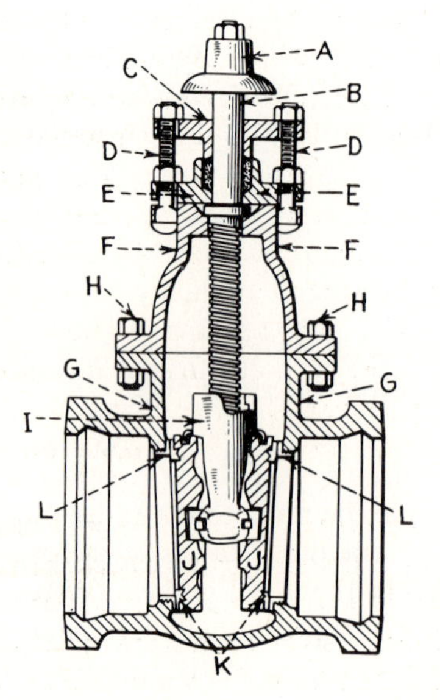

Figure 6-17 Taper-seat nonrising double-gate valve. *A*, operating nut; *B*, stem; *C*, follower; *D*, follower bolts; *E*, stuffing box; *F*, cover; *G*, body; *H*, body and cover bolts; *I*, ball; *J*, gate; *K*, gate ring; *L*, case ring.

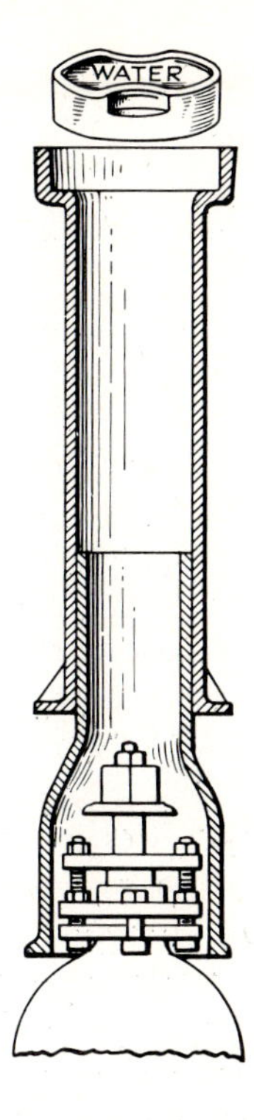

Figure 6-18 Telescope valve box in place.

box that is exposed to traffic loads resting upon the valve. It is good practice to place all valves in manholes, but this is usually done only with large valves. Valves can be obtained with threaded or flanged ends, bell-and-spigot ends, or combinations as needed. They are generally tested to 1000 kPa (150 psi).

Valves which are frequently operated, like those in filter plants, are selected upon a different basis than valves in distribution systems. Materials must be chosen to resist wear, while the pressures may be considerably less than those outside the plant. Most gate valves are designed for installation with the stem in a vertical position, and will not operate properly in other attitudes. If installation is required with the stem at an angle, special valves with disk tracks must be provided. Gate valves which are operated frequently may be actuated hydraulically or electrically as well as manually.

Large valves are subjected to large forces when in the closed position, and may be difficult to open manually. Gearing may be provided in such instances and a small bypass valve which will equalize the pressure on the disk is a desirable feature, both with respect to ease of opening and reduction of hammer when closing (Fig. 6-20).

Check valves permit water to flow in only one direction and are generally used to prevent reversal of flow when pumps are shut down. Check valves installed at the end of a suction line are called foot valves and prevent draining of the suction when the pump stops. Check valves are also installed on pump discharges to reduce hammer forces on the pump. Such valves may be

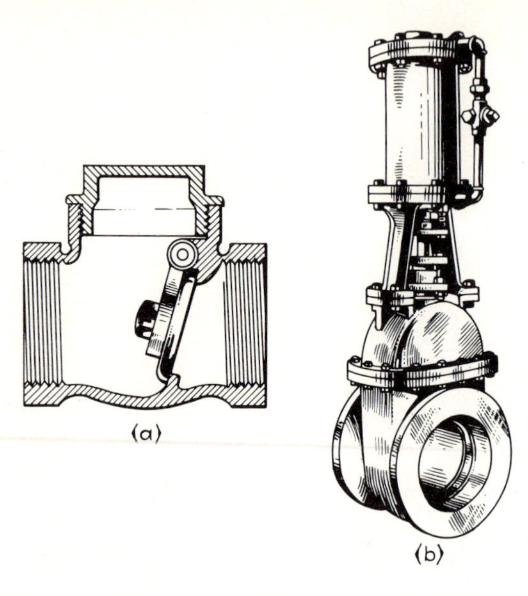

Figure 6-19 (*a*) Swing check. (*b*) Hydraulic cylinder valve with piping and four-way control valve.

Figure 6-20 Extra heavy pressure gate valve. *(Courtesy Clow Corporation.)*

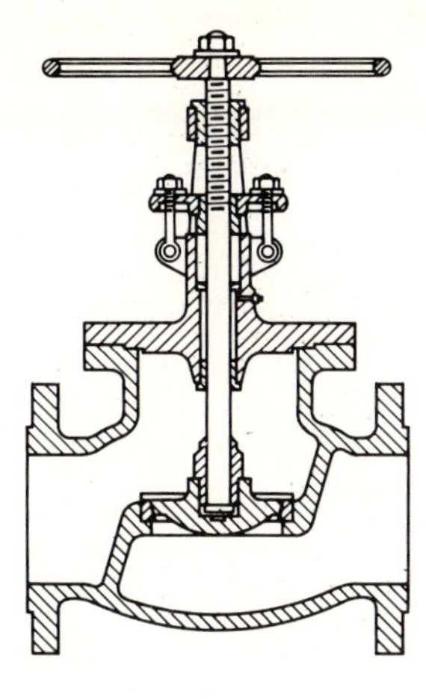

Figure 6-21 Globe valve closed.

simple swing-check or ball devices (Fig. 6-19) in small lines, but in large installations should be designed to close slowly, usually with discharge of some water through a bypass.

Globe and angle valves. Globe and angle valves are seldom used in water distribution systems because of their high headloss characteristics. The primary application of these valves is in household plumbing where their low cost outweighs their poor hydraulics.

Plug valves. These valves, also known as cone valves, have a tapered plug which turns in a tapered seat. A hole in the plug, when the valve is opened, coincides with ports in the seat, and these, in turn, are extensions of the pipe in which the valve is placed. Lubrication of the plug is accomplished by oil applied under pressure by a screw to grooves in the plug and a small chamber beneath it. This type of valve (Fig. 6-22), and its variation the cone plug, is adapted not

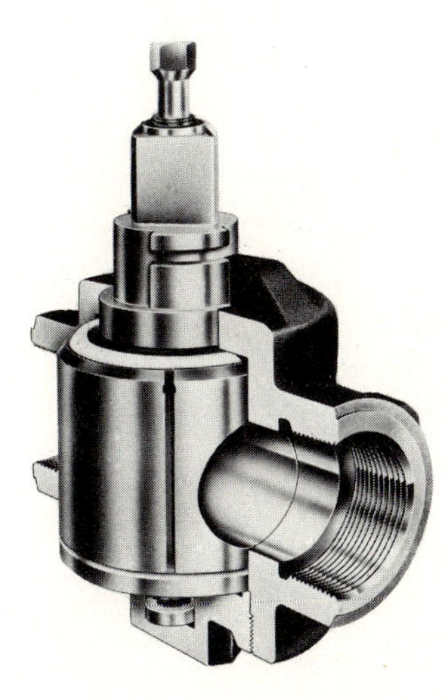

Figure 6-22 Cutaway view of round-port lubricated plug valve. *(Courtesy W-K-M Division, A.C.F. Industries.)*

only for water under high pressures but also for sewage, oil, abrasive liquids, and gases. It can also be obtained with multiple ports.

Butterfly valves. Butterfly valves are widely used in both low pressure applications in filter plants and in distribution systems where pressures may reach 860 kPa (125 psi).[15] They have numerous advantages over gate valves in large pipe lines, including lower cost, compactness, minimum friction wear, and ease of operation. They are not suitable for sewage and other fluids containing matter which might prevent their complete closure.

Pressure regulating valves. These automatically reduce pressure on the downstream side to any desired magnitude and are used on lines entering low areas of a city where, without such reduction, pressures would be too high. They function by using the upstream pressure to throttle the flow through an opening similar to that in a globe valve. An adjustable control permits setting the downstream pressure at the desired level and the valve will throttle itself until that pressure is attained.

Backflow preventers. Backflow preventers are automatic valves which are specifically designed to prevent contamination of water supplies by inadvertant unfavorable pressure gradients. The devices used may be either double check valves or reduced pressure valves. The former close when flow reverses, the latter when pressure drops. The type used depends upon the application.[17]

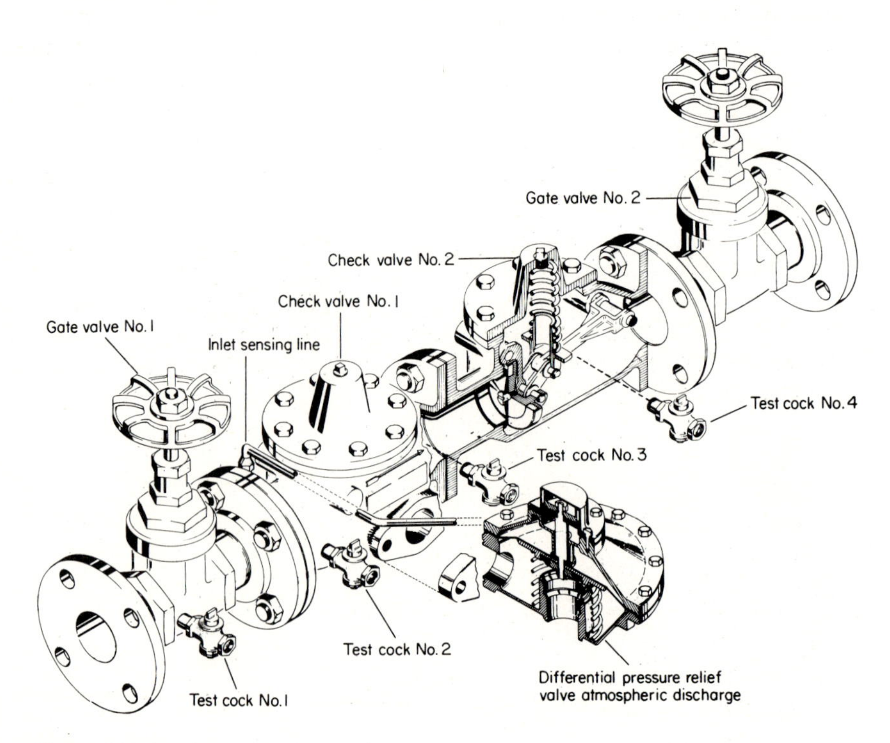

Figure 6-23 Reduced positive pressure backflow preventer. *(Courtesy Cla-Val Co.)*

Figure 6-24 Butterfly valve, 42 in (1050 mm), hand or motor operated. *(Courtesy S. Morgan Smith Company.)*

Altitude valves. These are used to close automatically a supply line to an elevated tank when the tank is full. Flow from the tank is permitted when low pressure below the valve indicates that water from the tank is required.

Air-and-vacuum and air-relief valves. In long pipe lines, air will accumulate in the high points of the line and may interfere with the flow. It is necessary, therefore, to place air valves at those points where trouble is expected. Manually operated valves have been used, but automatic types are preferred. Figure 6-25 is an automatic air-relief valve which will allow accumulated air to escape. As air fills the float chamber, the opening is cleared, and air escapes until the rising water again lifts the float. Automatic poppet air valves, also known as air-and-vacuum valves, allow air to enter the pipe when it is emptied, thus preventing collapsing pressure from the outside atmosphere. In the valve of Fig. 6-25, as the pressure decreases, the float drops and uncovers the opening. When the pipe is refilled air can escape. Clusters of small poppet air valves may be placed at pipe summits. Larger valves may also be used. The size required is a function of the pipe material, the pumping rate, and the rate at which the line drains when the pump is off.

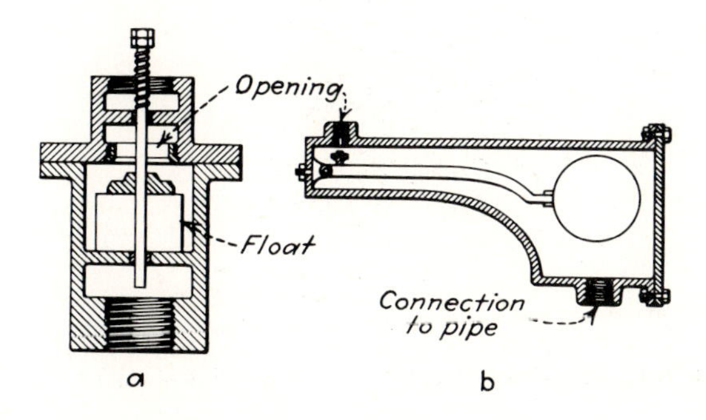

Figure 6-25 (*a*) Automatic poppet air valve. (*b*) Lever-and-float air-relief valve.

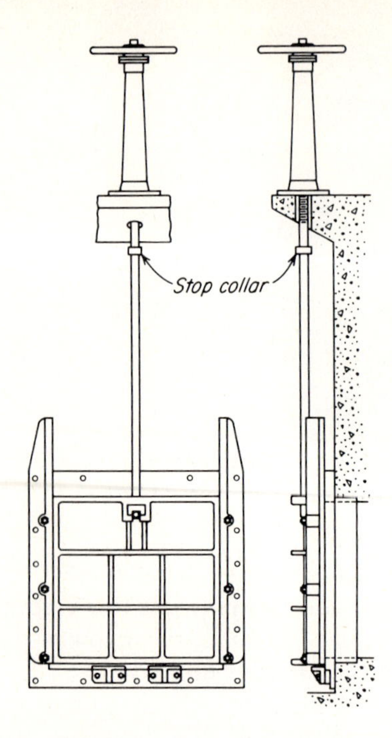

Figure 6-26 Sluice gate.

Sluice gates and shear gates. Sluice gates are vertically sliding valves which are used to open or close openings into walls. They can be obtained with any required length of stem and to be operated manually or by hydraulic cylinder. A light type is used to control flow in open conduits in water and sewage treatment plants. Shear gates are used for small circular openings.

6-24 Fire Hydrants

Hydrants consist of a cast-iron barrel with a bell or flange fitting at the bottom to connect to a branch from the water main, a valve of the gate or compression type, a long valve stem terminating in a nut above the barrel and one or more outlets.

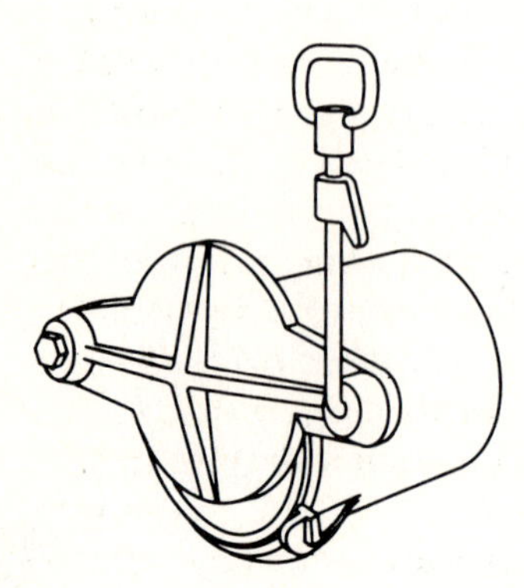

Figure 6-27 Shear gate.

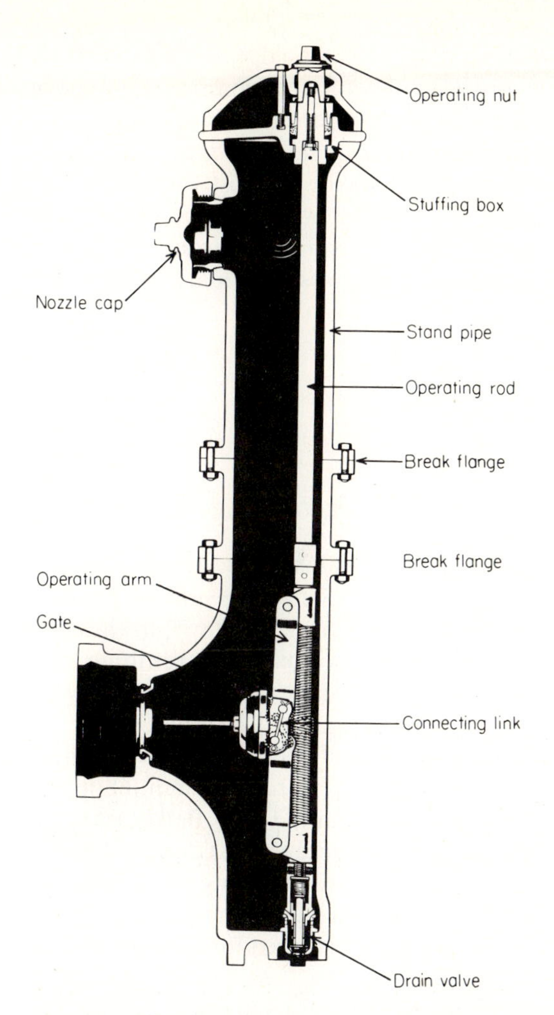

Figure 6-28 A fire hydrant.
(Courtesy Clow Corporation.)

Hydrants should be able to deliver 2.3 m^3/min (600 gpm) with a friction loss of not more than 20 kPa (3 psi) in the hydrant and not over 30 kPa (5 psi) from the street main to the outlet. They should have at least two 63.5-mm ($2\frac{1}{2}$-in) hose outlets and a large pumper outlet where needed. Hydrants are classified as one-way, two-way, three-way, or four-way according to the number of hose outlets provided. Specifications for hydrants have been formulated by the American Water Works Association. Briefly they include the following: (*a*) Pitch of stem thread shall be such that water hammer will not exceed working pressure when the latter is over 400 kPa (60 psi) or be above 400 kPa when working pressure is under this figure. (*b*) If the upper portion of the barrel is broken off, the hydrant must remain reasonably tight. (*c*) When a hydrant is discharging 0.95 m^3/min (250 gpm) from each 63.5-mm ($2\frac{1}{2}$-in) outlet, the total friction loss of the hydrant shall not exceed 10 for two-way, 20 for three-way, and 30 kPa (2 to 4 psi) for four-way hydrants. (*d*) To prevent freezing, a noncorrodable drip valve must be provided to drain the barrel when the main valve is closed. (*e*) The main valve must be faced with a yielding material, such as rubber for the compression type or a bronze ring for the gate type, with a seat of bronze or other noncorrodable material. (*f*) The outlet threads shall conform to those used in the system, preferably with those of the national standard, which for the hose outlets consists of 295 threads per meter ($7\frac{1}{2}$ per inch), with an outside diameter of 77.8 mm ($3\frac{1}{16}$in). This conformity is desirable as it permits fire equipment from a neighboring city to be used in emergencies. Article 6-28 gives information regarding methods of placing hydrants.

CONSTRUCTION AND MAINTENANCE OF DISTRIBUTION SYSTEMS

6-25 Excavation and Backfilling

Great care is not necessary in laying water pipe accurately to grade, but sufficient cover is necessary to give protection against traffic loads and to prevent freezing. For the latter, pipes are placed at depths varying from 2.4 m (8 ft) in the most Northern states to 0.75 m (2.5 ft) in the South. Future conditions resulting from grading and paving of streets should also be considered, or the mains may have to be lowered.

Trenches should be wide enough to allow good workmanship. Required widths range from 460 mm (18 in) for 50-mm (2-in) pipe to 1760 mm (68 in) for 1220-mm (48-in) pipe.[16] Extra excavation is necessary at the bells. Bell holes in the trench are required at each joint. These should be 150 mm (6 in) deeper than the maximum diameter of the bell, and extend 300 mm (12 in) along the pipe above the joint and 600 mm (24 in) along the spigot. An extra width of trench of 250 mm (10 in) on each side for 900 mm (36 in) will also be required for ease in making the joints.

Usually the trenches will not be deep enough to require bracing. Caving is less likely if the earth is piled on each side, but this cannot be done if the pipe is to be rolled into the trench. It is advisable to locate the trench at least 1.5 m (5 ft) from previously disturbed earth, as it is otherwise likely to cave. In rock excavation the rock should be removed so that it is at least 150 mm (6 in) away from the finished pipe line. A cushion of sand or earth, 150 mm (6 in) deep, should be placed between the rock and the pipe.

Backfill material should be free from cinders, refuse, or large stones. Stones up to 200 mm (8 in) in the largest dimension may be used from 300 mm (12 in) above the top of the pipe. Backfill from the trench bottom to the centerline of the pipe should be by hand with sand, gravel, shell, or other satisfactory material laid in 75-mm (3-in) layers and tamped. This material should extend to the trench sides. From the pipe centerline to 300 mm (12 in) above the pipe backfill should be by hand, or if by machine, very carefully done. Backfilling should not be done in freezing weather or with frozen material.

6-26 Pipe Handling and Laying

This includes removal from freight cars, conveying to the job, storage in a yard or on the street, and placing in the trench. Handling of smaller sizes of pipe is facilitated by use of a derrick mounted on a heavy truck. Each length is raised by a sling from the car, sounded for defects which may have occurred in transit, and then placed on the truck or trailer. The derrick truck also facilitates handling at the job. The pipe may be placed directly in the trench without the necessity of stringing it along the street. Without a derrick the pipe must be slid from the truck to the ground, and this cannot be done without possible breakage. Hand lowering into the trench is done by means of a rope at each end with a rolling hitch, one end of each rope being held under the feet of the men, and the other paid out as evenly as possible. When clear of the ground, the pipe should be given another sounding before lowering into the trench.

Several methods of bedding the pipe in the trench bottom are used, and they are shown in Fig. 6-29. The use of blocks, in methods 3 and 4, has the advantage of allowing easy leveling and alignment, but it produces greater crushing loads than methods 1, 2, and 5. When blocks are used they should be placed at two points along the length of the pipe. They should be placed on undisturbed earth and set in slots so that they project about 25 mm (1 in) above the trench bottom. Backfill should be carefully tamped around and under the pipe so that the blocks carry only a small part of the load. Type 2 bedding is generally specified.

6-27 Submerged Pipes

Occasionally, pipe lines must be run through streams or harbors, thus necessitating placing the pipe under water. Cast-iron pipes are generally used for this purpose, and a number of different procedures are employed in the laying. It is advisable to dredge out a trench in the channel bottom in which to lay the pipe. The channel will silt up and thereby give protection. A dredged bottom will be rather irregular, and it may be necessary to place blocking to furnish a suitable foundation. The use of highly flexible joints may make blocking unnecessary.

A number of lengths of pipe may be joined together on a barge and lowered to the trench, with divers making the underwater joints. In shallow water, A frames may be set up over the pipe location, and the connected pipe lowered into place. Pipe joined with flexible joints has been connected on a barge and lowered from the barge in the same manner as a cable would be lowered. Pipes, also having flexible joints, have been joined on one side of a stream and pulled into the previously dredged trench from the opposite side of the stream.

The joints used are frequently of the flexible or ball type shown in Fig. 5-2(*e*). A modified mechanical joint giving a considerable degree of flexibility has also been used. Divers may be employed to inspect and tighten all joints after the pipeline is in place. Joints are tested by compressing air in the pipe and noting the bubbles of escaping air.

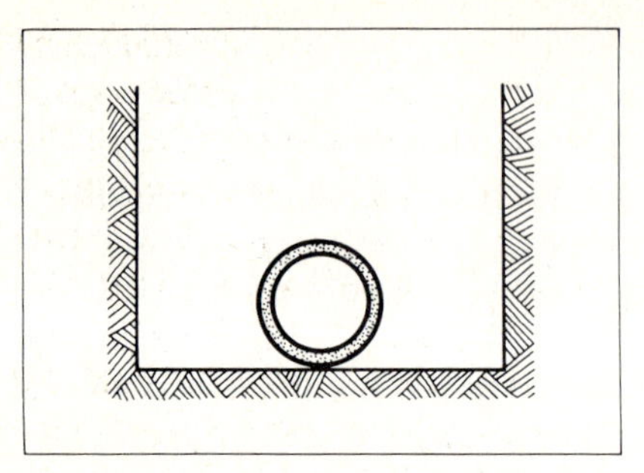

Type 1 Flat-bottom trench. Loose backfill.

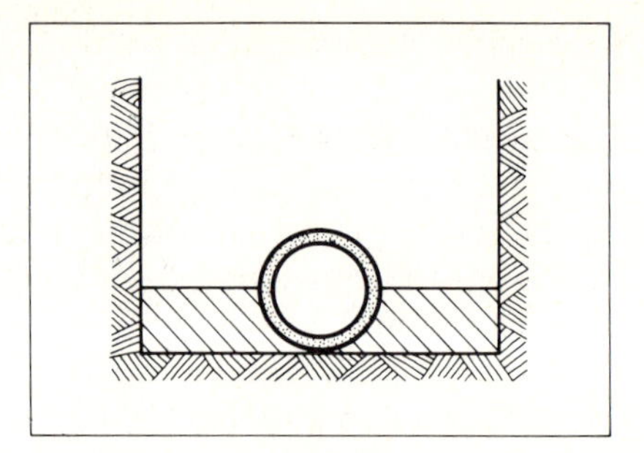

Type 2 Flat-bottom trench. Backfill lightly consolidated to centerline of pipe.

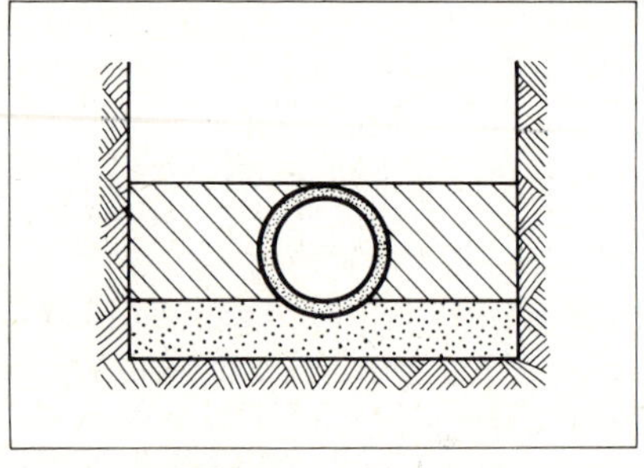

Type 3 Pipe bedded in 100-mm minimum loose soil. Backfill lightly consolidated to top of pipe.

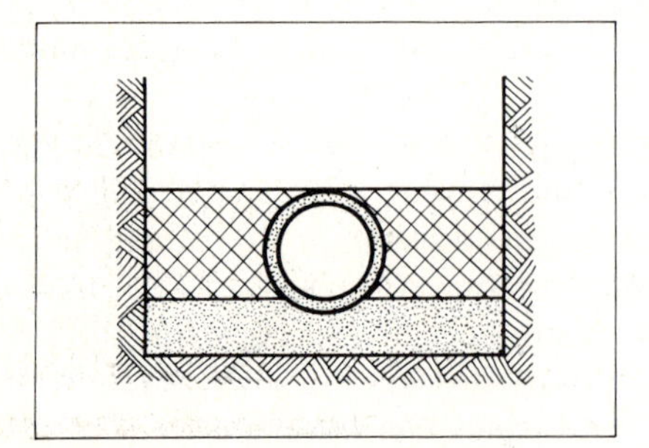

Type 4 Pipe bedded in sand, gravel, or crushed stone to depth of $\frac{1}{8}$ pipe diameter, 100-mm minimum. Backfill compacted to top of pipe. (Approx. 80 percent Standard Proctor, AASHTO T-99)

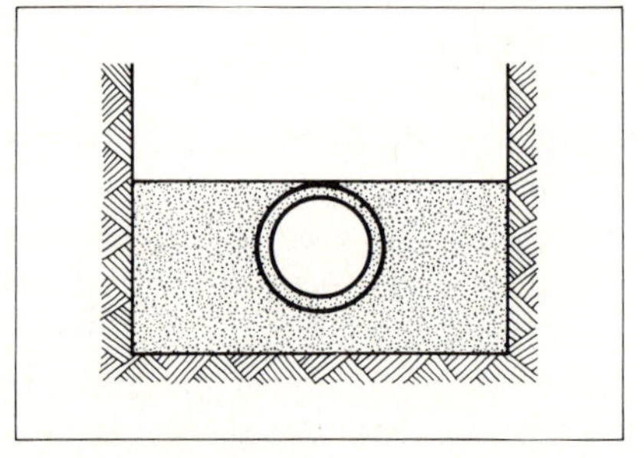

Type 5 Pipe bedded to its centerline in compacted granular material, 100-mm minimum under pipe. Compacted granular or select material to top of pipe. (Approx. 90 percent Standard Proctor, AASHTO T-99).

Figure 6-29 Standard pipe laying conditions.

6-28 Hydrant Placement

If summits and valleys are necessary in the distribution system, it is advisable to place hydrants near them to allow escape of air at proper time intervals and to allow blowing out of sediment. In general, of course, hydrant location will be dependent upon needs for fire protection. No portion of the pumper or hose nozzle cap should be more than 300 mm (12 in) or less than 150 mm (6 in) from the gutter face of the curb. If it is placed between sidewalk and curb or sidewalk and property line, no part should be within 150 mm (6 in) of the sidewalk. Placing the hydrant on a base of concrete will prevent settling. The branch from main to hydrant should not be less than 150 mm (6 in) in size. In order to prevent the hydrant from being displaced by water pressure, it should be braced on the side opposite to the branch entrance by concrete or brick extending to solid earth. Since hydrants have drain holes to prevent their freezing in winter, it is necessary to provide for escape of the drain water. This can be done by surrounding the base with about 0.1 m^3 (3 ft^3) of coarse gravel or crushed rock.

Care should be taken that the water does not drain beneath the street gutter. Frequent use of the hydrant for filling a street flusher and subsequent softening of earth beneath the gutter may cause damage by the flusher or other traffic.

Heavy trucks frequently break off hydrants, thereby causing loss of water before repairs are made. Placing valves in the supply branches is good practice and allows repairs to be made with minimum water losses. Hydrants are made that allow the barrel to break without allowing water to escape and are preferred in most installations (Fig. 6-28).

Hydrants also require maintenance. Observed leakage at the hose outlets means a damaged seat or valve. Since this requires pulling out the hydrant for shop repair, a new or reconditioned hydrant should be available for temporary placement. Pulled hydrants should be carefully inspected for worn or corroded parts that should be replaced. Parts that should be checked are the valve, valve seat, stuffing box packing, barrel opening, opening mechanism, and gaskets. In the field loose nozzles may require recalking and nozzle caps replacing, and their chains may have to be checked for freedom of action. The operating nut may be worn and require replacement. Only a special wrench should be used on such nuts. Street watering crews that use hydrants should be instructed on how to open and close a hydrant. Certain hydrants should be designated for street flushing purposes. Dead-end hydrants are ideal for this purpose.

Maintenance records are important, including a record card for each valve and hydrant in the distribution system.

6-29 Valve Placement

Valve spacing has already been discussed in Art. 6-23. In setting valves, a concrete base should be provided to prevent settling. Each valve must be provided with a valve box or manhole to allow quick access. Generally valves over 300 mm (12 in) in size are placed in manholes. At intersections of large mains there may be two, three, or four valves very close together. In this case the manholes may be dispensed with, and a vault constructed to accommodate all the valves.

6-30 Maintenance of Valves

Inspections made after a period of neglect have shown up to 20 percent of the gate valves to be defective; hence regular inspection is advisable, a practice that some cities follow on an annual or semiannual basis. Defects shown by inspections are of three types, the valves being inaccessible, inoperable, or closed.

Inaccessibility results from covering the valves with earth or paving, or from filling the boxes with earth or debris. Since valves must often be closed in the shortest possible time to prevent damage and loss of water from breaks, the importance of preventing this defect is apparent.

A valve may become inoperable through corrosion, especially if set in one position for many years, so that it cannot be closed, or the stem or some other part may break in closing. Pouring kerosene or diluted lubricating oil down the valve key will lubricate the joint between the valve stem and packing and may prevent trouble. Sediment frequently accumulates in the valve seat and prevents tight closure. This trouble can be corrected by carefully operating the gate up and down and allowing the increased velocity of the water to sweep out the sediment. Valves placed with stems horizontal do not give this trouble, but must be specially fabricated for this use. Valves may leak badly around the stem when operated, making it necessary to uncover the valve and repack the gland. If valves have not been standardized to either right or left turning, they should all be marked so that workmen will not break stems by turning them the wrong way.

Inspection will also show whether or not valves supposedly open are closed. Frequently, after repairs are made, workmen overlook opening one or more valves or do not open them completely, with resulting loss of pressure, a loss that may not be noticed until a fire occurs. It is good practice to require foremen to note on work sheets the valves closed and to check them when opened, the sheets being kept as a permanent valve record.

6-31 Disinfection of New and Old Water Mains

In the process of handling and placing, it is inevitable that newly laid water mains will be polluted. The mains may become polluted during storage on the street before the trench is opened, by the trampled mud in the bottom of the trench, by polluted water which may run into the trench, or by debris which workmen push into the open ends of the pipes and forget to remove in the morning.

Before disinfection the main should be flushed at a velocity of at least 0.76 m/s (2.5 ft/s). The use of a foam or rigid "pig" which is either driven through the line by the water pressure or pulled through by a cable is desirable. Mains have been satisfactorily disinfected with various chlorine compounds, potassium permanganate, copper sulfate, and a variety of heavy metals.

There is no satisfactory substitute for initial cleanliness of the mains. No disinfectant will kill bacteria which are sheltered by debris. Cleaning with "pigs" and flushing at a velocity of at least 0.76 m/s should be followed by filling with water containing a free residual chlorine concentration of at least 1.0 mg/l. A free residual of at least 0.5 mg/l must remain after 24 hours. Following this procedure bacteriological analyses of the water should be conducted to insure its suitability. If total bacterial counts exceed 500/ml or any coliform bacteria are found, the line should be filled with water containing 50 mg/l available chlorine which should not decrease below 25 mg/l in the 24-hour holding period.[18]

When it is necessary to repair or cut into an existing main, disinfection will also be necessary. No rule can be given as to methods, but by the use of fire hydrants for flushing and especially made taps the procedure given above may be followed.

Unsatisfactory bacteriological tests of water in new mains are sometimes caused by contaminated jute or hemp used in the joints. Sterilized material should be purchased and protected against contamination before it is used.

Growths of bacteria, crenothrix, etc., may cause tastes and odors in water passing through old mains. This nuisance has been lessened by killing or at least greatly reducing such growths by applying chlorine to those sections of the distribution system in which the trouble is occurring. While the chlorine is being injected into the mains, fire hydrants are opened until the water from them shows a deep yellow color when tested with orthotolidin.

6-32 Testing

Leakage allowed in new mains is frequently specified in contracts, varying from 5.5 to 23 l/mm diameter per km (60 to 250 gal/in per mi) per 24 hours at the working pressure. Since some jointing compounds, like Leadite, tighten up gradually, a specification calling for immediate testing may prevent the use of Leadite joints. Allowing the test to be made later—say in 30 days—may show the compound joints to be equal to or better than lead joints.

The specifications of the American Water Works Association require that no pipe installation be accepted until the leakage is less than that indicated by the formula

$$L = \frac{ND\sqrt{P}}{C} \tag{6-14}$$

in which L is the allowable leakage, N is the number of joints in the length of line tested, D is the nominal diameter of the pipe, P is the average test pressure during the leakage test, and C is a constant depending on units and equal to 326 (l/hour, mm, kPa) or 1850 (gal/hour, in, psi).

The testing is usually done upon lengths of pipe not exceeding 300 m (1000 ft). The pipe is filled with water, and pressure is applied and maintained by means of a hand pump. Care should be taken that no air has been retained in the pipe being tested. A test pressure of 50 percent above the normal operating pressure for at least 30 min is recommended. It should be recognized that leakage can also occur from the service connections as well as the joints.

6-33 Water Waste Surveys

Undiscovered breaks or blown joints in mains, unauthorized users of water, and unmetered customers who are wasting water may be discovered by waste surveys.[19] They are usually carried on by means of pitometers which are placed in the mains when flow is to be measured. Instead of using the pitometer it is possible to close all the valves on mains leading into a district and supply it through a hose connecting two fire hydrants and with a large water meter in the hose line. By closing valves on all mains but one entering a district, and measuring the flow in the one main at night when domestic use is low, the approximate location of sources of loss may be ascertained. Further restriction of the district by closing other valves and noting the effect upon the inflow rate in the supply main will locate leaks very closely. The approximate location of the leak having been found, the exact spot is located by methods described in the following article.

Waste surveys save enormous amounts of water, sometimes making it possible to defer distribution system enlargement for years. Large amounts of unaccounted-for water will justify a leak survey. Surveys make it possible, in a metered city, to account for at least 85 percent of the output. Pitometers are also useful in determining friction coefficients for large water mains and have sometimes indicated valves in distribution systems and supply mains that were supposed to be open but were partially or completely closed.

6-34 Leak and Pipe Location

The existence of a leak may be known or suspected, but its exact location may be difficult to discover. Various methods of investigation can be employed to advantage. Obviously, presence of melted ice or snow or green grass during a drought will indicate a leak. A steel rod may be thrust into the ground along the pipe line and withdrawn to determine whether or not its point is wet. A metal rod may be driven into the ground to make contact with the main, and the sound of escaping water may be discernible, either by placing the ear against the rod or by means of amplifying apparatus which is placed in contact with the rod. A serious leak in a long pipe line may be located with some accuracy by determining the hydraulic grade line, plotting it, and noting where the leak is causing a change in its slope.

It is sometimes necessary to locate lost underground mains and services. Various electrical magnetic devices are obtainable for discovering buried metallic structures. They are especially useful where there are no other interfering conduits or pipes but may be of little value in the streets of large cities where underground pipes are numerous.

6-35 Cleaning of Water Mains

Because of accumulations of sediment, rust, and growth of tubercles, the carrying power of water mains frequently drops to a very low figure. Cleaning will restore a large part of the capacity. Various types of scrapers are used for this purpose. At the lower end of the line to be cleaned the pipe is broken, and a 45° branch is used to bring the end to the street surface, where it is left open. A special sleeve is inserted at the upper end after a small float with a cable attached has been inserted. The upper valve is opened, and the float passes through. A large cable is then drawn through, and the scraper is inserted at the special sleeve. Again the water pressure is put on, and the scraper is pulled through by means of the cable, the dislodged clogging matter escaping from the open end with the flow of water. Pipes 20 to 50 years old have been cleaned and restored to a C of 120 in the Hazen-Williams formula, which is equivalent to 4-year-old pipe. Such high efficiencies are not retained, however, and recleaning may be necessary in 5 or 6 years. Cement lining of pipes in place (Art. 5-7) will result in permanent improvement.

6-36 Thawing Pipes

In most parts of the country waterworks departments must be prepared to thaw frozen mains and services, particularly the latter. At certain times many services may be frozen at once, necessitating very quick work. Mains, and occasionally services, are thawed by forcing a small steam pipe into the frozen pipes. Electrical thawing is widely used. Gasoline engine-generator sets are used for this purpose with the engine and generator mounted on a truck or with the generator driven by the truck motor. The maximum power required for the engine will probably be 60 kW, which will give 500 A at 100 V. Smaller apparatus may be used, but an engine power of at least 37 kW giving 500 A at 50 V would probably be the minimum required for a large city. In addition to the engine and generator an ammeter and rheostat must be used, so that the amount of current used can be measured and controlled. Flexible insulated copper cable, at least 150 m in 30-m lengths, will also be needed.

Alternating current is sometimes used for thawing by tapping the city power lines. It is, of course, necessary to step down the current from a voltage of 2200 or more to 110 to 55 V, depending upon the size of transformer used. The apparatus necessary will be a transformer, rheostat, ammeter, watt-hour meter if power is paid for, and the copper cable. The total transformer capacity should be 30 to 55 kW, the former for hydrants and short stretches of pipe, the latter for long stretches.

In thawing services the connection is made on the street side of the meter, the pipe being separated from the house piping to insure against damage to radio or other apparatus grounded on the water piping when the thawing current is turned

Table 6-5 Amperages for quick thawing of service pipes

Size, mm	Size, in	Wrought iron, steel, lead	Copper
12	$\frac{1}{2}$	200	500
19	$\frac{3}{4}$	250	625
25	1	300	750
31	$1\frac{1}{4}$	450	1000
38	$1\frac{1}{2}$	600	1500
50	2	800	2000

Table 6-6 Amperages for thawing cast-iron pipe

Cast-iron pipe size in	mm	Amperes
4	100	350–400
6	150	500–600
8	200	700–900
10	250	1000–1300

on. In no case should current be applied to pipes that have grounds attached to them. The other connection is made to the nearest fire hydrant located on the main supplying the service. Up to 50 V may be required for services, but usually far less will be needed. Mains are thawed by 200-m (600-ft) lengths if 110 V can be used; while if only 55 V can be maintained at the desired amperage, only 100-m (300-ft) lengths should be thawed at one time.

Tables 6-5 and 6-6 give amperages needed as suggested by Sheppard.[20] To thaw hydrants, use amperages of the size of pipe supplying the hydrant. Lower amperages will require longer times for thawing. According to Sheppard, a 12.5-mm ($\frac{1}{2}$-in) steel or iron service will be thawed in 10 min with 200 A, while with 100 A, 40 min will be needed. With cast-iron pipes the amperages of Table 6-6 will require $\frac{1}{2}$ to 2 hours. Use of lead substitutes, which are insulating in nature, for joints will retard thawing.

PROBLEMS

6-1 The pipe system in Fig. 6-11 is changed as follows: The input at A is 20 m^3/min and the outflow at G is 9 m^3/min. Line AB is 350 mm and AI is 350 mm in diameter. The elevation at A is 100 m and at G 97.5 m. Determine the distribution of flow and the headloss between A and G. If the pressure at A is 1000 kPa, what will be the pressure at G?

6-2 In Fig. 6-9 the dimensions are changed as follows:

Line	Length, m	Dia, mm
AB	450	250
BC	300	150
BD	600	250
CE	700	150
DE	400	200
EF	400	250

Find a single pipe, 2000 m long, which is hydraulically equivalent to the lines given.

6-3 From the following hourly demand figures find the uniform pumping rate which will just meet the demand and the volume of storage required to meet maximum hourly demand.

Time, A.M.	Rate, m^3/min	Time, P.M.	Rate, m^3/min
12 (midnight)	7.2	12 (noon)	23.9
1	7.1	1	24.5
2	7.0	2	24.4
3	7.1	3	24.2
4	7.0	4	24.3
5	7.3	5	25.6
6	7.5	6	26.9
7	12.2	7	34.0
8	18.9	8	33.0
9	21.0	9	20.0
10	22.8	10	8.4
11	23.5	11	7.6

6-4 Rework Prob. 6-3, assuming that water is only pumped from 6 A.M. to 6 P.M.

6-5 If the community in Prob. 6-3 must also provide a fire flow of 10 m^3/min for a period of 10 hours, what total storage should be provided for uniform 24-hour pumping?

6-6–8 In the pipe systems shown, find the distribution of flow and the pressure at point A. Assume all pipes not otherwise labeled are at identical elevations.

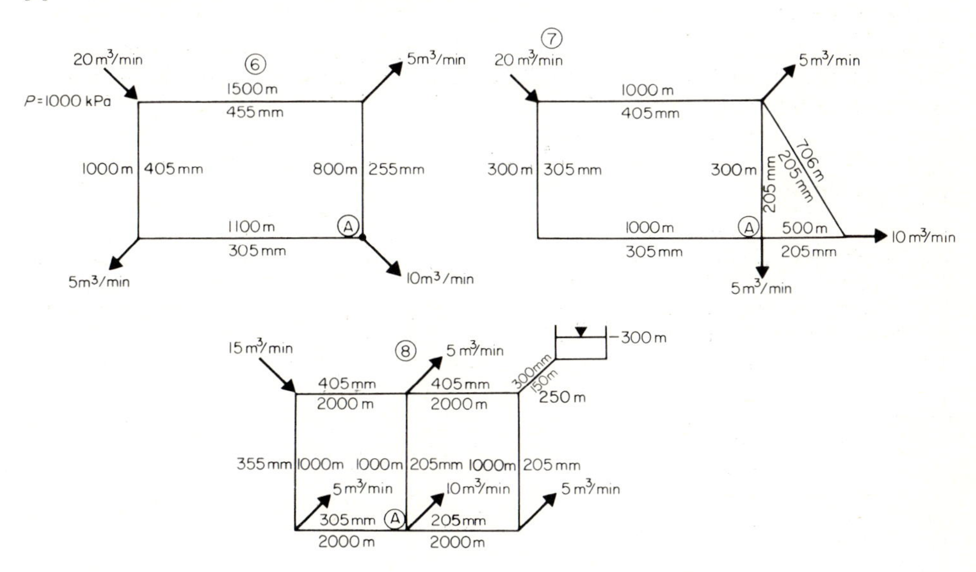

REFERENCES

1. Carl, K. J., R. A. Young, and G. C. Anderson, "Guidelines for Determining Fire Flow Requirements," *Journal American Water Works Association*, **65**:335, May, 1973.
2. Pentecost, Frank C., "Design Guidelines for Distribution Systems," *Journal American Water Works Association*, **66**:332, June, 1974.

3. Bankston, L. V., et al., "Form and Procedures for Fire Flow Tests," *Journal American Water Works Association*, **68**:264, May, 1976.
4. Cross, Hardy, "Analysis of Flow in Networks of Conduits or Conductors," *Bulletin 286*, University of Illinois Experiment Station, 1936.
5. McIlroy, M. S., "Direct Reading Electric Analyzer for Pipeline Networks," *Journal American Water Works Association*, **42**:347, April, 1950.
6. Appleyard, V. A., and F. P. Linaweaver, Jr., "The McIlroy Fluid Analyzer in Water Works Practice," *Journal American Water Works Association*, **49**:1, January, 1957.
7. Rich, Linvil G., "Environmental Systems Engineering," McGraw-Hill, New York, 1973.
8. Alexander, Stuart M., Norman L. Glenn, and Donald W. Bird, "Advanced Techniques in the Mathematical Modeling of Water Distribution Systems," *Journal American Water Works Association*, **67**:343, July, 1975.
9. Collins, Anthony G., and Robert L. Johnson, "Finite Element Method for Water Distribution Networks," *Journal American Water Works Association*, **67**:385, July, 1975.
10. Nakajima, Shigeki, "Improved Design of Distribution Systems by Minimum Route," *Journal American Water Works Association*, **67**:390, July, 1975.
11. AWWA C-708, "American Water Works Standard for Cold Water Meters, Multi-Jet Type for Customer Service," American Water Works Association, Denver, Colo., 1976.
12. Lacina, William V., and J. B. Coel, "Plastic Water Meters," *Journal American Water Works Association*, **68**:246, May, 1976.
13. AWWA C-706, "American Water Works Standard for Direct Reading Remote Registration Systems for Cold Water Meters," American Water Works Association, Denver, Colo., 1972.
14. AWWA C-707, "American Water Works Standard for Encoder Type Remote Registration Systems for Cold Water Meters," American Water Works Association, Denver, Colo., 1975.
15. Popalisky, J. R., "Experiences with Butterfly Valves," *Journal American Water Works Association*, **66**:349, June, 1974.
16. AWWA C-151, "American National Standard for Ductile-Iron Pipe, Centrifugally Cast in Metal Molds or Sand-Lined Molds, for Water or Other Liquids," American Water Works Association, Denver, Colo., 1976.
17. Angele, Gustave J., Sr., *Cross Connections and Backflow Prevention* 2d ed., American Water Works Association, Denver, Colo., 1974.
18. Buelow, Ralph W., et al., "Disinfection of New Water Mains," *Journal American Water Works Association*, **68**:283, June, 1976.
19. Case, E. D., "Water Waste Surveys and Unaccounted-for-Water," *Journal American Water Works Association*, **42**:3, March, 1950.
20. Sheppard, F., "Thawing Water Pipes Electrically," *Water Works Engineering*, **87**:11, May, 1934.

CHAPTER

SEVEN

PUMPS AND PUMPING STATIONS

7-1

A few cities which have supplies originating in mountainous areas can furnish water to consumers entirely by gravity. Usually, however, it is necessary to raise the water by means of pumps at one or more points in the system. Pumps may be needed, therefore, to lift water from a lake, reservoir, or river to a water treatment plant, and after treatment another lift will be needed to force the water into the mains and elevated storage. In the system, booster pumps may be needed at certain points to keep pressure at desirable levels. Where wells are the sources of supply, pumps will be needed to raise the water into a collecting basin, unless the wells are of the artesian or flowing type. From the collecting basin the main pump or pumps will force the water into the mains. Very small cities depending upon a single well may raise the water from the well and into the mains in one pumping operation. For pumping into the mains, standby or emergency pumps will be needed to operate when breakdowns occur or to take care of the great demand incident to a large fire.

The waterworks engineer is concerned with several factors when deciding upon the equipment required to pump water. These are the reliability of the service that will be given, the first cost of the equipment, and the operating cost. The total annual expense involved will include interest upon first cost, plus the annual depreciation and annual operating cost. Other factors being equal, the choice is made so that the sum of these three items is a minimum, although the necessity for reliability and uninterrupted operation is the most important consideration.

7-2 Classification of Pumps

Most pumping machinery used in public water supply systems may be broadly divided into three general classes, reciprocating, rotary, and centrifugal. The reciprocating class typically consists of a piston or plunger which alternately draws water into a cylinder on the intake stroke and then forces it out on the discharge stroke. The rotary type contains two rotating pistons or gears which interlock and draw water into the chamber and force it practically continuously into the discharge pipe. The centrifugal type has an impeller with radial vanes rotating swiftly to draw water into the center and discharge it by centrifugal force.

7-3 Work and Efficiency of Pumps

The work done by a pump is equal to the product of the mass flow and the total head against which the flow is moved.

Head is a common engineering term used to describe hydraulic energy, either kinetic or potential, as equivalent to the potential energy of a column of water of specified height. Head and pressure are interchangeable in that one may be expressed in terms of the other. If one considers a column of water of specific gravity 1.0, the pressure exerted upon a unit area will be a function solely of column height. In common engineering terms it is readily shown that one foot of water is equivalent to 0.433 psi, or that one meter of water is equivalent to 9.8 kPa. Pressure may be expressed in terms of either absolute or gauge readings. Either method may be used in considering pumped systems.

The *total dynamic head* (TDH) of a pump is the sum of the static suction lift, the static discharge head, the friction head, and the velocity head. In many water and wastewater applications the last is negligible. The sum of the static suction lift and static discharge head is called the *total static head*. If the pump inlet is submerged, i.e., if there is a positive head upon the intake, the static suction lift is

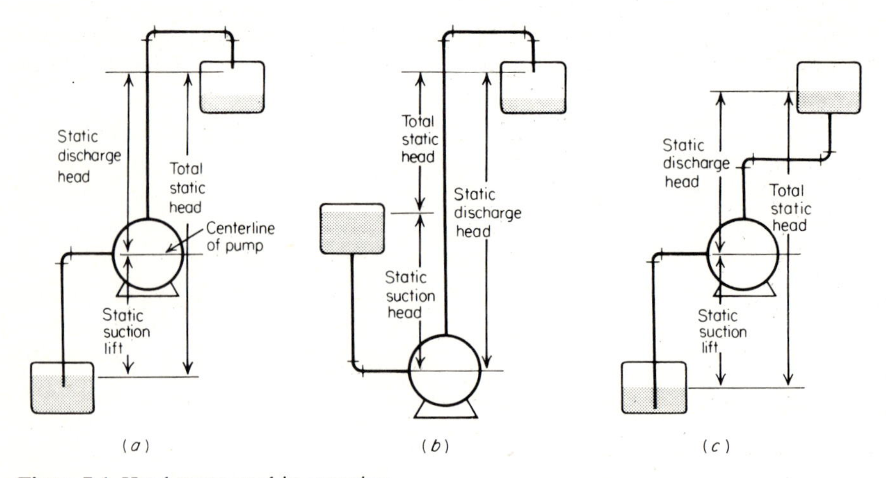

Figure 7-1 Head terms used in pumping.

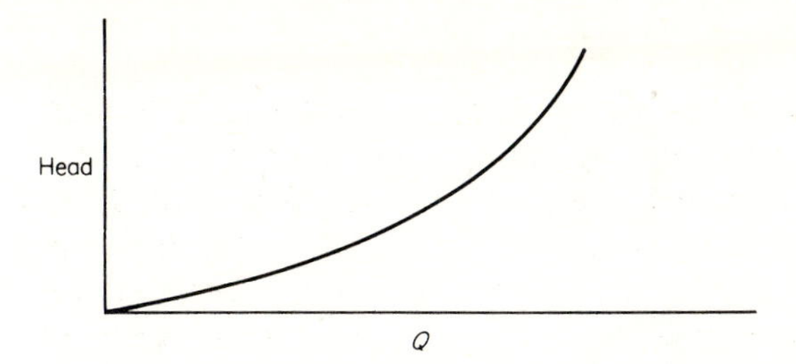

Figure 7-2 System headloss curve.

negative and reduces the total static head. These quantities are illustrated in Fig. 7-1.

The friction head consists of the loss in the pipelines as calculated from various formulas (Chaps. 6 and 15) plus the energy loss produced by flow through various fittings. The latter may be calculated as a function of velocity head ($h_L = KV^2/2g$), where K is related to fitting dimensions (Table 7-1), or may be expressed in terms of an equivalent length of straight pipe (Table 7-2).

Before selecting a pump it is necessary to evaluate the response of the piping system to variations in flow. The headloss in the system is a function of pipe diameter, length, material, and condition, and the number and type of fittings. Calculation of headloss for a number of flow rates permits plotting of a *system headloss curve* like that shown in Fig. 7-2. Such curves always pass through 0, 0 since there is no friction loss at zero flow, and are parabolic upward since friction losses vary approximately as velocity (or flow) to the second power. A *system head curve* is then obtained by adding the total static head to the system headloss curve. (Fig. 7-3).

The total head at any flow rate is available from a curve such as Fig. 7-3 and this permits calculation of the energy necessary to maintain specific flow rates.

The *water power* required is the net output of the pump and is equal to

$$P_w = kQH \tag{7-1}$$

in which P_w is the power, Q is the flow, H is the total head, and k is a constant dependant upon fluid density and units. For water at 20°C $k = 2.525 \times 10^{-4}$ (hp,

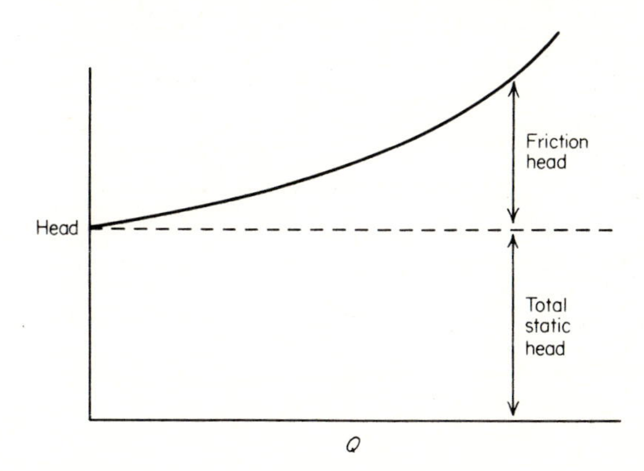

Figure 7-3 System head curve.

Table 7-1 Headloss of typical fittings

(From "Pump Application Engineering" by T. G. Hicks and T. W. Edwards. Copyright 1971 by McGraw-Hill Book Co., Inc. Used with permission of McGraw-Hill Book Company.)

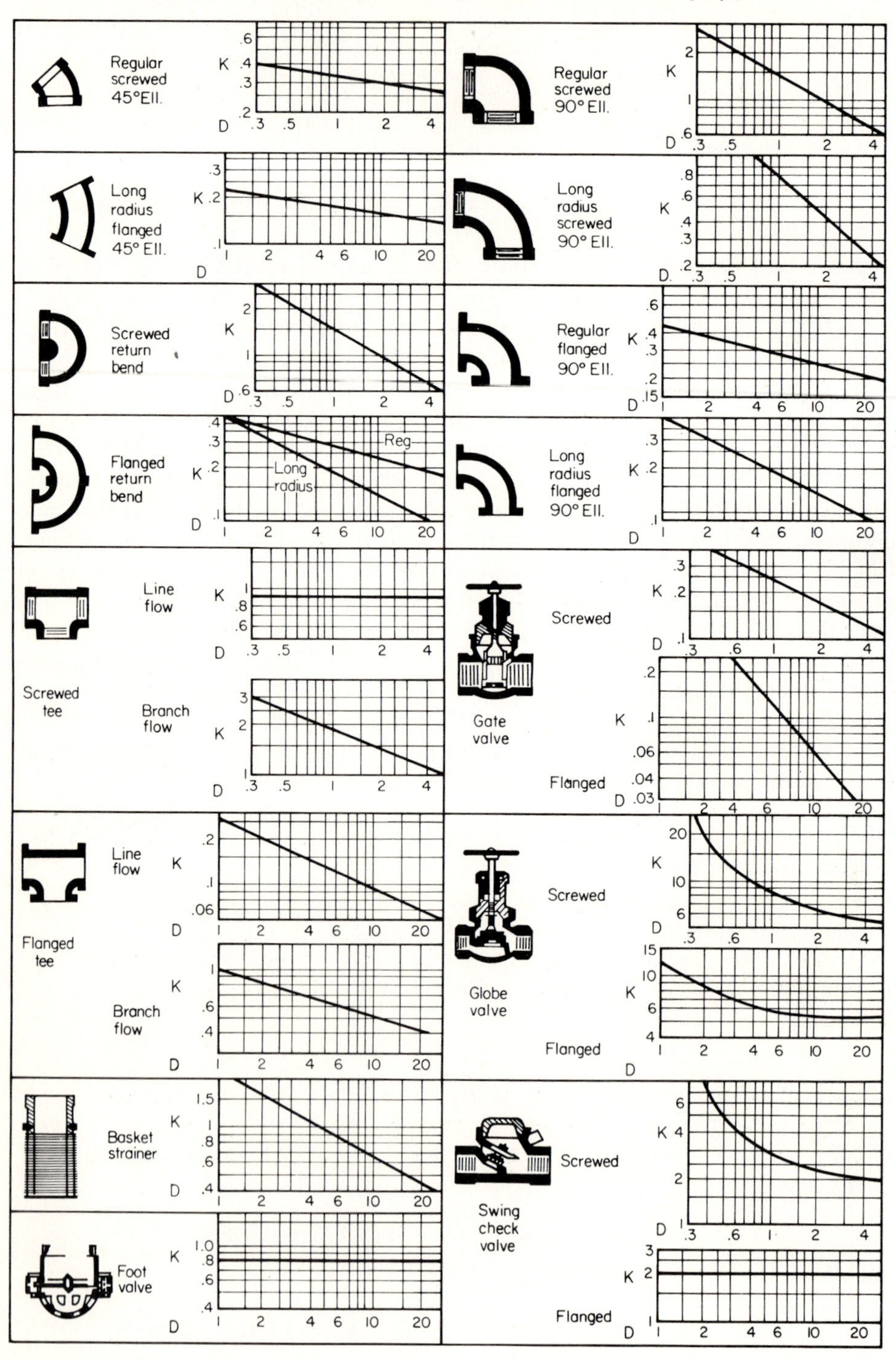

Table 7-2 Equivalent pipe lengths of various fittings

(Length of straight pipe, ft‡ giving equivalent resistance.)
(From "Pump Application Engineering" by T. G. Hicks and T. W. Edwards. Copyright 1971 by McGraw-Hill Book Co., Inc. Used with permission of McGraw-Hill Book Company.)

Pipe size, in†	Standard ell	Medium radius ell	Long-radius ell	45° ell	Tee	Gate valve, open	Globe valve, open	Swing check, open
1	2.7	2.3	1.7	1.3	5.8	0.6	27	6.7
2	5.5	4.6	3.5	2.5	11.0	1.2	57	13
3	8.1	6.8	5.1	3.8	17.0	1.7	85	20
4	11.0	9.1	7.0	5.0	22	2.3	110	27
5	14.0	12.0	8.9	6.1	27	2.9	140	33
6	16.0	14.0	11.0	7.7	33	3.5	160	40
8	21	18.0	14.0	10.0	43	4.5	220	53
10	26	22	17.0	13.0	56	5.7	290	67
12	32	26	20.0	15.0	66	6.7	340	80
14	36	31	23	17.0	76	8.0	390	93
16	42	35	27	19.0	87	9.0	430	107
18	46	40	30	21	100	10.2	500	120
20	52	43	34	23	110	12.0	560	134
24	63	53	40	28	140	14.0	680	160
36	94	79	60	43	200	20.0	1000	240

† in × 25.4 = mm
‡ ft × 3.28 = m

gpm, ft) or 0.163 (kW, m^3/min, m). The *power input* to the pump is a function of its efficiency and is equal to

$$P_p = P_w/E_p \tag{7-2}$$

in which E_p is the pump efficiency expressed as a decimal. The power required by the pump driver will also exceed its output, hence its efficiency must also be considered.

Example Determine the water power, pump power, and motor load for a pump system designed to deliver 1.89 m^3/min (500 gpm) against a total system head of 50 m (164 ft). Assume the efficiency of both pump and motor is 80 percent.

$$P_w = 0.163 \times 1.89 \times 50 = 15.4 \text{ kW (20.7 hp)}$$

$$P_p = 15.4/0.8 = 19.25 \text{ kW}$$

$$P_m = 19.25/0.8 = 24.06 \text{ kW}$$

Thus a total power demand of 24.06 kW (32.3 hp) would be experienced in pumping this flow.

The efficiency of pumps ranges from as little as 40 percent to as much as 90 percent depending upon the pump design, the fluid pumped, and the nicety with which pump and application are matched.

THE CENTRIFUGAL PUMP

7-4 Action of the Centrifugal Pump

If a vessel containing a liquid is rotated about a point (Fig. 7-4), centrifugal force will cause the liquid to rise to a point:

$$h = s^2/2g \tag{7-3}$$

where h is the height above the level at the center, s is the linear speed at the point where h is measured, and g is the acceleration of gravity. In practical centrifugal pumps water enters at the center of the casing and is accelerated by the rotating impeller which imparts to it both a radial and tangential velocity, the ratio of the two depending upon the design of the impeller (Fig. 7-5). The water leaving the impeller is slowed by the casing and a portion of its kinetic energy is converted to potential energy, that is, pressure. The conversion may be produced by the shape of the casing (volute pumps) or by internal diffusers or guide vanes.

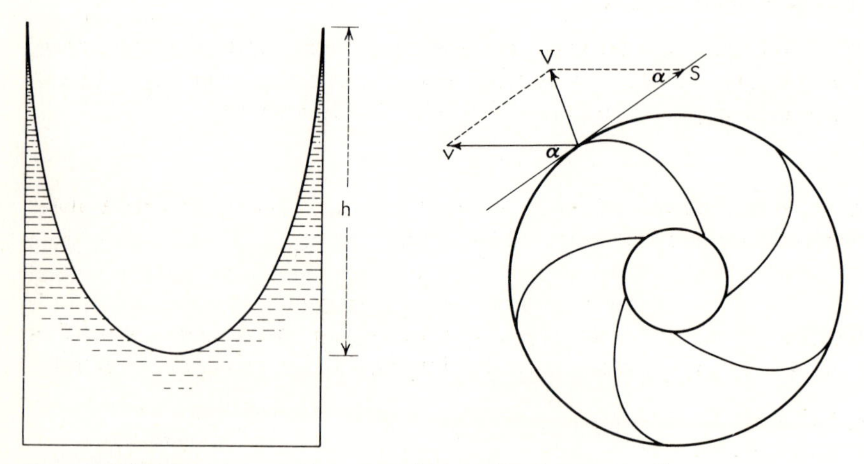

Figure 7-4 (*a*) Water in a rotating vessel. (*b*) Centrifugal pump impeller.

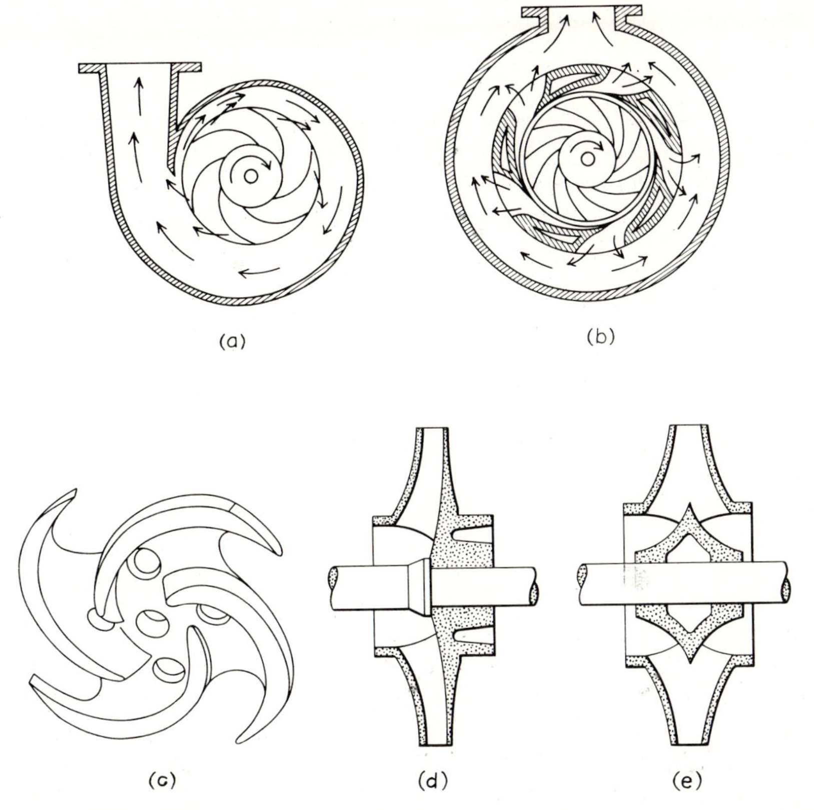

Figure 7-5 Pump details. (*a*) Volute pump. (*b*) Turbine pump. (*c*) Open impeller. (*d*) Closed impeller. (*e*) Closed impeller with double suction.

7-5 Effect of Varying Speed

From Eq. (7-3) it may be seen that, other variables being equal, the pressure (h) will vary as the square of the pump speed. Since the discharge from the pump can be likened to discharge through an orifice, $h = kV^2/2g$, whence

$$V = k'\sqrt{h} = k''\sqrt{s^2} = k''s \tag{7-4}$$

Thus the velocity of flow and hence the discharge will vary directly with pump speed. Since the power is a product of head and flow

$$P_w = k''' \cdot s^3 \tag{7-5}$$

From these relations the following observations can be drawn, noting that $s = D/2 \cdot \omega$ where D is the diameter of the impeller and ω its angular velocity.

1. Q varies with ω
2. h varies with ω^2
3. P_w varies with ω^3

These relationships neglect frictional and other losses and, more importantly, the fact that centrifugal pumps, unlike positive displacement pumps, do not deliver water only as a function of their speed, but also as a function of the pressure conditions under which they operate.

Pumps of similar design by a single manufacturer constitute a *series* and will have similar characteristics. For such a series of pumps

$$Q_1/\omega_1 D_1^3 = Q_2/\omega_2 D_2^3 \tag{7-6}$$

and

$$h_1/\omega_1^2 D_1^2 = h_2/\omega_2^2 D_2^2 \tag{7-7}$$

Thus h varies as D^2, Q varies as D^3, and P_w varies as D^5 where D is the impeller diameter. The *specific speed* of a pump series is defined as the speed at which a pump in the series will discharge a unit flow under a unit head at maximum efficiency. Thus, if the speed, discharge, and pressure are known for a single pump, one may solve Eq. (7-7), eliminating D, for

$$\omega_s = \omega Q^{1/2}/h^{3/4} \tag{7-8}$$

Specific speeds in English units are based upon a flow of 1 gpm at a head of 1 ft and do not correspond to metric units. To obtain equivalent values for ω_s Eq. (7-8) can be written

$$\omega_s = 6.67\omega Q^{1/2}/h^{3/4} \tag{7-9}$$

with Q in m^3/min, h in meters, and ω in r/min. As a general rule pumps with high specific speeds are more efficient than those with low. Between speeds of 1000 and 4000 centrifugal (radial) flow pumps perform well; from 4000 to 7000 mixed flow are most efficient: while above 7000 axial-flow (propeller) pumps have higher efficiencies. High specific speed is associated with high discharge at low head, hence axial-flow pumps are particularly applicable to storm water pumping (Chap. 17).

7-6 Characteristics of Centrifugal Pumps

As noted above, the discharge of centrifugal pumps is a function not only of their speed but also of the pressure conditions under which they operate. Characteristic curves like that of Fig. 7-6 permit the prediction of a pump's discharge under various head conditions. Such curves usually present head, power, and efficiency vs. flow at constant speed. The shape of the head-discharge curve is important in selecting pumps for specific applications. A pump which has a flat head-discharge curve, for example, will exhibit wide fluctuations in discharge with modest changes in pressure. This may be either desirable or undesirable depending upon the use which is intended.

Manufacturers frequently present pump data in the form shown in Fig. 7-7. A pump casing will accomodate impellers of several sizes which permits it to be used in different applications. This feature is often important when expansion of an

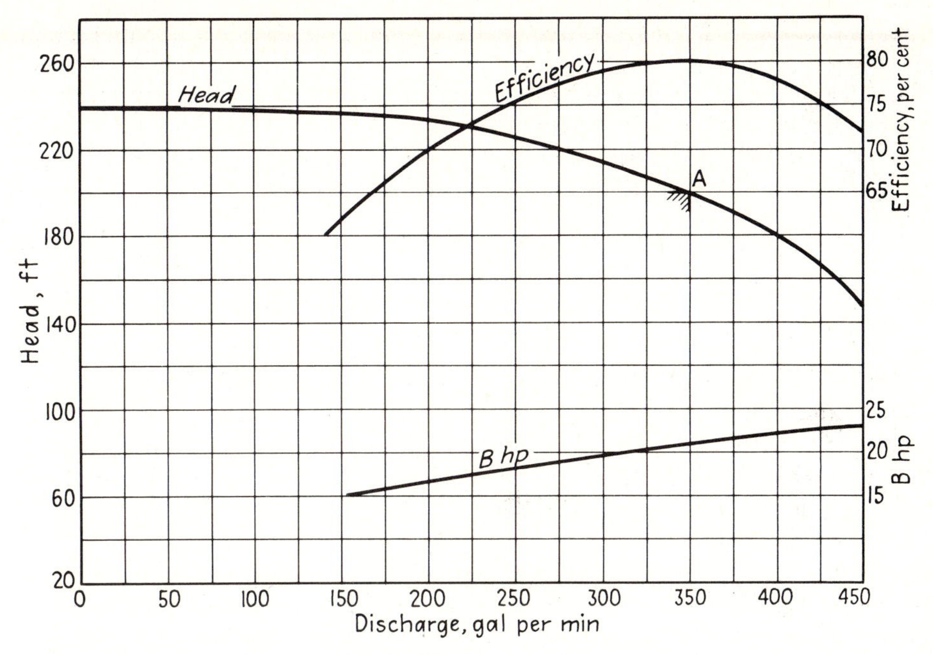

Figure 7-6 Characteristic curves of a centrifugal pump at constant speed.

existing system is being considered since replacement of an impeller and motor may permit an existing pump to continue in use. The cost of installation and alignment of large pumps may equal or exceed that of the pump itself.

Characteristic curves of centrifugal pumps may take various forms depending upon pump design. The major classifications of interest include *drooping head,*

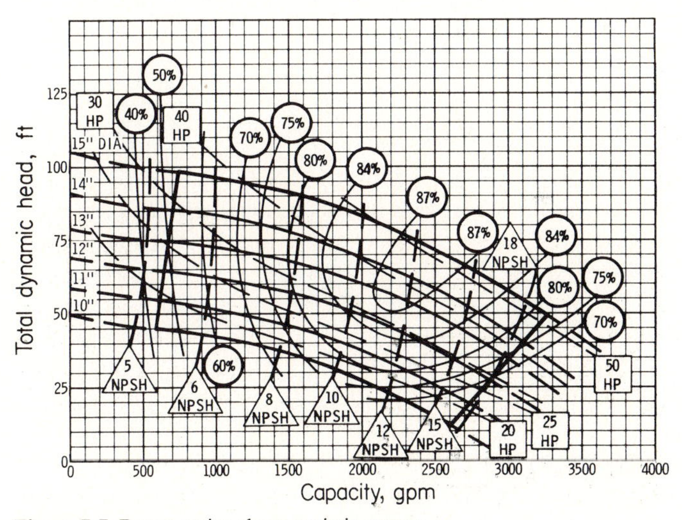

Figure 7-7 Pump series characteristic curves.

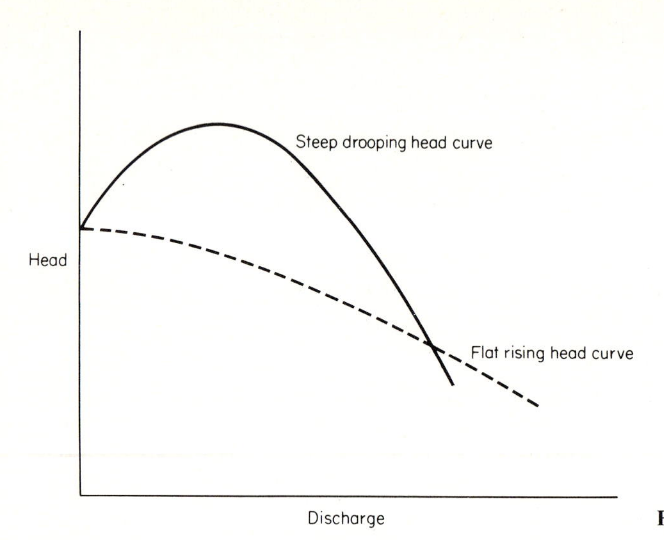

Figure 7-8 Characteristic curves.

rising head, *flat*, and *steep* characteristics. These are illustrated in Fig. 7-8. In rising head curves the pressure rises steadily to shutoff as flow decreases. Drooping head curves rise to a maximum head at an intermediate flow, then drop to a lower shutoff value. Such curves are typical of low specific speed pumps and may create operational difficulties under certain circumstances (Art. 7-10). Flat and steep curves reflect the variation in pressure with flow. In most sanitary engineering applications pumps with moderately steep characteristics are preferable.

7-7 Suction Lift

Net positive suction head (NPSH) is a function of the system design, that is, the pipe sizes, suction lift, flow, etc. Physically, NPSH is the force available to drive the flow into the pump. The *required* NPSH is a function of the pump design and

Table 7-3 Barometric pressure vs. altitude

Altitude		Pressure	
m	ft	kPa	ft H_2O
0	0	101	33.9
305	1,000	98	32.8
457	1,500	96	32.1
610	2,000	94	31.5
1220	4,000	88	29.2
1830	6,000	81	27.2
2439	8,000	75	25.2
3049	10,000	70	23.4
4573	15,000	57	19.2

Table 7-4 Vapor pressure of water vs. temperature

Temperature		Pressure	
°C	°F	kPa	ft H_2O
0	32	0.61	0.204
4.4	40	0.84	0.281
10.0	50	1.23	0.411
15.6	60	1.76	0.591
21.1	70	2.50	0.838
26.7	80	3.50	1.17
32.2	90	4.81	1.61
37.8	100	6.54	2.19
43.3	110	8.81	2.95
48.9	120	11.70	3.91
54.4	130	15.30	5.13
60.0	140	19.90	6.67

varies with flow, speed, and pump details. Characteristic curves (Fig. 7-7) may indicate the required NPSH or, alternately, the allowable suction lift at various flow rates.

Available NPSH is the sum of the barometric pressure and the static head on the pump inlet, less the losses in the pipe and fittings and the vapor pressure of the water. In calculating NPSH values, minimum anticipated barometric, and maximum anticipated vapor pressure should be used. Tables 7-3 and 7-4 present the variation of barometric and vapor pressure with altitude and temperature respectively. Barometric pressure may be further reduced by storm activity, and the values in Table 7-3 should be reduced by 1.2 ft of water or 3.5 kPa on that account.

Example Assume that a water pumping station at 500 m elevation uses pumps which require 30 kPa positive suction pressure (NPSH) when delivering water at 30°C. What is the allowable suction lift of these pumps if the entrance and friction losses are 15 kPa?

SOLUTION The barometric pressure, from Table 7-3, is 95.4 kPa. Reducing this to account for possible low pressure circumstances, $P_b = 95.4 - 3.5 = 91.9$ kPa. The vapor pressure of the water, from Table 7-4, is

$$P_v = 4.3 \text{ kPa}.$$

$$\text{NPSH}_{\text{avail}} = 91.9 - 15 - 4.3 - P_s = 72.6 - P_s$$

$$\text{NPSH}_{\text{reqd}} = 30 = 72.6 - P_s$$

$$P_s = 42.6 \text{ kPa}$$

Hence the allowable lift is 4.35 m or 14.26 ft. If the circumstances required a greater lift either another pump would be selected, the entrance conditions would be modified to reduce losses, or the pump would be moved to reduce the lift to that of which the pump was capable.

As the available NPSH decreases, the capacity of a pump decreases as indicated in Fig. 7-7.

7-8 Cavitation

If the NPSH drops below that required by the pump design, the pressure within the eye of the impeller may be reduced to the vapor pressure of the fluid. If this occurs the water will vaporize and a mixture of vapor and water will enter the pump. In the extreme, the pump may lose its prime, since pumps designed for liquids are poorly suited to pumping gas. A more common occurrence is continued flow at reduced capacity, with bubbles forming in the impeller eye and being passed out along the impeller. As the bubbles traverse the impeller they pass from a region of low pressure to one of high pressure, and eventually collapse permitting high pressure water to strike the impeller. This collapse frequently causes pitting of the impeller surface, and is always accompanied by a rattling or pinging noise.

Cavitation can be avoided by selecting pumps which operate at heads and capacities close to those corresponding to maximum efficiency, by following manufacturers' recommendations with regard to maximum suction lift and minimum discharge pressure, and by not altering pump speed from that of the original design.

7-9 Pump Selection

An estimate of pump requirements can be made from a knowledge of flow and head, Eq. (7-1). Selection of a specific unit requires examination of manufacturers' rating curves and comparison of these to the system curve (Fig. 7-3). If selection is

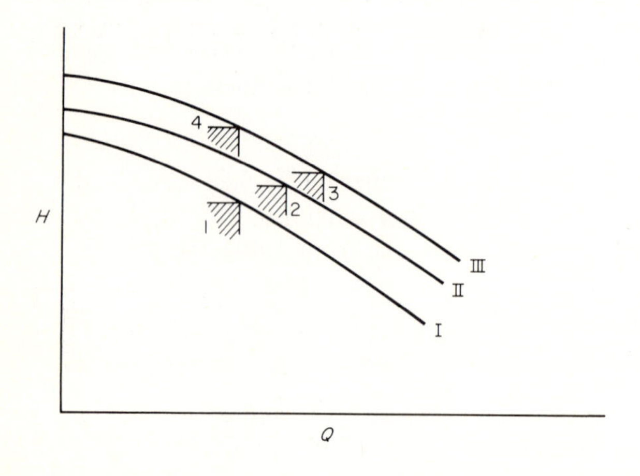

Figure 7-9 Effect of "conservative" pump selection.

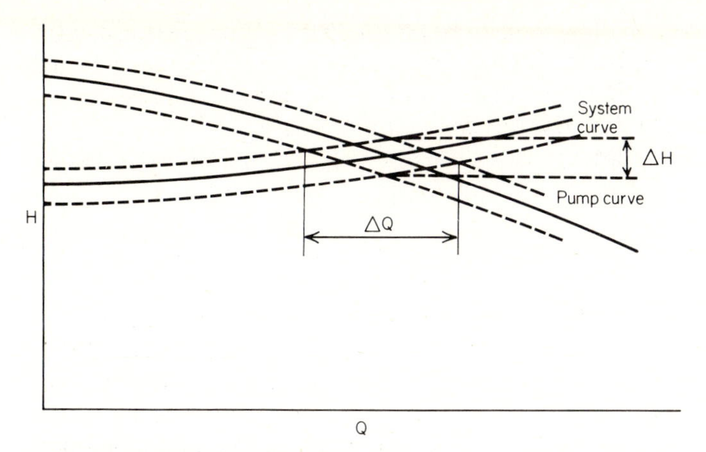

Figure 7-10 Flat system and pump curves.

left to the manufacturer, the engineer should check the proposed unit to insure that it is not excessively overdesigned. The manufacturer and designer may both increase the head and flow to insure satisfactory service. The actual design condition in Fig. 7-9 may be at point 1, with the engineer reporting point 2 to the manufacturer, and the manufacturer selecting a pump at point 3. In the series of pumps shown, the appropriate selection would be pump I, while that actually provided would be III. The actual operating point in a closed distribution system would be at point 4, the same flow as at point 1, but with a considerable penalty in head and efficiency.

If the engineer chooses to exercise his own judgment in pump selection it is desirable that he compare the system and pump curves or, at least, consider their shape.

System curves may be relatively flat or steep depending upon whether the head is primarily static or friction. Similarly, pump curves may be flat or steep, and drooping or rising depending upon impeller characteristics. If both curves are flat (Fig. 7-10) modest changes in head—which might occur as a result of pump

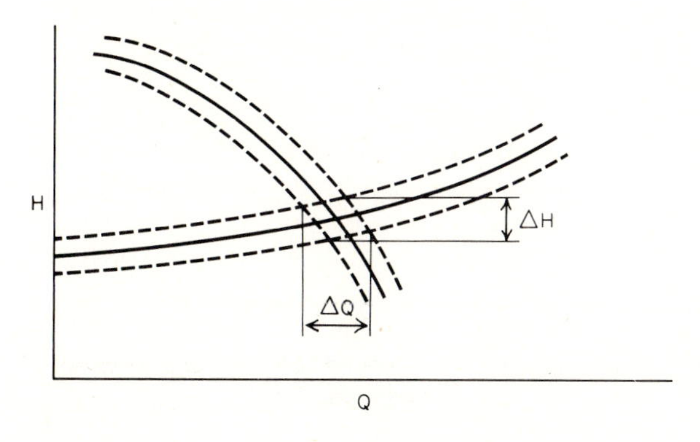

Figure 7-11 Flat system and steep pump curves.

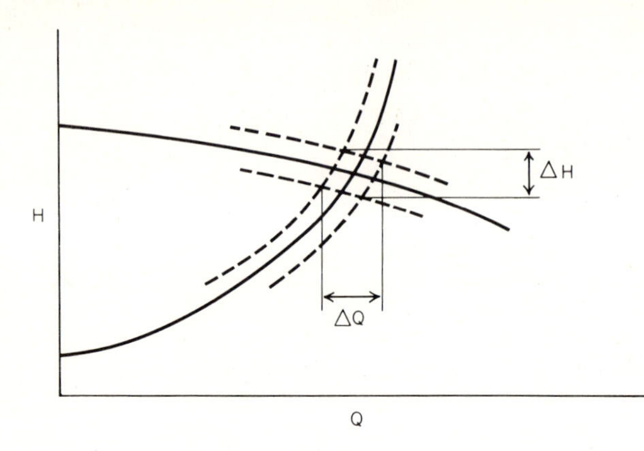

Figure 7-12 Steep system and flat pump curves.

wear or pipe aging—can produce marked changes in discharge, and might even shift the operating point to one outside the desirable range of the pump. If one of the curves is flat (Figs. 7-11 and 7-12) such variations are reduced, while if both are steep (Fig. 7-13), major changes in head are required to produce much variation in flow.

In most sanitary engineering pump applications it is not desirable that flow vary much with head, hence steep characteristic curves are generally preferable. In circumstances in which the system curve is steep a pump with a flat characteristic curve can be used with satisfactory results.

When specifying the head to be used in pump selection one should consider possible variations in static as well as friction head which might occur in operation. For example, the inlet water elevation used should be the minimum which can be reasonably expected, not the average or maximum.

7-10 Variation in Capacity

The variable demand which occurs in water distribution systems and the variable flow of wastewater makes it necessary to provide for variation in pumping capacity. The simplest technique is to size the pump for the maximum anticipated

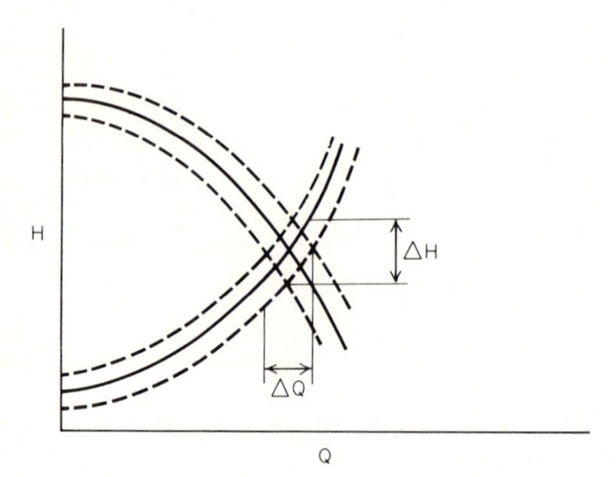

Figure 7-13 Steep system and pump curves.

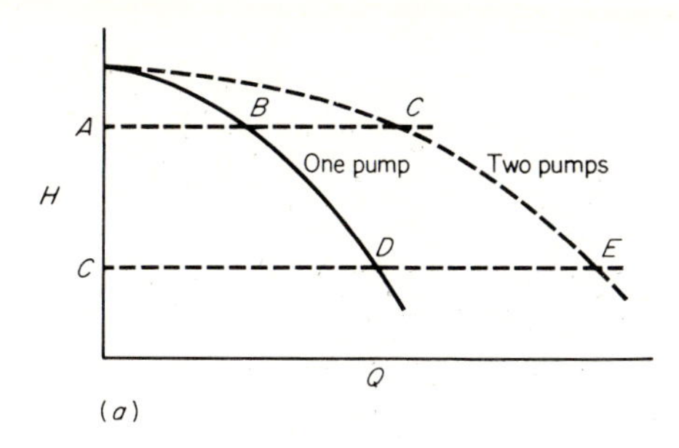

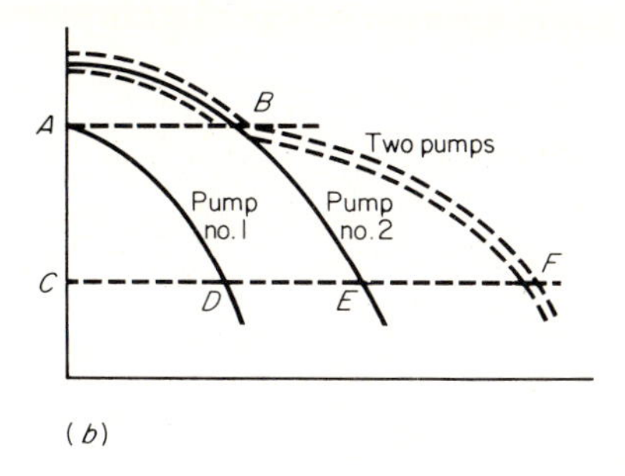

Figure 7-14 Combined characteristic curves. (*a*) Identical pumps. (*b*) Dissimilar pumps.

demand and let its capacity be reduced by throttling or by buildup of pressure in the system when demand is low. This results in low efficiency (Fig. 7-9) and may cause the pump to operate in a range for which it is not intended and, ultimately, may cause damage to the pump or motor.

Pumps may be installed in parallel to permit more variability in pumping capacity. This procedure is commonly used in water and wastewater pumping. The action of centrifugal pumps operating in parallel can be predicted by the addition of their characteristic curves. This is true whether the curves are identical or not, and is illustrated in Fig. 7-14. In Fig. 7-14 the combined curve for two pumps in parallel is obtained by adding the discharges of the pumps at each head. In Fig. 7-14(*a*), $AC = 2AB$, $CE = 2CD$. In Fig. 7-14(*b*) $CF = CD + DE$. At the head corresponding to points *A* and *B* note that pump no. 1 does not discharge and at that at this pressure and above, only pump no. 2 will provide any flow. This leads to the observation that in such systems low flows should be provided by the lower head pump (no. 1), since if they are provided by throttling pump no. 2, pump no. 1 will be unable to start. When pumps with drooping characteristics are employed, such as in high service water distribution lines, the combined characteristic curves appear as shown in Fig. 7-15.

More than one flow is possible at some values of head with pumps having drooping characteristics. This ordinarily presents no difficulty when a single pump

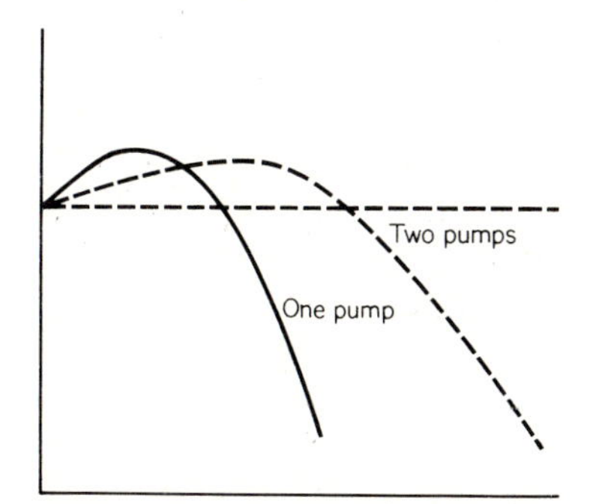

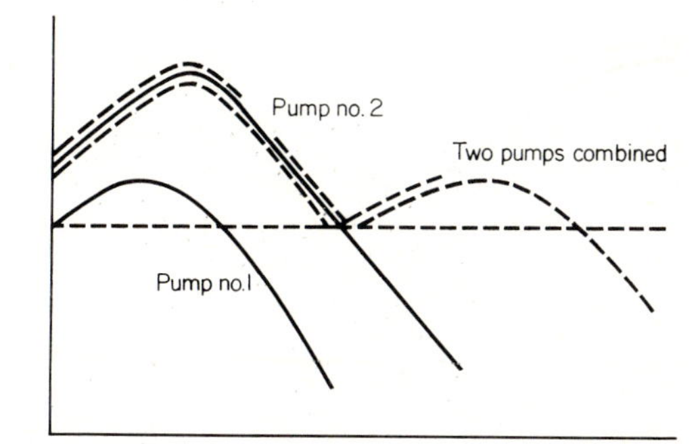

Figure 7-15 Combined characteristic curves—drooping head.

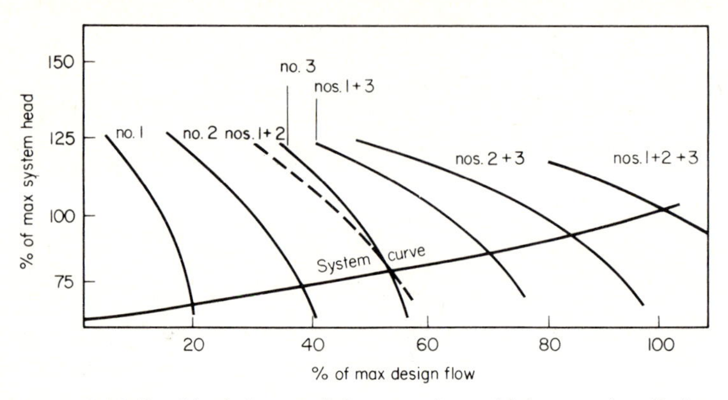

Figure 7-16 Combined characteristic curves for multiple pump installation.

is used, but with two or more the flow may divide in more than one way, leading to pulsing of individual pumps. As a general rule two pumps operated in parallel should have rising characteristic curves and provide some flow over the entire expected operating range. Pumps with drooping characteristics will divide flow precisely like other pumps provided they are operated at heads less than shutoff.

When more than two pumps are operated in parallel their flows are combined in the fashion above. Figure 7-16 presents the characteristic curves of three pumps, singly and in various combinations together with a system curve. Pump no. 1 alone might be used to provide a flow of 20 percent, no. 2 to provide 40 percent, no. 3 or nos. 1 and 2 together to provide 55 percent, nos. 1 and 3 to provide 72 percent, nos. 2 and 3 to provide 85 percent and nos. 1, 2, and 3 to provide 100 percent of maximum design flow. Intermediate flows would be provided by throttling or pressure increases in the system.

WELL PUMPS

7-11 Deep Wells

Deep wells are generally pumped by multistage diffuser centrifugal pumps, commonly called vertical turbines. These pumps develop high head by using a series of small impellers rather than a single large one. Vertical turbine pumps may be either oil or water lubricated. Water lubricated pumps are sometimes called open line shaft pumps since the drive shaft is exposed to the flow (Fig. 7-17). Deep well pumps may be driven either by a motor at the top connected to the pump by a line shaft, or by submersible motors below the pump at the bottom of the well, (Fig. 7-18).

The characteristic curves for such pumps depend upon the number of stages or impellers. Each impeller will have the same characteristic curve and the compos-

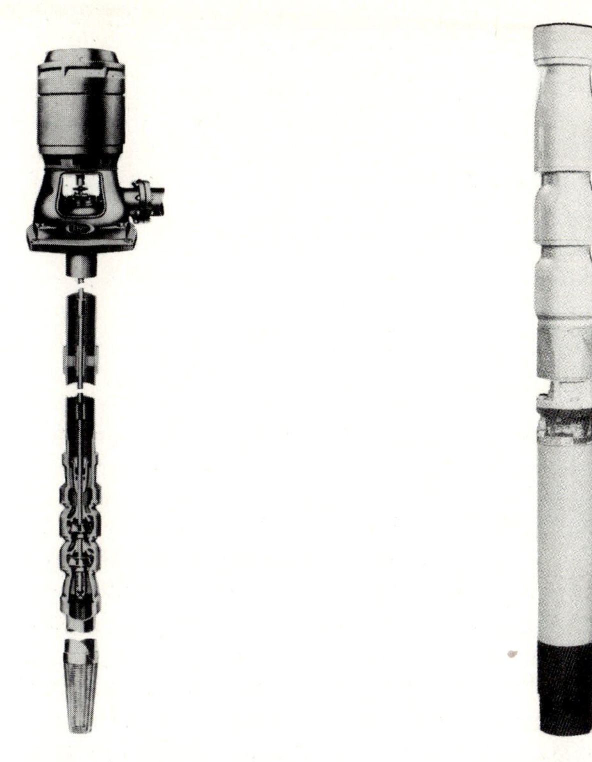

Figure 7-17 Vertical turbine pump. *(Courtesy Layne & Bowler, Inc.)*

Figure 7-18 Submersible well pump. *(Courtesy Crane Deming Pumps, Crane Co.)*

ite curve is obtained by adding them as shown in Fig. 7-19. Heads are added at a given discharge to determine the effect of series installation of centrifugal pumps. Deep well vertical turbines range from 100 mm (4 in) to 750 mm (30 in) in diameter, provide heads as high as 5000 kPa (1500 ft of water), and flows up to 100 m^3/min (30,000 gpm).

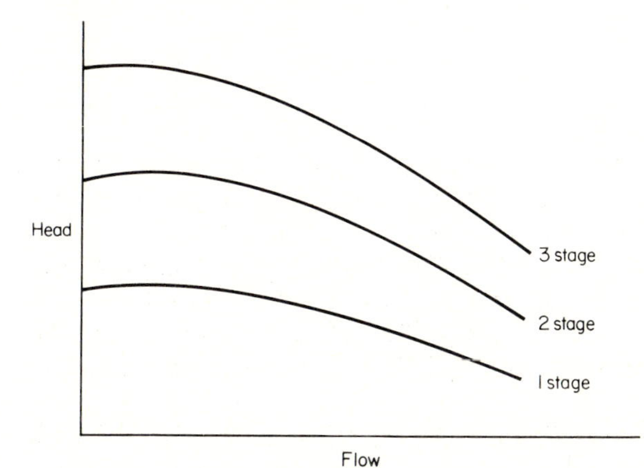

Figure 7-19 Series combination of pump stages.

7-12 Miscellaneous Pumps

Jet pumps employ a single stage centrifugal pump, installed at the ground surface, which directs a portion of its flow through an ejector and back to its own inlet. The ejector may be either at the ground surface if the well is shallow, or at the bottom if it is deep. This technique produces reduced pressure below the ejector and high pressure at the pump inlet, thus inducing a flow of water from the aquifer. The flow to the ejector ranges from 150 to 250 percent of the induced flow depending upon the head conditions.

Air lift pumps employ introduced air to reduce the weight of a column of water in a well, thus causing flow to occur under the hydrostatic pressure of the aquifer. Figure 7-20 illustrates an air lift system. The design of airlifts is largely empirical, with discharge being a function of submergence, eductor diameter, and air flow. The design procedure is described elsewhere.[1] The application of air lift in modern sanitary engineering practice is largely limited to solids recycle in small activated sludge systems in which its simplicity outweighs its inefficiency.

Helical rotor pumps consist of a helical impeller which rotates about its longitudinal axis in a double helix chamber. Water is driven through the depressions in the chamber by the action of the rotor. The drive is through a line shaft like that used with vertical turbine pumps. Helical rotor pumps are positive displacement pumps which may be used in deep wells. Capacities range up to 10 m^3/min (3000 gpm) at pressures of 3000 kPa (900 ft).

Rotary pumps are positive displacement devices which are particularly adapted to the pumping of viscous chemicals sometimes used in water and wastewater treatment. They are generally small, with some having adjustable capacity.

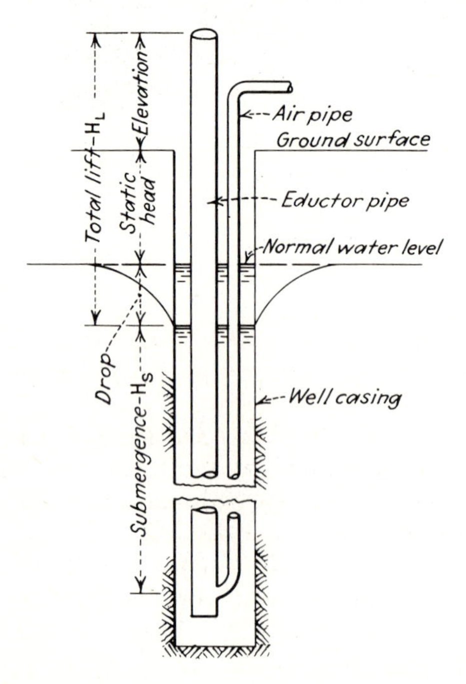

Figure 7-20 Diagram illustrating air lift.

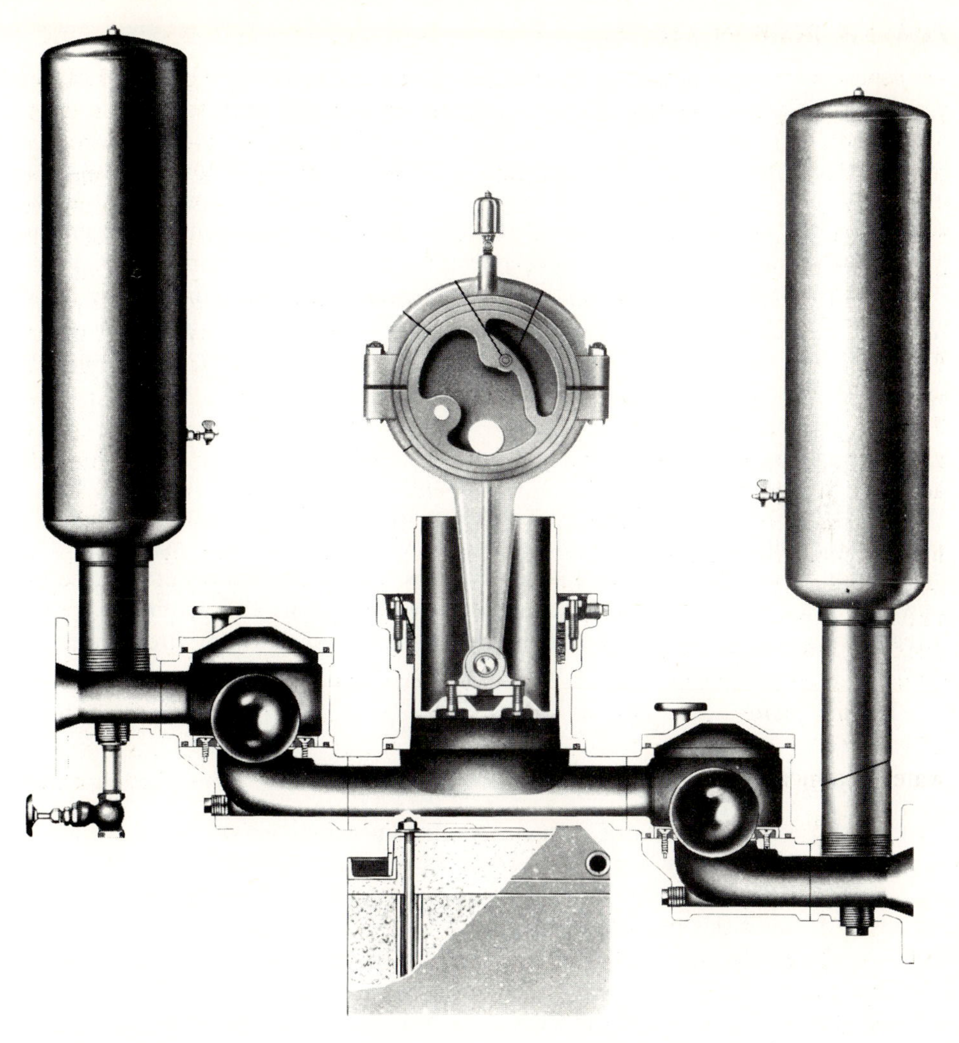

Figure 7-21 Positive displacement sludge pump. *(Courtesy ITT Marlow.)*

A variety of designs are available including cam and piston, external gear, internal gear, lobular, vane, shuttle block, flexible tube, and screw pumps like the helical rotor mentioned above.

Reciprocating pumps were widely used in individual wells and in municipal water supply in the past, but have been supplanted in this service by centrifugal designs. Their remaining application in sanitary engineering is primarily in sludge transport. A typical positive displacement sludge pump is shown in Fig. 7-21. In some designs the piston and cylinder are replaced by a disc and a flexible diaphragm. Pumps of this type require a moderate to high head for proper functioning. Their capacity ranges to 0.5 m^3/min (130 gpm) at a pressure of 200 kPa (30 psi). Screw pumps have also been adapted to sludge handling.

POWER FOR PUMPING

7-13 Choice of Prime Mover

Nearly all modern pumping stations rely upon electrical ac motors for their drives. Although variable speed drives are available, most pumps are driven at constant speed and the major power source is the squirrel-cage induction motor which starts at full voltage. Alternate power sources include wound rotor induction and synchronous motors, which permit greater flexibility in starting load, improve the power factor, and permit lower operating speeds at good efficiency. Motors of the latter type must justify their higher first cost through savings in power.

7-14 Steam Power

While electricity has largely replaced steam as the power for driving water supply pumps, large plants, particularly those located where fuel is cheap, may justify the installation of boilers and use of steam engines. The plant that generates electric power for city purposes may find it especially convenient and profitable to use direct-acting steam pumps and pumping engines, especially for fire service or where large amounts of water are to be pumped. The great reliability of steam also favors its use.

7-15 Steam Turbines

The steam turbine consists of a steam jet striking vanes upon a wheel which is caused to rotate rapidly. Impulse turbines are operated by the velocity of the moving jet of steam, while reaction turbines employ both the velocity and the expansive force of the steam. Both types are used to drive pumps. Because of its high speed it is necessary to place reducing gears between the shaft of the turbine and that of the centrifugal pumps. Use of condensers adds considerably to the efficiency of the turbine. The combined steam turbine and centrifugal pump has the advantages of small space, light foundations, simplicity, reliability, and automatic oiling. Efficiency of the turbine varies from 64 to 74 percent and of the reducing gears from 97.5 to 99 percent.

7-16 The Diesel Engine

The diesel is an internal-combustion engine which burns low-grade fuel oil. The fuel is atomized, forced into the cylinder, and ignited by the compression in the cylinder. The semidiesel burns more volatile fuel, such as kerosene or better grades of fuel oil. Compression in the cylinder is not so great as in the diesel, and ignition must, therefore, be brought about by a hot wire (glow-plug) or similar device. At full load the diesel will consume about 0.52 kg of fuel per MJ, and the semidiesel some 10 percent more. Neither type of engine should be operated at substantial overloads, but fuel economy is good at less than full load, say, 0.54 kg at 70 percent of full load. Diesel engines are especially adapted to plants of moderate size or where charges for electric current are high. The diesel plant has the advantage of being completely self-contained. It is, however, rather expensive in first cost and requires skilled attention. Gearing is necessary between engine and pump; it requires much floor space and is noisy in operation. The semidiesel is lower in first cost and operates at lower pressures but it is less efficient and harder to start.

7-17 The Gasoline Engine

Although gasoline engines are so high in operating cost that continuous operation is uneconomical, their low first cost makes them suitable for standby or emergency service, particularly for centrifugal pumps which are motor driven. The gasoline engine can be connected directly to the pump shaft on the other end from the motor. A reducing gear will not be needed, since the engine operates at a high speed which is adjustable at the carburetor. The gasoline engine chosen should have a capacity of 25 percent in excess of the need in order to allow for wear of valves. Provision must also be made for starting. Starters are operated by storage batteries which, because of infrequent operation of the engines, must generally be kept charged by the current used for daily operation.

7-18 Electric Motors

The ease with which electric current can be obtained from large power companies and its low price in most sections of the country have caused most cities to buy current for driving water supply pumps. Cities may be able to obtain very advantageous rates from power companies because power for pumping can, with proper system design, be consumed at off-peak periods, thereby improving the load factor of the power plant. The fact that motor drive is especially convenient for the centrifugal pump has also been a factor in the wide use of electric power in city waterworks. If possible the voltage should be selected for greatest economy in first cost of motors, control equipment, and wiring. There is no hard and fast rule, but some engineers assert that units of 37 to 45 kW are most economically served at 230 V, 45 to 150 kW at 460 V, and 150 kW and above at 2300 V. The cost of motor starting equipment at different voltages is an important factor especially where reduced voltage starters are required by the power company. In choosing motors for water pumping, the considerations involved are cost, simplicity in construction and operation, ruggedness, power factor, starting torque, and necessity or desirability for varying speed. Since dc motors are rarely used for water supply purposes, the discussion here given will apply only to ac motors. All motors should be manufactured in accordance with the standards of the American Institute of Electrical Engineers and of the National Electrical Manufacturers Association (NEMA).

The squirrel-cage induction motor is the type most widely used for driving pumps. It has the advantages of least cost, reliability, ruggedness, and simplicity of operation. It has the disadvantages of low starting torque, poor power factor, and high starting current. The current required for starting may be greatly reduced by using an autostarter or compensator. This type of motor can be used only for constant speeds.

The slip-ring induction motor does not take excessive current at starting, giving a starting torque about equal to full-load torque. It has the further advantage of allowing speed variation in a series of steps. Efficiency decreases markedly at the lower speeds. This type of motor costs 15 to 50 percent more than a squirrel-cage motor, the difference being inversely proportional to size. As with the squirrel-cage motor, the power factor is poor.

The synchronous motor has some features that make it preferable to the induction motor, particularly in larger installations. Its efficiency is generally higher than that of the induction motor. Its power factor may be made unity or leading at all loads by varying the exciting current. Its cost is higher than the induction motor except in larger slow-speed units. Its speed cannot be changed.

Motors are also employed to drive displacement pumps. For small installations of reciprocating pumps, squirrel-cage motors are used; but for larger pumps, and preferably for the smaller, slip-ring or synchronous motors should be used. For rotary pumps, squirrel-cage motors are used for constant-speed service, and induction motors for variable speed. Care should be taken, in making selections, that the motor is large enough to avoid overload and not so large that power is wasted. For preliminary calculations of pumping costs, motors may be considered as 85 percent efficient. Unless conditions are especially favorable as to interruption of current supply, emergency power sources, such as gasoline engines, should be provided.

PUMPING STATIONS

7-19 Location

The location of pumping stations has an important bearing on the pressure maintained in the distribution system. Uniformity of pressure would require a central location for the station, although the existence and sites of elevated storage tanks or reservoirs will also affect the pressures maintained. For small cities placing the station on one side of the high-use district and the elevated storage tank on the other gives good pressures. Large cities have a main pumping station as centrally located as circumstances will permit, with substations or booster stations at such points as experience indicates to be necessary.

7-20 Architecture

The building should be adequate in size and have space for placing additional units as needed. It should be fireproof, pleasing in appearance, and have well-kept grounds. Its architecture can and should be in harmony with the surroundings.

The interior should be well-lighted and designed so that it may be kept clean. This will necessitate numerous windows; light-colored walls; and floors of tile, terrazzo, or painted concrete. Since the public is likely to judge the quality of a water supply by the appearance and surroundings of the structures in which it is handled or treated, too much care cannot be given to these phases of design and operation.

Location of the equipment must be carefully planned so that unnecessary piping and wiring will be avoided. All units must be accessible, and a crane must be provided to permit moving and replacing heavy equipment. Space for office work and records will be needed, and provision must be made for heating, ventilation, and washrooms.

7-21 Capacity and Operation

In very small cities the pumping plant may operate for only 8 to 12 hours per day, with the operator quitting work when he has the elevated tank full enough to care for normal use during the night. Should there be a fire alarm, he hurries to the station and starts the pump. This may be an economical arrangement, since the cost of the larger pump needed to care for the whole day's supply in the working day is offset by the saving in labor cost. With very large storage and electrically operated pumps, night pumpage would reduce current costs. The larger city generally finds it necessary to have one or more attendants on hand at all times and to keep at least one pump operating all the time. Various methods of operation are employed: (*a*) There may be no elevated storage, with water pumped directly into the system with variations in consumption taken care of by turning on and turning off pumps. This method has the disadvantage of causing pressure fluctuations and presenting danger of interruption in service in case of breakdowns. Frequently the pumps operate at an underload and therefore inefficiently and uneconomically. Pumps with flat characteristic curves may be required to avoid excessive pressures during hours of minimum use. (*b*) All the water may be pumped to a large elevated reservoir and discharged therefrom by gravity to the distribution system. This allows very uniform pumping rates, permits use of pumps at rated capacities, with little head change, and provides safety against interruption. Usually, however, it is not feasible to provide so large an elevated reservoir or to locate it so that pressures will be favorable. (*c*) The water is pumped into the distribution system at a uniform or varying rate, with the excess over the consumption during part of the day going into an elevated tank or standpipe to be drawn upon during peak use or whenever the pumping rate is less than consumption. This method is most widely used, as it provides some water for emergencies, permits pumps to be operated at their maximum efficiencies, and also allows steep characteristic curves for the pumps.

Under method (*a*) the pumps must have sufficient capacity to care for the peak load, and the designer and operator must furnish and use such pump combinations as will result in as economical use as possible. Method (*b*) has the virtue of permitting constant pumping rates and therefore economical operation.

Method (*c*) is a compromise between the other methods and, with good engineering and operation, efficient pumping should be obtainable. Emergency pumps to replace units under repair will be needed. The very large rates required for fire fighting will necessitate special fire pumps. Since they operate only for relatively short periods, the importance of operation at most efficient loads is less important. When fire pumps operate, it is sometimes the practice to increase the pressure in the system. This may be accomplished by closing valves in the lines leading to elevated tanks. Such valves may be controlled by radio or electrically from the main pumping station or may be the "altitude" type, i.e., set to close automatically when the pressure on the inlet side reaches a certain amount and to reopen when the pressure falls to the usual level. Such valves are sometimes used in normal operation to prevent tanks from overflowing. (Art. 6-23)

7-22 Auxiliary Pumping Stations

At points in the distribution system where there is a high ground elevation, or where there are mains of inadequate size, undesirably low pressures may result. Such conditions are remedied by installing an auxiliary or booster pumping station. The pump takes water from the largest and most direct feeder main and discharges into a pipe which supplies the neighborhood, usually with the excess discharged into an elevated tank. Some stations pump only at periods of low use to fill the elevated tank, which is emptied during periods of high consumption and lower pressure.

The auxiliary station is frequently started and stopped automatically, the only attendance necessary being occasional visits for inspection and oiling. Automatic control is obtained in various ways: (*a*) A float rising and falling in the elevated tank may actuate an electric switch with wires leading to the pump room to start the pump when the water is low and stop it when full. (*b*) A pressure pipe leading from the tank to the pump station may actuate the switch. (*c*) An altitude valve in the discharge pipe from the tank closes when the pressure reaches a certain height, thereby permitting pressure to go still higher and finally to operate a switch at the pumping station. (*d*) The electrode type of control uses insulated wires or rods extending from the top of the tank and of such length that when submerged or exposed an electric circuit is closed or opened. Whatever method is used, care should be taken to prevent freezing from interfering with the action. This is most likely with the first method. For best operation of the control devices, the range of pressures between the start and cutoff should be large. In the pumping station, care should be taken to exclude seepage into the pump pit, and an automatic sump pump should be installed to keep it dry.

7-23 Valves and Piping Details

The foot valve on the suction line will hold priming water in the pump. A check valve is placed in the pipe on the discharge side to prevent backward surges of pressure which might injure the pump. With a centrifugal pump a check valve will prevent backward flow of the water, should the motor stop. On the outlet side of the check valve a gate valve is usually placed to permit repairs to the check and to allow throttling if the pump is the centrifugal type. If there is a positive suction head, a gate valve is also necessary on the suction side of the pump so that pump repairs can be made. A pressure-relief valve is sometimes placed on the discharge side of a reciprocating pump so that a stoppage or accidental closing of a valve in the discharge pipe will not cause harmful pressure to build up. The relief valve bypasses the water pumped back to the suction and thereby keeps the pressure down.

Piping, both suction and discharge, should be as free from bends as possible. The pipes should be of such size that friction head will be low and thereby keep down pumping costs. A computation of pumping costs with various sizes of pipe and the increased cost caused by the larger size will indicate the best combination.[2]

A pressure gauge should be installed so that the operator may be informed at all times of the pressure maintained. Large cities usually install recording gauges so that a permanent record will be available for study. It is advisable also that pressure records be kept for various strategic points throughout the distribution system. A meter measuring the output of each pump would be useful by permitting a check on its efficiency, but in any case a meter of the venturi or magnetic type should measure the total pumpage. A recording meter provides useful records. Accuracy in the venturi or magnetic meter will be assured if it is preceded by a length of straight pipe which is at least six times as long as the diameter of the meter tube. Daily log sheets kept by operators showing gauge and meter readings at hourly or more frequent intervals together with starting and stopping times for all units will supplement the gauge recording sheets.

PROBLEMS

7-1 A single-acting piston pump has a single cylinder 300 mm, in diameter. Its stroke is 300 mm, and there are 110 strokes per minute. The output is measured and found to be 2.2 m^3/min. Compute the percent slip.

7-2 A centrifugal pump operates at a speed of 1150 r/min and discharges 2.3 m^3/min against a head of 120 kPa. The power required is 8.2 kW. Compute (*a*) the efficiency of the pump, (*b*) the discharge, head, and power if the pump speed were changed to 1750 r/min.

7-3 A centrifugal pump is to operate against a total head of 360 kPa and is to discharge 1.1 m^3/min. Compute the water power and motor power assuming an efficiency of 60 percent. If the efficiency of an electric motor is 85 percent and the cost of current is 2.5 cents per kWh, compute the monthly cost of operation if the daily flow averages 750 m^3.

7-4 A pump must raise 2 m^3/min a vertical distance of 30 m. If the discharge and suction pipe are 200 mm in diameter and 275 m long, determine the total head against which the pump must act. Neglect minor losses.

7-5 A pump is to raise 1.9 m^3/min against a static head of 25 m through a pipe 365 m long. Determine the most economical pipe size, (that which gives minimum annual cost). Power costs 4 cents per kWh, the pump operates 13 hours per day, interest is 8 percent, annual depreciation is 2.5 percent, the cost of installed pipe is $30 for 150 mm, $36 for 200 mm, $48 for 250 mm, $54 for 300 mm, and $60 for 350 mm per meter of length. Assume overall pumping efficiency of 60 percent.

7-6 What is the maximum permissible difference in elevation between the water surface in an intake structure and the intake of a pump under the following circumstances:

Altitude—1000 m
Max water temp—25°C
Flow—2 m^3/min
Intake pipe—150 mm
Entrance loss—$V^2/2g$
$NPSH_{reqd}$ = 40 kPa

REFERENCES

1. Babbitt, H. E., J. J. Doland, and J. L. Cleasby, "Water Supply Engineering," 6th ed., McGraw-Hill, New York, 1962.
2. Hicks, T. G., and T. W. Edwards, "Pump Application Engineering," McGraw-Hill, New York, 1971.

CHAPTER

EIGHT

QUALITY OF WATER SUPPLIES

8-1 Impurities of Water

Water that is absolutely pure is not found in nature. Water vapor in the air condenses about particles and, as it falls, absorbs dust and dissolves oxygen, carbon dioxide, and other gases. At the ground surface it takes up silt and other inorganic matter. A few bacteria will enter the water from the air, but at the ground surface many more will be picked up as it runs off in streams or rivers. Small amounts of the products of the decomposition of organic matter, nitrites, nitrates, ammonia, and carbon dioxide will go into solution. Surface water will retain all these impurities for an indefinite period, but that part of the rainfall which percolates into the soil will lose the suspended silt and bacteria by filtration. The products of decomposition may also be lost in the filtration process by chemical combinations or by the action of plants. This loss of impurities, however, is offset by the solution of various salts which the percolating water will encounter in its path, the amount and character depending upon the length of underground travel and the chemical make-up of the strata traversed. Figure 8-1 shows the common impurities of water and their effects. These are discussed in more detail below, but it may be pointed out that certain microorganisms in water cause disease, certain algae cause disagreeable odors, certain salts in large amounts cause unpleasant tastes, hardness, or corrosiveness, and gases cause odors and corrosiveness. The chemicals produced by decomposition of organics ordinarily are not present in sufficient amounts to be of any concern.

There is no evidence that drinking hard waters or those containing salts in general causes kidney stones, gallstones, or other disorders. In fact, to the contrary, there is some evidence to the effect that soft water may be a contributing

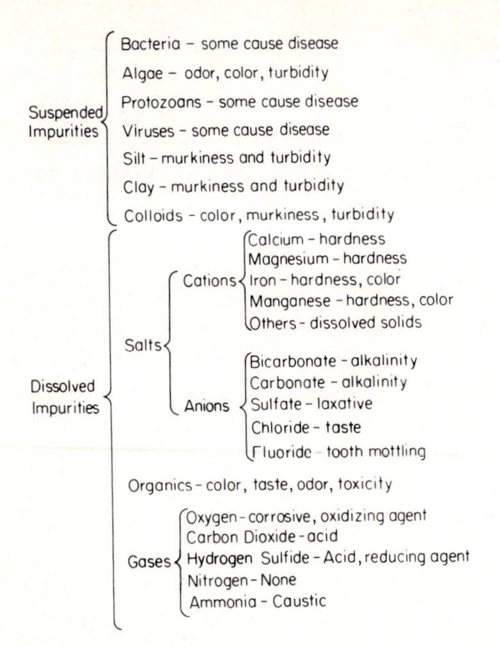

Figure 8-1 Common impurities in water.

factor in some ailments, particularly cardiovascular disease.[1] Drinking a water to which one is unaccustomed may have a laxative effect, particularly if it is high in sulfate, but this is only temporary and causes no injury.

A *potable* water is one that is safe to drink, pleasant to the taste, and usable for domestic purposes. A *contaminated* water is one that contains microorganisms, chemicals, industrial or other wastes, or sewage so that it is unfit for its intended use. The term *polluted* water is synonymous with contaminated water.

8-2 Waterborne Diseases

Communicable diseases which may be transmitted by water include bacterial, viral, and protozoal infections. The *bacterial diseases* include typhoid, paratyphoid, salmonellosis, shigellosis, bacillary dysentery, and asiatic cholera. *Viruses* cause infectious hepatitis and poliomyelitis, while *protozoans* can cause amoebic dysentery and Giardiasis. Schistosomiasis is caused by a worm which may be transmitted through water via a snail carrier. Other diseases are not generally transmitted by water.

Typhoid, paratyphoid, dysentery, gastroenteritis, and cholera are transmitted by the fecal and urinary discharges of sick persons and carriers. Through careless disposal of the discharges or inadequate treatment of city sewage, underground water may be contaminated to endanger well water supplies, or streams and lakes may be affected to contaminate surface supplies. Prevention of waterborne outbreaks of these diseases is primarily a matter of treating the water by methods described in Chaps. 9 to 11 and the placing of adequate safeguards around the water supply system that will prevent chance contamination.

Gastroenteritis is a clinically identifiable diarrheal disorder but the term is often used to describe outbreaks which are not directly attributable to specific organisms.[2] It resembles food poisoning, and apparently some of the organisms that are involved in food poisoning are sometimes waterborne after having reached water by means of bowel discharges. The role of bowel discharges is indicated by the fact that in the past, waterborne typhoid epidemics have been preceded by outbreaks of gastroenteritis, the incubation period of the latter apparently being only 12 to 48 hours while that of typhoid is usually about 10 days. On the other hand gastroenteritis has occurred, waterborne according to epidemiological evidence, while at the same time the water showed no evidence of bacterial pollution. There is now evidence that some gastroenteritis is caused by viral agents.[2] However, it is usual for an outbreak of gastroenteritis to be accompanied by the presence of coliform organisms in the water. Chemical irritants, the characteristics of which have not been determined, have been blamed for gastroenteritis in a few instances.

It should be recognized by the waterworks man that not all typhoid fever and dysentery is waterborne. The infection is carried by flies to foods, while carriers and the sick may infect foods that they handle. Infected milk has been responsible for many epidemics. Oysters and other shellfish grown in sewage-polluted water will themselves be polluted and highly dangerous. The characteristics of an outbreak of waterborne disease are described in Art. 8-27.

Infectious hepatitis, also known as epidemic hepatitis, is caused by a virus which has been responsible for several waterborne epidemics[2] but may also be transmitted by direct personal contact, food, and milk.

Poliomyelitis, also a virus disease, is found in the feces of infected persons and in sewage. Hence it has been assumed that the virus might occur in drinking water and it has, in fact, been detected there. It is possible that some cases may have occurred in this way, but there is no epidemiological evidence connecting outbreaks of poliomyelitis with drinking water or use of swimming pools.

Schistosomiasis organisms are animal parasites which pass part of their life cycle in certain species of water snails. After leaving the snails they are swimming forms known as cercariae, which may be drunk or, more likely, may enter the skin of a person who wades or swims in the water. Coagulation and filtration do not remove cercariae from water but chlorination, with a residual of 1 mg/l, will kill them in 10 min.

8-3 Methemoglobinemia

This disease causes "blue babies" and has caused some 262 reported cases and 29 deaths.[3] It is caused by nitrates in water consumed by infants under 2 months of age. In children of that age or under the nitrates are reduced in the body to nitrites which react with the oxygen receptor sites on the hemoglobin fraction of the blood and impair its oxygen-carrying capacity. This reaction does not occur in older children or adults. The water responsible has had nitrates in excess of 10 mg/l, and in all cases it has been groundwater. High nitrate waters are not uncommon and it is probable that subclinical cases in addition to more severe cases may occur that are not reported. One safeguard might be to use nitrate-free water for direct consumption by babies and in food formulas. Water quality standards, however, specify an upper bound for nitrate in public water supplies.

8-4 Lead Poisoning

Lead is sometimes found in water that has been in contact with lead pipes. As lead is a cumulative poison, habitual consumption of such water may result in lead poisoning. This may occur in waters containing over 0.3 to 0.5 mg/l by weight of lead. The U.S. Environmental Protection Agency Standards† for water quality, however, place the limit at 0.05 mg/l. The danger of lead poisoning has contributed to the reluctance of waterworks officials to use lead pipes for service lines. The waters likely to take up lead are soft or acid and include rainwater with its usual high carbon dioxide content, and swamp waters, which may contain humic acids and carbon dioxide. The interiors of metal water-storage tanks are not to be painted with lead paint. Zinc and copper are also liberated in water by corrosion of galvanized iron, brass, and copper pipes, but there is no indication that a sufficient quantity is freed to harm the water consumers.

8-5 Fluoride

Waters in certain sections of the country contain fluorides as impurities. When present in amounts greater than about 1.5 mg/l and especially over 3.0 mg/l as fluorine, and if the water is consumed during the period of formation of permanent teeth, a chemical combination occurs within the enamel, resulting in mottled and discolored teeth. The condition, which is incurable, occurs, with very few exceptions, only in the permanent teeth, which form in the jaw after birth. Therefore, the damage is done if the child consumes the fluoride bearing water from birth to 8 or 10 years of age. The teeth of adults moving to an area where the water contains fluoride are not affected. The known areas in which mottled enamel occurs include sections of Texas, New Mexico, Arizona, Colorado, Nevada, California, Utah, Oklahoma, Arkansas, Mississippi, Tennessee, Kansas, Iowa, Illinois, and the Dakotas. Scattered areas also occur in some of the Atlantic Coast states. Defluoridation of water is discussed in Art. 11-24.

It was observed that, where mottled enamel occurred, the incidence of dental caries was low. Extensive studies indicate that a moderate amount of fluoride in drinking water is important to good tooth development, the required amount being somewhere between 0.6 and 1.5 mg/l. The amount allowed under the U.S.

† The current water quality standards proposed by EPA under the provisions of the Safe Drinking Water Act (P.L. 93-523) are presented in App. I.

EPA standards is 1.4 to 2.4 mg/l depending upon the mean air temperature. Fluoridation of water as a means of preventing tooth decay is now practiced by many cities. Details are discussed in Art. 11-23.

8-6 Radioactivity in Water

The possibility of dangerous contamination of water by radioactive isotopes has received attention from the Environmental Protection Agency, which has proposed the following provisional maximum permissible concentrations (MPC):[4]

1. For gross alpha radiation, 15 pCi/l (picocuries/liter).
2. For combined Ra-226 and Ra-228, 5 pCi/l.
3. If gross alpha radiation is greater than 5 pCi/l, Ra-226 is to be monitored. If Ra-226 is greater than 3 pCi/l, Ra-228 is to be monitored.
4. Gross beta, Strontium-90, and H-3 in average annual concentration shall not produce an annual dose equivalent to the total body or any internal organ greater than 4 millirem (rem = roentgen equivalent, man). If gross beta is 50 pCi/l, Sr-90 8 pCi/l, or H-3 20,000 pCi/l the water should be monitored every 4 years.
5. Any community downstream from a nuclear facility must monitor gross beta and I-131 as well as Sr-90 and H-3.

One of the results of the detonation of a nuclear device is the dissemination of dust particles which have become radioactive by bombardment by neutrons or by adsorption of fission products. These are lifted to great heights above the ground surface and are carried long distances by air currents. They may serve as nuclei for formation of rain or snow, which will hasten their fall to the ground surface. Fallout from nuclear testing is minute when compared with radioactivity naturally present in the earth's crust, however, waterworks authorities have been concerned with the contamination which could result directly from fallout upon water surfaces, like impounding reservoirs, and also indirectly from surface runoff. Some of the radioactive isotopes in fallout are soluble, while others will settle to the bottom where decay will reduce their activity. Some, however, have long half-lives and the possibility of an accumulation sufficient to be injurious required investigation. Some investigations[5] have been made after the detonation of atomic and hydrogen bombs, but only small increases over background radioactivity were noted.

Waters may also be contaminated by liquid wastes from research laboratories, hospitals which use radioisotopes, laundries which serve such laboratories and hospitals, water-cooled nuclear reactors, plants that process reactor fuels, and uranium mining and preparation. The wastes from processing reactor fuels are highly radioactive, or "high-level" wastes and are not ordinarily discharged into streams. The radioactivity of low-level wastes may not be a problem after dilution, or they may be reduced in activity by water and sewage treatment (Art. 11-25). It is recognized, however, that they may be concentrated by algae or other water organisms, and these in turn may be consumed by fish, with further concentration. Present concern with the putative hazards of nuclear power production has led to stringent regulation of the discharges from such plants.

8-7 Water Bacteria

Many bacteria are found in water. Most of them are of no sanitary significance, some are indicators of pollution but are harmless; others, few in numbers, are pathogenic. These include bacteria causing typhoid fever, paratyphoid, dysentery, gastroenteritis, and cholera.

Groundwaters normally do not contain many bacteria since the effects of filtration, exposure to unfavorable environment, and time will eliminate most of them, including those of sanitary significance. Some shallow wells may contain considerable numbers, but these are frequently due to lack of safeguards in well construction. Soils or aquifers having cracks or crevices may allow insufficiently filtered waters to enter wells or springs. The waters of deep wells may have very few bacteria, but are rarely or never completely sterile.

A group of bacteria, the crenothrix, are found in groundwaters and may be troublesome. They thrive in waters containing minerals in solution. One type grows in iron-bearing waters and precipitates iron oxide or rust. It may occur in water mains and the rust produced will stain clothes and plumbing fixtures. After death crenothrix may cause unpleasant tastes and odors. Less common than the iron crenothrix are two other species, one that precipitates manganese as a dark-brown deposit and the other aluminum as a yellowish-white deposit. Growth of crenothrix can be prevented by iron and manganese removal. For control see Art. 11-13.

Untreated surface waters contain many bacteria. The sanitary engineer is not concerned with most of them. The coliform group is of great importance and includes a number of organisms.[6] *Aerobacter aerogenes* is widely distributed in nature, and normally found on plants and grains, in the soil, and to a varying degree in the feces of man and animals. *Aerobacter cloaceae* is found in the feces of man and animals, and also in soils. *Escherichia coli, (E. coli)* normally inhabits the intestinal tract of man and animals and is excreted with the feces. It is considered nonpathogenic but may cause infections of the genito-urinary tract.

The coliforms, therefore, are useful as indicators of pollution, since they show that the water has been in contact with soils or plants or has been polluted by sewage so recently that the bacteria have not died out naturally or been removed by natural filtration or artificial treatment. Coliforms do not increase in numbers in water.† They die at a logarithmic rate which may leave a few individuals

† Coliforms have been found in jute, hemp, cotton, etc., that is used as packing in pipe joints and they will multiply when the material is immersed in water. There is evidence that typhoid organisms will also multiply in jute. Leather washers used in pumps will also support growths of coliforms.

existing for weeks or months in freshwater. The death rate in saltwater is much higher.

Coliforms are of importance not only because they indicate pollution but also because their absence or presence, and their number, can be determined by routine laboratory tests. Tests for pathogens are not adapted to such routine work and are made only in special investigations. The routine membrane test differentiates between *Aerobacter* and *E. coli* and special procedures permit separation of coliforms of fecal and nonfecal origin.

8-8 Relationship between Pathogens and Coliforms

E. coli are excreted in enormous numbers. Fresh feces may contain from 5 million to 500 million per gram, and the average amount of feces is about 82 g daily per person. Coliforms in sewage vary widely according to its concentration, from 25,000 to 500,000 per ml, depending upon time of day of collection and also upon temperature. Greater numbers are found in summer than in winter. Pathogens are far less numerous and where exposed to light, unfavorable temperatures, effects of other bacteria, and other unfavorable environmental conditions will die out at least as rapidly as the coliforms. Water treatment methods will have the same effect, that is, more rapid removal of pathogens than coliforms. Chlorination, which is widely used in water treatment (Art. 11-2), is probably somewhat more efficient in killing pathogens than coliforms in the usual range of dosages, water temperature and pH. Kehr,[7] on the other hand, concludes that the ratio of typhoid organisms to coliforms in any water or sewage, remains constant even after natural purification or any artificial method of treatment. The ratio depends upon the typhoid fever case rate in the population which is polluting the water or producing the sewage. With usual case rates in the United States this would result in a ratio of 1 typhoid organism or less to 1 million coliforms. Such a ratio is, of course, extremely uncertain, but it must be concluded that to insure a safe water, methods of treatment used should reduce coliforms to as small a number as practicable and that frequent coliform tests should be made.

8-9 Determination of Bacterial Numbers[8]

The preferred method for bacterial enumeration is the *membrane filter technique.* In this procedure a measured volume of water is drawn through a cellulose acetate or glass filter with openings less than 0.5 microns (μm). The bacteria present in the sample will be retained upon the filter. The filter is rinsed with a sterile buffer solution, placed upon a pad saturated with a suitable nutrient medium, and incubated at an appropriate temperature. The bacteria which are able to grow upon the nutrient medium will produce visible colonies which can be counted, each colony representing one bacterium in the original sample. If the sample contains too many bacteria the filter may be overgrown, and counting impossible. In such circumstances smaller samples or diluted samples are used.

The medium used and the temperature of incubation depend upon the bacteria which are being enumerated. For *coliform bacteria* M-Endo Broth or LES Endo Agar are used with incubation at 35°C for 20 to 22 hours. Coliform bacteria will produce colonies which are pink to dark red with a golden metallic sheen, often with a greenish tint. Noncoliform bacteria which grow upon this medium will lack the characteristic sheen.

Fecal coliforms are assumed to be acclimated to a temperature corresponding to that in the intestines, while nonfecal coliforms would find such temperatures lethal or, at least, inhibiting. Membrane filter analysis for fecal coliforms is conducted at a temperature of 44.5°C for 22 hours. On the medium ordinarily used (M-FC Broth) fecal coliform colonies are blue. Other bacteria which grow upon this medium are grey to cream colored.

The presence of *fecal streptococci* specifically indicates fecal waste contamination by warm-blooded animals. Incubation is on M-Enterococcus Agar or KF Streptococcus Agar for 48 hours at 35°C. Colonies are light pink to dark red with pink margins.

8-10 Fecal Coliform/Fecal Strep Ratio[9]

It has been shown that waters which have a higher fecal coliform count than fecal strep are most likely to contain wastes of human origin. If fc/fs is greater than 4 it is nearly certain the pollution is of human origin, while if fc/fs is less than 0.7 it is nearly certain to be of animal origin. Intermediate values of the ratio are less certain. Above 2, human pollution is likely, while from 0.7 to 1.0 animal sources predominate. Values between 1.0 and 2.0 are of uncertain composition.

8-11 Multiple Tube Fermentation

Coliforms are defined, in part, as including all of the aerobic and facultative Gram negative non-spore forming bacilli which ferment lactose with gas formation within 48 hours at 35°C. The multiple tube fermentation technique is based upon the last property. Appropriate quantities of the water to be tested are placed in sterile tubes containing a nutrient medium which includes lactose. The tubes are incubated for 24 hours at 35°C, are then examined, and presence or absence of gas is noted and recorded. Coliforms will ferment lactose with formation of gas, and hence are sometimes called "gas formers." If no gas has formed in 24 hours, the tubes are again examined at the end of 48 hours. Gas in any amount in a tube is a *positive presumptive test.* If there is no gas it is a *negative test.* Tubes showing gas are subjected to a confirmatory test which will eliminate certain bacteria of no sanitary importance. Several procedures are used but usually a portion from the fermentation tube is transferred to a tube containing brilliant green bile broth. This is incubated for 48 hours, and if gas is formed, it constitutes a *confirmed test.* If no gas is formed, it is designated as a negative test. The tubes are examined at the end of 24 hours of the positive presumptive test and confirmation started because such tubes are more likely to be positive in the confirmed test than those which develop gas more slowly. A further procedure, not routine but sometimes done, is the *completed test*. Its purpose is to demonstrate with certainty that the organisms giving positive results for the confirmed test are really members of the coliform group as defined above. The confirmed test for coliforms, it will be noted, requires at least 48 hours for a negative test and at least 72 hours for a positive test. For sampling methods see Art. 8-23.

In order to arrive at the number of coliforms in a water it is necessary to note the positive results from various size portions of a sample. The *most probable number* (MPN) of coliforms in the water is obtained by applying the laws of statistics to the results of the tests. Tables which permit quick determination of the MPN are readily available.[8] The standard sample for a potable water is five 10-ml portions. If all of these are negative, the MPN is zero. If only one is positive, the MPN is 2.2 per 100 ml. If two are positive, the MPN will be 5.1 per 100 ml. In some cases the sample may include smaller portions also. If, for example, of a sample consisting of five 10-ml portions and one 1-ml portion, one of the larger portions and the single 1-ml portion is positive, the MPN is 4.4 per 100 ml. The MPN of a water should be obtained by averaging those of a number of samples.

8-12 Viruses in Water

Viruses are parasitic agents smaller in size than bacteria. Viral cells are essentially pure protein which is very specific in reaction, requiring a particular host cell which serves as a source of raw material for viral reproduction. Viruses will not grow on media other than that of their natural host, which makes their enumeration or even the ascertaining of their presence difficult. The presence of the enteric viruses reo, echo, and polio has been confirmed in raw water samples and in samples taken from distribution systems.[11]

Chlorination,[12] ozonation,[13] coagulation,[14] and diatomaceous earth filtration[15] have been demonstrated as reasonably effective techniques for reducing viral numbers. The various methods give removals of 95 to 99+ percent and thus such removals can be reasonably anticipated in water treatment plants employing these or similar methods. Since a single virus can produce infection, it may be objected that such removals are insufficient and this may well be true. Present epidemiological techniques and inadequacies in sampling, detection, and identification methods for viruses make it impossible to state with certainty whether viral contamination of public water supplies is rare or common. The absence of a large number of epidemics traceable to water is not considered conclusive evidence against the hypothesis of viral contamination.

8-13 Microscopic Organisms

Microscopic organisms other than bacteria often have undesirable effects upon water. The Sedgwick-Rafter test[8] has been devised to concentrate the organisms for ease of identification by the microscope. The test is not ordinarily a routine one but is made when conditions indicate that knowledge of the predominating

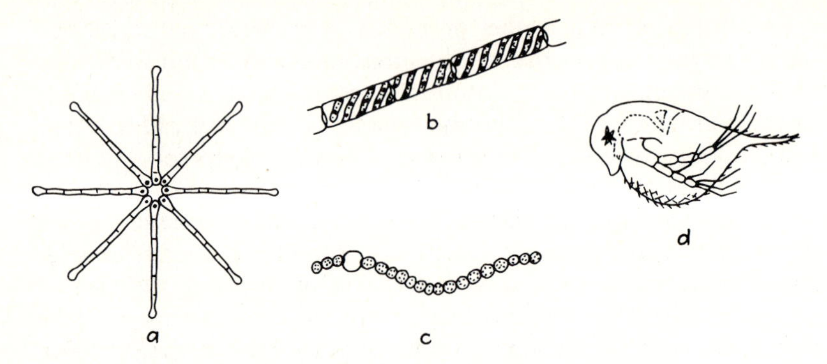

Figure 8-2 Some common organisms, greatly magnified. (*a*) *Asterionella*, a diatom. (*b*) *Spirogyra*, a green alga. (*c*) *Anabaena*, a blue-green alga. (*d*) *Daphnia*, a water flea.

type of organism is desirable. Several well-defined groups of freshwater plants and animals are of importance.

Algae. These are small, chlorophyll-bearing, generally one-celled plants of varying shapes and sizes which live in water. Certain types are filamentous and may form dense mats. When present in large numbers they may cause turbidity in the water and an apparent color. They sometimes cause trouble in waterworks by undue clogging of filters, but their most troublesome characteristic is the taste and odor that they may cause. The taste and odor are variable according to the particular algae involved, the most common being grassy, moldy, earthy, fishy, and aromatic odors. They are subdivided into the chlorophyta, or green algae; the cyanophyta, or blue-green algae; and the diatoms, which are distinguished by their beauty of shape and silicaceous structure. Other smaller groups are yellow-green, brown, and red. They must have light in order to live and propagate. Algae obtain their food from mineral matter in the water, nitrogenous compounds, and carbon dioxide. A product of their life processes is molecular oxygen which is taken into solution by the water. The tastes and odors are caused by certain oils which they secrete and which are released after death. Control methods are discussed in Art. 11-9.

Protists. One-celled animals are known as protozoans and are the lowest and simplest forms of animal life. The troublesome protists are similar to algae and are controlled in the same manner. From the protozoa, animal organisms in water rise in the scale of complexity in structure and life processes. The higher animals include rotifers, flatworms, and such crustacea (shellfish) as *Cyclops* and *Daphnia*, which are barely visible to the naked eye. They are rarely troublesome to the water plant operator.

One protozoan is of sanitary importance. It is *Endamoeba histolytica*, which causes amoebic dysentery. It forms cysts which are excreted in the bowel discharges of infected persons and which will live for long periods in water. They are removed by standard water treatment, coagulation followed by filtration, but are resistant to chlorination. However, breakpoint chlorination (Art. 11-3) with a free chlorine residual of 2 mg/l should kill them in a contact period of 30 min.

Fungi. These are non-chlorophyll-bearing plants and may therefore grow in the absence of light. Occasionally they will find conditions favorable in water mains and grow luxuriantly. As they die, the decomposition of their bodies will cause disagreeable tastes and odors. Certain fungi-like mosses, the bryozoa, which are really animals, may also grow in water mains and cause clogging or odors. They and the fungi can be controlled by treatment of the water with chlorine or copper sulfate.

Actinomycetes, which are related to both bacteria and fungi, are responsible for earthy, muddy, and musty odors in water. The troublesome types[16] are active in waters between 17 and 38°C.

8-14 Worms

Occasionally water consumers complain of small worms in the water. They may be red or grayish in color. The red blood worms are the larvae of *Chironomus* flies, and the others are larvae of the smaller but allied midge flies. The adult lays eggs on the surface of water in a basin or reservoir. The eggs hatch into the larvae, which attach themselves to concrete side walls or burrow in any sediment that has accumulated on the bottom. In each case the larvae make themselves characteristic shells or cases. Since they do not come to the water surface, a heavy infestation may occur before it is suspected. Remedial measures are draining and cleaning of basins to eliminate existing larvae and placing insect-tight covers over basins to prevent reinfestation. Chemicals in amounts usable for potable waters are of no use in killing the worms. They are not reported as ever causing tastes or odors in water.

8-15 Turbidity

A water is turbid when it contains visible material in suspension. While turbidity may result from living or dead algae or other organisms, it is generally caused by silt or clay. The amount and character of the turbidity will depend upon the type of soil over which the water has run and the velocity of the water. When the water becomes quiet, the heavier and larger suspended particles settle quickly, while the lighter and more finely divided ones settle very slowly. Very finely divided clay may require months of complete quiescence for settlement. Groundwaters are normally clear because the turbidity has been filtered out by slow movement through the soil. Lake waters are clearer than stream waters, and streams in dry weather are clearer than streams in flood because of the smaller velocity and because dry-weather flow is mainly groundwater seepage. Low inorganic turbidity (silt and clay) may result in a relatively high organic turbidity (color). The explanation of this is that low inorganic turbidity permits sunlight to penetrate freely into the water and stimulates a heavier growth of algae, and further, that organics tend to be adsorbed upon soil fractions suspended in water.[17] Turbidities are expressed in Turbidity Units (tu) and are measured by nephelometric comparison to a standard Formazin solution.[8] The EPA standard for drinking water is 1 tu monthly average with a maximum 2-day average of 5 tu. Sedimentation, with or without chemical coagulants, and filtration are employed to remove turbidity from water supplies.

8-16 Color

Color is caused by material in solution or a colloidal state and should be distinguished from turbidity which may cause an apparent color. True color in water supplies is generally caused by dyes derived from decomposing vegetation. Waters from swamps or weedy lakes and streams are therefore most likely to be troublesome in this respect. Colored water is not only undesirable because of consumers' objections to its appearance, but may discolor clothing and adversely affect industrial processes. Color is expressed in color units and is measured by comparison to a platinum-cobalt standard.[8]

8-17 Alkalinity and Acidity

Because the alkaline salts are very common in the ground, most waters are more or less alkaline. The carbonates and bicarbonates of calcium, sodium, and magnesium are the common impurities that cause alkalinity. In analyses, alkalinity is expressed in milligrams per liter in terms of equivalent calcium carbonate. Alkalinity is an important determination to the water treatment plant operator because some of the coagulants used to clarify water and prepare it for filtration require sufficient alkalinity to insure a proper reaction. If there are insufficient alkaline minerals in the water to react with the coagulants used or if, in other words, the alkalinity is too low, it may be increased by adding lime or sodium carbonate. Excessive alkalinity may, however, interfere with coagulation.

Caustic alkalinity is due to free hydroxyl ion and does not occur naturally in potable waters but sometimes results from treatment processes. In water-softening plants where lime is used there may be an excess of calcium hydroxide formed by the lime and water over that required to reduce the carbonate hardness. Caustic alkalinities are accompanied by high pH values, 9.5 or higher.

The most common cause of acidity in water is carbon dioxide, which may be present naturally or result from the reactions of coagulating chemicals used in water treatment. Actually, it is present as carbonic acid since it unites with water to form H_2CO_3. Acidity is measured in terms of the calcium carbonate needed to neutralize the carbonic acid and, as with alkalinity, it is expressed in milligrams per liter. When the two methods of determining alkalinity and acidity mentioned in this article are used a water may contain alkalinity, caused by bicarbonates, and at the same time acidity, caused by carbon dioxide.

8-18 pH

pH is a measure of the concentration of free hydrogen ion in water. Water, and other chemicals in solution therein, will ionize to a greater or lesser degree. The ionization reaction of water may be written:

$$HOH \rightleftharpoons H^+ + OH^- \tag{8-1}$$

This reaction has an equilibrium defined by the equation

$$\frac{(H)(OH)}{(HOH)} = K_w \tag{8-2}$$

in which (H) is the chemical activity of the hydrogen ion, (OH) the chemical activity of the hydroxyl ion, and (HOH) the chemical activity of the water. Since water is the solvent, its activity is defined as being unity. In dilute solution, molar concentrations are frequently substituted for activities yielding

$$[H][OH] = K_w(10^{-14} \text{ at } 20^\circ C) \tag{8-3}$$

Taking logs of both sides,

$$\log [H] + \log [OH] = -14 \tag{8-4}$$

$$-\log [H] - \log [OH] = 14$$

Defining $-\log = p$,

$$pH + pOH = 14 \tag{8-5}$$

In neutral solutions [OH] = [H], hence pH = pOH = 7. Increasing acidity leads to higher values of [H], thus to lower values of pH. Low pH is associated with high acidity, high pH with caustic alkalinity.

pH is important in the control of a number of water treatment and waste treatment processes and in control of corrosion. It may be readily measured potentiometrically by use of a pH meter.

8-19 Soluble Mineral Impurities

All natural waters have more or less dissolved mineral matter in them. As a matter of fact, we have become so accustomed to drinking water having such impurities that distilled water tastes flat and unpleasant. Its palatability can be increased by adding some common salt. Figure 8-1 shows the commoner impurities and their effects. It should be noted that divalent cations cause hardness, a quality that manifests itself in neutralization of soap. *Carbonate hardness* is caused by the carbonates and bicarbonates of calcium and magnesium. The carbonate of calcium is only slightly soluble in pure water; but when carbon dioxide is present, it will dissolve freely to form bicarbonates. This hardness is sometimes called temporary because it can be removed by boiling the water. The heat drives the carbon dioxide from the bicarbonates, leaving the insoluble carbonates to settle out as precipitate, forming scale in kettles or steam boilers. *Noncarbonate hardness* is caused by the sulfate and chloride salts of the divalent cations. It is not affected by boiling and is called permanent. The chlorides of magnesium and calcium are very corrosive to steam boilers and quickly cause pitting and grooving of boiler tubes. The sulfates of calcium and magnesium also cause scale in boilers.

The common sodium compounds are the carbonate, bicarbonate, sulfate, and chloride. They do not cause hardness. The carbonate and bicarbonate, on the contrary, will unite with grease and oil of the skin to form soap, thereby causing the slippery feel of soft alkaline water. In steam boilers the carbonate and bicarbonate will release carbon dioxide and cause corrosion of the tubes. Sodium sulfate may cause foaming in boilers if present in large amounts.

All the salts mentioned will cause tastes if present in sufficient amounts. Sodium chloride, common salt, for example, causes a perceptible taste when present in amounts greater than 200 mg/l, and water will be practically undrinkable if the amount is 3000 mg/l. Magnesium sulfate (Epsom salts) and sodium sulfate (Glauber salts) in large amounts, 500 mg/l or more, will both have noticeable laxative effects. The permissible limit of total dissolved solids has never been determined and in any case would depend upon the character of minerals present. A total of over 2000 mg/l would probably be objectionable for cooking and industrial uses even if the minerals were not hardness producers. The former Public Health Service Standards recommended 500 mg/l, with 1000 mg/l permissible if better water is unobtainable. The new standards of the EPA set no specific limit, but recommend 500 mg/l as a secondary standard.

8-20 Iron and Manganese

Iron in water causes hardness, but its important effect is that in very small amounts, about 0.3 mg/l and above, it will cause taste, discoloration of clothes and plumbing fixtures, and incrustations in water mains. Iron is very common in the earth, and water containing carbon dioxide which seeps through iron-bearing material dissolves it to form ferrous bicarbonate, $Fe(HCO_3)_2$. The ferrous bicarbonate, however, is easily oxidized into ferric hydroxide, $Fe(OH)_3$, which is precipitated as a rusty sediment. As previously indicated, iron in water is frequently accompanied by heavy growths of crenothrix which exaggerate the staining, pipe clogging, and other troubles. It should be recognized that iron in water may result from corrosion of the water mains, which will require appropriate corrective measures.

Manganese occurs in water less commonly than iron and generally in smaller amounts. It may be associated with iron, and trouble will probably result if the manganese concentration exceeds 0.05 mg/l or the total amount of iron plus manganese is more than 0.3 to 0.5 mg/l. Manganese is also oxidized into a sediment which clogs pipes, discolors fabrics, and stimulates organic growths. The color of the deposits and stains ranges from dark brown, if there is a mixture of iron, to black, if the manganic oxide is pure.

8-21 Chlorides

Occasionally special significance is given to the chloride content of a water, particularly sodium chloride. The salt used in preparation of food is excreted by the body, thereby making sewage higher in chlorides than the original water supply. Waters derived from the same sources, whether underground aquifers or surface areas, will have fairly uniform chloride contents. Should a particular well or stream show a higher chloride content, it might indicate sewage pollution. The test is useful only under special circumstances and is not recommended to replace bacterial tests for determining safety of water.

8-22 Gaseous Impurities

The soluble gases of the atmosphere, in addition to others which may be taken up at the ground surface or during percolation through the earth, are to be expected in water in greater or smaller amounts. They include nitrogen, methane, hydrogen sulfide, oxygen, and carbon dioxide. The last three are most important.

Hydrogen sulfide in groundwater is produced by reduction of sulfates, iron pyrites, or decomposition of organic matter and is likely to be found in water that has percolated through beds of lignite or other organic remains. Its disagreeable rotten-egg odor makes it highly objectionable in very small amounts. It is also corrosive to metals. Aeration removes it. It is poisonous when breathed, and atmospheres containing more than 0.005 percent should be avoided for exposures of over 1 hour, while concentrations of 0.06 to 0.08 percent will cause acute poisoning in a short time. Hence in aerating to remove this gas it should be dissipated into the open air with minimum risk of exposure to workmen.

Dissolved oxygen, O_2, is present in variable quantities in water. Its content in surface waters is dependent upon the amount and character of the unstable organic matter in the water and is an important factor in self-purification of polluted streams (see Art. 21-2). The amount of oxygen that water can hold is small and affected by the temperature; the higher the temperature, the smaller is the amount required for saturation. Corrosion of pipes is essentially an oxidation of the metal of the pipes, and free oxygen in solution is an agent of corrosion, particularly when carbon dioxide is also present.

Carbon dioxide, CO_2, is dissolved by water from the atmosphere, from decomposing organic matter at the earth's surface, or from underground sources. It unites with water to form carbonic acid, H_2CO_3, which is so easily decomposed that the carbon dioxide in the combination is still considered as "free." The presence in water of carbonic acid permits solution of calcium carbonate (limestone) and magnesium carbonate (dolomite) into the bicarbonates of those elements. The carbon dioxide in these combinations, which are $Ca(HCO_3)_2$ and $Mg(HCO_3)_2$, may be driven off by heat and is known as "half-bound" carbon dioxide. It will be remembered that the bicarbonates of divalent cations are the cause of temporary hardness. Carbon dioxide, particularly in the presence of free oxygen, causes corrosion of metals. In steam boilers the half-bound gas is also released to make matters worse. Free carbon dioxide can be removed by aeration or converted to bicarbonate (HCO_3) by use of lime, while the half-bound gas is removed by water softening. Aside from causing corrosion, carbon dioxide has no important effects in natural waters except that it is necessary for algal growth.

8-23 Sampling Methods

Methods of sample collection depend upon the analysis to be made. For bacteriological analysis only 120 ml of the water will be needed, but it must be collected in a sterilized bottle. If possible, the bottle and its stopper should be sterilized in the laboratory, although in an emergency both bottle and stopper may be boiled for 20 min. The standard procedure in sampling from a water faucet is as follows: *(a)* Flame the faucet briefly to kill clinging bacteria. This can be done with a piece of burning paper. *(b)* Turn on the water and allow it to run for 5 min. *(c)* Remove stopper from the bottle, being careful not to touch the inner portions of the stopper or bottle neck. *(d)* Fill bottle carefully, allowing no water to enter that has come in contact with the hands. It is sometimes necessary to collect a sample from a basin or reservoir. If the water can be reached, remove the stopper, plunge the bottle beneath the surface, and move the bottle while it is filling so that no water will enter that has been in contact with the hand. If the water is out of reach, in a well, the bottle can be lowered with a cord. It is advisable to give it a quick flaming, however, before it is lowered. Samples should be tested as soon as possible. If storage or transportation is necessary, they should be kept at a temperature between 0° and 10°C. If the sample has a chlorine residual, it should be dechlorinated immediately after collection.

Sampling for ordinary chemical analysis requires no other precaution than collection in a clean container. Not less than 4 l should be obtained.

8-24 EPA Standards[18]

Under the Safe Drinking Water Act of 1974 (Public Law 93-523) the Environmental Protection Agency has been charged with establishing Primary Drinking Water Standards for all "public water systems" which is defined as including all systems which serve 25 or more individuals or have 15 or more connections.

The present standards set *maximum contaminant levels* (MCLs) for certain inorganic and organic chemicals known to have toxic effects above a threshold concentration; for turbidity, since this is thought to be associated with difficulties in disinfection and, perhaps, with virus content; and for microbial population. The specific limits and the required sampling procedures and approved analytical techniques are presented in App. I.

States and cities are free to establish their own standards and most have done so in the past, generally conforming to the earlier Public Health Service Standards which applied by law only to waters used in interstate commerce. The major distinctions between the old and new standards are the inclusion in the new of limits on *chlorinated hydrocarbons*, *other pesticides*, *turbidity*, and *mercury* and the deletion of limits on iron, manganese, boron, chloride, sulfate, total dissolved solids, carbon chloroform extract, cyanide, detergents, and phenols. Most of the latter are primarily of aesthetic importance or are otherwise covered by the new standards.

8-25 Governmental Control of Water Supplies

The intention of the EPA is to work with the states to insure cooperation among the Agency, the states, and local governments. The states may receive grants from EPA for administration of the Act provided a supervisory program has been set up and the state has assumed or proposes to assume primary enforcement responsibility.[19] The Act is so worded that the maintenance of records, control of inspections, and other administrative actions appear to be under state jurisdiction unless the state abdicates its responsibility.

8-26 Liability for Unsafe Water

The seller of water to the public, whether a city, private corporation, or individual, does not guarantee the purity of the water furnished. Should a consumer of the water contract typhoid fever or other disease, he can obtain damages only if it is proved that the person, firm, or corporation supplying the water has been negligent in safeguarding the water and the disease was caused by said negligence. Instances of such negligence would be permitting the supply of chlorine to run out when this means of treatment was necessary to obtain safe water or allowing the existence of cross connections. Negligence would be definitely established if the danger of an existing condition had been pointed out by a state sanitary engineer to the water officials. A number of cities have had to pay heavy damages for sickness and death resulting from negligence. The seller of water may make the defense of contributory negligence upon the part of the complainant. For example, if it had been common knowledge that the water furnished was polluted and unfit for drinking unless boiled, then a person who drank raw water could not collect damages.

8-27 Characteristics of Waterborne Epidemics

The waterworks man should have an acquaintance with the epidemiology of typhoid fever and dysentery in order that he may investigate epidemics of those diseases. Such an acquaintance will enable him to find the cause if his water system is at fault and defend himself if the cause is elsewhere. The public usually leaps to the conclusion that contaminated water is the cause of any typhoid outbreak, in spite of the fact that it may also be caused by infected milk and other foods, particularly oysters, salads, and other uncooked foods. Oysters are contaminated by the water in which they are grown, but other foods are infected by carriers† or occasionally by flies. In the study of typhoid epidemics it is important that after taking infected food or water into the body the victim shows no signs of the disease for about 10 days. It may also be necessary to distinguish between primary cases, those persons infected by the original source, and secondary cases, those resulting from contact with the primary cases.

Whenever an epidemic occurs, the waterworks man should make certain that no defect exists or has existed in the system. If he has full records covering bacteriological tests, filter plant operating data, chlorine used, and residual tests, he is in a good position to show lack of negligence on his part. He should also see that the city health officer or his representative visits each typhoid case to obtain a complete history. Histories include date of onset of the disease; source of following foods, if used: milk, oysters, ice cream, raw vegetables; restaurants, banquets, or picnics attended; as well as journeys taken. Scrutiny of the questionnaires will generally show some common items that will give a clue to the investigators. If the only common item is the public water supply, then the water authorities may have to take the blame.

Waterborne epidemics have certain general characteristics. The cases are likely to be scattered over the city. Exceptions have been noted where water from several sources is pumped into the system and only one source is infected. Then the cases may be clustered around the point where the water enters the system. Similarly the cases may be grouped near a cross connection (see Art. 8-34) which has caused the trouble. All classes and all ages of people will be affected and if gastrointestinal disturbances and dysentery accompany or precede the typhoid, there is a strong likelihood that water is involved.

Water treatment and the application of sanitary science in general has much reduced waterborne epidemics. Past records indicate that between 1920 and 1945 there were 722 outbreaks of typhoid fever or an average of 28 per year. In comparison, in the period 1971 to 1974 there were a total of 99 waterborne outbreaks of which 4 were typhoid fever.[20]

The principal sanitary defects or poor practices which have caused waterborne outbreaks in the past can be classified as follows: *(a)* use of untreated surface waters; *(b)* use of untreated groundwater; *(c)* treatment deficiencies; and *(d)* distribution system deficiencies, particularly cross connections which permit polluted water to enter the system.

8-28 The Sanitary Survey

A survey of all surroundings and conditions that may affect the quality of a water supply is highly important. For state certification of a supply a favorable survey report is required, in addition to a satisfactory bacteriological test. The reasons for

† Carriers are persons who harbor disease germs and excrete them in body discharges but show no signs of disease. A considerable proportion of all persons having typhoid fever become temporary or permanent carriers. They can be discovered by a laboratory test. No one having had typhoid should be allowed to prepare food for the public unless tested and shown not to be a carrier.

surveys are that they will find conditions that are potential sources of contamination and waterborne epidemics, that when bacteriological testing of the water indicates pollution, a survey may find the danger, and that the survey is a necessity for proper interpretation of bacteriological tests. Difficulties frequently arise in connection with single water samples from small well supplies that show presence of coliforms. A sanitary survey may indicate that they are probably of nonfecal origin and that such drastic action as condemnation of the supply is not justified. Hasty action in condemnation is especially likely to cause adverse criticism if the water has been consumed regularly without causing disease. Sanitary surveys require judgment and technical knowledge. Operational procedures and techniques for correcting defects are discussed in the following articles.

8-29 Watershed and Reservoir Sanitation

When a city constructs an impounding reservoir, the question arises as to the sanitation of the watershed, especially the portion in close proximity to the reservoir, and the use that may be made of the reservoir for recreational purposes. The control to be exerted should depend upon the amount of treatment given to the water. A few cities use untreated reservoir water collected from uninhabited areas. In these cases no recreational use should be made of the reservoir, and use of the watershed should be strongly discouraged. If necessary, it should be purchased by the city. If the water is chlorinated, less vigilance may be needed on the watershed. When a complete treatment plant is constructed and properly operated, it appears that the only objection to using a reservoir for bathing, fishing, and hunting and the shores for camping and picnicking would be aesthetic. This, however, overlooks the possibility of a breakdown of the treatment plant or poor operation by unqualified personnel. The best solution as to the allowable use of reservoirs and watersheds is to allow decisions to be made by the state health department, which presumably will be impartial and not swayed by local interests or prejudice.

In the absence of state authority the following suggestions are made, and are applicable when an adequate treatment plant is being properly operated: *(a)* Recreational use of reservoirs should be permitted only when there is real need on the part of the public (not by interested persons) for such use and the need cannot be supplied by some other body or bodies of water. *(b)* Use should be regulated by caretakers having police authority. The pay of caretakers and cost of sanitary conveniences may be defrayed through a system of licenses or admissions paid by the persons who make the expense necessary. *(c)* Picnics should be restricted to certain areas where garbage cans and sanitary toilets are placed. The toilets should be the chemical type, although pit toilets are allowable if properly constructed and located. A sufficient number of toilets should also be installed to serve hunters and fishermen and should be kept in good condition. *(d)* Bathing should be permitted only in large reservoirs and then not within a distance of 2 km (1.2 mi) from the water intake. Warning signs should clearly indicate the limits. It is advisable to restrict the bathing, when allowed, to certain defined bathing beaches. *(e)* Boating, shooting, or fishing should not be allowed within 200 m (600 ft) of the water intake. The restricted area should be marked by a cable supported by buoys with warning signs displayed. *(f)* Farmhouses on the watershed in the immediate vicinity of the reservoir should have toilets that will not pollute surface water. If the watershed is large and includes cities, adequate treatment methods should be applied to their sewage.

The policy of the American Water Works Association[21] is that recreational use of reservoirs should not be permitted if they are a part of the water system and deliver water ready for consumption, or if they are terminal reservoirs, i.e., provide water at the end of storage and prior to treatment. Recreational use should not be permitted if the reservoir derives its water from an uninhabited or sparsely inhabited area at or near the point of rainfall or snow melt and if the water is clean and clear enough to be distributed to consumers with disinfection only. If the water is from an area not heavily inhabited and is allowed to flow from storage in a natural stream to the point of withdrawal and the water requires other treatment in addition to disinfection, limited recreational use may be permitted. Recreation is considered permissible under appropriate sanitary regulations if the water has flowed in a natural stream for a considerable distance before storage, has received polluting materials from cities, industries or agricultural areas, is confined in the reservoir primarily to supplement low stream flows, is allowed to flow in a natural stream with possible access of the public, and complete treatment is required. The sanitary regulations applied are to be determined solely by the waterworks system executive whose responsibility it is to provide a safe and potable water.

8-30 Lake Overturns

In deep lakes and impounding reservoirs the lower strata of water are cool in summer and change very little in temperature. The upper layers are warm and vary according to air temperature. The water temperature decreases slowly as depth increases, until at some level a quick change starts which in a relatively short vertical drop brings the water to the constant bottom temperature. The zone of quick temperature change is known as the *thermocline*. As the air temperature falls in the autumn, the upper water cools and sinks. This goes on until all the water in the lake is at the temperature normally below the thermocline. At this time all the lake water is at uniform density, and any disturbance, such as a heavy wind, will cause pronounced upward currents or an overturn of the stagnant bottom water to the top. A further lowering of the temperature of the upper water will have the same effect. The process goes on until the top water has reached a temperature of 4°C, at which point water is at its greatest density. A further lowering of temperature results in a decreased density, and the cold water stays at the surface. In the spring the top water, which has remained close to 0°C during the winter, is warmed to 4°C, and sinks, replacing the bottom water and causing another overturn.

Overturns are of importance to the waterworks man because the stagnant deep water when brought to the surface enters the water intake. Since it is low in dissolved oxygen and high in the products of decomposition, it frequently has a musty or moldy taste and odor and causes complaint among consumers. Water treatment plant operators should be prepared to use taste- and odor-preventive measures at the times when overturns are to be expected.

8-31 The Water Treatment Plant

The surveyor of a water system should carefully check the treatment plant. The investigation covers the competence and knowledge of the operator, the use of bacteriological and other tests, and keeping of adequate records. The danger of inadequate chlorination may be minimized by keeping on hand duplicate chlorinator parts and a reserve supply of chlorine. A check on the accuracy of the chlorinator should be obtained by placing the chlorine tank upon a scale and keeping a record of the weight, which should be systematically read.

Construction of the treatment plant should be noted. Bypasses around treatment units and auxiliary intakes which also bypass the plant may permit polluted water to enter treated water reservoirs. Raw-water basins and treated-water basins may be separated only by a concrete wall, and cracks, unsuspected pipes, or leaky valves may allow entrance of unsafe water. No precautions may have been taken against floods inundating basins with untreated water.

8-32 Groundwater Supplies

Springs must be safeguarded against pollution by surface or floodwaters. This is accomplished by curbing the spring to a sufficient height to prevent surface waters from entering. A cover is also necessary to prevent pollution from dust, animals, and insects.

Wells may also be polluted by flooding, and where possible they should not be located where this danger exists. The floodwater may enter wells insufficiently safeguarded at the top, and seepage downward outside the casing to the aquifer may occur.

Although shallow dug wells are seldom used for public supplies, they are much used for farm homes and may be grossly polluted. They should be lined with watertight concrete for at least 3 m (10 ft) below the ground surface to prevent surface water and insufficiently filtered water from entering. The well should be tightly covered, with a tight joint between the pump and the top.

Driven and drilled wells also require careful construction at the top. The annular opening between the casing and the delivery pipe should be tightly closed, and the casing should extend above the ground at least 300 mm (1 ft). An apron or other structure of concrete extended horizontally 600 mm (2 ft) or more from the casing at the ground surface will tend to prevent seepage of surface or shallow water down the outside of the casing. As a further safeguard the casing should be sealed into the first impervious stratum penetrated by means of cement or a metal collar. The well drillers should be required to do this and also close the openings that occur where the casing reduces in size. A metal collar expanded in the opening by means of a wedge will prevent leakage both inward and outward. Corroded casings sometimes permit contaminated water to enter wells and necessitate replacement or abandonment of the wells.

In older well installations it was sometimes found necessary to place the well pump in a pit so that the allowable suction lift would not be exceeded. In some cases the pits were drained into nearby sewers to care for seepage. Epidemics have resulted from a clogged sewer causing sewage to flood the pit and contaminate the well water. Pits should be avoided; but if used, seepage can be safely removed by means of a sump pump.

The surroundings of wells should be free from potential dangers. Sewers may break or leak and allow sewage to seep into the well. This is especially likely to happen where the soil or underground formation has cracks or crevices, which is frequently the case in limestone. Wells in limestone or those having the aquifer overlaid by limestone are frequently contaminated. Not only must sewers be guarded against, but also cesspools, septic tanks, privies, and abandoned wells. Abandoned wells may be open at the top so that unsafe water may enter the aquifer and contaminate nearby wells which are serving as sources of supply. Abandoned wells should always be plugged with concrete or capped. Most states have extensive regulations concerning closing of abandoned wells.

8-33 Pumping Plant, Pipes, and Reservoirs

It is dangerous practice to use clay or concrete sewer pipe to convey water not under pressure from wells to reservoirs. It is practically impossible to make such pipes watertight, and infiltration of groundwater may be expected. Sewers near such pipes have broken and permitted seepage of sewage into drinking water. Suction pipes leading from wells or reservoirs to pumps may leak and permit entrance of ground or surface water. They should preferably be placed well above the ground, and the joints should be carefully made. Presence of much air in the water beyond the pump is a symptom of leakage in the suction pipe.

Poorly located or poorly constructed basins are likely to permit contamination. They should be located so that floodwaters will not overflow or back up into them through drainpipes. The bottoms and walls should be free from cracks that will permit leakage inward or outward. Brick construction should be avoided, as it frequently permits leakage. Nearby sewers increase the hazard. The reservoir should be covered, if possible, to exclude dust, insects, and malicious pollution. Manholes should be locked and offset above the roof to prevent entrance of water from the roof or cover. Ventilators are needed to allow entrance of air as the water level fluctuates, but these should permit a minimum of dust and no insects to enter.

In the pumping plant, precaution should be taken against dangerous cross connections and use of unsafe priming water. If air lift is used for pumping from wells, the air intake should be located so that unpolluted air will be obtained.

8-34 Cross Connections

The greatest hazard in the distribution system is the cross connection. The AWWA defines cross connections as being "any connection or structural arrangement between a . . . potable water system and any nonpotable system through which backflow can occur."[22] Facilities that are particularly likely to have cross connections include hospitals, metal plating and chemical plants, car washes, laundries, and dye works.[23]

When cross connections to possibly contaminated systems are required for water supply, backflow preventers should be used to prevent contamination of the potable supply. Backflow preventers include a variety of devices. Single-check valves or double-check valves can be used in applications where the hazard is small, such as possible contamination with food products or other innocuous substances. Where the hazard is great, reduced pressure principle (RPP)[23,24] systems should be used (Fig. 6-23). These consist of compound spring-loaded-check valves which will positively close when the pressure on the downstream end rises to within about 10 kPa (2 psi) of that upstream, and which will vent the backflow to the atmosphere. RPP systems have a relatively high headloss but have been shown to provide positive control when in good condition.[23]

Regular inspection and testing of backflow prevention devices is essential to insure that they will operate when needed. Testing procedures which should be followed and an outline of a testing schedule are presented elsewhere.[23,25]

To protect against possible cross connections in relatively low risk situations such as might exist in private homes, air gaps and vacuum breakers are sometimes employed. The latter are not reliable and should not be used in critical locations.

Cross connections to auxiliary supplies may be surprisingly numerous in some large cities, and they will be discovered only by a careful survey of the water services connected to the public system. When found, they should be eliminated or remedied in one of several safe ways. One method is to require the auxiliary supply to be so treated or safeguarded that it will be at all times as safe as the public supply. Or the public supply is made available to the industrial or fire protection pipe systems only by first discharging it into a tank above the highest possible water level so that there will be an air break between the inlet pipe and the tank contents. Figure 8-3 shows an approved method. There is no physical connection and no dependence upon valves for safety. At all times the roof tank is kept full of water from the public supply by the float valve and is always available to supply the automatic sprinklers. When the fire pump is started, the check valve under the roof tank closes, and the sprinklers are served directly.

8-35 Plumbing Defects

Plumbing defects are special types of cross connections that are even more numerous than those already described. The danger arises from plumbing fixtures that have their water inlets below the rims. If the waste pipe is clogged and the fixture is full, a lowering of pressure in the water supply pipe will permit polluted

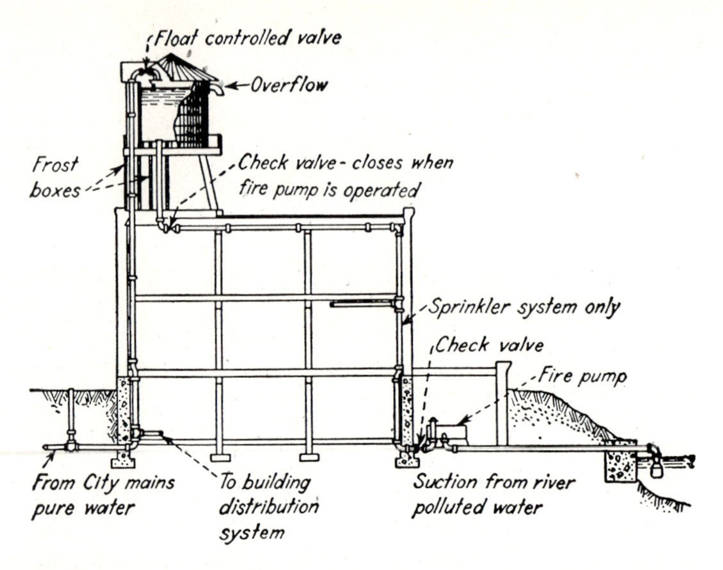

Figure 8-3 Safe method of obtaining dual water supplies. *(Courtesy Minnesota State Board of Health.)*

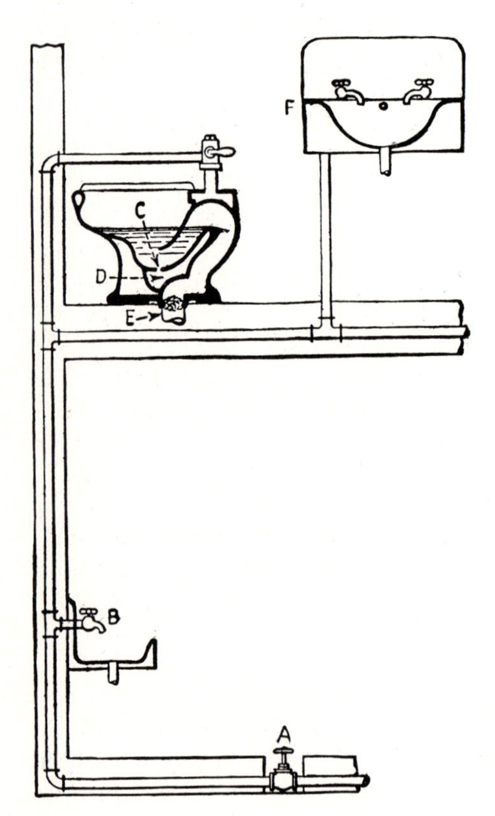

Figure 8-4 How back-siphonage occurs in water supply pipes. *(From "Municipal and Rural Sanitation" 5th ed., by V. M. Ehlers and E. W. Steel, Copyright 1958, McGraw-Hill Book Company, Inc. Used with permission of McGraw-Hill Book Company.)*

water to siphon back into the water line. Some fixtures, like bedpan washers, surgical-instrument sterilizers, and glass washers at soda fountains, usually have their water inlets in the bottom of the fixtures and are especially dangerous. Toilets and urinals of the flushometer type present a hazard. Some will permit siphonage of water from the bowl even when there is no clogging. Figure 8-4 shows the conditions that may permit water pollution. Valve *A* may be closed, or there may be a heavy draft on the supply outside the house. With these conditions and a clogged toilet or filled washbowl, opening valve *B* will siphon water into the piping system. Even without clogging, water from the toilet may be siphoned into the pipes through the siphon jet *C* and *D* and from there into the flush pipe. The remedy for cross connections of this type is prohibition of the use of underrim water inlets and the use of approved vacuum breakers in the water lines leading to flushometer toilets and urinals.

Other defects are also numerous. Actual cross connections are found between waste pipe lines and water lines in the complicated plumbing systems of large buildings. Leaky waste lines may be suspended above drinking-water tanks so that sewage can drip into them. Drain lines may be run from drinking-water tanks to waste lines without a break and may permit sewage to back up into the tanks. Such drainpipes should discharge into funnels that connect to the waste line, with the drainpipe terminating above the funnel's rim so that an air break is provided. Ingenuity must be employed by the inspector to discover all possible dangers. Cooperation between the city plumbing inspection force and the water department is of value.

8-36 The Piping System

Dangerous conditions sometimes result in the system of mains from placing sewers too close to the water pipes. An unsuspected leak in a water pipe may, while the water pressure is off, draw in sewage that has seeped into the soil. The practice of placing water pipes and sewers in the same trench should be discouraged. A break in the water main will necessitate excavation, and in the process the sewer may be broken, flood the hole with sewage, and contaminate the main. Should this occur, the main, after repairs, should be thoroughly disinfected before consumers resume use of water. This can be done by pumping a solution of chlorine into the main until a positive orthotolidin test can be obtained at the surrounding fire hydrants and from the services between the break and the nearest hydrants. Similar precautions are recommended after all breaks in mains where muddy water from the excavation made for repairs has entered the pipe. Methods of disinfecting water mains are given in Art. 6-31.

8-37 Drinking Fountains

Numerous types of drinking fountains are on the market, but many are not entirely satisfactory from the sanitary viewpoint. The vertical bubblers, it has been shown, may retain bacteria from the mouths of users. Others may form cross

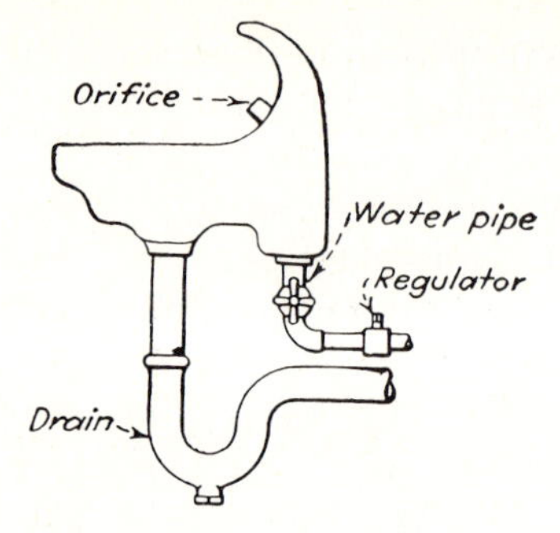

Figure 8-5 A sanitary drinking fountain.

connections. Fountains should conform to the following requirements: *(a)* The jet should rise at an angle from the orifice. *(b)* The orifice should be shielded so that the lips of the user cannot possibly come in contact with it or water drip from the mouth upon it. *(c)* The orifice should be above the rim of the bowl so that contaminated water cannot run back into the pipe, should the waste line be clogged. *(d)* The valve should be rugged and permit adjusting of the jet to the water pressure.

8-38 Tracing Pollution

Investigation of pollution in water supplies is sometimes furthered by use of dyes to trace the movement of water underground. Fluorescein (also called uranin) is a harmless organic dye which is only slightly removed by filtration through sand or earth and will impart a noticeable greenish color to water if present at a concentration of 0.002 mg/l. A solution is made of 125 g fluorescein and 125 g of caustic soda in 10 l of water. If surface pollution of a well is suspected, some of the solution is sprinkled on the ground surface, and the water is observed for color. If pollution of a well or spring by a cesspool or sinkhole is likely, enough of the solution is placed in the pool to give its contents a deep color. If pollution from a sewer is feared, several gallons of the solution should be poured into the sewer near the location of the leak or break.

A sample of the water should be taken before the test is started to compare with samples taken later. Large amounts of the dye are easily detected but to insure that minute amounts have not entered, comparison of the samples should be made visually by placing them in two deep colorless glass cylinders and observing the color in the full depth of the water, or spectrophotometrically.

REFERENCES

1. Craun, Gunther F., and Leland J. McCabe, "Problems Associated with Metals in Drinking Water," *Journal American Water Works Association*, **67**:593, November, 1975.
2. Sobsey, Mark D., "Enteric Viruses and Drinking Water Supplies," *Journal American Water Works Association*, **67**:414, August, 1975.

3. "Water Supply: Nitrate in Potable Waters and Methemoglobinemia," *Yearbook of the American Public Health Association*, New York, 1950.
4. Reid, Frank, "Drinking Water Quality Regulations: Measuring, Monitoring, and Managing," *Water and Sewage Works*, **123:**6:49, June, 1976.
5. Eliassen, R., and R. A. Lauderdale, "Radioactive Fallout in Water Supply at Portland, Me." *Journal American Water Works Association*, **48:**6, June, 1956.
6. McKinney, R. E., *Microbiology for Sanitary Engineers*, McGraw-Hill, New York, 1962.
7. Kehr, R. W., "Some Notes on the Relation Between Coliforms and Enteric Pathogens," *Public Health Reports*, **59:**589, 1943.
8. *Standard Methods for the Examination of Water and Wastewater*, 14th ed. American Public Health Association, New York, 1976.
9. Geldreich, Edwin E., and Bernard A. Kenner, "Concepts of Fecal Streptococci in Stream Pollution," *Journal Water Pollution Control Federation* **41:***Part 2:R336*, August, 1969.
10. Rivers, T.M., *Viral and Rickettsial Infections in Man*, J. B. Lippincott Co., Philadelphia, 1948.
11. McDermott, James H., "Virus Problems and Their Relation to Water Supplies," *Journal American Water Works Association*, **66:**693, December, 1974.
12. Culp, Russell L., "Breakpoint Chlorination for Virus Inactivation," *Journal American Water Works Association*, **66:**699, December, 1974.
13. Katzenelson, E., B. Kletter, and H. I. Shuval, "Inactivation Kinetics of Viruses and Bacteria in Water by Use of Ozone," *Journal American Water Works Association*, **66:**725, December, 1974.
14. York, David W., and William A. Drewry, "Virus Removal by Chemical Coagulation," *Journal American Water Works Association*, **66:**711, December, 1974.
15. Brown, Thomas S., Joseph F. Malina, Jr., and Barbara D. Moore, "Virus Removal by Diatomaceous Earth Filtration—Part 2," *Journal American Water Works Association*, **66:**735, December, 1974.
16. Silvey, J. K. G., J. C. Russell, D. R. Redden, and W. C. McCormick, "Actinomycetes and Common Tastes and Odors," *Journal American Water Works Association*, **47:**7, 1955.
17. McGhee, Terence J., and Reed A. Miller, "Soil Adsorption of Humic Color," *Journal Water Pollution Control Federation*, **48:**1970, August, 1976.
18. "National Primary Drinking Water Regulations," *Journal American Water Works Association*, **68:**57, February, 1976.
19. Barg, Raymond, "State Regulatory Agencies and the Safe Drinking Water Act," *Journal American Water Works Association*, **68:**70, February, 1976.
20. Craun, Gunther F., Leland J. McCabe, and James M. Hughes, "Waterborne Disease Outbreaks in the U.S.—1971–1974," *Journal American Water Works Association*, **68:**420, August, 1976.
21. "Recreational Use of Domestic Water Supply Reservoirs," *Journal American Water Works Association*, **50:**5, May, 1958.
22. American Water Works Association, "Backflow Prevention and Cross-Connection Control," *Manual M-14* American Water Works Association, Denver, Colo., 1966 and 1970.
23. Angele, Gustave J., Sr., *Cross Connections and Backflow Prevention* 2nd ed. American Water Works Association, Denver, Colo., 1974.
24. AWWA C-506, "AWWA Standard for Backflow Prevention Devices—Reduced Pressure Principle and Double Check Valve Types," American Water Works Association, Denver, Colo., 1969.
25. *Cross Connection Control Manual*, Environmental Protection Agency EPA-430/9-73-002, 1973.

CHAPTER

NINE

TREATMENT OF WATER—CLARIFICATION

9-1

It may be desirable to treat water for a number of reasons including the removal of pathogenic organisms, unpleasant tastes or odors, excessive color or turbidity, certain dissolved minerals, and a variety of unpleasant or potentially harmful chemical species. The product of water treatment may be suitable for general domestic and industrial purposes (meeting the criteria of Chap. 8) or may be produced to higher standards, such as those required for high or intermediate pressure boilers, manufacture of food and beverages, and other specialized industrial purposes.[1]

A number of treatment methods have been developed to produce water of the requisite quality. *Storage* and *plain sedimentation* alone have occasionally been used, but would not be incorporated in modern treatment works since more rapid sedimentation as well as removal of colloidal particles is obtained with *chemical coagulation* followed by *filtration*. *Aeration*, *chemical oxidizing agents*, and *adsorbents*, will reduce or remove tastes and odors. *Softening* (the removal of divalent cations—generally calcium and magnesium), *oxidation* and *precipitation* of iron and manganese, *stabilization* to prevent corrosion or deposition in distribution systems, and *disinfection* are all common treatment methods. More sophisticated techniques such as *ion exchange*, *reverse osmosis*, *electrodialysis*, and *distillation* are applied in those communities and industries in which the state of the raw water or the quality of the finished product require them.

The character and degree of treatment will depend upon the nature of the water and this depends largely upon its source. Surface waters are likely to be bacteriologically contaminated and more or less turbid. They will thus generally require coagulation, sedimentation, filtration, and disinfection. A few cities which obtain relatively clear water from lakes or reservoirs depend entirely upon disinfection, but no new surface water systems are constructed without provision for filtration. Groundwaters are usually clear and therefore may not require filtration if no chemical precipitates are formed by other processes. Filtration is desirable, however, following lime softening or iron and manganese removal—both of which may be necessary to produce a high quality supply from a groundwater source. Softening is also provided for many surface supplies. Taste and odor problems may occur in both surface and groundwaters—in the former as a result of organic contamination, in the latter as a result of high metallic content or dissolved hydrogen sulfide.

9-2 Storage

No recently constructed waterworks depend upon storage alone as a treatment technique although a number of large American cities still employ existing storage facilities in conjunction with other, more modern, methods. The desirable effects of storage are a reduction in turbidity and in bacterial numbers as indicated by the coliform test. The undesirable effects include the possible production of taste and odor by algal blooms and the likely fluctuation in water quality produced by overturns.

9-3 Screens

Screens may be employed both at surface water intakes and at inlets of sewage-treatment plants. They serve as a protective device for the remainder of the plant rather than as a treatment process. Large bulky objects can be removed by *bar screens*, which consist of vertically placed bars with openings generally on the order of 25 mm (1 in). Most such screens include mechanical cleaning devices similar to that shown in Fig. 9-1.

Fine screens are used at surface-water intakes, sometimes alone, sometimes following a bar screen. These devices have openings of approximately 6 mm ($\frac{1}{4}$ in) and thus remove leaves, twigs, and fish. A typical moving water screen is shown in Fig. 9-2, while a rotating sewage screen is shown in Fig. 9-3. The accumulated debris is carried from the flow by the screen and dislodged by gravity or by water sprays.

The headloss through unobstructed screens depends upon the nature of their construction (open area, blocked area, shape of the screen elements, etc.) as well as the approach velocity, which is normally 0.3 to 0.6 m/s (1 to 2 ft/s). The headloss can be calculated from a variety of empirical relationships such as that of Kirschmer[2] or from the basic equation for headloss through an orifice. Equipment manufacturers generally provide such information in their design brochures, but

Figure 9-1 Mechanically cleaned bar screen. *(Courtesy Envirex, A Rexnord Company.)*

the headloss through open screens is generally low—and insignificant in comparison to that during operation when the screen is partially blocked. Screen designs frequently incorporate provision for automatic presentation of a clean surface at a predetermined headloss. Simpler systems operate either continuously, or, more commonly, on an intermittent timed basis. Provision must be made for the development of significant headloss across intermittently cleaned screens. Some manufacturers recommend dropping the channel 150 to 300 mm (6 to 12 in) across a bar screen.

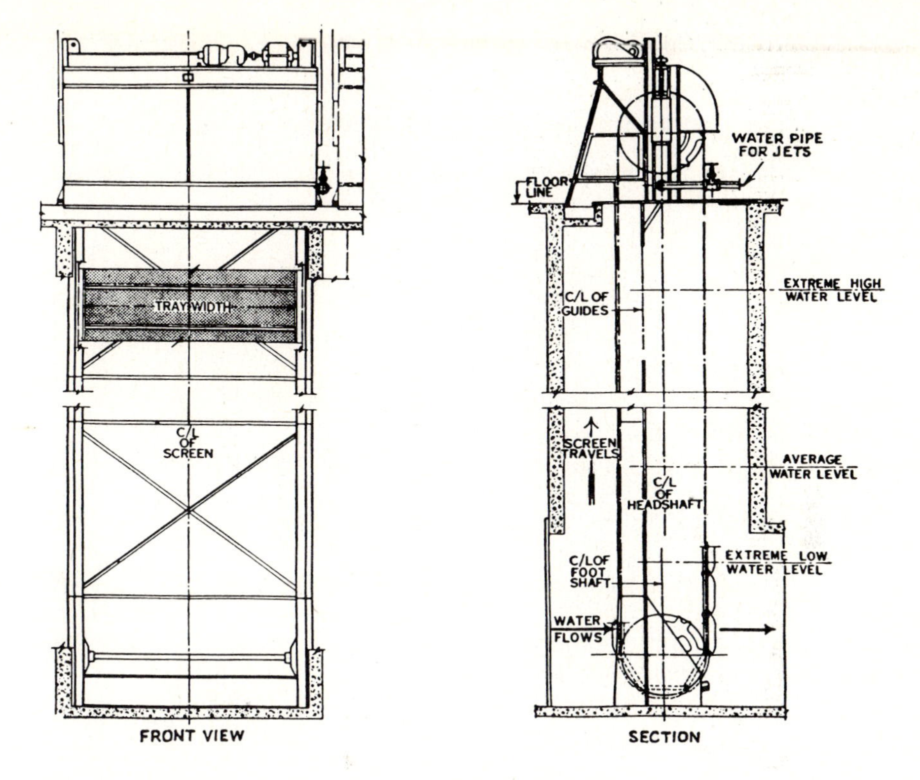

Figure 9-2 Traveling water screen. *(Courtesy FMC Corporation.)*

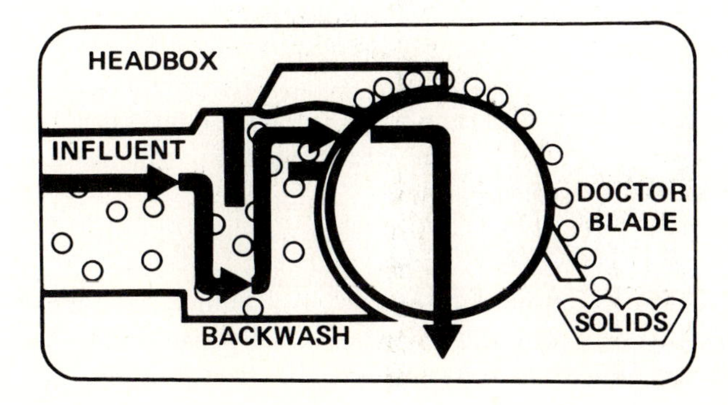

Figure 9-3 Rotating sewage screen. *(Courtesy Hycor Corporation.)*

PLAIN SEDIMENTATION

9-4 Principles of Sedimentation—Discrete Particles

The sedimentation of discrete spherical particles may be described by Newton's Law, from which the terminal settling velocity is found to be:

$$v = \left[\frac{4\mathbf{g}(\rho_s - \rho)d}{3C_D\rho}\right]^{1/2} \tag{9-1}$$

where v = the terminal settling velocity (L/T)
ρ_s = the mass density of the particle (M/L^3)
ρ = the mass density of the fluid (M/L^3)
$\mathbf{g}$ = the acceleration due to gravity (L/T^2)
d = the diameter of the particle (L)

and C_D is a dimensionless drag coefficient defined by

$$C_D = \frac{24}{N_R} + \frac{3}{\sqrt{N_R}} + 0.34 \tag{9-2}$$

in which N_R is the Reynolds number, $vd\rho/\mu$, where μ is the absolute viscosity of the fluid (M/LT) and the other terms are as given above. Equation (9-2) is applicable for Reynolds numbers up to 1000, which includes all situations of interest in sedimentation of water and wastewater. Where N_R is small (less than 0.5) the latter terms of Eq. (9-2) may be neglected to give

$$C_D = \frac{24}{N_R} = \frac{24\mu}{vd\rho} \tag{9-3}$$

which, when substituted in (9-1) yields

$$v = \frac{\mathbf{g}}{18\mu}(\rho_s - \rho)\, d^2 \tag{9-4}$$

which is Stokes' law.

Particles in water and sewage are not spherical. The effect of irregular shape is not pronounced at low settling velocities, and most sedimentation devices are designed to remove small particles which settle slowly. Larger particles which settle at higher velocities will be removed whether or not they follow Stokes' or Newton's laws.

The design of sedimentation basins is generally based upon the concept of the ideal sedimentation basin (Fig. 9-4). A particle entering the basin will have a horizontal velocity equal to the velocity of the fluid,

$$V = Q/A = Q/w \cdot h$$

and a vertical velocity, v_s, equal to its terminal settling velocity defined by Stokes' or Newton's law. If a particle is to be removed, its settling velocity and horizontal velocity must be such that their resultant, $\bar{V}$, will carry it to the bottom of the tank

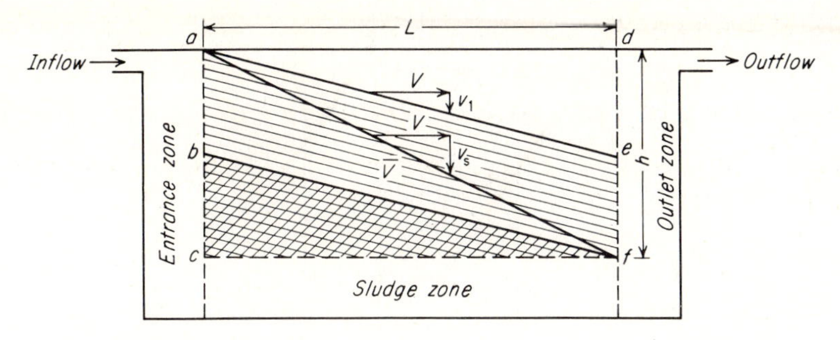

Figure 9-4 Diagram of ideal settling basin.

before the outlet zone is reached. If a particle entering at the top of the basin (point a) is so removed, all particles with the same settling velocity will be removed. Considering the slope of the velocity vector from a to f and the dimensions of the basin itself, one may write:

$$v_s/V = h/L, \text{ or}$$

$$v_s = Vh/L = h/L \cdot Q/w \cdot h = Q/wL \tag{9-5}$$

Equation (9-5) defines the *Surface Overflow Rate* (SOR) which is numerically equal to the flow divided by the plan area of the basin, but which physically represents the settling velocity of the slowest settling particle which is 100 percent removed. Those particles which settle at velocities equal to or greater than the SOR will be entirely removed, while those which settle at lower velocities will be removed in direct proportion to the ratio of their settling velocity to v_s, assuming that they are uniformly distributed upon entering the basin.

In Fig. 9-4 one may note that a particle with settling velocity v_1 which enters at the top (point a) will only have settled to point e and will thus enter the outlet zone in which higher velocities will carry it from the basin. An identical particle which enters at point b and settles at the same rate will be removed, as will all such particles which enter below point b. Again, considering the slopes of the velocity vectors and the geometry of the basin, one may equate the fraction of particles (X_r) with velocity v_1 removed to the vertical dimensions $b - c$ and $a - c$:

$$X_r = \frac{b - c}{a - c} = \frac{v_1/V \cdot L}{v_s/V \cdot L} = \frac{v_1}{v_s} \tag{9-6}$$

For actual suspensions of particles with a considerable variety of sizes and densities, prediction of the efficiency of a basin requires either a particle size distribution or a settling column analysis. From either technique a settling velocity cumulative frequency distribution curve similar to that of Fig. 9-5 may be obtained. As noted above, all particles with velocity greater than v_s will be removed. The fraction of all particles removed thus will be

$$F = (1 - X_s) + \int_0^{x_s} \frac{v}{v_s}\, dx \tag{9-7}$$

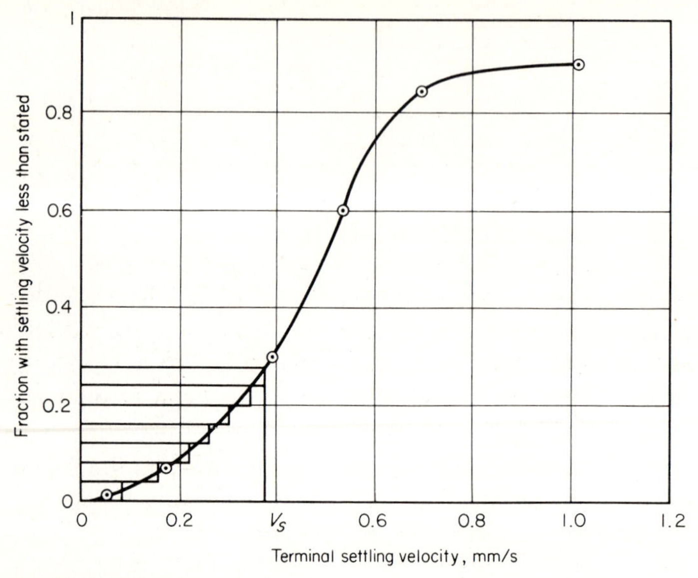

Figure 9-5 Cumulative distribution of particle settling velocity.

in which $(1 - X_s)$ is the fraction of particles with $v \geq v_s$ and the integral is the fraction of particles with $v < v_s$ which are removed in the basin.

The actual calculation is performed with a finite number of points on the distribution curve, so that, in practice, Eq. (9-7) is approximated by:

$$F = (1 - X_s) + \frac{1}{v_s} \Sigma v \, \Delta x \tag{9-8}$$

Example 9-1 Removal of discrete particles A settling basin is designed to have a surface overflow rate of 32.6 m/day (800 gal/ft^2 per day). Determine the overall removal obtained for a suspension with the size distribution given. The specific gravity of the particles is 1.2 and the water temperature is 20°C.

Particle size, mm	0.10	0.08	0.07	0.06	0.04	0.02	0.01
Weight fraction greater than size, (percent)	10	15	40	70	93	99	100

SOLUTION From App. IV, at 20°C $\mu = 1.027$ and $\rho = 0.997$. The settling velocities of the particles may be calculated from Stokes' law as follows:

$$v = \frac{\mathbf{g}}{18\mu}(\rho_s - \rho) \, d^2 = \frac{9800}{18(1.027)}(1.2 - 0.997) \, d^2$$

$$= 107.62 \, d^2$$

Weight Fraction, (percent)	10.0	15.0	40.0	70.0	83.0	99.0	100
v, mm/s	1.08	0.689	0.527	0.387	0.172	0.043	0.011
N_R	0.10	0.05	0.04	0.02	0.01	0.001	0.0001

Since the calculated Reynolds numbers are all less than 0.5, Stokes' law is applicable and the calculation of velocity is valid. From the calculated settling velocities the cumulative distribution curve of Fig. 9-5 is drawn. All particles with settling velocities greater than 0.37 mm/s (800 gal/ft^2 per day) will be removed. Thus, from the graph, the fraction $(1 - X_s)$ is equal to 0.73. The graphical determination of $\Sigma v \, \Delta x$ is tabulated below:

Δx	0.04	0.04	0.04	0.04	0.04	0.04	0.027
v	0.06	0.16	0.22	0.26	0.30	0.34	0.37
$v\Delta x$	0.0024	0.0064	0.0088	0.0104	0.0120	0.0136	0.0099

The overall removal is thus

$$F = (1 - X_s) + \frac{1}{v_s} \sum v \, \Delta x$$

$$= 0.73 + (1/0.37)(0.0635) = 0.898 \text{ or } 89.8\%$$

The preceding analysis indicates that, for sedimentation of discrete particles, only the plan area of the basin is important. This theoretical conclusion is subject to practical modifications dictated by wind currents, short circuiting and the need for uniform distribution of flow. Actual sedimentation tanks have depths ranging from less than 3 to more than 6 m (10 to 20 ft). This is generally taken into account by specifying a *detention time* for the basin, which is defined as the tank volume divided by the flow. If the surface overflow rate and detention time are selected, the plan area and volume may be calculated directly, and from these the depth is known.

The actual period individual particles of water remain in the basin depends upon the details of its design. Studies conducted using salt solutions, dyes, and radioactive tracers[3,4,5] indicate that the time elapsed to peak effluent concentration may be as little as 10 percent of the theoretical detention time, while the mean residence time may be 50 to 60 percent of theoretical. The observed phenomenon can be assessed by plotting the effluent concentration of the tracer vs. time and determining the time to the mode and to the mean. From

$$n = \frac{t_{\text{mean}}}{t_{\text{mean}} - t_{\text{mode}}} \tag{9-9}$$

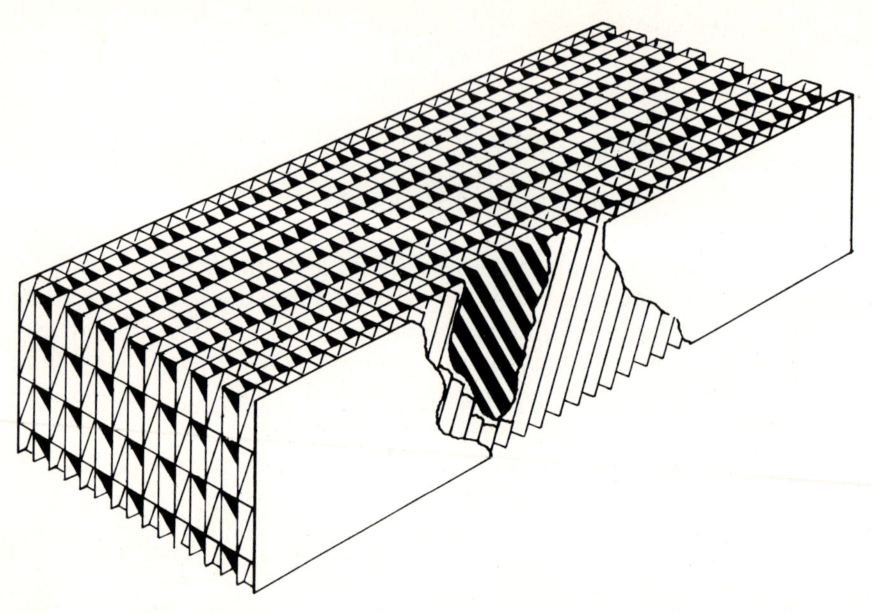

Figure 9-6 Tube settler. *(Courtesy Neptune Microflow, Inc., a subsidiary of Neptune International Corp.)*

one may calculate n, the number of hypothetical basins,[5] and evaluate the relative performance of the basin. Values of n in excess of 3 are considered to be satisfactory. One may also compare t_{mean} to the theoretical detention time and thus estimate what proportion of the basin is ineffective. The passage of water through the basin in less than the theoretical time results in removal efficiencies which are less than calculated.

Better utilization of the tank volume might be obtained by subdividing it vertically through the addition of horizontal trays. If, for example, a tray were placed in the basin of Fig. 9-4 from b to e, the slower settling particles shown would be removed. Such a scheme, in effect, increases the plan area of the tank and reduces the surface overflow rate. Although basins containing trays are not normally built, a number of proprietary devices such as the tube settlers of Fig. 9-6 and the Lamella clarifier of Fig. 9-7 take advantage of the same principle. These devices permit more effective utilization of existing clarifiers as well as higher design loadings on new systems.

Types of suspended solids The suspended matter in a turbid river water consists of finely divided silt, silica, and clay having specific gravities ranging from 2.65 for sand to 1.03 for flocculated mud particles containing 95 percent water. Suspended vegetable matter will have a specific gravity from 1.0 to 1.5. Water treated by coagulants will contain particles of precipitated floc of Al_2O_3 or Fe_2O_3. These, with adsorbed water, will have specific gravities of 1.18 and 1.34, respectively. Large amounts of entrained water in the floc, however, will reduce the specific

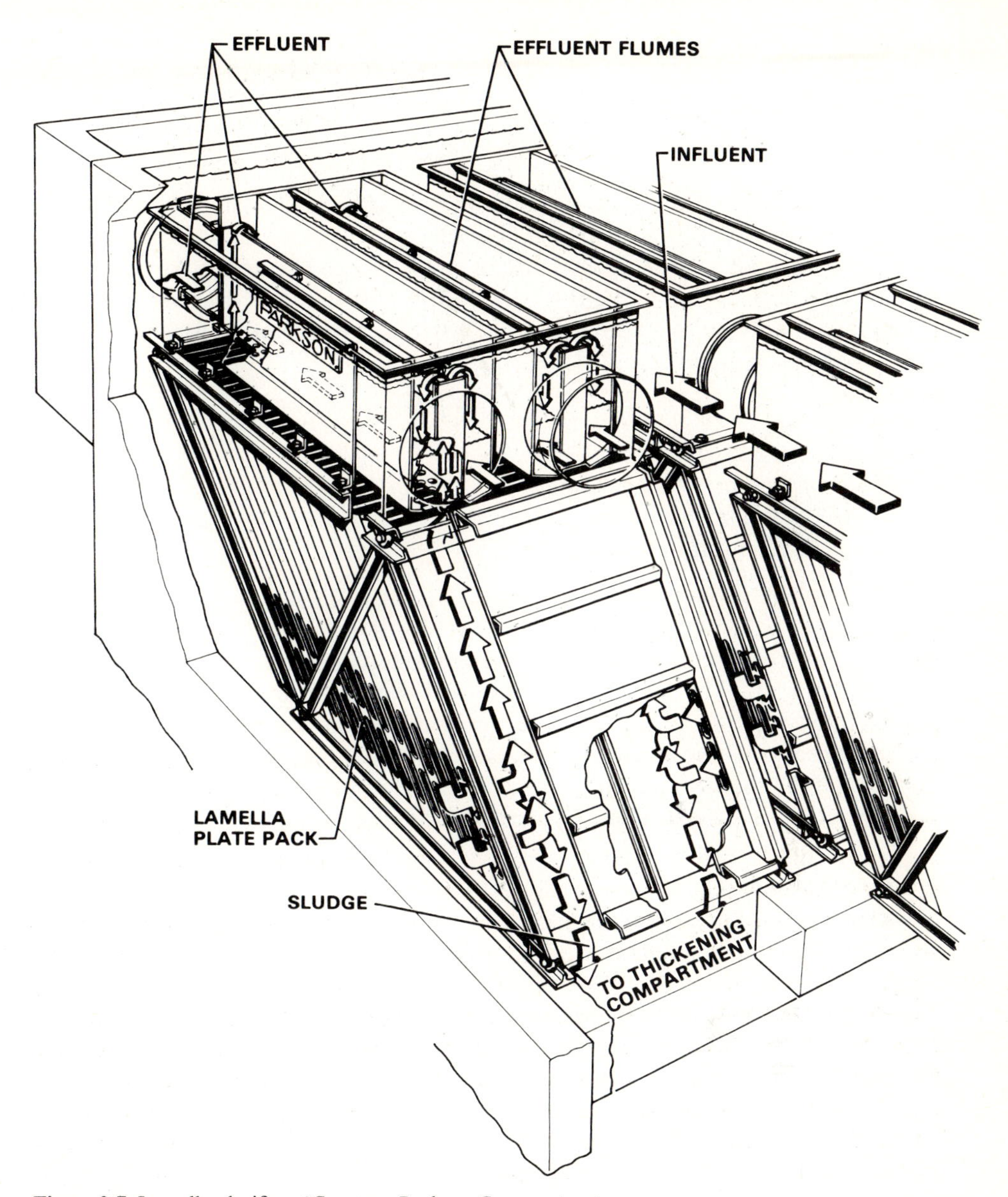

Figure 9-7 Lamella clarifier. *(Courtesy Parkson Corporation.)*

gravity to about 1.002. On the other hand, silt and clay enmeshed in the flocs will increase density. Floc particles range in size from submicroscopic to 1 mm or more.

In lime-soda water-softening plants the suspended particles are principally precipitated crystals of $CaCO_3$, having a specific gravity of 2.7 and particle size of 15 to 20 μm. Actually the crystals settle in clusters about 0.1 mm in size with about 75 percent water and a new specific gravity of about 1.2. If $Mg(OH)_2$ is also present, settling is less rapid.

9-5 Flocculent Suspensions

Flocculent particles such as those resulting from chemical precipitation or coagulation, or those found in biological treatment of wastewater will agglomerate while settling, with a resultant increase in particle size. The density of the composite particle may decrease due to the inclusion of water, but the overall result is generally an increase in settling velocity.[6] A suggested equation for the variation in floc density with particle size is:

$$\rho_s - \rho = kd^{-0.7} \tag{9-10}$$

where ρ_s = the density of the floc
ρ = the density of the water
d = the diameter of the particle (or the diameter of a sphere with equivalent surface area), and
k = a coefficient dependent upon the characteristics of the water and the chemicals involved.

Settling analyses of such suspensions are performed in columns at least 300 mm in diameter and equal in depth to the proposed clarifier. Samples are withdrawn at regular time intervals from multiple ports along the column and analyzed for percent removal. These percent removals are plotted as numbers vs. time and depth and a contour map of percent removal is plotted as in Fig. 9-8. From this plot percent removal is determined at selected time intervals as follows.

Example 9-2 Flocculent settling From the settling curves of Fig. 9-8 determine the efficiency of a sedimentation tank with a depth equal to that of the test cylinder and a detention time of 25 min.

SOLUTION At $t = 25$ min construct a vertical line as shown. In the volume of the basin corresponding to Δh_4 between 55 and 60 percent removal will occur. Similarly, in the volume corresponding to Δh_3 between 60 and 70 percent will be removed. In like fashion, the overall removal, assuming linear variation between contours, is:

$$F = \frac{55 + 60}{2} \cdot \frac{\Delta h_4}{h} + \frac{60 + 70}{2} \cdot \frac{\Delta h_3}{h} + \frac{70 + 80}{2} \cdot \frac{\Delta h_2}{h} + \frac{80 + 100}{2} \cdot \frac{\Delta h_1}{h}$$

Determining the Δh_i from Fig. 9-8,

$$F = 57.5(0.30) + 65(0.32) + 75(0.21) + 90(0.17) = 69\%$$

In order to obtain equivalent results in an actual tank the laboratory results should be modified to reduce the design SOR by a factor of 1.5 and increase the design detention time by a factor of 2.[7]

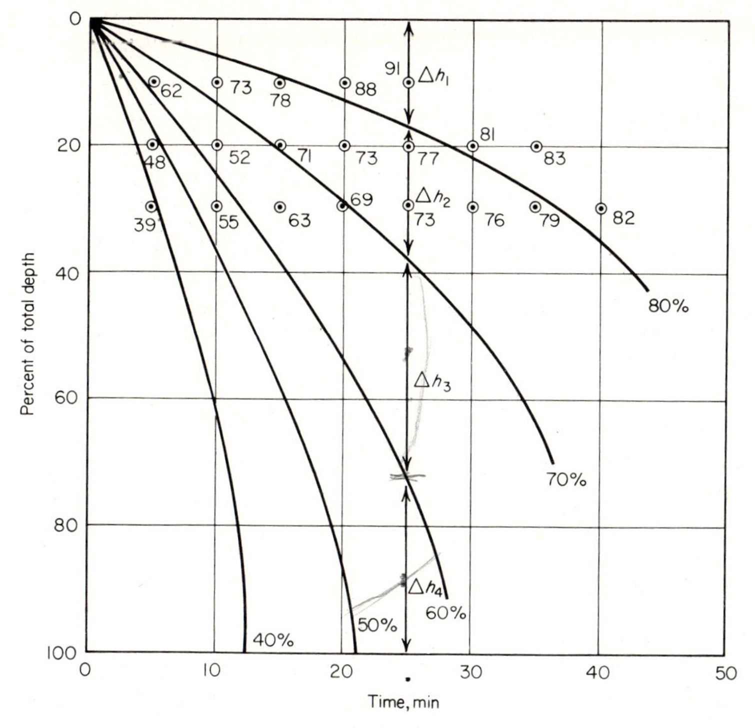

Figure 9-8 Settling of flocculent suspensions.

9-6 Hindered Settling

Hindered settling occurs when high densities produce particle interactions and individual particles are so close that the displacement of water by the settling of one affects the relative velocities of its neighbors. Such conditions occur in sludge thickeners and at the bottom of secondary clarifiers in biological treatment systems. An estimate of the extent to which settling is hindered may be obtained from Eq. (9-11).

$$v_h/v = (1 - C_v)^{4.65} \tag{9-11}$$

where v_h = the hindered settling velocity (L/T)
v = the free settling velocity as calculated from Eqs. (9-1) or (9-4)
C_v = the volume of particles divided by the total volume of the suspension.

Equation (9-11) is valid for Reynolds numbers less than 0.2, which is generally the situation in hindered settling.

The design of clarifiers intended for hindered settling is based upon three considerations: the area required for discrete settling of particles at the top of the clarifier, the area required for settling of the interface between the discrete and hindered settling zones, and the rate of sludge withdrawal. In general the settling rate of the interface is the controlling factor.

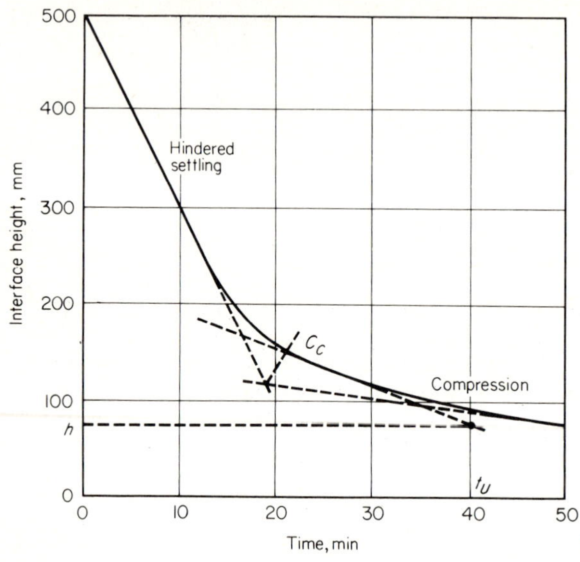

Figure 9-9 Hindered settling analysis.

The area required for hindered settling may be determined by column settling tests in which the height of the interface is plotted vs. time as in Fig. 9-9. The rate of subsidence at any time is equal to the slope of the curve. The critical area is given by:[8]

$$A = \frac{Qt_u}{h_0} \tag{9-12}$$

where A = the area of the basin (L^2)
Q = the flow (L^3/T)
h_0 = the depth of the column (L)
t_u = the time required to reach the desired underflow solids concentration (T)

The critical concentration is determined by extending tangents from the hindered and compressive settling lines to their point of intersection and bisecting the angle formed. The bisector will intersect the subsidence curve at C_c, the critical concentration. A tangent is constructed to the subsidence curve at C_c and the intersection of this tangent with the depth, h, required for a desired underflow concentration, C, will yield the required retention time, t_u.

Example 9-3 Hindered settling The subsidence curve shown in Fig. 9-9 was obtained for a sludge with an initial concentration of 3000 mg/l. Determine the surface overflow rate required to obtain a thickened sludge with a solids concentration of 2 percent by weight.

SOLUTION In order to obtain a solids concentration of 20,000 mg/l (2 percent), $h_0 C_0 = hC$ must be satisfied.

$$h = \frac{h_0 C_0}{C} = \frac{500(3000)}{20{,}000} = 75 \text{ mm}$$

From Fig. 9-9, $t_u = 40$ min

$$A = \frac{Qt_u}{h_0} = \frac{Q(40)}{500} = 0.080Q$$

$$\text{SOR} = Q/A = Q/0.08Q = 12.5 \text{ mm/min } (442 \text{ gal/ft}^2 \text{ per day})$$

From the slope of the subsidence curve in the hindered settling zone the surface overflow rate necessary for hindered settling is

$$\text{SOR} = 15.0 \text{ mm/min}$$

therefore, thickening governs.

9-7 Scour

Horizontal velocity in sedimentation basins must be limited to a value less than that which will resuspend the particles. The horizontal velocity just sufficient to cause scour has been defined as:

$$V = \left[\frac{8\beta(s-1)\mathbf{g}d}{f}\right]^{1/2} \tag{9-13}$$

where V = the horizontal velocity of flow (L/T)
s = the specific gravity of the particle (dimensionless)
$\mathbf{g}$ = the acceleration due to gravity (L/T^2)
d = the particle diameter (L)
β = a dimensionless constant ranging from 0.04 to 0.06
f = the dimensionless Darcy-Weisbach friction factor, usually 0.02 to 0.03.

In most sedimentation tanks the horizontal velocity is well below that required to produce scour. In grit chambers, (Art. 22-5) scour is an important design parameter.

9-8 Sedimentation Tank Details

Sedimentation tanks may be rectangular, circular, or square. *Rectangular basins* employ a horizontal flow pattern with the flow along the long axis. Such a flow pattern minimizes the effect of inlet and outlet disturbances. Sludge removal equipment in such basins consists of horizontal scrapers which drag the solids to a hopper at one end from which they are removed intermittently or continuously by gravity, auguring, or buckets. Typical designs are shown in Figs. 9-10 and 9-11.

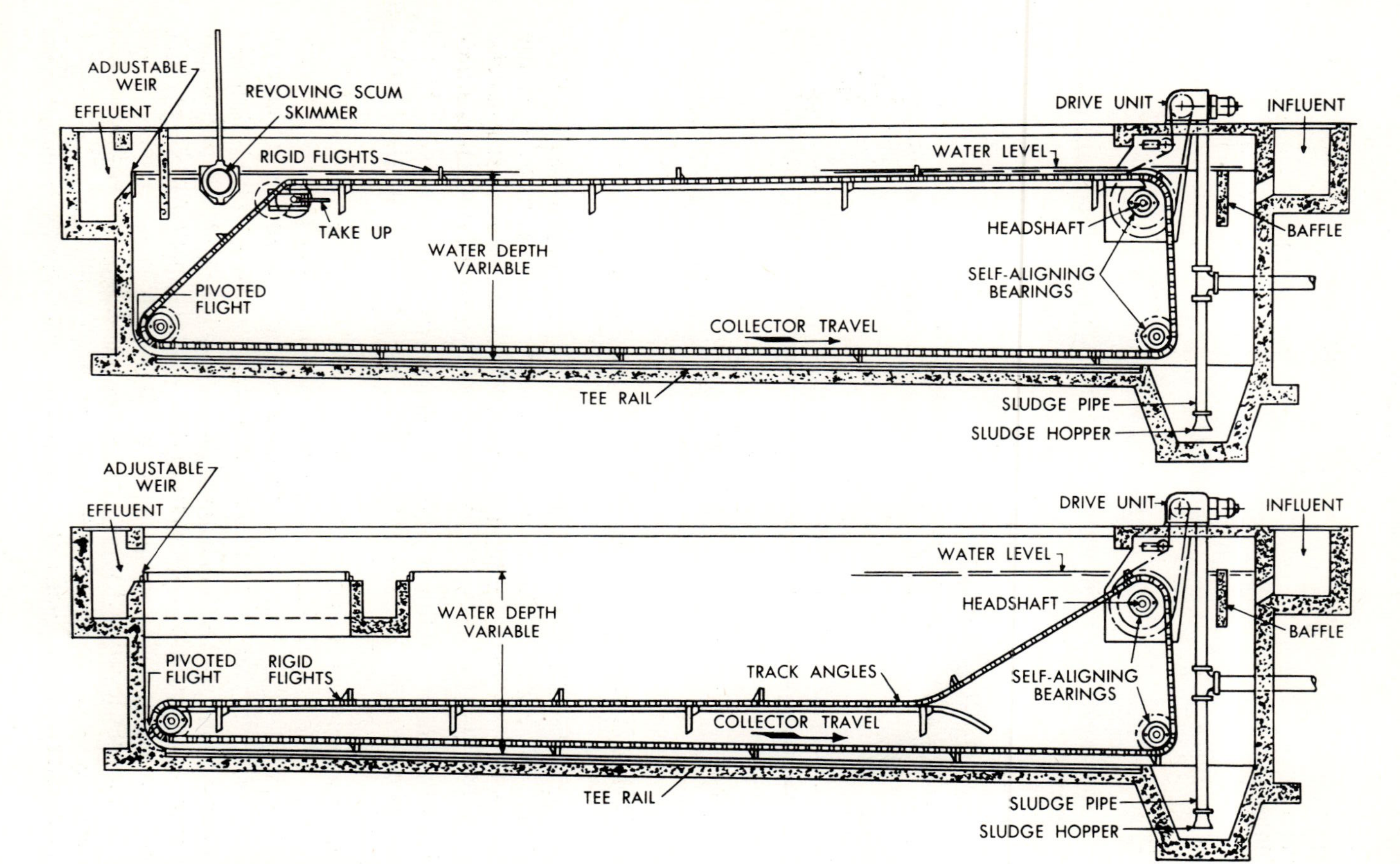

Figure 9-10 Rectangular clarifiers. *(Courtesy Envirex, a Rexnord Company.)*

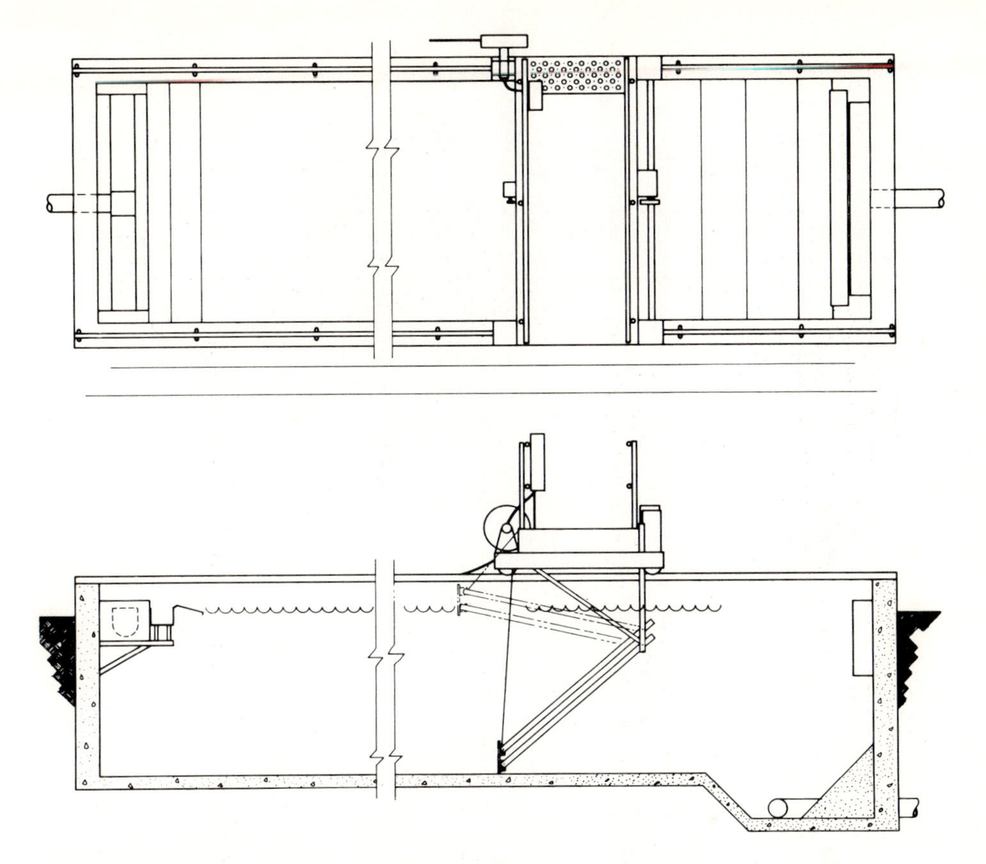

Figure 9-11 Travelling bridge sludge collector. *(Courtesy Aqua-Aerobic systems, Inc.)*

In *circular basins* the flow may enter either at the periphery as in Fig. 9-12, or at the center as in Fig. 9-13. Recent studies[3] have demonstrated that the average detention time is greater in peripheral feed basins. Cleaning equipment in circular basins usually consists of scraper blades mounted on radial arms. These windrow the sludge into a spiral which is gradually pushed to a central hopper from which it is removed by gravity. Very light sludges may be vacuumed from the bottom by devices such as that shown in Fig. 9-14.

Square basins may be used in situations in which land area is limited. The equipment installed in such basins is generally similar to that in circular designs save for the addition of a corner sweep mechanism like that shown in Fig. 9-15.

Careful design of inlets and outlets is important to assure reasonable performance of sedimentation tanks. The *ideal inlet* reduces the entrance velocity to prevent pronounced currents toward the outlet, distributes the water as uniformly as possible across the width and depth of the tank, and mixes it with the water already in the tank to prevent density currents. Figure 9-16 shows some of the inlets used. To assure uniform distribution the headloss through the multiple openings should be large in comparison to the difference in head available at them.

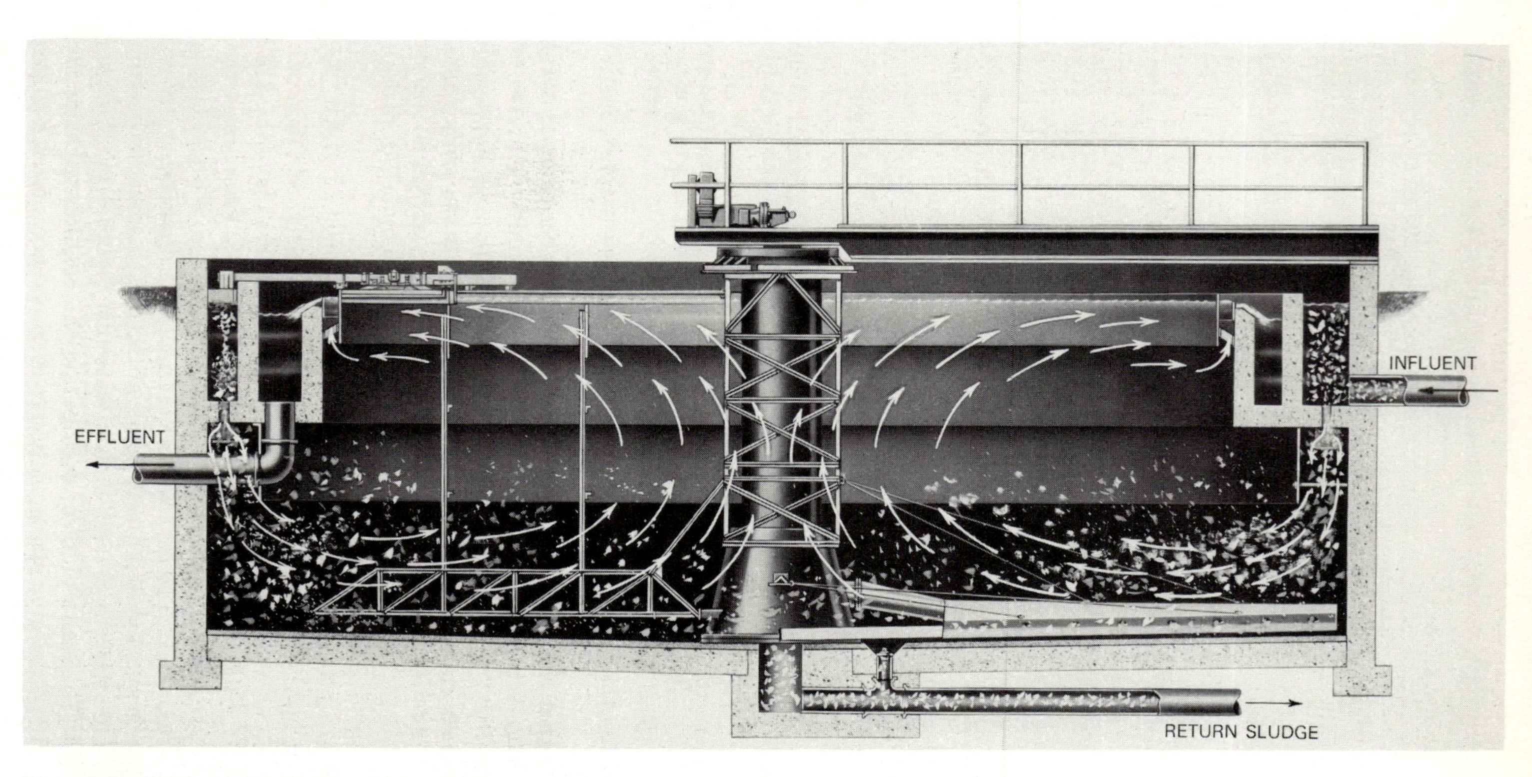

Figure 9-12 Peripheral flow clarifier. (*Courtesy Envirex, a Rexnord Company.*)

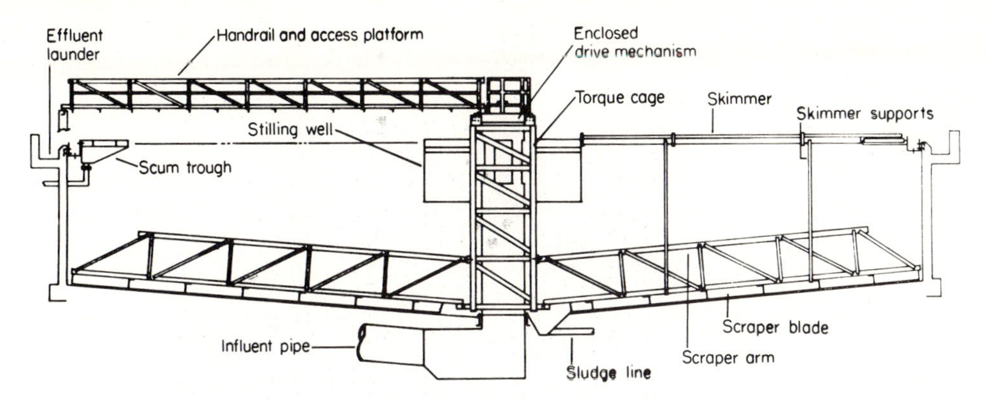

Figure 9-13 Center flow clarifier. *(Courtesy Environmental Elements Corporation.)*

Outlets frequently consist of free-falling weirs discharging into effluent launders like those of Fig. 9-17. Weir loading rates are limited to prevent high approach velocities near the outlet. Loading rates are generally specified in volume/unit length per day, which permits the ready calculation of the total weir length required. Outlets are placed as far from the inlet as possible—at the end of rectangular, at the periphery of center fed, and toward the center and along the radii of peripherally fed tanks. The weirs may cover a significant portion of the area of the basin. The length calculated from the weir overflow rate is the total weir length, not the length over which flow occurs. Weirs frequently consist of V-notches approximately 50 mm (2 in) in depth placed 300 mm (12 in) on centers.

Figure 9-14 Vacuum sludge withdrawal system.

Figure 9-15 Corner sweep mechanism. *(Courtesy Envirex, a Rexnord Company.)*

A free-falling weir which discharges into an effluent launder creates a flow condition which does not have a precise mathematical solution. A solution based upon the assumption that the shape of the curve defined by the water surface is parabolic is:

$$H = \left[h^2 + \frac{2q^2x^2}{\mathbf{g}b^2h}\right]^{1/2} \tag{9-14}$$

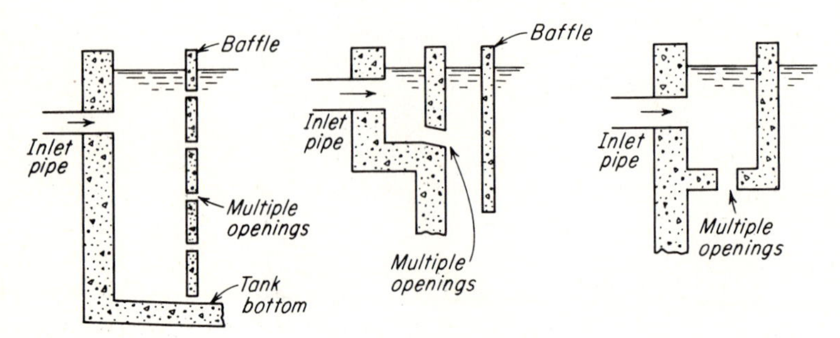

Figure 9-16 Three types of inlets for settling tanks.

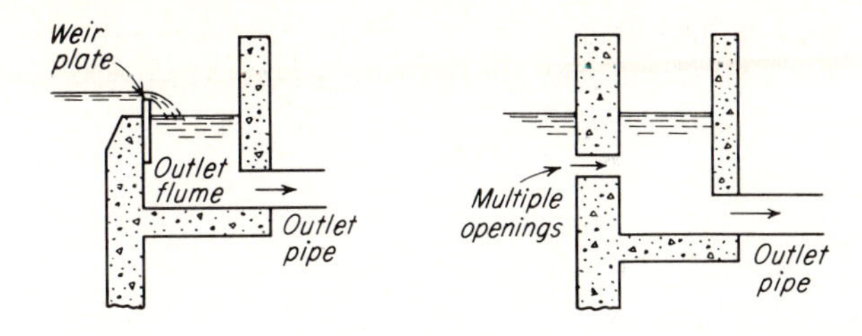

Figure 9-17 Outlet details of settling tanks.

in which H is the water depth in the launder at the upstream end, h is the depth at some distance x from the upstream end, q is the discharge per unit length, b is the launder width, and $\mathbf{g}$ is the acceleration due to gravity. The equation is derived for flumes with level inverts and parallel sides and does not include frictional losses. These can be calculated from the velocity in the channel. If no other control is built into the system, it is reasonable to assume that the flow at the lower end of the channel will be at its critical depth.

Example 9-4 Effluent launder design Calculate suitable dimensions for an effluent launder around the periphery of a 15 m (50 ft) radius tank. The design flow for the basin is 11,000 m^3/day (3 mgd). Half of the flow is assumed to go about each side of the basin.

SOLUTION A number of initial assumptions are necessary in solving a problem of this nature. If the initial assumptions prove to be unsuitable they are corrected and the problem is reworked. Assuming that the channel width is 300 mm, the depth at the lower end can be calculated for the assumption of critical depth from:

$$h = \left[\frac{Q^2}{b^2\mathbf{g}}\right]^{1/3} \tag{9-15}$$

whence,

$$h = \left[\frac{0.0637^2}{0.30^2(9.8)}\right]^{1/3} = 0.166 \text{ m} = 166 \text{ mm}$$

The depth at the upper end can now be calculated,

$$H = \sqrt{0.166^2 + \frac{2(0.00135)^2(47.1)^2}{9.8(0.166)(0.30)^2}}$$

$$H = 0.288 \text{ m} = 288 \text{ mm}$$

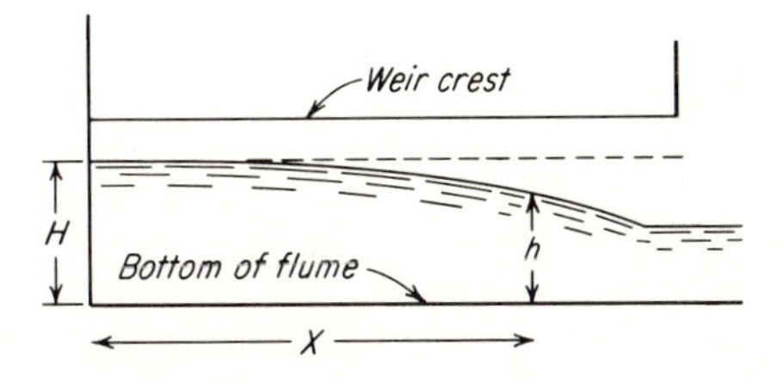

Figure 9-18 Drawdown curve in flume receiving weir discharge.

To this one must add the frictional headloss. This is precisely calculable only by integrating over the length of the flume. This calculation is complicated by the constant change in flow, depth, and velocity. The frictional headloss just above the outlet can be calculated from the Chezy-Manning equation (Art. 15-2) to be

$$s = \left[\frac{Qn}{Ar^{2/3}}\right]^2$$

$$= \left[\frac{0.0637(0.013)}{(0.30)(0.166)(0.0788)^{2/3}}\right]^2$$

$$= 0.008 \text{ m/m}$$

while that at the midpoint of the flume is

$$s = \left[\frac{0.0318(0.013)}{(0.277)(0.30)(0.0973)^{2/3}}\right]^2$$

$$= 0.0005 \text{ m/m}$$

from these values an overall frictional headloss in the flume is estimated to be 0.10 m. The depth of the launder below the bottom of the weirs discharging into it is thus 0.288 + 0.10 or about 400 mm. If the weirs are 50 mm deep V-notches the resulting channel dimensions are 450 mm deep by 300 mm wide (18 × 12 in).

9-9 Presedimentation

Some cities obtain water from rivers which are highly turbid at times. The Mississippi, its tributaries, and some rivers of the southwest have turbidities which frequently, but temporarily, reach 10,000 to 40,000 mg/l. Much of this suspended matter is coarse and will settle rapidly without chemical treatment. Additionally, it is desirable to prevent this high concentration of material from entering and overloading coagulation and filtration systems. For these reasons some cities employ preliminary sedimentation with detention times of 3 to 8 hours before coagulation.

COAGULATION—SEDIMENTATION

9-10 Purposes and Action of Coagulants

A large portion of the suspended particles in water and wastewater are sufficiently small that their removal in a sedimentation tank is impossible at reasonable surface overflow rates and detention times.

Colloidal particles, as a result of their small size, have a very large ratio of surface area to volume. For example, 1 cm^3 of material, if divided into cubes 0.1 mm on a side, (the size of fine sand), will have a surface area of 0.06 m^2, while if

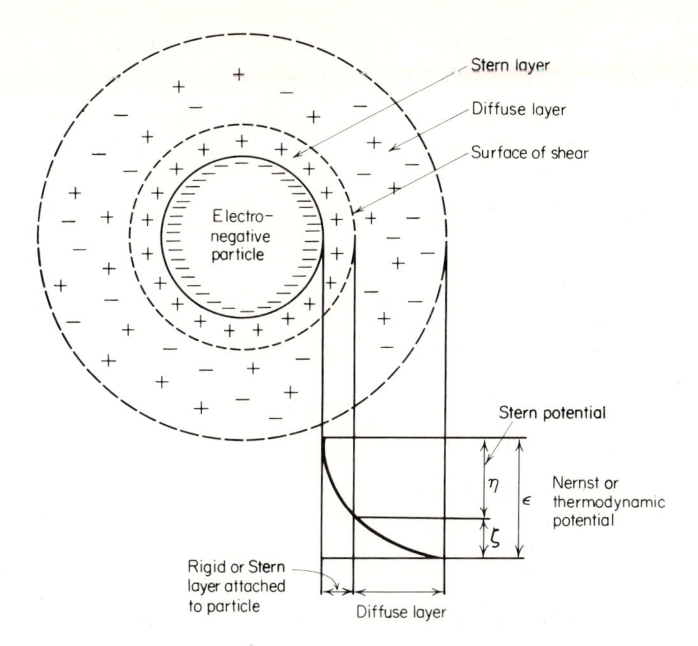

Figure 9-19 Guoy-Stern colloidal model.

divided into cubes 10^{-5} mm on a side, (the midpoint of the colloidal range), will have a surface area of 600 m^2. As a result of this immense area, surface chemical phenomena predominate. Adsorption results, with a tendency for other substances to concentrate on the particle surface. Particle charges result from preferential adsorption or from ionization of chemical groups on the surface. A schematic representation of the resulting colloidal state is presented in Fig. 9-19. Most colloidal particles in water and wastewater are negatively charged as shown. The stationary charged layer on the surface is surrounded by a bound layer of water in which ions of opposite charge drawn from the bulk solution produce a rapid drop in potential. This drop within the bound water layer is called the Stern potential. A more gradual drop, called the zeta potential, occurs between the shear surface of the bound water layer and the point of electroneutrality in the solution.

The surface charge on colloidal particles is the major contributor to their long-term stability. Particles which might otherwise settle or coalesce are mutually repelled by their like charge. *Coagulation* is a chemical technique directed toward destabilization of colloidal particles. *Flocculation*, in engineering usage, is a slow mixing technique which promotes the agglomeration of the destabilized particles.

Although other techniques are possible, the coagulation of water and wastewater generally involves the addition of chemicals—either hydrolyzing electrolytes or polymers. The action of metallic coagulants is complex, involving the dissolution of the salt (which may reduce the zeta potential by altering the ionic concentration in the bound layer), the formation of complex hydroxy-oxides of the metal which may be highly charged, and the entrapment of individual particles in the chemical precipitate formed. The processes involved are very complex[9] but may be described as follows:

1. Dissolution:

$$Al_2(SO_4)_3 \rightarrow 2Al(H_2O)_6^{+3} + 3SO_4^{-2} \tag{9-16}$$

2. Hydrolysis:

$$\begin{aligned} Al(H_2O)_6^{+3} + H_2O &\rightarrow Al(H_2O)_5OH^{+2} + H^+ \\ Al(H_2O)_5OH^{+2} + H_2O &\rightarrow Al(H_2O)_4(OH)_2^{+1} + H^+ \\ Al(H_2O)_4(OH)_2^{+1} + H_2O &\rightarrow Al(H_2O)_3(OH)_3 + H^+ \\ Al(H_2O)_3(OH)_3 + H_2O &\rightarrow Al(H_2O)_2(OH)_4^- + H^+ \end{aligned} \tag{9-17}$$

3. Polymerization:
   The products of the hydrolysis combine to form a variety of molecules including:

$$Al_6(OH)_{15}^{+3}$$
$$Al_7(OH)_{17}^{+4}$$
$$Al_8(OH)_{20}^{+4}$$
$$Al_{13}(OH)_{34}^{+5}$$

with a molecular structure of the form,

```
              H            H            H
              O    H2O     O    H2O     O    H2O
          H   |   H   |   H   |   H   |   H   |   H   |   H
          O   |   O   |   O   |   O   |   O   |   O   |   O
(H2O)4Al    Al    Al    Al    Al    Al    Al    Al(H2O)4
          O   |   O   |   O   |   O   |   O   |   O   |   O
          H   |   H   |   H   |   H   |   H   |   H   |   H
            H2O   O    H2O    O    H2O    O
                  H           H           H
```

Hydrolysis of iron salts is somewhat different than that of aluminum, but results in the formation of similar polymeric species. The net effect of addition of a metallic coagulant is seen to be the formation of large, insoluble, positively charged particles and production of free hydrogen ion from the water involved in the hydrolysis. This complex process is frequently represented by the simplified equation

$$Al_2(SO_4)_3 + 6H_2O \rightarrow 2Al(OH)_3 + 3H_2SO_4 \tag{9-18}$$

The polymeric species formed and the effectiveness of coagulation are dependent upon both pH and the applied concentration of coagulant. For any water there is an optimum pH range and optimum coagulant concentration. Curves

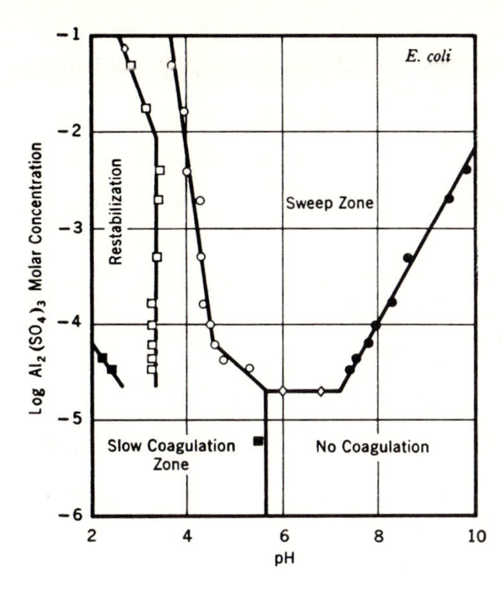

Figure 9-20 Aluminum sulfate domain of stability. *(Reprinted from Journal American Water Works Association* **62**, *by permission of the Association. Copyright 1970 by the American Water Works Association, Inc. 6666 W. Quincy Avenue, Denver, Colorado 80235.)*

such as that presented in Fig. 9-20[10] can be obtained for particular waters and coagulants by means of laboratory jar tests. The practical control of coagulation is dependent upon such analyses.

The chemicals commonly used in coagulation include alum (aluminum sulfate), ferric chloride, ferric sulfate, sodium aluminate, ferrous sulfate and lime, and chlorinated copperas. The choice is dictated by relative cost and effectiveness in particular areas. Alum is by far the most commonly used.

9-11 Simplified Reactions

Most of the metallic coagulants react with water to produce free hydrogen ion as shown in Eqs. (9-17) and (9-18). Since nearly all natural waters contain significant alkalinity, the hydrogen ion released will react with this, reducing the variation in pH. If a water contains insufficient alkalinity, the addition of a coagulant may depress the pH below the range in which the particular salt is effective. In such a circumstance an alkaline salt must be added to increase the buffer capacity of the solution. The adequacy of the natural alkalinity may be estimated from the following equations, which represent the approximate overall reactions involved.

Aluminum sulfate (alum)

$$Al_2(SO_4)_3 \cdot 18H_2O + 3Ca(HCO_3)_2 \rightarrow 2Al(OH)_3 + 3CaSO_4 + 18H_2O + 6CO_2$$

Ferric chloride

$$2FeCl_3 + 3Ca(HCO_3)_2 \rightarrow 2Fe(OH)_3 + 3CaCl_2 + 6CO_2$$

Ferric sulfate

$$Fe_2(SO_4)_3 + 3Ca(HCO_3)_2 \rightarrow 2Fe(OH)_3 + 3CaSO_4 + 6CO_2$$

Ferrous sulfate and lime

$$FeSO_4 \cdot 7H_2O + Ca(OH)_2 \rightarrow Fe(OH)_2 + CaSO_4 + 7H_2O$$

followed by, in the presence of dissolved oxygen

$$4Fe(OH)_2 + O_2 + 2H_2O \rightarrow 4Fe(OH)_3$$

Chlorinated copperas

$$3FeSO_4 \cdot 7H_2O + 1.5Cl_2 \rightarrow Fe_2(SO_4)_3 + FeCl_3 + 21H_2O$$

followed by

$$Fe_2(SO_4)_3 + 3Ca(HCO_3)_2 \rightarrow 2Fe(OH)_3 + 3CaSO_4 + 6CO_2$$

and

$$2FeCl_3 + 3Ca(HCO_3)_2 \rightarrow 2Fe(OH)_3 + 3CaCl_2 + 6CO_2$$

The optimum pH range for each of the coagulants is:

	pH
Alum	4.0 to 7.0
Ferrous sulfate	8.5 and above
Chlorinated copperas	3.5 to 6.5 and above 8.5
Ferric chloride	3.5 to 6.5 and above 8.5
Ferric sulfate	3.5 to 7.0 and above 9.0

9-12 Polymeric Coagulants

Polymeric coagulants, or polyelectrolytes, are long chain high molecular weight molecules which bear a large number of charged groups. The net charge on the molecule may be positive, negative, or neutral. Representative molecular structures are:

Anionic

$$-CH_2-\underset{\displaystyle COO^-}{\underset{|}{CH}}-CH_2-\underset{\displaystyle COO^-}{\underset{|}{CH}}-CH_2-\underset{\displaystyle COO^-}{\underset{|}{CH}}-$$

Cationic

$$-CH_2-\underset{\displaystyle \underset{\displaystyle \underset{\displaystyle R}{N^+}}{\underset{|}{C_6H_4}}}{\underset{|}{CH}}-CH_2-\underset{\displaystyle \underset{\displaystyle \underset{\displaystyle R}{N^+}}{\underset{|}{C_6H_4}}}{\underset{|}{CH}}-CH_2-\underset{\displaystyle \underset{\displaystyle \underset{\displaystyle R}{N^+}}{\underset{|}{C_6H_4}}}{\underset{|}{CH}}-$$

Ampholytic

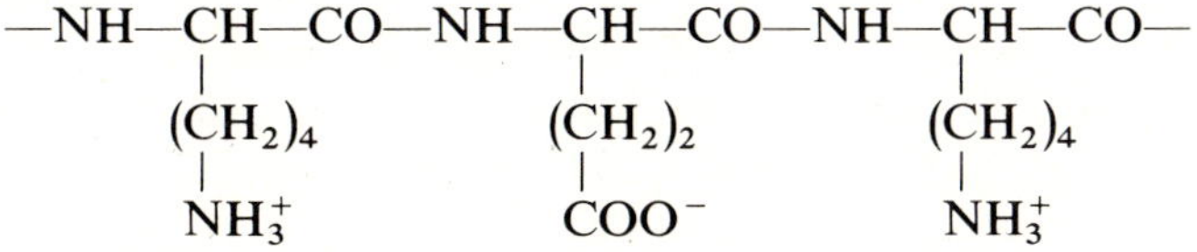

$$
\begin{array}{ccc}
-\mathrm{NH{-}CH{-}CO{-}} & \mathrm{NH{-}CH{-}CO{-}} & \mathrm{NH{-}CH{-}CO{-}} \\
| & | & | \\
(\mathrm{CH_2})_4 & (\mathrm{CH_2})_2 & (\mathrm{CH_2})_4 \\
| & | & | \\
\mathrm{NH_3^+} & \mathrm{COO^-} & \mathrm{NH_3^+}
\end{array}
$$

Although it would appear that cationic polymers should be most effective in the coagulation of the negatively charged colloids found in water and wastewater, this is not always the case.[9] The chemical groups on the polymer are thought to combine with active sites on the colloid. Such interaction of a single molecule with a large number of particles would produce a bridging effect combining them into a larger particle which might settle under the action of gravity.

Polyelectrolytes are excellent coagulants which may be used alone, or in conjunction with metallic coagulants. 237 coagulant aids are presently approved by EPA for use in treating public water supplies.[11] Permissible dosages range from 1 to 150 mg/l, with most being less than 10 mg/l. The selection of an appropriate coagulant requires determination of necessary dosage by jar tests and comparison of all relevant costs, including subsequent solids handling.

9-13 Coagulant Aids

Coagulant aids, properly speaking, do not in general aid in coagulation, but rather in flocculation. The agents used include oxidizing agents such as chlorine, weighting agents such as clay, activated silica, and polyelectrolytes.

Oxidizing agents are thought to improve coagulation-flocculation by the oxidation of organic compounds which might otherwise interfere. When chlorine is used the dosage is usually that required to reach the breakpoint. (See Art. 11-3).

Weighting agents are used in the coagulation of waters of low initial turbidity. It is a curious phenomenon in water treatment that highly turbid waters are more easily clarified than those which are relatively limpid. The addition of materials like bentonite clay increases the particle density, the average weight of the suspension, and provides a considerable surface for the adsorption of organic compounds. Dosages of clay typically range from 10 to 50 mg/l.[12] Other weighting agents/adsorbents which have been used include activated carbon, powdered silica, and limestone. These agents may have other effects in addition to those associated with clay.

Activated silica consists of a preparation of colloidal sodium silicate which can act as a coagulant itself, as a coagulant aid in association with alum, or as a flocculating (bridging) agent. The preparation of activated silica is necessarily a continuous process since it ages rapidly and may gel within a matter of hours. Its preparation involves the neutralization of approximately 80 percent of the alkalinity of a 1.5 percent solution with any available acid. Chlorine is frequently used.

The solution is then diluted after being aged for about 10 min and is fed to the water being treated. The details of the process are available elsewhere.[12]

Under optimum conditions activated silica will increase the rate of coagulation and flocculation, reduce the coagulant dosage, broaden the pH range of effective coagulation, produce larger and tougher floc particles, and increase the removal of both color and colloidal material. Dosages are on the order of 10 percent of the alum dose, with the optimum being determined by jar test. When used as a filter aid dosages are 1 mg/l or less.

Polyelectrolytes, as well as being used as coagulants, have applications as coagulant or flocculant aids. The bridging phenomenon described above is particularly important since it can produce agglomerated particles as much as 100 times larger than those produced by metallic coagulants alone. Their use will usually result in a decrease in primary coagulant dosage as well as an increase in flocculation and sedimentation rate. The role of polyelectrolytes as a coagulant aid is similar to that of activated silica. The choice of an appropriate agent is best made by experimental and economic comparison.

9-14 Mixing

Mixing provides for the rapid dispersion of chemical additives in the raw water and requires a high degree of turbulence and power dissipation. A variety of static (baffled piping, baffled channel, hydraulic jump) and mechanical (paddle, turbine, propeller) mixing devices are commercially available. Detention times in flash or rapid mixers of 10 to 20 s are usually adequate, but some regulatory agencies may employ other standards. Power required is 2 to 5 kW per m^3/min (1 to 2 hp per ft^3/s). Power input in mixing and flocculation is frequently expressed in terms of the mean velocity gradient G $(1/T)$:

$$G = (P/\mu V)^{1/2} \tag{9-19}$$

where P = the power dissipated (FL/T)
μ = the absolute viscosity (M/LT)
V = the volume to which P is applied (L^3)

For rapid mixing the product GT should be 30,000 to 60,000, with T generally being 60 to 120 s.

9-15 Flocculation

Flocculation is a slow mixing process in which destabilized colloidal particles are brought into intimate contact in order to promote their agglomeration. The rate of agglomeration or flocculation is dependent upon the number of particles present, the relative volume which they occupy, and the velocity gradient (G) in the basin.

The theory of flocculation is complex and is detailed elsewhere.[4,5] Practical design factors include the use of detention times of 20 to 30 min and values of G

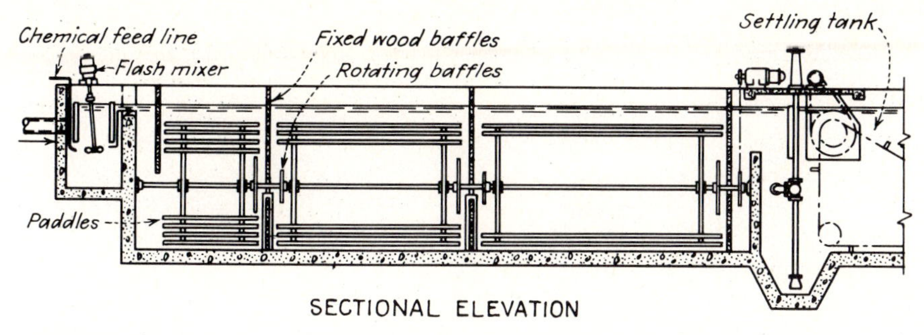

Figure 9-21 Flash mixer and flocculator. *(Courtesy Envirex, a Rexnord Company.)*

between 25 and 65 s^{-1}. The flocculation technique most commonly used involves mechanical agitation with rotating paddle wheels or vertically mounted turbines. Typical paddle flocculation units are illustrated in Figs. 9-21, 9-22, and 9-23.

9-16 Suspended Solids Contact Clarifiers

Contact or upflow clarifiers combine chemical addition, mixing, flocculation, and sedimentation in a single unit. Systems of this type are shown in Figs. 9-24, 11-6, and 11-7. The water enters in the center where the chemical addition and rapid mixing occur, then flows downward through the central area under the skirt where flocculation occurs, and thence upward through the bulk of the basin which serves as a sedimentation tank. As the water moves upward its velocity will decrease since the flow area increases. Particles borne by the water will be carried upward until they reach the point at which their settling velocity is equal to the upward velocity of the fluid. As the number of particles so suspended increases, a sludge blanket will be formed which will act in a sense as a filter—straining out or flocculating particles which might otherwise be carried out by their upward velocity. As the particle density in the sludge blanket increases, the water velocity through the blanket will increase resulting in upward movement of the suspended

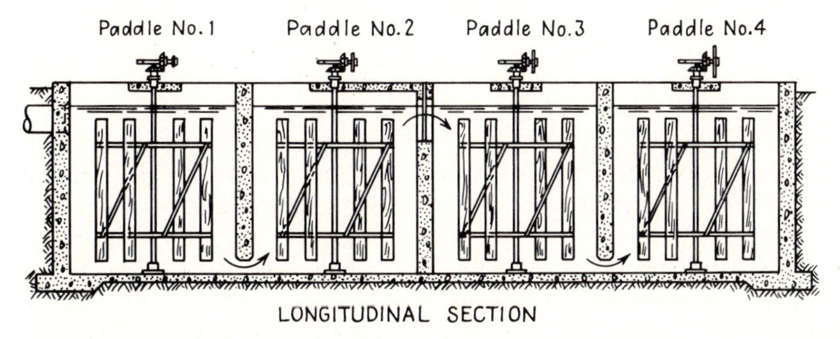

Figure9-22 Vertical flocculator. *(Courtesy FMC Corporation.)*

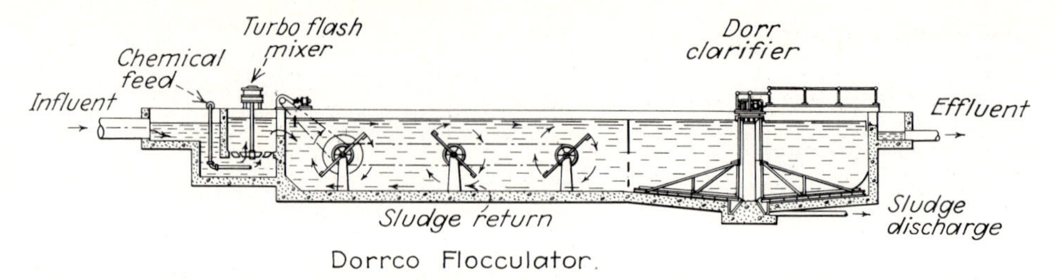

Figure 9-23 Flash mix, flocculation, and sedimentation. *(Courtesy Dorr-Oliver, Inc.)*

mass to a new equilibrium point. In time, the sludge would be carried over the weirs, however a portion of the blanket is wasted from time to time to maintain a reasonable elevation.

9-17 Design Criteria

The design criteria for clarifiers consist of specification of the surface overflow rate and the detention time or depth. The values used are dependent upon the type of clarifier used and the characteristics of the liquid being treated. Typical design values are presented below, but may not satisfy individual state standards.

Type of basin	Detention time, h	Weir overflow rate		Surface overflow rate	
		$m^3/m/day$	gal/ft/day	m/day	$gal/ft^2/day$
	Water treatment				
Presedimentation	3–8				
Standard basin following					
Coagulation and flocculation	2–8	250	20,000	20–33	500–800
Softening	4–8	250	20,000	20–40	500–1000
Upflow clarifier following					
Coagulation and flocculation	2	175	14,000	55	1400
Softening	1	350	28,000	100	2500
Tube settler following					
Coagulation and flocculation	0.2				
Softening	0.2				
	Sewage treatment				
Primary before					
Activated sludge	0.75–1.0	125	10,000	60	1500
Trickling filter	2–2.5	125	10,000	20–37	500–900
Intermediate between filters	2	125	10,000	40	1000
Final after					
Standard trickling filter	2	125	10,000	40	1000
High-rate trickling filter	2	125	10,000	33	800
Conventional activated sludge	2–3	125	10,000	24–33	600–800
Contact stabilization	2.5–3.5	125	10,000	20–28	500–700
Extended aeration	3–4	125	10,000	12–24	300–600

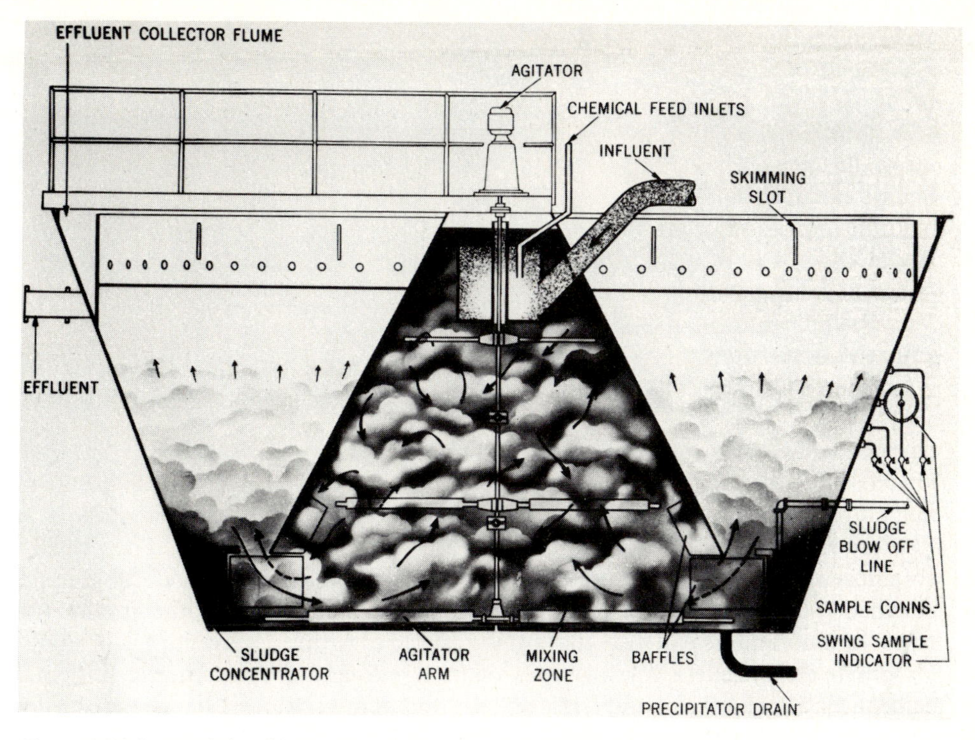

Figure 9-24 Suspended solids contact clarifier. *(Courtesy Permutit Co., Inc.)*

9-18 Chemical Feeding Methods

The feeding of chemicals includes handling, storage, proportioning, and conveying of the proper amounts to the water. Elaboration of the handling devices will depend upon the size of the plant. Ordinarily the chemicals are stored in the second floor of the filter building or separate chemical house, thus requiring hoists or elevators. In the floor of the storage room, openings are placed immediately above the proportioning devices. Such openings are screened with mesh wire cloth to hold large lumps or foreign matter. Storage spaces should, of course, always be dry and well ventilated.

The older method of feeding chemicals was by solutions of known strength which were made up and stored in vats. Generally such vats were supplied in duplicate so that the operator on each shift could mix up enough solution to supply his successor. The feeding was done by means of a constant-head orifice box, as pictured in Fig. 9-25. The float valve keeps the solution in the box at a fixed depth over the adjustable orifice, thus insuring a constant rate of flow independent of the depth in the supply vat. Such feeding methods are still to be found in the older and smaller plants. Some chemicals, however, as ferric chloride and sodium aluminate, must be fed in solution form.

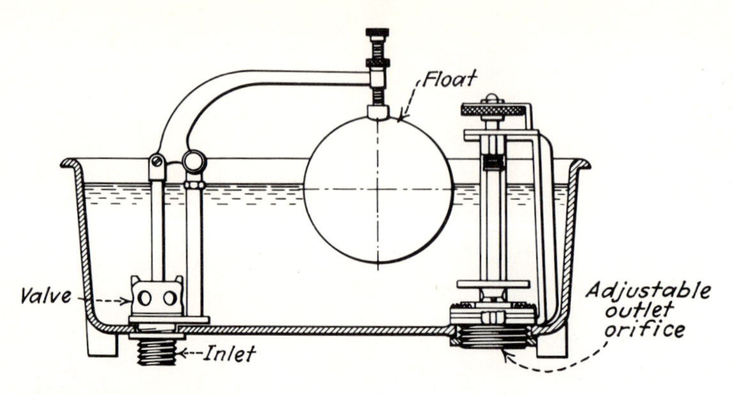

Figure 9-25 Constant-head orifice box for solution feeds.

Dry feed machines of various types are used in most plants. They can be obtained with various capacities, and each machine itself has considerable range in capacity, at least 40 : 1 for volumetric machines and 100 : 1 for the gravimetric types. Manufacturers can also guarantee accuracies within 1 percent plus or minus for gravimetric feeders for all types of materials and for volumetric feeders 3 percent for alum, ferrous and ferric sulfate, and dense soda ash, and 5 percent for lime, activated carbon, and light soda ash.

Feeders include the following: (*a*) A hopper which is an integral part of the machine usually with an agitating device. (*b*) The proportioning mechanism which can be set at the feed rate required and which varies in the different machines. In the volumetric types it may be a conveyor screw; a rotating horizontal disk which receives the chemical from the hopper and which has a fixed blade that pushes the chemical into a funnel; an oscillating hopper and tray or pan at the bottom of the main hopper as in Fig. 9-26, or a long, slightly sloping metal trough which vibrates, the slope and speed of the vibration controlling the flow. Gravimetric types may use a weighing belt over which the chemical moves or a container is filled and the feed is controlled by the loss in weight. (*c*) The solution or suspension chamber in which the chemical is mixed with water. Specifications may require a detention period of 3 min when 1 l of water is used per 120 g of chemical fed. (*d*) The extension hopper is specified as high enough to extend from the machine hopper to the ceiling so that it can be filled from the floor above. It also acts as a storage bin. (*e*) Other features are desirable. A meter for measuring the water used. An alarm, red light, or buzzer, to give notice if the feed is more than 5 percent from the setting. It is also possible to get automatic controls that will change the chemical feed with variations in the amount of water to be treated. With activated carbon and light soda ash it is advisable to have a locking device on the feed that will prevent the material from "flooding" through the mechanism. (*f*) For quicklime feeders some type of dry feeding mechanism is necessary. This measures and discharges the lime into the slaking chamber where lime and water are agitated,

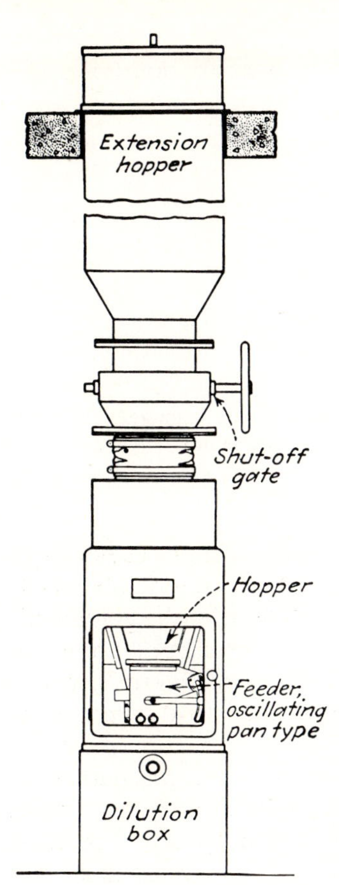

Figure 9-26 Dry chemical feeder with hopper extended to charging floor above. *(Courtesy Infilco-Degremont, Inc.)*

preferably for 30 min, to form milk of lime which is then discharged into the water to be treated. Some provision is made for removal of grit and dust, and vapors must be cared for. For good efficiency the water during slaking should not be less than 77°C, and to attain that temperature it may require preheating of the feed water, which may be done by using heat of the milk of lime. In some cases the slaking chambers are insulated.

Proper selection of piping to convey the chemical solutions to the water is of importance, or much trouble will result from clogging or corrosion. Rubber hose is suitable for either acid or alkaline chemicals, but it must be so supported that air pockets will not form. Plastic is satisfactory for alum, ferrous sulfate, and acid solutions in general. Lined steel pipe is more rigid than plastic and is equally suitable for carrying acid solutions of iron sulfate and alum. Pipe lined with hard rubber is also used for solutions of coagulants and chlorine. Iron and steel are suitable for the alkaline chemicals, but in any case the piping should be completely accessible and easily taken down for cleaning or with cleanouts for rodding.

PROBLEMS

9-1 Kirschmer's formula[2] for the headloss through bar screens is:

$$h_1 = \beta\left(\frac{w}{b}\right)^{4/3} h_v \sin\theta$$

where h_1 = the headloss (L)
w = the width of the bars (L)
b = the clear space between the bars (L)
h_v = the velocity head upstream (L)
θ = the angle which the screen makes with the horizontal, and
β = a dimensionless coefficient which is a function of bar geometry, ranging from 2.42 for rectangular bars to 0.76 for teardrop shapes.

Calculate the headloss through a bar screen consisting of rectangular bars 5 mm wide, 20 mm on centers placed in a 600 mm deep by 1 m wide channel which carries 20 m^3/min.

9-2 The evaluation of peripheral feed and center feed clarifiers by Dague and Baumann[3] yielded the following results:

	Center flow	Rim flow
Theoretical detention		
Time, min	48.7	48.7
T_{mean}, min	30.4	37.2
T_{mode}, min	5.4	24.0

Determine n for each basin and state which, if either, you would select.

9-3 A rectangular sedimentation basin is to treat 3000 m^3 of water per day with a detention time of 6 hours. It is to be hand cleaned of sludge at 6-week intervals. The raw water has an average suspended solids content of 250 mg/l and this is reduced to 5 mg/l with the aid of coagulants. The settled sludge, which includes 40 mg/l (based on raw water flow) of metallic precipitates, has a moisture content of 85 percent and a specific gravity of 1.24. If the tank length is to be twice its width, and if the depth just before cleaning is to be 3 m, determine the basin dimensions.

9-4 A water plant treating 12,000 m^3/day uses alum at a rate of 20 mg/l. Determine the daily usage and the storage volume required if a minimum of one-month's supply is desired and deliveries are bi-weekly. Alum has a density (bulk) approximately equal to that of water.

9-5 A plant treating 30,000 m^3/day uses 13 mg/l ferrous sulfate. Calculate the daily chemical consumption and the required dosage of lime as CaO.

9-6 A circular sedimentation tank 30 m in diameter has an overflow rate of 28.5 m/day (approximately 700 gpd/ft^2). The discharge is over a free-falling peripheral weir to a channel 600 mm wide which has a level invert, parallel sides, and a single outlet. The water depth at the outlet is at critical depth and the channel length may be assumed to be equal to the basin circumference. What will be the water depth in the channel at a point diametrically opposite the outlet?

REFERENCES

1. Nordell, E.: *Water Treatment for Industrial and Other Uses* 2d ed., Reinhold, New York, 1961.
2. Kirschmer, O.: "Untersuchungen Uber den Gefallsverlust an Rechen," (Investigations Concerning the Headloss of Racks) cited in: Parker, H. W., *Wastewater Systems Engineering*, Prentice-Hall, Inc., Englewood Cliffs, N.J., 1975.

3. Dague, R. R. and E. R. Baumann: "Hydraulics of Circular Settling Tanks Determined by Models," Annual Meeting, Iowa Water Pollution Control Association, Lake Okoboji, Iowa, 1961. Reprinted in *Spiraflo Clarifier Hydraulic Studies*, Lakeside Engineering Corp.
4. Fair, G. M., J. C. Geyer, and D. A. Okun: *Water and Wastewater Engineering, Vol. 2*, John Wiley and Sons, New York, 1968.
5. Gemmel, R. S.: "Mixing and Sedimentation," in *Water Quality and Treatment* 3d ed., McGraw-Hill, New York, 1971.
6. Lagvankar, A. L. and R. S. Gemmel: "A Size-Density Relationship for Flocs," *Journal American Water Works Assoc.* 60:1040, 1968.
7. Metcalf and Eddy, Inc.: *Wastewater Engineering*, McGraw-Hill, New York, 1972.
8. Talmadge, W. P. and E. B. Fitch: "Determining Thickener Unit Areas," *Industrial Engineering Chemistry*, **47**, 1955.
9. Weber, W. J. Jr.: *Physicochemical Processes for Water Quality Control*, Wiley-Interscience, New York, 1972.
10. Hanna, G. P. and A. J. Rubin: "Effect of Sulfate and Other Irons in Coagulation with Aluminum (III)". *Journal American Water Works Association* **62:** 315, 1970.
11. "Coagulant Aids for Water Treatment—EPA Report," *Journal American Water Works Association* **67:**468, August, 1975.
12. Cohen, J. M. and S. A. Hannah: "Coagulation and Flocculation," in *Water Quality and Treatment* 3d ed., McGraw-Hill, New York, 1971.

CHAPTER

TEN

TREATMENT OF WATER—FILTRATION

10-1

Sedimentation, with or without coagulation, will not ordinarily give adequate treatment to water. The production of clear and sparkling water, with the utmost safety so far as disease is concerned, requires the use of a filter. Filtration also aids in the removal of color, tastes, odors, iron, and manganese.

Filters of two general types have been used. The slow sand filter was first used in Great Britain early in the nineteenth century, and a number of plants of this type were constructed in the United States in the years 1890 to 1910. The rapid filter, also known as the American and mechanical filter, was developed in the United States during the period 1900 to 1910; and, because of its greater adaptability to our more turbid waters, it has largely displaced the slow filter in this country. Some of our large cities, however, still retain their slow sand filters, and slow filters have modern applications in waste treatment (Chap. 24).

10-2 The Rapid Filter

Rapid filtration generally implies a process which includes coagulation, flocculation, clarification, filtration, and disinfection. In treating waters of low turbidity and industrial process waters the clarification step may be omitted[1,2] while other processes may be added if the raw water requires them.

Figure 10-1 is a diagrammatic sketch of a standard rapid filter plant showing the path of the water through the various units. Figure 10-2 shows a section of a

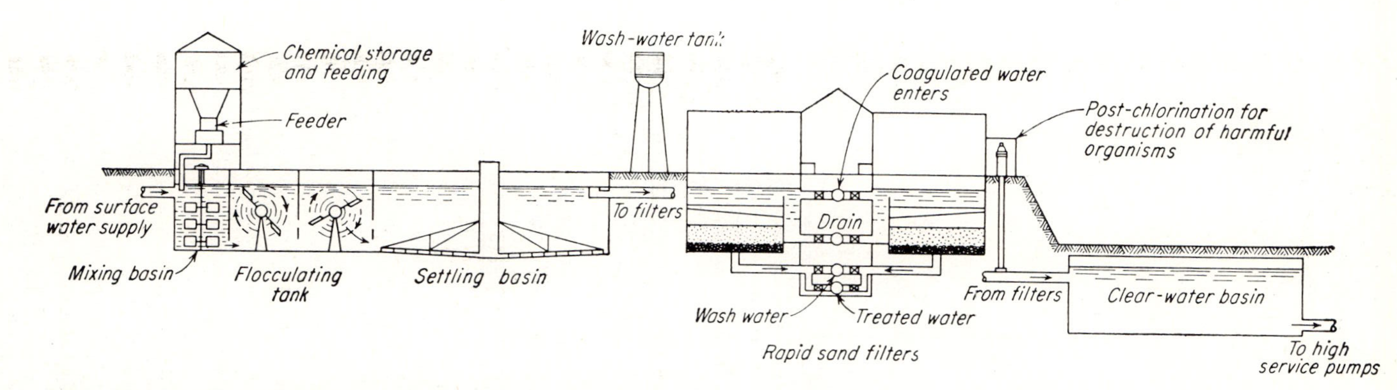

Figure 10-1 Filtration plant, including coagulation, settling, filtration, postchlorination and clear-water storage. Prechlorination, i.e., addition of chlorine at the mixing basin, is also common practice.

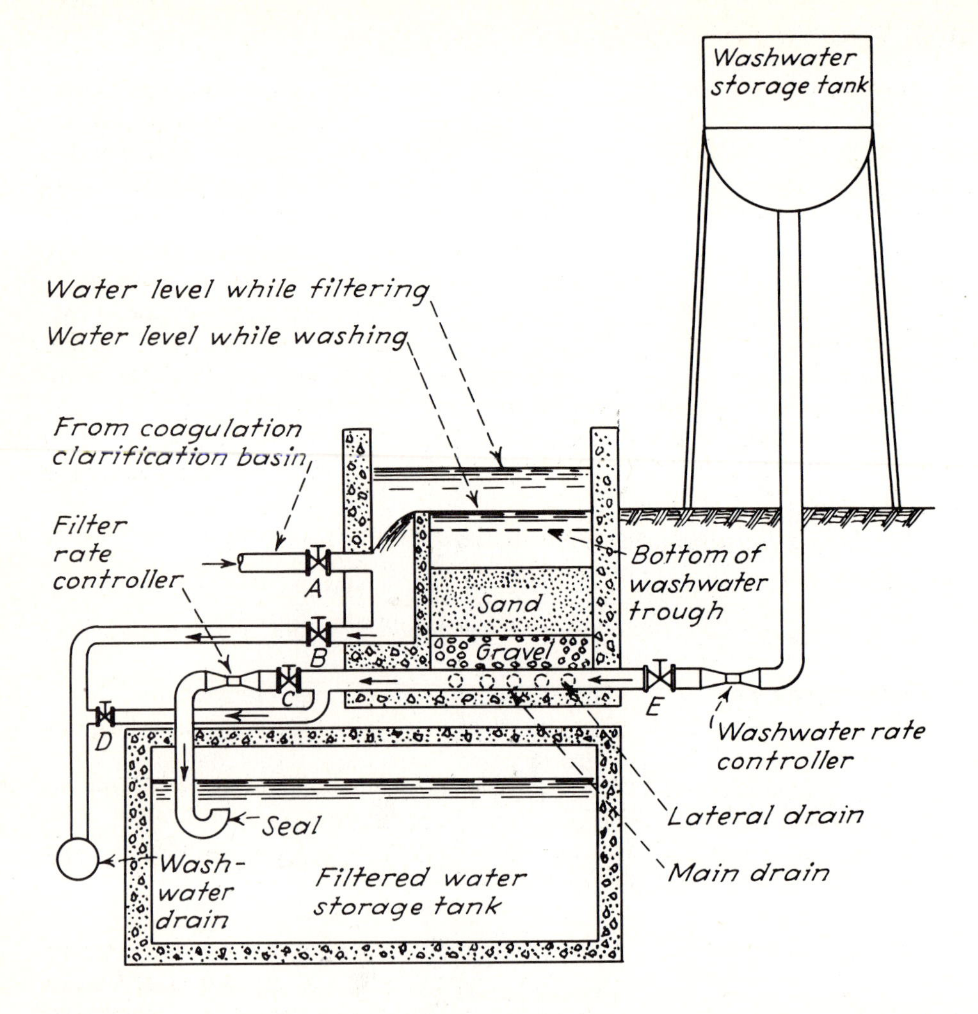

Figure 10-2 Diagrammatic section of a rapid sand gravity filter. A, B, C, D, and E are valves, which may be hydraulically or pneumatically actuated. Valve D permits wasting filtered water. The seal in the effluent pipe keeps the pipe full at all times so that the rate controller will function.

single standard filter, the valves, rate controller, and relation to the wash water tank and filtered water storage tank.

The essential characteristics of a rapid filter are three in number: (*a*) careful pretreatment of the water in preparation for filtration, according to the methods discussed in Chap. 9—turbidity of the water applied to the filters should not exceed 10 units and preferably not 5 units; (*b*) high rate of filtration, 120 to 240 m/day (2 to 4 gal/ft^2 per day) or more; (*c*) washing the filter units by reversing flow of filtered water upward through the filter to remove mud and other impurities which have lodged in the sand.

10-3 Theory of Filtration Through Coarse Media

A rapid filter consists of a bed of coarse media, such as sand, ranging in depth from 300 mm to several meters. The kinetics of removal of particles smaller in size than the pore openings in the bed is described as consisting of a transport and an attachment step.[3,4] The *transport* to the surface of the filtration medium may be produced by diffusion, interception, settling, impingement, or hydrodynamic carriage. The transport process is aided by flocculation in the interstices of the filter and by the relatively short travel required for removal by sedimentation.

Attachment of the particles after their contact with the medium is chemical in nature and is influenced by pH, ionic composition of the water, age of the floc, nature and dosage of polymer, and the composition and surface condition of the medium.[4,5,6] Both the filter medium and the suspended particles in the influent may bear significant charges which can either aid or inhibit attachment. These surface charges can be affected both by changes in pH and by the addition of chemical coagulants.

The removal of particulate matter in a granular filter occurs in the following manner:[7]

1. The quantity of particles removed by a layer of filter medium is proportional to the concentration entering that layer.
2. Filter performance varies with time, first increasing, then decreasing.
3. The quantity of particles removed by a layer of filter medium equals the quantity accumulated in the filter pores.
4. A layer eventually reaches a point at which it no longer effectively clarifies the suspension and the concentration leaving equals that entering.
5. The equilibrium condition is reached first at the inlet layer and proceeds progressively through the filter in the direction of flow.

There is some question whether the equilibrium condition results from no retention of entering particles or from a combination of deposition and scour. Both theories are supported by some experimental evidence and it is conceivable that different mechanisms may exist with different suspensions.

As the successive layers of the filter become saturated with removed material, the headloss through the bed will increase due to the constriction of the flow. If the headloss becomes excessive a partial vacuum may be created within the medium causing air bubbles to be formed from gases drawn from solution. Such "air-binding" further restricts the area of flow, increases velocity and headloss, and may cause particles to be carried from the filter.

Large particles which are strongly bound will tend to be removed in the upper layers of the filter, producing high headloss with little penetration of floc. This phenomenon will be particularly pronounced if the filter medium is fine. Finer suspended particles, particularly upon coarse filters, will tend to penetrate further into the bed, distributing the reduction in flow capacity and producing lower headlosses for equivalent removal over equal time periods. Since deeper penetration is expected, coarse beds should logically be deeper,[8] and this has been shown to be necessary.[9] Consideration of the theory of filtration leads to the conclusion that it is the area of the filter medium which is important,[10] but since this is a function of particle size and bed depth, it follows that increasing the size of the medium will require an increase in depth.[11]

Pretreatment is required prior to filtration and thus coagulation, flocculation, and filtration cannot properly be considered as separate unit processes. Conventional rapid filtration using sand as a medium requires low influent turbidity and has typically employed coagulation with alum or lime, flocculation, and sedimentation. These processes may be incorporated in a single basin (Chap. 9).

Filters employing *mixed media*, which give an approximation of reverse gradation, or those with an *upflow* filtering pattern are less sensitive to influent turbidity and may be preceded only by a flash mix of chemicals, with coagulation and flocculation occurring within the filter. In such applications it is important that the floc be small but tough so that deep penetration will be obtained without shearing of the floc particles. This can be achieved by using short flocculation periods (10 min) and relatively high gradients in the mixing basins.[12]

10-4 Filter Media[13]

The choice of a filter medium is dictated by the durability required, the desired degree of purification, and the length of filter run and ease of backwash sought. The ideal medium should have such a size and be of such material that it will provide a satisfactory effluent, retain a maximum quantity of solids, and be readily cleaned with a minimum of washwater.[5]

The size of filter media is specified by the *effective size*, which is the sieve size in millimeters that permits 10 percent by weight to pass. Uniformity in size is specified by the *uniformity coefficient*, which is the ratio between the sieve size that will pass 60 percent by weight and the effective size. Fine materials will provide a better effluent, but will produce high headlosses in the upper layers of the bed, thus yielding short filter runs. Coarse media permit deeper penetration of the floc, better utilization of the storage capacity of the filter, longer filter runs, and easier cleaning upon backwash. Fine media have been found to contribute to the formation of mudballs.

Sand is the cheapest filter medium and has been widely used. The sand used in rapid filters should be free from dirt, hard and resistant, and preferably of quartz or quartzite. It should not lose more than 5 percent by weight after being placed in 40 percent hydrochloric acid for 24 hours. The sand depth, when used alone ranges from 600 to 700 mm in most applications. In present practice sands with effective sizes of 0.45 to 0.55 mm are most widely used. Uniformity coefficients do not exceed 1.70 and may be required to be not less than 1.20. Usually there is also a requirement that no sand grains over a certain size be permitted.

Anthracite has been used as a substitute for sand in some filter plants and may be used in conjunction with sand and other materials in mixed media filters. Crushed anthracite for filters has an effective size of 0.70 mm or more and a uniformity coefficient of 1.75 or less. Bed depths when anthracite is used alone are similar to those in sand filters and, in such cases, the smaller effective sizes are used.

Garnet sand or *ilmenite*, a particularly dense material (s.g. = 4.2), may be used as a constituent in mixed media filters. Its relatively high cost and limited availability as well as its high density make its use as a sole filter material impractical.

Other materials which may be locally available such as crushed glass, slag, metallic ores, and even shredded coconut husks and burned rice husks[14] have been used as filter media. The major materials of engineering interest are however limited to sand and anthracite.

10-5 Mixed Media

Ordinary granular filters, upon being backwashed, will settle with the finest particles on top and the coarsest on the bottom. This gradation is unfavorable in that suspended particles not removed in the upper layers will tend to pass through the bed, and the bulk of the medium is thus unused. Removal of the influent solids in the top layers leads to higher headloss than would result if they were distributed through the filter. The unfavorable gradation can be reversed, to a degree, by employing two or more media of different density, so selected that the coarser particles will settle more slowly than the finer. Mixed media filters usually employ anthracite (s.g. ~ 1.5) and silica sand (s.g. ~ 2.6) and may include garnet or ilmenite sand (s.g. ~ 4.2). It can be shown[5] that for equal settling velocities the required particle sizes for media of different density can be calculated from

$$\frac{d_1}{d_2} = \left[\frac{\rho_2 - \rho_w}{\rho_1 - \rho_w}\right]^{2/3} \tag{10-1}$$

Example Determine the particle sizes of anthracite and ilmenite which have settling velocities equal to that of sand 0.5 mm in diameter.

SOLUTION For the anthracite:

$$d_1 = (0.5)\left[\frac{2.6 - 1}{1.5 - 1}\right]^{2/3} = 1.1 \text{ mm}$$

For the ilmenite:

$$d_1 = (0.5)\left[\frac{2.6 - 1}{4.2 - 1}\right]^{2/3} = 0.3 \text{ mm}$$

Thus anthracite smaller than 1.1 mm would remain above 0.5 mm sand, and grains of ilmenite larger than 0.3 mm would remain below it.

Mixed media filters are not true depth filters but provide two or three filter surfaces with progressively smaller openings and thus permit effective use of a larger portion of the volume. Filter runs are proportionately longer and headlosses lower than those in single medium filters. Since solids concentrations are to be expected on planes within the filter as well as upon the surface, backwash systems should be selected to provide for these accumulations. Scour systems (Art. 10-10) which have proven satisfactory include air-water systems and rotary wash designs (Fig. 10-10) with a second rotor located about 10 mm above the unexpanded level of the denser medium.

10-6 Gravel

In standard rapid filters the filter medium may be underlain by 400 to 600 mm of gravel which serves to support the sand, permit the filtered water to move freely toward the underdrains, and also allows the washwater to move more or less uniformly upward to the sand. It is placed in five or six layers with the finest size on top. It is also specified that it should be hard, rounded, durable, weigh approximately 1600 kg/m^3; be free from flat, thin, or long pieces; and contain no loam, sand, clay, shells, or other foreign material. A common grading and layer thickness are as follows:

40 to 60 mm	120 to 200 mm
20 to 40 mm	80 to 120 mm
10 to 20 mm	80 to 120 mm
5 to 10 mm	60 to 80 mm
$2\frac{1}{2}$ to 5 mm	60 to 80 mm
Total depth	400 to 600 mm

Not uncommonly, unequal distribution of the washwater causes a jet action and mounding of gravel at the sand-gravel interface. To overcome this difficulty, at some plants an 80-mm layer of torpedo sand is placed between the gravel and the filter sand. The torpedo sand should have an effective size between 0.8 and 2.0 mm and a uniformity coefficient of not over 1.70.

10-7 Filtration Rates

For many years the standard rate of filtration has been about 120 m/day (2 gal/ft^2 per min). Some state regulatory agencies still adhere to this criterion despite overwhelming evidence that it is unreasonably conservative. Rates of 300 to 360 m/day have been used recently[10] and 240 m/day is standard practice in some states. Conservative practice might consist of sizing the filters for a nominal rate of 120 m/day, while providing hydraulic capacity permitting flows up to 240 m/day. Later "overloads" could thus be handled without difficulty. Filtration rates of 600 to 1200 m/day have been reported to be successful[15] but such rates are not recommended for domestic water supply systems. High filtration rates tend to carry floc deeper into the bed and require that the floc be strong if it is not to be sheared and carried through the filter. Polyelectrolytes have made a significant contribution to the increase in filter rates and length of filter runs.

10-8 The Underdrain System

The underdrains collect the filtered water from the gravel and also distribute the washwater during the washing process. To be satisfactory, the underdrains should collect and distribute water as evenly as possible, although this cannot always be done exactly because of the slight difference in head occurring at various points in the system. Hydraulic systems which provide for uniform distribution of water to or uniform collection of water from an extensive area require that the headloss through the orifices be large in comparison to that in the main carriage system. A widely used type is the perforated-pipe system. In it, head differences over the bed are minimized by keeping velocities in the pipes or conduits of the system at proper value and the orifices of proper size, number, and distribution. The combined area of the orifices should be from 0.2 to 0.33 percent of the plan area of the filter unit. A simple type of underdrain system consists of a central cast-iron manifold or header having openings into which cast-iron laterals can be attached. Asbestos-cement and cement-lined cast iron have been used for manifolds and laterals, and stainless steel for laterals. The laterals are generally 150 or 200 mm (6 or 8 in) on centers and are perforated on the underside with 5 to 10 mm (0.2 to 0.4 in) openings. The perforations are sometimes placed alternately on the underside but 30° off center. Placing openings on the underside requires supporting the laterals on concrete blocks 40 mm ($1\frac{1}{2}$ in) above the filter bottom. The advantage is that jet action of the washwater is discouraged. The perforations are sometimes lined with bronze bushings to prevent corrosion. H. N. Jenks at Sacramento established the following data covering design of pipe underdrain systems:

1. Ratio of length of lateral to its diameter should not exceed 60.
2. Diameter of perforations in the lateral should be from 6 to 12.5 mm ($\frac{1}{4}$ to $\frac{1}{2}$ in).
3. Spacing of perforations along the lateral may vary from 75 mm (3 in) for the smaller holes to 200 mm (8 in) for the larger.
4. Ratio of total area of perforations in the underdrain system to total cross-sectional area of laterals should not exceed 0.5 for the larger holes and should decrease to 0.25 for the smaller.
5. Ratio of total area of perforations in the underdrain system to the entire filter area may be as low as 0.002.
6. Spacing of laterals may be as great as 300 mm (12 in) for satisfactory diffusion but is limited by total head available.
7. Rate of washing may be varied from 0.15 to 0.90 m (3.5 to 22 gal/ft^2) per min, provided that the foregoing factors are used in design.

Manufacturers of filter equipment have devised various shapes of manifolds and nozzles or strainers for the laterals. They are of various designs, but are usually of brass, bronze or plastic and made in umbrella shape or as a cylinder having small perforations. Their principal function is to distribute washwater without jet action. Figure 10-4 shows a nozzle having a special air orifice which is used in those plants employing an auxiliary air wash.

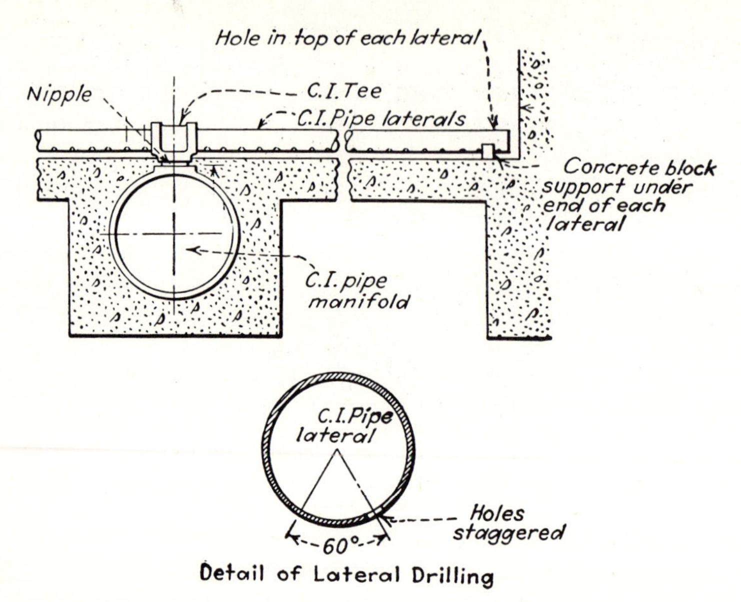

Figure 10-3 A perforated pipe underdrain. *(Courtesy Black and Veatch.)*

A number of proprietary filter bottoms are available which may permit the elimination of all or part of the gravel support. The Leopold bottom which employs a double chamber, the lower acting as a carriage system and the upper as a distributor, has a lower headloss than pipe lateral systems and requires only a very shallow fine gravel support. The Camp filter bottom (Fig. 10-6) may include air diffusers to aid in backwash, while the M-block filter bottom is able to introduce air without nozzles.

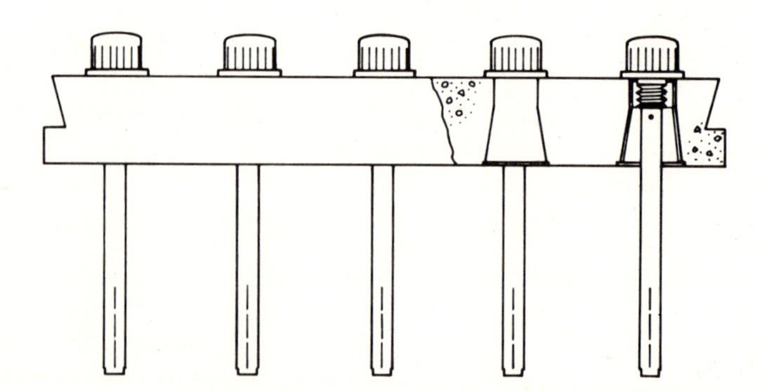

Figure 10-4 Special distributing nozzle for combined air and water wash. *(Courtesy Infilco-Degremont, Inc.)*

Figure 10-5 Glazed tile filter block used in the Leopold filter bottom. *(Courtesy Leopold Company, Division of Sybron Corporation.)*

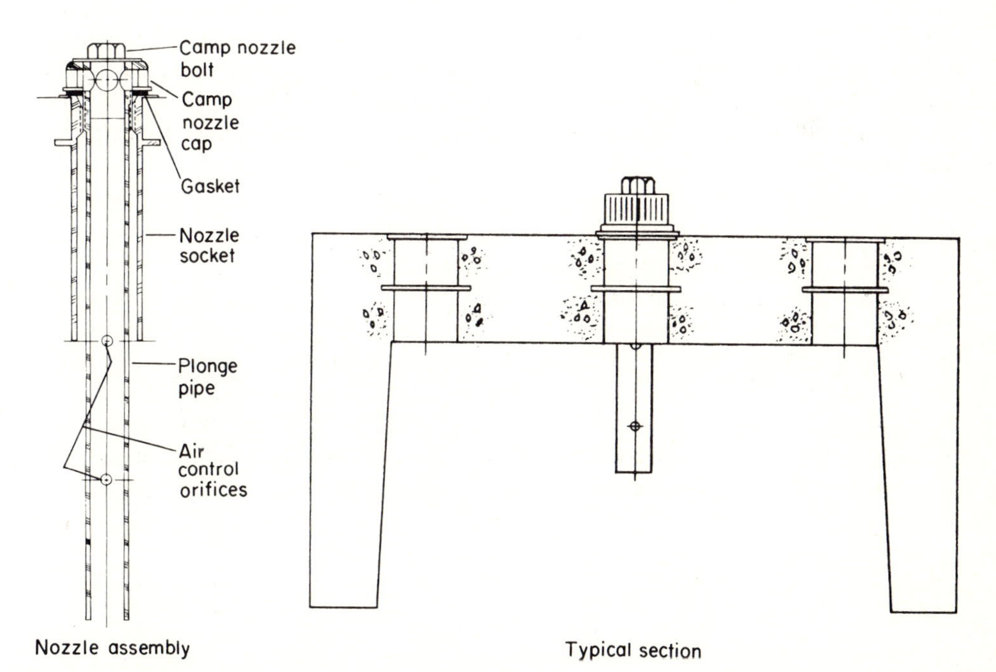

Figure 10-6 Camp filter bottom. *(Courtesy Walker Process Division, Chicago Bridge and Iron Company.)*

More modern filter designs employ backwash from a large chamber which eliminates the variation in head typical of standard filters and permits the use of low headloss nozzles and diffusers.

10-9 The Filter Unit and Washwater Troughs

A filter consists of two or more units of sizes depending upon the capacity of the plant. Some state authorities require that plants producing no more than 380 l/min (545 m^3/day) should have the complete plant capacity in each of two units. Units of 2, 4, 8, 12, 16, 20 and 24 thousand m^3/day capacity are commonly used, the smaller units being used in the smaller plants. Usually all units are of the same capacity. They may be placed in one or two rows, usually the latter in large plants. A unit is essentially a rectangular concrete box open at the top, with a depth of 3 m (10 ft) or more. The depth must be adequate to care for the underdrain system, gravel, sand, and a water depth of 1 to 1.5 m (3 to 5 ft) while filtering, plus an additional height according to conditions. The side walls are usually roughened by guniting or otherwise to discourage streaming of water between walls and the sand. The influent pipe discharges behind a baffle wall so that currents will not disturb the sand. The washwater also leaves the unit from behind the baffle or through a separate trough.

The rising washwater, after passing through the media, flows into washwater troughs. The lips of the troughs are horizontal and are all placed at the same height, usually at a distance equal to the rate of washwater rise per minute, 600 to 900 mm (2 to 3 ft), above the sand level. The trough arrangement depends upon the size of the unit. In small units the troughs discharge to a gullet on one side, while in large units they discharge into a central gullet which divides the unit into two sections (see Fig. 10-7).

In European practice, washwater is generally discharged to the sides and no troughs are placed above the filter. Designs of this sort require narrow filters but lessen the chance of washing filter media from the box.

Since washwater should not travel more than 1 m (40 in) laterally in order to minimize pressure differences on the filter, the troughs are never spaced more than 2 m (7 ft) apart from lip to lip. They are constructed of concrete, fiberglass or steel, and are of various cross sections. Concrete troughs have vertical sides and a V-shaped bottom. They must be made of such capacity that the sides will act as free-falling weirs for the washwater. A freeboard of 50 to 100 mm is provided at the upper end. Trough bottoms may slope toward the outlets but usually are horizontal.

The dimensions of washwater troughs, when the bottom is horizontal and the flowing water can be allowed to assume its own slope, can be obtained from the following formula

$$y = 1.73 \sqrt[3]{\frac{Q^2}{gb^2}} \qquad (10\text{-}2)$$

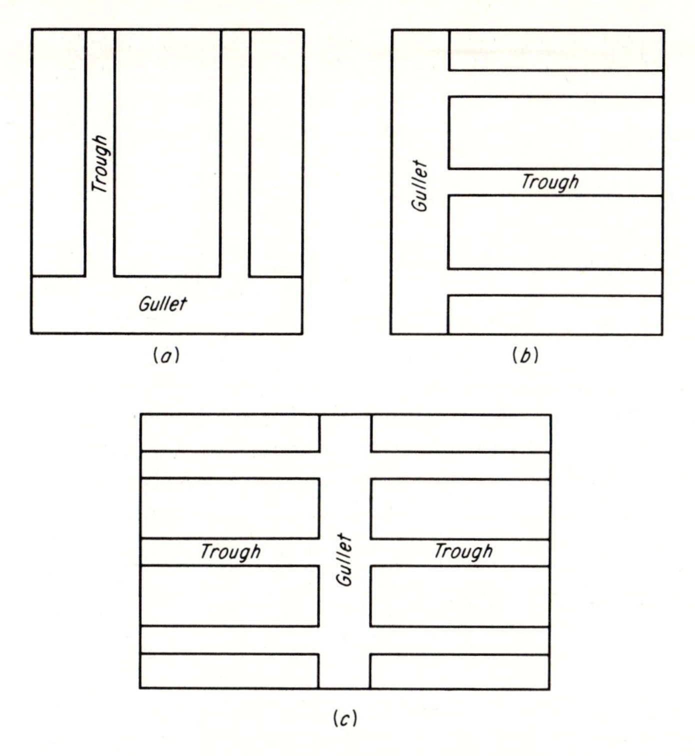

Figure 10-7 Arrangements of troughs and gullets in rapid sand filter plants. Arrangement (c) may have two separate underdrain systems, with one-half the bed being washed at one time.

in which Q is the total amount of water received by the trough, b is the width of the trough, y is the depth of the water at the upper end of the trough, and $\mathbf{g}$ is the acceleration of gravity. Fifty mm freeboard is added to the y, and since the formula is based upon a flat-bottomed trough, the V-shaped bottom if used must be compensated for by obtaining an area equivalent to the rectangular cross section. If the troughs are long an allowance for frictional losses must also be made.

Example A V-bottom trough with a horizontal bottom is to receive washwater from a section of filter bed 2 m wide and 3 m long. The washwater rate is 0.60 m/min. The trough width is approximately equal to its depth.

Solution The maximum flow in the trough is $Q = 0.6 \times 3 \times 2 = 3.6\ \text{m}^3/\text{min}$. If b and y are assumed equal,

$$y^5 = 1.73^3 \times \frac{0.06^2}{9.8}$$

$$y = 0.29\ \text{m}$$

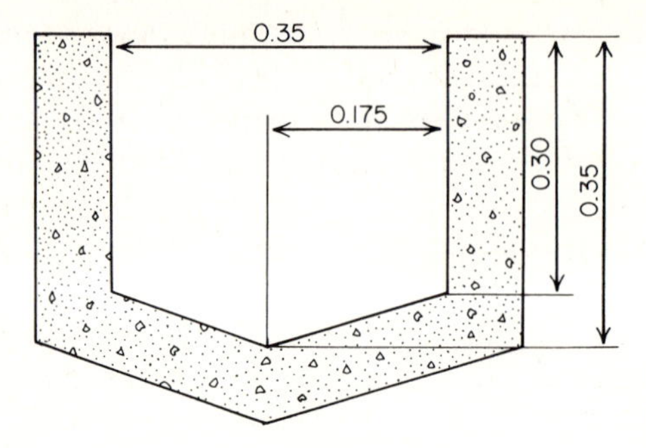

Figure 10-8 Cross section of a concrete wash-water trough.

The actual channel dimensions might be as shown in Fig. 10-8. This provides equivalent area of flow (0.084 m^2) and a freeboard of 50 mm at the upper end.

The depth at the lower end of the channel, if it were rectangular, would be equal to $y/1.73 = 0.17$ m, assuming critical velocity. The velocity, for a rectangular channel, would be $0.06/0.30 \times 0.17 = 1.18$ m/s, hence the hydraulic gradient at the lower end is given by

$$s = \left(\frac{nV}{r^{2/3}}\right)^2 = \left(\frac{0.013 \times 1.18}{0.0797^{2/3}}\right)^2 = 0.007 \text{ m/m}$$

Thus the maximum possible friction loss would be $3 \times 0.007 = 0.02$ m or 20 mm, but the actual loss is negligible since the velocity and hydraulic radius are not constant. If the channel were longer the headloss might be significant and would be calculated by determining sequential depths, velocities, and slopes at a series of points along the channel. Equation (10-2) is based upon the assumption of a parabolic water surface in the channel, hence the depth at any point x is given by

$$y_x = y\left(1 - 0.422\,\frac{x^2}{L^2}\right)$$

where y is the depth at the upper end and L is the total channel length. The flow at any point x may be calculated from

$$Q_x = Q(x/L)$$

and ranges from zero at the upper end to Q at the free discharge. The frictional headloss could then be approximated from a series of calculations averaged along the length of the trough.

10-10 The Washing Process

Washing consists of passing filtered water upward through the bed at such a velocity that it causes the sand bed to expand until its thickness is 25 to 40 percent greater than during filtering, depending upon the media.[16] The grains move through the rising water, rub against each other, and are cleaned of deposits. A bed is usually washed when the gauge provided for the purpose shows that the friction head through the bed has reached 2 to 3 m (7 to 10 ft), or whatever amount the operator deems the advisable limit. If the bed is not operating continuously, washing may be done at the end of the run or at the beginning of the new run regardless of the headloss. Necessity for washing, however, is indicated not only by loss of head but by the presence of floc in the effluent. Turbidity may break through at headlosses as little as 1.2 m (4 ft). Operators may backwash at 1.5 m (5 ft) headloss or 48 hours, whichever occurs first.

While the filtering action of a granular bed is primarily a function of the effective size of the particles (D_{10}), its action when expanded is governed by the larger grains and is best expressed as a function of its 60 percent size (D_{60}). The expansion of a granular bed during backwash can be calculated as follows.[17] The velocity which just begins to expand the bed is

$$V_b = \frac{0.3682(D_{60})^{1.82}[\rho(\rho_s - \rho)]^{0.94}}{\mu^{0.88}} \tag{10-3}$$

in which V_b is the backwash velocity in meters per minute, D_{60} the product of the effective size and the uniformity coefficient in millimeters, ρ_s the density of the medium, ρ the density of water, and μ the viscosity of water in centipoises (cP). Since ρ and μ are temperature dependent it is obvious that V_b will vary with temperature and provision should be made for varying backwash velocity with season.

The backwash process may consist of simple fluidization with or without auxiliary scour or surface wash, air scour and partial fluidization, or combinations of the first two. The backwash velocity must be sufficiently great to carry off the suspended matter removed by the filter, yet not so great as to wash out the filter medium. As a practical matter this means, for most suspensions, that the backwash rate must exceed 0.3 m/min (7 gal/ft^2 per min), but be less than 10 D_{60} m/min for sand and less than 4.7 D_{60} m/min for anthracite where D_{60} is, as above, the 60 percent size in millimeters.[10] The velocities given are approximations for 20°C and specific gravities of 2.65 and 1.55 respectively for sand and coal.

The bed is completely fluidized when the frictional force exerted by the washwater equals the particle weight and when the superficial approach velocity of the backwash exceeds:[10]

$$V_b = V_t \times f^{4.5} \tag{10-4}$$

where f is the porosity of the medium.

Example Calculate the terminal velocity and fluidization velocity of filter sand with an effective size of 0.55 mm, a uniformity coefficient of 1.5 and a specific gravity of 2.65. The porosity of the sand is 0.45.

SOLUTION

$$V_t = 10 \times (1.5 \times 0.55) = 8.25 \text{ m/min}$$

$$V_b = 8.25 \times 0.45^{4.5} = 0.23 \text{ m/min}$$

Thus a backwash velocity of 8.25 m/min would carry the medium from the bed and a velocity of 0.23 m/min would fluidize but not expand the sand.

The cleansing of a granular bed during backwash is a result of the shear produced by the rising water and of the abrasion resulting from contacts between particles in the fluidized bed. The maximum abrasion occurs, as shown by Kawamuru[10] when the bed is 10 percent expanded or,

$$V_b = 0.1V_t \tag{10-5}$$

Thus for sand beds

$$V_b = D_{60} \tag{10-6}$$

and for anthracite,

$$V_b = 0.47D_{60} \tag{10-7}$$

in which V_b is in meters/min and D_{60} is in millimeters. The rates above are for a temperature of 20°C but can be corrected for other temperatures by

$$V_{b(T)} = V_{b(20)} \times \mu_T^{-1/3} \tag{10-8}$$

in which μ_T is the viscosity in centipoises at the temperature in question.

Example Determine the appropriate backwash rate for a sand medium with an effective size of 0.5 mm and a uniformity coefficient of 1.5 at 5 and 35°C.

SOLUTION

$$D_{60} = 0.5 \times 1.5 = 0.75 \text{ mm}$$

$$V_{b(20)} = D_{60} = 0.75 \text{ m/min}$$

$$V_{b(5)} = 0.75 \times (1.52)^{-1/3} = 0.65 \text{ m/min}$$

$$V_{b(35)} = 0.75 \times (0.71)^{-1/3} = 0.84 \text{ m/min}$$

In designing a backwash system one must consider not only the size and density of the medium, but also the type and arrangement of underdrains, the number, size, and location of wash troughs, the head needed, either pumped or elevated, the control system, and the type and capacity of auxilliary scour or surface wash.

The head lost on backwash is the sum of the heads lost in the expanded bed, the gravel, the underdrains, and the pipe, valves, and rate controller. This may be written as

$$H = h_f + h_g + h_u + h_p \tag{10-9}$$

h_f is the head lost in the expanded bed and is equal to

$$h_f = L(1 - f)(\rho_s - \rho) \tag{10-10}$$

in which L is the depth of the unexpanded bed, f is the porosity, ρ the density of water, and ρ_s the density of the medium.

h_g is the head lost in the gravel and is equal to

$$h_g = 200L_g \frac{V_b \mu}{\rho g \phi^2 D_{60}^2} \times \frac{(1-f)^2}{f^3} \tag{10-11}$$

in which L_g is the depth of the gravel, ϕ is a shape factor

$$\left(\phi = \frac{\text{Area of sphere of equal volume}}{\text{Area of particle}}\right),$$

and the other terms are as previously defined. As a first approximation the headloss through a gravel layer may be taken as

$$h_g = 0.03L_g V_b \tag{10-12}$$

h_u is the head lost in the underdrains, variable with design and given by

$$h_u = \frac{1}{2g}\left(\frac{V_b}{\alpha\beta}\right)^2 \tag{10-13}$$

in which α is an orifice coefficient and β is the ratio of the orifice area to the filter bed area (normally 0.2 to 0.7 percent).[10]

h_p is the headloss in the pipe, valves, and controller. If these are replaced by an equivalent pipe (Art. 6-15) the headloss is calculable from

$$h_p = F\frac{L}{d}\frac{1}{2g}\left(\frac{4AV_b}{\pi d^2}\right)^2 \tag{10-14}$$

in which L is the length, d the diameter, and F the friction factor of the equivalent pipe and A is the filter bed area.

When backwash is provided from an elevated tank the controller is selected so that the head lost therein is large with respect to the change in head in the tank. Such a controller will provide minimum change in backwash flow with change in storage head. Backwash may also be provided by pumping from storage, or directly from the distribution system through a pressure reducing valve (Art. 6-23). When pumped backwash is employed the pumps should be axial or mixed flow in order that a steep characteristic curve be obtained (Art. 7-9). A pump of this sort can provide the variation in flow required at different times of year through throttling of its discharge.

The amount of water required for washing varies from 1 to 5 percent of that produced. Backwash generally continues for about 5 min, hence the volume required is readily calculable once the backwash rate is known.

Surface wash or *air wash* should be incorporated in all filter designs to help assure thorough cleaning of the media and to prevent the formation of mudballs (Art. 10-14). Figure 10-9 illustrates the pipe arrangement necessary for the Bayliss surface wash. It consists of 25-mm (1-in) vertical pipes on 600-mm (2-ft) centers each way, terminating in brass caps having four holes 6 mm ($\frac{1}{4}$ in) in diameter

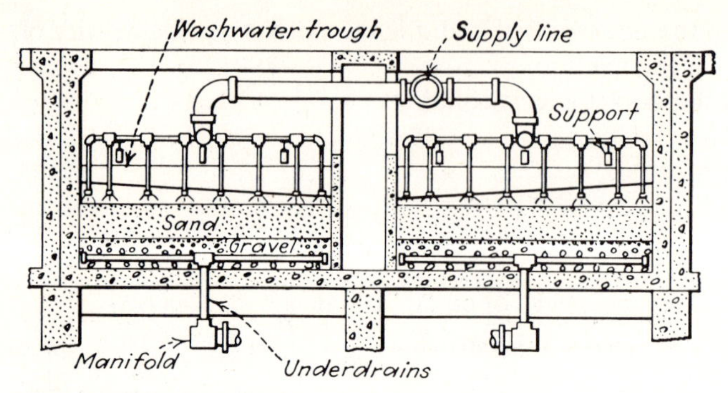

Figure 10-9 Section of filter with surface wash.

bored at a downward angle of about 38° to the horizontal and one hole directed vertically downward. The pipes end 100 mm (4 in) above the undisturbed sand and apply water at 70 to 100 kPa (10 to 15 psi) at the nozzles at the rate of 0.2 m/min (5 gal/ft^2 per min) to the total bed area. Use of the surface wash with a moderate upward flow at the same time, 0.48 to 0.60 m/min (12 to 15 gal/ft^2 per min), will keep the bed in good shape and prevent formation of mudballs (see Art. 10-14).

A more widely used surface wash is the Palmer Filter Bed Agitator, Fig. 10-10. It consists of rotating pipe arms of proper number and length to cover the whole bed. The arms are 50 mm above the bed and beneath the troughs and have nozzles which direct water jets at an angle to the sand surface, with one or more at the ends to stir the sand in the corners. The pressure differential set up by the nozzles causes the arms to rotate. The water requirements for the surface wash are 0.02 to 0.04 m/min ($\frac{1}{2}$ to 1 gal/ft^2 per min) to the total bed area at not less than 340 kPa (50 psi). In washing, the water in the filter is brought to the washing level, and the agitator is operated for 1 min. Then the washwater valve is opened and washing continues as long as necessary. The agitator is then turned off and the

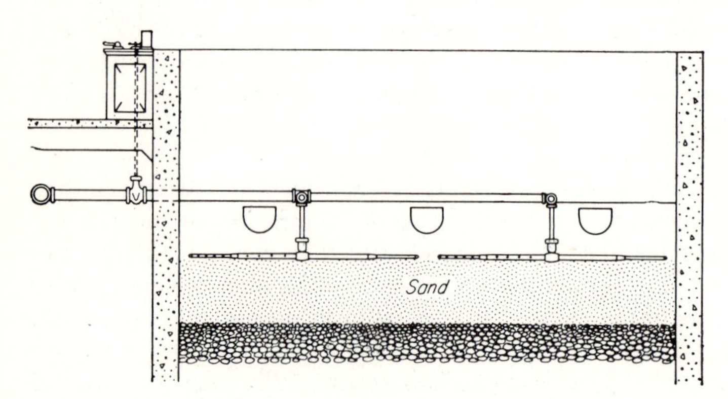

Figure 10-10 Palmer filter bed agitator.

upward wash is continued for 2 min. This apparatus, by violent agitation of the upper layers of sand, prevents formation of mudballs and allows lower washwater rates and saving of washwater. A vacuum breaker should be placed in the water supply line to prevent backsiphonage to potable water.

Air wash systems employing the M-Block, Camp filter bottom, or similar schemes permit agitation of the unexpanded bed. Air is introduced through the underdrains at rates ranging from 0.3 to 1.5 m/min (commonly 0.9 to 1.2 m/min) for a few minutes, followed by water at 0.3 to 0.5 m/min (7 to 12 gal/ft^2 per min). The air is left on in some designs until the washwater reaches the bottom of the troughs. Air agitation may also help to restratify dual media beds (Art. 10-5) and is particularly useful in cleaning the interface between coal and sand. Air flow through underdrains is generally held to velocities less than 50 m/min (150 ft/min) while backwash water velocities range from 0.3 to 0.6 m/s (2 to 3 ft/s).

10-11 The Control System

It has been customary in the past to provide rate of flow controllers to maintain constant filtration rates despite variations in headloss within the filter. Such controllers, while still available, are not necessarily desirable. Filters equipped with such devices produce neither equal quality[18] nor equal volumes[19] of water between backwash cycles when compared to declining rate filters.

The types of filter controls are discussed individually below, but may generally be classed as constant rate, declining rate, constant level, equal loading, or constant pressure.[10,20]

1. Constant headloss, constant head The standard rapid filter with rate controller is generally intended to function as a constant headloss device. In these filters a variable controller (Fig. 10-11) maintains a constant headloss by opening gradually as the filter bed becomes dirty. By maintaining both constant headloss and constant water level in the filter, a uniform flow rate should be obtained.

2. Variable headloss, variable head—single filter A rate of flow controller like that in Fig. 10-11 can, within certain limits, maintain a constant velocity of flow. In variable head filters the water depth on the bed is allowed to vary. The controller must then compensate for both changes in headloss in the media and changes in water surface elevation.

3. Variable headloss, variable head—multiple filters In these designs the total head on the filter is automatically made equal to the headloss. If the inlet divides the flow equally among the filters the head on each will vary freely with constant flow—from 10 to 20 mm, to the total available within the filter box. Such filters may employ either a restricted discharge or discharge above the sand level to insure water cover on the media.[21] The restricted or elevated discharge requires that more head be available on the filter. Standard filters with broken rate controllers may operate in this mode.

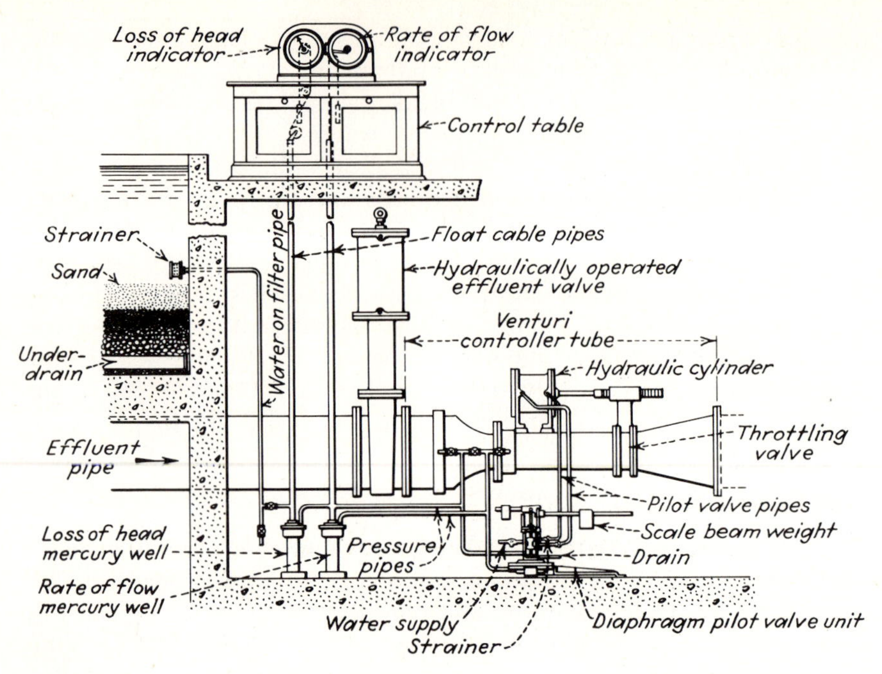

Figure 10-11 Control apparatus for a rapid sand filter unit. Influent and washwater pipes and pressure pipes from valves to control tables not shown. *(Courtesy Builders-Providence, Inc.)*

4. Variable headloss, variable head—unrestricted declining flow If water enters a filter with no significant influent headloss and the flow decreases with time, the increase in headloss in the bed will match the decrease in headloss in the underdrains. In practice the water level in the filter will increase since the headloss in the bed increases more rapidly than that in the underdrains decreases. Designs of this sort are similar to the Greenleaf filter (Fig. 10-12) and are sometimes called variable declining rate filters.[22] The cleanest filter in such a battery will carry the highest flow.

Variable declining rate filters must consist of multiple units, have a total production with one unit out of service which is sufficient to backwash one unit, and must have a low headloss underdrain system. The advantages of this filter include:[20]

1. Minimum mechanization: No controllers, headloss gauges, wash equipment or pipe gallery.
2. Slowly starting backwash which prevents possible bed upsets.
3. No possibility of air binding (Art. 10-14).
4. Underdrain is accessible.
5. Filtered water is visible.

5. Variable headloss, variable head, restricted declining flow Flow into the filter can be restricted at either the inlet or outlet. Restriction at the inlet prevents the

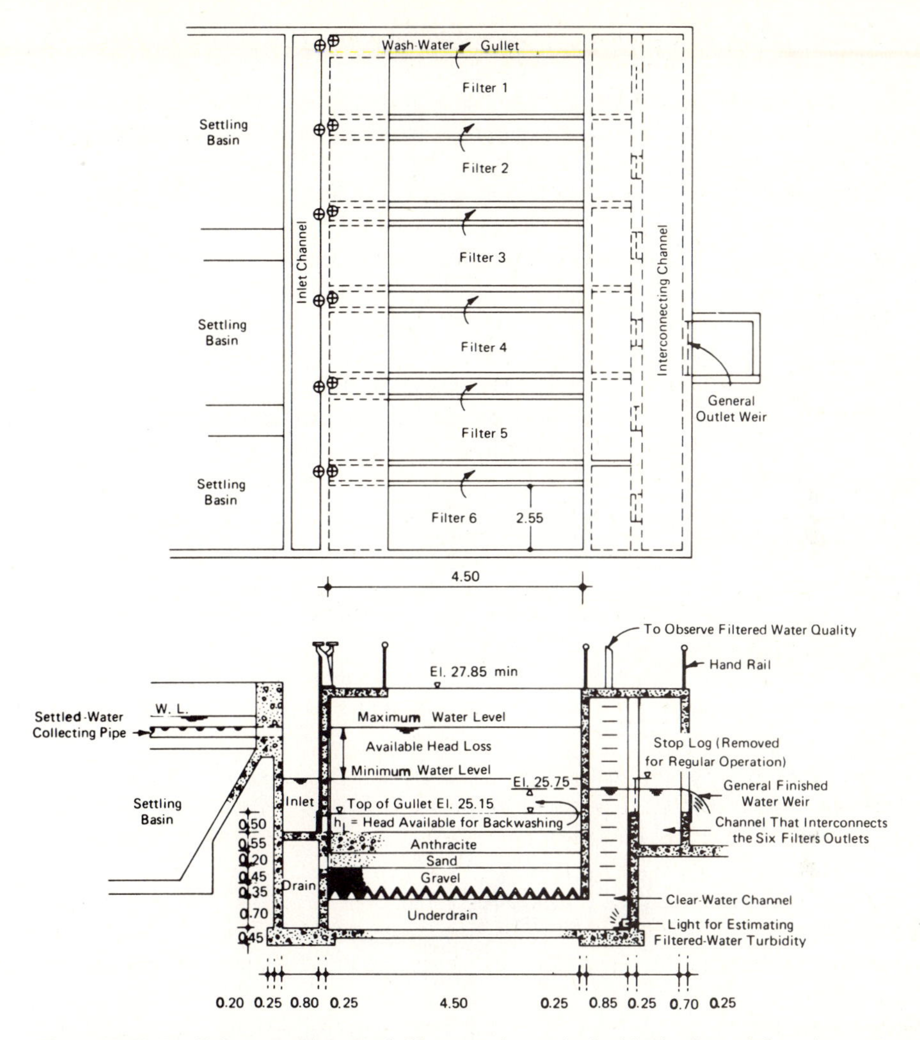

Figure 10-12 Declining-rate filter. Each filter may be washed with the flow of the others. *(Reprinted from Journal American Water Works Association* **66**, *by permission of the Association. Copyright 1974 by the American Water Works Association, Inc., 6666 W. Quincy Avenue, Denver, Colorado 80235.)*

possibility of initially high flows which might carry suspended matter through the bed.[19,22] The operation is otherwise similar to unrestricted declining flow systems.

6. Variable headloss, constant head Hydraulic devices such as a siphon or a float operated butterfly valve permit maintenance of constant head on a filter. An orifice or other restriction in the outlet is required to prevent initial high flow rates. The headloss in such a restriction must equal the difference between the headloss in the clean filter at maximum design flow and the head on the filter at the established level. Mechanically pressurized filters (Art. 10-18) may also operate in this mode.

The various filter controls above have a common intent, which is the minimization of surges and high filtration velocities which might cause breakthrough. The least successful, but most commonly used device is the mechanical rate controller which may hunt, leading to surges, and is unlikely to receive the maintenance which it requires.

Declining rate designs are more logical in that the higher rates are applied to clean filters, making breakthrough less likely. Additionally, surging is unlikely since velocity can increase only if the head upon the filter increases, and this requires measurable time. Mechanical rate controllers are expensive and their elimination in a treatment plant may save considerable money.

Controllers like that in Fig. 10-13 may also be used in backwash systems. As noted in Art. 10-10 a high headloss orifice can serve the same purpose when

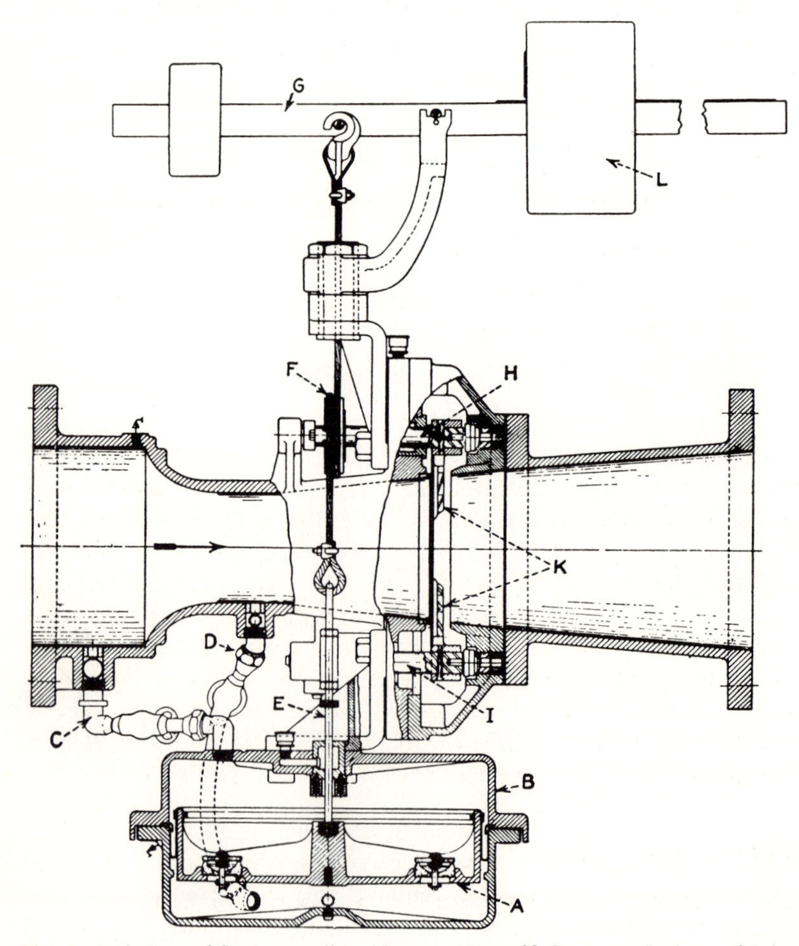

Figure 10-13 Rate-of-flow controller. *(Courtesy Leopold Company, Division of Sybron Corporation.)*

coupled with either a pump or an elevated storage tank. A delivery system with a high initial headloss will suffer less change in flow with change in head than a low headloss system.

Example A filter, without controller, has a total headloss in meters equal to $1.0 + 10\ V_b + 5000\ V_b^2$ through the sand, gravel, and underdrains respectively. Find the static head required to wash the filter and the variation in rate if the level in the storage tank drops 2 m during backwash. Rework for the same system with an orifice producing a headloss of $100{,}000\ V_b^2$.

SOLUTION The backwash rate is taken as 0.6 m/min (0.01 m/s).

$$H = 1 + 10(0.01) + 5000(0.01)^2 = 1.60 \text{ m}$$

The initial head must be 2 m greater to provide the desired velocity at the end of backwash. The initial head is thus 3.6 m and the initial velocity given by

$$H = 1 + 10(V_b) + 5000(V_b)^2 = 3.6$$

$$V_b = 0.0218 \text{ m/s}$$

for an increase of 118 percent. With the high loss orifice

$$H = 1 + 10(0.01) + 105{,}000(0.01)^2 = 11.60 \text{ m}$$

The initial head must be 2 m greater, as above, thus

$$H = 1 + 10(V_b) + 105{,}000(V_b)^2 = 13.60 \text{ m}$$

$$V_b = 0.0113 \text{ m/s}$$

for an increase of 13 percent.

A large array of sensing instruments such as *headloss gauges*, *turbidity sensors*, *flow meters*, *sand expansion gauges*, and *sand leakage detectors* are available. Their necessity or desirability depends upon the plant control system. Filters designed to operate without controllers and with low headloss backwash systems do not require anything other than a turbidity sensor to warn of breakthrough. Standard designs employing mechanical rate of flow controllers should be provided with headloss gauges and turbidity sensors as a minimum. If excessive backwash velocities are possible a sand expansion gauge is desirable. Flow meters on individual filters are helpful in assessing plant performance but are not critically important. If air scour is used, air control and measurement devices similar to those used in waste aeration systems (Chap. 24) are necessary.

10-12 Pipe Galleries and Piping

Pipe galleries are unnecessary in declining rate or variable head filter designs such as that illustrated in Fig. 10-12. When multiple filters with rate controllers and pumped or elevated backwash are employed, a pipe gallery provides a common location for pipes, controls, valves, and other fittings. Galleries should be uncrowded to facilitate maintenance and should be well lighted, ventilated, and drained. Some leakage will occur even with good maintenance, thus a drainage system and possibly a sump pump will be required.

Sampling faucets and small cocks to allow air to escape at high points in the piping should be provided. Flanged joints are generally used for piping and valves. Pipe sizes are based upon velocities which result in reasonable losses of head. The piping is usually of cast or ductile iron, although the wastewater pipe or sewer, if underground, may be of clay tile, either plain or encased in concrete. In recently designed plants cement-lined steel is used for some of the large pipes. Condensation on pipes can be prevented by plastic or other coatings. Influent lines are occasionally constructed under the operating floor as rectangular concrete conduits and sometimes are placed on the side of the bed opposite the pipe gallery with a hydraulically operated sluice gate to control the flow into each unit (see Fig. 10-14). These lines are designed for low velocities, 0.3 to 0.6 m/s (1 to 2 ft/s), since it is usually not desirable to lose much head in this conduit. Velocity in the washwater line is usually about 3 m/s (10 ft/s), and in the sewer 2.5 to 3.5 m/s (8 to 12 ft/s). The velocity in the filtered effluent line to the clear well is kept in the range of 1.2 to 1.5 m/s (4 to 5 ft/s). This pipe discharges below the water level in the clear well or is trapped, so that the outlet is always submerged and thus insures proper action of the controller. Velocity in the filtered-waste or rewash pipe is usually about 4.5 m/s (15 ft/s). Lower available heads may necessitate lower velocities and larger pipes and valves.

Valves may be pneumatically, electrically, hydraulically, or manually operated. If the latter, they are equipped with long stems which pass to stands on the operating floor. The valves may be either double disk gate or butterfly types. The latter are far more commonly used in new construction (Art. 6-23).

10-13 Negative Head

When a filter is clean, there is some slight loss of head, 150 to 300 mm (6 to 12 in), in the sand, gravel, and underdrains. As clogging occurs, the friction losses increase greatly, mostly, however, in the top of the filter sand. When the loss in the top layers becomes greater than the head of water above the sand, the column of water below acts as a draft tube, and a partial vacuum results. This condition is known as negative head and, when excessive, allows air to escape from solution in the water and lodge in the sand. This is known as air binding, and it may interfere considerably with filtration. Also, a mass of air may, at the beginning of the backwash, escape before the whole surface of the sand is broken. This will allow high local velocity of the washwater and may displace the gravel. In Fig. 10-15 it is assumed that c is 3 m below the water level of the filter and that the total loss of head is 2.55 m. The three tubes with their water levels at A, B, and C show the pressure heads at the water level, at b, which is 0.15 m below the sand surface, and at c, which is in the effluent pipe, respectively. With the total friction loss 2.55 m it

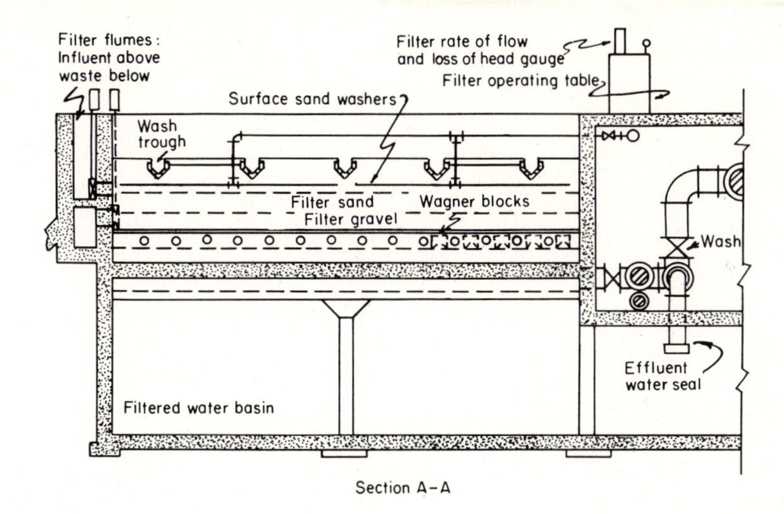

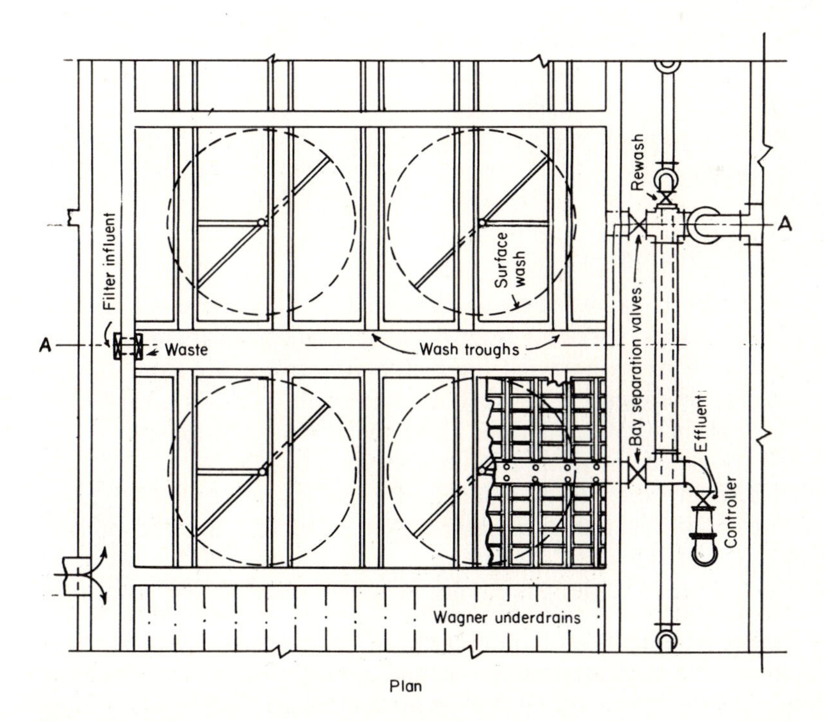

Figure 10-14 Plan and section of a two-bay filter bed, with separate underdrain systems but one rate controller. *(Courtesy Infilco-Degremont, Inc.)*

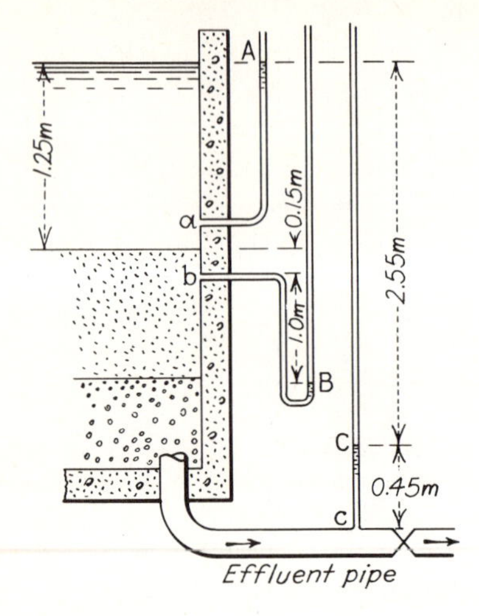

Figure 10-15 Diagram illustrating negative head.

is reasonable to assume that 2.25 m occurs in the top 150 mm of sand. By applying Bernoulli's equation between A and b, with the datum at c, by equating pressure head and potential head at b plus friction head to the total head, neglecting velocity heads, then

$$p_b = 3 - (3 - 1.4) - 2.25$$
$$= -0.85 \text{ m}$$

in which p_b is the pressure at b in meters of water. At b, therefore, there is a partial vacuum or negative head of 0.85 m.

10-14 Operating Difficulties

Air Binding This condition is caused by air dissolved in the water escaping into the media to form bubbles. It results from negative head, from increase of the water temperature as it passes through the bed, and occasionally from oxygen released by algae. Trouble is especially likely to occur if the water is saturated or supersaturated with air in solution. Serious interference with filtration may result, and when the bed is washed, the rising air bubbles may cause uneven washing and a loss of media. The remedies are avoidance of excessive negative head, algae control, remedying conditions that supersaturate the water with air, and precautions, if possible, against the water's warming as it passes through the plant. As noted in Art. 10-11, some filter designs cannot produce negative head within the media.

Mud accumulations Mud may accumulate on the filter surface to form a dense mat. As washing begins, there may be some lateral pressure between the points where the water breaks through, and the sticky mud is pushed into lumps or balls. If the lumps are largely mud and floc, they may stay near the surface and increase in size. If they contain much sand, they will sink to the gravel during washing.

Mudballs are also formed when cracks appear in the bed surface. In the process of crack formation a dense blanket of mud, floc, and organic matter first appears on the bed. Then, by shrinkage of the blanket, settlement of the sand, or pressure of the water above, the blanket pulls apart to form cracks either at the sides of the filter or elsewhere. In any case they extend a distance into the media. Water enters the crack, and it soon fills with mud and floc. During the washing process the rising water compacts the material in the crack, and a mudball is formed. The ball sinks sooner or later to the gravel surface where it will interfere with the rising washwater and cause excessive velocities around the edge of the mudball. This causes the gravel to move and form mounds, while above the ball the sand is improperly washed and more mud accumulates. This may, and frequently does, progress until a clogged mass extends from the gravel to the sand surface. The following remedies have been successfully applied. The mudballs may be removed from the sand with rakes while the bed is washed. Clogged masses in the bed are broken up with a 10-mm ($\frac{3}{8}$-in) pipe having a pointed closed end and two perforations a bit above the point. The pipe is connected to a rubber hose furnishing water under good pressure. Caustic soda has also been used. The bed is first washed; then the water is drawn off to 100 mm (4 in) above the bed; and the caustic is applied. After 12 hours of soaking, the sand is thoroughly agitated, preferably with air wash, and 8 hours later it is washed until the water runs clear. The use of surface wash and air scour has been very successful in reducing this trouble. Filters which have been thoroughly fouled with mud accumulations must have the medium replaced or cleaned by passage through an ejector.

Sand incrustation Where heavy lime treatment of water is practiced, a delayed crystallization of calcium carbonate may occur in the filter and enlarge the sand grains. This can be prevented by carbonating the water passing through the filter or by sequestering the calcium with polyphosphate. These techniques are discussed in Art. 11-22. Sand sometimes accumulates a dark-brown coating of manganese. If the manganese is largely mineral, soaking with a sulfuric acid solution will improve matters. If the manganese is in the organic state, caustic soda should be used in the manner suggested above for clogging.

10-15 Laboratory Control and Records

A water treatment plant should always include a laboratory adequate in size and equipped to make bacteriological, chemical, and physical tests. Daily tests should include water temperature, turbidity, alkalinity, color, hardness, residual chlorine, and coliform bacteria.

The tests should be made on the raw water and partially and completely treated water in order to check the efficiency of all parts of the plant. For example, coliforms and total bacterial count could be obtained for raw water, filter influent, filter effluent, and chlorinated water. Residual chlorine, particularly at large plants, should be tested every 2 hours and continuous recording of potentiometric chlorine analysis is now common.

The daily record should show the test results, amount of water filtered, hours of operation of various filters, filters washed and amount of washwater, and the chemicals used in total amount and in milligrams per liter. Such a record is usually kept on a large sheet covering a month with one horizontal line for each day. Good records are of value to the operator by allowing him to increase efficiency and decrease costs, to the waterworks superintendent by giving him cost data, and to health authorities and the operator when epidemics of water-borne disease occur and the public supply is under suspicion. It is good practice to have bacteriological tests made by other laboratories, often that of the city or state health department, as a check and as a further safeguard against criticism.

10-16 The Clear Well and Plant Capacity

The maximum capacity of a filter plant depends upon the water consumption and the storage available for filtered water. The greater the storage, the nearer may the plant capacity be to the annual average consumption. It will often be economical, therefore, to add storage in order to reduce the size of the filter plant needed. In any case a clear well is constructed in connection with the filters as reserve storage and also to allow the plant to be operated without too frequent variation of its output rate. The operator need work only a sufficient number of units to keep the clear well nearly full. There is much variation in the sizes of clear wells but one-fourth to one-third of the daily capacity of the plant is most common. Clear wells must be covered, and all possible precautions should be taken to prevent pollution from entering. At some plants they are located beneath the filter units, while at others they are separate structures.

The number and size of the filter units affect the cost of the plant and its flexibility in operation. A small number of large units is economical because less control equipment is necessary, but flexibility of operation is reduced. There is no standard relationship between total capacity and size of units except that, in general, larger plants tend to use larger units. Even the smallest plant, however, should have at least two filter units. Because adding units is comparatively easy, the design period of filter plants is short. At some plants, idle capital is reduced by constructing the filter boxes without installing the mechanism, underdrains, and filtering medium until they are needed.

10-17 Filter Galleries

These are also known as infiltration galleries and have been used by a number of cities in the United States and abroad for water supplies. They are horizontal conduits or pipes with perforations or open joints so that the groundwater can enter by gravity flow. The water is then pumped from a well either at one end or at the middle. The pipe should preferably be of long-lived material, like concrete or clay tile, of such size that headloss will be small for the amount of water to be collected and with entrance openings small enough so that excessive amounts of sand will not enter. Galleries are usually located in shallow aquifers of sand or

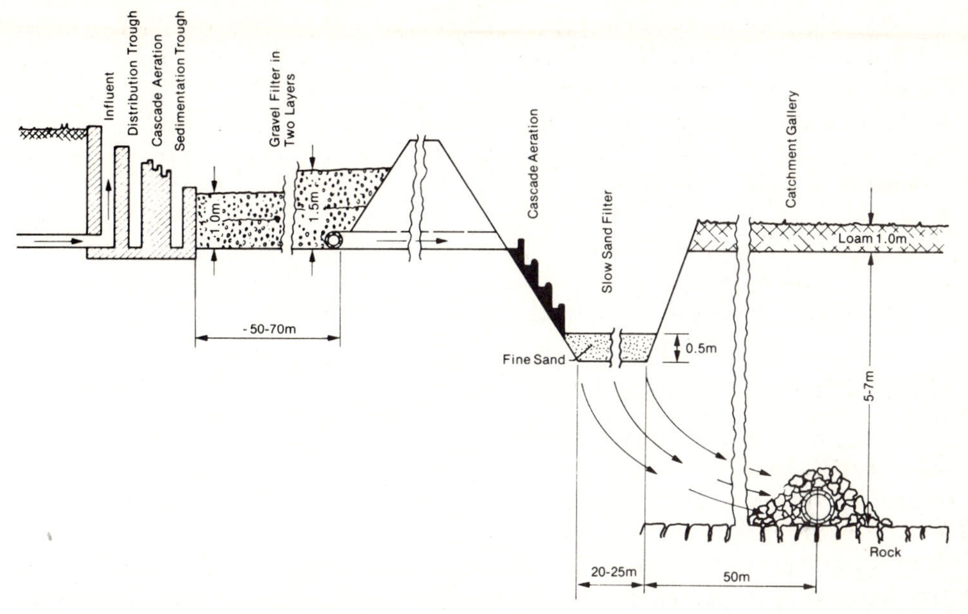

Figure 10-16 Filter beds and galleries used along the Ruhr River, Germany. *(Reprinted from Journal American Water Works Association* **68,** *by permission of the Association. Copyright 1976 by the American Water Works Association, Inc., 6666 W. Quincy Avenue, Denver, Colorado 80235.)*

gravel or in fissured rock. Frequently the gallery is adjacent to a stream or it may tap a natural underground reservoir. If it is an aquifer of sand or gravel, the possible yield can be predicted by the method indicated in Art. 4-5. At some existing filter galleries yields up to 12.5 m^3/m per day (1000 gal/ft per day) have been obtained.[23]

Galleries have the advantage of low first cost and operating cost compared with the rapid sand filter. First cost will probably be more than that of a shallow well field, but operating cost will be lower. Use of galleries may be combined with recharge of the aquifer with floodwaters or reclaimed wastewaters.

Figure 10-16 shows a type of installation used along the Ruhr River in Germany[24] where, as a result of intense industrial activity and population density, the river water is re-used a number of times. Here additional water was obtained in the galleries by pumping from the river to prefiltration basins constructed on the river bank and containing extremely coarse grained material. The water then passes through slow sand filters and then to the infiltration gallery.

10-18 Pressure Filters

The pressure filter is a type of rapid filter which is in a closed container and through which the water passes under pressure. Figure 10-17 shows a section of a pressure filter. Pressure filters are also obtainable with the axis of the cylindrical tank placed horizontally. The sand bed is generally 450 to 600 mm (18 to 24 in) thick with effective sizes and uniformity coefficients of the sand following those of

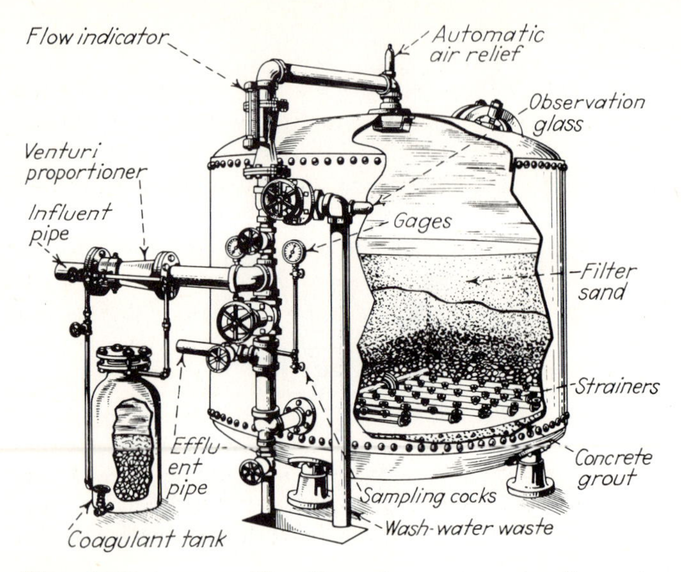

Figure 10-17 A pressure filter. The washwater enters the effluent pipe. *(Courtesy Infilco-Degremont, Inc.)*

open gravity filters. Gravel layers follow the same practice as in ordinary filters, and the underdrains are pipe grids or false bottoms. Washing is accomplished through reversing flow by manipulating the valves in the piping. A loss-of-head gauge indicates when washing is needed. The water is given a small dose of coagulant before it reaches the filter, dosing being frequently accomplished by passing small portions of the flow through small chemical tanks or pots in which it takes up the required amounts of alum and alkali. Ammonia alum or potash alum in lump form is used for pot feeds, and sal soda, also in lumps, is used for an alkali. A more efficient method is to inject a solution from a separate tank or tanks. In this case less expensive filter alum and lime or polymers can be used as coagulants. Rate controllers should also be a part of the equipment. Vertical units are obtainable in diameters from 400 to 2500 mm (16 to 96 in), and horizontal units in 2 to $2\frac{1}{2}$ m (7 to 8 ft) diameters and $2\frac{1}{2}$ to $7\frac{1}{2}$ m (8 to 25 ft) long. An installation may include a battery of filters all receiving raw water from and discharging filtered water into common header pipes.

Because they are considered unreliable in removal of bacteria, pressure filters are rarely used for treating municipal water supplies where filtration is necessary for safety, although they may be used in connection with softening of and iron removal from groundwaters. They are much used in clarifying softened water at industrial plants and in treating swimming-pool water that is recirculated. Filtering rates range from 120 to 300 m/day (2 to 5 gal/ft^2 per min). The swimming-pool rate is usually 180 m/day (3 gal/ft^2 per min). Since swimming-pool water is recirculated, its alkalinity must be renewed if metallic coagulants are used, and accordingly arrangements are generally made for dosing with both alum and lime or some other alkali. Alternately, polymeric coagulants can be employed.

10-19 Diatomaceous Earth Filter

This type of filter was developed by the Army for field use to remove chlorine-resistant dysentery cysts and the cercariae of schistosomiasis. The latter will pass through sand filters. This type of filter requires very little space and for this reason was adapted to swimming-pool use. In recent years many such installations have been made and, with good operation, they have been satisfactory.

The filter consists of rigid elements which are covered by closely wound noncorrosive wire or fabric of metal, plastic, or other material, or they may be cylinders of fused alumina or porous stone. In operation a precoat of diatomite in the amount of 0.3 kg/m^2 of filter area is provided by placing it in the first water applied to the filter after washing. Water must be wasted until the coating is established on the filter elements. Thereafter diatomite is added continuously as a slurry to the raw water to form the body coat. This keeps the mixture of turbidity, etc., removed from the water in porous condition and allows longer runs between backwashings. Body coat feed is based upon turbidity, about 2 to 3 mg/l per mg/l of turbidity of the applied water. The filter is backwashed, using filtered water,

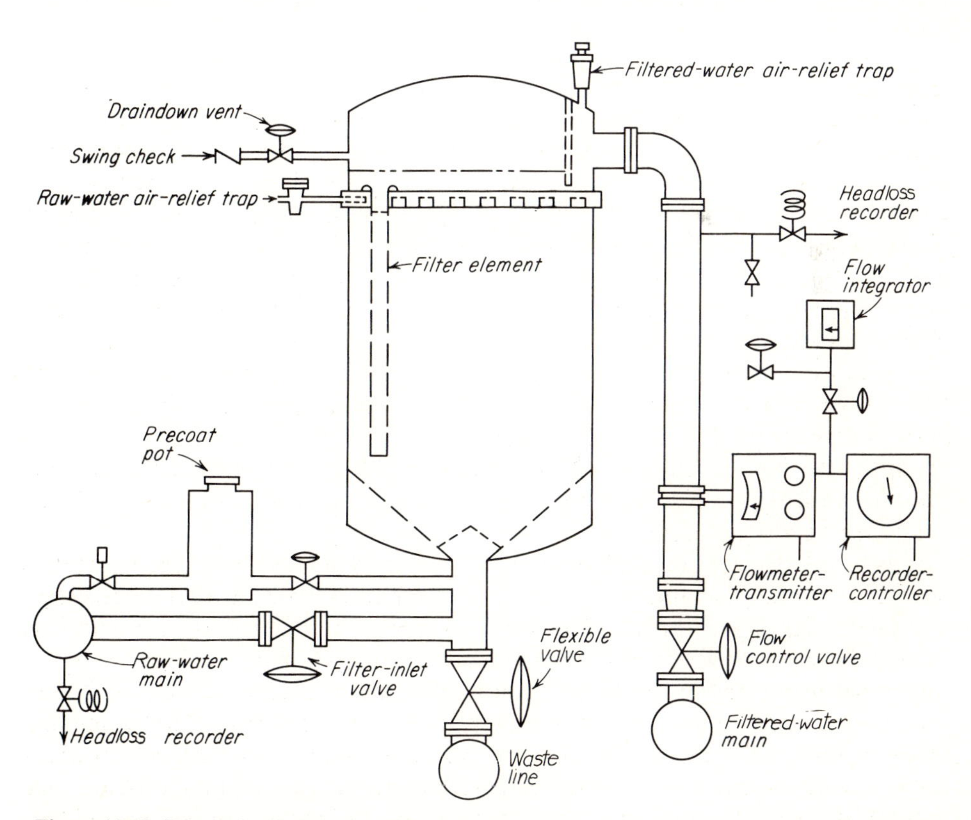

Figure 10-18 Schematic diagram of a diatomite filter. Precoat is placed by passing water through the precoat pot. Body coat is fed into the raw water main. Only one filter element is shown.

when the difference in pressure between the influent and effluent sides of the filter is about 70 kPa (10 psi). Backwashing is done by reversing flow through the filter elements with or without use of auxiliary water jets. Some manufacturers of filters provide for an "air bump" to dislodge material from the elements. Runs between washings may vary from a few hours to several days, depending upon the water characteristics and skill in operation. Filtration rates are about 60 m/day (1 gal/ft^2 per min).

Diatomite filters are not adapted to waters of high turbidity. The New York State Health Department places an upper limit of 30 mg/l on raw waters to be treated for municipalities, although higher turbidities have been successfully treated. The raw water should be chlorinated before application to the filter.

A few states now approve the use of diatomite filters but most do not approve them or do not have a fixed policy. This reluctance results from several considerations. Only a thin layer of diatomite is on the filter element and raw water might pass through. Only a metal wall separates raw and filtered water. Also, carelessness in washing might permit mixing of raw and filtered water.

Some industrial diatomite filters utilize vertical plate structures with water being applied from either side. As the filter plugs, the flow is reversed, washing the filter and then filtering in the direction of the wash until the headloss increases again. The backwash water may be collected and the diatomite recovered in designs of this type.

10-20 Upflow Filters

The unfavorable gradation occurring in standard filters is only partially overcome in mixed media designs. In *upflow filters* the direction of filtration is the same as that of the backwash, and thus the water is filtered by progressively finer layers as it passes through the bed. Such a design permits maximum utilization of the storage capacity of the filter and thus, longer filter runs. In order to prevent undesired expansion during filtration at rates above the minimum fluidization velocity, the medium (usually sand) is restrained by a grid which produces arching in the fine grains. Expansion is provided during backwash by breaking the arches with air. Once the arches are broken the filter bed will expand as if the grid were absent.

Upflow filters have been loaded at rates equaling or exceeding the maxima used on standard filters and have the potential of producing potable waters[2] although their use in the United States is restricted to industrial water production. Their application is generally as a single treatment unit preceded by flash mixing with an appropriate polymer. Coagulation, flocculation, filtration, and storage of suspended solids all occur within the filter medium.

10-21 Other Filtration Processes

A process developed in the USSR employs a modification of upflow filtration in which the expansion of the bed is restrained during filtering by directing a portion of the flow downward. Effluent is collected through a screened pipe a few centimeters below the upper surface of the sand. In this *biflow system* approximately 80 percent of the flow is directed upward and 20 percent downward.[25]

Continuously cleaned filters employ *moving bed* systems in which the filter medium is moved against the direction of flow of water with the top layer automatically falling from the path of motion.[26] The fallen sand is washed in an ejector and returned to the process. The advantage of this system is that water production is continuous with no interruptions for backwashing.

Slow sand filters are still used by some large cities where installations were made before the rapid filter was developed and established. New construction of slow filters is largely confined to additions to existing slow filter plants, although even such additions are often of the rapid filter type.

The characteristic features of the slow sand filter are the slow rate of operation, 3 to 6 m/day (3 to 6 gal/ft^2 per hour); lack of pretreatment of the water in the original plants; and cleaning of beds by scraping and sand removal. Without preliminary clarification, slow filters cannot cope with the highly turbid waters of many American rivers. This consideration, together with the great flexibility of the rapid filter, has placed the slow filter in the background. In European cities, however, use of rapid filters as pretreatment with slow filters for final treatment is not uncommon. A few American cities have followed the same policy, using preliminary rapid filters which are roughing filters or scrubbers. Others have added plain sedimentation basins and coagulation-sedimentation tanks.

As with the rapid filter, a clear well is needed at the plant, and this is generally made large enough to equalize filter output and demand over a day. The action of the slow filter is much the same as that of the rapid filter. The slower filtration rate gives the bed a greater efficiency in removal of bacteria than is attained by the rapid filter, considering the filter alone and not the auxiliary treatment. The absence of a gelatinous floc in the applied water requires the sticky coating of the sand grains to be acquired from the mud, silt, and organic matter removed. Probably slimes made up of bacterial growths play a larger part in removing organic matter from the water than is the case with the rapid filter. To give time for "ripening," a slow filter must be operated for some days when first completed before its effluent can be used. Even after cleaning, the filtered water may be wasted for several days, or the filter may be operated temporarily at a low rate.

The sand bed thickness varies from 300 to 600 mm (12 to 24 in), is less carefully specified than with the rapid filter, and the tendency is to use finer material. A common specification used places the effective size at 0.25 to 0.35 mm, and the uniformity coefficient at not over 3.0. The sand should also contain not more than 2 percent of calcium and magnesium computed as carbonates. The gravel is generally placed in three layers, the largest being 20 to 50 mm (1 to 2 in) in size and about 180 mm (7 in) thick. The other layers are about 50 mm (2 in) thick and are, respectively, 10 to 20 mm ($\frac{3}{8}$ to $\frac{7}{8}$ in) and 5 to 10 mm ($\frac{3}{16}$ to $\frac{3}{8}$ in). When the loss of head reaches 1 to 2 m (3 to 7 ft) slow sand filters are usually cleaned. The period between cleanings depends upon the character of the water and rate of filtration. A bed operating at a rate of 5 m/day should run for at least 10 and usually 40 days. Beds are cleaned by scraping off the top layer of sand. The amount removed depends upon the depth of the clogging; clay-carrying waters, for example, give greater penetration of solids.

The washing is done in a device which agitates the sand in a stream of water. About 0.5 percent of the filter output is required for washing purposes. After washing, the sand is stored and replaced on the bed when, by successive cleanings, the thickness has been reduced to about 300 mm. Replacement is done by hand or by hydraulic means. Raking the surface of the sand will lengthen the time between scrapings.

After cleaning or emptying for any purpose beds are refilled with filtered water from below until the sand is completely covered. This prevents entrapping of air in the sand. Thereafter the inlet valve is opened slowly to prevent disturbance of the sand.

PROBLEMS

10-1 A rapid filter plant is to treat 23,000 m^3 per day at a rate of 120 m/day. Determine the size and number of units required if the filtration rate is not to exceed 180 m/day with one filter being backwashed, nor 240 m/day with one filter out of service and one filter being backwashed. How much water would be required to backwash one filter if the wash is at 1 m/min and continues for 10 min?

10-2 What would be the area of a rapid filter with a capacity of 4000 m^3/day, if it is to operate at 120 m/day? With a washwater rate of 0.6 m/min compute the rate of washwater application (m^3/min) and the total amount used in a 5 min wash. If the Baylis surface wash is used (Art. 10-10) the upflow velocity can be reduced to 0.48 m/min. How much water will this save in a 5-min wash?

10-3 Calculate the backwash velocity necessary to expand a sand filter with an effective size of 0.6 mm and a uniformity coefficient of 1.3. This medium (or any other, for that matter) can be effectively cleaned at a backwash velocity of 0.3 m/min if fluidization is provided by air scour. Calculate the saving in dollars each time a filter 10 m by 10 m is washed, if the net cost of water production is $0.10 per m^3. Assume the backwash duration is 10 min.

10-4 A rapid filter plant has units with a capacity of 8000 m^3/day at a rate of 120 m/day. What should be the capacity of the elevated storage tank if it is to hold sufficient for two consecutive 5-min washes at a rate of 0.8 m/min? What commercial size pipe would you use in the filter gallery?

10-5 A valveless filter contains a sand medium with an effective size of 0.6 mm and a uniformity coefficient of 1.2. Determine the number of cells into which the filter must be divided if one unit is to be backwashed by the flow through the others and the design filtration rate is 120 m/day based upon the entire filter area. The flow to the entire filter array is not reduced during backwash—the loading upon the units in service increases to match the incoming flow.

10-6 Rework Prob. 10-5 assuming that one filter unit might be out of service, and that it must be possible to backwash a single bed with the flow produced by the total number less two.

10-7 Rework Prob. 10-5 using an anthracite medium with an effective size of 0.95 mm and a uniformity coefficient of 1.4.

10-8 Determine the variation in backwash velocity required to expand an anthracite medium with an effective size of 1 mm and a uniformity coefficient of 1.3 if the water temperature varies from 3 to 30°C.

10-9 Determine the headloss through the medium of Prob. 10.8. If the backwash line is equivalent to a 500-mm, cast-iron line 150 m long with a velocity of 3 m/s and the minimum headloss through the controller and underdrains is 4.5 m, what total pressure must be provided at the source?

REFERENCES

1. Hutchinson, W. R.: "High Rate Direct Filtration," *Journal American Water Works Association*, **68:**292, June, 1976.
2. Haney, Ben J. and Stephen E. Steimle: "Potable Water Supply by Means of Upflow Filtration (L'Eau Claire Process)," *Journal American Water Works Association*, **66:**117, February, 1974.
3. O'Melia, C. R. and W. Stumm: "Theory of Water Filtration," *Journal American Water Works Association*, **59:**1393, November, 1967.
4. Logonathan, P. and W. J. Maier: "Some Surface Chemical Aspects in Turbidity Removal by Sand Filtration," *Journal American Water Works Association*, **67:**336, June, 1975.
5. Kawamura, Susuma: "Design and Operation of High Rate Filters—Part 1," *Journal American Water Works Association*, **67:**535, October, 1975.
6. Boyd, Russell H. and Mriganka M. Ghosh: "An Investigation of the Influences of Some Physicochemical Variables on Porous Media Filtration," *Journal American Water Works Association*, **66:**94, February, 1974.
7. "Water Filtration—The Mints-Ives Controversy 1960–1973," *Filtration and Separation* (Brit.), **13:**131, March/April, 1976.
8. *Water Treatment Plant Design*, American Water Works Association, Denver, Colo., 1969.
9. Stanley, D. R.: "Sand Filtration Studied with Radiotracers," *Proceedings American Society of Civil Engineers*, **81:** Separate No. 592, January, 1955.
10. Kawamura, Susuma: "Design and Operation of High Rate Filters—Part 2," *Journal American Water Works Association*, **67:**653, November, 1975.
11. Jung, Helmut and E. S. Savage: "Deep Bed Filtration," *Journal American Water Works Association*, **66:**73, February, 1974.
12. Hutchinson, W. and P. D. Foley: "Operational and Experimental Results of Direct Filtration," *Journal American Water Works Association*, **66:**79, February, 1974.
13. AWWA B-100, "AWWA Standard for Filtering Materials," American Water Works Association, Denver, Colo., 1972.
14. Frankel, Richard J.: "Series Filtration Using Local Media," *Journal American Water Works Association*, **66:**124, February, 1974.
15. Harris, V. L.: "High Rate Filter Efficiency," *Journal American Water Works Association*, **62:**515, August, 1970.
16. Camp, Thomas R., S. David Graber, and Gerard F. Conklin: "Backwashing of Granular Filters," *Journal Sanitary Engineering Division, Proceedings American Society of Civil Engineers*, **97:SA6**:903, December, 1971.
17. Amirtharajah, A. and John L. Cleasby: "Predicting Expansion of Filters During Backwash," *Journal American Water Works Association*, **64:**52, January, 1972.
18. Hudson, H. E. Jr.: "Declining Rate Filtration," *Journal American Water Works Association*, **51:**1455, November, 1959.
19. Hudson, H. E. Jr.: "Functional Design of Rapid Sand Filters," *Journal Sanitary Engineering Division Proceedings American Society of Civil Engineers*, **89:SA1**:17, 1963.
20. Arboleda, Jorge: "Hydraulic Control Systems of Constant and Declining Flow Rate in Filtration," *Journal American Water Works Association*, **66:**87, February, 1974.
21. Aultman, W. W.: "Valve Operating Devices and Rate of Flow Controllers," *Journal American Water Works Association*, **51:**1467, November, 1959.
22. Cleasby, J. L.: "Filter Rate Control Without Rate Controllers," *Journal American Water Works Association*, **61:**181, April, 1969.
23. Stone, R.: "Infiltration Galleries," *Proceedings American Society of Civil Engineers*, **80** Separate No. 472, August, 1954.
24. Kuntschik, Otto R.: "Optimization of Surface Water Treatment by a Special Filtration Technique," *Journal American Water Works Association*, **68:**546, October, 1976.
25. Ives, Kenneth J.: "Progress in Filtration," *Journal American Water Works Association*, **56:**1229, 1964.
26. Allanson, J. T. and E. P. Austin: "Development of a Continuous Inclined Sand Bed Filter," *Filtration and Separation* (Brit.), **13:**165, March/April, 1976.

CHAPTER

ELEVEN

MISCELLANEOUS WATER TREATMENT METHODS

11-1

The public now demands that the waterworks operator do more than furnish a water that is clear and free from disease-causing organisms. It desires a water that is soft, free from tastes and odors, and does not discolor plumbing fixtures or corrode metals. Industry also requires water that will not interfere with its processes. Recently there has been increasing concern about the presence of minute quantities of organic material, particularly chlorinated hydrocarbons, which are thought to be causative agents of a variety of diseases. Such contaminants are known to be present in many water supplies[1,2] although their effect upon health is unknown. Standard disinfection practice using chlorine contributes to the production of these compounds.[3]

11-2 Chlorine in Water

Disinfection of water is the killing of disease-causing microorganisms that it may contain. In the process, coliforms will also be killed and other bacteria will be reduced in numbers. Complete sterilization, however, is not ordinarily obtained, nor is it necessary. Chlorine, in its various forms, has been widely used in disinfecting water. It is cheap, reliable, and presents no great difficulty in handling. Chlorine and water react according to the following equation:

$$Cl_2 + H_2O \rightleftharpoons HOCl + HCl \qquad (11\text{-}1)$$

which is nearly complete to the right. The hypochlorous acid, HOCl, ionizes or dissociates into hydrogen (H^+) ions and hypochlorite (OCl^-) ions in another reversible reaction

$$HOCl \rightleftharpoons H^+ + OCl^- \tag{11-2}$$

It is the hypochlorous acid and the hypochlorite ions which accomplish the disinfection. The hypochlorous acid is the better disinfectant and since the relative concentration of the two species is a function of the hydrogen ion concentration, the efficiency of disinfection with chlorine is affected by pH. As a general rule chlorine and its products are most effective at low pH. At pH values under 3.0 there may be some molecular chlorine. The chlorine existing in water as hypochlorous acid, hypochlorite ions, and molecular chlorine is defined as *free available chlorine.*

Hypochlorites, such as calcium hypochlorite, chloride of lime, and sodium hypochlorite, act in the same fashion. For example, when calcium hypochlorite is dissolved in water it ionizes as follows:

$$Ca(OCl)_2 \rightarrow Ca^{++} + 2OCl^- \tag{11-2}$$

The hypochlorite ions then combine with hydrogen ions to form hypochlorous acid, the reverse of the action following solution of chlorine in water.

$$H^+ + OCl^- \rightleftharpoons HOCl \tag{11-3}$$

The ratio of hypochlorous acid to hypochlorite ion, however, is again dependent upon pH, and accordingly disinfection efficiency is also affected by hydrogen ion concentration.

Chlorine is a very active element and when added to water as free chlorine it will combine with organic and inorganic matter and oxidize some organic and inorganic compounds. Free available chlorine reacts with ammonia and many organic amines to form chloramines. The chloramines, while oxidizing agents, are less active than hypochlorous acid and their disinfecting efficiency, therefore, is considerably reduced. The chlorine in water in chemical combination with ammonia or other nitrogenous compounds which modify its rate of bactericidal action is known as the *combined available chlorine.*

The *chlorine demand* of a water is the difference between the amount of chlorine added and the amount of chlorine present as a residual, either free or combined, after some designated period.

Chlorine is used in water treatment for disinfection, prevention and destruction of odors, iron removal, and color removal. While its principal use is as a disinfectant the mechanism of its bactericidal action is uncertain. It is likely that the chlorine destroys the extracellular enzymes of the bacterial cells, and possible that it actually passes through thc cell wall to attack intracellular systems. The bactericidal efficiency of chlorine is reduced by increased pH values and low water temperatures.

11-3 Chlorination

Chlorination of water is practiced for the purposes listed above and the various needs may be satisfied simultaneously. Chlorination is classified according to its point of application and its end result.

Plain Chlorination. In some cities surface waters are used with no other treatment than chlorination, although in some of these cities long storage is also given. In such cases chlorination is extremely important as the principal if not the only safeguard against disease. Such otherwise untreated waters are likely to be rather high in organic matter and require high dosages and long contact periods for maximum safety. The chlorine may be added to the water in the pipe leading from an impounding reservoir to the city. For disinfection alone a dose of 0.5 mg/l or more may be required to obtain a combined available residual in the city distribution system.

Prechlorination. This is the application of chlorine before any other treatment. The chlorine may be added in the suction pipes of raw-water pumps or to the water as it enters the mixing chamber. Its use in this manner has several advantages. It may improve coagulation and will reduce tastes and odors caused by organic sludge in the sedimentation tank. By reducing algae and other organisms it may keep the filter sand cleaner and increase the length of filter runs. Its range of effective action will, of course, depend upon the maintenance of a residual through the units of the plant. Frequently the dosage is such that a combined available residual of 0.1 to 0.5 mg/l goes to the filters. The combination of prechlorination with postchlorination may be advisable or even necessary if the raw water is so highly polluted that the bacterial load on the filters must be reduced in order that a satisfactory coliform count or MPN be obtained in the final effluent.

Postchlorination. This usually refers to the addition of chlorine to the water after all other treatments. It has been standard treatment at rapid sand filter plants, and when used without prechlorination and with low residuals it is sometimes called marginal chlorination. The chlorine may be added in the suction line of the service pump, but it is preferable to add it in the filter effluent pipe or in the clear well so that an adequate contact time will be assured. This should be at least 30 min before any of the water is consumed if only postchlorination is given. Dosage will depend upon the character of the water and may be 0.25 to 0.5 mg/l in order to obtain a combined available residual of 0.1 to 0.2 mg/l as the water leaves the plant. Greater residuals will probably be needed if it is desired to hold a disinfecting effect throughout the distribution system. This is considered desirable since it affords protection against contamination from cross connections and prevents organic growths in mains and their resulting odors. In some cases additional chlorine is injected into feeder mains at strategic points in the distribution system to maintain the residual in the mains. Residuals tend to disappear in the pipes because of the combination of the chlorine or chloramine with living organic matter, products of decomposition, or the pipes themselves.

Breakpoint Chlorination. As previously indicated, when chlorine is added to water it unites with organic and inorganic matter. Some of the resulting chlorine combinations, like chlorophenols, may be odorous and undesirable. If the chlorine dosage is increased, the combined available residual will also increase with the chlorinous odors increasing as well. This will continue until, with most waters, the residual will exhibit a drop in a curve of residuals plotted against chlorine dosage, and then with a further increase in dosage the residual will resume its increase. The dosage at which the residual begins to increase again is the breakpoint. Apparently the breakpoint indicates complete oxidation of the chloramines and other chlorine combinations, and the residual above the breakpoint is mostly free available chlorine. Usually the chlorinous and other odors will disappear at or before the breakpoint. Dosages are likely to be 7 to 10 mg/l in order to obtain a free available residual of about 0.5 mg/l or more. The chlorine, when the breakpoint procedure is applied, usually, but not always, is added at the influent to the plant. In some cases ammonia has been added to waters lacking in it in order to form a more pronounced breakpoint.

Free Residual Chlorination vs. Combined Residual Chlorination. Most prechlorination and postchlorination is to obtain a combined residual, although there may be some free chlorine residual also present. Combined residuals, as indicated by bacterial results, are generally adequate. For greater safety, free residual chlorination can be used by adding sufficient chlorine to destroy the ammonia, and the residual so obtained will be a very efficient disinfectant. In the process of obtaining the free residual, odorous materials will probably have been eliminated, and a residual can be maintained in the distribution system once its chlorine demand is satisfied if the water is not exposed to sunlight. The cost of the extra chlorine required by the process is the only disadvantage other than the possible formation of additional halogenated hydrocarbons.

Dechlorination. Heavy chlorination required for quick disinfectant action or destruction of tastes and odors may produce high residuals which are undesirable. This process has been called *superchlorination* and it may be followed by dechlorination. Aeration will remove chlorine, hypochlorous acid, and dichloramine by shifting the equilibria of the several ionization reactions as chlorine gas is stripped. Sulfur dioxide, sodium bisulfite and sodium sulfite can be used as reducing agents to convert chlorine and its ionization products to chloride. Activated carbon (Art. 11-12) is also useful as an adsorbent. Samples of chlorinated water collected for bacteriological tests are dechlorinated with sodium thiosulfate.

11-4 Chloramines

The bactericidal qualities of the chloramines have already been mentioned. If an excess of ammonia is present in the water, they will be formed in relatively large amounts, and there will be little or no combination of chlorine with organic matter that may produce undesirable odors. Hence chloramines have been used as an odor *preventive* with satisfactory bactericidal effects if the contact period is long enough. They will not remove or reduce odors present before the ammonia-chlorine treatment is applied. The combined residuals so obtained are persistent in the distribution system and are useful in obtaining good bacteriological tests of the water throughout the city. They should be of value for disinfection of contaminating matter which may enter the mains by a cross connection and should also prevent the growth of crenothrix or other organisms which may cause odors.

The following reactions occur in the formation of chloramines:

$$NH_3 + HOCl \rightarrow H_2O + NH_2Cl \quad \text{(monochloramine)} \qquad (11\text{-}4)$$

$$NH_2Cl + HOCl \rightarrow H_2O + NHCl_2 \quad \text{(dichloramine)} \qquad (11\text{-}5)$$

$$NHCl_2 + HOCl \rightarrow H_2O + NCl_3 \quad \text{(nitrogen trichloride)} \qquad (11\text{-}6)$$

The monochloramine predominates at pH over 7.5; the dichloramine will predominate at pH from 5 to 6.5; below pH 4.4 odorous nitrogen trichloride, which is not a disinfectant, will be produced. It, however, is sufficiently volatile that it will disappear if there is a detention period of a few hours in the plant.

Recent research[4] indicates that there is a transient presence of HOCl and OCl^- at sub-breakpoint levels. The concentration of these species is a function of pH, temperature, chlorine, and ammonia dosage. The maximum transient concentration of HOCl, the most effective species, occurs at pH 7.5 which should thus be the optimum pH for sub-breakpoint chlorination in the presence of ammonia. As noted above, in the absence of ammonia, lower pH is better—in fact, the predominant species at pH 7.5 in the absence of ammonia is OCl^-.

Ammonia is usually added in the ratio of 1 part of ammonia to 4 parts of chlorine, although experiment may indicate the desirability of a higher or lower proportion. It is used as the gas, as a solution of the gas in water, or as ammonium sulfate or ammonium chloride. Ammoniators much resembling chlorinators are used to feed the gas. The ammonium sulfate used is the chemical fertilizer, which is sufficiently pure to be used for this purpose.

11-5 Use of Chlorine Gas

The widespread use of chlorine gas for water treatment has led to the development of a variety of equipment designed for its application. Chlorine is obtained in pressurized cylinders ranging from 45 to 1000 kg (100 to 2200 lb) and in tank cars which may be 15, 30, or 50 thousand kilograms (16 to 55 tons) in capacity. Small plants commonly use the 45-kg cylinders, while large ones requiring 75 to 100 kg/day generally use 1000-kg containers as a matter of convenience and economy.

A satisfactory chlorinator must feed the gas into the water at an adjustable rate, and it must do this although the pressure in the gas container changes as the temperature changes. While some chlorinators apply the measured amount of gas to the water through a porous porcelain diffuser, most types dissolve the gas in

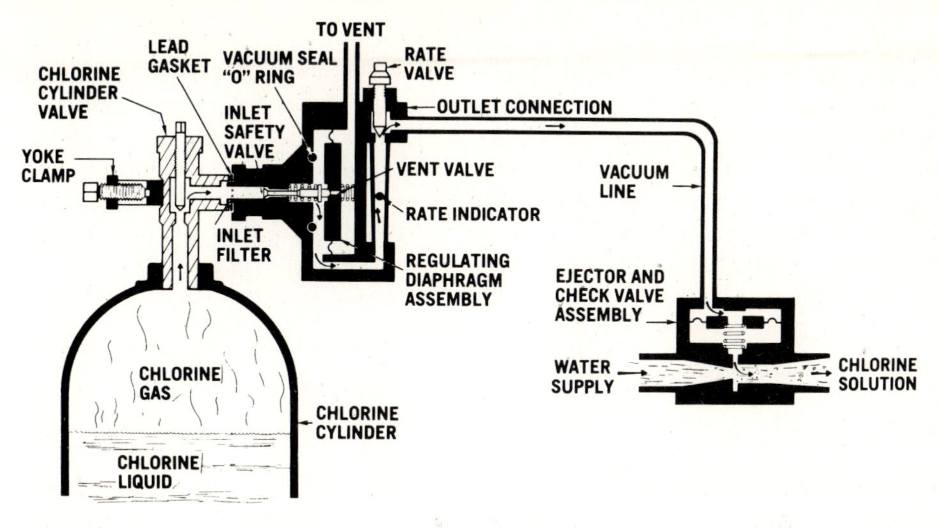

Figure 11-1 Chlorination apparatus. *(Courtesy Capital Controls Company.)*

water and feed the solution. The type to be chosen from those manufactured will largely depend upon the capacity required. Figure 11-1 shows a chlorinator in which a water operated aspirator draws chlorine through a regulator and measuring device. The flow can be set to provide a predetermined dosage of chlorine. The quantity of water required and the pressure at which it must be delivered are a function of the size and design of the system. At the injector the chlorine is dissolved in the water and is conveyed to the point of application. In some instances the injector may be at the point of application. The inlet valve closes if the vacuum fails. If it should not close, the chlorine pressure forces the seal of the vent, and the chlorine is discharged outside the building.

Figure 11-2 is a simplified diagram of the V-notch chlorinator. It also operates under a vacuum generated by an injector, with the chlorine flow controlled by a variable V-notch orifice. The amount of chlorine fed is shown by a feed rate indicator of the rotameter type. The differential valve and the chlorine pressure-regulating valve maintain a constant vacuum differential across the V notch. The feed rate is changed by changing the area of the V-notch orifice.

Certain precautions are needed in chlorination practice. As a check on the accuracy of the chlorinator, the cylinder or cylinders should be placed on scales, the loss of weight checked at least daily, and a permanent record kept. Chlorinators should be in duplicate so that service will never be interrupted by breakdowns, and a 2-week supply of chlorine should always be on hand. If chlorine gas, after leaving the cylinder, strikes a colder part, it may reliquefy and partially clog openings and cause irregular flow of gas. This can be avoided by keeping the air temperature around the chlorinators at 24°C or higher. If the water is below 10°C, gas will combine with moisture to form a yellow crystalline chlorine hydrate, commonly called chlorine ice. It may cause trouble at the chlorine pressure-

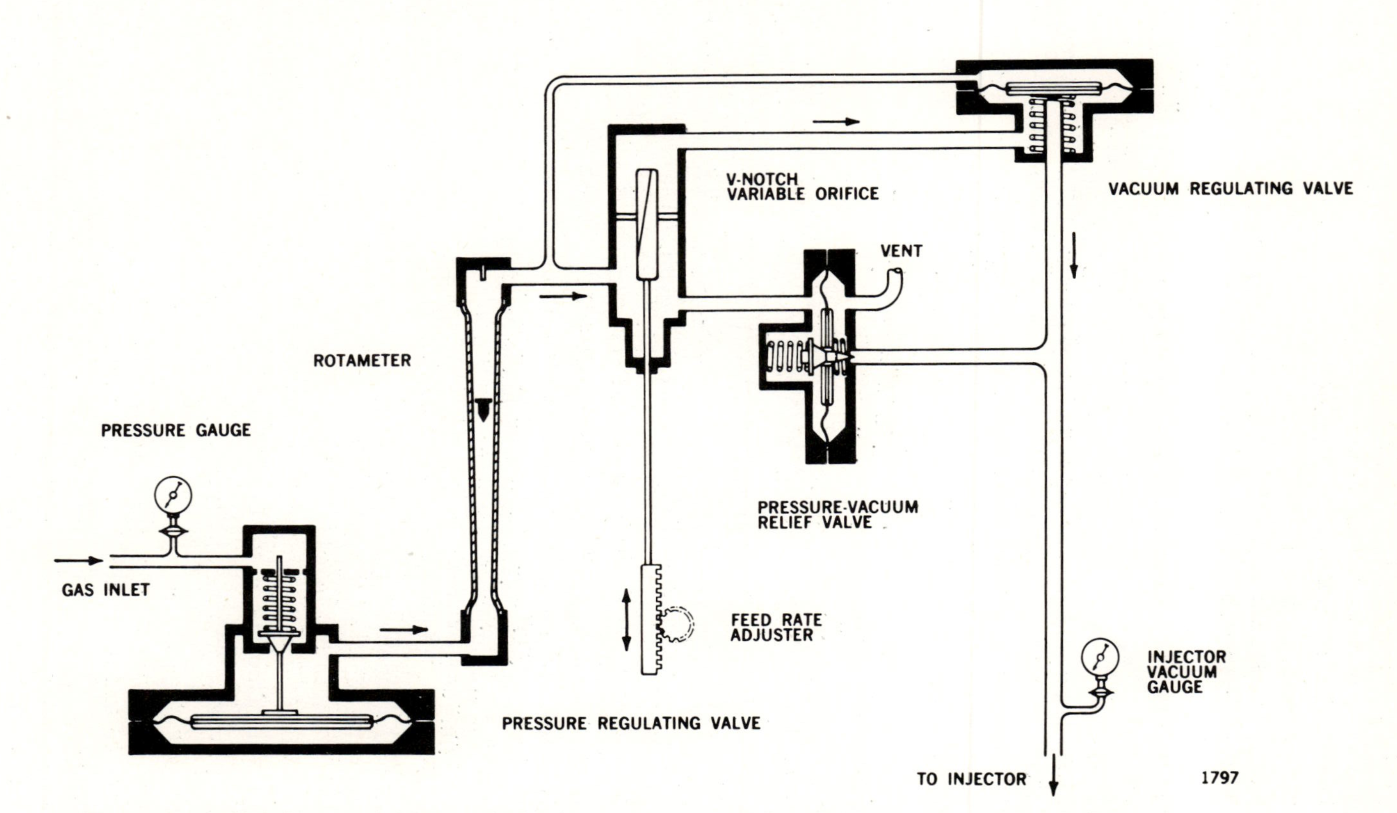

Figure 11-2 Simplified diagram of the V-notch chlorinator. *(Courtesy Wallace and Tiernan, Div., Pennwalt Corp.)*

reducing valve and completely or partially clog the orifice. Leaks in piping can be found by moving an open bottle of ammonia (ammonium hydroxide) along the piping. When white fumes appear, it is an indication of reaction between chlorine and ammonia to form ammonium chloride. It is necessary to have a gas mask stored outside the chlorinator room available for use if major leaks occur, and to employ a detector and alarm system to warn of dangerous chlorine concentrations within the chlorinator room.

Chlorinators and cylinders of chlorine should be housed separately and not in rooms used for other purposes. Ventilation should be provided to give a complete air change each minute and the air outlet should be near the floor since chlorine is heavier than air. Switches for fans and lights should be outside the room and near the entrance. The entrance door should have a clear-glass window to allow observation from outside. Chlorinator rooms should be heated to 15°C with protection against excessive heat. Cylinders should be protected against temperatures greater than that of the equipment.

11-6 Hypochlorination

The oldest method of water chlorination employed chlorinated lime which releases chlorine when placed in water, but has the disadvantages of deteriorating with age and being disagreeable to handle. Its maximum chlorine content is 30 to 35 percent, and careful control over its use is necessary because of the chlorine content variability. Chlorinated lime has been largely displaced, not only by chlorine gas, but also by improved commercial compounds of sodium and calcium hypochlorite. Sodium hypochlorite, NaOCl, sometimes known as laundry bleach, is a liquid obtainable in carboys and large containers at a strength of 12 to 15 percent available chlorine. Calcium hypochlorite, $Ca(OCl)_2 \cdot 4H_2O$, is a powder obtainable in cans, drums, and large containers under such trade names as HTH, Perchloron, and Pittchlor. Calcium hypochlorite has a chlorine content of 70 percent, does not lose its chlorine content with storage, and can be applied with dry-feed machines. Solution-feed apparatus is also used, for both sodium and calcium hypochlorite, generally of the type that feeds at a constant rate through an adjustable orifice having a constant head (Fig. 9-25). Small positive displacement diaphragm pumps, with speed and stroke adjustments, are available to pump hypochlorite solution into water to be treated. Duplex and triplex types can also be obtained to apply two or more chemicals, such as hypochlorite for disinfection, and alum and sodium carbonate for coagulation or softening. They may be operated by a motor or a turbine-type water meter. In the latter case there can be automatic proportioning.

Hypochlorination is especially applicable to emergency use where supplies are endangered and there would be considerable delay in obtaining chlorine gas and chlorinators. Many small plants, however, use this method continuously. An emergency hypochlorite dosing apparatus can be improvised from one or more barrels and a constant head orifice box equipped with a float valve of the type used in the ordinary toilet flush tank. The barrels serve as mixing and storage tanks, and pipes with valves allow a solution of known strength to run to the constant head feed box. Solutions of not over 0.65 percent of chlorine by weight are used as this is near the solubility limit at ordinary temperatures. In calculating chlorine dosage, consideration must be given to the chlorine content of the chemical.

Hypochlorites are also useful in disinfecting waters of reservoirs and swimming pools or in controlling algae. For reservoirs the water surface may be sprayed from a boat, or a sack of chemical may be dragged through the water as suggested for copper sulfate in algae control (Art. 11-9).

11-7 Ozonation

Chlorination has been shown to be responsible for increases in the concentration of volatile halogenated organics in many water supplies.[5] Chloroform, bromodichloromethane, dibromochloromethane, bromoform, 1, 2 dichloroethane, and carbon tetrachloride, particularly the first four, are commonly found in chlorinated water.[3]

Of the alternate disinfection techniques available,[5] ozonation is most likely to be readily applicable in American practice. Ozonation is widely practiced in Europe[6] where it has been successful both in disinfection and reduction of tastes and odors. Since it is a strong oxidizing agent, ozone may be applied in any situation where chlorine has been used. It is effective as a germicide, in destruction of organic matter which might produce tastes or odors, and in oxidation of iron and manganese.[7]

Unlike chlorine, which exhibits a distinct time-concentration relationship in disinfection, with higher dosages shortening the required contact time, ozone is an all-or-nothing disinfectant. Before disinfection can be obtained the ozone demand of the water must be met, i.e., sufficient ozone must be added to oxidize the various impurities present. Dosages range from 0.25 mg/l for high quality groundwaters to 5 mg/l following filtration for poor quality surface waters.[8] Effective ozone dosages for viruses range from 0.25 to 1.5 mg/l at contact times of 45 s to 2 min. As a general rule ozone is far more rapid in its action than other oxidizing agents. Ozone demand is usually satisfied in 30 s or less.[6]

Ozone, unlike chlorine and the other halogens, is not particularly sensitive to pH within the range of pH 5–8, but is significantly affected by temperature. As temperature increases both the rate and degree of completion of oxidation reactions is likely to increase, hence ozone demand may increase with temperature. The spontaneous breakdown of ozone to molecular oxygen likewise increases with temperature while the solubility of the gas in water decreases. The combination of these factors may require large increases in dosage at high temperatures.

The disadvantages of ozonation, which have restricted its use, are its cost relative to chlorine, the need to generate it at the point of use, and its spontaneous decay which prevents maintenance of a residual in the distribution system. The demonstrated production of halogenated hydrocarbons by present chlorination practice is likely to make these disadvantages of lesser importance.

Ozone is manufactured by electrical discharge into cooled dried air. Approximately 1 percent of the atmospheric oxygen is converted to O_3 at an energy consumption of 0.025 kWh per gram O_3. The mixture of air and ozone is transferred into the water either by bubbling it through the bulk solution or by permitting droplets of water to fall through a rising column of gas. Manufacture at locations remote from the point of use is impractical due to spontaneous decay of the stored gas.

The decay phenomenon which requires on-site generation also prevents maintenance of a residual ozone concentration which could guard against contamination of the distribution system by cross-connections or biological growths. A possible technique for providing protection within the distribution system involves both ozonation and chlorination.[9] Ozonation can provide both disinfection and, possibly, oxidation of the organic precursors of the halogenated hydrocarbons. Subsequent chlorination in modest dosages could provide residual protection without formation of troublesome by-products. It is possible that the problem of halogenated organic formation may be complicated by the presence of bromide in water. Bromide may be oxidized to bromine by either ozone or chlorine with subsequent possible formation of bromoform, etc. The actual degree of oxidation of organics by ozone is small, the major mechanism appears to be a change in chemical structure to organics which will not combine to form haloforms.

11-8 Other Disinfection Techniques

The other halogens, bromine and iodine, are also effective germicidal agents. The chemistry of these compounds is similar to that of chlorine, although the ionization constants involved, and thus the optimum pH, are different. Hypochlorous acid predominates at pH 7.4 and below, hypobromous acid at pH 8.7 and below, while hypoiodite is found only at very high pH.

Iodine, unlike bromine and chlorine, does not react with ammonia or organic nitrogen compounds to form amines, and thus persists as hypoiodous acid and molecular iodine. It is an effective disinfectant which has been applied in swimming pools, but is unlikely to be widely used in water treatment due to its possible physiological effects upon thyroid activity and its relatively high cost.[10]

A mixture of halogens, *monochloramine and iodide*, has been studied as a technique for disinfection.[11] The combination exhibited faster disinfection than chloramine alone, presumably from formation of hypoiodous acid by the chloramine and iodide. The iodine-iodate system is a better disinfectant at neutral pH values than are chloramines. The combination proved to be more effective than either chloramine or iodine used alone.

Bromine, while an effective disinfectant, is more expensive than chlorine and is implicated in the formation of halogenated hydrocarbons. For these reasons and lack of broad experience in its use,[10] bromine is not expected to be commonly used as a disinfectant.

Chlorine dioxide and *bromine chloride* are halogen compounds which have been applied to the disinfection of water and wastes. Bromine chloride ionizes to HCl and hypobromous acid which may combine with ammonia to form bromamines.[12] It is not used in municipal water treatment, but has had applications in wastewater disinfection. Chlorine dioxide is particularly useful in taste and odor removal as well as in disinfection.[5,6] It is produced by the chlorination of sodium chlorite in a ratio of 1 mole chlorine to 2 moles chlorite. The resulting compound, ClO_2, does not react with ammonia, is unaffected by pH within the normal range encountered in water, and is a powerful oxidizing agent and germicide. Like ozone, it requires on-site generation and appears to be able to oxidize organics without formation of halogenated hydrocarbons,[6] by blocking active sites more quickly than halogen substitutions can occur. Unlike ozone, it is relatively persistant at ordinary temperatures and can provide residual protection.

Ultraviolet irradiation is effective in killing all types of bacteria and viruses. The mechanism is thought to be destruction of nucleic acids by the rays generated in mercury vapor-quartz lamps. The wavelength of the light should be 253.7 nm and the intensity 50 uv W/m^2 at a distance of 50 mm.[5] Minimum retention times are on the order of 15 s with water films less than 120 mm thick. The advantages of uv disinfection include ready automation, no chemical handling, short retention time, no effect upon chemical characteristics and taste, low maintenance, and no ill effect from overdoses. The disadvantages lie primarily in the lack of residual protection, relatively high cost, and need for low turbidity in the water to insure penetration of the rays. The process is used primarily in industrial applications and in small private water systems. Efficiency of uv production in modern lamps is about 30 percent.

Extreme values of pH, either high or low, can provide good bacterial kills. Precipitation of magnesium (associated with a pH of 11) can give coliform reductions of more than 99.9 percent. Mineral acids, if present, can produce similar reductions but, like lime, are not added for purposes of disinfection. Bacterial reduction by pH variation is incidental to other processes.

Heat can be used to disinfect water but the method is impractical on a large scale. Continuous flow pasteurization has been successfully used on small scale systems[13] but is high in cost and provides no residual protection.

Ultrasonic waves at frequencies of 20 to 400 kHz have been demonstrated to provide complete sterilization of water at retention times of 60 min and very high percentage reductions at retention times as low as 2 s. Costs have been excessive, but combinations of short-term sonation and ultraviolet light might be economically attractive.[14]

Metallic ions such as silver, copper, and mercury exhibit disinfecting action. Silver is effective in concentrations less than those known to be harmful to human health (0.05 mg/l). The advantages of silver include its low effective dose, ease of application, and residual protection. The disadvantages include the possibility of silver adsorption on colloidal materials, possibility of inhibition or precipitation by other chemicals, reduction in efficiency with temperature, and relatively high cost.

Quaternary ammonium compounds are potent disinfectants but have not been accepted for potable waters due to cost, possible toxic effects, and their objectionable taste.[13]

Other oxidizing agents such as potassium permanganate, which are used in iron and manganese removal or for taste and odor control exhibit germicidal activity, but are seldom used as primary disinfectants.

11-9 Algae Control

The characteristics of algae have been discussed in Art. 8-13. Control of algae in small basins may be obtained by covering them to exclude sunlight. Many reservoirs and basins are so large that this is impractical, necessitating the use of chemicals. Copper sulfate is widely used for this purpose. The amounts necessary to kill various types of algae are shown in Table 11-1. Some waterworks men, however, use a standard dose, such as 1 mg/l for all types. Smaller amounts are often successful as prophylactic doses to prevent trouble. Prevention is to be preferred since killing a heavy algae growth with copper sulfate is likely to be followed by a temporary intensification of tastes and odors. Care must also be taken to prevent overdosage in bodies of water that support fish life, or many will be killed. Table 11-2 shows the lethal doses of copper sulfate for various fish. In large reservoirs the copper sulfate is applied from boats by means of sacks or screened hoppers which contain the crystals and are dragged through the water. The boat systematically follows parallel paths some 5 m (15 ft) apart, and wave action is depended upon for further dispersion. The amount of chemical needed may be based upon a treatment of the top 2 m (6 ft) of the water, although some troublesome diatoms may flourish as low as 6 to 10 m (20 to 33 ft) below the surface. Another method of application consists of blowing the crystals onto the water surface from a boat, using material passing a 1-mm screen. The largest of such particles penetrated as far as 5 m (15 ft) before completely dissolving.

In waters high in alkalinity copper may be precipitated as the carbonate. Addition of sodium citrate will complex the copper as the citrate which is soluble in waters of high alkalinity.[15]

Chlorine is also useful in controlling algae. Table 11-1 shows the amounts required by various organisms. If the trouble is occurring in a small collecting basin or reservoir, the water may be chlorinated before it enters. Prechlorination at rapid filter plants may be necessary from the standpoint of algae control in the plant. The entire contents of a reservoir may also be chlorinated using a portable gas chlorinator installed near the inlet. Combination of chlorination and dosing with copper sulfate has reportedly given excellent results.[16]

11-10 Aeration

Aeration of water has a number of useful functions. It readily removes hydrogen sulfide. Odors caused by the gases due to organic decomposition are partly removed, but odors and tastes from volatile oil secretions of living organisms and

Table 11-1 Quantity of copper sulfate and chlorine required for different organisms

Organisms	Odor	Copper sulfate, mg/l	Chlorine, mg/l
Chrysophyta:			
Asterionella	Aromatic, fishy, geranium	0.10	0.5–1.0
Melosira	....	0.30	2.0
Synedra	Earthy	1.00	1.0
Navicula	....	0.07	
Chlorophyta:			
Conferva	....	1.00	
Scenedesmus	....	0.30	
Spirogyra	....	0.20	0.7–1.5
Ulothrix	....	0.20	
Volvox	Fishy	0.25	0.3–1.0
Xygnema	....	0.70	
Coelastrum	....	0.30	
Cyanophyta:			
Anabaena	Moldy, grassy, vile	0.10	0.5–1.0
Clathrocystis	Grassy, vile	0.10	0.5–1.0
Oscillaria	....	0.20	1.1
Aphanizomenon	Moldy, grassy, vile	0.15	0.5–1.0
Protozoa:			
Euglena	....	0.50	
Uroglena	Fishy, oily	0.05	0.3–1.0
Peridinium	Fishy	2.00	
Chlamydomonas	....	0.50	
Dinobryon	Aromatic, violet, fishy	0.30	0.3–1.0
Synura	Cucumber, fishy, bitter	0.10	0.3–1.0
Schizomycetes:			
Beggiatoa	Putrefactive	5.00	
Crenothrix	Putrefactive	0.30	0.5

solution of decomposed organic matter or industrial wastes are only slightly affected. Simple chlorine odor can be removed, but odors due to combinations of chlorine and organic matter are not significantly removed. Carbon dioxide can be removed to the extent of 70 percent or more, and corrosiveness will be remedied to some degree. When used for these purposes, aeration is primarily considered as a physical or mechanical action by which the undesirable matter is swept out of the water and replaced with oxygen and other gases of the air. It is also used in removal of iron and manganese, in which case the oxygen made available from the air unites chemically with the metals to oxidize and precipitate

Table 11-2 Fish tolerance to copper sulfate

Fish	mg/l
Trout	0.14
Carp	0.30
Suckers	0.30
Catfish	0.40
Pickerel	0.40
Goldfish	0.50
Perch	0.75
Sunfish	1.20
Black bass	2.10

them. It is also used for mixing chemicals with water and for flocculation with the use of diffused compressed air. The aeration methods used depend upon the material to be removed, efficiency desired, and local conditions as to head, etc. Efficiency will depend directly upon the concentration in the water of the gas to be removed, the ratio of the exposed area to the volume of the water, the time of aeration, the diffusivity of the gas, and the water temperature.

Aerators, where the water is not filtered later, should be in insect-tight enclosures to prevent midge flies laying eggs which later hatch into worms. Coliform contamination has also been noted after aeration. The various aeration methods will be discussed briefly.

Spray nozzles provide a large air-water surface area but exposure time is short, 2 s or less. They require considerable head and so much space that housing is difficult and cold weather operation may be impossible. They may reduce carbon dioxide by as much as 90 percent. They are usually operated at heads of 70 to 140 kPa (10 to 20 psi). The amount of water discharged per nozzle will depend upon nozzle design and head used. A nozzle of good design having a 25-mm (1-in) orifice will discharge 0.3 m^3/min (80 gpm) to a height of 2 m (6 ft) at a pressure of 70 kPa (10 psi).

Cascades consist of a flight of three or four concrete or metal steps over which the water tumbles in a thin sheet. Some types have low weirs at the periphery of the metal steps. The cascade is less used than formerly. It is of little or no use in reducing algae odors but may reduce CO_2 from 20 to 45 percent. Head required will be 1 to 3 m (3 to 10 ft).

Multiple-tray aerators consist of a series of trays, each containing 200 to 300 mm (8 to 12 in) of coke, slag, stone, or ceramic balls 50 to 150 mm (2 to 6 in) in size, supported by slats, perforated plates, or screens. Some aerators, however, use no contact media and depend only upon the slats, etc. Usually three beds are used 500 mm (20 in) apart, and application is by means of spray nozzles to the top tray or, more commonly, by means of a shallow pan with a perforated bottom, with perforations 5 to 12.5 mm ($\frac{3}{16}$ to $\frac{1}{2}$ in) in diameter 75 mm (3 in) on centers. Application rates are from 0.04 to 0.20 m^3/min per m^2 of total tray area (1 to 5 gal/min per ft^2), depending upon the amount and kind of the dissolved gases to be removed. Low rates and fine material, say 40 to 60 mm in size, should be used where the gas is highly concentrated. The removal of carbon dioxide can be approximated by the following formula:

$$C_n = C_o e^{-Kn} \tag{11-7}$$

in which C_n is the concentration of carbon dioxide in milligrams per liter after passing through n trays, including the distribution pan, K is a coefficient and C_o is the carbon dioxide in milligrams per liter in the original water. K varies from 0.28 to 0.37, the larger value being applicable with good ventilation.

Diffused-air aerators consist of concrete tanks with depth of 3 to 5 m (10 to 18 ft), width 3 to 10 m (10 to 33 ft), and length to give the desired detention period, which is from 5 to 30 min. The air is applied by means of diffusers of the same type as used in treating sewage (Chap. 24), which may be placed in the tank bottom or mounted on the tank walls. In either case the air is broken up into fine bubbles, and the diffusers are placed on one side of the tank so that a spiral flow is set up that prolongs contact of the bubbles and the water. Amounts of air used vary from 0.04 to 1.5 m^3/m^3 of water. This type of aerator has the advantage of providing a long contact, eliminates freezing troubles, and may also combine mixing and flocculation.

11-11 Prevention of Odors

Tastes in water, except those caused by salts, are practically indistinguishable from odors. Odors in water are caused by (*a*) dissolved gases, like hydrogen sulfide; (*b*) organic matter derived from algae and other microorganisms, either living or in process of decay; (*c*) decomposing organic matter in general; (*d*) industrial wastes of which phenol is commonly the most troublesome; (*e*) chlorine, either as residual chlorine or in combination with other substances like phenol or decomposing organic matter.

Aeration, as indicated above, is useful in removing or reducing hydrogen sulfide but is far less so for algal or other organic odors. It will remove residual chlorine, though it is rarely used for this purpose, but has less effect upon combination odors. These and the residual odor may be eliminated, however, by using breakpoint chlorination or may be prevented by using chloramines.

Odors, except those caused by free chlorine alone, can be removed by an oxidizing agent. The small amounts of chlorine generally used for disinfecting will not oxidize odor-producing organic matter but only cause unpleasant combination odors. The operator, therefore, is tempted to reduce the disinfecting chlorine dose to prevent complaints. The alternative is to increase the chlorine to the breakpoint level (Art. 11-3).

Potassium permanganate has been used with some success to oxidize odor-producing material. Usually only 0.05 to 0.10 mg/l of permanganate is required, and it may be added to the raw, filtered, or filtered chlorinated water. Ozone also has an odor oxidizing effect, and may be more widely used in the future due to its lesser role in the formation of halogenated hydrocarbons.

11-12 Activated Carbon

This has become the most important of the available agents for overcoming odor troubles. Activated carbon is made from lignite, paper-mill waste, sawdust, and similar materials by heating in a closed retort and oxidizing, or "activating," by means of air or steam to remove the hydrocarbons which would interfere with adsorption of organic matter. It is very porous and has many carbon atoms with free valences. It is obtainable under several trade names, in granular form of various sizes or as a fine powder. The carbon accomplishes its purpose by its adsorptive properties, the highly porous structure and large surface area permitting the carbon to remove and hold many of the impurities of water. Molecules in solution are attracted to, and may be held by, a surface in contact with the solution. The forces holding a molecule against the surface may result from chemical bonding or van der Waals attraction. Adsorption will remove gases, liquids, and solids from solution with the rate of reaction and completeness of removal being dependent upon pH, temperature, initial concentration, molecular size, molecular weight, and molecular structure.

Adsorption is greatest at low pH since at low pH activated carbon is positively charged due to adsorption of hydrogen ion, while most colloids and all ionized polar groups on organic molecules are negatively charged.

Although adsorption is an exothermic reaction it proceeds more rapidly at elevated temperature due to increased diffusion of molecules into the fine pores of the carbon. The amount removed at equilibrium, if equilibrium is attained, will, however, be lower at high temperatures.

The rate of removal increases with increasing concentration of pollutant but decreases with increasing molecular size, molecular weight, and complexity of molecular structure. In mixtures of pollutants a degree of preferential adsorption may occur with smaller lighter molecules being more completely removed.

The relative capacity of different carbons is best assessed by trial upon the water to be treated. For screening purposes many manufacturers will test their product upon standard solutions to yield iodine, molasses, or phenol numbers. These values indicate the ability of a particular carbon to adsorb, respectively, small, large, and complex molecules. The number represents the number of milligrams of a particular compound adsorbed per gram of carbon at a specified equilibrium concentration. Values of these indices generally are greater than 1000.

Activated carbon has been used in water processing primarily as a short-term treatment to correct seasonal taste and odor problems. Fixed-bed systems containing granular carbon could be used for this purpose, as they are in tertiary treatment of wastewater, but the capital investment is seldom justifiable. Attempts to combine filtration and adsorption within a single bed are not satisfactory, since activated carbon changes significantly in effective specific gravity as its adsorptive capacity is exhausted. Spent carbon thus migrates to the bottom of the bed where it may release contaminants, and from which it is difficult to remove. Fixed-bed granular activated carbon systems are discussed in Chap. 26.

Powdered activated carbon is generally less than 0.075 mm in size and thus has an extremely high ratio of area to volume. Since adsorption is a surface phenomenon, this increases its effectiveness, but also makes it slow to settle and difficult to remove once added. Points of application include the raw water as early as possible in the plant, the mixing basin, split feed: with a portion in the mixing basin and the balance just ahead of the filters, and ahead of the filters, either at constant rate or at a heavy rate immediately after the filter is washed, followed by a light rate. In a few instances carbon has been used in impounding reservoirs to reduce algae or other odors. In this case it is applied as a slurry and sprayed over the water surface at 1 to 10 g/m^2 (10 to 100 lb/acre).

Dosing is accomplished by the ordinary dry feed machines used for coagulants with a special device using compressed air to prevent "flooding," which is an uncontrolled running of the carbon through the feeding device. The feeder should be in a separate room to segregate the dust, and it should have a hood into which sacks of carbon can be emptied with a minimum escape of dust. The dosages used vary from 0.25 to 8 mg/l with 1 to 2 mg/l most common. When added before sedimentation, the activated carbon not only has the opportunity of removing matter from the water, but that portion which settles will continue to adsorb products of decomposition in the sludge and prevent their appearance in the water above. Some will go over into the sand of the filters where it will act upon the water passing through. During washing, the carbon will be removed, and the lessening effect upon filter runs is negligible.

The required dosage of activated carbon is frequently controlled by means of the threshold odor test. It consists essentially of comparing varying dilutions of an odor-bearing water (dilutions made with odor-free water) to an odor-free standard. The dilution at which the odor can just be detected is called the *threshold point*. This is expressed quantitatively by the threshold number, which is the number of times that the odor-bearing sample is diluted with odor-free water. The standard amount tested is 250 ml; and if 50 ml of the odor-bearing sample, when diluted with odor-free water to 250 ml, produces a barely perceptible odor, the threshold number is 5. As a result of observations of the water supply it is decided what threshold number should not be exceeded; and if the tap samples approach or exceed the maximum threshold desired, activated carbon is used or, if already being used, the dose is increased.

Although activated carbon can be regenerated this is not commonly done in water plants adding carbon on a seasonal basis. If the carbon is to be recovered from the sludge, polyelectrolytes must be used as the sole coagulant to prevent accumulation of inerts. Wet oxidation systems (Chap. 26) have been adapted to the regeneration of such powdered carbon slurries.

11-13 Removal of Iron and Manganese

Both well and surface waters, but particularly the former, are likely to have iron dissolved in them as the result of carbon dioxide coming in contact with iron salts to form soluble ferrous bicarbonate. Manganese may be associated with the iron and in this case removal is more difficult. When they are present in amounts greater than 0.3 mg/l either alone or as a total, the following effects may be noted: an unpleasant taste, a reddening of the water which may cause staining of plumbing fixtures and clothing, accumulations of precipitated iron in the water mains, growths of crenothrix in the mains, and accompanying odors in the water. Manganese stains are dark brown rather than red. Iron may also enter the water through corrosion of the water pipes, in which case prevention methods must be aimed at stabilization of the water (Art. 11-22).

Iron and manganese are both removed in water softening, but it is not always desirable to soften water solely for removal of relatively small quantities of these metals. Their removal can be facilitated by their oxidation to a higher valance state in which their solubility is reduced. The basic reactions of interest are:

$$Fe^{++} \rightarrow Fe^{3+} + e^- \qquad (11\text{-}8)$$

and

$$Mn^{++} \rightarrow Mn^{4+} + 2e^- \qquad (11\text{-}9)$$

The oxidizing agent used to effect the reactions can be atmospheric oxygen, chlorine, ozone, permanganate, or any other strong oxidant which will not leave an unwanted residue.

Iron alone, in groundwaters containing no organic matter, can be removed by simple aeration followed by sedimentation and filtration. The aeration can be by any of the techniques discussed in Art. 11-10, but is most effective in towers containing slats or trays which may be either seeded with catalytic material such as pyrolusite (MnO_2) or which are permitted to accumulate deposits of Fe_2O_3 which also serve to catalyze the oxidation reaction. The sedimentation basin serves primarily as a retention device which permits further oxidation of iron by the oxygen in the aerated water. Precipitated iron is removed by the filter.

Bacterial growth in filters used to remove iron and manganese can lead to reduction of the oxidized metals and their return to solution. If sufficient biological activity exists to produce anaerobic conditions, the oxidized metals may be used as electron acceptors. This difficulty can be avoided in part by more frequent backwash and by maintenance of a chlorine residual through the filter. If a filter has developed heavy biological growth this can be removed by heavy dosage with chlorine or permanganate.[17]

If *both iron and manganese* are present, or if an iron-bearing water contains organic matter, aeration will be sufficiently rapid only when catalyzed by pyrolusite or by an accumulation of oxidation products (Fe_2O_3 and MnO_2) upon a porous bed such as coke. The other simple aeration processes will not provide oxidation within a reasonable time, although elevation of the pH by lime dosage will increase the rate of reaction. Manganese is much more slowly oxidized than iron, the rate being negligibly slow at pH levels below 9.[17] Removal upon oxide deposits in aeration towers is by adsorption followed by slow oxidation.

Organic materials in water may interfere with iron removal by peptization, i.e., by complexing without complete neutralization of molecular charge, by forming soluble complexes with both oxidized and reduced iron, by reduction of ferric to ferrous iron, or by a combination of these.[7]

The application of strong oxidizing agents such as chlorine, ozone, chlorine dioxide, or potassium permanganate can serve to oxidize manganese more rapidly and to destroy organic molecules which might interfere with the metallic oxidation. Permanganate will function satisfactorily at neutral pH in oxidation of manganese, the others require a pH above 8.5 for reasonable rates of manganese removal. Iron is removed adequately at neutral pH but its removal is enhanced at higher pH.

Manganese and iron can be removed by adsorption upon a bed of pyrolusite. This process is sometimes called the *manganese zeolite* process although it does not involve an ion exchange reaction. The adsorbed iron and manganese is periodically oxidized by dosing the bed with permanganate. Systems employing this process are widely used in individual home treatment units.

The aesthetic problems associated with iron and manganese in water can be reduced by sequestering these metals with polyphosphates. The phosphates must be added prior to any other treatment and are not effective at total Mn plus Fe concentrations above 1 mg/l. Dosages are approximately 5 mg/l sodium hexametaphosphate per milligram Mn plus Fe.

11-14 Water Softening

A description of hardness producing compounds is presented in Art. 8-19. In effect, hardness consists of divalent metallic cations, which, while not undesirable from a standpoint of ingestion, may contribute properties which interfere with nonpotable uses, both industrial and domestic.

In addition to reducing the concentration of calcium and magnesium, standard softening processes contribute to the removal of iron and manganese, organic material, bacteria, viruses, and other suspended solids. The latter substances can, of course, be removed by other techniques.

Municipal softening provides consumer savings due to reduction in soap consumption, longer life of water heaters, and less scale accumulation in pipes. Industries receiving softened water will realize savings associated with their own treatment processes since softened water is more easily demineralized if this is required, and may be directly usable for cooling purposes.[18] The economy of scale possible in municiple treatment may be attractive to industrial users.

From a health standpoint there is increasing evidence that the presence of hardness, particularly calcium, is desirable. In Great Britain the government does not recommend softening and will not participate in the financing of water-softening plants.[19] Ordinary lime-soda softening cannot yield total removal of hardness and is generally operated to produce a water with a total hardness of about 100 mg/l. Greater reductions are uneconomic for public water supplies and may be unhealthy as well.

11-15 The Lime-soda Method

The chemical dosages required to precipitate calcium and magnesium are best calculated by consideration of the ionic balance of a water presented on a bar diagram like that of Fig. 11-3.[20] The concentrations of the various chemical species determined from a water analysis are expressed in meq/1 (milliequivalents per liter) and drawn to scale as shown. The species which are important to the analysis include Ca^{++}, Mg^{++}, HCO_3^-, $CO_3^=$, and CO_2. The remaining ions may be shown as "other species."

Calcium is precipitated as the carbonate in accord with the equation

$$Ca^{++} + CO_3^= \rightarrow CaCO_3 \tag{11-10}$$

Provision of a carbonate ion concentration equal to that of the calcium will permit nearly complete combination. The theoretical solubility of $CaCO_3$ in waters containing equal calcium and carbonate concentrations is approximately 7 mg/l, however, in practice, residual concentrations are generally 35 mg/l or more.

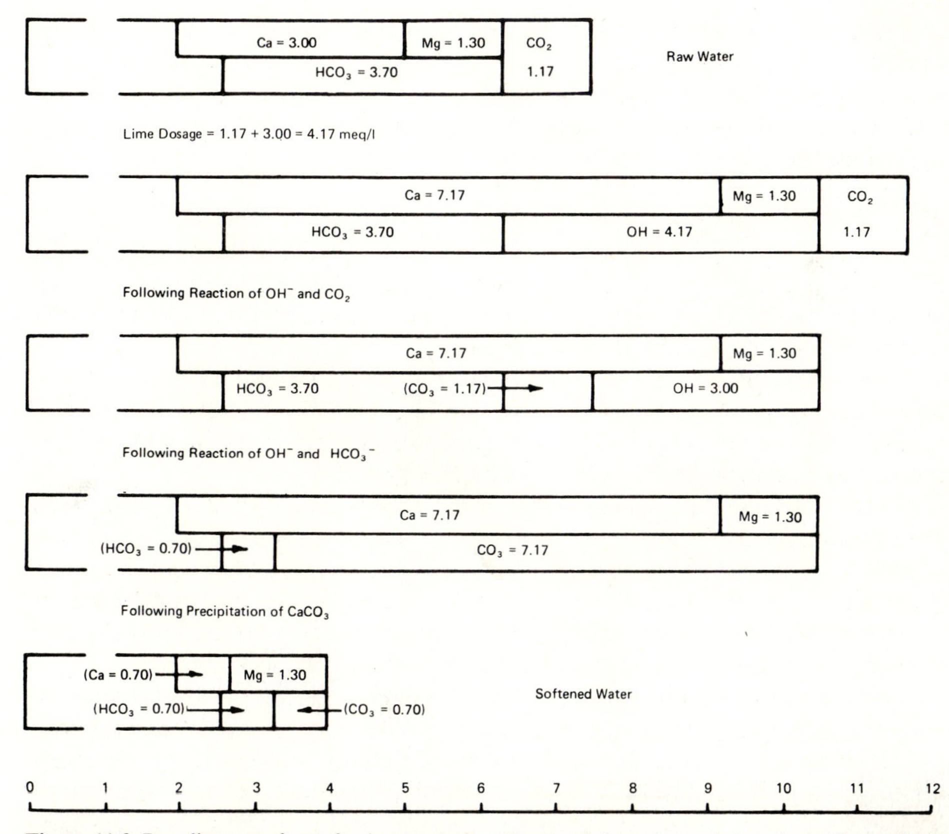

Figure 11-3 Bar diagrams for softening analysis. *(Reprinted from Journal American Water Works Association* **67**, *by permission of the Association. Copyright 1975 by the American Water Works Association, Inc. 6666 W. Quincy Avenue, Denver, Colorado 80235.)*

The carbonate ion required can be added directly if necessary, but it is generally more economical to produce it by reaction of lime with the bicarbonate alkalinity. If the quantity available from this reaction is inadequate, soda-ash (Na_2CO_3) is used to supply the deficit. If a final calcium concentration greater than 35 mg/l (as $CaCO_3$) is desired this can be achieved by providing a carbonate concentration 35 mg/l as $CaCO_3$ greater than the required removal of calcium. This is not precisely correct theoretically or practically but will give a first order approximation to the dosage—which requires adjustment in any event.

The chemical equilibria involved in waters containing the carbonate-bicarbonate buffer system are complex. The pH of maximum carbonate concentration is a function of the total alkalinity and temperature as shown in Fig. 11-4, and also of the total dissolved solids concentration.[21] The total alkalinity and dissolved solids, it should be noted, are increased in proportion to the quantity of lime added and change during the softening reactions. A first order approximation

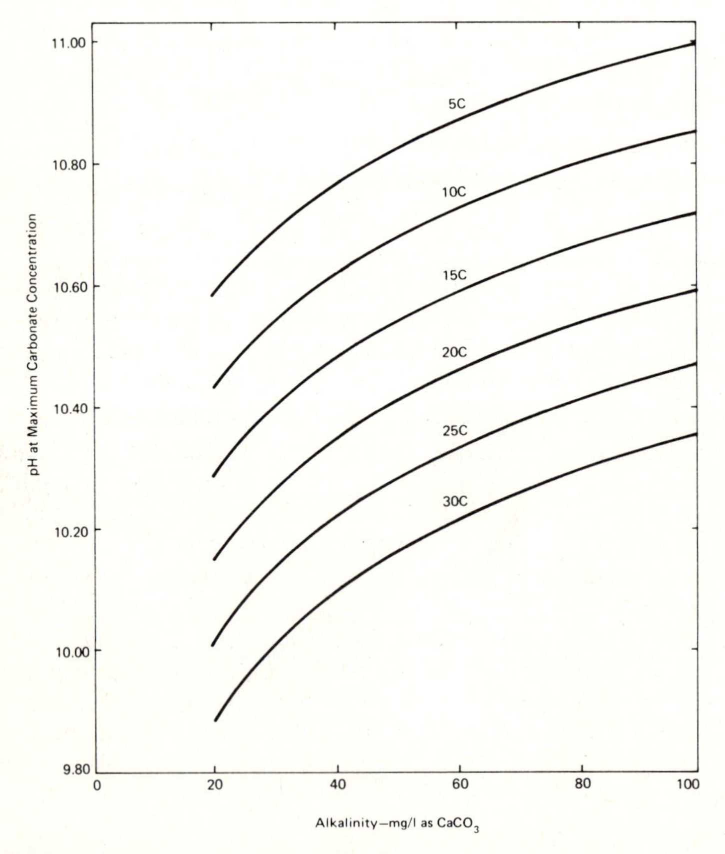

Figure 11-4 Alkalinity vs. pH at maximum carbonate concentration at various temperatures for TDS = 200 mg/l. *(Reprinted from Journal American Water Works Association* **68,** *by permission of the Association. Copyright 1976 by the American Water Works Association. Inc., 6666 W. Quincy Avenue, Denver, Colorado 80235.)*

to the required dosages can be calculated from the bar diagram, with the dosage then being adjusted in the laboratory to yield the optimum pH based upon alkalinity and total dissolved solids.

Example Calculate the necessary amounts of lime and soda-ash to soften a water of the following composition: Total hardness as $CaCO_3$, 215 mg/l; Alkalinity as $CaCO_3$, 185 mg/l; Mg as Mg, 15.8 mg/l; Sodium as Na, 8 mg/l; Sulfate as SO_4, 28.6 mg/l; Chloride as Cl, 10.0 mg/l; Nitrate as N, 1.0 mg/l; Carbon Dioxide as CO_2, 25.8 mg/l; pH 7.07.

SOLUTION A large portion of the data given are not required. The pertinent values are:

$$\text{Total Hardness} = 215 \text{ mg/l as } CaCO_3 = 4.30 \text{ meq/l}$$

$$\text{Magnesium} = 15.8 \text{ mg/l as Mg} = 1.30 \text{ meq/l}$$

$$\text{Calcium} = \text{Total Hardness} - \text{Mg} = 3.00 \text{ meq/l}$$

$$CO_2 = 25.8 \text{ mg/l as } CO_2 = 1.17 \text{ meq/l}$$

$$\text{Alkalinity} = 185 \text{ mg/l as } CaCO_3 = 3.70 \text{ meq/l}$$

Since the pH is close to 7, it may be concluded that the alkalinity is entirely bicarbonate [at pH 7, $CO_3 \approx 0.001\ (HCO_3)$]. The bar diagram can then be constructed as shown in Fig. 11-3.

If the final hardness desired is 100 mg/l as $CaCO_3$ (2 meq/l) it is evident that Mg removal will not be required. This may be concluded since the calcium concentration can be reduced to 35 mg/l as $CaCO_3$ (0.7 meq/l) and this, plus the influent magnesium will meet the required standard. Lime will be required to react with the CO_2 and to convert the bicarbonate to carbonate to whatever degree is necessary. The carbonate concentration needed for removal of 2.3 mg/l Ca is 3.0 meq/l (the required removal plus a 35 mg/l excess). This can be provided from the bicarbonate conversion since HCO_3 exceeds 3.0. If it did not, the excess would be supplied by soda-ash.

The required lime dosage is thus equal to the CO_2 concentration plus the necessary CO_3 production or, $1.17 + 3.00 = 4.17$ meq/l. The effect of the addition of this dosage is shown in the later bar diagrams of Fig. 11-3. The first reactions occur rapidly while the precipitation is slow. The concentrations shown in the last diagram are those which would be obtained after mixing and sedimentation. The effluent pH, based upon the carbonate and bicarbonate concentrations shown, may be calculated to be 10.3.

Magnesium is precipitated as the hydroxide in accord with the equation

$$Mg^{++} + 2OH^- \rightarrow Mg(OH)_2 \qquad (11\text{-}11)$$

in order to obtain effective removal, excess hydroxyl ion must be added, usually about 1 meq/l. As with calcium, exact doses must be determined empirically. With an excess hydroxyl ion concentration of 1 meq/l the practical solubility of

$Mg(OH)_2$ is 10 mg/l as $CaCO_3$ (0.2 meq/l). If final magnesium concentrations higher than 10 mg/l are desired they are obtained by treating only a portion of the flow and bypassing the remainder. The portion to be treated depends upon the influent and effluent magnesium concentrations and may be calculated from

$$X = \frac{Mg_i - Mg_e}{Mg_i - 0.2} \tag{11-12}$$

in which Mg_i and Mg_e are the influent and effluent concentrations of magnesium in meq/l and X is the fraction of the flow to be treated.

Example Assume that the water of the example above is to be softened to a final hardness of 2 meq/l with the additional restriction that the final magnesium concentration does not exceed 0.6 meq/l (30 mg/l as $CaCO_3$).

SOLUTION The proportion of the flow which must be treated for magnesium removal is:

$$X = \frac{1.3 - 0.6}{1.3 - 0.2} = 0.636$$

The lime dosage required will be that required to react with the carbon dioxide, to convert the HCO_3 to CO_3, to combine with the magnesium, plus an excess.

$$\text{Lime} = 1.17 + 3.70 + 1.30 + 1.00 = 7.17 \text{ meq/l}$$

This dosage is added only to the 63.6 percent of the flow being treated for magnesium removal. The ionic distribution in the water following the addition of lime, and the subsequent reactions are shown in Fig. 11-5. The initial chemical reactions are rapid while the precipitation is slow. The effluent from the magnesium-softening process will have the constituents shown in the fifth diagram of Fig. 11-5. This flow is then mixed with the bypass flow which still has the makeup shown in the first diagram. The constituents of the mixed flow are calculated as follows:

Calculation of ionic strength in the mixed flow

	Concentration in		
Ion	Bypass flow	Mg flow	Net concentration in mixed flow
Mg^{++}	1.30	0.20	1.30(0.364) + 0.20(0.636)
Ca^{++}	3.00	10.17	3.00(0.364) + 10.17(0.636)
CO_2	1.17	0	1.17(0.364) + 0
HCO_3^-	3.70	0	3.70(0.364) + 0
CO_3^{--}	0	8.57	0 + 8.57(0.636)
OH^-	0	1.20	0 + 1.20(0.636)

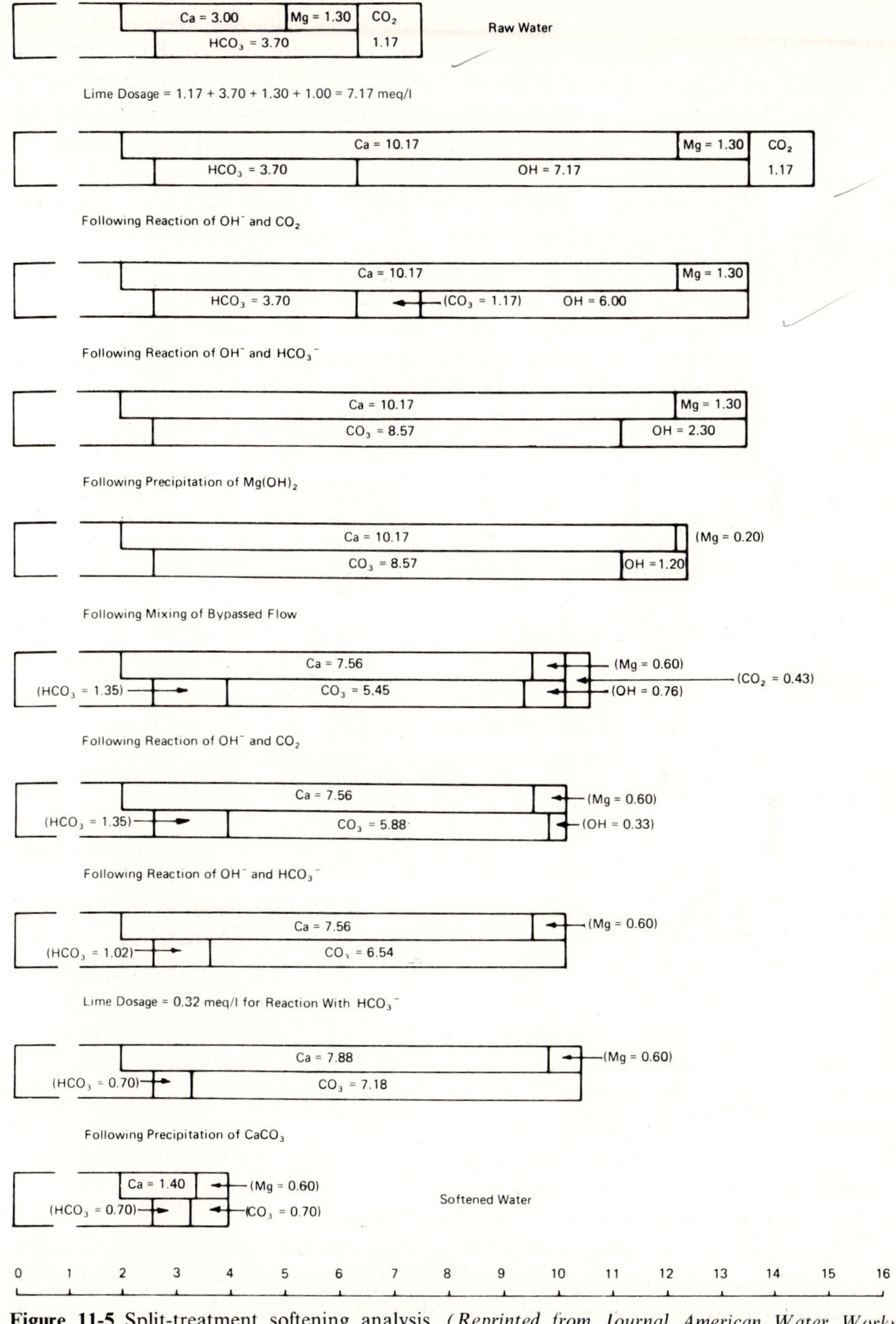

Figure 11-5 Split-treatment softening analysis. *(Reprinted from Journal American Water Works Association* **67,** *by permission of the Association. Copyright 1975 by the American Water Works Association, Inc., 6666 W. Quincy Avenue, Denver Colorado 80235.)*

The concentration in the mixed flow is shown in the sixth diagram of Fig. 11-5, with subsequent reactions below, as the hydroxyl ion from the treated flow combines with the bypassed flow. The resulting water in the eighth diagram must then be considered as an additional problem in calcium removal. In order to reduce the calcium concentration to 1.4 meq/l (for a final hardness of 2.0 meq/l) the CO_3 concentration must be 0.7 meq/l greater than the desired removal. The desired removal is $7.56 - 1.4 = 6.16$, thus the required CO_3 concentration is $6.16 + 0.7 = 6.86$ meq/l. Since 6.54 is present (diagram 8) 0.32 must be added and this can be produced from the HCO_3 present by adding 0.32 meq/l of lime. If no HCO_3 remained, the CO_3 could be provided by adding soda-ash. The effect of the lime addition and the subsequent reactions are shown. The resulting water has a final hardness of 2.0 meq/l as in the first example but the proportions of calcium and magnesium are changed. The total limc dosage is 7.17 meq/l to 63.6 percent of the flow plus 0.32 meq/l to the total flow. The net dosage is thus

$$7.17(0.636) + 0.32 = 4.87 \text{ meq/l}$$

This may be compared to the 4.17 meq/l required to obtain the same final hardness by calcium removal alone, and illustrates the economy possible by not removing magnesium.

Lime-soda softening plants usually include the following units:

1. *Feeding and mixing apparatus.* These are similar in design and operation to the feeding machines and mixing and flocculation basins used in water coagulation. A 40-min detention period, at least, is recommended.
2. *A settling basin.* The settling basins used resemble those employed for coagulation-sedimentation in rapid filter plants, but the detention periods are generally somewhat longer. Mechanical sludge-removal apparatus is usually employed.
3. *A recarbonation or stabilization unit.* Lime softening leaves water supersaturated with calcium carbonate. Unless prevented, this will precipitate later to cause enlargement of the sand grains in the filters, incrustation of mains, and stoppage of water meters. In the settling basin a dose of alum may be used to assist sedimentation, and the carbon dioxide generated will react with the excess calcium carbonate to hold it in solution as the bicarbonate. Alternately, after sedimentation, carbon dioxide generated from burning coke, gas, or oil, or delivered to the plant as a pressurized liquid, may be blown into the water. One type of recarbonator consists of an underwater burner or burners which burn a proper mixture of gas and air forced to them by a compressor. Other recarbonators include a combustion chamber; a washer or scrubber, in which the products of combustion are cleaned and cooled; a drier and trap which remove moisture coming over from the scrubber; a blower to raise the pressure above the hydrostatic head in the carbonation chamber; and a carbonation chamber of 15- to 30-min detention period, usually with perforated pipes in the bottom to obtain diffusion. Diffusers employ about 300 holes 1.5 mm in diameter per m^3 of gas per minute. Carbonation must be carefully controlled with relation to

the alkalinity and pH of the water to produce a water which is neither corrosive nor depositing. Alum or carbonation with fuel gas may be used to remove incrustation already upon sand or pipes.

4. *Filters.* While sedimentation will allow most of the precipitated material to settle, filtration is applied to insure complete clarification. Pressure filters are used in industrial water-softening plants and in some municipal plants softening groundwater. It is, of course, feasible to soften a water in the conventional rapid filter plant in which case the gravity filters will serve for final clarification. In a plant which is softening a groundwater and in which filtration is unnecessary from a health standpoint, recarbonation may make the filters unnecessary. Without recarbonation the filters, if used, will require more frequent washing and perhaps sand replacement.

Return of sludge to the mixing basin, or filtration of the mixed water through a blanket of sludge suspended in the coagulation tank increases the efficiency of crystallization and flocculation of the precipitates. The use of a sludge blanket requires tanks of special design with upward flow of the water through the slurry or suspended sludge (Art. 11-16). Excess lime treatment can be supplemented by the use of hexametaphosphate for stabilization. Sodium hexametaphosphate, which is sold under various trade names like Calgon and Micromet, has been used to prevent incrustation of pipes by lime-soda softened water. The chemical has the property of sequestering divalent cations and thus preventing the formation of crystals of calcium carbonate and loosening scale already in the distribution system. It is used in amounts varying from 0.5 to 5.00 mg/l and is applied after the softening process is completed. Where there are filters, it can be added before, in which case it should prevent incrustation of the sand grains, or it may be added after the filters. Other polyphosphates besides sodium hexametaphosphate have also been found useful in this field.

The chief advantages of lime-soda softening are the following: The total mineral content of the water is reduced, the pH of the treated water is increased, which reduces the corrosiveness of the water upon the distribution system, if the magnesium content of the water is high, there is more efficient coagulation of the water and better removal of suspended material thereby reducing the amount of coagulant needed, pathogenic bacteria may be reduced in numbers and iron and manganese are removed, and in many cases color caused by iron and organic matter is adsorbed.

There are, however, the following disadvantages: Large amounts of sludge are produced and must be disposed of, the operation requires close and skillful supervision, and unless recarbonation or stabilization is practiced, filtering media and the surfaces of the distribution system, meters, etc., will accumulate a coating of calcium carbonate.

Difficulties with sludge may require, or at least encourage, calcining it and obtaining lime to use in the plant. This is impractical if the sludge has turbidity from the water incorporated with it and accordingly the process is applicable only to groundwaters or to surface waters which have been separately clarified before softening. The process is presently used at only a few plants (see Art. 11-26).

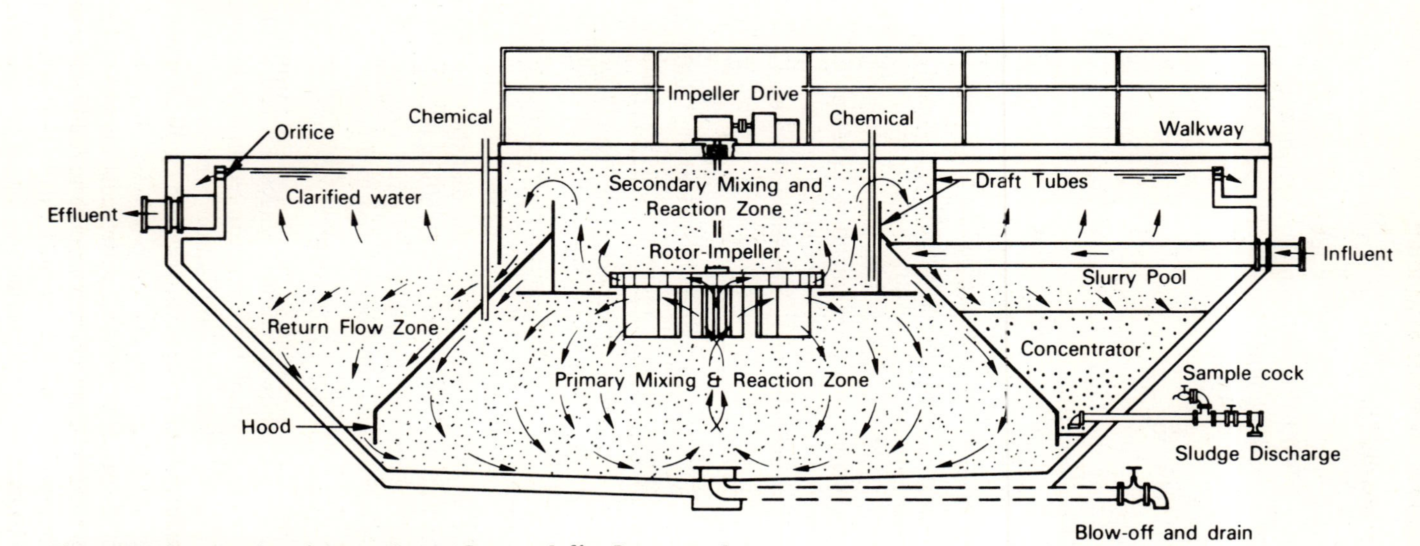

Figure 11-6 Cross section of the accelator. *(Courtesy Infilco-Degremont, Inc.)*

11-16 Suspended Solids Contact Units

These are now widely used at lime-soda filter plants and combine mixing, flocculation, and clarification in the same structure. The raw water enters a central chamber which also receives the chemicals. In this chamber there is provision for agitating and circulating the chemicals and the water and, as in Fig. 11-6 mixing and reaction zones may be formed. The water, precipitated chemicals, floc, and turbidity pass to a clarification chamber. In most types the effluent must pass upward through a blanket of suspended floc and other solids, known as slurry, before it can leave the tank at effluent weirs or launders. There may be provision for some of the slurry to recirculate to the mixing chamber. One or more sludge hoppers or concentrators provide a location from which the sludge is discharged continuously or upon a schedule either manually or automatically.

There are a number of distinctive features to this type of plant. The chemicals are introduced, not into the raw water, but into water which already has a slurry of previously formed precipitates. Precipitation of the newly arrived chemicals will take place on the surface of old particles, and the resulting flocculent material, aided by agitation, quickly forms larger masses. The raw water is introduced into this larger amount of previously formed slurry and is further agitated. The result is that the reactions with the water impurities and their incorporation in the floc masses are quickly accomplished. After this contact the water-slurry mixture goes to the clarifying portion of the tank. As the water rises, its vertical velocity decreases, and suspended material falls. The rising water passes, therefore, through a suspended filter and clarification is thus aided.

Costs are less for this type of structure than for the conventional type of plant. Treatment rates for softening are approximately 120 m/day (2 gal/min per ft^2) at the slurry separation level with detention periods, from 1 to 2 hours. The weir loading should not exceed 0.125 m^3/min per meter of weir length (10 gal/min per ft) and if orifices are used, they should produce uniform rising rates in the tank. Since this unit should not be overloaded, multiple units may be necessary to give safety and flexibility in operation.

Records from the many industrial plants and increasing number of cities using these units indicate that they are generally satisfactory. They require careful and skilled operation, however, particularly if surface water is being treated and turbidity increases greatly in a very short time. The short detention period requires quick change in chemical dosages and control of the slurry level in the sedimentation chamber, or large amounts of solid material may go to the filters.

11-17 The Cation-exchange Method

Zeolites used in water softening are complex compounds of sodium, aluminum, and silica which have the faculty of exchanging bases. They may be obtained and refined from natural deposits or made synthetically. Since the natural material is green in color, it is often known as greensand. The artificial product is lighter in color and usually has larger grains. It varies in size from 0.25 to 2.5 mm, with an effective size of 0.42 mm.

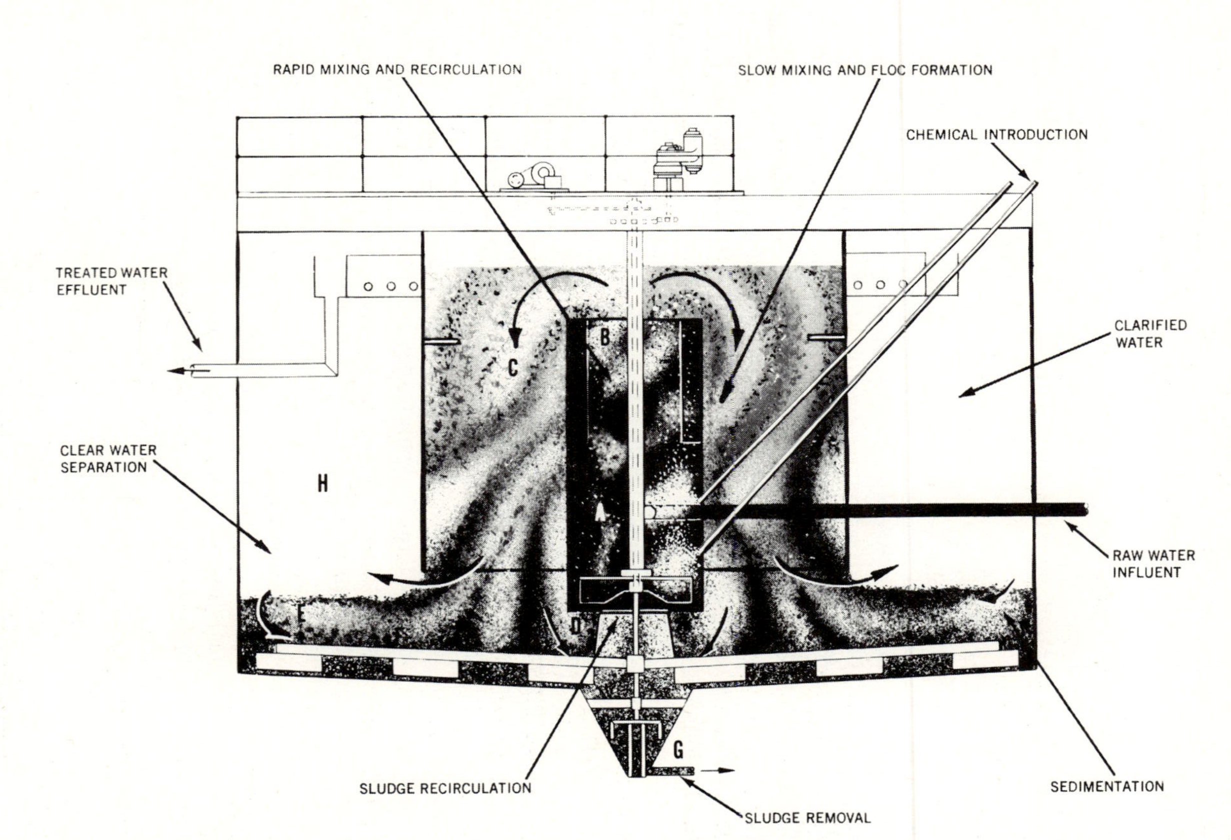

Figure 11-7 Ecodyne reactivator®. *(Courtesy Ecodyne Corporation-Graver Water Division.)*

SOFTENING

$$\underset{\text{Hard water}}{\begin{matrix}Ca\\Mg\end{matrix}\left\{\begin{matrix}(HCO_3)_2\\SO_4\\Cl_2\end{matrix}\right.} + \underset{\text{Sodium exchange bed}}{Na_2X} = \underset{\text{Soft water}}{Na_2\left\{\begin{matrix}(HCO_3)_2\\SO_4\\Cl_2\end{matrix}\right.} + \underset{\text{Exhausted exchange bed}}{\left.\begin{matrix}Ca\\Mg\end{matrix}\right\}X}$$

REGENERATION

$$\underset{\text{Exhausted exchange bed}}{\left.\begin{matrix}Ca\\Mg\end{matrix}\right\}X} + \underset{\text{Salt solution}}{2NaCl} = \underset{\text{Sodium exchange bed}}{Na_2X} + \underset{\text{Waste water}}{\left.\begin{matrix}Ca\\Mg\end{matrix}\right\}Cl_2}$$

Figure 11-8 Water softening by cation exchange and regeneration. X is the zeolite.

When water containing calcium and magnesium compounds is passed over the zeolite, the calcium and magnesium cations are removed, and the sodium is given up in exchange, thereby softening the water and at the same time increasing its sodium content. When the sodium of the zeolite is exhausted, it is regenerated by applying a solution of sodium chloride. An exchange is again effected, and the brine is withdrawn as a calcium and magnesium chloride solution (see Fig. 11-8).

Zeolite water softeners are occasionally arranged as gravity units but usually are constructed in the same manner as pressure sand filters. A unit consists of a closed steel cylinder containing a bed of zeolite which may be from 0.75 to 2 m (30 to 75 in) thick. The thicker bed has a greater capacity per unit area, takes up less space, and requires less water when washing is needed. It may operate with an upward or downward flow. The advantage claimed for upward flow is that compaction of the zeolite is prevented, greater efficiency is obtained, and washing is unnecessary. The unit must have a hard-water inlet, soft-water outlet, inlet and outlet for washwater (for downflow unit), salt-solution inlet and outlet, rinse-water outlet, rate-of-flow controller, sampling cocks, a support system for the

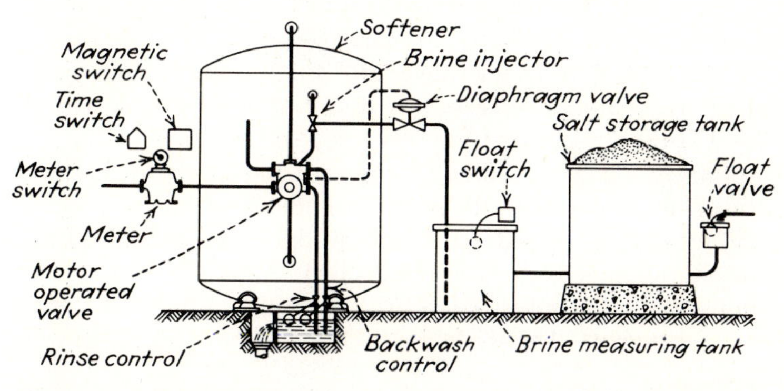

Figure 11-9 Diagram of Permutit Company's automatically controlled zeolite softening plant.

zeolite, and an underdrain system. The plant must also include provisions for salt storage and measurement of the brine as it is pumped to the softening unit.

The rate of water application varies directly as the thickness of the bed. For a bed 0.75 m (30 in) thick a rate equal to that of the rapid sand filter, 120 m/day (2 gal/ft^2 per min) is used. The length of time between regenerations will depend upon the character of the water and the rate of application.

The capacity of ion-exchange media has been historically expressed in grains or kilograins of $CaCO_3$ per cubic foot. The conversion to equivalents per m^3 involves multiplication by 45, thus a zeolite with a reported capacity of 4000 grains per cubic foot would exchange 4000/1000 × 45 or 180 equivalents per cubic meter. Natural greensands will remove 135 equivalents per cubic meter between regenerations, high-capacity greensands 230 to 250, and synthetic types between 275 and 550 equivalents. The need for regeneration can be noted when hardness begins to appear in the softened water. It is not advisable to exhaust the zeolite entirely, since this requires a relatively greater amount of salt for regeneration. The economical salt use apparently is from 125 to 175 grams per equivalent of hardness removed. Regeneration is accomplished by running a 5 to 10 percent salt solution into the unit from below, either continuously for greensand or on a batch basis for 5 to 15 min with synthetic zeolite. The amount of salt required is determined by estimating or calculating the total hardness removed and using salt in the amount suggested above. After regeneration, water is wasted to rinse out the salt, the wastage continuing until the water has a hardness of not more than 20 mg/l. Washing of zeolite beds is necessary only in the downflow types and is done at rates of 250 to 500 m/day (4 to 8 gal/ft^2 per min). It should precede regeneration, and at other times may be needed to remove precipitated iron and other suspended matter held in the bed or to overcome compaction. Hard or softened water may be used for this purpose.

Zeolite softening has a number of advantages over the lime-soda method: There is no sludge to dispose of, but the discharge of brine may be subject to governmental regulation in municipal or industrial plants. It permits a compact and easily operated plant and water of nearly zero hardness is obtainable. Water of zero hardness is not desired for municipal water supply but is useful in textile processes and for boiler waters. The hardness of the water can be closely regulated by mixing the treated water with bypassed untreated water.

If the sodium zeolite is used for iron removal the iron must be in solution. If it has been precipitated by aeration the zeolite will act as a filter and partially remove it. Clogging will then occur and the efficiency of the zeolite will be impaired. Beds so clogged can be improved by dosing with 10 percent hydrochloric acid. Proper dosage of the raw water with polyphosphate (Calgon) will prevent retention of iron and fouling of the beds. Iron removal by aeration and filtration before softening on ion exchange beds is recommended.

Zeolite softening is used by industrial plants and cities. Where salt is not costly it may be cheaper than the lime-soda method. It is not adapted to a water that is high in turbidity.

11-18 Hydrogen Exchange and Demineralizing

The zeolite previously described exchanges sodium for the calcium, magnesium, and iron in the water. The sodium of the zeolite is renewed by the use of sodium chloride. Other zeolites are available which exchange all the cations, including sodium, for hydrogen. These are sometimes called carbonaceous zeolites since they are made from such materials as coal and lignite. They are regenerated by the use of sulfuric acid or hydrochloric acid. The water treated in the hydrogen zeolite has the sulfates, chlorides, and bicarbonates transformed into sulfuric, hydrochloric, and carbonic acids. This acidity in the treated water can be neutralized by mixing with it the proper proportion of alkaline raw water to obtain the desired characteristics in the treated water. The hydrogen zeolite also has the advantage of containing no silica, and hence the treated water does not have its silica content increased, which is especially desirable for boiler waters.

The carbonaceous zeolites can also be operated on the sodium cycle, i.e., regeneration being accomplished by sodium chloride. This has the advantage of producing a water low in silica, and the process is used for treating low-silica waters where it is desired to keep the silica content low. Otherwise, results are as with the common zeolite.

For certain industrial uses the acids of the effluent from the hydrogen zeolite may be undesirable. The sulfuric and hydrochloric acids can be removed by passing the water through a weakly basic exchange material, which removes the sulfate, chloride, and nitrate anions and leaves the water practically demineralized. The carbonic acid must be removed by aeration. A strongly basic exchanger is available that will also remove such weak acids as H_2CO_3 and H_2SiO_3, but its regeneration is more costly (see Fig. 11-11). It is regenerated by means of sodium carbonate or caustic soda. The effluent from the two treatment steps

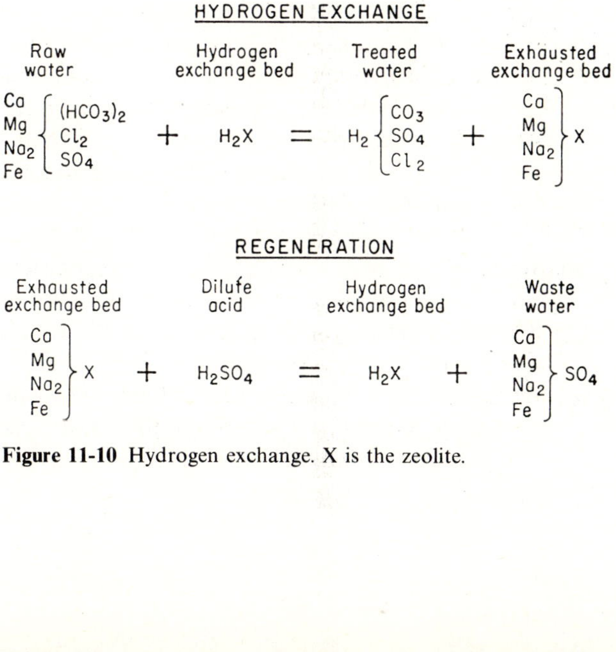

HYDROGEN EXCHANGE

Raw water		Hydrogen exchange bed		Treated water		Exhausted exchange bed
$\left.\begin{matrix}Ca\\Mg\\Na_2\\Fe\end{matrix}\right\}\left\{\begin{matrix}(HCO_3)_2\\Cl_2\\SO_4\end{matrix}\right.$	+	H_2X	=	$H_2\left\{\begin{matrix}CO_3\\SO_4\\Cl_2\end{matrix}\right.$	+	$\left.\begin{matrix}Ca\\Mg\\Na_2\\Fe\end{matrix}\right\}X$

REGENERATION

Exhausted exchange bed		Dilute acid		Hydrogen exchange bed		Waste water
$\left.\begin{matrix}Ca\\Mg\\Na_2\\Fe\end{matrix}\right\}X$	+	H_2SO_4	=	H_2X	+	$\left.\begin{matrix}Ca\\Mg\\Na_2\\Fe\end{matrix}\right\}SO_4$

Figure 11-10 Hydrogen exchange. X is the zeolite.

STRONG BASE ANION EXCHANGE

Cation exchange effluent		Anion exchange bed		Treated water		Exhausted exchange bed
$H_2 \begin{cases} SO_4 \\ Cl_2 \\ (NO_3)_2 \\ SiO_3 \\ CO_3 \end{cases}$	+	$A(OH)_2$	=	H_2O	+	$A \begin{cases} SO_4 \\ Cl_2 \\ (NO_3)_2 \\ SiO_3 \\ CO_3 \end{cases}$

REGENERATION

Exhausted exchange bed		Caustic soda		Regenerated exchange bed		Waste water
$A \begin{cases} SO_4 \\ Cl_2 \\ (NO_3)_2 \\ SiO_3 \\ CO_3 \end{cases}$	+	$2NaOH$	=	$A(OH)_2$	+	$Na_2 \begin{cases} SO_4 \\ Cl_2 \\ (NO_3)_2 \\ SiO_3 \\ CO_3 \end{cases}$

Figure 11-11 Strong base anion exchange. A is the exchange material.

mentioned, first through the hydrogen zeolite and second through the anion exchanger, compares with distilled water in quality and should cost less, although the process is too costly for municipal use and is not desirable from a health standpoint. It has been found useful, however, for manufacturing soft drinks, in high-pressure boilers, and in other industrial processes where waters free from mineral salts are required.

11-19 Desalination

Increasing water consumption and depletion of existing water resources has led to considerable interest in conversion of saline or brackish waters. The Office of Saline Waters has financed some research and other processes have been developed privately. The cost of the systems discussed below is generally not competitive with treatment of freshwater, but where adequate freshwater is not available their use is feasible. Continued development is projected to reduce the cost of desalination to levels comparable to that of freshwater treatment by the end of the century,[22] although recent increases in energy costs make such a development unlikely. Desalination systems can be separated into those which employ a phase change, like distillation or freezing, and those which separate water and dissolved minerals within the aqueous phase, like ion exchange, electrodialysis, and reverse osmosis.

Evaporators, generally multiple-effect systems, can be used to distill freshwater from salty or brackish water. Problems associated with evaporators include accelerated corrosion and scaling due to the high temperatures involved. First cost and operating cost are high. The cost of water production by this technique ranged from about $0.25 to 0.55 per m^3 ($1 to 2/1000 gal) in 1972 in plants with a capacity of 4700 to 10,000 m^3 per day (1.25 to 2.62 mgd).

Solar stills employ shallow basins covered with glass or plastic upon which vaporized water condenses. The technique requires a climate with a high percentage of sunlight. Costs are comparable to those of multiple-effect evaporators but energy consumption is considerably less. Production is uncertain since it is dependent upon the weather and will vary with the time of year and intensity of incident sunlight. The use of geothermal power to operate multiple-effect stills has also been studied.[23]

Freezing for desalination is effected by application of vacuum processes in which evaporation of a portion of the water or of a miscible secondary refrigerant cools the flow. The influent to the process is cooled in a heat exchanger by product and waste streams to close to the freezing point. The evaporator withdraws the refrigerant, dropping the temperature and producing ice crystals which are removed and washed with a portion of the product water. The ice crystals are used to condense the vapor stream from the evaporator, or cool it following its compression, and this, in turn, melts the ice. A portion of the product is used to rinse the crystals and the rinse water, product, and brine are used to chill the incoming flow.

Freezing, due to the low temperature and lack of direct surface heat exchange, minimizes corrosion and scaling and is generally more economical than distillation.[24]

Ion exchange has been applied to desalting using systems like that illustrated in Figs. 11-10 and 11-11. Ion-exchange processes are simple to operate, have moderate capital costs and few operating problems, but require costly regenerants and produce troublesome waste streams. The net production ranges from 50 to 80 percent of the influent flow depending upon the total dissolved-solids concentration. A number of novel industrial systems[25] which reduce regenerant requirements have been developed. One, the Sul-biSul process, employs an anionic bed in which sulfates are converted to bisulfates which may be regenerated by the raw water. Ion exchange is more applicable to brackish (TDS < 5000 mg/l) than to saltwater. Costs of production in 1972 ranged from $0.05 to 0.40 per cubic meter ($0.20 to 1.60 per 1000 gal) depending upon flow and dissolved solids content.

Electrodialysis employs electrical energy to drive dissolved ionized solids across semipermeable membranes. The system consists of cathodic and anodic semipermeable membranes and two electrodes. Cations and anions are driven across their respective membranes (which reject oppositely charged or neutral molecules) leaving demineralized water behind. The yield is approximately 85 to 90 percent with the remaining flow containing the brine rejected from the system. The cost in 1972 ranged from $0.05 to 0.45 per cubic meter ($0.20 to 1.80 per 1000 gal) depending upon flow and dissolved solids content in the raw water.

Reverse osmosis is perhaps the best demonstrated technology for saline water conversion. If a semipermeable membrane separates two bodies of water with differing salt concentrations the water will flow from the less to the more concentrated. The flow can be halted by applying pressure equal to the osmotic pressure of the concentrated solution, and reversed by exceeding that pressure. In practice the pressure ranges from 5 to 50 times the osmotic pressure of the water.[26] Reverse osmosis systems include the membrane, a support structure, a pressure vessel and a pump. The optimum membrane configuration is the hollow fiber, which has an area to volume ratio of up to 30,000 m^2/m^3 (compared to 300 to 1000 m^2/m^3 for other designs), requires no separate support, and has satisfactory flux rates. Modern membranes are made of aromatic polyamides which have more desirable mechanical and chemical properties than earlier cellulose acetate designs. Reverse osmosis is not easily applied to seawater although development of new membrane materials and configurations and use of multistage processes shows promise.[26] The cost of treating brackish water by this technique ranged from $0.06 to 0.25 per cubic meter ($0.25 to 1.25 per 1000 gal) in 1972.

11-20 Color Removal

The most commonly used method of color removal is coagulation followed by sedimentation. In most waters the negatively charged color colloid is precipitated by a trivalent ion of opposite charge, usually by dosage with alum, occasionally with ferric salts. In colored waters of high alkalinity heavy dosages of the coagulant may be required to precipitate the color. Not all waters of the same chemical characteristics respond in the same way to color removal measures. Some waters of high color and low alkalinity will require alkalinity in the form of lime or sodium carbonate. Many Florida cities, cited by Black, combine softening with color removal although during periods of low hardness they may resort to coagulation. Free residual chlorination was found effective at Miami. Experimentation is required before adoption of a color removal method and the problem is complicated if the water varies in alkalinity or hardness.

11-21 Treatment of Boiler Waters

The quality of boiler feed water is of great importance in many industries. The requirements for a good boiler water are entire freedom from turbidity and noncarbonate hardness, practical freedom from oil and oxygen, and total hardness not to exceed 35 mg/l. For very high pressures all hardness should be removed. Boiler troubles produced by water include scale formation, corrosion, priming, and foaming.

Boiler scale interferes with heat movement to the water, causing fuel loss and overheating of the metal. Softening by the methods already described will greatly reduce scale. The hot process of softening[27] is also used. It employs lime and soda, but the water is first heated, thereby stripping carbon dioxide, increasing the efficiency and speed of action of the chemicals, and reducing the solubility of the precipitates.

Boiler compounds are used to prevent scale formation. They are made of many organic and inorganic materials, including sugar, glycerin, tannin, sodium silicate, soda ash, etc. Many of them accomplish little, but where beneficial they probably prevent scale by interfering with the formation of the crystals that adhere to the metal. The scale is therefore loose and can be removed by flushing.

Boiler corrosion is caused by mineral acids and salts, organic acids in water from swamps and ponds, free carbonic acid, and oxygen. Magnesium chloride is considered injurious, as it breaks up in the boiler to form hydrochloric acid. Free carbonic acid and other acids can be overcome by alkalies. Oxygen is very important in boiler corrosion, and it is often necessary to deoxygenate feed water. Deoxygenators are tanks containing steel filings over which the water flows. A high caustic alkalinity of the feed water, pH 9.5 or higher, is also desirable.

Foaming is the formation of bubbles on the surface of the water in the boiler. They may be carried off in the steam pipe to cause wet steam. If the bubbles are produced suddenly and explosively, the result is priming. The causes are variously given: improper design of the boiler so that there is little steam space; poor operation, including variations in load, irregular firing, and improper height at which the water level is carried; water condition, including excessive concentration of salts and presence of organic matter. Aside from operation and design of boilers, prevention of foaming, according to present knowledge, is a matter of keeping concentration of foreign matter to a minimum, whether in suspension or solution, organic or inorganic.

11-22 Stabilization of Water

This term is applied to correction of a water so that it will neither corrode the pipes through which it passes nor deposit incrusting films of calcium carbonate. Some waters are naturally corrosive because of the presence of oxygen and free carbon dioxide while others, after coagulation, may be quite soft, with low pH values and will also be aggressive, that is corrosive. Other waters, which have been softened by the lime-soda process, will be soft but have high pH values and will be saturated to a greater or lesser extent with calcium carbonate. The situation is that with a low pH value corrosion may occur and with a high pH value incrustation will result. The latter may be prevented by recarbonation of the water, as discussed in Art. 11-15 and the former by addition of lime to increase the pH to a value at which it is in equilibrium with calcium carbonate or at which a thin film of calcium carbonate will be established that will protect against corrosion without the deposition of excessive amounts.

The pH of stability for a water may be approximated from Langlier's Index:

$$\mathrm{SI} = \mathrm{pH} - (\mathrm{p}K_2' - \mathrm{p}K_s' + \mathrm{pCa} + \mathrm{pAlk}) \tag{11-13}$$

in which pH is the negative log of the hydrogen ion concentration, pCa and pAlk are, similarly, the negative logs of the molar concentrations of calcium and alkalinity, $\mathrm{p}K_2'$ and $\mathrm{p}K_s'$ are empirical constants dependent upon temperature and ionic strength, and SI is Langlier's Index. If the index is greater than zero the water will be depositing, if less than zero, corrosive, if zero, stable. The magnitude of SI does not indicate the rate of corrosion or deposition nor is the prediction always accurate in this abbreviated form.

The marble test for measuring stability illustrates the conditions desired. The water of known alkalinity is held in contact with powdered calcium carbonate for

24 hours, and its alkalinity is then determined. If there has been a decrease of alkalinity, the water is supersaturated with calcium carbonate and incrustation will occur. If there has been an increase in alkalinity, the water will dissolve a calcium carbonate deposit, and corrosion may result. If the alkalinity remains unchanged, the water is stable, and the accompanying pH will be that of equilibrium.

Corrosion and incrustation can, therefore, be prevented by adjustment of the pH of the water, alkalinity being increased by adding lime. The use of coatings to prevent corrosion has been discussed in Art. 5-17. Another method of preventing incrustation is the use of sodium hexametaphosphate at doses of 1 to 2 mg/l. It will also prevent corrosion at higher doses, amounts of which will depend upon the pH and hardness of the water. Protection apparently results from formation of a protective film.

11-23 Fluoridation

It has been established that fluoride is helpful in the development of teeth free from decay and that presence in the drinking water of modest amounts of fluoride will provide the protection desired. Sodium fluoride is commonly used, but there is considerable usage of sodium silico-fluoride and hydrofluosilicic acid. Sodium fluoride is a white powder but can be obtained colored blue to distinguish it from other chemicals, containing 90 to 95 percent NaF. It is soluble to the extent of 4 percent in water, and $2\frac{1}{2}$ percent solutions have been fed from solution pots by means of aspirators. Hypochlorinators have also been used successfully.

The therapeutic dosage of fluoride is thought to be about 1 mg/l as F, while dosages slightly higher have been associated with mottling of teeth and tooth loss. The Environmental Protection Agency Standards specify variable upper limits for fluoride depending upon air temperature, presumably upon the basis that more water is drunk in warm climates. The range of concentration is 0.6 to 2.4 mg/l. The variation with temperature is presented in App. I.

11-24 Defluoridation of Water[28]

Although over a million people in the United States are reported to use water containing excessive amounts of fluoride, there are only a few plants designed for its removal, probably because of the high cost of such treatment. Various methods of removal have been tried with varying success and costs. Dolomitic lime, magnesia, or magnesium sulfate may be dissolved, converted to magnesium hydroxide and settled. This will remove the fluoride, but 45 to 65 mg/l of magnesium will be required for removal of 1 mg/l of fluoride. This is too expensive for water containing more than 3 to 4 mg/l of fluoride. Bone char and an artificial bone char consisting of calcium hydroxy phosphate salts have been used as specific ion exchange systems for fluoride. Aluminum sulfate has a relatively high capacity for fluoride absorption, and its capacity can be increased with such filter aids as activated silica and clays. In some waters, however, interfering anions may so adversely affect its capacity that expense will be too great. Plants at Bartlett,

Tex. and Clovis, N.Mex. use activated alumina as an ion exchanger. Regeneration is by means of a strong caustic solution, followed by a wash with 15 percent sulfuric acid solution, and a final rinse. Raw water may also be used to rinse the residual caustic from the bed.

11-25 Radioisotope Removal by Water Treatment

The sources of nuclear radiation in water are given in Art. 8-6. Wastewaters in nuclear facilities are treated primarily by coagulation with lime, alum, and iron. Coagulation and filtration of any type will remove particulate matter which is radioactive, while radioactive phosphorus can be removed, like other phosphorus, with lime, alum, iron, or other metallic coagulants.

Strontium is chemically similar to calcium and it is removable by softening although, like calcium, it has some residual solubility in ordinary softening.

Radium, for which EPA proposed a standard in 1976, is removable only by ion exchange and no resin specific for radium is available. The removal of radium by ion exchange requires the removal of hardness as well, and this total softening process is expensive and conceivably harmful to the water consumers' health.[29]

11-26 Water Treatment Wastes

The manufacture of potable water from surface or ground supplies usually results in the production of a variety of waste streams. These flows, depending upon circumstances, may not be suitable for discharge to surface waters and may thus require treatment prior to their disposal.

The major waste streams arising from surface water treatment include primary solids (largely sand and silt), coagulation sludges (which may include significant quantities of metallic coagulants, minor amounts of polymer, or both), and filter backwash. Groundwater treatment may produce minor quantities of fine sand and precipitated iron and manganese which is found primarily in the filter backwash. Miscellaneous wastes which could be produced in either surface or groundwater treatment include softening sludges, waste activated carbon or its regeneration wastes, brines from ion exchange, reverse osmosis, electrodialysis, freezing or distillation and the various impurities found in commercial chemicals.

Wastes which are essentially liquid such as brines and activated carbon regenerant streams have been disposed of by lagooning to obtain evaporation if the climate permits and by discharge to deep wells or saline surface waters. Discharge to sewers simply moves the problem, since wastes of this sort do not profit from sewage treatment and may interfere with some sewage-treatment processes. $CaCl_2$ brines resulting from ion-exchange softening can be converted to useful regenerant by dosing with soda-ash. The resulting $CaCO_3$ sludge requires further handling.

Filter backwash water may be passed through a surge tank which will equalize the flow and provide some sedimentation, and then back to the head of the plant where the solids agglomerated on the filter may be removed by the coagulation processes. Polymers are sometimes added in the surge tank.[30]

The *sludges* found in the bottom of the clarifiers may vary from 96 percent moisture content for filter backwash to 40- to 70-percent moisture content for primary (grit) removal systems dosed with iron, lime, or polymers.[31] Average solids content of softening and coagulation sludges is 8 to 10 percent but varies widely from plant to plant. In order to handle these wastes economically it is necessary to reduce their moisture content. This can be done by lagooning, dewatering on sand drying beds, gravity thickening, vacuum or pressure filtration, centrifugation, solvent extraction, or freezing. The solids may be pretreated prior to dewatering by either polyelectrolyte, chemical, heat, or pressure conditioning. Once the sludge is reduced to a manageable volume it may be disposed of by direct landfill, drying followed by landfill, or by a variety of chemical recovery techniques. The particular process or series of processes is dependent upon the nature of the raw water and the purification techniques employed, since these dictate the nature of the sludge.

Alum sludges can be concentrated to 20- to 36-percent solids in centrifuges or to 40-percent solids in pressure filters.[32] Polymer doses are approximately 0.1 to 0.8 percent of the dry solid mass. *Freeze treatment* requires no chemicals, is independent of solids character, and produces a coarse slurry which is readily dewatered upon vacuum filters or drying beds.[33] The slurry will settle to about 20 percent solids content in 1 to 5 hours. The use of *aliphatic amine solvents*[34] which are miscible with water at low temperature (18°C) but separate when warmed (55°C), offers a sophisticated technique of providing solids streams which are entirely water free. The process has not yet been applied to large-scale systems, but has been demonstrated on a pilot scale. The cost of this technique reportedly compares favorably with other dewatering methods. Dewatered alum sludge does not rewet and will support plant growth.

Alum is recoverable from coagulation sludges by dissolution in sulfuric acid. The liquid alum solution gradually becomes colored but the color reportedly does not interfere with its reuse.[30]

Lime softening sludges can be dewatered by centrifuges to 50 to 60 percent solids. Centrifugation is desirable for this sludge since it also classifies the waste to a degree, leaving the bulk of the Mg, Si, Fe, and Al in the centrate.[32] Calcium carbonate sludge can be recovered by *recalcining*. The process generally requires a series of steps to remove any magnesium which may have been precipitated. Addition of CO_2 will dissolve $Mg(OH)_2$ without dissolving $CaCO_3$ provided the pH is kept above 7.5.[32] The $CaCO_3$ sludge can then be heated to 900 to 1200°C at which temperature it will be converted to quicklime.

$$CaCO_3 \xrightarrow{\Delta} CaO + CO_2 \tag{11-14}$$

The process is practical in some instances and has been applied at some large plants. The cost of the lime produced by this system is about the same as if it were purchased, but the saving in waste handling may make the method attractive.

Water treatment sludges may in some cases be handled by discharge to sewers. Their mixture with sewage solids may facilitate dewatering the latter and their mixture with the raw flow can aid in its clarification and precipitate phosphorus. The solids handling problem is simply transferred by this expedient but this may result in savings in overall capital and operating costs.

Considerable research upon the handling of water treatment sludges is now in progress. Systems which have proven satisfactory in sewage sludge dewatering are generally applicable in principle, but loadings, pretreatment required, and final moisture content may be significantly different. The dewatered sludges appear to be disposable in properly operated landfills.

PROBLEMS

11-1 The ionization constant for hypochlorous acid is $K = 3 \times 10^{-8}$. Determine the percentage of hypochlorous acid present at pH values of 5, 7, and 9 using the equilibrium expression

$$\frac{(H)(OCl^-)}{(HOCl)} = K$$

11-2 The following data was obtained in a chlorination experiment. Plot the data and determine the breakpoint dosage.

Dosage (mg/l)	1.00	2.00	3.00	4.00	5.00	6.00	7.00
Residual (mg/l)	0.80	1.55	1.95	1.25	0.50	0.85	1.95

What dosage is required to provide a free residual of 1 mg/l?

11-3 If the breakpoint chlorine dosage at a water plant is 5 mg/l, in what form would you obtain the required chemical for plants with capacities of 2, 20, and 100 thousand m^3/day? Assume that 2-weeks supply must be kept on hand at all times.

11-4 Determine the power required for generation of ozone for each of the plants of Prob. 11-3 assuming that the ozone and breakpoint chlorine dosages are the same. If power cost 3 cents per kWh what will be the annual cost of disinfection with ozone for power alone?

11-5 Calculate the cost of disinfecting each of the flows above using ultraviolet radiation. Assume that the water film is 100 mm thick and that the retention time is 15 s.

11-6 Determine the theoretical number of trays required in an aeration tower to provide 95 percent removal of CO_2. Assume that K for the tower is 0.35.

11-7 A water contains 200 mg/l Ca as $CaCO_3$, 75 mg/l Mg as $CaCO_3$, and 180 mg/l HCO_3^- as $CaCO_3$. The total concentration of free CO_2 and carbonic acid is 150 mg/l as $CaCO_3$. Determine the required chemical dosages to soften this water as much as possible without removing magnesium. What will be the final hardness of the water?

11-8 If the water of Prob. 11-7 is to be softened to a final hardness of 85 mg/l as $CaCO_3$ and is not to contain more than 25 mg/l Mg as $CaCO_3$, find the required chemical dosages. Show the condition of the water at various stages through the process on a series of bar diagrams.

11-9 Present a diagram of an ion-exchange process which could be used to meet the minimum effluent criteria of Prob. 11-8. What would be the total annual salt consumption for a plant with a capacity of 5000 m^3/day?

11-10 A commercial ion-exchange resin is reported to have a capacity of 12,500 grains per cubic foot. What quantity of this resin would be exhausted per day in a plant which reduces the hardness of a 7000 m^3/day flow from 250 mg/l as $CaCO_3$ to 100 mg/l as $CaCO_3$? What total bed area and volume would you specify, and how often would individual beds be regenerated? Assume that the regeneration cycle takes 1 hour.

REFERENCES

1. Grigoropoulas, S. G., et al.: "Organic Contaminants in Water Supplies," *Journal American Water Works Association*, **67**:418, August, 1975.
2. Jenkins, David, et al.: "Organic Contaminants in Water," *Journal American Water Works Association*, **66**:682, November, 1974.
3. Symons, James M., et al.: "National Organics Reconnaissance Survey for Halogenated Organics," *Journal American Water Works Association*, **67**:634, November, 1975.
4. Cleasby, John L.: "Research Achievement—Existing and Expected," *Journal American Water Works Association*, **68**:272, June, 1976.
5. Hoehn, Robert C.: "Comparative Disinfection Methods," *Journal American Water Works Association*, **68**:302, June, 1976.
6. Rook, Johannes J.: "Developments in Europe," *Journal American Water Works Association*, **68**:279, June, 1976.
7. Cromley, J. Timothy and John T. O'Connor: "Effect of Ozonation on the Removal of Iron from Ground Water," *Journal American Water Works Association*, **68**:315, June, 1976.
8. Harris, Weldon C. "Ozone Disinfection," *Journal American Water Works Association*, **64**:182, March, 1972.
9. Gomella, Cyril: "Ozone Practices in France," *Journal American Water Works Association*, **64**:39, January, 1972.
10. Carrell, Morris J.: "Chlorination and Disinfection: State of the Art," *Journal American Water Works Association*, **63**:769, December, 1971.
11. Kinman, Riley N. and Ronald F. Layton: "New Method for Water Disinfection," *Journal American Water Works Association*, **68**:298, June, 1976.
12. Johnson, J. D. and R. Overby: "Bromine and Bromamine Disinfection Chemistry," *Journal Sanitary Engineering Division, Proceedings American Society of Civil Engineers*, **99:SA 3**:373, 1973.
13. Goldstein, Melvin, L. J. McCabe, and Richard L. Woodward: "Continuous-Flow Water Pasteurizer for Small Supplies," *Journal American Water Works Association*, **52**:247, 1960.
14. Laubusch, Edmund J.: "Chlorination and Other Disinfection Processes," in *Water Quality and Treatment* 3d ed., McGraw-Hill, New York, 1971.
15. Headstream, Marcia, Dan M. Wells, and Robert M. Sweazy: "The Canyon Lakes Project," *Journal American Water Works Association*, **67**:125, March, 1975.
16. Courchene, John E. and James D. Chapman: "Algae Control in Northwest Reservoirs," *Journal American Water Works Association*, **67**:127, March, 1975.
17. O'Connor, John T.: "Iron and Manganese," in *Water Quality and Treatment* 3d ed., McGraw-Hill, New York, 1971.
18. Cecil, Lawrence K.: "Industry's Benefit from Municipal Water Softening," *Journal American Water Works Association*, **67**:80, February, 1975.
19. Ingols, Robert S. and T. F. Craft: "Analytical Notes—Hard vs. Soft Water Effects on the Transfer of Metallic Ions from Intestine," *Journal American Water Works Association*, **68**:209, April, 1976.
20. McGhee, Terence J.: "Heuristic Analysis of Lime-Soda Softening Processes," *Journal American Water Works Association*, **67**:626, November, 1975.
21. Schierholz, Paul M., John D. Stevens, and John L. Cleasby: "Optimum Calcium Removal in Lime Softening," *Journal American Water Works Association*, **68**:112, February, 1976.
22. Stobel, Joseph J.: "Growth of Desalting Technology—Studies and Applications," *Journal American Water Works Association*, **64**:701, November, 1972.
23. O'Brien, James J.: "Geothermal Resources as a Source of Water Supply," *Journal American Water Works Association*, **64**:649, November, 1972.
24. Frazer, James H. and Walter E. Gibson: "Secondary Refrigerant Desalination," *Journal American Water Works Association*, **64**:746, November, 1972.
25. Faber, Harry A., Sidney A. Bresler, and Graham Walton: "Improving Community Water Supplies With Desalting Technology," *Journal American Water Works Association*, **64**:705, November, 1972.

26. Lynch, Maurice A. Jr. and Milton S. Mintz: "Membrane and Ion-Exchange Processes—A Review," *Journal American Water Works Association*, **64:**711, November, 1972.
27. Nordell, Eskel: *Water Treatment for Industrial and Other Uses*, 2d ed., Reinhold, New York, 1961.
28. Maier, F. J.: "Fluorides in Water," in *Water Quality and Treatment* 3d ed., McGraw-Hill, New York, 1971.
29. Larsen, T. E.: "The Proposed Drinking Water Standards for Radium," *Journal American Water Works Association*, **68:**501, October, 1976.
30. Nielsen, Hubert L., Keith E. Carns, and John N. DeBoice: "Alum Sludge Thickening and Disposal," *Journal American Water Works Association*, **65:**385, June, 1973.
31. Calkins, Ronald J. and John T. Novak: "Characterization of Chemical Sludges," *Journal American Water Works Association*, **65:**423, June, 1973.
32. Burris, Michael A., Kenneth W. Cosens, and David M. Mair: "Softening and Coagulation Sludge—Disposal Studies for a Surface Water Supply," *Journal American Water Works Association*, **68:**247, May, 1976.
33. Wilhelm, J. H. and C. E. Silverblatt: "Freeze Treatment of Alum Sludge," *Journal American Water Works Association*, **68:**312, June, 1976.
34. Olson, Richard L.: "Alum Sludge Drying with Basic Extractive Treatment," *Journal American Water Works Association*, **68:**321, June, 1976.

CHAPTER
TWELVE

SEWERAGE—GENERAL CONSIDERATIONS

12-1

As cities have grown, the more primitive methods of excreta disposal have given place to the water-carried sewerage system. Even in small towns the greater safety of sewerage, its convenience, and freedom from nuisance have caused it to be adopted wherever finances permit.

Sewerage implies the collecting of wastewaters from occupied areas and conveying them to some point of disposal. The liquid wastes will require treatment before they can be discharged into a body of water or otherwise disposed of without endangering the public health or causing offensive conditions.

12-2 Definitions

Sewage is the liquid conveyed by a sewer. It may consist of any one or a mixture of liquid wastes which will be separately defined. *Sanitary sewage*, also known as domestic sewage, is that which originates in the sanitary conveniences of a dwelling, business building, factory, or institution. *Industrial waste* is a liquid waste from an industrial process, such as dyeing, brewing, or papermaking. *Storm sewage* is liquid flowing in sewers during or following a period of rainfall and resulting therefrom. *Infiltration* is the water that has leaked into sewers from the ground. *Inflow* is water which enters sewers from surface sources such as cracks in manholes, open cleanouts, perforated manhole covers, and roof drains or basement sumps connected to the sewers. Inflow occurs only during runoff events.

A *sewer* is a pipe or conduit, generally closed but normally not flowing full, for carrying sewage. A *common sewer* is one in which all abutting properties have equal rights of use. A *sanitary sewer* is one that carries sanitary sewage and is designed to exclude storm sewage, surface water, and groundwater. Usually it will also carry whatever industrial wastes are produced in the area that it serves. It is occasionally, although improperly, called a separate sewer. A *storm sewer* carries storm sewage, including surface runoff and street wash. A *combined sewer* is designed to carry domestic sewage, industrial waste, and storm sewage. A sewer system composed of combined sewers is known as a *combined system*, but if storm sewage is carried separately from the domestic and industrial wastes, it is said to be a *separate system*. The term *sewerage* is applied to the art of collecting, treating, and disposing of sewage. *Sewerage works* or *sewage works* are comprehensive terms covering all the structures and procedures required for collecting, treating, and disposing of sewage.

A *house sewer* is a pipe conveying sewage from the plumbing system of a single building to a common sewer or point of immediate disposal. A *lateral sewer* has no other common sewer discharging into it. A *submain sewer* is one that receives the discharge of a number of lateral sewers. A *main sewer*, also known as a trunk sewer, receives the discharge of one or more submain sewers. A *sewer outfall* receives the discharge from the collecting system and conducts it to a treatment plant or point of final disposal. An *intercepting sewer* is one that cuts transversely a number of other sewers to intercept dry-weather flow, with or without a determined quantity of storm water, if used in a combined system. A *relief sewer* is one that has been built to relieve an existing sewer of inadequate capacity.

Sewage treatment covers any process to which sewage is subjected in order to remove or alter its objectionable constituents so as to render it less dangerous or offensive. *Sewage disposal* applies to the act of disposing of sewage by any method. It may be done with or without previous treatment of the sewage.

12-3 General Considerations

Providing adequate sewerage for an urban area requires careful engineering. The sewers must be adequate in size or they will overflow and cause property damage, danger to health, and nuisances. Adequacy in size calls for estimation of the amount of sewage and use of hydraulics to determine proper sizes and grades of the sewers. Another important consideration is the velocity of flow in the sewers. If not great enough, deposits of solids will occur with accompanying odors and stoppages. After the sewage is collected, it becomes a liability to the city because of its potential danger to health and possible production of nuisance in streams.

The degree of treatment required depends upon the water quality standards applicable to the receiving stream and the flow and quality of both the stream and waste. In no case is treatment which produces an effluent containing more than 30 mg/l BOD_5 and 30 mg/l suspended solids (see Chap. 20) considered adequate for new construction.

The required effluent quality can be achieved by combining a variety of different treatment processes. No single process will suit all circumstances and the engineer must select the combination of systems which will provide the desired treatment at minimum cost and with maximum reliability.

The following chapters cover the collection, pumping, treatment, and disposal of liquid wastes. Estimation of waste quantities is treated in Chap. 2.

12-4 Combined vs. Separate Sewers

Present-day construction of sewers is largely confined to the separate system except in those cities where combined systems were constructed many years ago. In newly developing urban areas the first need is for collection of sanitary sewage, and, since sanitary sewers are relatively small and inexpensive, they can usually be constructed without long delay. For years storm water will be cared for by street gutters and natural watercourses. As the city grows, however, underground conveyance of storm-water runoff may be needed. Many of the cities having combined systems were highly developed before the establishment of water-carried sewerage and already had storm sewer systems. It is interesting to note that in the early 1800s in some cities the discharge of household wastes into the sewers was actually forbidden, but later the storm sewers received all liquid wastes and became combined sewers. Further extensions of such systems were then specially designed as combined sewers, often with provision for separating the dry-weather flow, which is largely sanitary sewage, from the large wet-weather flow.

In new construction, combined sewers might be used where the storm water flow is sufficiently contaminated to require treatment equal to that given the sanitary sewage. While the "first-flush" of storm sewage has been shown to be heavily polluted, the flows involved are very large and the receiving streams will be otherwise contaminated with non-urban runoff. It appears to be economically impractical to treat storm sewage although some work has been done on development of physical-chemical systems intended for that purpose.[1]

As a general rule, new sewer systems must be designed to separate sanitary and storm flows—even to the extent of providing strict limitations upon infiltration at sewer joints. The Environmental Protection Agency will not finance treatment works unless it has been demonstrated that excessive infiltration and/or inflow does not exist in the collection system. This regulation effectively prevents the construction of new combined sewers and tends to make replacement of existing ones desirable.

It should also be noted that sanitary sewer systems must be carefully supervised to insure that illegal connections of roof and basement drains are not made after initial construction is complete.

12-5 Liability for Damages Caused by Sewage

The courts of the various states have differed considerably in decisions governing damages resulting from installation of sewerage works. A few principles may be discerned, however, through the legal fog. City officials cannot be required to furnish sewerage service to the citizens, nor can they or the city be held liable should it not be furnished. Once sewerage works have been provided, however, the city assumes some definite responsibilities for damages which may result to health or property from inadequacy or faulty design of the sewers and other works, poor maintenance and operation, contamination of water supplies, and pollution of streams.

Should a sewer system be designed by an engineer presumably competent to do such work, and the construction be carried out in practical accordance with his designs, a city is usually not held liable for damages resulting from their inadequacy. Hence, property owners whose basements or ground floors are flooded would have no redress. A storm sewer may discharge surface water in such a manner that it damages property. If the sewer does not discharge upon the property that normally received the drainage before the sewer was built, or if, normally receiving it, damage now results which did not occur previously, the city is liable.

For damages resulting from poor maintenance and negligence in operation the city is always liable. For example, if stoppage of a sewer is reported to the proper authorities and they do not respond promptly, and sewage backs up and overflows from house fixtures, the city would be liable. Leaky sewers have contaminated water supplies. If the city officials had knowledge of the situation and did not remedy it, negligence could be established, and damages could be collected. At sewage treatment plants odors may occur, with or without negligent operation. Property owners may collect temporary or permanent damages for depreciation of property values so caused. As intimated, negligence may not have to be proved in this case. For this reason sewage treatment plants should not be located where odors are likely to cause depreciation of property values. In preparation of environmental assessments for new construction, the impact of both noise and odor from the sewage works upon adjacent property must be evaluated.

Discharge of untreated or partially treated sewage may so pollute streams that the water is offensively odorous, resulting in nuisances and possibly depreciated property values. Riparian owners, under the common law, are entitled to reasonable use of water in streams; and should excessive pollution prevent their using such water for washing, bathing, irrigation, or watering stock, they may demand and receive compensation. In addition, fines may be assessed against cities and industries which violate the terms of their NPDES permits (Art. 1-3).

REFERENCES

1. Agnew, Robert W.: "Some Schemes for the Treatment of Combined Sewer Overflows" *Transactions 25th Annual Sanitary Engineering Conference*, University of Kansas, Lawrence, Kansas, February, 1975.

CHAPTER

THIRTEEN

AMOUNT OF STORM SEWAGE

13-1

The first step in design of sewers is the estimation of the flow which they will receive. Chapter 2 considers the variation in flow in sanitary sewers, while this chapter presents methods of estimating storm flow from urban areas.

The primary source of storm flow is rainfall. The sources of rainfall data, descriptions of rain gauges, etc., are presented in Chap. 3.

13-2 The Rational Method

All presently used techniques for estimating storm flow are based upon the use of rainfall data—either implicitly or explicitly. The rational method relates the flow to the rainfall intensity, the tributary area, and a coefficient which represents the combined effects of ponding, percolation, and evaporation. The total volume which falls upon an area, A, per unit time under a rainfall of intensity, i, is:

$$Q_{\text{prec}} = iA \tag{13-1}$$

Of this total, a portion will be lost by evaporation, percolation, and ponding. The portion lost is not constant, but may be determined for differing conditions of temperature, soil moisture, and rainfall duration. The actual amount which appears as runoff may then be calculated from:

$$Q = CiA \tag{13-2}$$

in which C is the runoff coefficient, i.e., the fraction of the incident precipitation which appears as surface flow.

Table 13-1 Runoff coefficients for various surfaces[1]

Type of surface	C
Watertight roofs	0.70–0.95
Asphaltic cement streets	0.85–0.90
Portland cement streets	0.80–0.95
Paved driveways and walks	0.75–0.85
Gravel driveways and walks	0.15–0.30
Lawns, sandy soil	
2% slope	0.05–0.10
2–7% slope	0.10–0.15
> 7% slope	0.15–0.20
Lawns, heavy soil	
2% slope	0.13–0.17
2–7% slope	0.18–0.22
> 7% slope	0.25–0.35

13-3 Runoff Coefficients

C for an area is not invarient, but tends to increase as the rainfall continues. Research conducted in the early part of the twentieth century led to the development of formulas (13-3) through (13-5). For impervious surfaces:

$$C = 0.175t^{1/3} \tag{13-3}$$

or

$$C = t/(8 + t) \tag{13-4}$$

For improved pervious surfaces:

$$C = 0.3t/(20 + t) \tag{13-5}$$

in which t is the duration of the storm in minutes.

Average values of C commonly used for various surfaces are presented in Table 13-1.

An effective runoff coefficient for a composite drainage area can be obtained by estimating the percentage of the total which is covered by roofs, paving, lawns, etc., and multiplying each fraction by the appropriate coefficient and then summing the products.

Example Determine the runoff coefficient for an area of 0.20 km^2. 3000 m^2 is covered by buildings, 5000 m^2 by paved driveways and walks, and 2000 m^2 by Portland cement streets. The remaining area is flat, heavy soil, covered by grass lawn.

SOLUTION From Table 13-1 obtain values for C for each area. Calculate the percentage of land area in each category.

$$\text{Roofs} - C \times A/A_{\text{total}} = 0.70 \text{ to } 0.95 \times \frac{3000}{200{,}000}$$

$$= 0.01050 \text{ to } 0.01425$$

$$\text{Driveways and walks} = 0.75 \text{ to } 0.85 \times \frac{5000}{200{,}000}$$

$$= 0.01875 \text{ to } 0.02125$$

$$\text{Street} = 0.80 \text{ to } 0.95 \times \frac{2000}{200{,}000}$$

$$= 0.0080 \text{ to } 0.0095$$

$$\text{Lawn} = 0.13 \text{ to } 0.17 \times \frac{190{,}000}{200{,}000}$$

$$= 0.1235 \text{ to } 0.1615$$

$$C_{\text{avg}} = 0.16 \text{ to } 0.21$$

Conservative practice, in the absence of local knowledge, would be to use the higher values of those given in the table. Some engineers use values of 0.7 to 0.9 for densely built areas, 0.5 to 0.7 for well built-up areas adjacent to densely built zones, 0.25 to 0.50 for residential areas with detached houses, and 0.15 to 0.25 for suburban sections with few buildings. Typical values of C for areas of different character are presented in Table 13-2.

As noted in Chap. 3, the rational method should not be used for large areas nor for regions incorporating significant surface storage such as ponds or swamps. The technique is used for urban areas, generally not exceeding 5 km^2 in area. For areas larger than 0.5 km^2 other techniques may be more suitable (see Chap. 3).

13-4 Time of Concentration

When a rainfall event occurs upon an area served by a storm sewer, the runoff will flow over roofs, yards, and pavement to the gutter and eventually to the sewer inlet. This travel will require measurable time and, while the areas immediately adjacent to the inlet will contribute flow quickly, areas which are distant will not. If one assumes that 15 minutes will be required for flow from the most distant point to reach the inlet and that the rainfall event lasts 5 min, the flow at the inlet will increase for 5 min and then decrease for the next 15 min. The decrease occurs because flow from the area near the inlet ceases although flow from more remote areas is still arriving.

Table 13-2 Runoff coefficients for different areas[1]

Description of area	C
Business	
Downtown area	0.70–0.95
Neighborhood area	0.50–0.70
Residential (urban)	
Single family area	0.30–0.50
Multi-units, detached	0.40–0.60
Multi-units, attached	0.60–0.75
Residential (suburban)	0.25–0.40
Apartment areas	0.50–0.70
Industrial	
Light	0.50–0.80
Heavy	0.60–0.90
Parks, cemeteries	0.10–0.25
Playgrounds	0.20–0.35
Railroad yards	0.20–0.40
Unimproved areas	0.10–0.30

The maximum rate of runoff for a given rainfall intensity will occur when the rainfall has continued for a period sufficient to permit flow to reach the inlet from the most remote point of the drainage area. The time required for the maximum runoff rate to develop is called the time of concentration. Figure 13-1 illustrates a rectangular watershed discharging into an inlet I. It is assumed that it takes 5 min for water to run from the boundary of one zone to the next, or to the inlet in zone

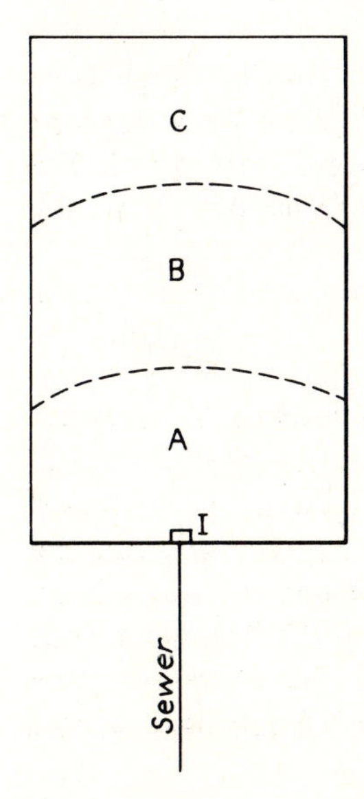

Figure 13-1 Figure illustrating inlet time.

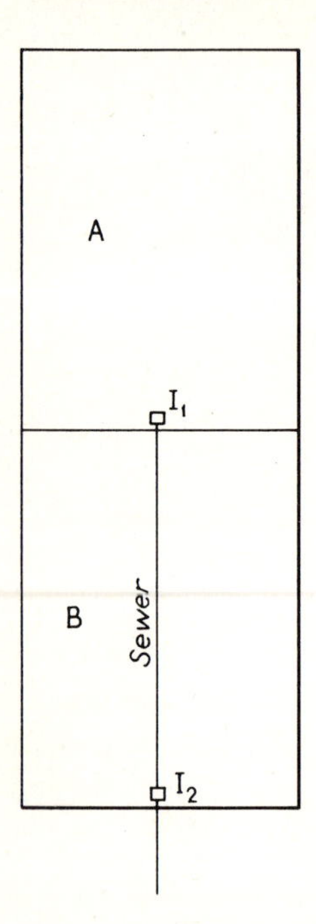

Figure 13-2 Figure illustrating inlet time and flow time.

A. Only zone A will contribute flow after 5 min, only A and B after 10 min, and all three after 15 min or more. If the rain lasted only 10 min, the water arriving at I from zone C during the period 10 to 15 min after beginning would be offset by the decreasing runoff from zone A.

The same logic may be applied to the watershed of Fig. 13-2. The water from A enters the sewer at I_1, and that from B at I_2. The time of concentration at I_2 is either the time of concentration for area B or the *inlet time* plus the *time of flow* from I_1 to I_2, whichever is greater. The *inlet time* is the time of concentration at I_1, while the *time of flow* is a function of the velocity in line $I_1 - I_2$ and its length.

The time of concentration for each sewer line is determined in a similar fashion. At each point the inlet time to the sewer most remote in time is added to the time of flow in the sewer. When branches join, the longest time of concentration for any branch is used as the basis for subsequent design. The time of concentration will depend largely upon the slope of the ground surface and the resulting slope of the sewer, which generally parallels the ground surface.

Inlet times may simply be assumed, frequently being taken as 5 to 10 min. They may also be calculated using Izzard's overland flow technique (Art. 3-26). The calculation requires that the rainfall intensity be known, and this, in turn, requires a knowledge of time of concentration. A suitable procedure is to assume a time of concentration, determine intensity as illustrated in Art. 13-7, and calculate a new time of concentration.

An alternate technique involves use of a nomogram like that of Fig. 13-3. In the example shown on the figure, a flow distance of 60 m over an ordinary grass

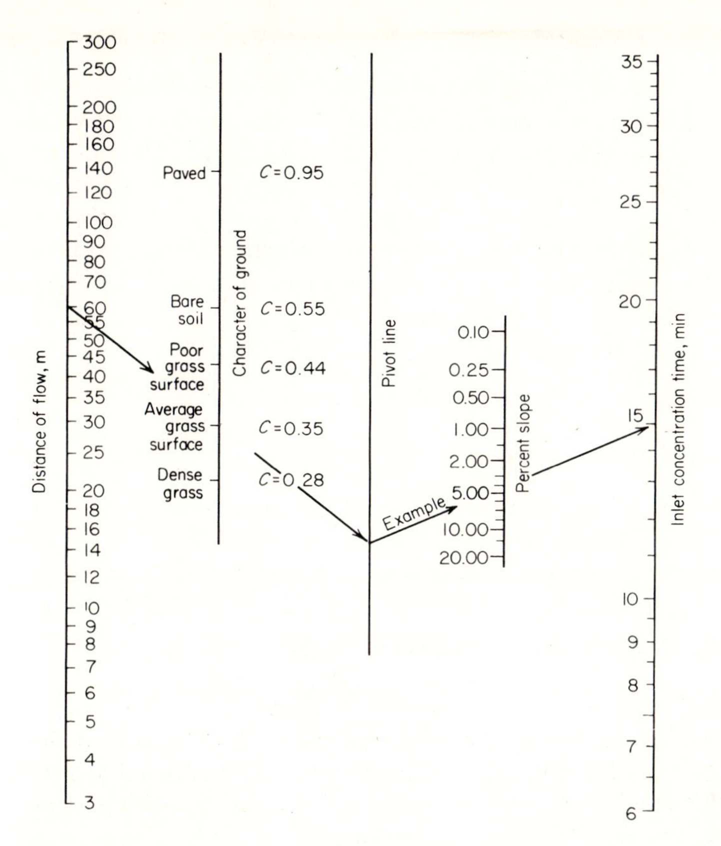

Figure 13-3 Overland flow time. (*Modified from a figure in "Data Book for Civil Engineers, Volume 1 Design" 3d ed., by Elwyn E. Seelye. Copyright 1960. With permission of John Wiley & Sons, Inc.*)

surface with a slope of 4 percent yields a time of concentration of 15 min. This procedure neglects the effect of rainfall intensity, but is adequate for most urban drainage projects.

13-5 Rainfall Intensity

In determining rainfall intensity for use in the rational formula it must be recognized that the shorter the duration, the greater the expected average intensity will be. The critical duration of rainfall will be that which produces maximum runoff, and this will be that which is sufficient to produce flow from the entire drainage area. Shorter periods will provide lower flows since the total area is not involved, and longer periods will produce lower average intensities. The storm sewer designer thus requires some relationship between duration and expected intensity. Intensities vary in different parts of the country and curves or equations are specific for the areas for which they were developed.

13-6 Intensity Curves and Formulas

The data obtained at rainfall gauging stations (Art. 3-4) can be used to develop intensity-duration-frequency curves such as those shown in Fig. 13-4. Curve A represents the rainfall intensity-duration which will be equaled or exceeded once in 30 years on average. The other curves present the intensities to be expected, on average, once in 20, 15, 10, or 5 years. Wide variations in point rainfall occur in relatively short distances, hence data from two stations which are separated by a reasonable distance may be combined to obtain what is, effectively, a longer period of record.

The equations of intensity-duration curves are often more convenient. The equations are typically of the form:

$$i = \frac{A}{t + B} \tag{13-6}$$

in which i is the precipitation rate (usually per hour), t is the duration (usually in minutes), and A and B are constants. Table 13-3 presents formulas of this type prepared from data collected from broad sections of the country (Fig. 13-5). These equations should be used with caution and only in the absence of better local information since wide variations are to be expected in such large sections of the country.

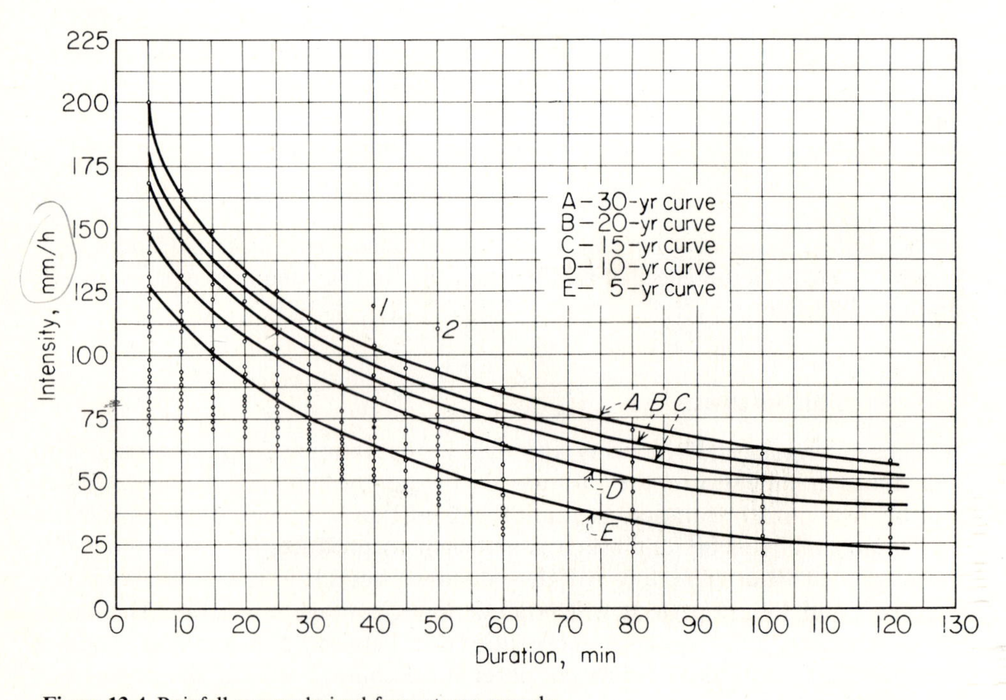

Figure 13-4 Rainfall curves derived from storm records.

Table 13-3 Precipitation formulas for various parts of the United States (i, mm/h; t, min)

Frequency, years	Area 1	Area 2	Area 3	Area 4	Area 5	Area 6	Area 7
2	$i = \frac{5230}{t + 30}$	$i = \frac{3550}{t + 21}$	$i = \frac{2590}{t + 17}$	$i = \frac{1780}{t + 13}$	$i = \frac{1780}{t + 16}$	$i = \frac{1730}{t + 14}$	$i = \frac{810}{t + 11}$
5	$i = \frac{6270}{t + 29}$	$i = \frac{4830}{t + 25}$	$i = \frac{3330}{t + 19}$	$i = \frac{2460}{t + 16}$	$i = \frac{2060}{t + 13}$	$i = \frac{1900}{t + 12}$	$i = \frac{1220}{t + 12}$
10	$i = \frac{7620}{t + 36}$	$i = \frac{5840}{t + 29}$	$i = \frac{4320}{t + 23}$	$i = \frac{2820}{t + 16}$	$i = \frac{2820}{t + 17}$	$i = \frac{3100}{t + 23}$	$i = \frac{1520}{t + 13}$
25	$i = \frac{8300}{t + 33}$	$i = \frac{6600}{t + 32}$	$i = \frac{5840}{t + 30}$	$i = \frac{4320}{t + 27}$	$i = \frac{3300}{t + 17}$	$i = \frac{3940}{t + 26}$	$i = \frac{1700}{t + 10}$
50	$i = \frac{8000}{t + 28}$	$i = \frac{8890}{t + 38}$	$i = \frac{6350}{t + 27}$	$i = \frac{4750}{t + 24}$	$i = \frac{4750}{t + 25}$	$i = \frac{4060}{t + 21}$	$i = \frac{1650}{t + 8}$
100	$i = \frac{9320}{t + 33}$	$i = \frac{9520}{t + 36}$	$i = \frac{7370}{t + 31}$	$i = \frac{5590}{t + 28}$	$i = \frac{6100}{t + 29}$	$i = \frac{5330}{t + 26}$	$i = \frac{1960}{t + 10}$

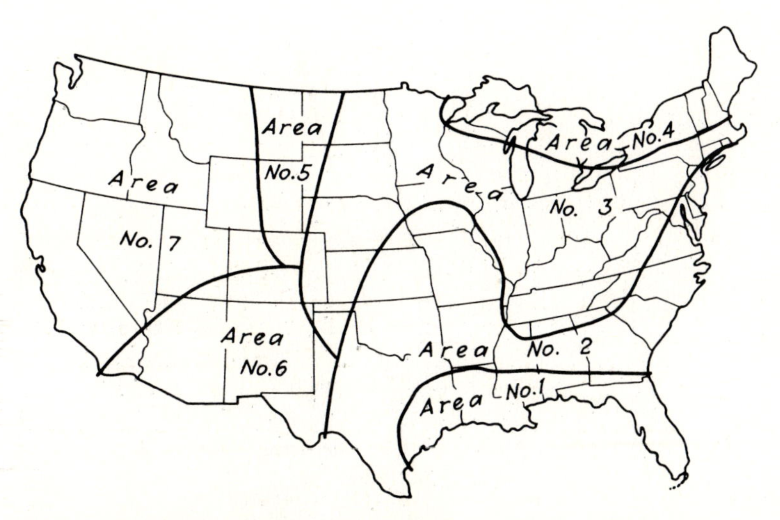

Figure 13-5 Map showing areas of approximately similar rainfall characteristics.

Table 13-4 Rainfall intensity at New Orleans and Shreveport, Louisiana
(mm/h)

Duration, min	New Orleans			Shreveport		
	2 yr	5 yr	10 yr	2 yr	5 yr	10 yr
5	157	185	206	132	168	193
6	150	178	198	127	160	183
7	145	173	191	122	152	178
8	140	165	185	117	147	170
9	135	160	180	112	142	165
10	130	155	173	109	137	157
11	127	150	168	104	132	152
12	122	147	163	102	127	150
13	119	142	160	99	124	145
14	114	140	155	94	122	140
15	112	135	152	91	117	137
16	109	132	147	89	114	132
17	107	130	145	86	112	130
18	104	124	140	84	109	127
19	102	122	137	84	107	124
20	99	119	135	81	104	119
21	97	117	132	79	102	117
22	94	114	130	76	99	114
23	91	112	127	76	97	112
24	91	112	124	74	94	112
25	89	109	122	74	94	109
26	86	107	119	71	91	107
27	86	104	117	69	89	104
28	84	102	117	69	86	102
29	81	102	114	66	86	102
30	81	99	112	66	84	99
31	79	97	109	64	84	97
32	79	97	109	64	81	97
33	76	94	107	64	81	94
34	76	94	107	61	79	91
35	74	94	104	61	79	91
—	—	—	—	—	—	—
40	69	86	97	56	71	84
45	64	81	91	51	69	79
50	61	76	86	48	64	74
55	56	71	81	46	61	71
60	53	69	79	43	56	66

(mm × 0.04 = in)

Actual data for New Orleans and Shreveport, Louisiana are presented in Table 13-4 for return periods of 2, 5, and 10 years. Data which are applicable to specific areas are generally available and should be used whenever possible. It should be observed that both New Orleans and Shreveport lie within Area 1 of Fig. 13-5, but have significantly different recorded intensity-duration-frequency relations.

13-7 Use of Intensity-Duration-Frequency Data

In Art. 13-5 it was pointed out that the critical rainfall duration was equal to the time of concentration. Design of storm sewers therefore requires estimation of the time of concentration using the techniques of Art. 13-4. This time of concentration is then used as the rainfall duration in determining intensity. The intensity is then used in the rational formula, Eq. (13-2), to determine the flow in the storm sewer segment being designed.

Example A sewer line drains a single-family residential area with $C = 0.35$. The distance of flow from the most remote point is 60 m over ordinary grass with a slope of 4 percent. The area drained is 100,000 m^2 and the intensity-duration formula is:

$$i = \frac{5230}{t + 30} \text{ mm/h}$$

from Fig. 13-3, the time of concentration is found to be 15 min, and,

$$i = \frac{5230}{45} = 116 \text{ mm/h}$$

Substituting in the rational formula,

$$\begin{aligned} Q &= CiA \\ &= 0.35 \times 0.116 \text{ m/h} \times 100{,}000 \text{ m}^2 \\ &= 4060 \text{ m}^3/\text{h} \\ &= 1.13 \text{ m}^3/\text{s} \end{aligned}$$

The engineer must select the appropriate return period for the design storm. This establishes the frequency with which the collection system will be overloaded, on average. It is too expensive to design a sewer to carry the largest flow which could ever occur, but, on the other hand, flood damage to property should be avoided when possible. A formal calculation of minimum expected cost is seldom made in storm drainage design for urban areas, but consideration of these costs is implicit in selection of 2- to 5-year frequencies for residential areas and 10- to 15-year frequencies for business and commercial districts. An example of the application of the rational method is presented in Chap. 17.

13-8 Other Techniques

The *unit hydrograph* procedure described in Art. 3-27 may be applied to urban areas if sufficient rainfall-runoff data have been accumulated. The hydrograph will reflect the effects of depression storage and infiltration more accurately than the rational technique. Gutter storage may be quite significant in design level storms, and this is also included in derived hydrographs.

The application of such techniques to specific urban areas has been reported.[2, 3, 4] Storm rainfall and physical parameters of the drainage area may be used to predict runoff from areas of different characteristics and degrees of imperviousness.[5, 6] The U.S. Soil Conservation Service technique for establishing peak runoff discharges[4] is particularly useful for areas less than 8 km^2.

PROBLEMS

13-1 A residential urban area has the following proportions of various areas: Roofs, 25 percent; asphalt pavement, 14 percent; concrete sidewalk, 5 percent; gravel driveways, 7 percent; grassy lawns with average soil and little slope, 49 percent. Compute the runoff coefficient for the whole area using the values of Table 13-1.

13-2 An urban area has an area of 100,000 m^2 and a runoff coefficient of 0.45. Using a duration of 25 min and the curves of Fig. 13-4 compute the runoff resulting from point rainfalls with a recurrence interval of 5 and 15 years.

13-3 Using a runoff coefficient of 0.40 and the rainfall intensities of the 10-year curve of Fig. 13-4, plot a new curve showing runoff in mm/h as ordinate and durations up to 2 hours as abscissa for the area of Prob. 13-2.

13-4 Using Fig. 13-3 determine the time of concentration for a flow which travels 50 m over average grass on a slope of 5 percent and 100 m over concrete pavement with a slope of 1 percent.

REFERENCES

1. *Design and Construction of Sanitary and Storm Sewers*, American Society of Civil Engineers, *Manual of Practice*, **37**, New York, 1960.
2. Tholin, A. L. and C. J. Keifer: "Hydrology of Urban Runoff," *Journal Sanitary Engineering Division, Proceedings American Society of Civil Engineers*, **85 SA-2**, March, 1959.
3. McPherson, M. B.: "Urban Hydrological Modeling and Catchment Research in the U.S.A.," *American Society of Civil Engineers Urban Water Resources Research Program*, American Society of Civil Engineers, New York, 1975.
4. "Urban Hydrology for Small Watersheds," Soil Conservation Service, *Technical Release*, No. **55**, 1975.
5. Tersfriep, M. L. and J. B. Stall: "The Illinois Urban Drainage Area Simulator, ILLUDAS," Illinois State Water Survey, *Bulletin*, **58**, Urbana, Illinois, 1974.
6. Bras, R. L. and F. E. Perkins: "Effects of Urbanization on Catchment Response," *Journal Hydraulics Division Proceedings American Society of Civil Engineers*, **101 H-3**, 1975.

CHAPTER

FOURTEEN

SEWER PIPES

14-1

The pipe materials used to transport water (Chap. 5) may be used for sewage carriage as well. It is common, however, to employ less expensive materials such as clay, concrete, or plastic, depending upon the particular application. Cast-iron or steel pipes are used only under unusual loading conditions or for force mains in which the sewage is under pressure.

PRECAST SEWERS

14-2 Clay Sewer Pipe

Clay pipe is made of clay or shale which has been ground, wet, molded, dried, and burned in a kiln. Near the end of the burning process, sodium chloride is added to the kiln and vaporizes to form a hard waterproof glaze by reacting with the pipe surface. The burning itself produces a fusion, or vitrification, of the clay, making it very hard and dense. Vitrified glazed clay pipe is not subject to mineral or bacterial corrosion (Art. 14-12).

Clay pipe is manufactured with integral bell and spigot ends fitted with polymeric rings of various designs on the spigot end,[1] with fitted fiberglass polyester sockets, and in a plain end configuration in which pipe are joined using a sleeve which fits over two abutting ends.[2] Clay pipe is manufactured to standard specifications of the ASTM[3] in diameters from 100 to 1070 mm (4 to 42 in). Tables 14-1 and 14-2 present the strength and dimensions of standard and extra-strength clay pipe.

Table 14-1 Minimum crushing strength of clay pipe[3]
(*Reprinted by Permission of the American Society for Testing and Materials, Copyright 1977.*)

Nominal size, in (mm)	Extra strength clay pipe		Standard strength clay pipe	
	lbf/linear ft	kgf/linear m	lbf/linear ft	kgf/linear m
4 (100)	2,000	2,980	1,200	1,790
6 (150)	2,000	2,980	1,200	1,790
8 (200)	2,200	3,270	1,400	2,080
10 (250)	2,400	3,570	1,600	2,380
12 (305)	2,600	3,870	1,800	2,680
15 (380)	2,900	4,320	2,000	2,980
18 (460)	3,300	4,910	2,200	3,270
21 (530)	3,850	5,730	2,400	3,570
24 (610)	4,400	6,550	2,600	3,870
27 (690)	4,700	6,990	2,800	4,170
30 (760)	5,000	7,440	3,300	4,910
33 (840)	5,500	8,190	3,600	5,360
36 (915)	6,000	8,930	4,000	5,950
39 (990)	6,600	9,820		
42 (1,070)	7,000	10,410		

Table 14-2 Dimensions of clay pipe[3]
(*Reprinted by Permission of the American Society for Testing and Materials, Copyright, 1977.*)

Nominal size, in (mm)	Laying length: Min, ft (m)	Laying length: Limit of minus variation, in/ft (mm/m)	Difference in length of two opposite sides, max, in (mm)	Outside diameter of barrel, in (mm): Min	Outside diameter of barrel, in (mm): Max	Inside diameter of socket at $\frac{1}{2}$ in (13 mm) above base, min, in (mm)[a, b]
4 (100)	2 (0.61)	$\frac{1}{4}$ (20)	$\frac{5}{16}$ (8)	$4\frac{7}{8}$ (124)	$5\frac{1}{8}$ (130)	$5\frac{3}{4}$ (146)
6 (150)	2 (0.61)	$\frac{1}{4}$ (20)	$\frac{3}{8}$ (9)	$7\frac{1}{16}$ (179)	$7\frac{7}{16}$ (189)	$8\frac{3}{16}$ (208)
8 (200)	2 (0.61)	$\frac{1}{4}$ (20)	$\frac{7}{16}$ (11)	$9\frac{1}{4}$ (235)	$9\frac{3}{4}$ (248)	$10\frac{1}{2}$ (267)
10 (250)	2 (0.61)	$\frac{1}{4}$ (20)	$\frac{7}{16}$ (11)	$11\frac{1}{2}$ (292)	12 (305)	$12\frac{3}{4}$ (324)
12 (305)	2 (0.61)	$\frac{1}{4}$ (20)	$\frac{7}{16}$ (11)	$13\frac{3}{4}$ (349)	$14\frac{5}{16}$ (364)	$15\frac{1}{8}$ (384)
15 (380)	3 (0.91)	$\frac{1}{4}$ (20)	$\frac{1}{2}$ (13)	$17\frac{3}{16}$ (437)	$17\frac{13}{16}$ (452)	$18\frac{5}{8}$ (473)
18 (460)	3 (0.91)	$\frac{1}{4}$ (20)	$\frac{1}{2}$ (13)	$20\frac{5}{8}$ (524)	$21\frac{7}{16}$ (545)	$22\frac{1}{4}$ (565)
21 (530)	3 (0.91)	$\frac{1}{4}$ (20)	$\frac{9}{16}$ (14)	$24\frac{1}{8}$ (613)	25 (635)	$25\frac{7}{8}$ (657)
24 (610)	3 (0.91)	$\frac{3}{8}$ (30)	$\frac{9}{16}$ (14)	$27\frac{1}{2}$ (699)	$28\frac{1}{2}$ (724)	$29\frac{3}{8}$ (746)
27 (690)	3 (0.91)	$\frac{3}{8}$ (30)	$\frac{5}{8}$ (16)	31 (787)	$32\frac{1}{8}$ (816)	33 (838)
30 (760)	3 (0.91)	$\frac{3}{8}$ (30)	$\frac{5}{8}$ (16)	$34\frac{3}{8}$ (873)	$35\frac{5}{8}$ (905)	$36\frac{1}{2}$ (927)
33 (840)	3 (0.91)	$\frac{3}{8}$ (30)	$\frac{5}{8}$ (16)	$37\frac{5}{8}$ (956)	$38\frac{15}{16}$ (989)	$39\frac{7}{8}$ (1013)
36 (915)	3 (0.91)	$\frac{3}{8}$ (30)	$\frac{11}{16}$ (17)	$40\frac{3}{4}$ (1035)	$42\frac{1}{4}$ (1073)	$43\frac{1}{4}$ (1099)
39 (990)	5 (1.52)	$\frac{1}{4}$ (20)	$\frac{3}{4}$ (19)	$45\frac{3}{8}$ (1152)	$47\frac{1}{4}$ (1200)	$48\frac{1}{2}$ (1232)
42 (1070)	5 (1.52)	$\frac{3}{8}$ (30)	$\frac{7}{8}$ (23)	$48\frac{1}{2}$ (1232)	51 (1295)	$52\frac{1}{2}$ (1333)

Table 14-2 (*continued*)

Nominal size, in (mm)	Depth of socket[a, b]		Thickness of barrel: Extra strength		Thickness of barrel: Standard strength		Thickness of socket at $\frac{1}{2}$ in (13 mm) from outer end[b]	
	Nominal, in (mm)	Min, in (mm)	Nominal, in (mm)	Min, in (mm)	Nominal, in (mm)	Min, in (mm)	Nominal, in (mm)	Min, in (mm)
4 (100)	$1\frac{3}{4}$ (44)	$1\frac{1}{2}$ (38)	$\frac{5}{8}$ (16)	$\frac{9}{16}$ (14)	$\frac{1}{2}$ (13)	$\frac{7}{16}$ (11)	$\frac{7}{16}$ (11)	$\frac{3}{8}$ (9)
6 (150)	$2\frac{1}{4}$ (57)	2 (51)	$\frac{11}{16}$ (17)	$\frac{9}{16}$ (14)	$\frac{5}{8}$ (16)	$\frac{9}{16}$ (14)	$\frac{1}{2}$ (13)	$\frac{7}{16}$ (11)
8 (200)	$2\frac{1}{2}$ (64)	$2\frac{1}{4}$ (57)	$\frac{7}{8}$ (22)	$\frac{3}{4}$ (19)	$\frac{3}{4}$ (19)	$\frac{11}{16}$ (17)	$\frac{9}{16}$ (14)	$\frac{1}{2}$ (13)
10 (250)	$2\frac{5}{8}$ (67)	$2\frac{3}{8}$ (60)	1 (25)	$\frac{7}{8}$ (22)	$\frac{7}{8}$ (22)	$\frac{13}{16}$ (21)	$\frac{5}{8}$ (16)	$\frac{9}{16}$ (14)
12 (305)	$2\frac{3}{4}$ (70)	$2\frac{1}{2}$ (64)	$1\frac{3}{16}$ (30)	$1\frac{1}{16}$ (27)	1 (25)	$\frac{15}{16}$ (24)	$\frac{3}{4}$ (19)	$\frac{11}{16}$ (17)
15 (380)	$2\frac{7}{8}$ (73)	$2\frac{5}{8}$ (67)	$1\frac{1}{2}$ (38)	$1\frac{3}{8}$ (35)	$1\frac{1}{4}$ (31)	$1\frac{1}{8}$ (29)	$\frac{15}{16}$ (24)	$\frac{7}{8}$ (22)
18 (460)	3 (76)	$2\frac{3}{4}$ (70)	$1\frac{7}{8}$ (48)	$1\frac{3}{4}$ (44)	$1\frac{1}{2}$ (38)	$1\frac{3}{8}$ (35)	$1\frac{1}{8}$ (29)	$1\frac{1}{16}$ (27)
21 (530)	$3\frac{1}{4}$ (83)	3 (76)	$2\frac{1}{4}$ (57)	2 (51)	$1\frac{3}{4}$ (44)	$1\frac{5}{8}$ (41)	$1\frac{5}{16}$ (33)	$1\frac{3}{16}$ (30)
24 (610)	$3\frac{3}{8}$ (86)	$3\frac{1}{8}$ (79)	$2\frac{1}{2}$ (64)	$2\frac{1}{4}$ (57)	2 (51)	$1\frac{7}{8}$ (48)	$1\frac{1}{2}$ (38)	$1\frac{3}{8}$ (35)
27 (690)	$3\frac{1}{2}$ (89)	$3\frac{1}{4}$ (83)	$2\frac{3}{4}$ (70)	$2\frac{1}{2}$ (64)	$2\frac{1}{4}$ (57)	$2\frac{1}{8}$ (54)	$1\frac{11}{16}$ (43)	$1\frac{9}{16}$ (40)
30 (760)	$3\frac{5}{8}$ (92)	$3\frac{3}{8}$ (86)	3 (76)	$2\frac{3}{4}$ (70)	$2\frac{1}{2}$ (64)	$2\frac{3}{8}$ (60)	$1\frac{7}{8}$ (48)	$1\frac{3}{4}$ (44)
33 (840)	$3\frac{3}{4}$ (95)	$3\frac{1}{2}$ (89)	$3\frac{1}{4}$ (83)	3 (76)	$2\frac{5}{8}$ (67)	$2\frac{1}{2}$ (64)	2 (51)	$1\frac{3}{4}$ (44)
36 (915)	4 (102)	$3\frac{3}{4}$ (95)	$3\frac{1}{2}$ (89)	$3\frac{1}{4}$ (83)	$2\frac{3}{4}$ (70)	$2\frac{5}{8}$ (67)	$2\frac{1}{16}$ (52)	$1\frac{7}{8}$ (48)
39 (990)	$4\frac{1}{8}$ (105)	$3\frac{7}{8}$ (98)	$3\frac{3}{4}$ (95)	$3\frac{3}{8}$ (86)	... ...	... ...	$2\frac{3}{4}$ (70)	$2\frac{5}{8}$ (67)
42 (1070)	$4\frac{1}{8}$ (105)	$3\frac{7}{8}$ (98)	4 (102)	$3\frac{1}{2}$ (89)	... ...	... ...	$2\frac{3}{4}$ (70)	$2\frac{5}{8}$ (67)

[a] The minimums for inside diameter of socket and depth of socket may be waived where such dimensions are conducive to the proper application of the joint.

[b] Plain-end pipe shall conform to the dimensions in Table 14-2, except those dimensions pertaining to sockets.

Pipe fittings are cast in the forms illustrated in Fig. 14-1. Other shapes may be made on special order. Wyes and tees are used for joining house sewers to common sewers. Stoppers are used in bell ends of pipe or fittings and fastened with mortar to close the pipe until connections are made later. A saddle is used when a hole is broken into the top of a sewer to allow a vertical connection to be made. Such a connection is also used when the common sewer is very deep. Slants are used when a hole is broken in the side of the sewer and the branch comes in at an angle and particularly if the sewer is of concrete or brick. Concrete or mortar is used liberally around the joint to prevent weakness and leakage. The trap pictured is a type sometimes used for main traps in house sewers.

14-3 Strength and Loading of Vitrified Clay Pipe

The static load produced on buried pipe may be calculated using an equation of the form:

$$W = CwB^2 \tag{14-1}$$

in which W is the load on the pipe per unit length, w is the weight of the fill material per unit volume, B is the width of the trench just below the top of the pipe

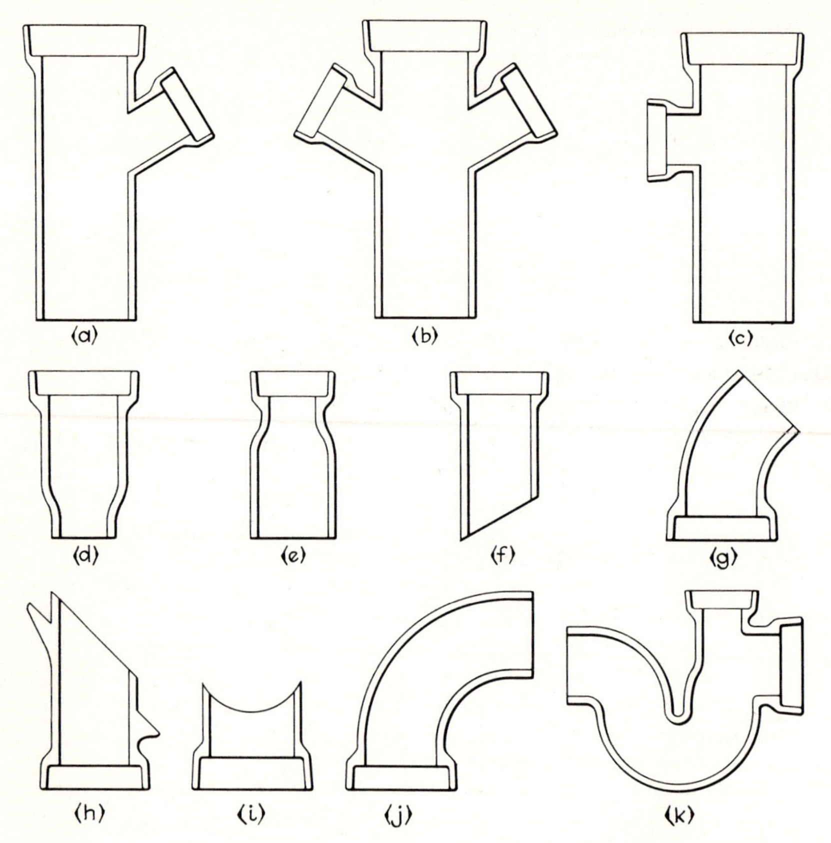

Figure 14-1 Sections of bell-and-spigot fittings. (*a*) Y branch. (*b*) Double Y branch. (*c*) T branch. (*d*) Reducer. (*e*) Increaser. (*f*) Slant, used for making connections to brick and concrete sewers. (*g*) Short-radius $\frac{1}{8}$ bend. Also obtainable as long-radius and $\frac{1}{16}$ bend. (*h*) Y saddle. (*i*) T saddle. Saddles are used where use of standard branches is impracticable. (*j*) Long-radius elbow or $\frac{1}{4}$ bend. Also obtainable in short radius. (*k*) Running trap.

(which for ease of joining is generally not less than one and one half pipe diameters plus 300 mm), and C is a coefficient which depends upon the depth of the trench, the character of the construction, and the fill material. For ordinary trench construction the value of C can be calculated from:

$$C = \frac{1 - e^{-2K\mu' H/B}}{2K\mu'} \tag{14-2}$$

in which H is the depth of fill above the pipe, B is as identified above, K is the ratio of active lateral pressure to vertical pressure, and μ' is the coefficient of sliding friction between the fill material and the sides of the trench. The product $K\mu'$ ranges from 0.1 to 0.16 for most soils (Table 14-3).

Graphical solutions of Eqs. (14-1) and (14-2) and of similar equations for other construction conditions are presented in references 5 and 6. Tabular listings

Table 14-3 Value of the product $K\mu'$

Soil type	Maximum value of $K\mu'$
Cohesionless granular material	0.192
Sand and gravel	0.165
Saturated top soil	0.150
Clay	0.130
Saturated clay	0.110

of allowable loadings or depths of fill for the bedding conditions shown in Figs. 14-2 and 14-4 are also available in manufacturers' publications.

The weights of the materials commonly used for backfill are presented in Table 14-4. From these values the backfill load can be calculated using Eqs. (14-1) and (14-2).

Table 14-4 Unit weight of backfill material

Material	Unit weight kg/m³	Unit weight lb/ft³
Dry sand	1600	100
Ordinary sand	1840	115
Wet sand	1920	120
Damp clay	1920	120
Saturated clay	2080	130
Saturated topsoil	1840	115
Sand and damp topsoil	1600	100

Example A 610-mm (24-in) sewer is to be placed in an ordinary trench 3.66 m (12 ft) deep, 1.22 m (4 ft) wide which will be backfilled with wet clay weighing 1920 kg/m^3 (120 lb/ft^3). Determine the load upon the pipe.

SOLUTION From Eq. (14-2),

$$C = \frac{1 - e^{-0.110(2)(3.66/1.22)}}{(2)(0.110)}$$

$$= 2.20$$

From Eq. (14-1),

$$W = 2.20(1920)(1.22)^2$$

$$= 6290 \text{ kg/m } (4220 \text{ lb/ft})$$

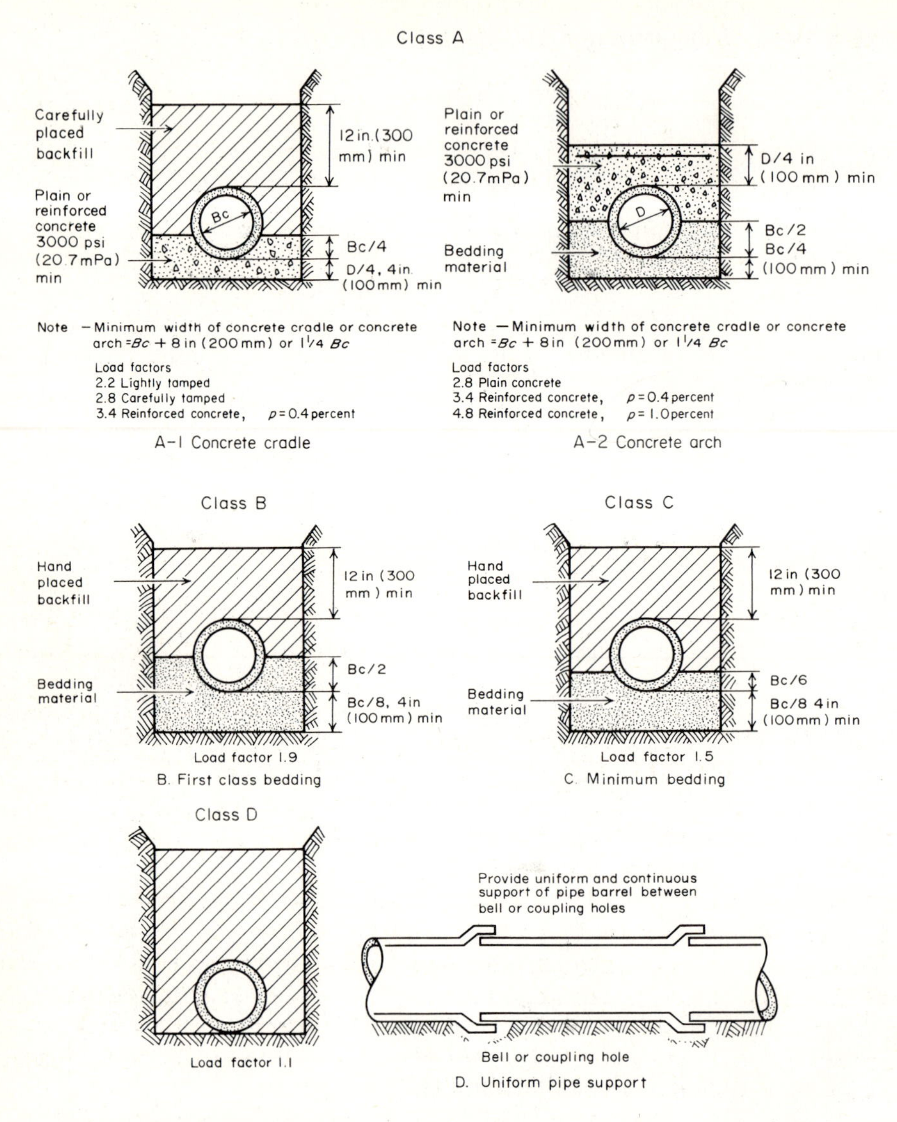

Figure 14-2 Methods of bedding clay pipe and load factors applicable to strength.[4] (*Reprinted by Permission of the American Society for Testing and Materials, Copyright 1977.*)

The type of bedding required may then be determined from Table 14-1 and Fig. 14-2. Standard strength clay pipe has a minimum crushing strength of 3870 kg/m. Applying a safety factor of 1.5 yields an allowable load of 2580 kg/m. The load factor for the pipe is thus 6290/2580 = 2.44. From Fig. 14-2 one may observe that only a concrete encasement (Class A) will provide this load factor. Alternately, extra strength pipe could be used with a crushing strength of 6550 kg/m which, through a similar calculation, provides

Table 14-5 Proportion of "long" superficial loads reaching pipe in trenches

Ratio of depth to width	Sand and damp topsoil	Saturated topsoil	Damp yellow clay	Saturated yellow clay
0.0	1.00	1.00	1.00	1.00
0.5	0.85	0.86	0.88	0.89
1.0	0.72	0.75	0.77	0.80
1.5	0.61	0.64	0.67	0.72
2.0	0.52	0.55	0.59	0.64
2.5	0.44	0.48	0.52	0.57
3.0	0.37	0.41	0.45	0.51
4.0	0.27	0.31	0.35	0.41
5.0	0.19	0.23	0.27	0.33
6.0	0.14	0.17	0.20	0.26
8.0	0.07	0.09	0.12	0.17
10.0	0.04	0.05	0.07	0.11

a load factor of 1.4. This, from Fig. 14-2, may be provided by any bedding other than Class D.

The method of bedding the pipe is important in developing its strength, in assuring it is laid to grade, and in preventing subsequent settlement. In areas with poor soil conditions (heavy clays, peats, etc.) bedding is particularly critical. In some circumstances trenches may require sheeting and flooring or even pile support to prevent excessive settling or damage to the pipe.[7]

In addition to the loads imposed by backfill, loads produced by building foundations, stockpiled bulk materials, traffic, and machinery may also reach buried sewers. The proportion of the load reaching the sewer may be estimated from Tables 14-5 and 14-6. Long loads (Table 14-5) are those longer than the trench width, such as stockpiled bulk materials. Short loads (Table 14-6) are those resulting from traffic, foundations at right angles to the trench, etc. In Table 14-6 the maximum values are for a length of load equal to the trench width. The minima are for a load length one-tenth the trench width.

Example A sewer trench in damp clay, 1.22 m wide is crossed by a concrete tunnel carrying steam pipes. The tunnel is 0.91 m wide and weighs 1340 kg/m. Its bottom is 1.83 m above the top of the sewer. What load will be transmitted to the pipe?

SOLUTION The weight of the tunnel per meter of pipe length covered is $(1340/0.91) \times 1.22 = 1796$ kg/m. The ratio of depth to width is $1.83/1.22 = 1.5$. From Table 14-6, the maximum proportion of the load reaching the pipe will be 0.51. Therefore the load reaching the pipe will be $0.51 \times 1796 = 916$ kg/m.

Table 14-6 Proportion of "short" superficial loads reaching pipe in trenches

Ratio of depth to width	Sand and damp topsoil		Saturated topsoil		Damp clay		Saturated clay	
	Max	Min	Max	Min	Max	Min	Max	Min
0.0	1.00	1.00	1.00	1.00	1.00	1.00	1.00	1.00
0.5	0.77	0.12	0.78	0.13	0.79	0.13	0.81	0.13
1.0	0.59	0.02	0.61	0.02	0.63	0.02	0.66	0.02
1.5	0.46	.....	0.48	.....	0.51	.....	0.54	
2.0	0.35	.....	0.38	.....	0.40	.....	0.44	
2.5	0.27	.....	0.29	.....	0.32	.....	0.35	
3.0	0.21	.....	0.23	.....	0.25	.....	0.29	
4.0	0.12	.....	0.14	.....	0.16	.....	0.19	
5.0	0.07	.....	0.09	.....	0.10	.....	0.13	
6.0	0.04	.....	0.05	.....	0.06	.....	0.08	
8.0	0.02	.....	0.02	.....	0.03	.....	0.04	
10.0	0.01	.....	0.01	.....	0.01	.....	0.02	

14-4 Plain Concrete Sewer Pipe

Precast concrete pipe may be used for small storm drains and for sanitary sewers in locations where grades, temperatures, or sewage characteristics prevent corrosion (Art. 14-12). It should not be used for sanitary sewers in the southern United States where high temperatures and flat sewer grades are common.

Precast concrete pipe is manufactured to specifications of the ASTM.[8] For use in sanitary sewers, joints should be constructed using rubber gaskets

Table 14-7 Physical and dimensional requirements for nonreinforced concrete pipe

(Reprinted by Permission of the American Society for Testing and Materials, Copyright 1977.)

Internal diameter, mm (in)	Class 1		Class 2		Class 3	
	Minimum thickness of wall, mm	Minimum strength, kN/linear m, three-edge bearing	Minimum thickness of wall, mm	Minimum Strength, kN/linear m, three-edge bearing	Minimum thickness of wall, mm	Minimum strength, kN/linear m, three-edge bearing
100 (4)	15.9	21.9	19.0	29.2	22.2	35.0
150 (6)	15.9	21.9	19.0	29.2	25.4	35.0
200 (8)	19.0	21.9	22.2	29.2	28.6	35.0
250 (10)	22.2	23.3	25.4	29.2	31.8	35.0
310 (12)	25.4	26.3	34.9	32.8	44.5	37.9
380 (15)	31.8	29.2	41.3	37.9	47.6	42.2
460 (18)	38.1	32.1	50.8	43.8	57.2	48.1
530 (21)	44.5	35.0	57.2	48.1	69.9	56.2
610 (24)	54.0	37.9	76.2	52.5	95.3	64.2

(Fig. 14-3). Pipe manufactured for rubber gasket joints must be held to closer tolerances than ordinary concrete pipe,[9] and must be more carefully placed.

Loads upon concrete pipe are calculated using the techniques of Art. 14-3. The dimensions and load-bearing capacity of concrete pipe are presented in Table 14-7. The pipe is manufactured in diameters from 100 to 610 mm (4 to 24 in) in three wall thickness classes and two strengths.

Bedding classes for concrete pipe are shown in Fig. 14-4. These, it should be noted, while similar, are not identical to those shown in Fig. 14-2 for clay pipe.

Allowable loads for various sizes of concrete pipe as a function of strength and bedding are presented in Table 14-8.

Joints for use in storm sewers may be made as in sanitary sewers, of bituminous material, or of cement in the form of mortar, neat cement, or grout (Fig. 14-3). Procedures for forming these joints are presented in reference 6.

14-5 Reinforced Concrete Sewer Pipe

Precast concrete pipe in diameters larger than 610 mm (24 in) is reinforced. Reinforcing can also be obtained in smaller sizes. The pipe is manufactured in sizes from 305 to 4570 mm (12 to 180 in). Joints are either bell and spigot or tongue and groove from 305 to 760 mm and tongue and groove above that size.

Table 14-8 Supporting strength of concrete pipe[a] [5] per linear foot of pipe in thousands of pounds (kips)

ASTM spec. no.	Standard strength concrete sewer pipe C 14 Safety factor = 1.5				Extra strength concrete sewer pipe C 14 Safety factor = 1.5			
Bedding class	D	C	B	A	D	C	B	A
Load factor	1.1	1.5	1.9	3.0	1.1	1.5	1.9	3.0
Internal diameter of pipe, in								
6	0.8	1.1	1.4	2.2	1.5	2.0	2.5	4.0
8	0.9	1.3	1.6	2.6	1.5	2.0	2.5	4.0
10	1.0	1.4	1.8	2.8	1.5	2.0	2.5	4.0
12	1.1	1.5	1.9	3.0	1.6	2.2	2.8	4.5
15	1.2	1.7	2.2	3.5	2.0	2.8	3.5	5.5
18	1.4	2.0	2.5	4.0	2.4	3.3	4.2	6.6
21	1.6	2.2	2.8	4.4	2.8	3.8	4.9	7.8
24	1.7	2.4	3.0	4.8	2.9	4.0	5.1	8.0

[a] Supporting strengths shown in table are for concrete pipe meeting ASTM specifications (3-edge bearing test) and include safety and bedding factors as indicated. (kips/ft × 14.6 = kN/m, in × 25.4 = mm)

A —Typical cross sections of joints with mortar or mastic packing

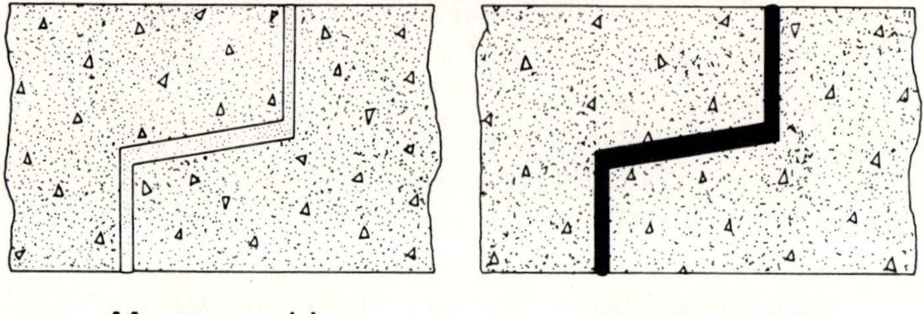

B —Typical cross sections of basic compression type rubber gasket joints

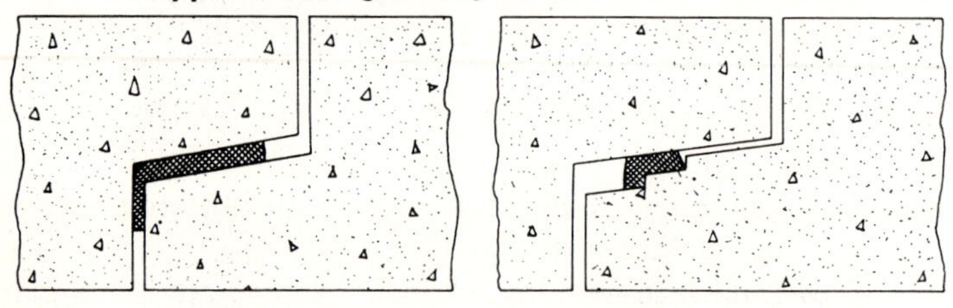

C —Typical cross sections of opposing shoulder type joint with O-ring gasket

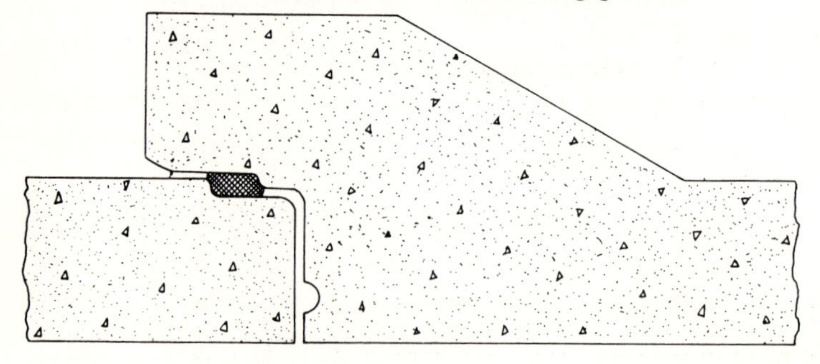

D —Typical cross section of spigot groove type joint with O-ring gasket

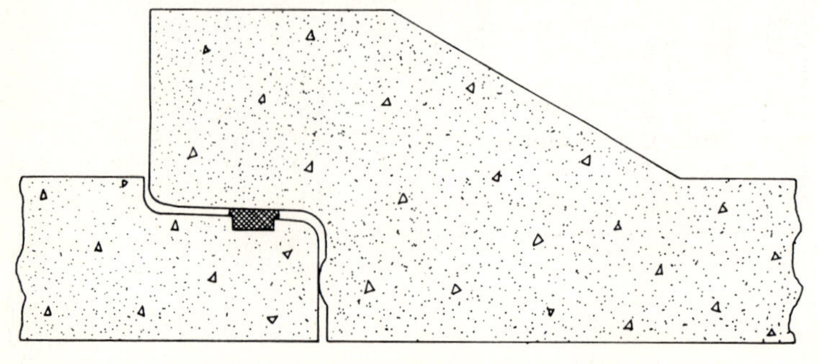

Figure 14-3 Typical concrete pipe joints.[5] (*Courtesy American Concrete Pipe Association.*)

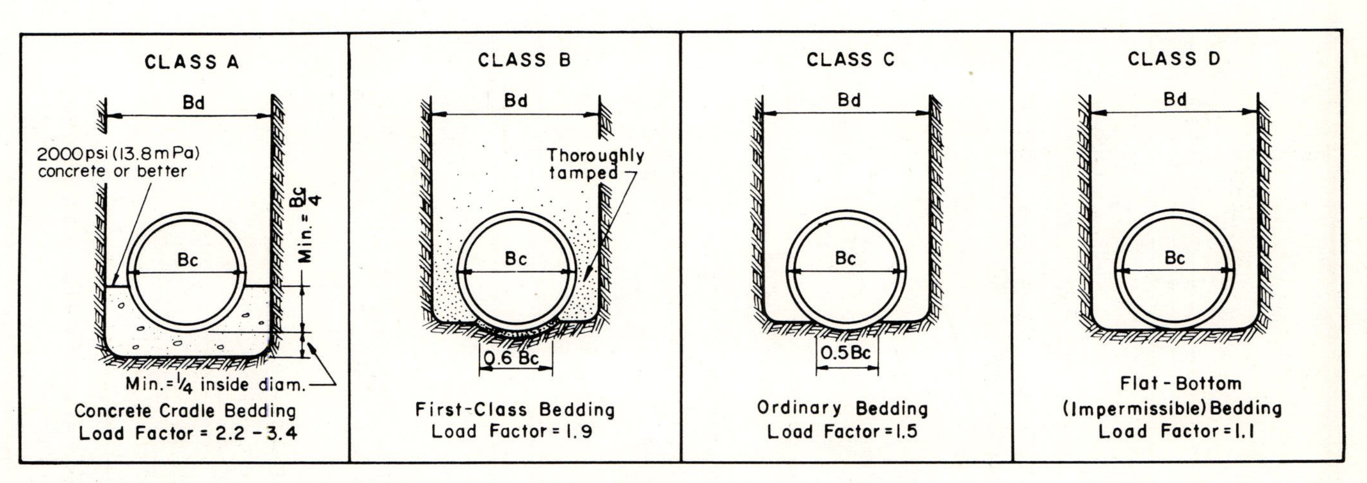

Figure 14-4 Bedding methods for concrete pipe.[6] (*Courtesy Portland Cement Association.*)

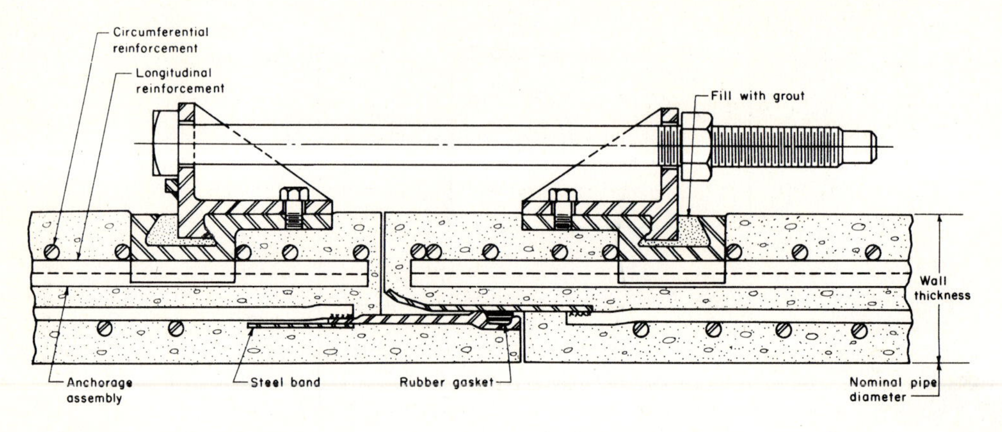

A. Subaqueous joint for concrete pipe outfall

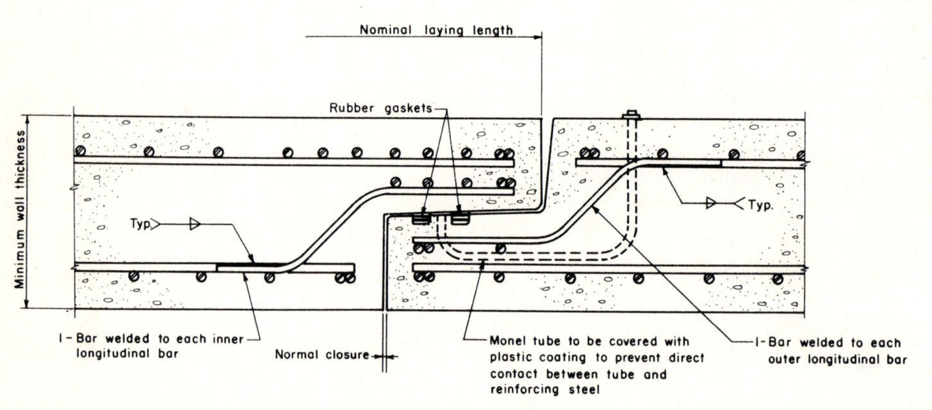

B. Double rubber gasket joint for large-diameter concrete pipe outfall

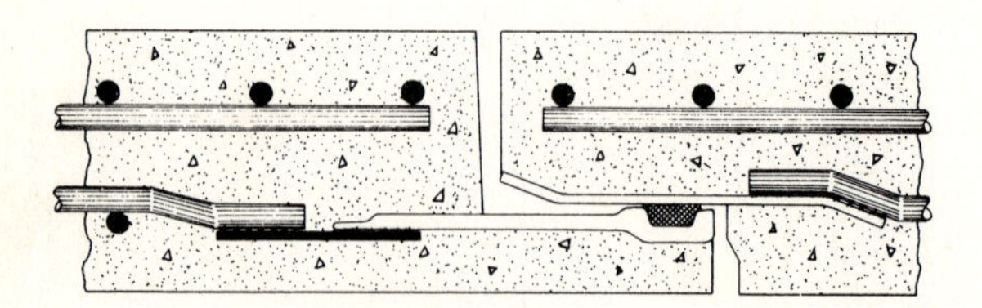

C. Typical cross section of steel end ring joint with spigot groove and O-ring gasket

Figure 14-5 Reinforced concrete pipe joints. (*Courtesy Portland Cement Association and American Concrete Pipe Association.*)

Table 14-9 Design loads for reinforced concrete pipe[10]

(Reprinted by Permission of the American Society for Testing and Materials, Copyright 1977.)

	Design load			
	To produce a 0.25-mm crack		Ultimate	
Class	N/m per mm dia	lb/ft per in dia	N/m per mm dia	lb/ft per in dia
I	38.3	800	57.4	1200
II	47.9	1000	71.8	1500
III	64.6	1350	95.8	2000
IV	95.8	2000	144.0	3000
V	144.0	3000	180.0	3750

Table 14-10 Dimensions of reinforced concrete pipe[a]

Internal diameter, mm (in)	Wall thickness, mm (in)		
	Wall A	Wall B	Wall C
310 (12)	44 ($1\frac{3}{4}$)	51 (2)	
380 (15)	47 ($1\frac{7}{8}$)	57 ($2\frac{1}{4}$)	
460 (18)	51 (2)	63 ($2\frac{1}{2}$)	
530 (21)	57 ($2\frac{1}{4}$)	70 ($2\frac{3}{4}$)	
610 (24)	63 ($2\frac{1}{2}$)	76 (3)	95 ($3\frac{3}{4}$)
690 (27)	66 ($2\frac{5}{8}$)	83 ($3\frac{1}{4}$)	101 (4)
760 (30)	70 ($2\frac{3}{4}$)	89 ($3\frac{1}{2}$)	108 ($4\frac{1}{4}$)
840 (33)	73 ($2\frac{7}{8}$)	95 ($3\frac{3}{4}$)	114 ($4\frac{1}{2}$)
910 (36)	76 (3)	101 (4)	120 ($4\frac{3}{4}$)
1070 (42)	89 ($3\frac{1}{2}$)	114 ($4\frac{1}{2}$)	130 ($5\frac{1}{4}$)
1220 (48)	101 (4)	127 (5)	146 ($5\frac{3}{4}$)
1370 (54)	114 ($4\frac{1}{2}$)	140 ($5\frac{1}{2}$)	159 ($6\frac{1}{4}$)
1520 (60)	127 (5)	152 (6)	171 ($6\frac{3}{4}$)
1680 (66)	140 ($5\frac{1}{2}$)	165 ($6\frac{1}{2}$)	184 ($7\frac{1}{4}$)
1830 (72)	152 (6)	178 (7)	197 ($7\frac{3}{4}$)
1980 (78)	165 ($6\frac{1}{2}$)	190 ($7\frac{1}{2}$)	209 ($8\frac{1}{4}$)
2130 (84)	178 (7)	203 (8)	222 ($8\frac{3}{4}$)
2290 (90)	190 ($7\frac{1}{2}$)	216 ($8\frac{1}{2}$)	235 ($9\frac{1}{4}$)
2440 (96)	203 (8)	229 (9)	248 ($9\frac{3}{4}$)
2590 (102)	216 ($8\frac{1}{2}$)	241 ($9\frac{1}{2}$)	260 ($10\frac{1}{4}$)
2740 (108)	229 (9)	254 (10)	273 ($10\frac{3}{4}$)
2900 (114)	241 ($9\frac{1}{2}$)		
3050 (120)	254 (10)		
3200 (126)	267 ($10\frac{1}{2}$)		
3350 (132)	279 (11)		
3500 (138)	292 ($11\frac{1}{2}$)		
3650 (144)	305 (12)		
3800 (150)	318 ($12\frac{1}{2}$)		
3960 (156)	330 (13)		
4110 (162)	343 ($13\frac{1}{2}$)		
4270 (168)	356 (14)		
4420 (174)	368 ($14\frac{1}{2}$)		
4570 (180)	381 (15)		

[a] Not all sizes are available in all classes. (See Table 14-11 and references 5, 6, and 10).

Table 14-11 Supporting strength of concrete pipe[a] [5] per linear foot of pipe in thousands of pounds (kips)

Reinforced concrete culvert, storm drain, and sewer pipe C76
Safety factor = 1.0. Based on 0.01-in (0.25-mm) crack

ASTM spec. no.	Class I				Class II				Class III				Class IV				Class V			
Bedding class	D	C	B	A	D	C	B	A	D	C	B	A	D	C	B	A	D	C	B	A
Load factor	1.1	1.5	1.9	3.0	1.1	1.5	1.9	3.0	1.1	1.5	1.9	3.0	1.1	1.5	1.9	3.0	1.1	1.5	1.9	3.0
Internal diameter of pipe, in																				
12					1.1	1.5	1.9	3.0	1.5	2.0	2.6	4.0	2.2	3.0	3.8	6.0	3.3	4.5	5.7	9.0
15					1.4	1.9	2.4	3.8	1.8	2.5	3.2	5.0	2.8	3.7	4.8	7.5	4.1	5.6	7.1	11.3
18					1.6	2.2	2.8	4.5	2.2	3.0	3.8	6.1	3.3	4.5	5.7	9.0	5.0	6.8	8.6	13.5
21					1.9	2.6	3.3	5.2	2.6	3.5	4.5	7.1	3.9	5.3	6.7	10.5	5.8	7.9	10.0	15.8
24					2.2	3.0	3.8	6.0	3.0	4.0	5.1	8.1	4.4	6.0	7.6	12.0	6.6	9.0	11.4	18.0
27					2.5	3.4	4.3	6.8	3.3	4.6	5.8	9.1	4.9	6.7	8.5	13.5	7.4	10.1	12.9	20.2
30					2.7	3.7	4.7	7.5	3.7	5.1	6.4	10.1	5.5	7.5	9.5	15.0	8.2	11.2	14.2	22.5
33					3.0	4.1	5.2	8.2	4.0	5.6	7.0	11.1	6.0	8.2	10.4	16.5	9.0	12.4	15.7	25.0
36					3.3	4.5	5.7	9.0	4.4	6.1	7.7	12.2	6.6	9.0	11.4	18.0	9.9	13.5	17.1	27.0
42					3.8	5.2	6.6	10.5	5.2	7.1	8.9	14.2	7.7	10.5	13.3	21.0	11.5	15.7	20.0	31.5
48					4.4	6.0	7.5	12.0	6.0	8.1	10.2	16.2	8.8	12.0	15.2	24.0	13.2	18.0	22.8	36.0
54					4.9	6.7	8.5	13.5	6.7	9.1	11.5	18.2	9.9	13.5	17.1	27.0	14.8	20.2	25.7	40.5
60	4.4	6.0	7.6	12.0	5.5	7.5	9.5	15.0	7.4	10.1	12.8	20.2	11.0	15.0	19.0	30.0	16.5	22.5	28.5	45.0
66	4.8	6.6	8.3	13.2	6.0	8.2	10.4	16.5	8.1	11.1	14.1	22.3	12.1	16.5	21.0	33.0	18.1	24.8	31.3	49.5
72	5.3	7.2	9.1	14.4	6.6	9.0	11.4	18.0	8.9	12.1	15.4	24.3	13.2	18.0	22.8	36.0	19.8	27.0	34.2	54.0
78	5.7	7.8	9.9	15.6	7.1	9.7	12.3	19.5	9.6	13.2	16.7	26.3	14.3	19.5	24.7	39.0				
84	6.1	8.4	10.6	16.8	7.7	10.5	13.3	21.0	10.4	14.2	18.0	28.4	15.4	21.0	26.6	42.0				
90	6.6	9.0	11.4	18.0	8.2	11.2	14.2	22.5	11.1	15.2	19.2	30.4								
96	7.0	9.6	12.2	19.2	8.3	12.0	15.2	24.0	11.9	16.2	20.5	32.4								
102	7.5	10.2	12.9	20.4	9.3	12.7	16.1	25.5	12.6	17.2	21.8	34.4								
108	7.9	10.8	13.7	21.6	9.9	13.5	17.1	27.0	13.4	18.2	23.1	37.0								

[a] Supporting strengths shown in table are for concrete pipe meeting ASTM specifications (3-edge bearing test) and include safety and bedding factors as indicated. (kips/ft × 14.6 = kN/m, in × 25.4 = mm)

Joints are made by mortaring the cleaned and wetted tongue and groove before assembly or by mechanical or O-ring joints like those shown in Fig. 14-5.

Reinforced concrete pipe is made in five classes with two wall thicknesses in class I and three wall thicknesses in the other four classes. Strength is based upon either the load which produces a 0.25-mm (0.1-in) crack, or the ultimate load. The design load for the different classes is presented in Table 14-9. Dimensions are given in Table 14-10. Allowable design loads are shown in Table 14-11.

Special fittings are seldom used with reinforced concrete pipe. In large sizes curves are made by deflecting the joints and then filling the resulting opening with poured concrete. Connections are made by cutting an opening and mortaring the smaller line to the larger.

If reinforced concrete pipe is used to convey untreated sanitary sewage it may be subject to corrosion in the same fashion as plain concrete pipe. Large concrete sewers may be lined in place with corrosion-resistant material (Art. 14-12). Outfall lines carrying treated sewage are unlikely to be subject to corrosion save in very unusual circumstances.

14-6 Asbestos Cement Pipe

Asbestos cement pipe is manufactured in the sizes and classes shown in Table 14-12.[11] The load upon this pipe may be calculated as for other rigid pipe (Art. 14-3). Joints are made using a sleeve and ring arrangement which slips over

Table 14-12 Minimum crushing loads for asbestos cement pipe[11]

(Reprinted by Permission of the American Society for Testing and Materials, Copyright 1977.)

Nominal size in (mm)	Crushing strength per lineal foot, lbf (kN/m)		
	Class 100	Class 150	Class 200
4 (102)	4,100 (59.8)	5,400 (78.8)	8,700 (126.9)
6 (152)	4,000 (58.4)	5,400 (78.8)	9,000 (131.3)
8 (203)	4,000 (58.4)	5,500 (80.2)	9,300 (135.8)
10 (254)	4,400 (64.2)	7,000 (102.1)	11,000 (160.5)
12 (304)	5,200 (75.8)	7,600 (110.8)	11,800 (172.3)
14 (356)	5,200 (75.8)	8,600 (125.5)	13,500 (197.1)
16 (406)	5,800 (84.6)	9,200 (134.2)	15,400 (224.8)
18 (457)	6,500 (94.8)	10,100 (147.4)	17,400 (254.0)
20 (508)	7,100 (103.6)	10,900 (159.0)	19,400 (283.2)
24 (610)	8,100 (118.2)	12,700 (185.3)	22,600 (329.9)
30 (762)	9,700 (141.5)	15,900 (231.9)	28,400 (414.6)
36 (914)	11,200 (163.4)	19,600 (285.9)	33,800 (493.5)

the ends of the pipe which are somewhat reduced in diameter and grooved to retain the rubber ring.[12] Wyes and tees are made.

Asbestos cement is subject to the corrosion mechanism common to sanitary sewers and may be undesirable for that reason.

14-7 Plastic Truss Pipe

Plastic truss pipe (Fig. 14-6) consists of an extruded shell with integral diagonal stiffeners between the inner and outer membranes. The space between the inner and outer surfaces is filled with lightweight concrete, which increases the pipe stiffness. The pipe sections are joined by chemical welding of the parent ABS material which yields, in effect, a continuous pipe. Miscellaneous fittings are available to permit connection to household sewers made of clay, solid plastic, asbestos cement, impregnated fiber, etc. The pipe is manufactured only in sizes of 200 to 380 mm (8 to 15 in).[13]

Plastic truss pipe is not rated in the same fashion as concrete or clay pipe. The manufacturer claims that its flexibility makes the three-point bearing test meaningless, since it can deform sufficiently to develop lateral support when in place. Design procedures are available in the manufacturer's literature. The major

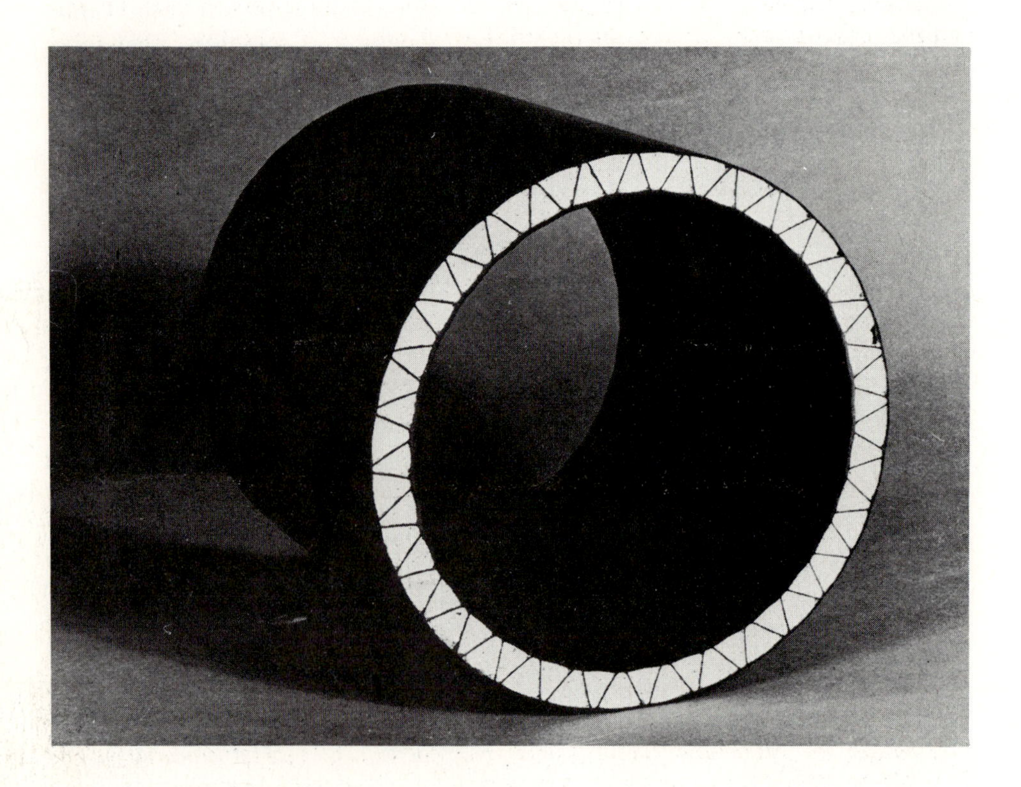

advantages of this pipe are its watertightness and its ability to undergo differential settling without failure.[14] Its long-term durability and resistance to corrosion if the inner surface were to be damaged, can only be established by the passage of time. Truss pipe has been in use since 1965.

14-8 Other Sewer Materials

Bituminized fiber pipe is manufactured in small sizes and is primarily used for house sewers. *Solid wall plastic* pipe is manufactured in sizes from 100 to 305 mm (4 to 12 in) under one specification,[15] and in sizes from 100 to 380 mm (4 to 15 in) under another.[16] The pipe is made of polyvinyl chloride (PVC) in either case. The difference between the two specifications is dimensional—wall thickness, etc.

Solid wall plastic pipe may be used alone, in conjunction with plastic truss pipe, and in vacuum and pressurized collection systems (Art. 16-17, 16-18). Like plastic truss pipe, the cemented joints tend to be virtually watertight.

Cast or *ductile iron* is used in force mains and small outfall lines. It may also be employed where sewers are constructed above the ground surface and in inverted siphons. Lines within sewage treatment plants are often constructed of iron. Joints are made in the same fashion as in water distribution systems.

Corrugated metal pipe is sometimes used for storm sewers. It may be galvanized, coated with bituminous material, or coated with asbestos. Corrugated metal pipe sections are available in many thicknesses and cross-sections. Their primary use is in roadway drainage.

14-9 Infiltration and Sewer Joints

All water which enters a sewer is likely to remain in it and pass through whatever pumping and treatment units are incorporated in the system. Since both pumping and treatment are expensive it may be economical to spend more upon the sewer to reduce the possibility of infiltration.

The cost of treating wastewater depends upon the plant size and the actual processes employed, and ranges from \$0.013 to 0.066/m^3 (\$0.05 to 0.25/1000 gal). The annual cost of treating infiltration can be calculated from:

$$\text{A.C.} = I \times d \times L \times 365 \times \text{U.C.} \tag{14-3}$$

where A.C. = the annual cost of treating infiltration
d = the diameter of the line
L = its length
I = infiltration per unit diameter per unit length per day, and
U.C. = unit cost of waste treatment

The annual cost per kilometer of sewer could be as much as \$1000. The elimination of this flow could justify an additional capital investment of \$10,000/km (\$16,000/mi) or more, depending upon the cost of money.

Pumping costs depend upon the total flow, the head against which the flow is pumped, and the cost of power. For typical pumping conditions the cost of transporting infiltration might be as much as $300/km per year, justifying a capital investment of $3000/km ($4800/mi). The costs listed above are probable maxima, but indicate the savings possible through careful design, construction, and maintenance of a sewer system.

SEWERS BUILT IN PLACE

14-10

When the required sewer size exceeds that which can be economically precast and transported to the site, cast in place sewers are used. Specially designed sections may also be required to clear obstacles in the sewer path.

14-11 Design of Concrete Sewers

Large concrete sewers may be analyzed as closed rings or fixed arches using the techniques of structural design.[17] For small diameters or spans, empirical designs may be used since theoretical calculations lead to thicknesses too small for ordinary construction techniques. Reinforced concrete arches are often constructed with the thickness of the crown equal to $\frac{1}{12}$ the span with a minimum of 125 mm (5 in). For plain concrete the thickness might be $\frac{1}{10}$ the span with the same minimum. The thickness of the invert is 25 mm (1 in) greater than the crown, and the haunches two to three times the crown thickness.

The shape of the sewer depends upon hydraulic considerations, construction conditions, and available space. The lower surface is generally curved to concentrate low flows and maintain self-cleansing velocities. Figure 14-7 shows some typical sewer shapes.

The sewer must be designed to conform to the bearing capacity of the foundation material. This may require placing a sub-base of crushed stone or gravel or, in some cases, piles. The bottom should be excavated to conform to the invert of the sewer, which is usually placed immediately following excavation. The invert should be consolidated by vibration and troweled to a smooth finish.

The concrete is usually placed in two or more lifts with waterstops at construction joints. For large sections the invert, walls, and arch may be constructed in separate pours. When a constant sewer cross section is maintained for some distance, collapsible steel forms may be used to form the arch.

14-12 Corrosion of Sewers

Organic matter is likely to accumulate in sanitary sewer lines as a result of deposition at low flow velocities or coagulation of grease at the water surface. This material will undergo decomposition and, under the conditions existing in most

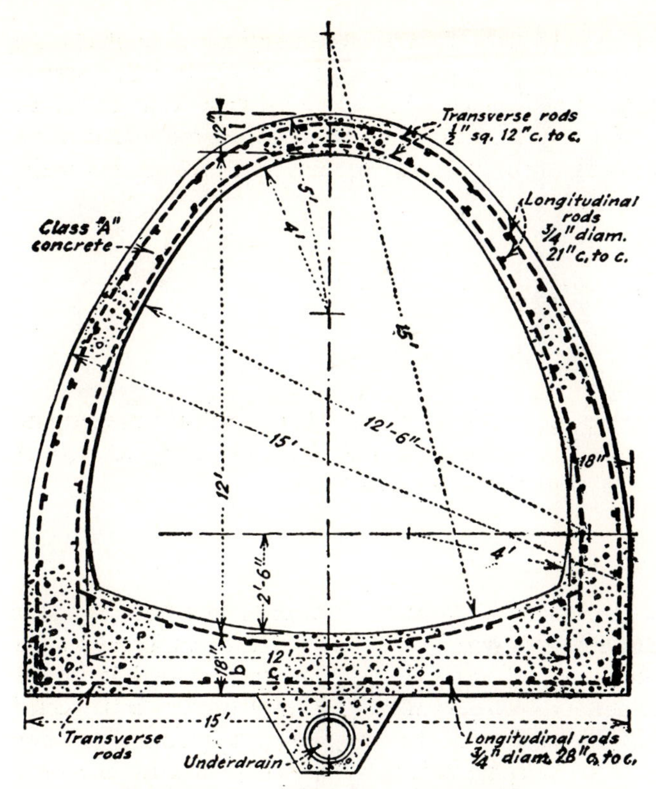

Semielliptical, with underdrain, Newark.

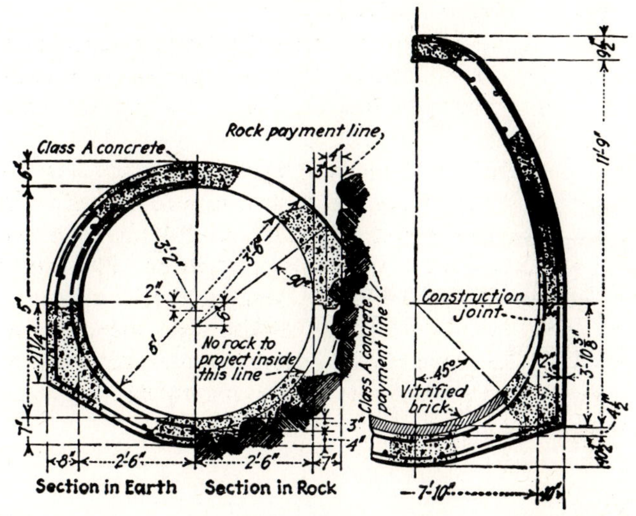

Circular and inverted egg, Louisville. Circular section in earth is suitable for soft material.

Figure 14-7 Sewer shapes. (ft × 0.305 = m, in × 25.4 = mm.)

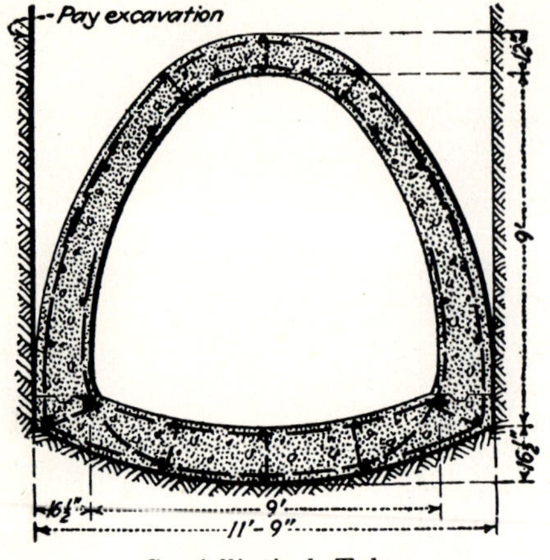

Semielliptical, Tulsa.

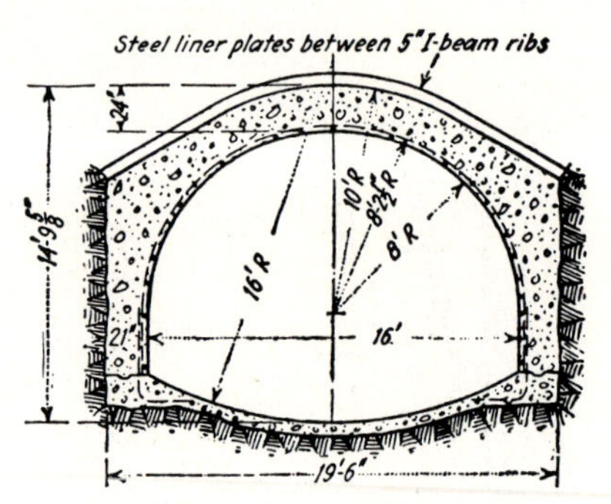

Horseshoe, Dallas.

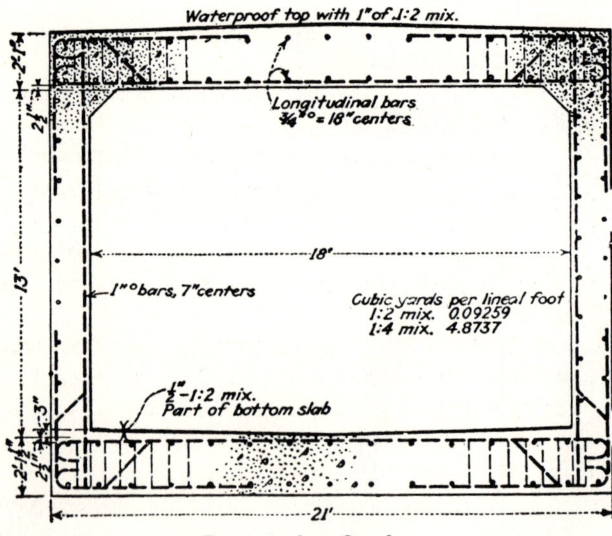

Rectangular, Omaha.

Figure 14-7 (*continued*)

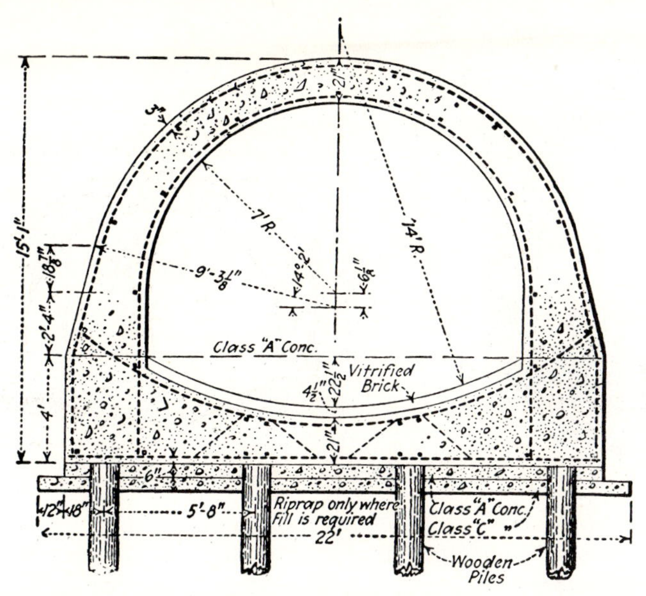

Horseshoe shape on piles. Pile support is necessary in very soft material.

Figure 14-7 *(continued)*

sanitary sewers, the breakdown of organic matter will be accompanied by bacterial reduction of sulfates originally present in the waste matter or water.

The anaerobic oxidation of complex organics is accompanied by formation of short chain fatty acids which may depress the pH in the sewer. The combination of sulfate reduction and low pH can produce free hydrogen sulfide in the air space of the sewer. The hydrogen sulfide in the sewer atmosphere may redissolve in condensed moisture at the sewer crown (Fig. 14-8) where the bacterium *thiobacillus*[18] can reoxidize it to sulfuric acid.

In sewers made of acid soluble materials such as concrete, iron, or steel, this acid formation can lead to destruction of the crown and failure of the sewer. The problem is aggravated by warm temperatures and flat sewer grades which produce low velocities and long retention times. Extensive damage to concrete sewers has occurred in the Gulf Coast region where both conditions exist.

Sewer corrosion has been combatted by chlorination, forced ventilation, and lining with inert materials. Chlorination halts biological action, at least temporarily. Forced ventilation reduces crown condensation, strips H_2S from the atmosphere of the sewer, and may provide sufficient oxygen in solution to halt anaerobic action.

Sewers which flow full and outfall lines carrying sewage treated to secondary standards do not provide the necessary conditions for crown corrosion. Ordinary sewer lines are normally made of vitrified clay since this is the only material which has been proven by long service to be resistant to corrosion. In new construction, particularly where foundation conditions are poor or the groundwater table high,

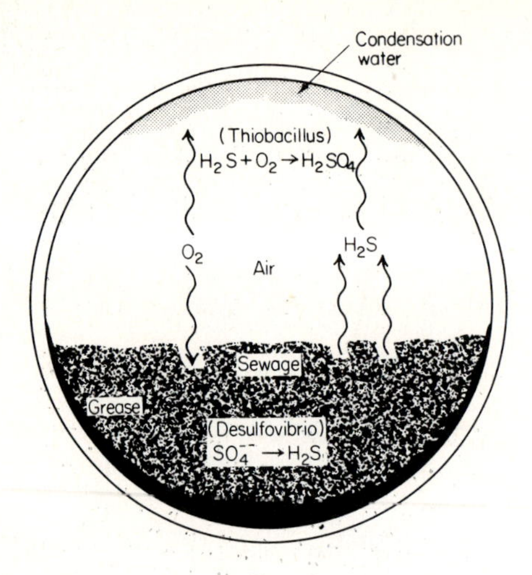

Figure 14-8 Schematic diagram of sewer corrosion.

solid wall plastic or plastic truss pipe should be considered for pipe 380 mm (15 in) or less in diameter. Iron, concrete, and asbestos cement pipe are used for force mains and outfall lines.

Where sewer sizes larger than 1070 mm (42 in) are required, concrete is used. If corrosion is anticipated the sewer may be cast with an integral lining or be lined in place with plastic, clay tile, or asphaltic compounds. Joints are made with plastic, hot tar, or mastic (Fig. 14-3).

PROBLEMS

14-1 A 250-mm clay sewer is to be placed in a trench 5 m deep. The backfill material is damp clay weighing 1920 kg/m^3. Determine the minimum required width of the trench at the top of the pipe and the total load per unit length. Using a factor of safety of 1.5, what type of bedding is required?

14-2 A 200-mm clay sewer is to be buried 3 m deep in damp topsoil weighing 1600 kg/m^3 in a trench of minimum width. A building foundation 300 mm wide crosses the trench at right angles. It has a load of 3000 kg/m and bears 2.5 m above the pipe. Determine the load per unit length of sewer in the ditch and under the foundation.

14-3 A precast reinforced concrete sewer 1220 mm in dia. is buried under a 5 m cover (to the top of the pipe) in a trench 2 m wide. Consider the safe load to be that which produces a 0.25-mm crack modified by a factor of safety of 1.25. Determine the type of bedding and class of pipe to be specified.

REFERENCES

1. "Standard Specification for Compression Joints for Vitrified Clay Bell-and-Spigot Pipe," (C-425), American Society for Testing Materials, Philadelphia.
2. "Standard Specification for Compression Couplings for Vitrified Clay Plain End Pipe," (C-594), American Society for Testing Materials, Philadelphia.
3. "Standard Specification for Extra Strength and Standard Strength Clay Pipe and Perforated Clay Pipe," (C-700), American Society for Testing Materials, Philadelphia.

4. "Standard Recommended Practice for Installing Vitrified Clay Sewer Pipe," (C-12), American Society for Testing Materials, Philadelphia.
5. *Concrete Pipe Design Manual*, American Concrete Pipe Association, Arlington, Va., 1970.
6. *Design and Construction of Concrete Sewers*, Portland Cement Association, Chicago, 1968.
7. Mayer, John K., et al.: *Sewer Bedding and Infiltration—Gulf Coast Area*, U.S. Environmental Protection Agency, 1972.
8. "Standard Specification for Concrete Sewer, Storm Drain, and Culvert Pipe," (C-14), American Society for Testing Materials, Philadelphia.
9. "Standard Specification for Joints for Circular Concrete Sewer and Culvert Pipe, Using Flexible Watertight, Rubber-Type Gaskets," (C-443), American Society for Testing Materials, Philadelphia.
10. "Standard Specification for Reinforced Concrete Culvert, Storm Drain, and Sewer Pipe," (C-76), American Society for Testing Materials, Philadelphia.
11. "Standard Specification for Asbestos-Cement Pressure Pipe," (C-296), American Society for Testing Materials, Philadelphia.
12. "Standard Specification for Rubber Rings for Asbestos Cement Pipe," (D-1869), American Society for Testing Materials, Philadelphia.
13. "Standard Specification for Acrylonitrile-Butadiene-Styrene (ABS) Composite Sewer Pipe," (D-2680), American Society for Testing Materials, Philadelphia.
14. "Standard Recommended Practice for Underground Installation of Flexible Thermoplastic Sewer Pipe," (D-2321), American Society for Testing Materials, Philadelphia.
15. "Standard Specification for Type PSP Poly (VinylChloride) (PVC) Sewer Pipe and Fittings," (D-3033), American Society for Testing Materials, Philadelphia.
16. "Standard Specification for Type PSM Poly (VinylChloride) (PVC) Sewer Pipe and Fittings," (D-3034), American Society for Testing Materials, Philadelphia.
17. *Analysis of Arches, Rigid Frames, and Sewer Sections*, Portland Cement Association, Chicago.
18. McKinney, R. E.: *Microbiology for Sanitary Engineers*, McGraw-Hill, New York, 1962.

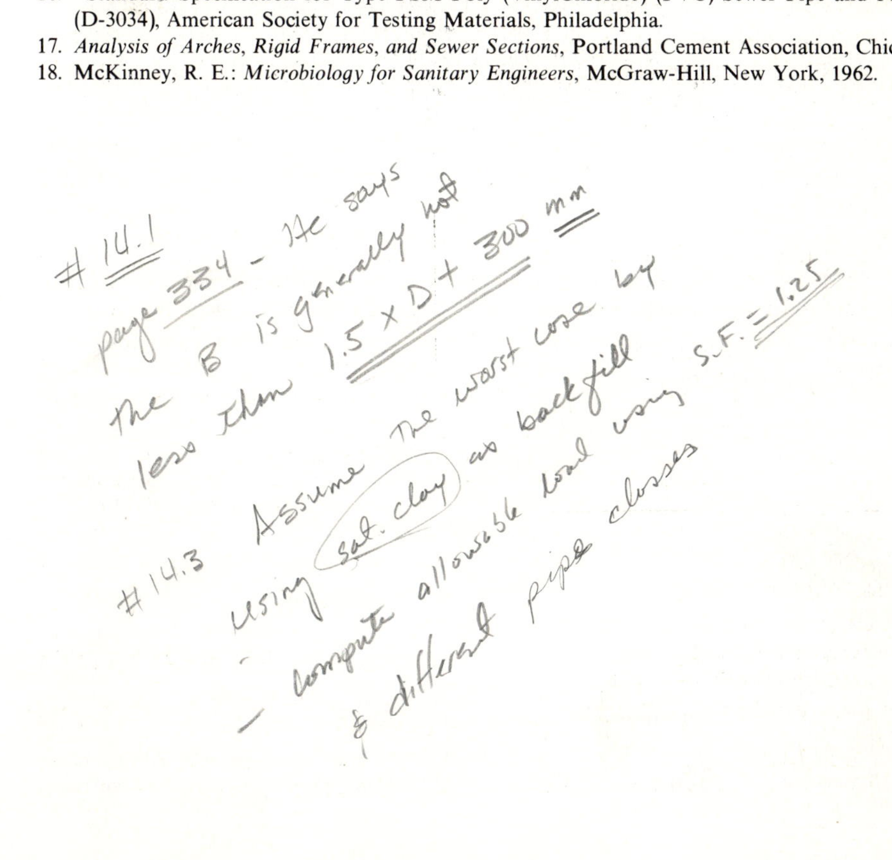

CHAPTER

FIFTEEN

FLOW IN SEWERS

15-1

Most sewers are designed to flow as open channels and not under pressure, even though they may flow full at times. There are exceptions, such as inverted siphons and discharge lines from sewage pumping stations which are always under pressure. Occasionally the capacity of storm sewers will be overtaxed, inlets will be full and overflowing, and water will rise in manholes. Sewers in such condition are said to be surcharged. Sanitary sewers may also be surcharged by excessive inflow during storms, by stoppages in the lines, or by flows greater than those designed for. Vacuum and pressurized collection systems are designed to flow full.

15-2 Flow Formulas

When water enters a pipe or channel at a constant rate, and escapes freely at the lower end, steady uniform flow will soon be established. Steady flow is that in which the same volume of liquid flows past any given point in each unit of time. Uniform flow is that which is free from changes in velocity along the course of the conduit or stream. In the usual sewer design problem, steady flow may be assumed. Uniform flow may be expected in straight sewer lines, but velocity changes may be expected at obstacles and changes in cross section of the pipes or channels and must in some cases be considered in making hydraulic computations.

Water moves downstream in a pipe or channel impelled by the force of gravity. It will move at such a velocity that the available head or fall will be used up in overcoming friction and, in small part, in attaining kinetic energy or velocity head. The amount of friction or resistance that must be overcome varies directly with the roughness of the surface of the pipe or channel, directly with the area of the contact surface, approximately with the square of the velocity, and directly with the density of the liquid. The contact surface will be the wetted perimeter of the conduit multiplied by its length. These relationships can be expressed as a formula

$$V = C\sqrt{Rs} \tag{15-1}$$

which is known as the Chezy formula. In it V is the mean velocity, R is the hydraulic radius, or area of the stream divided by the wetted perimeter, s is the slope of the hydraulic grade line or, in open channels, the slope of the water surface for uniform flow, and C is an experimental coefficient. Since the effects of roughness, velocity, and other factors are only approximate, the value of C is not constant but varies with V, R, and s according to the Kutter formula, which is, in metric units,

$$C = \frac{(23 + 0.00155/s) + 1/n}{1 + (n/\sqrt{R})(23 + 0.00155/s)} \tag{15-2}$$

the equation may be converted to English units by multiplying the right side by 1.81.[1]

In this formula a new quantity n is introduced. It depends upon the roughness of the pipe or channel surface and affects the velocity inversely as its value. Table 15-1 gives the values of n as determined by observation of stream and pipe flow. For sewers of clay pipe or concrete, n is sometimes considered to be 0.015. With good construction methods, careful aligning of the pipe, and smooth joints, n may be taken as 0.013, and this is standard practice. For large concrete pipes, with careful construction, n may be even less. Some state regulatory agencies permit values of n equal to 0.0125 for any smooth and durable pipe with lengths of 1.5 to 3 m (5 to 10 ft) and 0.012 for lengths of 3.3 m (11 ft) or more.

Table 15-1 Values of n in the Kutter and Manning formulas

n	Character of surface
0.009	Plastic pipe
0.009	Well-planed timber evenly laid
0.010	Neat cement. Very smooth pipe
0.012	Unplaned timber. Cast-iron pipe of ordinary roughness
0.013	Well-laid brickwork. Good concrete. Riveted steel pipe. Well-laid vitrified clay pipe
0.015	Vitrified tile and concrete pipe poorly jointed and unevenly settled. Average brickwork
0.017	Rough brick. Tuberculated iron pipe
0.020	Smooth earth or firm gravel
0.030	Ditches and rivers in good order, some stones and weeds
0.040	Ditches and rivers with rough bottoms and much vegetation

The Manning formula is widely used for open channel flow calculations. It is:

$$V = \frac{1}{n} R^{2/3} s^{1/2} \tag{15-3}$$

If English units are used (ft, ft/s), the right side of the equation is multiplied by 1.486.

The n of this formula is the same as the n of the Kutter formula, and the values of Table 15-1 can be used. Use of the Manning formula will give results that are as satisfactory for sewer problems as those of the Chezy formula with Kutter's expression for C.

Manning's "n" is not constant since, as may be observed from Eqs. (15-2) and (15-3) it serves to replace Kutter's "C" which varies with hydraulic radius, and, therefore, depth of flow. Camp[1] has presented a relation between n and depth of flow which is shown as a part of Fig. 15-5. The formula is also applicable to flow in closed pipes with the values of n given in Table 15-1.

Design of sewers requires many determinations of velocities, pipe sizes, and slopes, and, therefore, convenience and speed of solution are of importance. Diagrams have been constructed to simplify the design procedure, and these are used when they are available. The Chezy formula is particularly clumsy to use without such short-cut methods. Article 15-5 describes the use of nomographic diagrams based upon the Manning formula.

15-3 The Hydraulic Grade Line, or Piezometric Head Line

In open-channel flow the hydraulic grade line, or piezometric head line will be the water surface, and the gradient or slope will be the fall or the grade per unit of length. Under ordinary conditions the slope of a sewer is considered as the slope or gradient of its invert. This implies that the hydraulic grade line or water surface will parallel the invert. In most sewer lines this will be the case, but it must be remembered that any conditions that will change the slope of the water surface will change the carrying capacity of the sewer irrespective of the invert slope. It may be necessary to consider this possibility where large sewers intersect or where changes in velocity affect the hydraulic grade line through conversion of velocity head to pressure head or the reverse. The outlets of sewers discharging into streams or lakes may be submerged at times, thereby causing a hydraulic gradient which is less than that of the invert and materially reducing carrying capacity.

Very large sewers are sometimes curved on long radii. This will increase the loss of head and therefore affect the hydraulic grade line. Little experimental work is available on which to base calculations of headloss in such circumstances. One procedure is to increase the roughness coefficient in the curved portions by adding 0.003 to 0.005 to the value of n.

Changes of direction in small sewers are made in manholes in a curved channel of very short radius. Little is known as to the headloss in open channels under such conditions, but it may be considered to approximate the loss in a closed elbow. Such losses are expressed as a portion of the velocity head. In this case it would be approximately $1.25(V^2/2g)$. With usual velocities this would justify a fall in the invert in the manhole of about 30 mm (0.1 ft). It is common practice, therefore, to drop inverts this amount in manholes where there is a considerable change in direction. Some engineers drop the invert elevation at manholes when the sewer increases in size an amount equal to the difference in size; i.e., the tops or crowns of the two sewer pipes remain on the same line. Some state regulatory agencies permit the placing of the 0.8 depth points of the two sewers at the same elevation. This expedient overcomes some of the headlosses caused by changes in size. Since size increases are likely to occur where branches come into the mains, the invert drop lessens the danger of sewage backing up and surcharging the branches. If sewers are designed to run full, the hydraulic grade line will not parallel the invert at the junction, and surcharging may occur in parts of the sewer unless such a drop is made.

15-4 Required Velocities

An important consideration in sewer design is the velocity obtained in sewers. Experience indicates that a velocity of not less than 0.6 m/s (2 ft/s) is required in sanitary sewers in order to prevent settlement of the sewage solids. The minimum allowable slopes, therefore, are those which will give this velocity when the sewer is flowing full, and greater slopes should be used if they are compatible with existing topography. The regulations of state agencies frequently include minimum allowable slopes for sewers of various sizes. They are the slopes which will give a velocity in the pipes of 0.6 m/s when flowing full with n equal to 0.013. Under special conditions some state regulatory agencies will permit slopes slightly less than those which provide a velocity of 0.6 m/s, but in this case the engineer must show by his computations that the depth of the sewage at design average flow will be at least 0.3 of the diameter.

In storm sewers greater velocities are required than in sanitary sewers because of the heavy sand and grit which is washed into them. The minimum allowable velocity is 0.75 m/s (2.5 ft/s), and 0.9 m/s (3 ft/s) is desirable. Because of the abrasive character of the solids, excessively high velocities should be avoided, 2.4 m/s (8 ft/s) being considered the desirable upper limit. Even within that limit, large and important sewers are frequently lined in the lower portions with vitrified-tile blocks or other hard materials to prevent damage.

In very flat areas where there is difficulty in obtaining the minimum grades, there is a temptation to use larger pipes because they will provide a velocity of 0.6 m/s at lower grades. It should be recognized, however, that a cleaning velocity of 0.6 m/s will be reached only when the pipes flow full or 78 percent of full. As may be seen from Fig. 15-5, pipes flowing less than 78 percent full will have velocities less than pipes flowing full. Therefore, using larger pipes for low flows may make matters worse. The actual flow that is anticipated and the actual velocity produced in the partially filled sewer should be considered in selecting an appropriate slope. In storm sewer designs it may be necessary to use grades that are too low for self-cleansing velocities. In such a case provision should be made for convenient flushing to prevent stoppages or to remove them should they occur.

15-5 Flow Diagrams

Figures 15-1 to 15-3 are nomograms for the solution of the Manning formula for various ranges of quantities and pipe sizes and with *n* equal to 0.013. Use of the diagrams can best be explained by means of an example. It is desired to determine the size of pipe required to carry 3.4 m^3/min, with an available slope of 0.003. Figure 15-1 is used. A straightedge is placed on 3.4 of the quantity scale and also on 0.003 of the slope ratio scale. It will be seen that the straightedge cuts the diameter scale at 305 mm and the velocity scale at 0.77, indicating that a 305-mm pipe will be required and the velocity will be 0.77 m/s when flowing full.

In a similar manner, with any two of the hydraulic quantities known the other two may be obtained. Should the necessary size not fall upon a commercial size, the next larger pipe will be used.

Some engineers design sewer pipes to run half full when carrying the expected quantity. This practice has much in its favor when designing laterals and submains, as a factor of safety, although it is not justified for mains and outfall sewers. Figures 15-1 to 15-3 can also be used in designing sewers to run half full. It is then necessary, however, to double the expected quantity before using the diagrams.

Figure 15-4 is a nomogram based upon the Manning formula which introduces various values of *n*. It can be used for pipes and conduits of all shapes and for open channels. It is used as follows:

A circular pipe 1.83 m in diameter is of very old and rough brickwork. The value of *n* is therefore assumed to be 0.017. The slope of the pipe is 0.003. What will it carry when flowing full, and what will be the velocity? The hydraulic radius when flowing full will be 0.46 m. With *n* as 0.017 and *s* as 0.003, lay a straightedge to find the intersection with the pivot scale, as shown by the dashed line. From the intersection with the pivot scale place a straightedge to intersect the hydraulic radius scale at 0.46 m. This line crosses the velocity scale at 1.83 m/s. With this velocity and the pipe flowing full the quantity carried will be 4.8 m^3/s.

15-6 Partial-flow Diagrams

It is frequently necessary to determine the velocity and depth of sewage in a pipe which is flowing only partially full. Use of Fig. 15-5 will allow quick computation of the hydraulic elements of partially filled circular sewers. In using the diagram it is first necessary to use Figs. 15-1 to 15-3 in order to find conditions when the sewer is flowing full. Then by calculating the ratio of any two known hydraulic elements the others can be found. As an example:

A 915-mm (36-in) circular sewer is laid on a slope of 0.003. *n* is 0.013 when the sewer is full. What will the velocity and depth of flow be when the sewer is carrying 8.5 m^3/min?

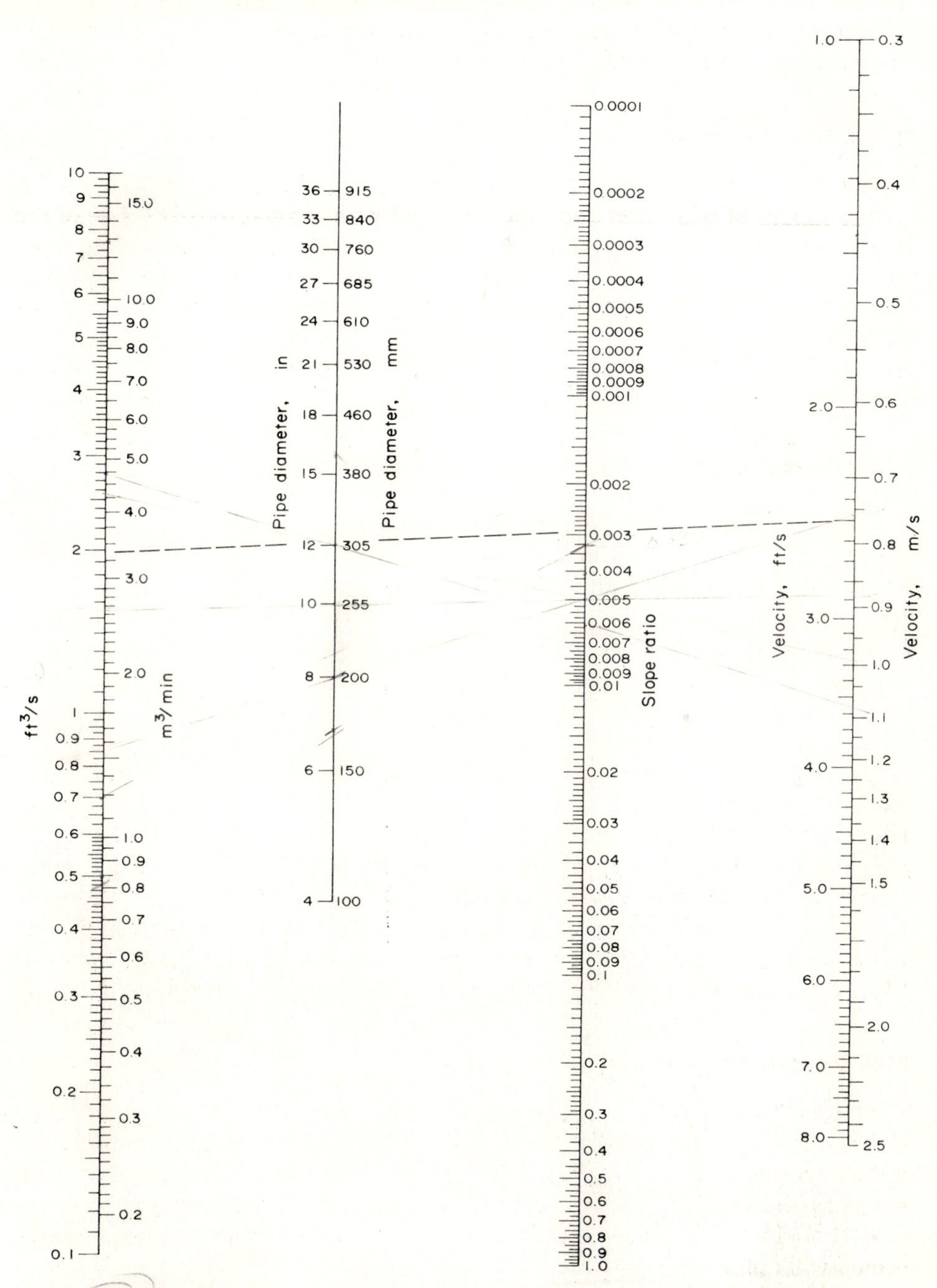

Figure 15-1 Diagram for solution of Manning formula for circular pipes flowing full. $n = 0.013$.

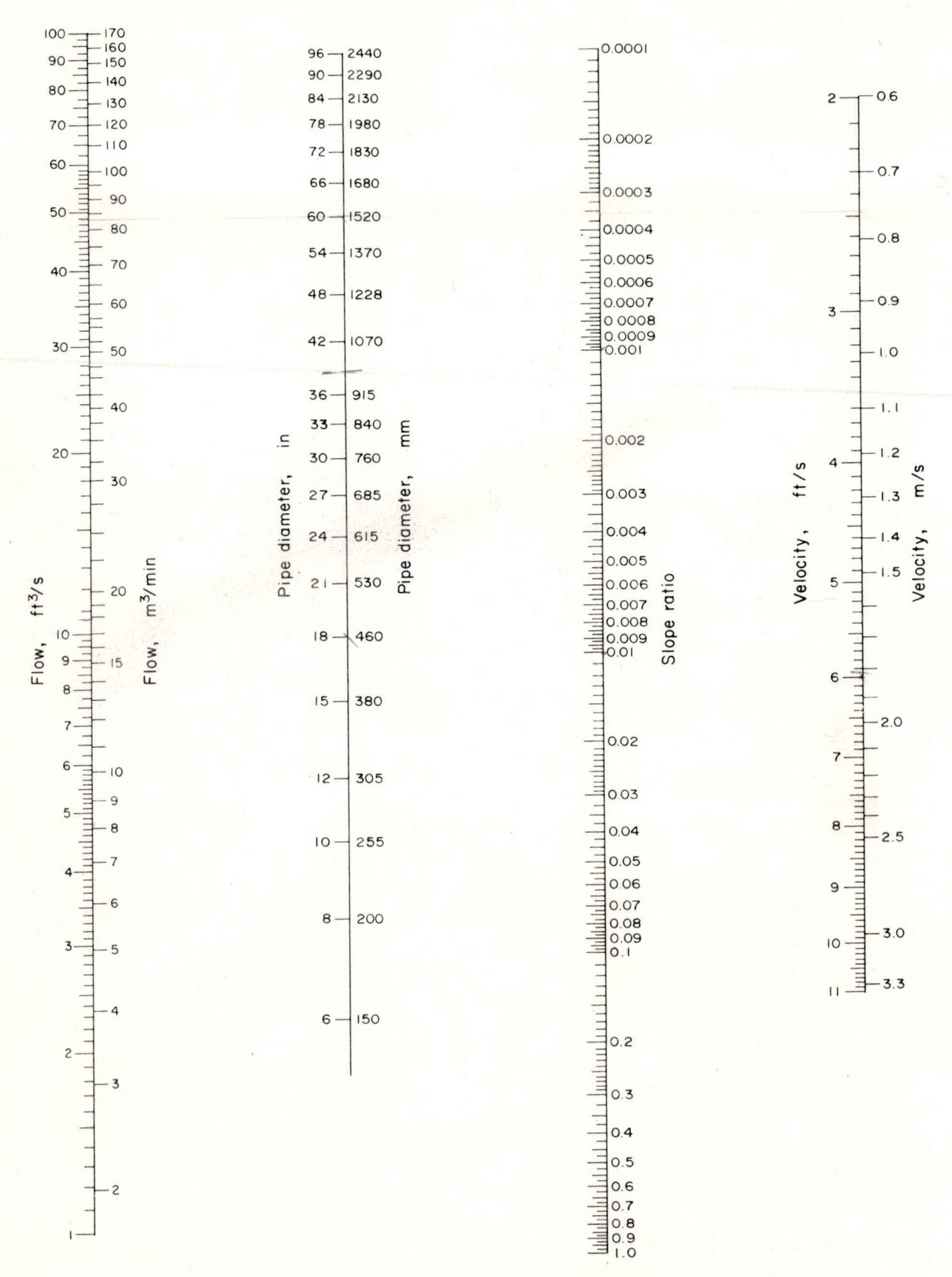

Figure 15-2 Diagram for solution of Manning formula for circular pipes flowing full. $n = 0.013$.

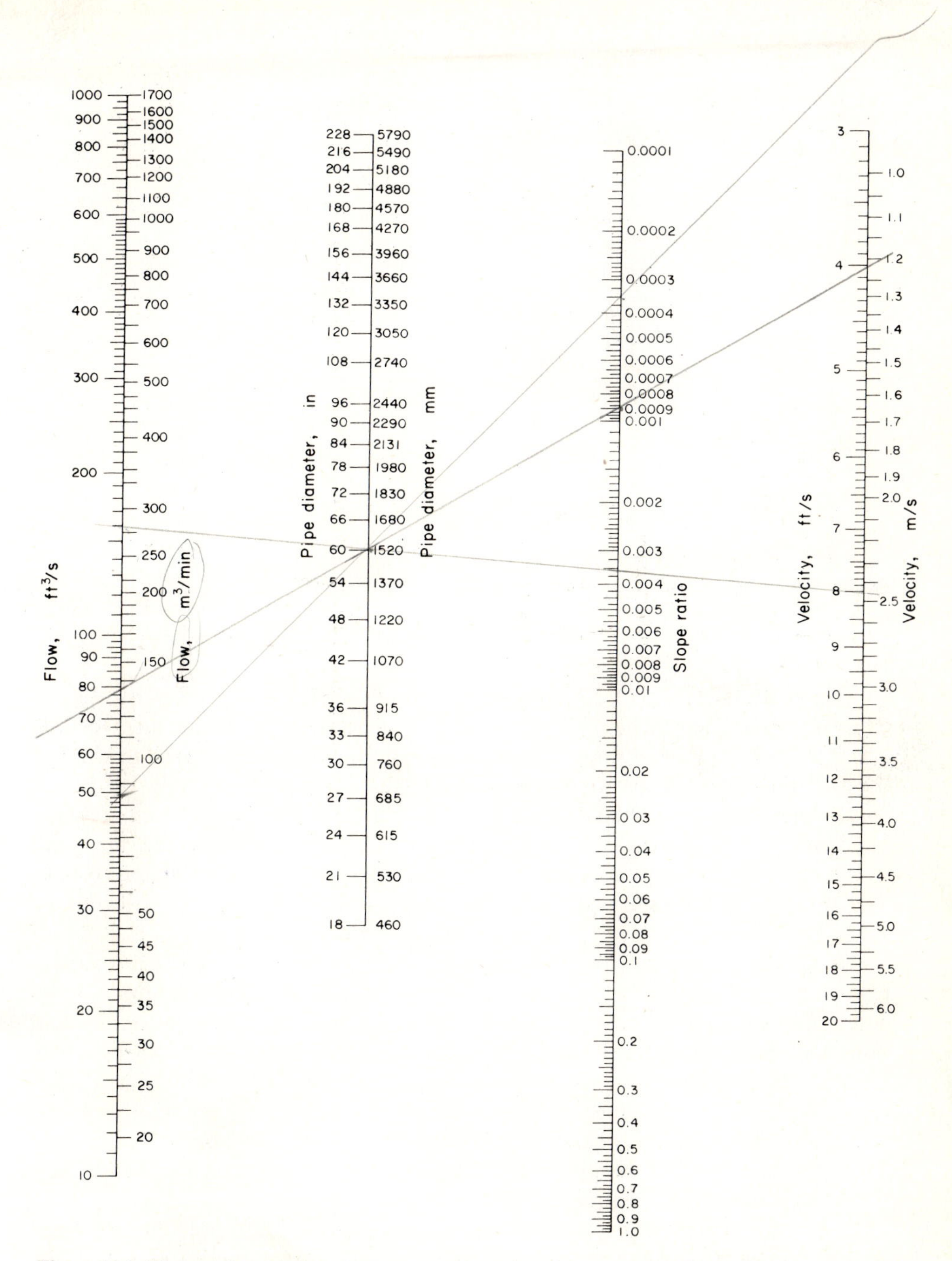

Figure 15-3 Diagram for solution of Manning formula for circular pipes flowing full. $n = 0.013$.

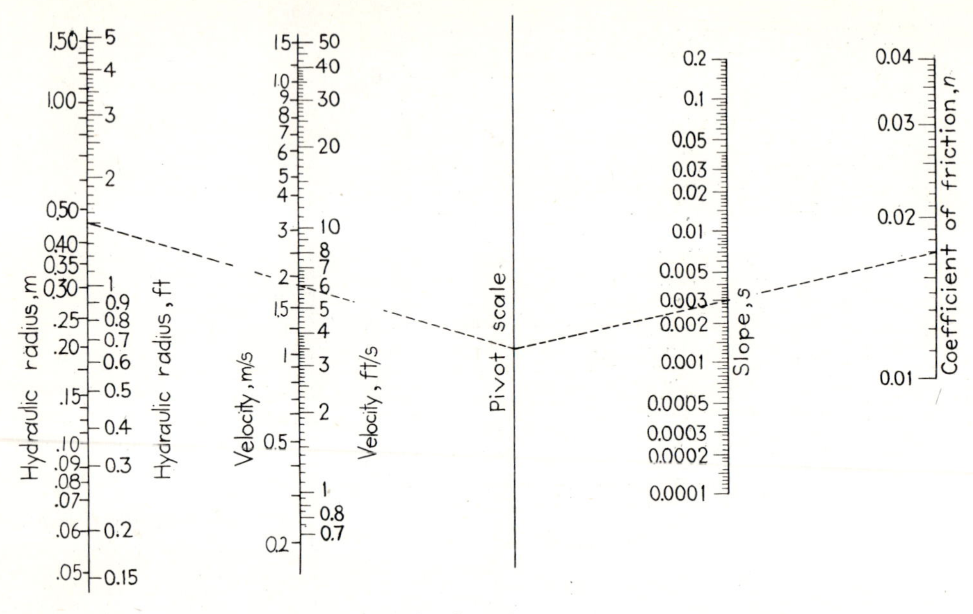

Figure 15-4 Nomogram based upon Manning formula.

Reference to Fig. 15-2 shows that the discharge when flowing full is 62 m^3/min and the velocity is 1.57 m/s. The ratio of the actual to the full discharge is $q/Q = 8.5/62 = 0.14$. Using this ratio as abcissa in Fig. 15-5, follow it upward to the discharge curve and from the point of intersection read the ordinate on the left hand scale, which is 0.30. The depth of flow is thus $0.3 \times 915 = 275$ mm at a flow of 8.5 m^3/min. To obtain the velocity, follow the depth line (0.30) to the right until it intersects the velocity curve and read the abcissa to obtain v/V equal to 0.60. The velocity in the pipe when it is carrying a flow of 8.5 m^3/min will thus be $0.6 \times 1.57 = 0.94$ m/s.

It should be recognized that partial-flow diagrams give only approximate results, particularly for high velocities. The discrepancies between computed and actual flow conditions may be caused by wave formation, surface resistance, and other factors of which our knowledge, so far as partly filled channels is concerned, is very scanty. This is particularly the case in circular sewers flowing nearly full at high velocities. Consequently in practical design a factor of safety may be necessary. Note that velocities below 78 percent of total depth decrease and above that point first increase and then decrease. Actually a pipe will not flow full unless it is surcharged or under pressure.

Conditions during partial flow must frequently be determined in combined sewers. Velocities during dry-weather flow must be calculated to eliminate the possibility of deposits occurring. Knowledge of the depth of flow is of value in designing sewer intersections. Large sewers should be brought together at elevations so that water from one will not back up into the other. (Art. 15-3.)

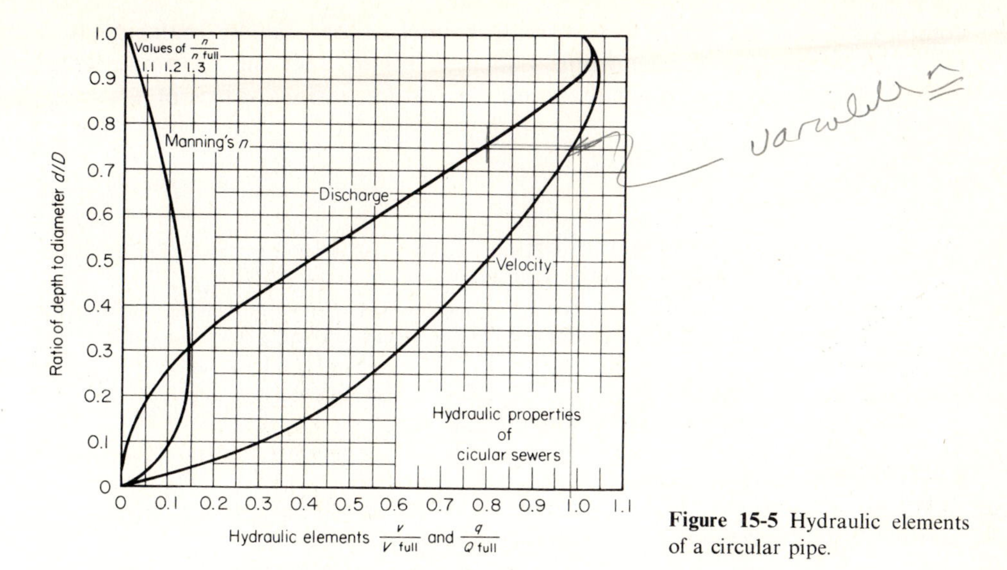

Figure 15-5 Hydraulic elements of a circular pipe.

15-7 Sewer Shapes

Most sewers are circular in cross section. This shape has the advantages of giving the maximum cross-sectional area for the amount of material in the wall, convenience in manufacturing of tile and precast concrete pipe, and fairly good hydraulic qualities. It is fairly stable in place, but excavations must be shaped to the barrel of the pipe or some type of bedding used. (Art. 14-3.)

Egg-shaped sewers were formerly used more than at present, especially for combined sewers. Their principal advantage is the slightly higher velocity for low flows over the circular sewer of equal capacity. Since the small end of the egg is down, they present difficulty in building and are somewhat unstable. They are always built of brick or constructed of concrete in place. A commonly used egg-shaped cross section has for its upper portion a semi-circle of radius equal to one-third of the depth or the long diameter and an invert curved on a radius of one-sixth of the long diameter, with the sides having a radius equal to the long diameter.

Other shapes of sewers used are shown in Fig. 14-7. The rectangular section is popular for storm sewers of moderate or large size. It is easy to design and construct. Its hydraulic qualities are fair. It should be recognized that the hydraulic radius decreases in value very sharply when the flow increases from nearly full to completely full so that the roof is a part of the wetted perimeter. Discharge will then fall off some 30 percent from the maximum. A curved invert is generally used in rectangular sewers to concentrate very small flows. Other shapes used are the semi-elliptical, horseshoe, and basket handle. They are adopted for ease of construction or reasons of economy rather than for hydraulic advantages. Large combined sewers sometimes have a small channel, called a *cunette*, in the invert to carry dry-weather flow at favorable velocities. Figure 15-4 can be conveniently used in designing shapes other than circular and for various values of n.

PROBLEMS

15-1 A circular sewer is to have a slope of 0.0022 and is to carry 0.05 m^3/s when flowing full. $n = 0.013$. What size will be used and what will be the velocity?

15-2 A circular sanitary sewer is to carry 0.07 m^3/s when flowing full, and conditions are such that the minimum allowable grade must be adopted. $n = 0.013$. Determine the commercial size pipe that would be used and the grade.

15-3 A circular sewer is to carry 0.07 m^3/s when flowing half full, and the slope is 0.006. With $n = 0.013$, what will be the pipe size and velocity?

15-4 A rough brick circular sewer 1.83 m in diameter has a slope of 0.0009 with $n = 0.020$. Compute the capacity when full and the velocity.

15-5 A circular sewer 1.52 m in diameter is laid on a slope of 0.00036 and n is 0.013. Determine the velocity and depth of flow when the minimum flow of 17 m^3/min occurs.

15-6 Two circular sewers join. One is 2.13 m in diameter and at maximum flow carries 93.5 m^3/min on a slope of 0.0002. The branch sewer is 0.76 m in diameter, has a slope of 0.001, and a maximum flow of 16 m^3/min. At what height above the invert of the larger sewer should the smaller one enter so that during maximum flow there will be no back-up into the smaller? Let $n = 0.013$.

REFERENCES

1. Addison, Herbert: *A Treatise on Applied Hydraulics* 4th ed., John Wiley & Sons, Inc., New York, 1954.
2. Camp, T. R.: "Design of Sewers to Facilitate Flow," *Sewage Works Journal*, **18**:3, 1946.

CHAPTER

SIXTEEN

SEWER APPURTENANCES

16-1

Sewer systems require a variety of appurtenances to insure proper operation. These include manholes, inlets, inverted siphons, pumping stations, and regulators.

16-2 Manholes

Manholes are used as a means of access for inspection and cleaning. They are placed at intervals of 90 to 150 m (300 to 500 ft) and at points where there is a change in direction, change in pipe size, or considerable change in grade. Large sewers, 1520 mm (60 in) or more in diameter, can be entered for inspection and need fewer manholes. Manhole design has become largely standardized, and most of the larger cities have standard plans which they have adopted for general use. The manhole has a cast-iron frame and cover with a 500 to 600 mm (20 to 24 in) clear opening. The frame rests upon brickwork which is corbeled to form an opening not less than 1 m (40 in) and usually 1.25 m (48 in) in diameter in a distance of 0.9 to 1.5 m (3 to 5 ft) below the manhole. The 1.25-m cylinder is continued downward until it reaches the sewer. If the total depth is less than 4 m (12 ft), the walls are made 200 mm (8 in) thick. For each additional 2 m of depth, an extra 100 mm of thickness should be added. Concrete walls are often used.

The bottom of the manhole is usually made of concrete slightly sloped on the upper surface toward the open channel or channels, which are continuations of the sewer pipes. The channels are sometimes lined with half-round or split sections of sewer pipes. In any case the channel depth should be nearly equal to the pipe diameter to prevent sewage from spreading over the manhole bottom. Should this

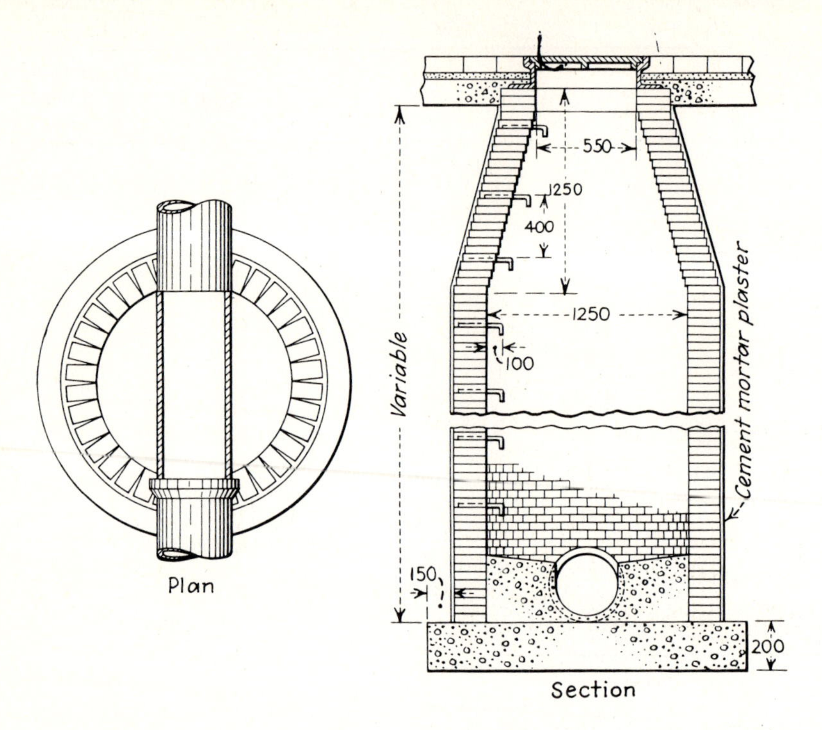

Figure 16-1 A brick manhole. (Dimensions in mm.)

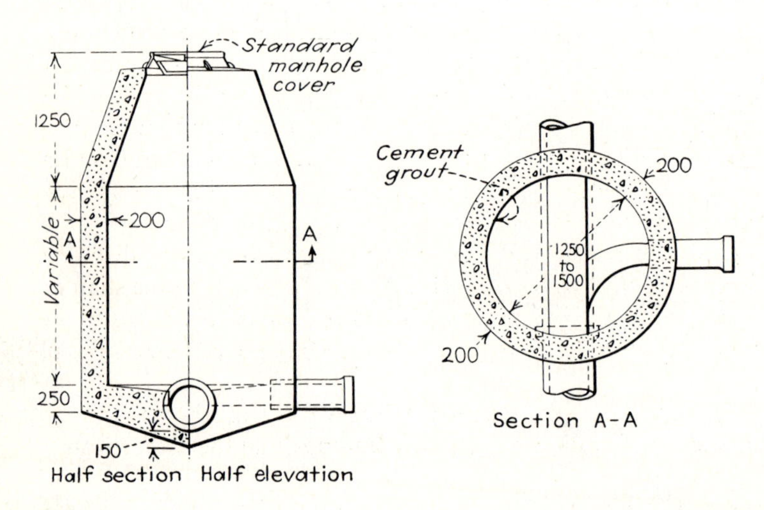

Figure 16-2 A concrete manhole with junction of branch sewer. (Dimensions in mm.)

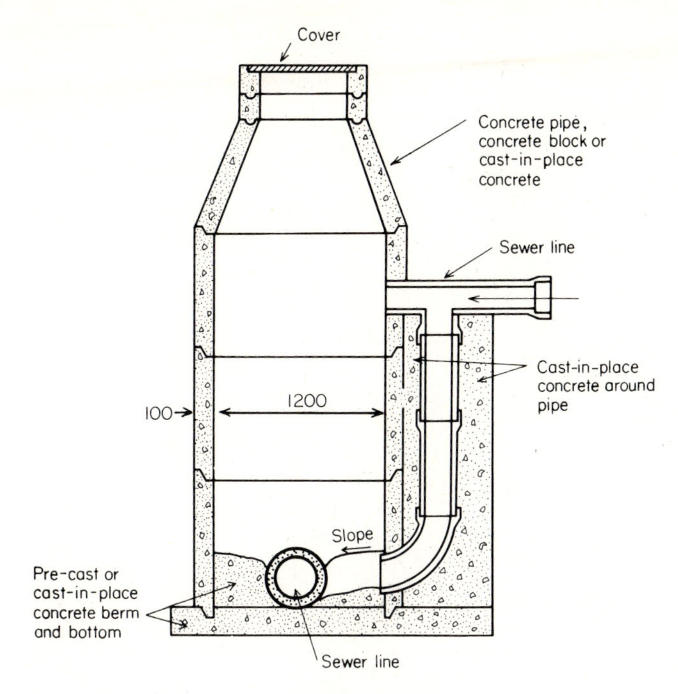

Figure 16-3 Section of drop manhole. (Dimensions in mm.)

happen, solids may be stranded, and odors result, although the slope on the bottom tends to discourage such nuisance. Changes of direction are made in the channels.

Where laterals or submains join in a deeper sewer, excavation will be saved by keeping the upper sewer at a reasonable grade and making a vertical drop at the manhole. This is known as a drop manhole, and the method of construction is shown in Fig. 16-3. Note that, while the sewage drops in the vertical pipe, the sewer line intersects the manhole wall so that the branch can be rodded for cleaning. If the drop is less than 0.6 m (2 ft), it is usually cared for by increasing the sewer grade instead of using a drop manhole. If a drop of less than 0.6 m is made in the manhole, the bottom is so arranged that the incoming sewage falls into a sloping channel without splashing or possible deposition of solids.

A wellhole is a shaft in which a large amount of sewage drops a long distance. The force of the fall may be broken by staggered horizontal plates in the shaft or by means of a well or sump at the bottom from which the sewage overflows to the low-level sewer. In a flight sewer the sewage runs down a flight of concrete steps to break the fall.

Manholes in large sewers are constructed as shown in Fig. 16-4 or may be constructed as a separate shaft with a passageway leading from the shaft through the sewer wall.

Manhole covers and frames are obtainable from foundries in different standard weights. For the heaviest city traffic a cover and frame weighing about

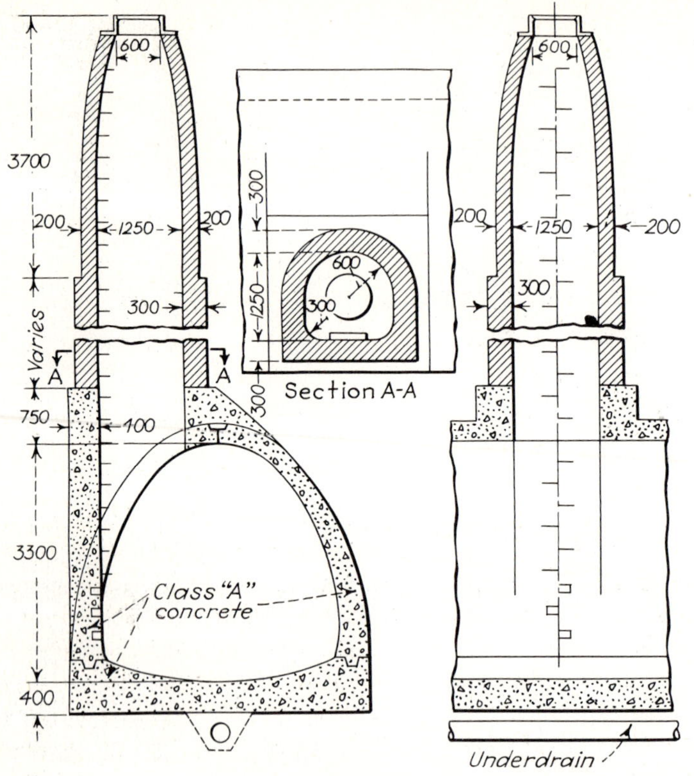

Figure 16-4 Manhole giving access to a large sewer. (Dimensions in mm.)

340 kg (750 lb) may be specified, while for lighter city traffic a 245-kg (540-lb) weight may be considered strong enough. Suburban traffic requires less strength, and weights from 150 to 180 kg (325 to 400 lb) should suffice. Where only foot traffic will be encountered, frames and covers weighing only about 70 kg (150 lb) may be used. Figure 16-5(*a*) shows a heavy frame and cover. The lighter types are not only thinner, but are smaller in the dimension from top to web, as in Fig. 16-5(*c*). The covers should be roughened to prevent excessive slipperiness, but perforated covers should not be used for sanitary sewers. It was formerly supposed that the openings ventilated the sewers, but this is more efficiently accomplished by the stacks of building plumbing systems. The openings have the disadvantage of permitting rainwater, sand, and grit to enter the sewers.

In deeper manholes, steps are needed to allow workmen to descend. Rungs of steel imbedded in the walls have been used for this purpose, but their life is short, and they may be dangerous. A better step, obtainable from foundries, is constructed of cast iron which is inserted in brick joints or concrete.

As an economy measure some cities use cleanouts instead of manholes. The cleanout used by Dallas is shown in Fig. 16-6. It consists of a light cast-iron frame and cover, weighing not less than 36 kg (80 lb), and a line of pipe to the sewer. Connection to the sewer is made by means of a Y which has the side outlet at an

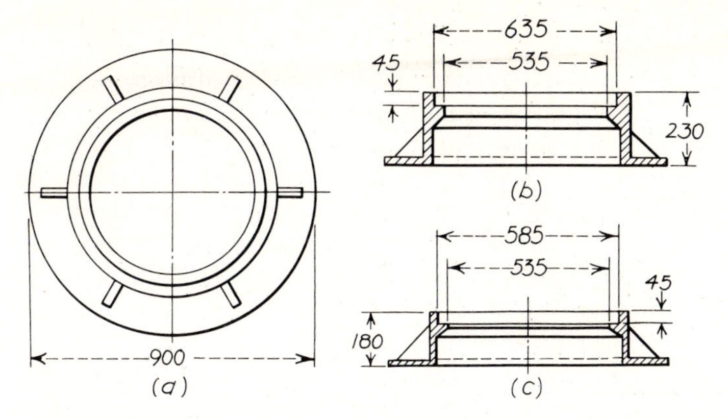

Figure 16-5 Manhole frame and cover. (*a*) Plan. (*b*) Section of heavy frame. (*c*) Section of light frame. (Dimensions in mm.)

angle of about 27° instead of the standard 60°. If the sewer is very deep, it may be advisable to insert a $\frac{1}{16}$ bend just above the Y so the cleanout can come to the surface at about a 45° angle. This construction permits flushing the sewer with a fire hose, and rods may also be inserted to clear heavy obstructions. Dallas uses cleanouts on 150-mm and 200-mm (6- and 8-in) sewers at 75- to 90-m (250- to 300-ft) intervals, except at junctions, and at the upper end of laterals. Cleanouts are not permitted as a substitute for manholes in many cities. If used, they should be used with caution. Some engineers use cleanouts at the head end of small sewers, but nowhere else.

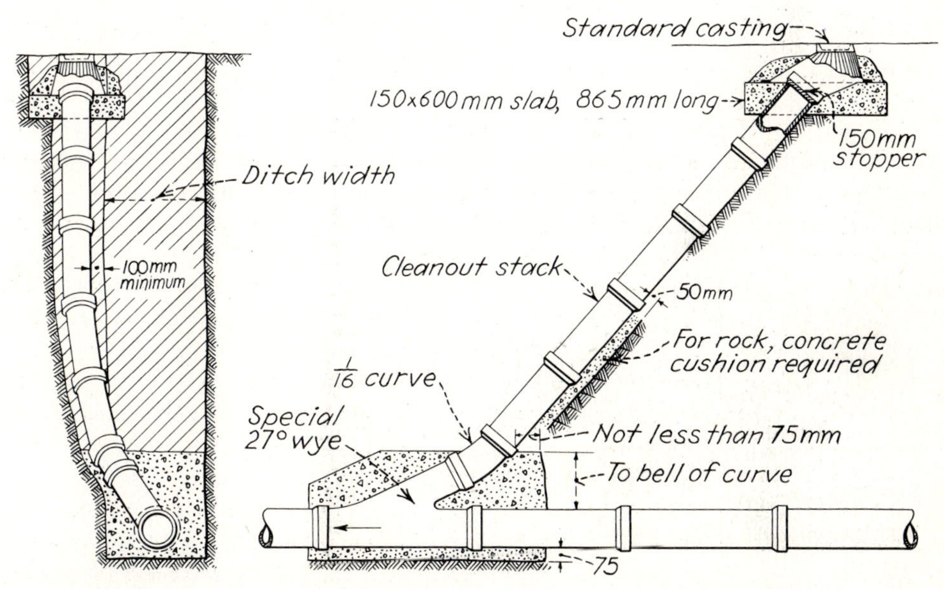

Figure 16-6 Standard cleanout used by Dallas, Tex. The $\frac{1}{16}$ curve is sometimes omitted.

16-3 Inlets

An inlet is an opening into a storm or combined sewer for entrance of storm runoff. Inlets are placed at the gutters, usually at street intersections but occasionally at midpoints of the blocks, if they are more than 150 m (500 ft) long. The usual location at an intersection is shown in Fig. 16-7. With the streets sloping as indicated by the arrows, note that the inlets are placed so that crosswalks will not be flooded. Placing of inlets at the corners not only requires the pedestrians to step across flooded gutters but also subjects the inlets to considerable traffic wear and damage. The short branches needed to connect the inlets to the sewer may all enter at a manhole or may enter through Ys at the nearest points.

The curb-opening inlet (Fig. 16-8) is used widely and by some cities exclusively. Determination of the capacity of such inlets has received considerable study by various investigators,[1] but the methods of analysis are too lengthy for discussion here. The length of opening depends upon the amount of storm water, the depth of water in the gutter as it reaches the inlet, and the depression that is given to the gutter. Depressions are up to 130 mm (5 in) in depth, extend the length of the inlet and usually 900 mm (36 in) outward from the curb line. Table 16-1 shows inlet capacities used by the Texas Highway Department. The depth of flow in the gutter can be obtained by the Manning formula applied to the usual street-gutter cross section. It is

$$Q = K\frac{z}{n}s^{1/2}y^{8/3} \tag{16-1}$$

where Q = the gutter flow
z = the reciprocal of the transverse slope of the bottom of the gutter
n = the roughness coefficient, usually 0.015 for smooth concrete gutters
s = the slope of the gutter
y = the depth of the water in the gutter at the curb line, and
K = a constant dependant upon units and equal to 22.61 (m^3/min, m) or 0.56 (ft^3/s, ft).

The value of y and z, taken together, permit calculation of the distance which water will extend into the street. Ponding can be reduced by placing inlets closer together, thereby reducing Q and y.

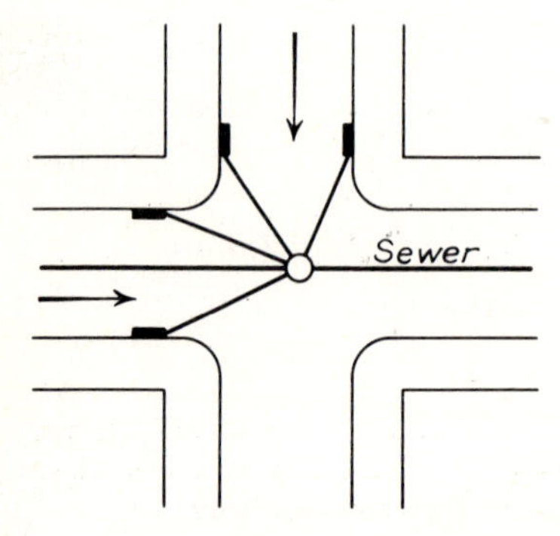

Figure 16-7 Street intersection showing location of storm sewer inlets. Branch lines enter at manholes. The arrows show the direction of surface flow in the streets.

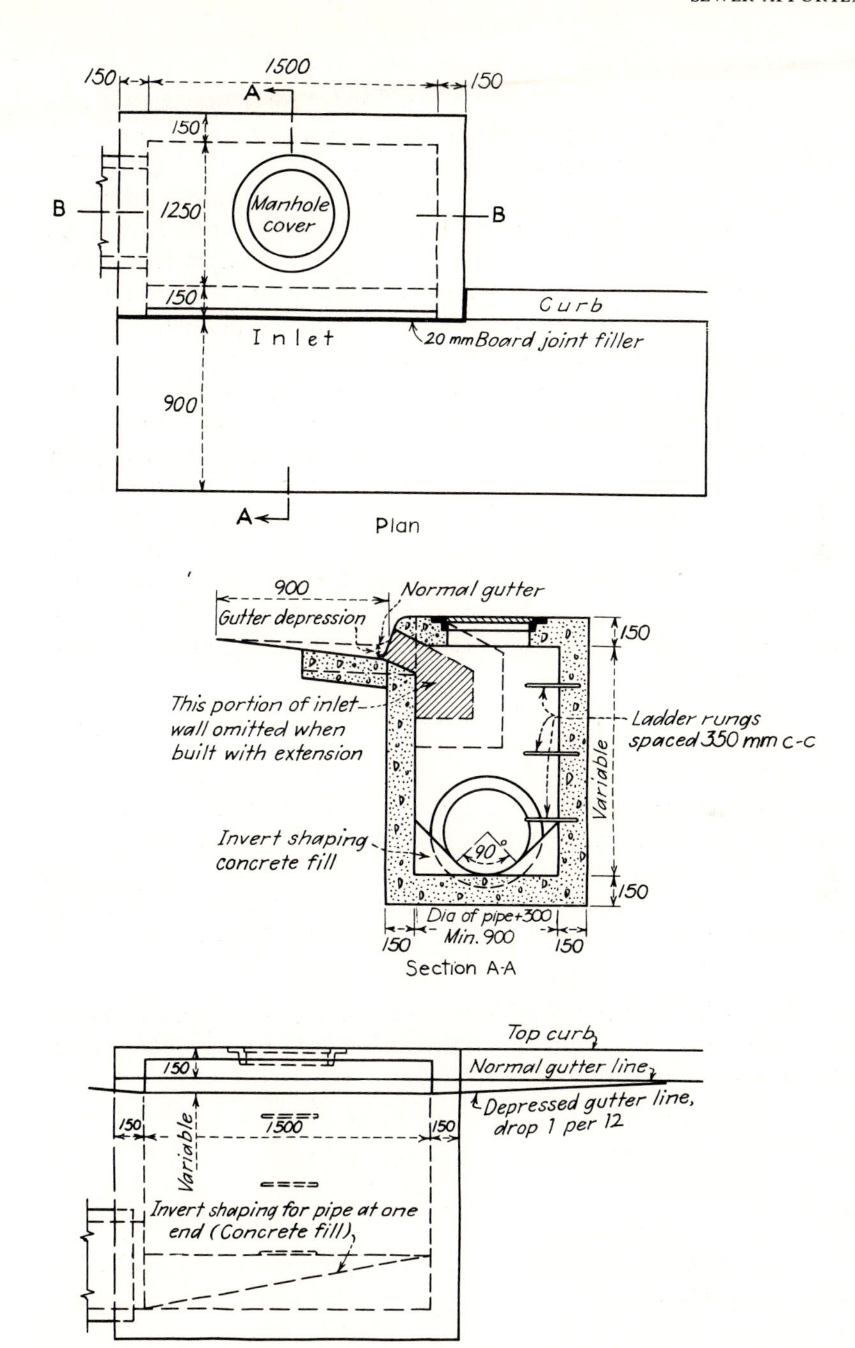

Figure 16-8 Curb-opening storm-water inlet. The crosshatched area of section A-A shows an extension which is used when a greater length than 1500 mm is required. The bottom of the extension slopes toward the main chamber of the inlet. *(Texas Highway Department.)* (Dimensions in mm.)

Table 16-1 Capacities of storm-flow inlets per meter of length

Depth of flow in gutter, mm	Depth of depression, mm	Average capacity per m of length, m^3/min	Depth of flow in gutter, mm	Depth of depression, mm	Average capacity per m of length, m^3/min
15	0		90	0	0.641
	25	0.351		25	1.143
	50	0.825		50	1.784
	75	1.394		75	2.508
	100	1.990		100	3.289
	125	2.843		125	4.225
30	0	0.123	120	0	0.975
	25	0.479		25	1.544
	50	1.003		50	2.230
	75	1.617		75	3.010
	100	2.341		100	3.802
	125	3.094		125	4.794
60	0	0.346	150	0	1.377
	25	0.786		25	2.018
	50	1.366		50	2.731
	75	1.996		75	3.567
	100	2.787		100	4.437
	125	3.623		125	5.295

(m^3/min × 0.588 = ft^3/s, mm × 0.04 = in, m × 3.28 = ft.)

The procedure involves determining how much runoff can be allowed to accumulate in the gutter before it must be intercepted by an inlet. On very flat grades or for the first inlet on a grade, this value is equal to the gutter capacity. Inlets on a moderate grade, however, can be expected to bypass a certain proportion of the flow. The quantity bypassed is a function of the inlet design, and can be expected to vary from zero on grades less than 0.5 percent to as much as 60 percent of the gutter flow on grades greater than 3 percent. The permissible accumulation between inlets is thus the gutter capacity less the residual gutter flow (or bypass) from the inlet next upstream.

Example The storm-water runoff in a gutter at an inlet is 2.55 m^3/min. The transverse slope of the gutter is 0.0156, and its longitudinal slope is 0.03. The value of n is 0.015. Determine the value of y and the inlet capacity per foot of length.

Solution The value of $z = 1/0.0156 = 64$. Substituting in the equation and solving for y shows its value to be 0.037 m. Referring to Table 16-1, with a depression of 75 mm, by interpolation between $y = 30$ and 60, it is found that the average capacity per meter length will be 1.705 m^3/min. The necessary inlet length will be

$$\frac{2.55}{1.705} = 1.5 \text{ m}$$

If a 75-mm depression is used, a 1.5-m length inlet will be ample. With a flow of 2.55 m^3/min the water will extend from the curb line for a distance of $0.037 \times 64 = 2.37$ m. This should not be undesirable, since the flow would be based upon a storm of high intensity.

It is sometimes advisable to permit some of the water to pass the inlet and to be intercepted by the next inlet below, particularly where standard lengths or multiples are in general use. For proportions of the water intercepted by given inlet length, the student is referred to reference 1.

Combined curb openings and gratings are used by some cities and gutter gratings alone in others. Gratings to be effective should have openings parallel to the flow.

16-4 Catch Basins

The catch basin is an inlet with a basin which allows debris to settle out. The outlet pipe is usually trapped in order to prevent escape of odors from the sewer, a provision that also causes retention of floating matter. Catch basins were formerly considered necessary to prevent stoppage of storm and combined sewers with sand and grit. Present practice, however, emphasizes good sewer grades and careful construction, and simple inlets are preferred. The water held in catch basins frequently produces mosquitoes and may itself be a source of odors. If they are to function properly, they must be cleaned frequently, and this is expensive. The most economical method is pumping the basin contents into trucks by means of a specially designed portable centrifugal pump. Figure 16-9 shows two types of catch basins. Usually a curb opening is used.

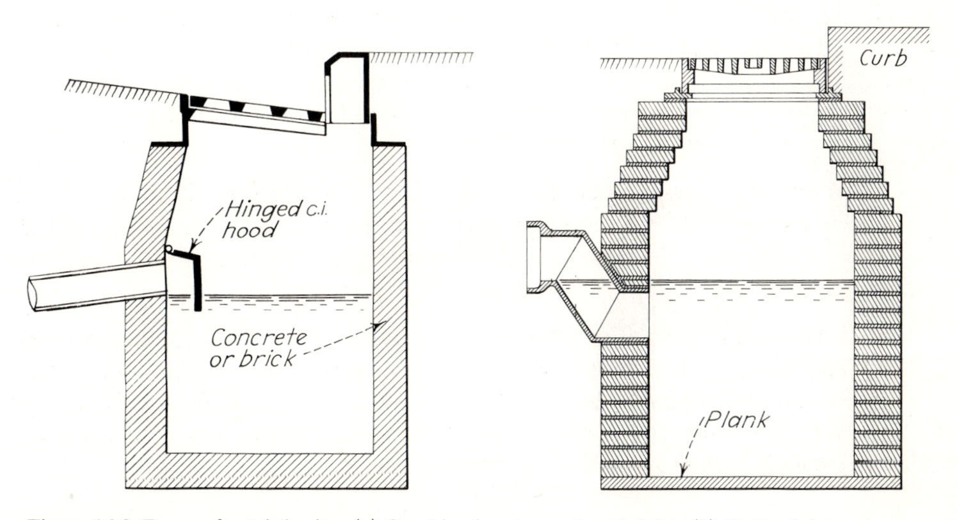

Figure 16-9 Types of catch basins. (*a*) Combined gutter and curb inlet. (*b*) Gutter inlet.

16-5 Flushing Devices

Formerly it was the practice to place automatic flush tanks at the upper ends of laterals where grades are low. They are seldom used in present practice. Figure 16-10 shows such a tank. It resembles a manhole but is equipped with a siphon placed in the bottom. A connection to the water supply system provides a small but constant supply of water so regulated that the tank will fill at least once daily. When the tank is full, the siphon goes into operation and quickly discharges the water down the sewer. The usual flush is about 750 l (200 gal) of water.

The siphon works as follows: Immediately after the siphon has emptied the tank, the water level in both legs of the U-shaped trap stands at the same level as the outlet A. In the tank it is at the level of the small sniff hole B. As the tank fills, the sniff hole is covered, and water rises under the bell C, compressing the air under the bell and depressing the water level in the longer leg of the U trap. This continues until the water in the tank is at the discharge line and in the trap is at the level shown. A little more water in the tank further compresses the air under the bell and causes some of it to escape at the lip A with considerable violence. This sudden release of air pressure causes water from the tank to rush under the bell and down the pipe and thus starts the siphon discharging. Discharge continues until the water level in the tank reaches the sniff hole. Air at atmospheric pressure can then enter the bell and break the vacuum so that the siphon will stop discharging. The cycle is then repeated. The outlet A must protrude above the outlet bottom, or the siphon will merely trickle and never empty the tank. Stoppage of

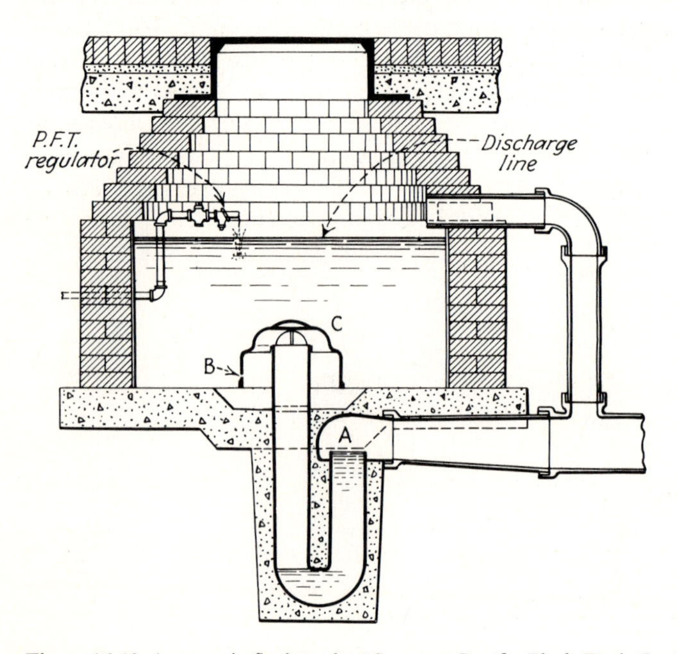

Figure 16-10 Automatic flush tank. *(Courtesy Pacific Flush Tank Co.)*

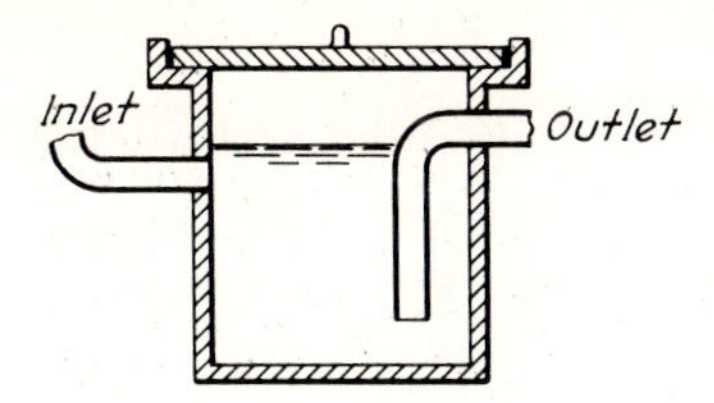

Figure 16-11 Grease trap.

the sniff hole will have the same effect. The overflow pipe will prevent a stoppage of the siphon from causing overflow at the street. It should be recognized that a flush tank with the water supply as arranged is a dangerous cross connection (see Art. 8-34) with the water supply.

A nonautomatic flushing manhole may be constructed by placing slots at the outlet side of an ordinary manhole to receive a wooden stop gate. Closing the gate will permit sewage to accumulate, and opening it will allow a rush of sewage to pass down the pipe. The manhole may also be filled with water from a hose to a fire hydrant, or a permanent water connection may be installed.

16-6 Sand, Grease, and Oil Traps

The sewage from kitchens of hotels and restaurants contains grease which tends to accumulate on sewer walls and cause clogging. Some cities, therefore, require grease traps in the waste lines from the sinks of such kitchens. Figure 16-11 shows a section of a grease trap. Sewage from garages, particularly from the floor drains and wash racks, contains oil, mud, and sand. Figure 16-12 shows a combination sand and oil trap required by some cities. Floor drains could also be connected to the same trap. Such a trap would also prevent gasoline from entering the sewers and causing an explosion hazard. It is important, however, that the traps be regularly cleaned, or they are of no value. Regular inspections are necessary to insure this.

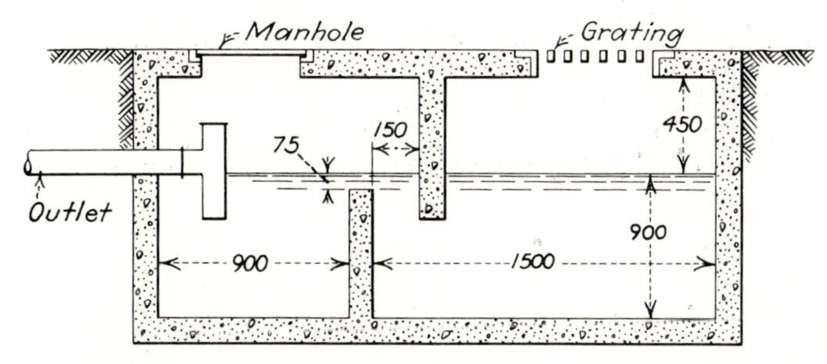

Figure 16-12 Sand and oil trap for garages and car washes. (Dimensions in mm.)

16-7 Regulators

A regulator is a device that diverts sewage flow from one sewer into another. The regulator usually goes into action when the sewage flow reaches a predetermined amount. It may then divert all the sewage into another channel or only that part above the predetermined flow at which it begins to function. Regulators are mostly used where combined sewers discharge into interceptors. The interceptors take the dry-weather flow, but the storm water is diverted into a sewer which flows to the nearest watercourse. Since the first washings from streets may contain much organic matter, and in any case what domestic sewage is diverted should be well diluted, regulators are sometimes designed to operate only after the flow has reached three times the maximum dry-weather flow. Numerous ingenious devices have been used for this purpose, but these have little or no application in modern sewers which are designed to exclude storm-water flows.

16-8 Junctions

The junction of small sewers is made in manholes. The junction of a small sewer with one large enough to be entered is usually a Y branch or slant penetrating the sewer wall. Where two large sewers join, either a bellmouthed or a flat-topped junction may be used. In each case, care should be taken that there is no decrease in velocity and a minimum of water disturbance in either sewer or in the junction

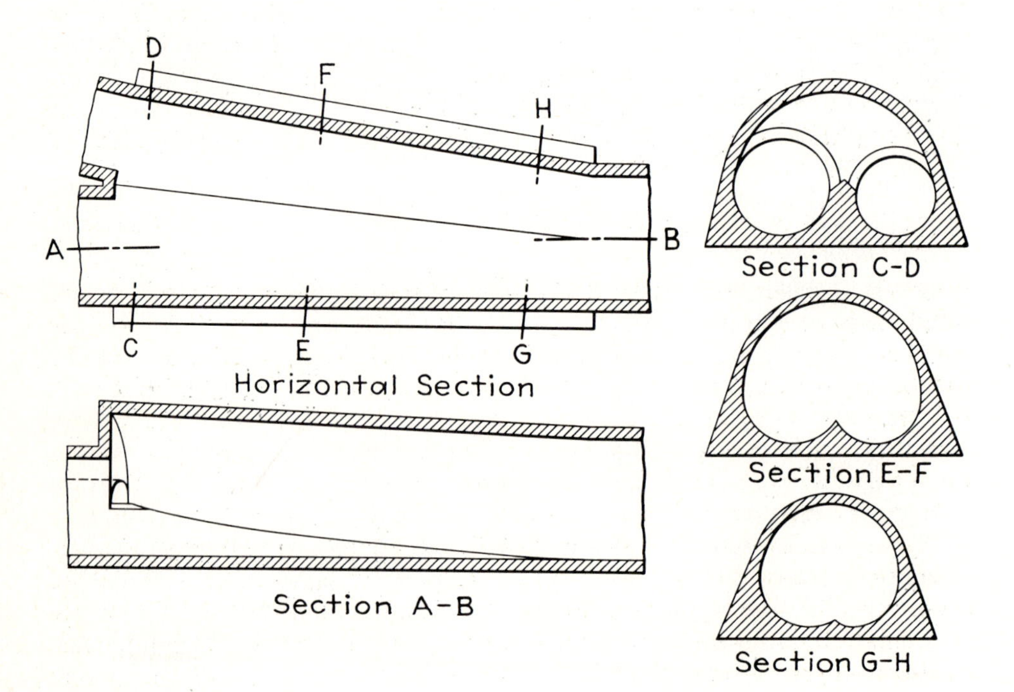

Figure 16-13 Bellmouthed sewer junction.

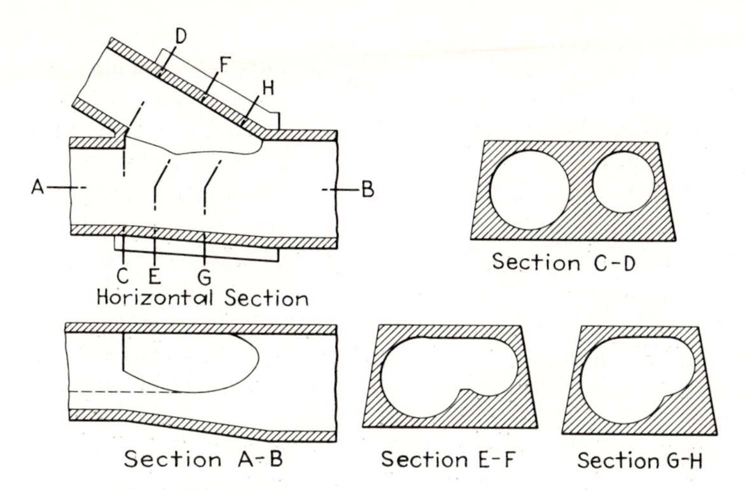

Figure 16-14 Flat-topped sewer junction.

which would cause eddies and deposition of solids. The normal flow lines of the intersecting sewers should coincide at the junction. In the bellmouthed junction the arches of the two sewers are discontinued as soon as their adjoining springing lines intersect. A single arch is then thrown over both sewers from each outer wall. This arch is reduced in span and height until it assumes the size of the sewer formed by the junction. Figure 16-13 shows a bellmouthed junction. In the flat-topped junction both sewers terminate where their walls intersect, and the junction chamber has a flat top and no arch, as shown in Fig. 16-14. The inverts and side walls make the transition to the cross section of the sewer beyond the junction.

16-9 Sewer Outlets

If the sewer outlet discharges treated sewage into a small stream, nothing more elaborate is needed than a simple concrete head wall, similar to those used for highway culverts, to prevent undercutting of the pipe by the sewage and the stream. An apron of concrete should be provided on the bank below the head wall. It is advisable to place a flap valve or automatically closing gate on the outlet discharging treated sewage to prevent muddy flood water from backing up into the sewage treatment plant when the stream level is high. Filters may be badly damaged by deposition of mud in them in this fashion.

Sewers discharging into harbors or other large bodies of water are frequently extended long distances beyond the banks into deep waters or where currents will cause rapid mixing of the sewage and the diluting water and thereby reduce environmental problems. Such pipes may be laid on the bottom with joints made by divers or may be connected on barges and lowered in large sections. Precast reinforced concrete pipes are most widely used for submarine outfalls 915 mm (36 in) in size or greater. Cast iron is used for sizes 610 mm (24 in) or less.

Corrugated metal, ductile iron, and steel have been used to some extent. They may be protected from waves by placing them in a dredged trench or by a row of piles on each side with cross timbers placed just above the pipes. Outfalls may have multiple openings at the outlet to aid in diffusion of the sewage in the water.

16-10 Inverted Siphons

In sewerage works the term inverted siphon is applied to a portion of a sewer which dips below the hydraulic grade line to avoid such an obstruction as a railway cut, a subway, or a stream. The pipe used must be able to withstand internal pressure. As high velocities as possible should be maintained, with 0.9 m/s (3 ft/s) minimum. If there is sufficient head to permit good velocities, single pipes 305 to 610 mm (12 to 24 in) in size have been used with little trouble. Where very little head and a widely varying flow make it difficult to obtain favorable velocities, a pipe large enough to carry the maximum flow will have the sewage barely moving through it during low flow. Several pipes in parallel will in large part overcome this difficulty. A sewer, for example, may have one of the pipes of the inverted siphon large enough to take the minimum flow at a good velocity with a second pipe to take the difference between the minimum and average flow, and a third to carry peak flow. Inverted siphons are constructed of cast-iron, concrete, or tile pipe, depending upon the construction conditions that are encountered. Concrete or tile pipe must be encased in concrete to prevent leakage. Where a stream is crossed, cast-iron pipe is normally used and is placed by the same methods used for water mains (Art. 6-27). Design of an inverted siphon can be shown by an example (Fig. 16-15).

A sewer 1070 mm in diameter carries 42.5 m^3/min when flowing full. The design requires that the head loss be kept to a minimum. The low flow is 5.5 m^3/min, and the average flow 20.4 m^3/min. Three pipes will be used, one to carry the minimum flow, the second to carry flows from minimum to average, and the third to carry all flows above the average. Manning's n is 0.013. The invert

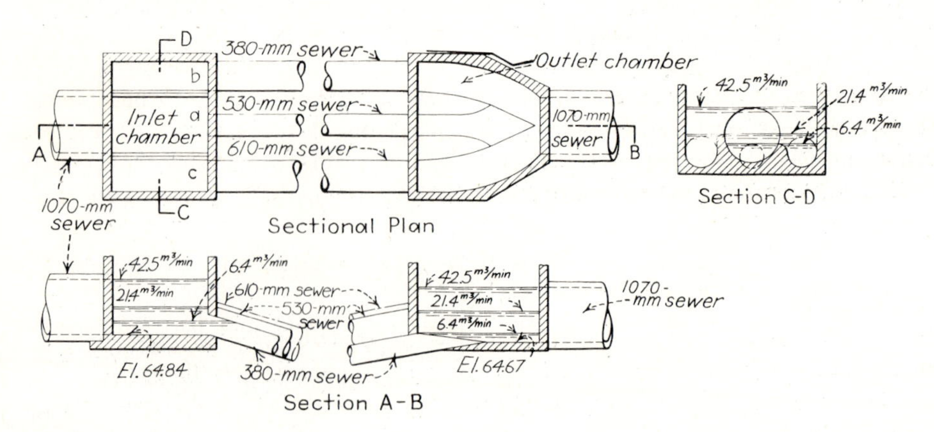

Figure 16-15 An inverted siphon.

elevation at the inlet to the inverted siphon is 64.84 m, and the pipe will be 33 m long. The inlet chamber, as illustrated in the figure, consists of a channel "a" which is a continuation of the sewer and discharges into the smallest of the three pipes. The weir on one side is set at an elevation equal to the depth in the sewer at the flow rate which is carried by the smallest pipe. When the flow exceeds this value, the depth will increase, and the excess flow will spill over into the channel which feeds the second pipe. The weir on the other side is set at the depth which will exist when both channel "a" and "b" are full. Flows above the capacity of the first two lines are carried by the third.

Since the headloss is to be minimized, the pipe will be designed for a velocity of 0.9 m/s when flowing full. The headloss at the entrance will be at least equal to the velocity and an allowance must be made for loss over the weir as well. For purposes of calculation, the entrance loss will be assumed to be 0.06 m. The losses due to bends will be neglected.

The small pipe must carry 5.5 m^3/min which requires a 380-mm (15-in) pipe on a grade of 0.0032. This size is the nearest commercial size and will actually carry 6.4 m^3/min when flowing full at a velocity of 0.9 m/s. The total fall which must be provided is $33(0.0032) + 0.06 = 0.17$ m.

The second pipe must carry 20.4 − 6.4, or 14.0 m^3/min. The commercial pipe size required to carry this amount on the grade previously established is 530 mm (21 in). The capacity of this pipe is 15.0 m^3/s.

The last pipe must carry $42.5 - 6.4 - 15.0 = 21.1$ m^3/min. The size required will be 610 mm (24 in). It should be observed that with equal hydraulic gradients the larger pipe will have higher velocities than the 380-mm line.

It is now possible to set the side weir elevations. When the main sewer is carrying 6.4 m^3/min the water depth will be 277 mm, giving a side weir elevation of 65.12 m. When the flow has reached 15.0 + 6.4, or 21.4 m^3/min, the depth will be 536 mm, and the side weir elevation 65.38 m.

The weir length will be taken to be 2 m. The velocity over the weir to the 530-mm sewer will be

$$\frac{15.0}{(0.536 - 0.277) \times 2} = 28.95 \text{ m/min} = 0.483 \text{ m/s}$$

The velocity head over the weir will be 12 mm. The velocity in the new direction in the pipe is assumed to be 0.9 m/s, giving a velocity head of 41 mm. The total 53 mm is less than the assumed 0.06 m.

The velocity over the weir into the 610-mm pipe will be

$$\frac{21.1}{(1.070 - 0.536) \times 2} = 19.76 \text{ m/min} = 0.329 \text{ m/s}$$

The velocity head over the weir is thus 6 mm, and the total, assuming a pipe velocity of 0.9 m/s, is 47 mm—which is less than the assumed 0.06 m.

The outlet chamber is so arranged that the inverts of the three pipes merge into a single channel which becomes the invert of the outlet sewer. The elevation

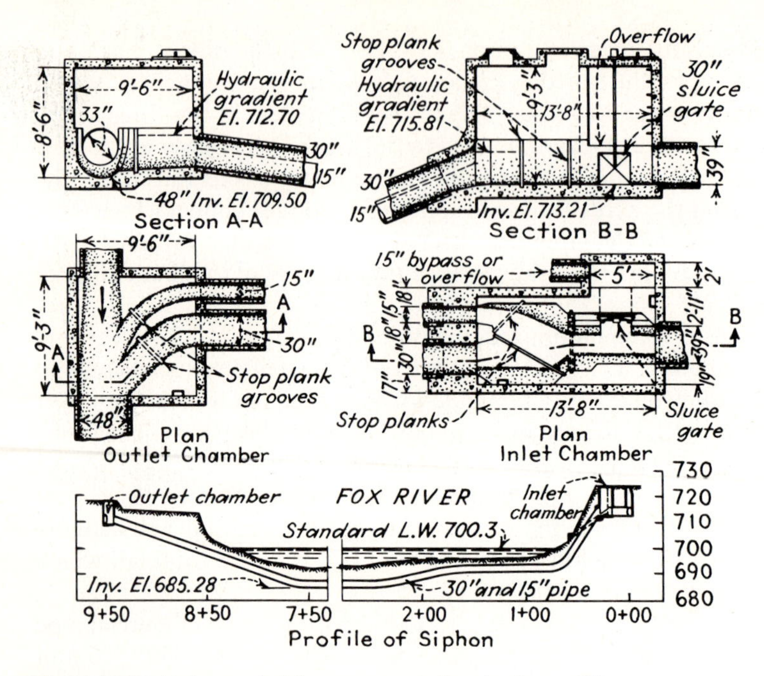

Figure 16-16 An inverted siphon constructed at Appleton, Wis.

of the outlet will be 64.84 − 0.17 = 64.67. Because of the greater velocities in the two larger pipes, their outlets should be higher than that of the smallest, as shown, to avoid eddies and possible accumulation of solids during periods when they are not in operation.

Inverted siphons are sometimes constructed with a simpler type of inlet chamber. Simple stop boards may be placed in the inlet box so that, when the smallest pipe is flowing full, the sewage overflows to the next pipe, as in Fig. 16-16. A bypass, or overflow pipe, may also be installed. Generally, inlet and outlet chambers are placed in manholes or in closed structures which may be entered by manholes in the roofs.

16-11 Sewer Crossings

It is frequently possible to cross depressions or small watercourses without resorting to inverted siphons. Occasionally an existing bridge may be used by suspending the pipe from a truss. Vibration will occur, however, and cast or ductile iron pipe should be used with special couplings, rather than standard lead joints, to avoid leakage. A more common method of crossing shallow depressions is to support the pipe on piers or bents of concrete. Here, cast or ductile iron pipe should also be used, with a support at each joint, but ordinary lead joints may be used under these conditions. If the exposed pipe is long, provision should be made for thermal expansion.

Figure 16-17 Sewer supported on piers.

PUMPING OF SEWAGE

16-12 Need for Pumping

Most large cities find it necessary to pump at least a part of their sanitary sewage. Cities protected by levees, like New Orleans, must pump all sewage, even storm water, and may need to pump it more than once. Pumping is required when basements are deep, terrain is flat, obstacles lie in the path of the sewer, the receiving stream is higher than the sewer outlet, or when gravity flow is desired through an above-ground treatment plant.

16-13 Pumps for Sewage

Sewage is commonly pumped using specially designed centrifugal units. When large volumes of flow must be moved against low to moderate heads, as in pumping storm sewage from a leveed area, mixed or axial flow pumps are used. Speeds in this service are generally in the range[2] of 200 to 1200 r/min.

Nonclog pumps have impellers which are usually closed and have, at most, two or three vanes. The clearance between the vanes is generally sufficiently large that anything which will clear the pump suction will pass the pump. A bladeless impeller, sometimes used as a fish pump, has also been applied to pumping sewage. For a given capacity, bladeless impellers are larger than vaned designs.

Manufacturers of sewage pumps specify the sphere size which the pump will pass. A pump with a 100-mm (4-in) discharge might pass a 75-mm (3-in) sphere, while a 250-mm discharge would pass a 200-mm sphere. Pump suctions are usually larger than the discharge by about 25 percent. The smallest discharge size which is commonly used is 75 mm (3 in).

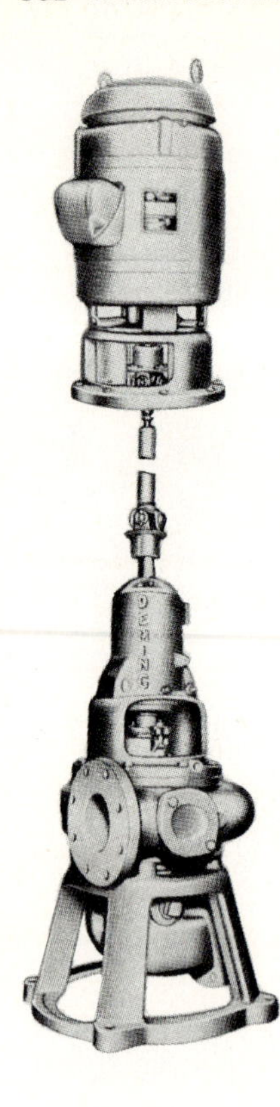

Figure 16-18 Typical vertical sewage pump installation. *(Courtesy Crane Deming Pumps, Crane Co.)*

Nonclog pumps are manufactured in both vertical and horizontal configurations. Vertical pumps are, by far, the more popular. They require less floor space, eliminate high points in suction lines, and permit vertical separation of pump and motor. A typical installation is shown in Fig. 16-18.

Large pumping stations may incorporate screens and even grit removal equipment ahead of the pumps. These devices are particularly likely to be incorporated at pumping stations located at treatment plants. The screens and grit removal serve to protect the pumps and may prolong their life.

Pumps are generally provided in duplicate, even in small lift stations. The pumps alternate in ordinary service and if one should fail, the other will serve as a standby. Emergency power is provided only in large stations, such as those at treatment plants. In these locations a generator may be installed.

Submersible pumps which have been specially modified to provide easier servicing are available from several manufacturers. These units are close-coupled vertical centrifugal pumps which may be removed from their piping connections without entering the sump. A system of this type is illustrated in Fig. 16-19.

Sewage pumps operate against variable head conditions and the location of the system curve relative to the pump curve must be considered for the total range of suction and discharge conditions. A flat head-discharge curve is not desirable in this service (Art. 7-6).

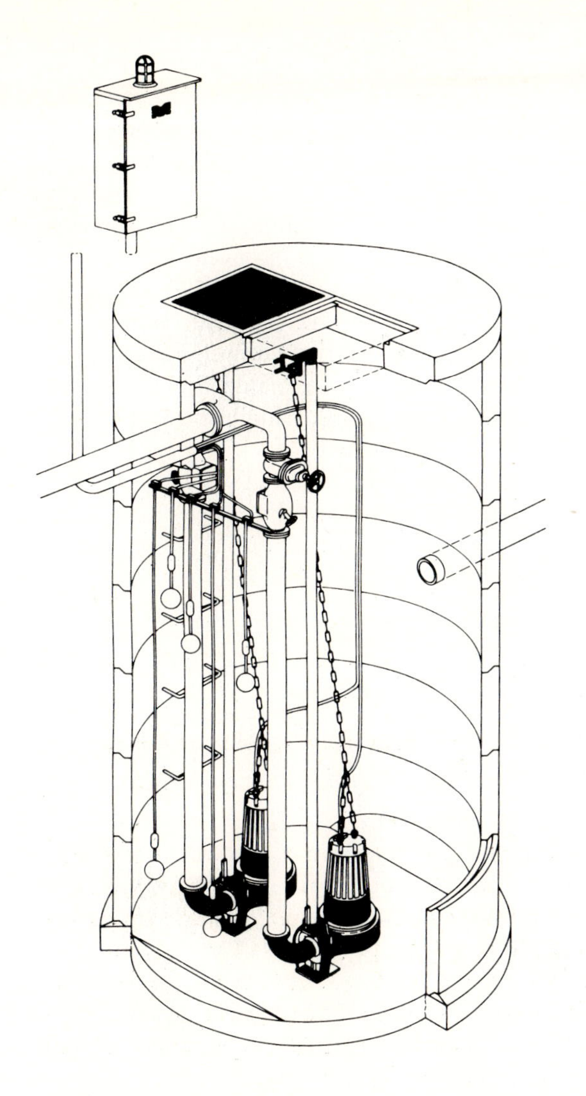

Figure 16-19 Submersible sewage pump installation. *(Courtesy LFE Corp., Fluids Control Div.)*

16-14 Pumping Stations

In *dry pit* pumping stations the pumps may be either vertical (Fig. 16-18) or horizontal (Fig. 16-21). The pump takes suction through a wall pipe to an adjoining wet well, is accessible for maintenance and service at all times, and is less subject to corrosion than if it were submerged.

The pumps and the wet well must be sized together in order to achieve a satisfactory design. The smallest capacity pump which can be used depends upon the size of the line to which it discharges since self-cleansing velocities (about 0.6 m/s) must be maintained. The smallest pump which should be connected to a 100-mm (4-in) force main, for example, would have a capacity of about 280 l/min (75 gpm).

Since sewage flow is variable (Chap. 2) the pumping station must be able to adjust to these variations. In small pump stations the pump is sized to meet the peak flow and lower flows accumulate in the wet well until sufficient liquid is present to permit the pump to run for at least 2 min. In order that the pump not cycle too frequently, it is standard practice to permit sufficient storage in the wet well to insure that the pump will not start more than once in 5 min. The application of these considerations will be illustrated in the following example.

A small subdivision produces an average wastewater flow of 120,000 l/day. The minimum hourly flow is estimated to be 15,000 l/day, and the maximum, 420,000 l/day. Determine the design pumping rate and the wet well capacity.

The pump must deliver 420,000/1440 = 292 l/min to match the peak flow rate. If the discharge were to a large pipe, a higher capacity might be required to maintain self-cleansing velocities.

The pump running time may be calculated from:

$$t_r = \frac{V}{D - Q} \tag{16-2}$$

in which t_r is the running time, V the storage volume, D the pump discharge, and Q the influent flow. The filling time, with the pump off, is:

$$t_f = \frac{V}{Q} \tag{16-3}$$

and the total cycle time is:

$$t = t_r + t_f = \frac{V}{D - Q} + \frac{V}{Q} \tag{16-4}$$

To assure 2-min running time, the minimum volume is:

$$V = t_r \times D = 2 \times 292 = 584 \text{ l}$$

To assure a 5-min cycle, at average flow,

$$t = \frac{V}{D - Q} + \frac{V}{Q} = 5$$

$$1/V = \tfrac{1}{5}(292 - 83.3) + \tfrac{1}{5}(83.3)$$

$$V = 298 \text{ l}$$

Therefore the running time governs, the volume between starting and stopping elevations will be about 600 l, and the actual cycle time at average flow will be:

$$t = \frac{600}{292 - 83.3} + \frac{600}{83.3} = 10.1 \text{ min}$$

A minimum depth, variable with inlet features,[2,3] must be maintained over the pump suction. At an intake velocity of 0.6 m/s (2 ft/s) a submergence of about 300 mm (1 ft) is required. It is normal to provide some freeboard above anti-

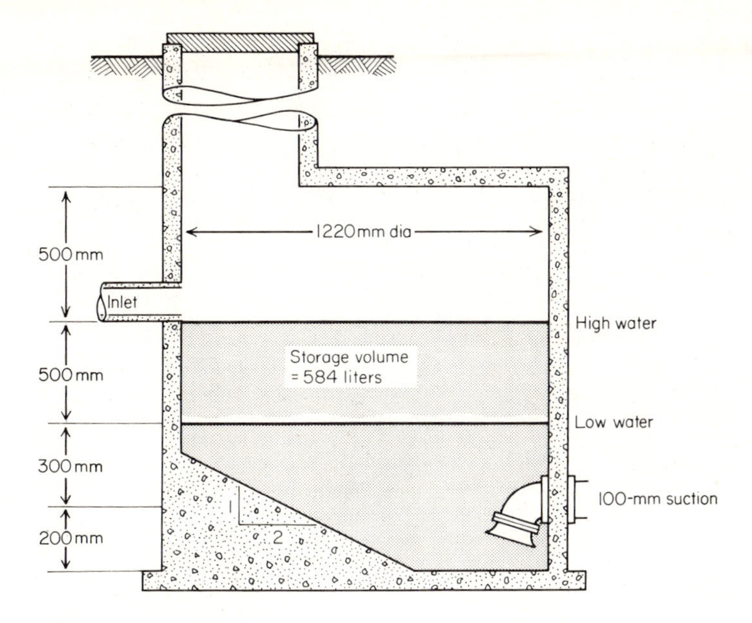

Figure 16-20 Wet pit details.

cipated maximum water level. Typical installations allow 600 mm (2 ft) for this purpose. A wet pit suitable for the example above is illustrated in Fig. 16-20.

Larger pump stations can employ pumps of different capacities to match effluent to influent flow. The different pumps are set to come on at different water elevations in the wet well. Such stations have a minimum of three pumps so sized that the maximum flow can be pumped with the largest unit out of service.

Switching of pumps can be effected by floats on chains or rods, by air bubblers, or by mercury switches encased in rubber floats which hang free in the wet well. Modern pumping stations use the latter two devices. If mercury switches in hanging floats are used, no air supply is required.

Gate valves are needed in the suction line and in the discharge line. They should be all-iron wedge type with rising stems to avoid corrosion. A check valve should be placed between the pump and the gate in the discharge pipe. It should be an all-iron swing type with outside lever and weight to reduce slam. A sump in the dry-well floor is needed to collect seepage. A small sump pump should be installed to transfer this to the wet well (Fig. 16-21). A trolley rail should be provided to assist in lifting the pumps and motors.

The size of the wet well is kept small in order to forestall septic action in the sewage, and the bottom is sloped toward the pump inlets to prevent accumulation of solids. In large installations the wet well may be divided into two compartments connected with an opening that can be closed by a shear gate or sluice gate, thus permitting cleaning or repair. A sump pump in the dry well may have a suction pipe that will permit drainage of the wet well. Ventilation of the wet well should be

Figure 16-21 Small two-pump underground sewage pumping station. *(Courtesy Marolf, Inc.)*

provided to prevent excessive condensation and accumulation of odors. A manhole or other means of entrance must be provided. If gasoline and oils are to be expected in the incoming sewage in large amounts, explosion-proof electric motors and starters should be used.

Small pumping stations must frequently be placed in residential sections. To forestall criticism the building should be of pleasing architectural design. The stations should be properly lighted, ventilated, and heated if necessary.

Outfall lines from sewage pumping stations may be so long that hydrogen sulfide is generated in them. This produces no trouble in the force mains themselves, since they flow full, but may cause odors and corrosion of concrete at outlet structures. Houston has prevented this difficulty by injecting air into a 2400-m (1.5-mi) force main at the rate of 13 percent by volume at STP. Without the air, the sulfide increased from 0.7 to 2.5 mg/l from wet well to outlet. With air, only a trace was found.

16-15 Sump Pumps

Basements of buildings are sometimes below sewers, and sewage produced in them must be raised. The sewage is collected in a closed metal tank which is emptied by a centrifugal pump of nonclogging submerged type; i.e., pump casing, volute, and suction pipe are all placed on a vertical shaft below the water or sewage level. A motor is placed above the pump, and a float operates a switch. A high-water alarm may be installed. For unscreened sewage, pumps with suction diameters of less than 75 mm (3 in) should not be used.

Figure 16-22 Sump pump installation. *(Courtesy Crane Deming Pumps, Crane Co.)*

16-16 Sewage Ejectors

The sewage ejector works with compressed air. Figure 16-23 shows a well-known type. It consists of a tank or receiver into which the sewage runs, the air escaping through a vent, until the tank is full. The upper bell then has air entrapped under it, and its buoyancy raises the rod, closing the air exhaust, opening the compressed air inlet, and thus forcing the sewage up the discharge pipe. When the receiver is emptied, the weight of the lower bell becomes great enough to pull down the rod, which closes the compressed air valve and opens the exhaust. Check valves are placed in the inlet and discharge pipes. Ejectors have been used for pumping large quantities of sewage, but centrifugal pumps have largely replaced them. They are still used as sump pumps in buildings where compressed air is available and are obtainable in capacities of 100 to 3800 l/min (30 to 1000 gpm). They have the important advantage of presenting little difficulty with clogging and can easily handle unscreened sewage. When clogging occurs it is at the valves.

16-17 Vacuum Collection Systems

For ease of cleaning, gravity sewers are generally at least 200 mm (8 in) in diameter. In order to maintain self-cleansing velocities at low flow the sewer grade must be steepened, hence gravity sewers installed in areas where population density is low may be inordinately expensive.

Vacuum systems utilize 50- to 150-mm (2- to 6-in) plastic pipe which is buried deep enough to prevent freezing. Each customer's residence is connected to the vacuum main through a special entry valve which opens automatically when a

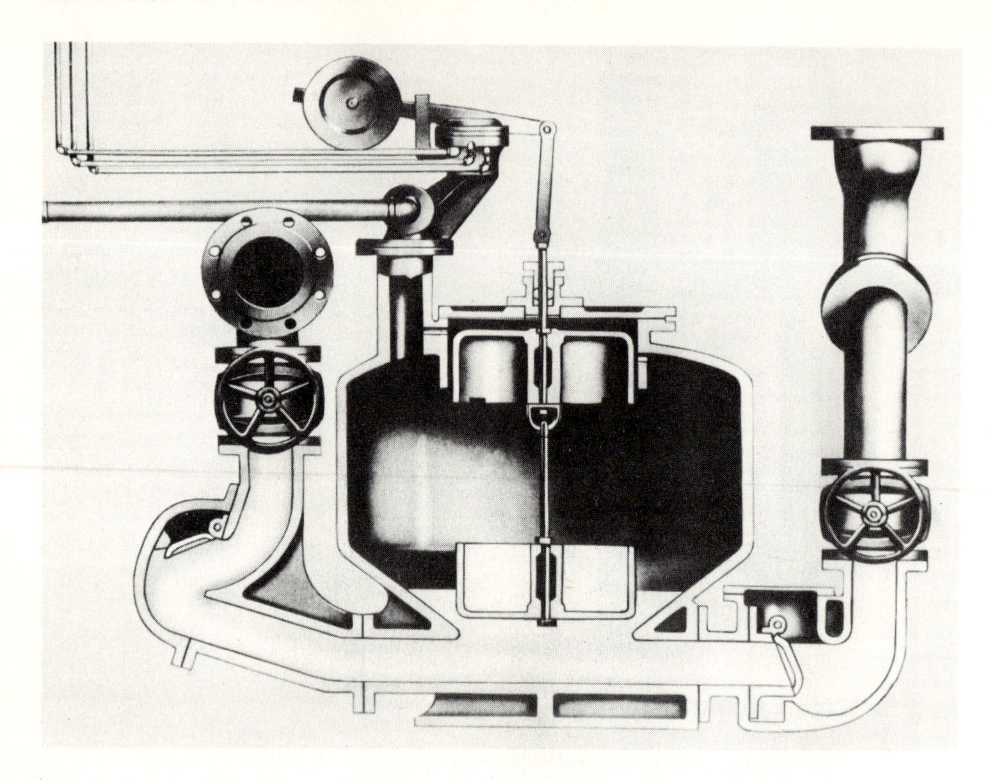

Figure 16-23 Shone sewage ejector. *(Courtesy Clow Corporation.)*

sufficient volume of sewage has collected in the gravity house lateral. The valve closes again after a period of time sufficient to withdraw the accumulated waste. The vacuum in the collection system is maintained at a central collecting tank to which a number of mains may be run and from which the flow is then pumped in the usual fashion.

Vacuum systems are applicable in special circumstances, but are not a satisfactory substitute for gravity systems in populated areas.

16-18 Grinder Pumps

Another technique which has been used to avoid the deep sewers and expensive construction typical of areas with low population densities is the grinding of sewage to reduce the particle size. If the largest particle is reduced from 50 to 5 mm there is no reason why it cannot be pumped through a small line using a small pump and a small wet well.

Pumps which will both macerate and pump sewage have been developed and are available for this purpose (Fig. 16-25). Like vacuum systems, pressure systems employing grinder pumps are useful primarily as a supplement to, rather than a substitute for, standard gravity systems.[4]

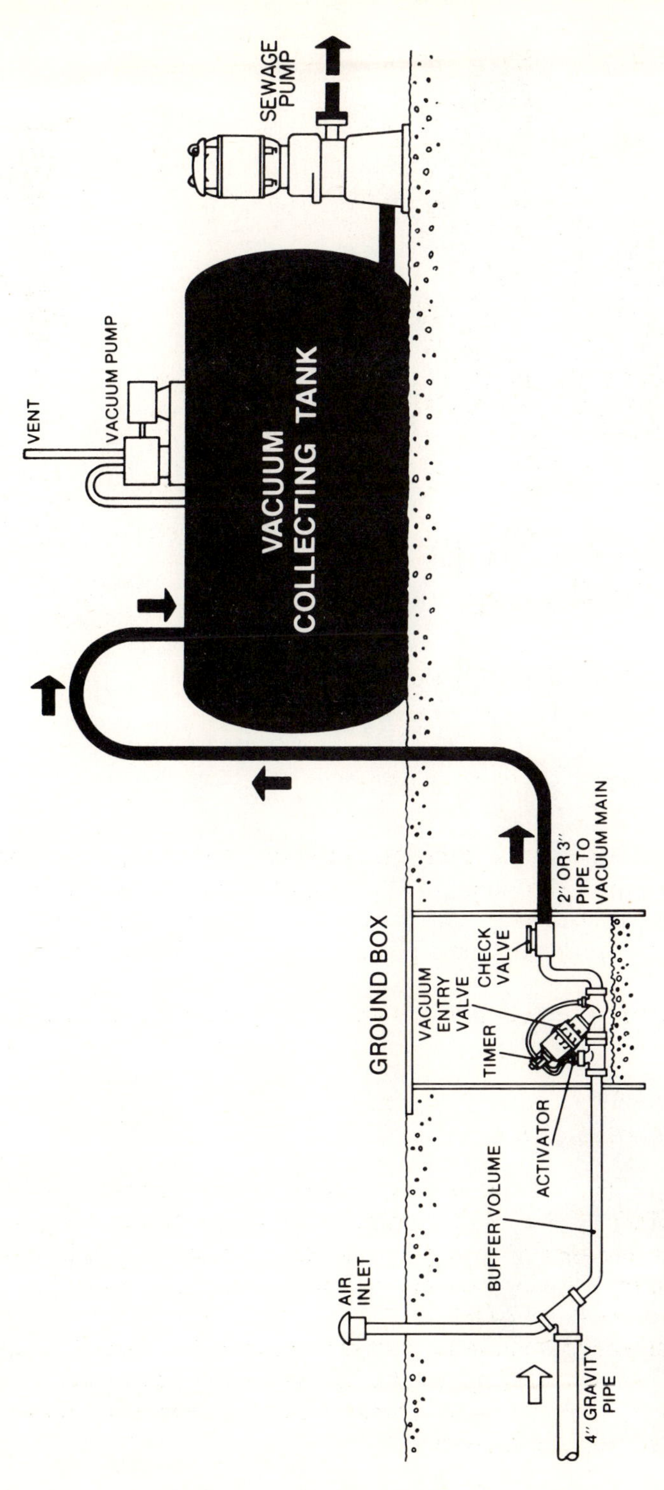

Figure 16-24 Vacuum collection system. (*Courtesy Colt Industries.*)

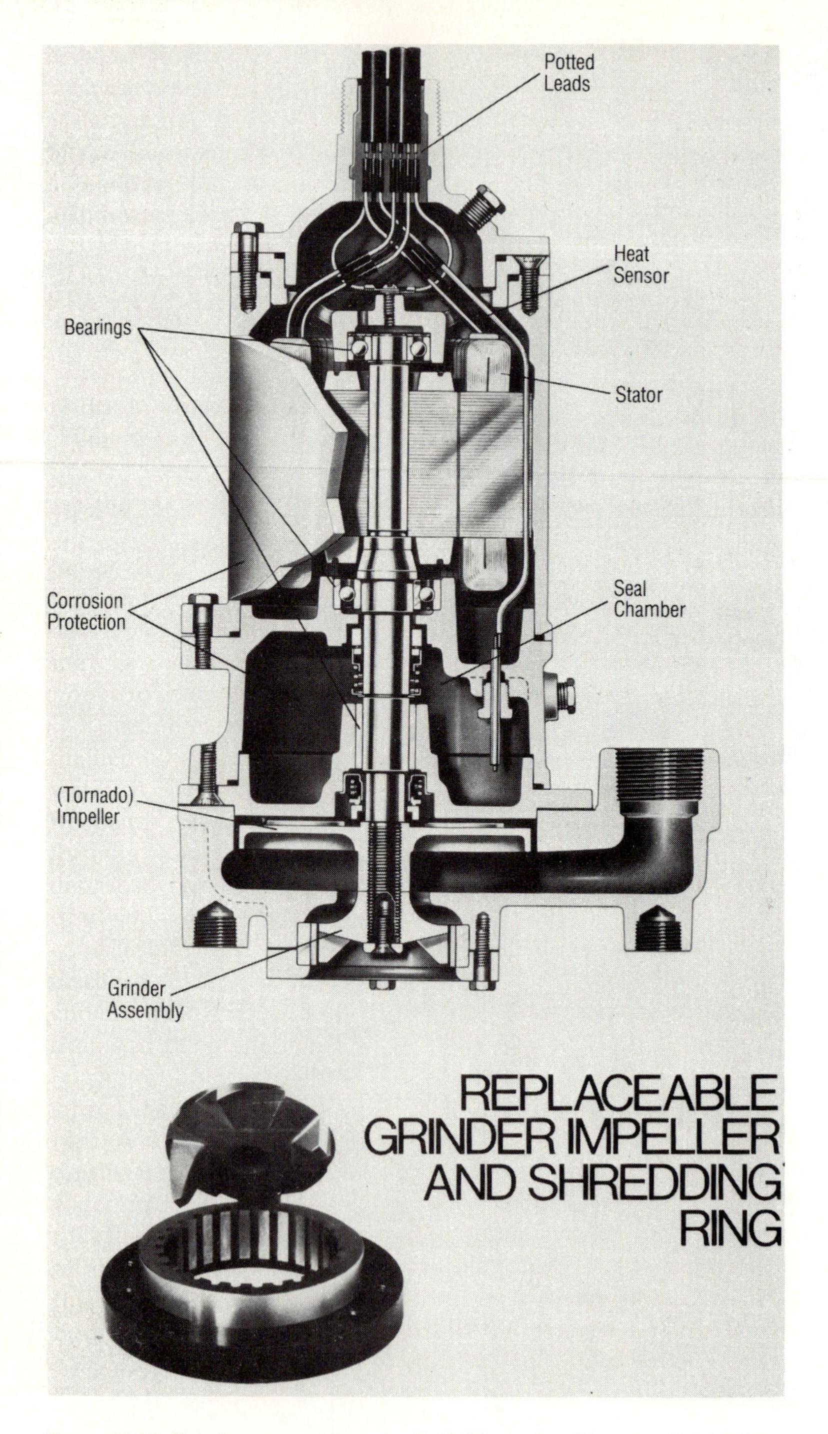

Figure 16-25 Grinder pump. *(Courtesy F. E. Myers Co., Division of McNeil Corporation.)*

PROBLEMS

16-1 A circular sewer 1220 mm in diameter has a slope of 0.001. The minimum sewage flow is 14 m^3/min, and average flow is 40 m^3/min. Design an inverted siphon which will give the minimum loss of head and will consist of three pipes, one for the minimum flow, one to carry the flow from minimum to average, and the third to carry flows up to the capacity of the 1220-mm sewer. The inverted siphon will be 220 m long with $n = 0.013$.

16-2 A small sewage pumping station is to have only one pump operating at a time. It is expected that the average flow will be 450,000 l/day. The minimum rate of flow is one-half the average, and the maximum is 175 percent of average. Determine the size of the wet well, the capacity of the pump, and the cycle time at maximum, minimum, and average flow.

REFERENCES

1. "Drainage of Highway Pavements," *Hydraulic Engineering Circular No. 12*, U.S. Department of Transportation, Federal Highway Administration, Washington, D.C., 1969.
2. Hicks, Tyler G. and Theodore W. Edwards: *Pump Application Engineering*, McGraw-Hill, New York, 1971.
3. Metcalf and Eddy, Inc., *Wastewater Engineering: Collection, Treatment, Disposal*, McGraw-Hill, New York, 1972.
4. Carcich, I. G. et al.: "Pressure Sewer Demonstration," *Journal Environmental Engineering Division, Proceedings American Society of Civil Engineers*, **100 EE 1**:25, 1974.

CHAPTER

SEVENTEEN

DESIGN OF SEWER SYSTEMS

17-1

The design of a sewer system entails preliminary investigations, a detailed survey, the actual design, preparation of final drawings, and correction of plans to conform to changes made during construction.

17-2 Preliminary Investigations

Preliminary investigations are required to arrive at an estimate of cost which will serve as the basis for bond issues, assessments, or other fund raising. Preliminary designs from maps are the basis for facility plans made under Section 201 of Public Law 92–500. Generally, maps are available which show streets and the more important topographic features. Small towns which have no official maps may have been mapped by assessors, insurance companies, or public utilities who will usually permit their maps to be copied. If no maps are available, aerial strip photography is probably the least expensive method of obtaining a map with the necessary detail.

The anticipated population, its density, and its waste production must be estimated for the planning period which is normally 20 years or more. Prospective disposal sites are selected and their suitability is evaluated with regard to collection of sewage and effects of its disposal. Site selection will affect the cost of collection and can affect the degree of treatment required.

Cost estimates are made for the alternatives identified as being physically practicable and environmentally acceptable. Costs should be based, when possible, upon local bid tabulations for recent similar construction. When such information is not available, the use of average curves such as those published by the Environmental Protection Agency,[1] corrected using the local construction cost index, may be satisfactory for preliminary purposes.

17-3 The Underground Survey

Before final lines and grades are selected for sewers, the designer should be aware of the location of all underground obstacles such as existing sewers, water or gas lines, electrical or telephone wires, tunnels, foundations, or other construction. Many city engineering departments maintain maps showing all underground structures. In the absence of such maps the engineer must compile information from as-built drawings obtained from the various utilities.

The presence of rock or high groundwater will have a significant effect upon the cost of construction, hence soil borings or soundings may be desirable. Borings may be made by driving a sharpened steel rod into the ground until rock is struck. For depths in excess of 5 to 6 m (15 to 20 ft) augers or coring bits are necessary. Sufficient borings should be made to establish the location of the rock surface. The actual number required will depend upon the geological character of the area.

17-4 The Survey and Map

Preparation of construction drawings requires knowledge of paving characteristics of the streets, the location of all existing underground structures, the location and basement elevations of all buildings (usually estimated for residences), the profiles of all streets through which the sewers will run, the elevations of all streams, culverts, and ditches, and maximum water elevations therein. Permanent benchmarks should be established for use in construction.

The map scale is usually 1 : 1000 to 1 : 3000, depending upon the detail desired. Contours should be shown unless surface relief is negligible. Contour intervals range from 250 mm to 3 m (1 to 10 ft). Elevations of street intersections and any abrupt changes in grade are shown. The elevations of any existing structures (sewers, lift stations, outfalls, etc.) which the new construction must meet are noted on the map.

17-5 Layout of the System

A tentative layout is made by drawing lines along the streets or utility easements. Arrows show the direction of flow, which is generally in the direction in which the ground slopes. The result of this layout will be a main sewer leaving the area at its lowest point with smaller laterals and submains radiating to the outlying areas. The sewers will follow the natural surface drainage as closely as the layout of the streets and easements permits. Ridges may require either pumping or construction of separate sewer systems draining to different points. In flat terrain a central location may be selected to which all lines will drain for pumping to a gravity main or to the treatment plant.

Location in backlot utility easements will reduce pavement damage during construction, but may make access to the sewer difficult. Generally, water, gas, and sewer lines are laid within the street right of way. It is desirable that the sewer and water lines be separated—preferably by the width of the street. On very wide streets sewers may be placed on either side to reduce the length of connections.

The vertical layout is dictated by the need to provide minimum cover and the desirability of minimum excavation. In northern states 3 m (10 ft) of cover may be required to prevent freezing. In the south, minimum cover is dictated by traffic loads and ranges upward from 0.75 m (2.5 ft), depending upon pipe size and expected loads (Art. 14-3).

Manholes are located at all sewer intersections, changes in horizontal direction, major changes in slope, changes in size, and at intervals along straight runs. Manhole spacings generally do not exceed 100 m (300 ft) and should never exceed 150 m (500 ft). Manholes are numbered and the numbers identify the individual sewer lines as well, since these run from manhole to manhole.

The area tributary to each sewer is sketched on the map, considering the location of lots and buildings. Some lines will have no additional tributary area but will merely carry sewage collected by lines upstream. For storm sewers a similar procedure is used except the lines are considered to run from inlet to inlet. Since inlets are generally at corners, storm sewers usually run from corner to corner. The area tributary to each inlet is sketched according to the ground contours on the map.

17-6 The Profile

A vertical profile is drawn from survey notes for each sewer line. The horizontal scale ranges from 1 : 500 to 1 : 1000 depending upon the detail to be shown. The vertical scale is usually 10 times the horizontal. The profile shows the ground surface, tentative manhole locations, rock indicated by borings, location of bore

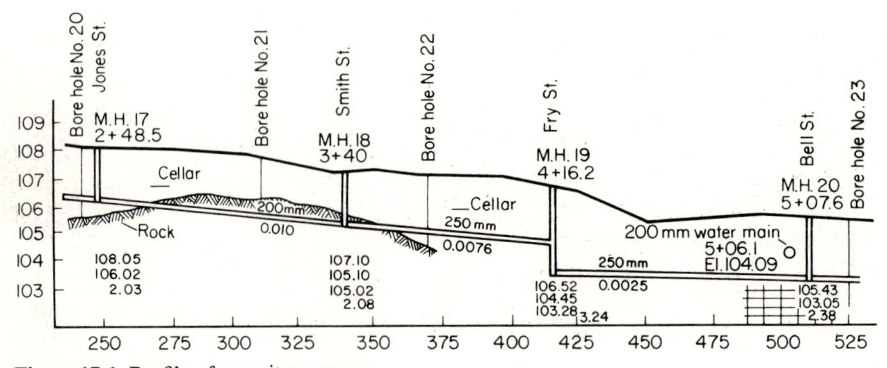

Figure 17-1 Profile of a sanitary sewer.

holes, all underground structures, cellar elevations, and cross streets. A map of the street may be shown above or below the profile.

The profile is used to assist in design and serves as the basis for construction drawings. When the design is completed the lines are shown as in Fig. 17-1 with grades, sizes, and elevations. At each manhole the surface elevation, the elevation of the sewer invert entering and leaving the manhole, and the cut to the leaving invert are listed.

The sheet may also contain a tabulation of lengths and sizes of pipe, their cuts, and the number and depth of manholes. Such a tabulation is useful in estimating construction costs.

Manhole 19 of Fig. 17-1 is a drop manhole used to obtain sufficient cover at the low point without resorting to a steep grade which would cause a deep cut at manhole 20.

17-7 Design of a Sanitary Sewer System

The designer should familiarize himself with the requirements of the state regulatory agency. These will cover such matters as minimum flows per capita, infiltration, and manhole spacing. The following design illustrates a method which must be varied according to local and state requirements.

Figure 17-2 shows a portion of a city that it is proposed to sewer. The sewers for the area north of Maple Ave. are already designed, and the sewer leading from a portion of that area flows south on 15th St. to manhole 22. The illustrative design here given covers the area west of 12th St. and that portion east of 12th St. and south of Beech Ave. The topography is such that the trunk sewer will flow

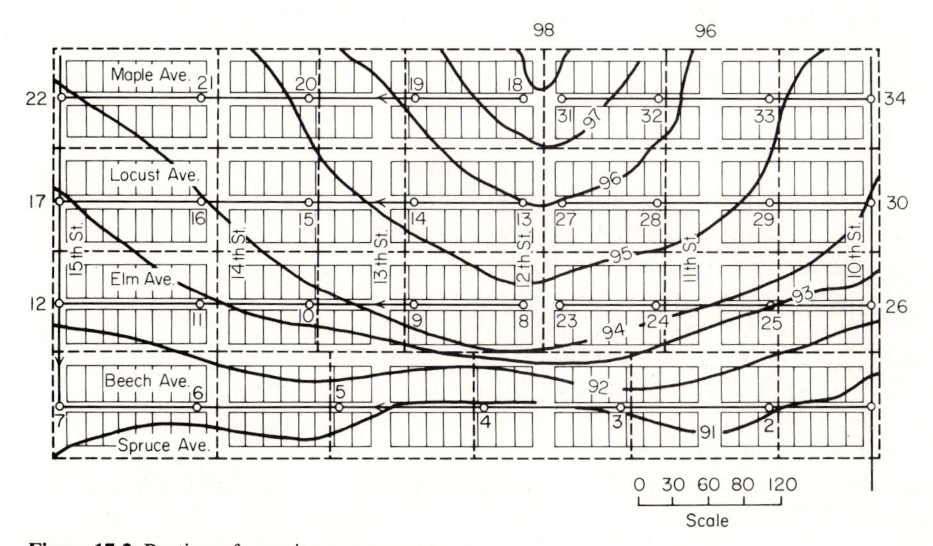

Figure 17-2 Portion of a sanitary sewer system.

south on 10th St. from manhole 1. The area east of 12th St. and north of Beech Ave. is disregarded in this design.

It is assumed that the maximum density of population in the area will be 10,000 persons/km^2. It is also assumed that the maximum rate of sewage flow, including infiltration, will be 1500 l/day per capita (Art. 2-18). This might be reduced to 1000 l per capita for mains and this figure might have been applied to design of the sewer on 15th St. It was assumed, however, that the uncertainties of the development of the whole area served were such that the 1500-l rate should be used for all the sewers. The method of tabulation to obtain amounts is shown in Table 17-1. Note that if the 1000-l rate were applied to the 15th St. sewer the amount of sewage of line 5, for example, would be 1000×6085 or 6,085,000 l instead of 9,127,500 l. The minimum cover over the sewers is to be 2 m above the vertical inside diameter of the sewer. The minimum permissible size of sewer is 200 mm. The value of n is 0.013.

Guided by contours, the lines are placed on the map in the streets and alleys, so that all buildings will be served, with arrows indicating the assumed directions of flow. Manholes are placed at intersections, changes of direction, and intervals not exceeding 120 m along the laterals. In some cases manholes are placed at street intersections regardless of block lengths, but here they are placed so as to obtain line lengths of 90 m in the upper ends of laterals where stoppages are most likely to occur. The manholes have been numbered consecutively from the outlet of the system upward until an intersection is reached, whereupon the numbering is continued from the upper end of the tributary lateral. Each line from manhole to manhole is given a number for use in Tables 17-1 and 17-2. The area tributary to each line is then sketched as shown by the dashed lines on Fig. 17-2. The area boundaries are determined therefore, as on 15th St., by the location of the building lots. Some lines have no tributary areas, although they have an increment of groundwater infiltration, if this is separately considered. Design proceeds from the most remote point of the system downward. Where a branch joins a line already designed a new design is started at the upper end of the branch and is completed to the intersection.

Table 17-1 shows the procedure followed to compute the sewage flow the various lines must carry. Line O is the sewer flowing south on 15th St., and terminating at manhole 22. It has already been designed, and in the table the total tributary population and flow are given. If infiltration is separately computed, extra columns can be inserted in the table.

The procedure in arriving at the sanitary sewage flow is simple. Note that for line 1 the tributary area is 10,000 m^2, which has a population of 100. This, multiplying by 1500 gives a sewage flow of 150,000 l/day. For line 2, which has an area of 7000 m^2, the population increment is 70, and this is added to 100 to give a total tributary population of 170. For lines 3 and 4 a similar procedure is followed. At the end of the lateral its total tributary population of 360 is added to the 5725 population of line O to give a total of 6085 and a flow of 9,127,500 l/day for line 5.

After the flow for line 5 has been obtained, the lateral in the alley between Locust and Elm Aves. is investigated in the same way, and its total is added to that

Table 17-1 Data for flow in a sanitary sewer system

Line no. (1)	On street (2)	From man-hole (3)	To man-hole (4)	Length of line, m (5)	Increment of area, m^2 (6)	Increment of population (7)	Total tributary population (8)	Sewage flow, l/day (9)	Sewage flow, m^3/min (10)
0	15th	...	22	....	....	....	5725	8,587,500	5.96
1	Alley between Maple and Locust	18	19	90	10,000	100	100	150,000	0.10
2	Alley between Maple and Locust	19	20	90	7,000	70	170	255,000	0.18
3	Alley between Maple and Locust	20	21	90	7,000	70	240	360,000	0.25
4	Alley between Maple and Locust	21	22	120	12,000	120	360	540,000	0.38
5	15th	22	17	87	....	....	6085	9,127,500	6.34
6	Alley between Locust and Elm	13	14	90	10,000	100	100	150,000	0.10
7	Alley between Locust and Elm	14	15	90	7,000	70	170	255,000	0.18
8	Alley between Locust and Elm	15	16	90	7,000	70	240	360,000	0.25
9	Alley between Locust and Elm	16	17	120	12,000	120	360	540,000	0.38
10	15th	17	12	87	....	....	6445	9,667,500	6.71
11	Alley between Elm and Beech	8	9	90	10,000	100	100	150,000	0.10
12	Alley between Elm and Beech	9	10	90	7,000	70	170	255,000	0.18
13	Alley between Elm and Beech	10	11	90	7,000	70	240	360,000	0.25
14	Alley between Elm and Beech	11	12	120	12,000	120	360	540,000	0.38
15	15th	12	7	87	....	....	6805	10,207,500	7.09
16	Alley between Beech and Spruce	7	6	120	12,000	120	6925	10,387,500	7.21
17	Alley between Beech and Spruce	6	5	120	9,000	90	7015	10,522,500	7.31
18	Alley between Beech and Spruce	5	4	120	12,000	120	7135	10,702,500	7.43
19	Alley between Beech and Spruce	4	3	120	11,000	110	7245	10,867,500	7.55
20	Alley between Beech and Spruce	3	2	120	11,000	110	7355	11,032,500	7.66
21	Alley between Beech and Spruce	2	1	90	7,000	70	7425	11,137,500	7.73

Table 17-2 Design of a sanitary sewer system

Line no. (1)	To street (2)	From man-hole (3)	To man-hole (4)	Length of line, m (5)	Sewage flow, m^3/min (6)	Ground elevations: Upper man-hole (7)	Ground elevations: Lower man-hole (8)	Diam. of pipe, mm (9)	Grade of sewer (10)	Fall of sewer, m (11)	Velocity flowing full, m/sec (12)	Capacity flowing full, m^3/min (13)	Q/Q_{full} (14)	V/V_{full} (15)	V, m/sec (16)	Invert elevations: Upper man-hole (17)	Invert elevations: Lower man-hole (18)
0	15th	...	22	....	5.96	......	93.69	305	....	.....	.....	.....	.....	.....	.....	.....	91.23
1	Alley between Maple and Locust	18	19	90	0.10	97.74	96.40	200	0.0180	1.62	1.42	2.72	0.04	0.44	0.62	95.54	93.92
2	Alley between Maple and Locust	19	20	90	0.18	96.40	95.27	200	0.0130	1.17	1.20	2.30	0.08	0.53	0.64	93.92	92.75
3	Alley between Maple and Locust	20	21	90	0.25	95.27	93.93	200	0.0113	1.02	1.10	2.10	0.12	0.58	0.64	92.75	91.73
4	Alley between Maple and Locust	21	22	120	0.38	93.93	93.69	200	0.0070	0.84	0.89	1.70	0.22	0.68	0.61	91.73	90.89
5	15th	22	17	87	6.34	93.69	92.99	380	0.0040	0.35	1.04	6.95	0.91	1.02	1.06	90.71	90.36
6	Alley between Locust and Elm	13	14	90	0.10	96.04	95.37	200	0.0180	1.62	1.42	2.72	0.04	0.44	0.62	93.84	92.22
7	Alley between Locust and Elm	14	15	90	0.18	95.37	94.57	200	0.0130	1.17	1.20	2.30	0.08	0.53	0.64	92.22	91.05
8	Alley between Locust and Elm	15	16	90	0.25	94.57	93.81	200	0.0113	1.02	1.10	2.10	0.12	0.58	0.64	91.05	90.03
9	Alley between Locust and Elm	16	17	120	0.38	93.81	92.99	200	0.0070	0.84	0.89	1.70	0.22	0.68	0.61	90.03	89.14
10	15th	17	12	87	6.71	92.99	92.32	460	0.0015	0.13	0.71	6.97	0.96	1.03	0.73	88.88	88.75
11	Alley between Elm and Beech	8	9	90	0.10	94.85	94.30	200	0.0180	1.62	1.42	2.72	0.04	0.44	0.62	92.65	91.03
12	Alley between Elm and Beech	9	10	90	0.18	94.30	93.48	200	0.0130	1.17	1.20	2.30	0.08	0.53	0.64	91.03	89.86
13	Alley between Elm and Beech	10	11	90	0.25	93.48	92.90	200	0.0113	1.02	1.10	2.10	0.12	0.58	0.64	89.86	88.84
14	Alley between Elm and Beech	11	12	120	0.38	92.90	92.32	200	0.0070	0.84	0.89	1.70	0.22	0.68	0.61	88.84	88.00
15	15th	12	7	87	7.09	92.32	91.92	530	0.00092	0.08	0.62	8.18	0.87	1.02	0.63	87.67	87.59
16	Alley between Beech and Spruce	7	6	120	7.21	91.92	91.74	530	0.00092	0.11	0.62	8.18	0.88	1.02	0.63	87.56	87.45
17	Alley between Beech and Spruce	6	5	120	7.31	91.74	91.71	530	0.00092	0.11	0.62	8.18	0.89	1.02	0.63	87.45	87.34
18	Alley between Beech and Spruce	5	4	120	7.43	91.71	91.40	530	0.00092	0.11	0.62	8.18	0.91	1.02	0.63	87.34	87.23
19	Alley between Beech and Spruce	4	3	120	7.55	91.40	91.43	530	0.00092	0.11	0.62	8.18	0.92	1.02	0.63	87.23	87.12
20	Alley between Beech and Spruce	3	2	120	7.66	91.43	91.40	530	0.00092	0.11	0.62	8.18	0.94	1.03	0.64	87.12	87.01
21	Alley between Beech and Spruce	2	1	90	7.73	91.40	90.61	530	0.00092	0.08	0.62	8.18	0.94	1.03	0.64	87.01	86.93

of line 5 to obtain the flow for line 10. Similarly the flow of the lateral between Elm and Beech Aves. is added to that of line 10 to obtain that of line 15. Lines 15 to 21 between Beech and Spruce Aves. become a main sewer and have tributary areas.

Table 17-2 is a tabulation showing the design of the system. In practice this would be combined with Table 17-1. Also a profile of the streets and alleys would be available and sewer elevations would be placed on it. In the design the following requirements are to be complied with. At design flow the velocity is to be at least 0.6 m/s. If a sewer changes direction in a manhole without change of size, a drop of 30 mm is to be provided in the manhole. If the sewer changes size, the crowns of the inlet and outlet sewers are to be at the same elevation. Branches coming into a manhole shall have their crowns at the same elevation as that of the large sewer. Drop manholes will be used only if the invert of the branch would be 0.6 m or more above what its location would be when following the rule just stated.

Design starts with line 1. The ground elevations at the upper and lower manholes are noted from column (7) and (8), and an invert elevation at the upper manhole is set at $97.74 - 2 - 0.2 = 95.54$, which provides for the 2 m of cover and the diameter of the sewer. A tentative elevation at the lower manhole is obtained in the same way and is 94.20. These elevations are placed in columns (17) and (18). This provides a fall of 1.34 m and grade of 0.0146 which, for a 200-mm pipe, results, from Fig. 15-1, in a capacity, when flowing full, of 2.45 m^3/min and a velocity of 1.28 m/s.

Upon consideration of the partial flow diagram of Fig. 15-5, it is found that the actual velocity at a flow of 0.1 m^3/min is only 0.44 m/s on this slope. Therefore, the slope is increased as noted to provide a velocity of at least 0.6 m/s at design flow.

A number of slopes may be tried before a satisfactory velocity is obtained. The slope is then entered in column 10, the drop is calculated from the product of columns 10 and 5, and the lower invert elevation in column 18 is obtained by subtracting column 11 from column 17. Note that the invert elevation at the lower end of a line is the same as that at the upper end of the next except where a change in size or direction occurs.

Manhole 22 is the lower end of lines 4 and 0 and the upper end of line 5. Above the manhole line 0 is 305 mm in diameter. Trial indicates that a 305-mm pipe would require a slope steeper than the ground to carry the flow of 6.34 m^3/min. Since the sewer invert of line 4 is already lower than dictated by minimum cover, it is preferable to use a larger sewer on a lower slope. A 380-mm sewer is required. Since a change in size occurs the crowns of the intersecting sewers must be matched. The crown of line 4 is at 91.09, that of line 0 at 91.54, therefore the crown of line 5 must be at 91.09, giving an invert elevation of 90.71. The drop provided by matching crowns (0.18 m) exceeds that dictated by the change in direction. Manholes 17 and 12 are handled in a similar fashion. At manhole 7 there is a change in direction with no change in pipe size. The invert is thus dropped 30 mm or 0.03 m across the manhole.

17-8 Design of a Storm Sewer System

Figure 17-3 shows a section of a city that is to be sewered. It is assumed that it is a residential area and that the coefficient of runoff (C of the formula) will be 0.40 at the time of maximum development. In residential areas, high frequency rainfall data is used, thus the two year formula of Area 3 (Table 13-3) is used. From this, $i = 2590/(t + 17)$. The minimum sewer size is 305 mm, n is 0.013, the minimum allowable velocity is 0.75 m/s with the sewer flowing full, and the minimum allowable cover over the crown is 1.5 m. In practice, the location of all existing underground structures, including sanitary sewers, would need to be known, and interference avoided. In this design it is assumed that no interference exists.

As with sanitary sewers, the first step is to place the lines tentatively, in the streets in this case, following the natural slopes. Since no house connections are required, the sewers need not abut upon the individual lots, and this may permit them to be run by shorter routes than the sanitary sewers must follow. The drainage districts tributary to each line are then sketched. They are laid out after noting the contours and considering also that the front portions of lots, including the house roofs, will drain toward the street even though the general lay of the land indicates otherwise. Likewise the corner lots will be graded to drain in part to the front and in part to the side streets. The areas of the drainage districts are then computed, after scaling dimensions from the map, and are placed on the map; or the districts may be given distinguishing marks on the map, and the area noted on the design tabulation.

In Fig. 17-3 laterals were placed to run south on 12th, 13th, 14th, and 15th Sts., with a main running east on Spruce Ave. Inlets were placed at street intersections, and the area tributary to each line was drawn with sufficient accuracy. Uncertainties as to rainfall intensity and concentration time do not warrant ex-

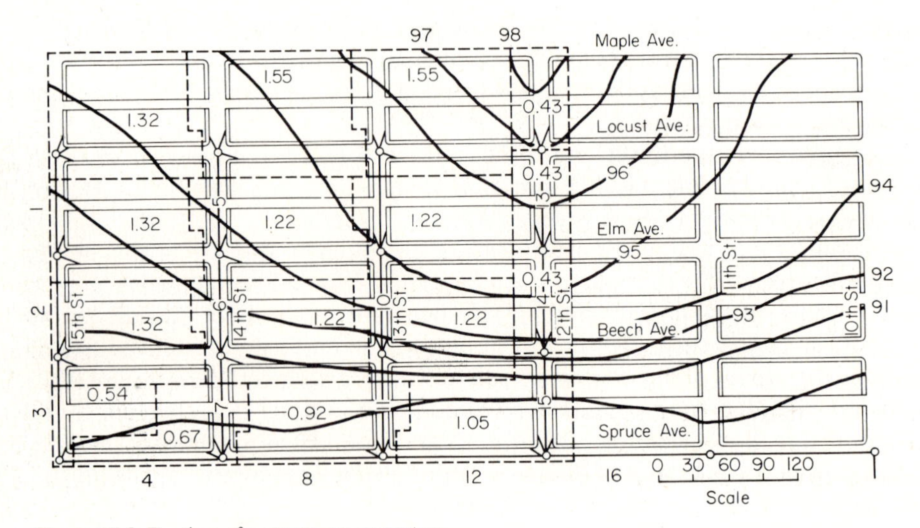

Figure 17-3 Portion of a storm sewer system.

treme care in tracing drainage districts that will produce highly complicated shapes. Since 12th St. is nearly on the crest of a watershed, it is open to question whether it requires a sewer. Neighboring lots will drain toward it, however, and it is possible that the flow in the gutters near its lower end might cause trouble in heavy storms. If extreme economy is necessary, it may be permissible to dispense with these lines provided gutter details at the street corners are such that the water will be diverted to the cross streets. Whether or not the laterals should be extended north to Maple Ave. is also debatable. The relatively small amount of water that will be collected by the gutters on Maple Ave. may be conducted around the corners finally to reach the inlets at Locust Ave. All lines have been numbered consecutively, starting at the one farthest from the last line of the system until a lateral is reached, then resuming the numbering at the farthest end of the lateral until line 16, on Spruce Ave. from 12th to 11th Sts., is reached. The inlets are placed to intercept the flow in the gutters before it reaches the corners.

A tabulation is useful in making the design. The first four columns of Table 17-3 show the line numbers and their location. The tributary area of each line is given in column 5. The runoff coefficient is given in column 6. Changes in surface characteristics may occur in different portions of the section of the city, and these can be shown in this column. Column 7 is the area of each district multiplied by the runoff coefficient to give an equivalent area that is 100 percent impervious. Column 8 is the total area tributary to each line, i.e., the sum of the increments of column 7.

The inlet time must be obtained by the techniques of Chap. 13 or assumed for the first line of each lateral. Here it is taken as 10 min in each case except for line 13, which, since it serves a small area, is assumed to have an inlet time of only 6 min. Column 9 gives time of concentration, which is made up of inlet time plus the time of flow in the preceding lines. Column 10 is the rainfall intensity in millimeters per hour and is obtained from the rainfall formula, using the time of concentration for the duration t. The amount of sewage that must be cared for is found by multiplying the rainfall and area of columns 10 and 8.

After determination of the sewage flow, the lines can be designed with the help of the flow charts. Using Fig. 15-2 permits diameter of the pipe, slope, and velocity to be found. Care must be taken that the slope conforms to the topography so that excessive cuts are avoided and the necessary cover is obtained. For this reason it is convenient to find a tentative grade of the pipe based upon the ground surface and check the velocity. If the velocity is too low, the slope must be increased. This may have the advantage of permitting a smaller sewer to be used. Columns 12, 13, 14, and 17 give information derived from the chart. The capacity as shown in column 17 must be greater than the expected flow of column 11. From the length of line and the velocity in the pipe, the flow time is found and placed in column 16. This is added to the time of concentration for preceding lines to obtain the time of concentration for the following lines. Columns 18 and 19 show ground elevations, and columns 20 and 21 give the invert elevations at the upper and lower ends of the lines. The difference in elevation is found by multiplying the length of the line by its slope.

Table 17-3 Design of a storm sewer system

Line no. (1)	Location (2)	From street (3)	To street (4)	Increment of area $m^2 \times 10^{-4}$ (5)	C (6)	Equivalent area Ca 100 percent $m^2 \times 10^{-4}$ (7)	Total area $\sum Ca$ $m^2 \times 10^{-4}$ (8)	Time of concentration, min (9)	i mm/h (10)	Q m^3/min (11)	Grade (12)	Dia. of pipe, mm (13)	Velocity flowing full, m/s (14)	Length of line, m (15)	Time of flow, min (16)	Capacity of sewer, m^3/min (17)	Ground elevations Upper end (18)	Ground elevations Lower end (19)	Invert elevations Upper end (20)	Invert elevations Lower end (21)
1	15th St.	Locust	Elm	1.32	0.40	0.53	0.53	10.0	96	8.5	0.0077	380	1.4	87	1.0	9.3	93.29	92.62	91.41	90.74
2	15th St.	Elm	Beech	1.32	0.40	0.53	1.06	11.0	93	16.4	0.0070	530	1.7	87	0.9	22.6	92.62	92.01	90.59	89.98
3	15th St.	Beech	Spruce	1.32	0.40	0.53	1.59	11.9	90	23.9	0.0074	610	1.9	87	0.8	32.3	92.01	91.37	89.90	89.26
4	Spruce Ave.	15th	14th	0.54	0.40	0.22	1.81	12.7	87	26.3	0.0011	840	0.9	140	2.8	28.9	91.37	91.22	89.03	88.88
5	14th St.	Locust	Elm	1.55	0.40	0.62	0.62	10.0	96	9.9	0.0052	460	1.3	87	1.0	12.7	94.05	93.60	92.09	91.64
6	14th St.	Elm	Beech	1.22	0.40	0.49	1.11	11.0	93	17.2	0.0147	460	2.2	87	0.7	22.1	93.60	92.32	91.64	90.36
7	14th St.	Beech	Spruce	1.22	0.40	0.49	1.60	11.7	90	24.0	0.0126	530	2.2	87	0.6	28.9	92.32	91.22	90.29	89.19
8	Spruce Ave.	14th	13th	0.67	0.40	0.27	3.68	15.5	79	48.5	0.0020	915	1.3	140	1.8	51.0	91.22	91.10	88.81	88.53
9	13th St.	Locust	Elm	1.55	0.40	0.62	0.62	10.0	96	9.9	0.0084	380	1.5	87	1.0	10.2	95.46	94.73	93.58	92.85
10	13th St.	Elm	Beech	1.22	0.40	0.49	1.11	11.0	93	17.2	0.0243	460	2.8	87	0.5	27.2	94.73	92.62	92.77	90.66
11	13th St.	Beech	Spruce	1.22	0.40	0.49	1.60	11.5	91	24.3	0.0193	460	2.5	87	0.6	25.0	92.62	91.10	90.66	88.98
12	Spruce Ave.	13th	12th	0.92	0.40	0.37	5.65	17.3	76	71.6	0.0009	1220	1.0	140	2.3	72.4	91.10	91.07	88.22	88.09
13	12th St.	Locust	Elm	0.43	0.40	0.17	0.17	6.0	113	3.2	0.0168	305	1.8	87	0.8	7.8	96.92	95.46	95.12	93.66
14	12th St.	Elm	Beech	0.43	0.40	0.17	0.34	6.8	109	6.2	0.0256	305	2.2	87	0.7	9.5	95.46	93.23	93.66	91.43
15	12th St.	Beech	Spruce	0.43	0.40	0.17	0.51	7.5	106	9.0	0.0248	305	2.3	87	0.6	9.9	93.23	91.07	91.43	89.27
16	Spruce Ave.	12th	11th	1.05	0.40	0.42	6.58	19.6	71	77.9	0.0011	1220	1.2	140	2.0	79.9	91.10	91.19	88.35	88.20

The design is started at the most remote point from the outlet of the system and proceeds until the intersection with the first branch line. Design of the branch is then begun, starting from its most remote point. Branches make connections at manholes. Where they intersect main sewers the elevation of the crown of the branch was made the same as that of the main sewer. Great differences in elevation would justify the use of drop manholes, but none were necessary. Where sewers changed in size the invert elevation of the outlet was dropped an amount equal to the increase in size. The time of concentration, for the line below the junction, is that of the longer branch, or, more correctly, the longer of the two times of concentration. For example: The time of concentration for line 8 is that of line 4 (which is longer than that of line 7) plus the time of flow in line 4, or $12.7 + 2.8 = 15.5$ min. On Spruce Ave. where the sewer becomes large in diameter, the grades were increased slightly to prevent increases in size. This was economical because the slight extra excavation was more than offset by the saving in pipe size.

PROBLEMS

17-1 Design the sewers required to serve the area east of 12th St. and north of Beech Ave. of Fig. 17-2. Assume that the line from the north of 10th St., ending at manhole 34, is 305 mm in diameter and has a tributary population of 5300, with a total sewage flow of 5.52 m^3/min based upon a flow of 1500 l/day per capita. The ground elevation at this manhole is 94.27, and its invert elevation is 92.13. The line south from manhole 26 must enter manhole 1 no lower than elevation 88.31. Assumptions as to per capita sewage, required cover, and minimum size are the same as in the illustrative design. Assume ground-surface elevations from the contours of the map. Make tabulations as shown in Tables 17-1 and 17-2.

17-2 Using the assumptions of the illustrative design of Fig. 17-3, lay out a system of storm sewers to serve the balance of the district, and make a tabulation similar to that of Table 17-3. Design the line on Spruce Ave. from 11th to 10th Sts., and find the total storm-water flow of the district. Assume ground elevation at 10th St. and Spruce Ave. as 90.5.

REFERENCES

1. "A Guide to the Selection of Cost-Effective Wastewater Treatment Systems," United States Environmental Protection Agency, Office of Water Program Operations, Washington, D.C., 1975.

CHAPTER

EIGHTEEN

SEWER CONSTRUCTION

18-1

The engineer is responsible for insuring that the plans and specifications for sewerage works are complied with, and for authorizing any changes which may become necessary during the course of construction. While methods and scheduling are the responsibility of the contractor, the engineer should exercise sufficient control to prevent foreseeable trouble or delays. Location of lines and grades may be made by either the engineer or the contractor, but responsibility for compliance with the plans lies with the former. Actual construction of sewers consists of excavation, bracing and dewatering of trenches, pipe installation, backfilling, and construction of appurtenances.

18-2 Lines and Grades

Sewer lines in large cities are carefully located during planning to avoid existing underground structures. In smaller cities and in sparsely settled residential districts less care is needed, and generally only water mains need be avoided. In new construction, paving will not ordinarily have been done, and the middle of the street will be chosen so that house sewers, which are paid for by the property owners, will be equal in length to properties on each side. An offset line is located where it will not be disturbed or covered. The contractor then measures from the offset line and lays out the trench on the ground. When excavation is nearly completed, batter boards are placed across the trench at 10- or 15-m (30- or 50-ft) intervals, with their station marked indelibly. They may be supported as shown in the figure or may be held in place by piles of earth. The center line of the sewer is

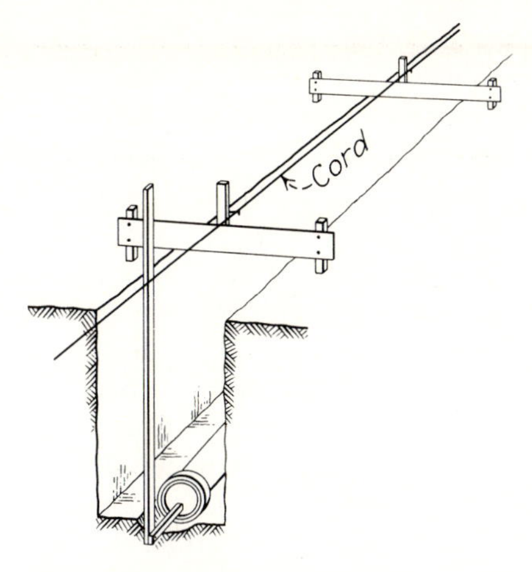

Figure 18-1 Batter boards used in transferring line and grade to trench. Stick for checking grade is shown in position.

indicated on the batter board by a nail or, better, by the edge of an upright cleat. The same edge of all cleats should be used to indicate line. Elevations are then run, and a mark is placed on each cleat at some even distance above the invert of the sewer. A nail is driven in the edge of each cleat at the grade mark, and a cord is stretched from cleat to cleat. Line is transferred from the cord to the bottom of the trench by means of a plumb bob. Grade is transferred by means of a stick marked in even increments and having a short piece fastened at a right angle to its lower end. Grade is checked by placing the short piece on the invert of each length of sewer pipe and noting whether the proper mark touches the cord.

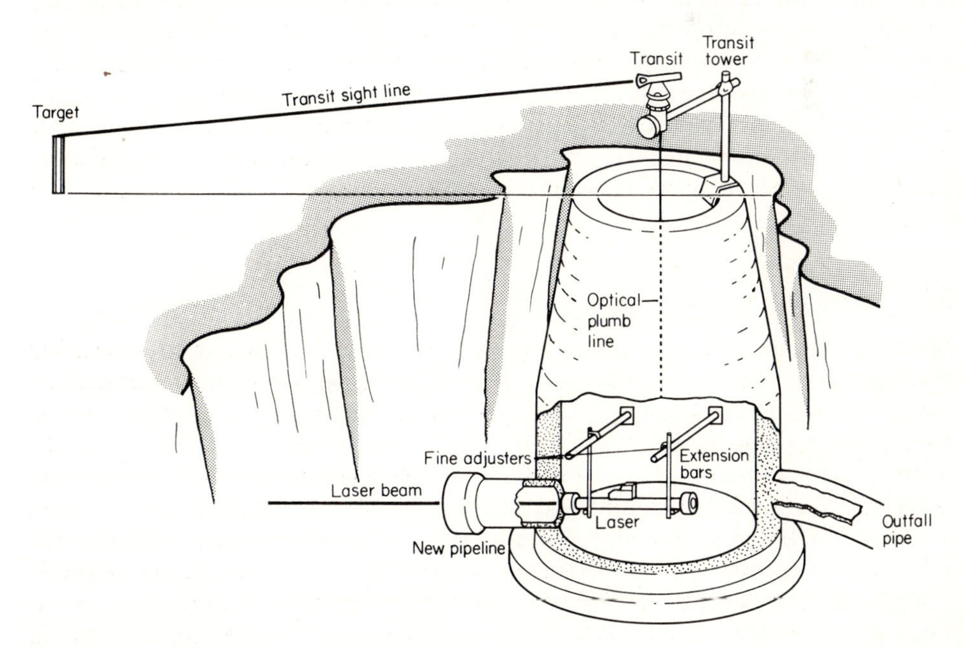

Figure 18-2 Laser alignment setup.

A newer technique for maintaining both line and grade employs a laser beam generated at a manhole and directed down the pipe as it is placed. Accuracy of line and grade can be held to within 0.01 percent over a range of 300 m (1000 ft).

18-3 Classification of Excavation

In contracts, excavation is frequently divided into classes A, B, and C. Class A consists of solid rock in the original bed or in well-defined ledges removable only by blasting and of all boulders over 0.25-m^3 (8-ft^3) volume. Another definition of rock is that it shall mean any material geologically in place, cemented into a mass by natural causes and having a hardness, when first exposed, of 3 or more in the mineral scale of hardness. Frozen material is not classified as rock. Class B excavation comprises disintegrated limestone, shale, soapstone, slate, hardpan, fire-clay, cemented gravel, macadam pavement, and boulders less than 0.25 m^3 (8 ft^3) and more than 0.03 m^3 (1 ft^3) in volume. This material can be removed by hand or mechanically with minor difficulty. Class C includes all other materials. Where quicksand (Art. 18-9) is known to occur, it should also be included as a definite item in the contract.

18-4 Hand Excavation

Hand excavation is held to an absolute minimum in sewer construction. Sewer lines should be located to permit maximum use of mechanical equipment with hand excavation being limited to intersections with existing structures and minor excavation at pipe joints.

When hand-dug trenches exceed 1.5 m (5 ft) in depth it is necessary to rehandle the material excavated at the bottom to move it further from the trench edge. The minimum shelf space should be over 500 mm (18 in). In trenches over 2.5 m (8 ft) deep it is necessary to construct relay platforms at about 2-m (6-ft) intervals.

18-5 Machine Excavation

Machine excavation is much less expensive than hand work and should be employed wherever possible. Trenches can be excavated by machine, the pipe laid, and the trench backfilled in less time than hand excavation alone would require. When gas, water, or electrical conduits are encountered, hand excavation is required for short distances.

Trenches are excavated using specialized equipment or standard construction machinery such as back hoes, clamshells, or draglines. Special equipment which is sometimes used includes continuous chain drives upon which are mounted buckets with cutting teeth. The buckets discharge the earth to a conveyor which deposits it to the side of the trench. Side cutters added to the buckets permit excavation of trench widths up to 1.3 m (4 ft). Depths of 9 to 10 m (30 to 33 ft) can be excavated with such equipment. The trenching machine is mounted upon a tractor and can excavate a shallow trench at a rate up to 10 m/min (30 ft/min) under the most favorable conditions.

Excavation of trenches should be carried below the final elevation of the pipe so that suitable granular backfill can be placed to provide good bedding (Chap. 14). When bell and spigot pipe is used, hand excavation is required for each bell after the bedding material has been placed. The excavation should be carried a minimum of 200 mm (8 in) below grade for cuts up to 5 m (16 ft) and 40 mm more for each additional meter of excavation.

18-6 Rock Excavation

Excavation and trenches in rock should be made to a depth of one-fourth the pipe diameter below the bottom of the pipe, but in no case less than 100 mm (4 in). The space between the trench bottom and the pipe should be filled with granular material. When small amounts of rock are encountered, they can frequently be broken by means of picks, bars or wedges, bull points, and sledges. These methods are applicable to soft and seamy rock. Harder rock may be broken by drilling a row of holes and splitting off a mass of rock by driving in plugs and feathers. This is essentially a quarrying method but may be used where a limited amount of rock does not justify the use of explosives.

Explosives must be used with care in sewer trenches or damage may be done to nearby buildings. Their use involves drilling holes of proper depth and spacing, loading with the explosive, and detonating the charges. The common explosives used in trench construction are blasting powder and dynamite. The former must be tamped into the drill hole and then detonated by the flame, or heat of a fuse, or by explosion of a cap. Dynamite is obtained in cylindrical sticks which are exploded by means of caps. A cap or detonator is a small copper cylinder containing fulminate of mercury and potassium chlorate. The cap is set off by means of a fuse or an electric spark generated by a small electric apparatus.

For information concerning spacing of holes and amount of explosive required for various types of rock, the reader is referred to works on excavation and to the publications of manufacturers of explosives.

18-7 Sheeting and Bracing[1]

Trenches in unstable materials require sheeting and bracing to prevent caving or collapse of the side walls. The danger to workmen and increase of construction cost justify all possible precautions to prevent extensive cave-ins. Practical experience with various types of earth is necessary in order to determine whether extensive, light, or no bracing will be necessary. If there is doubt as to its necessity, however, the safe course should be followed. Some soils will stand without bracing for a time, while long exposure or rainy weather will necessitate bracing. The sides of trenches that are more than 1.5 m (5 ft) deep and 2.5 m (8 ft) or more in length should be held securely by shoring and bracing, or sloped to the angle of repose of

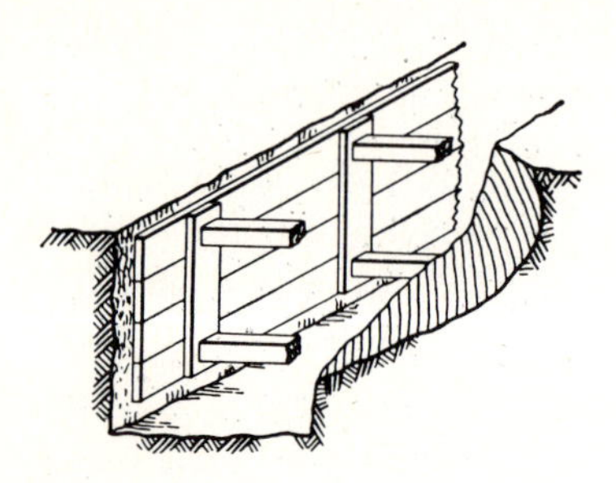

Figure 18-3 Box sheeting.

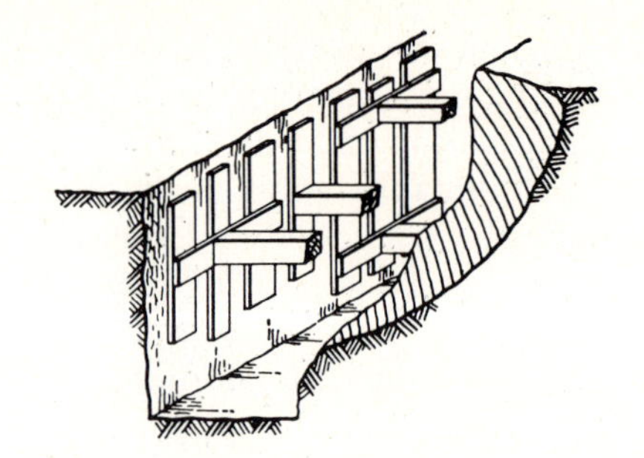

Figure 18-4 Poling boards.

the materials being excavated. Trenches which are constructed without sheeting are dug vertically at the bottom to reduce the load upon the pipe.

Sheeting is the term applied to the planks in actual contact with the trench sides. Braces are the crosspieces extending from one side to the other. Rangers, or wales, are the timbers that transfer or convey the load on the sheeting boards to the cross braces. Various systems of trench protection are used, and those in common use will be described.

Stay Bracing Stay braces consist of a pair of vertical boards placed on opposite sides of the trench with two cross braces to hold them in place. The intervals between the stay braces will depend upon the type of soil. The vertical planks should be 50 by 100 mm (2 × 4 in), and the cross braces not less than 100 by 100 mm (4 × 4 in). Stay braces should be used only in shallow trenches in stable materials. The security they give is not very great, and the sides of the trench should be carefully watched by the foreman in charge.

Poling Boards These are short pieces of board placed vertically against the sides of the trench with short rangers and braces completing the system. The boards may be nonuniform in length and frequently have gaps between them, as shown in the illustration. This method of trench protection is used in materials that will stand for 1 to 1.5 m (3 to 5 ft) without collapsing. It has the advantage of not requiring driving of the sheeting and does not necessitate boards extending above the street level. It is well adapted to use in the first stages of excavation where more elaborate vertical sheeting will be required for lower tiers. The boards are usually 50 mm (2 in) thick, the rangers 100 by 150 or 200 mm (4 × 6 or 4 × 8 in) and the cross braces 150 by 150 or 200 by 200 mm (6 × 6 or 8 × 8 in) depending upon the trench width.

Box Sheeting This method uses horizontal sheeting boards and vertical rangers with one or more cross braces for each pair of rangers. In stable material the sheeting may have gaps between the tiers and between the individual boards. In loose material the sheeting must be close and strong, with 50-mm (2-in) sheeting boards and 50 by 200-mm (2 × 8-in) rangers if they are not longer than 1 m (40 in), and 150 by 150-mm (6 × 6-in) cross braces. Box sheeting is well adapted to an unconsolidated soil. The excavation can proceed for a depth equal to the width of a single board; the two opposite boards are temporarily braced until three or four are in position; and then rangers with permanent cross bracing can be placed. This method is also used for the first 1 to 1.5 m (3 to 5 ft) of an excavation where vertical sheeting will be used for the lower portion.

Vertical Sheeting Vertical sheeting is the strongest, most elaborate, and most expensive method used. It is employed in deep excavations in soft materials and where groundwater is expected. In applying the method the trench is first excavated as far as possible without endangering the banks. The first set of rangers is then placed about 300 mm (12 in) beneath the ground surface against three planks placed vertically against the squared and trimmed sides of the trench, one flush at each end and one at the middle of each ranger. If possible, another set of rangers will also be used and placed 1 m (3 ft) below the first pair. The rangers are then braced so that the vertical planks are pushed slightly into the earth. Other planks can then be driven down in the opening between the rangers and the trench sides.

After the sheeting is in position behind the rangers and driven as far down as the bottom of the excavation, heavy driving begins. Mauls or air hammers are used to drive the boards in turn into the earth, and excavation proceeds as they go down. At 1-m (3 ft) intervals, rangers and cross braces are placed. Generally, sheeting boards 3 m (10 ft) in length are used for the top 2.5 m (8 ft) of excavation, requiring three sets of rangers. This has the advantage of not requiring the sheeting to project far above the surface during driving. Should the trench be 5 m (16 ft) or less in depth, only one tier will be used. Rangers are usually 5 m (16 ft) long.

A second tier of sheeting is started after the lowest rangers and braces of the upper tier are in place. These rangers are used as guides for the lower sheeting, and the rangers for the new tier are set and braced some 600 mm

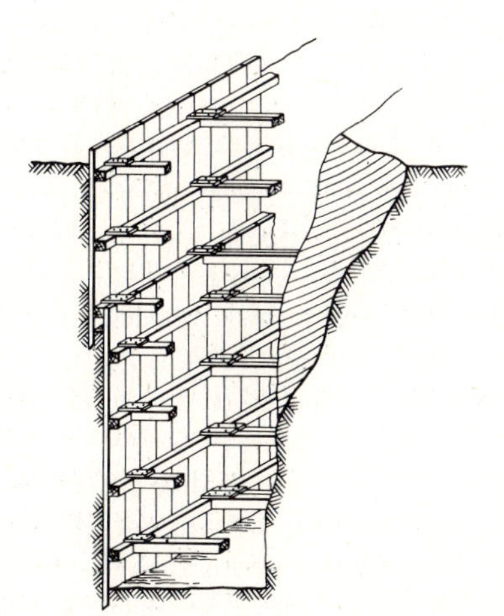

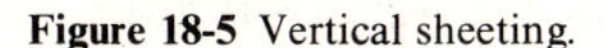

Figure 18-5 Vertical sheeting.

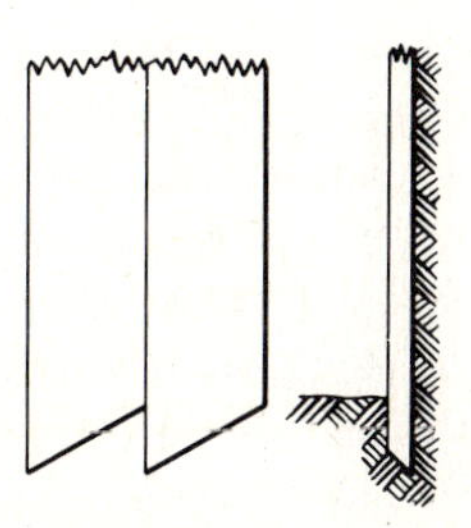

Figure 18-6 Beveling of vertical sheeting.

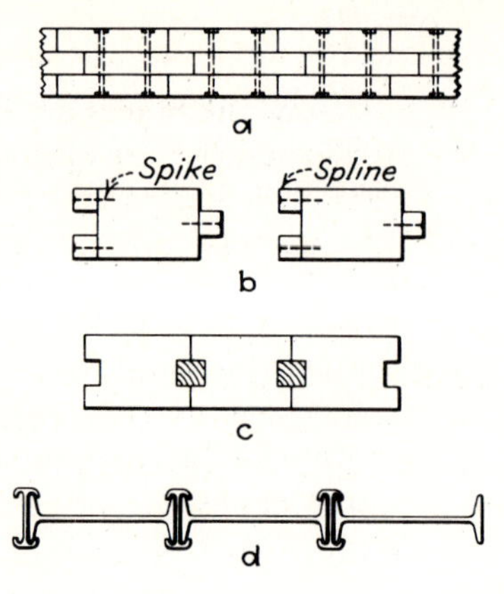

Figure 18-7 Sheet piling. *(a)* Wakefield (timber). *(b)* Tongue and groove (timber). *(c)* Groove and spline (timber). *(d)* Steel.

(2 ft) below them. Driving of the new sheeting can then proceed in the same manner as in the first tier. The cross braces of the higher tier, however, prevent driving the sheeting boards that would come immediately below them. These leave openings which must be closed with boards nailed in place.

The sheeting planks are usually 50 by 250 mm (2 × 10 in). The rangers and cross braces may be varied in size as the excavation becomes deeper, or they may be placed at closer intervals. Theoretical computations based upon Rankine's formula for earth pressures can also be made. Rangers may be 150 by 150 mm (6 × 6 in) or, better, 150 by 200 mm (6 × 8 in) at the top of the excavation, increasing to 200 by 250 mm (8 × 10 in) with an excavation 9 m (30 ft) deep. Cross braces may vary from 100 by 150 mm (4 × 6 in) to 100 by 200 mm (4 × 8 in).

It is advisable to have a cross brace for each ranger at the ends instead of placing a single brace over the point of contact. This precaution makes each section of trench independent, and, should one section collapse, others will not be affected. The sheeting boards are beveled in two directions, so that they will tend to fit snugly against each other as they are driven and also against the trench side. Their tops are frequently protected by malleable iron caps to prevent brooming as they are driven. Braces are tightened by driving wedges between their ends and the rangers. Should they become loose, they may fall unless cleats are nailed on their ends. When vertical sheeting is used, care must be taken to allow for the loss in width caused by each tier of sheeting.

Where much water is anticipated, wood sheet piling may be used instead of planks. Steel sheet piling is used on especially important jobs. It is stronger than wood piling, more nearly watertight, and can be pulled and reused more often.

Skeleton (Open) Sheeting This type of protection includes rangers and cross bracing as in vertical sheeting, but the sheeting planks are placed only at the ends and middle points of the rangers. It is used where some protection is needed and more elaborate protection is anticipated. Should conditions warrant, skeleton sheeting can quickly be transformed into vertical sheeting by driving planks between the previously placed rangers and the sides of the trench.

18-8 Removal of Sheeting and Bracing

Salvage of timber bracing and sheeting is usually practiced. The braces are first loosened and removed, a process requiring caution and experience, or workmen may be caught in cave-ins. If a hoisting engine is available, a loop of rope or cable is thrown around one or more sheeting planks, and they are pulled out. Steel piles have a hole in their upper ends in which a hook can be placed. Hand pulling of timber sheeting is done with a bar and the clamp shown in Fig. 18-8.

The technique of backfilling and removing sheeting must be carefully specified to avoid damaging the pipe. As noted in Art. 14-3, the soil load upon pipe is reduced by friction developed between the fill material and the walls of the trench. If the sheeting is pulled after the trench is backfilled, the friction force will not be developed and the load upon the pipe will be higher than originally calculated. A satisfactory procedure is to pull the sheeting in increments as the trench is backfilled or to pull it part way and cut it off. In particularly difficult soil conditions, where the danger to workers or structures requires it, specifications may require that sheeting be left in place.

18-9 Dewatering of Trenches

If the groundwater table is above the bottom of the trench, water will flow into it. Such a condition requires careful sheeting and bracing and also removal of the water by pumping. If the trench is in sandy material, and the water flow is large, a quicksand results, and considerable difficulty is encountered. Quicksand may be

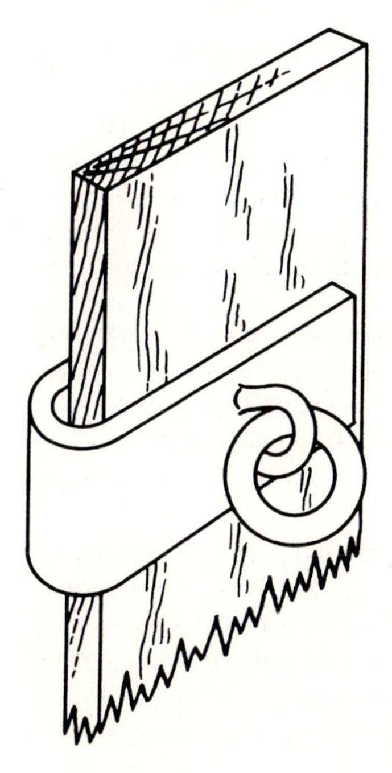

Figure 18-8 Steel clamp used to pull timber sheeting.

defined as a loose, granular material through which water is moving with sufficient velocity to separate the grains and keep them in partial suspension. Unless sheeting of trenches is driven well below the excavation in quicksand, it will flow into the trench and undermine the sides. Placing hay or straw in the trench bottom, weighted down with brickbats or other heavy rubbish, will stabilize the sand and permit water to rise without endangering side walls until work in the trench is finished.

Quicksand can best be combated by using well points. These are pipes 50 to 75 mm (2 to 3 in) in diameter, pointed at the lower ends, with a length of screen just above the point. In effect they are small driven wells (Art. 4-14). The pipes are driven or jetted into the ground in a line on one or both sides of the trench, some 2 m (6 ft) from the centerline, 1 m (3 ft) apart, and well below the water table. A 150- to 200-mm (6- to 8-in) header pipe parallels the trench and has valved branches connected to a pump. This arrangement lowers the level of the groundwater table below the trench bottom and keeps it dry.

Where water is encountered but there is no trouble with sand, the usual procedure is to allow it to run along the trench bottom to a sump from which it is raised and discharged by a pump. Since this water will contain gritty material, only centrifugal pumps, diaphragm pumps, jet pumps, and vacuum pumps are practical (see Chap. 7).

In a trench for a large sewer, an open-jointed tile underdrain surrounded with gravel may be placed below the grade of the sewer. An underdrain generally discharges into a sump, but its advantage is that there is no water running in the trench to damage the bottom. In order to keep the drain open it may be necessary to draw a rope or chain back and forth through it to break up silt accumulations. The drain is, of course, left in place when the sewer is completed, but should be closed to prevent continuous drainage of the soil.

18-10 Pipe Laying

Before placing pipe the grade of the bottom of the trench should be checked. Specifications may require that the grade be held within 10 mm ($\frac{1}{2}$ in) of that on the plans.

The pipe should be inspected to insure that it has no cracks or defects, with particular attention being paid to the joints. Bell-and-spigot pipe is sometimes placed using a pipe hook, as shown in Fig. 18-9, but gasketed pipe is better handled with a sling. Pipe with premolded joint rings or attached couplings should be handled so that no load, including the weight of the pipe itself, will be supported by the jointing material.[2]

Pipe lengths are placed on line and grade after the trench has been completely dewatered. Lengths are joined by pressing them together with a lever or winch. The pipe is more easily moved if a portion of the weight remains on the sling. Small diameter rubber gasketed pipe may tend to separate slightly unless backfill is tamped about it before it is released.

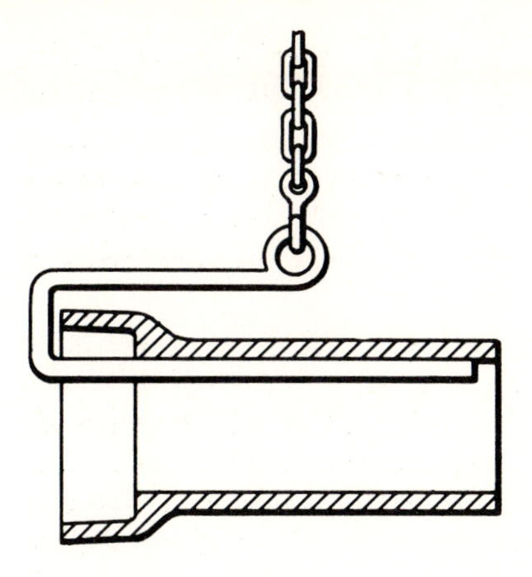

Figure 18-9 Hook and chain for lowering sewer pipe into trench.

Plastic, plain end clay, or asbestos cement pipe may have the joining ring placed upon one end before it is lowered into the trench. The specific recommendations of the manufacturer should be followed. Depending upon the jointing technique, placement of additional sections may be delayed until joints already made have set up.

18-11 Jointing

Joints in ordinary bell-and-spigot pipe are made with Portland cement mortar or bituminous materials. In small pipe, centering is provided by wrapping the spigot with an oakum cord of appropriate thickness. The gasket is driven into the bell with a calking tool and the joint is filled with mortar or bitumen. The inside of the pipe is smoothed with a swab or drag.

Cement-mortar joints have the disadvantage of being rigid, and settlement of the pipe will cause cracks which permit infiltration. Special precautions are necessary where the groundwater table is above the sewer. Under such conditions the tighter and more flexible, though more expensive, bituminous materials can be used. They are poured hot, at about 200°C, and require a jute or oakum gasket calked in place to center the joint and prevent the material from running into the pipe; they also require a joint runner on the outside as in the making of lead joints in water pipe. It is permissible to make alternate joints before the pipe is lowered into the trench. These, of course, are poured with the pipes in a vertical position. A difficulty with this type of joint is overheating of the bituminous material. This results in an inflexible joint which soon leaks.

Bituminous joints, while providing some flexibility, are not elastic. A better type of joint uses plastic or rubber materials which provide resilience as well as resistance to root penetration and infiltration. These joints allow a deflection of up to 4 percent in any direction without leakage at an internal pressure of 30 kPa (5 psi). The joints are of two types. One has collars of resilient material placed at

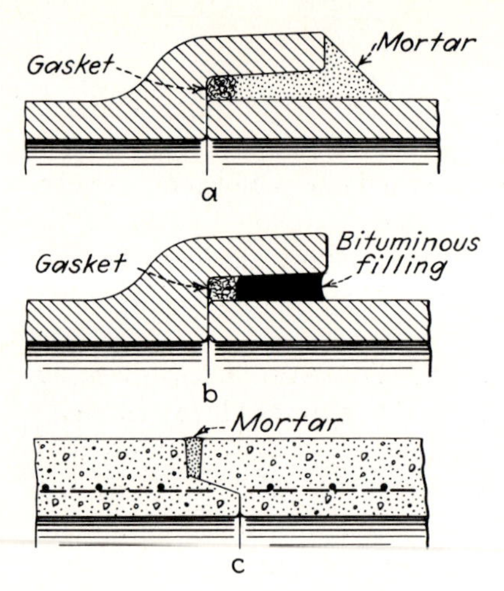

Figure 18-10 Joints. *(a)* Mortar joint in tile or small concrete pipe. *(b)* Bituminous joint. *(c)* Joint in large concrete sewer pipe.

the factory on the spigots and in the bells. A combined lubricant and cementing compound is brushed on the spigot which is then forced into the bell. A second type of resilient joint uses a gasket or compression ring which will be compressed between plastic material placed on the spigot and in the bell. When the spigot is forced into the bell the ring fills the annular space to make the seal.

Y and T branches for house sewer connections are placed wherever the plans indicate or the engineer directs. When not immediately used they are closed by clay disks or stoppers held in place by mortar. Junctions should be placed so that no load will endanger the sewer. They should also be carefully located in the construction field notes so that they can be found when a connection is to be made later.

18-12 Jacking and Boring

Installation of sewers under highways, railroads, airfields, etc., may be accomplished without disruption of surface traffic by jacking and boring.

In this technique, pipes are driven by hydraulic jacks mounted in a jacking pit at the point of beginning. The lead-pipe section may have a cutting ring installed upon it, and lubrication fittings may be placed upon the pipe. A cutter head operating ahead of the pipe opens the hole into which the pipe is jacked, while an auger operating on the same shaft as the cutter draws the excavated material through the pipe to the jacking pit.

Jacking is usually done upgrade to keep the cutting face and pipe dry. The pipe should be kept in motion once the installation is begun. If motion ceases, particularly under surfaces subject to vibratory loads, the soil may consolidate around the pipe, freezing it in position. It may then be necessary to jack and bore from the opposite side to meet the frozen pipe.

18-13 Backfilling

Trenches should be backfilled immediately after the pipe is laid unless Class A bedding is used (Fig. 14-2), in which case backfill is delayed until the concrete has set up sufficiently to support the backfill. No water should be permitted to rise in unbackfilled trenches.

Fill material should be free of brush, debris, frozen material, large rocks, and junk. No rock should be placed in the upper 400 to 500 mm (15 to 20 in) of the trench, nor within 900 mm (36 in) of the top of the pipe.

The fill should be tamped in layers 150 mm (6 in) deep around, under, and over the pipe to a depth of 600 mm (2 ft). Earth should be dropped into the trench carefully until 600 mm (2 ft) of cover is in place. Thereafter backfill may proceed more rapidly. Fill beneath streets and other surface construction must be bedding material, sand, or tamped earth placed in uniform layers at a moisture content assuring maximum density with the compaction technique used.

Puddling or flooding with water to consolidate the backfill is permissible only in sandy or gravelly material. If this method is used, the first flooding should be applied only after the first 600 mm (2 ft) of fill above the pipe has been placed and compacted by tamping, and the second flooding during or after the subsequent filling of the trench. An excess of water should be avoided to prevent disturbance of the earth under and around the pipes and excess pressure upon them. Where trenches are in fields, the backfilling above the 600-mm level is not tamped. All the earth is replaced, and the resulting mound is allowed to settle naturally.

CONSTRUCTION OF SEWERS IN PLACE

18-14 Concrete Sewers

If the trench is dry, and the bottom is firm, it will be shaped so that it will constitute the outside form of the invert. In soft material a flat bottom or cradle must be built, in some cases with pile supports. If the invert is comparatively flat, it is built complete in one operation for a length of 5 to 6 m (15 to 20 ft). The previously built portion of the invert serves as a template at one end of the section, and a metal template acts as a bulkhead at the other end. With a straightedge or screeds, the newly placed concrete is worked into the invert shape. Construction joints between the invert and the arch are keyed. Joints between the lengths are vertical and also keyed.

The arch forms are supported by the invert and consist of collapsible ribs or centers made of steel. The centers are spaced according to the loads they must carry and are in immediate contact with tongue-and-groove planking or lagging which supports the concrete. All-metal collapsible forms are also obtainable from manufacturers who specialize in this type of equipment. If the trench sides will permit, outside forms may be avoided, and the outside of the top of the arch is shaped by troweling. The inside forms, after collapsing, may be arranged to rest upon a truck which is moved on rails laid on the invert to the new sections to be

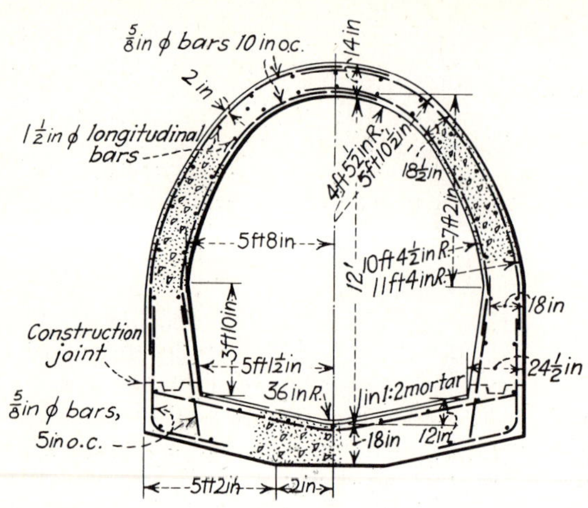

Figure 18-11 Horseshoe-shaped sewer constructed in two operations. (ft × 3.28 = m, in × 25.4 = mm.)

poured. Very large sewers are sometimes poured in three stages: the invert, a portion of the sides, and finally the arch.

Circular sewers are generally built in two operations. A collapsible inside form for the whole circumference is placed so that the lower half can be poured between the form and the trench. After the concrete has hardened, an outside form is placed resting upon the concrete sides of the lower half. An opening at the top of the outside form permits the concrete to be poured in and tamped. The work is done in sections, and bulkheads are needed at the open ends of the forms.

Forms should always be tight, clean, and oiled before being used. It is good practice not to remove the inside forms until the sewer is covered with backfill to a depth of 600 mm (2 ft). Reinforcing must be carefully placed and checked to see that it complies with the plans. Care must also be taken that the concrete is properly tamped, so that it will be uniformly dense and smooth inside.

If the invert is to be lined with brick, the sewer bottom is shaped up to a line 10 mm ($\frac{1}{2}$ in) below the bottom of the bricks and allowed to set before bricklaying begins. The bricks are bedded in 1:3 mortar on edge as stretchers, and the end joints are broken by a lap of at least 50 mm (2 in). The joints, also a 1:3 mix, should not exceed 5 mm ($\frac{1}{4}$ in) in thickness. Where liner plates are used, they should not exceed 300 by 450 mm (12 by 18 in) in surface dimension and be at least 25 mm (1 in) thick. They should be bedded in 10 mm ($\frac{1}{2}$ in) of wet mortar. Unlined inverts may be finished by floating and troweling. Within 2 hours after the concrete is placed, a dry mix of 2 parts Portland cement to 1 part sand, free from dust and particles over 3 mm, should be sprinkled over it, floated, and troweled.

18-15 Tunneling

Where cuts exceed 8 m (25 ft), the relative cost of tunneling compared with ordinary construction should be considered. If tunneling is advisable or necessary, the method used will depend upon the material. In solid rock, no shoring or bracing will be required; the excavation will be made by blasting; and the tunnel cross section will be the same shape as the outside of the sewer. Seamy rock may require bracing. Rock tunnels should be carefully ventilated to remove fumes and the dust caused by drilling holes for the explosives.

Where caving is likely, timbering of the tunnel is necessary. If the material will stand for a short time, frames are used to support lagging or poling boards that are in contact with the earth. In soft materials the poling boards are beveled and are driven ahead into the soft material as it is removed. When the safe limit of the unsupported poling boards is reached, another frame is placed, and another set of poling boards is driven ahead. If the tunnel is large, it is excavated in sections, the top usually first. It should be noted that the inverted egg-shaped sewer fits into the timbering with least loss of space. In long tunnels materials are transported in and out on small railway cars. Concrete is mixed on the surface and taken into the tunnel when it is needed. Usually the entire space between the sewer and the sides of the excavation is filled with concrete. Shafts are needed to give access to the tunnel, for removal of excavated earth and delivery of construction materials. They must be braced and shored.

In soft materials the shield method is frequently used. This is a steel cylinder having a cutting edge on the forward end. It is driven forward by means of hydraulic jacks backed against the already constructed walls of the sewer or a strong tunnel lining. Excavation proceeds at the forward edge of the shield.

In wet material the shield is enclosed at the end away from the face to form a caisson, and compressed air is supplied to keep out the water. This requires an auxiliary chamber or air lock through which materials and workmen pass as they enter or leave the caisson. When workmen leave, they are held in the air lock for a period depending upon the air pressure under which they have been working. During this time the pressure is slowly decreased. This precaution is necessary to prevent caisson disease, also known as the bends, diver's palsy, etc.

REFERENCES

1. "Trench Excavation," *Data Sheet 254* (revised), National Safety Council, Washington, D.C., 1959.
2. "Standard Recommended Practice for Installing Vitrified Clay Sewer Pipe," (C-12) American Society for Testing Materials, Philadelphia.

CHAPTER
NINETEEN

MAINTENANCE OF SEWERS

19-1

Maintenance of sewers consists principally of the removal or prevention of stoppages, cleaning of sewers, some repair work, and cleaning of catch basins, if any. It will not be costly unless there are many sewers with flat grades, or conditions are such that tree roots find easy entrance through the joints. Good maintenance requires adequate knowledge as to sewer locations, and a competent working force that is properly equipped and on call at all times.

19-2 Protective Ordinances

Cities find it necessary to pass ordinances designed to protect the sewers against injury and stoppage. Some cities do not permit plumbers to connect house sewers with the common sewers, such work being done by city forces to insure a workmanlike job. Should plumbers be permitted to make connections, their work should always be inspected before it is covered, and the plumber may forfeit a previously filed bond if he does not leave the sewer in good condition. Ordinances also forbid discharge into sewers of steam or corrosive, flammable, and explosive liquids, gases and vapors, and of garbage and dead animals. Wastes from kitchen sinks and floor drains of restaurants, hotels, and boarding houses and from packing houses, creameries, bakeries, cleaning establishments, laundries, garages, and stables may be required by ordinance to pass through grease traps or tanks similar to those described in Art. 16-6 before entering the sewers.

SEWER MAINTENANCE

19-3 Equipment

The equipment required will depend upon the size and type of sewers serviced. A small city will have less and simpler equipment than a large city. If sanitary sewers are to be maintained the troubles encountered will be largely roots and grease. In storm and combined sewers grit and sand will accumulate and a wide variety of miscellaneous rubbish will be encountered. Manufacturers' catalogs are helpful in choosing the appropriate equipment. A typical list of items carried by a full-time large-city maintenance crew working on all types of sewers is as follows:

Major equipment:

Medium size truck
Power winch
Portable manually operated winch
300 m (1000 ft) flexible steel cable
300 m (1000 ft) fire hose
200 to 250 m (600 to 750 ft) flexible steel rods (power drive desirable)
150 m (500 ft) interlocking wood sewer rods
Root cutters of assorted sizes
Sewer brushes of assorted sizes
Sand buckets, scoops, and drags of assorted sizes
Turbine flushing heads
Steel sewer tapes and heavy wire (for small sewers)
Sewer flushing bags

Minor equipment:

Shovels, picks, and mattocks
Assorted wrenches
Hydrant and manhole tools
Flashlights (explosion-proof)
Rubber boots, coats, and gloves
Buckets and rope

Safety equipment:

Hydrogen sulfide detector
Carbon monoxide detector
Combustible gas detector
Wolf safety lamp
Hose mask (double) with safety harness
Safety belts (2 or 3) with 7.5 m (25 ft) of 20-mm ($\frac{3}{4}$-in) rope

Complete first-aid kit
Manhole guard rails
Traffic signs and flags
Oil lamps and flares

Sewer cleaning rods may be inserted and operated by hand but are more efficient if pushed or pulled and turned by power as in Fig. 19-1. Up to about 75 m (250 ft) of wooden rods can be manipulated by hand. Figure 19-2 shows the tools used with the rods. The scoop shown in the figure will be more effective if it has a flap gate. The turbine cleaner (Fig. 19-3) is operated at the end of a hose and is pulled through the sewer.

19-4 Stoppage Clearing

A stoppage is a condition which partially or completely prevents flow of sewage through the sewer line. Stoppages may be caused by large objects which have been

Figure 19-1 Power-operated steel flexible rods. *(Courtesy Flexible Rod Equipment Company.)*

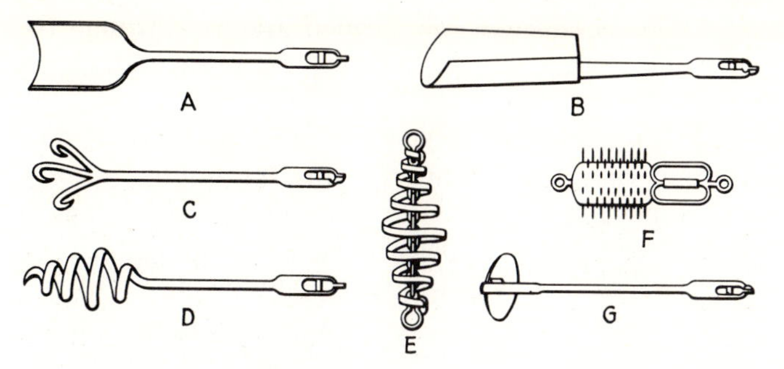

Figure 19-2 Sewer cleaning tools. A, gouge for cutting obstructions; B, scoop for removing sand and loose material; a flapgate on the end is desirable; C and D, claw and screw for removing paper and rags; E, root cutter; F, wire brush for removing grease; G, scraper.

placed in the sewers at manholes, sand or grit, grease, or miscellaneous materials. Clearing calls for choice of the proper tool. In many instances a stoppage can be cleared by forcing a rod headed by a pointed tool through the obstruction and then allowing the velocity of the released sewage to scour the pipe. Large objects may require excavation and opening the line. Sand and grit in large amounts, if not resulting from a bypassed sand trap, may mean a broken pipe and necessitate repairs.

1. *Root removal.* In sewers up to 380 mm (15 in) in diameter roots are frequently removed by using flexible-type rods with an auger-like cutter on the end. The cutter is rotated by twisting the rods either by power or manually as they are pushed into the sewer. In larger sewers or for particularly obstinate cases in small sewers, cutting drags are drawn or pulled through with cables and winches. In some cities much effort is given to preventing recurrence of root troubles. Defective joints which admit roots are uncovered and repaired. Especially troublesome lines have been relaid with iron pipe. Permanent control in house sewers has been obtained by requiring cast-iron pipe with leaded joints from the sewer to the property line and not less than vitrified pipe with asphalt joints inside the property.
2. *Sand and grit removal.* If punching a hole through the deposit and manual flushing does not suffice, buckets or scoops are pulled through by the cable and winch. Where the deposit is not too extensive the turbine cleaner, with its

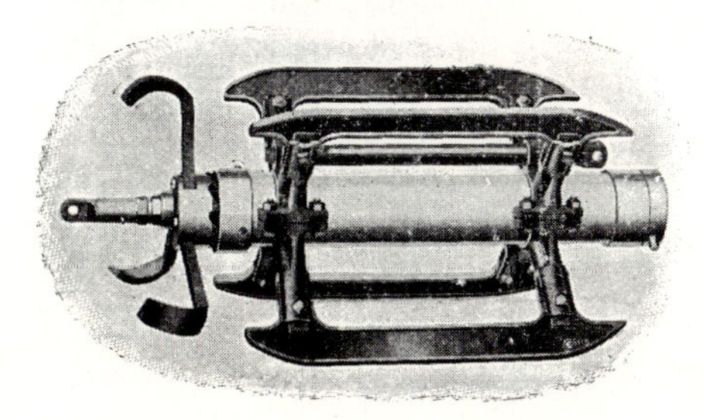

Figure 19-3 Turbine sewer cleaner. *(Courtesy Turbine Sewer Machine Company.)*

water-powered, rotating cutter, is effective. This, when dragged through the sewer by means of a winch and cable, combines flushing, cutting, and dragging in one operation.

3. *Grease removal.* Stoppages by grease are largely confined to house sewers although industrial plants may cause important problems. It is not uncommon, however, for large sewers to have their capacities reduced by grease coating the sewer pipe. In house sewers flat steel sewer tapes with various tools on the ends have been used. A 6-mm ($\frac{1}{4}$-in) steel wire 15 m (50 ft) long having a 150-mm (6-in) bend on the end and rotated by turning the end as it is fed into the sewer has been used successfully. Grease stoppages in the common sewers are removed by cutting and flushing with a final cleaning with a power-rotated brush. Some maintenance crews always use a brush drag to finish cleaning a line that has been dragged with other tools.

19-5 Sewer Cleaning

Sewer cleaning is a prevention measure applied to certain lines where experience indicates that stoppages may occur or if inspection indicates that cleaning is advisable. Flushing is widely used for cleaning but brushes, scrapers, or scoops may be required for removal of grease or miscellaneous debris and cutters or hooks for encroaching roots.

Flushing is of value because it temporarily increases velocity of flow and also the depth of the liquid, both conditions increasing the transporting power of the water. It is in the lateral sewers, where the flows are rarely great in volume and frequently cease entirely, that deposits are especially likely to occur and where flushing is most frequently required. Automatic flush tanks have already been described (Art. 16-5). Though little used now they were formerly placed at the dead ends of laterals having grades less than the minimum permissible.

Flushing is also done by use of fire hose connected to a fire hydrant and discharging into a manhole. Flushing must always be used with caution, or enough head may be built up in the sewer to cause backing up into basement drains or plumbing fixtures.

Some cities use rubber balls to clean sewers if serious clogging has not occurred. The ball is of soft rubber and it is inflated to slightly less diameter than the pipe. A rope is attached to the ball and it is allowed to travel from a manhole downstream in the pipe. The ball adjusts itself to irregularities in the pipe, but water held behind it escapes in a strong stream between the ball and the pipe wall thus flushing loose grease and other debris. Roots up to pencil size will be broken off. Large material dislodged from the sewer is removed at the lower manhole. If the ball cannot go through the sewer the clogging material can be removed with a bucket or scoop (Art. 19-4). Some of the balls have spiraled ribs which impart a rotary motion to the water on the downstream side. Others are enclosed in a light chain network. Operation is most effective at a head of 0.3 to 1.2 m (1 to 4 ft) above the ball, but where there are basements or other low connections good but slower operation is possible with a head of 50 mm (2 in).

Mention was made above that flushed material should be intercepted at the downstream manhole. This is accomplished by some cities by placing an L temporarily in the outlet of the manhole, with one leg vertical. The outflowing sewage must rise to the open end of the L and the bottom of the manhole acts as a trap.

19-6 Inspection Practice

Up-to-date maps of the sewer system are important in attaining good maintenance. They should show exact locations of manholes, directions of flow, grades, house sewers, etc. Inspection practices differ. In some cities inspections are made and work done only when difficulties arise. In other cities routine inspections are made. Here also practice differs as to frequency of inspections. Sewers on flat grades or previously troubled by roots may be examined every 3 months; sewers in which no trouble is recorded, once or twice yearly; intercepting sewers, one to four times per month; flush tanks, monthly; inverted siphons monthly to weekly; storm-water overflows, during and following heavy rains. In some cities trained crews follow an inspection routine, make repairs or cleanings according to a prearranged schedule, and make records of the conditions found, the work done, and the costs. Inspections of industrial plants may be necessary to discover conditions which are causing undesirable accumulations in the sewers and are otherwise required in conjunction with industrial waste sampling programs required for industrial cost recovery programs.

Inspections are made by looking through a sewer toward a flashlight (explosion-proof) during low flows. In some cities laterals and other sewers are inspected whenever a house connection is made. Observations should not be confined to periods of low flows. In main sewers surcharging in certain sections may indicate partial stoppages that will not appear during low-flow periods.

Sewers may also be inspected by passing small television cameras through them and observing the actual conditions. Such inspection is often a part of Sewer System Evaluation Surveys (SSES) required under PL 92-500 for sewerage systems which exhibit excessive inflow and infiltration. Television inspection permits precise location of leaks, intrusions, failures, etc., and accurate estimation of the repair required.

19-7 Making Repairs and Connections

Large brick sewers will require some repairs, pointing up of joints, and replacement of arch bricks that have fallen out. Guniting of large sewers requiring extensive repairs is sometimes done. Manholes may need similar treatment. Manhole frames and covers break or wear and become noisy. This nuisance can be eliminated by placing a gasket or cushion made for the purpose between the rim of the lid and the seat. Cracked or crushed lengths of pipe must be replaced. This is done by chipping off the upper portion of the bell of the length below the gap and a corresponding portion from the length that is to be inserted. The new length can then be laid in place and turned until the gap in its bell is also at the top of the pipe. The joints are made in the usual way, and the gaps in the bells are carefully filled with mortar. When a house connection is made, a Y-branch should be inserted. A poorer method consists of chipping a hole in the pipe and inserting the spigot of the house sewer pipe into the hole. A slant may be used for this connection. Concrete should be liberally used to hold the pipe in place, and care should be taken that the spigot does not protrude into the sewer. A saddle over the hole in the sewer will make a more satisfactory connection. Pipe repairs may also be made by slip-lining damaged sewers from manhole to manhole with plastic pipe made specifically for this purpose. Lining cuts off the household connections, which must be remade, but avoids excavation of the entire length of pipe.

19-8 Cleaning Catch Basins

Several cleaning procedures are used. Hand methods are employed mainly where catch basins are not accessible to machines. A three-leg derrick with a pulley, rope, and bucket is set up over the basin. The water is first removed, and the solids are shoveled into the bucket by a man in the basin. The bucket is then raised and emptied into a truck. Another method employs a specially designed eductor which consists of a centrifugal trash pump which discharges into a tank truck. Heavy material and sticks are first removed by hand, and the eductor then pumps out water and dirt. A third method employs a small orange-peel bucket.

19-9 Gases in Sewers

Explosions in sewers are not uncommon. Manhole covers may be blown into the air with danger to traffic, and occasionally the sewer lines themselves are destroyed or damaged. Asphyxiation of workmen, although rare, has also occurred. Combustible gases in sewers are due to the following: (*a*) gasoline from garages, filling stations, and clothes-cleaning establishments; (*b*) gas that has leaked into sewers from nearby gas mains or service pipes; (*c*) miscellaneous chemicals, like calcium carbide, discharged by accident or design into sewers; (*d*) gaseous products of decomposition, notably methane. Of these, gasoline is by far the most important. Fuel gas is less likely to cause trouble because its light weight allows it to escape more easily than gasoline vapor from the sewers. Products of decomposition are probably of little importance. Present-day sewers usually do not retain sewage long enough to permit any considerable accumulation of methane.

Prevention of accumulations of explosive mixtures in sewers can be accomplished in part by the following: requiring traps in the waste lines for garages, cleaning establishments, etc.; encouraging gas companies to trace leaks; inspecting and investigating the sources of all combustible materials that appear in sewers.

Accidents may be prevented by testing the air of sewers before bringing an open flame into them. Manufacturers of first-aid apparatus have devised appliances that will test air for its explosiveness and also for its toxic and asphyxiating properties.[1] These should be available for use under suspicious circumstances. The tester should wear a hose mask or if this is not available he should wear a safety belt with a rope attached, with other workmen outside the manhole to pull him out if he is overcome. Normally there is no danger in entering sewers for ordinary sewer air is not injurious to health. The term "sewer gas" is meaningless. Odors of gasoline and natural gas are indicative of danger, but putrefactive odors are of little significance.

In addition to the explosive gases certain toxic gases may occur in sewers. These are carbon monoxide, hydrogen sulfide, and hydrogen cyanide. The last named may occur in industrial sewers.

REFERENCES

1. "Occupational Hazards in the Operation of Sewage Works," *Manual of Practice No. 1*, Water Pollution Control Federation, Washington, D.C.

CHAPTER

TWENTY

CHARACTERISTICS OF SEWAGE

20-1

Sewer systems bring the wastewater of a city to a limited number of locations for treatment and disposal. The techniques used following collection depend, to a degree, upon the nature of the waste and the location of the discharge. A multitude of analyses could be performed upon wastewater to determine its characteristics, but normally only a few are used. The analyses necessary are those required to demonstrate compliance with the NPDES permit (Art. 1-3) and those required for proper operation of plant processes.

20-2 Physical Characteristics

Sewage is over 99.9 percent water, but the remaining material has very significant effects. Fresh domestic sewage has a slightly soapy or oily odor, is cloudy, and contains recognizable solids, often of considerable size. Stale sewage has a pronounced odor of hydrogen sulfide, is dark grey, and contains smaller, but occasionally recognizable, suspended solids.

At temperatures of about 20°C sewage will change from fresh to stale in 2 to 6 hours, with the time depending primarily upon the concentration of organic matter. The latter varies with per capita water consumption, infiltration, and the quantity of industrial waste discharged to the collection system. The quantity of waste produced per person on a dry solids basis is relatively invariant, but the quantity of carriage water used is not.

20-3 Solids Determinations[1]

The solids in sewage may be suspended or in solution. *Total solids* includes both and is determined by evaporating a known volume or weight of sample and weighing the residue. Results are expressed in mg/l.

Suspended solids and *dissolved solids* determinations require filtration of the sample. The filtration is made through a membrane filter similar to those used in bacteriological analyses (Art. 8-9). If suspended solids are to be determined the filter is dried and preweighed, a measured volume of sample is drawn through it, and it is dried and reweighed. The change in weight divided by the sample volume yields the suspended solids concentration. A measured volume or weight of the filtrate is evaporated to dryness and the residue weighed to determine the dissolved solids. If any two of the three concentrations are measured, the third may be calculated as either the sum or difference of the others.

Volatile solids are those solids ignitable at 550°C. The concentration of these materials is considered to be a rough measure of organic content, or, in some instances, of the concentration of biological solids such as bacteria and protozoa. Volatile solids may be measured upon the total sample (total volatile solids), the suspended fraction (volatile suspended solids), or the filtrate (volatile dissolved solids). The determination is made by ignition of the residue from the total solids test in a muffle furnace. In the case of volatile suspended solids either a glass filter (which will undergo only a slight weight loss which is corrected with a blank) or a cellulose acetate filter (which will leave no ash) must be used. The volatile fraction is determined by difference between the residue following drying and that following ignition. The residue following ignition is called *nonvolatile solids* or *ash* and is a rough measure of the mineral content of the wastewater.

20-4 Chemical Characteristics

Sewage contains both organic and inorganic chemicals. The *inorganic constituents* are present in the carriage water but increase due to water use. Ordinary sewage treatment is not directed toward altering the concentration of inorganic contaminants. Tertiary treatment, which may be required in some cases to maintain water quality, employs techniques similar to those used in water treatment (Chaps. 11 and 26).

The *organic constituents* include those present in the wastes discharged to the sewers and their degradation products. These may be separated analytically into fats, proteins, carbohydrates, acids, etc., but the effort is seldom worthwhile.

Nitrogen and *phosphorus* may be present either as a part of the organic fraction or as inorganic chemicals. Their concentration is important both from a standpoint of possible water pollution, and because they are necessary in moderate concentration in biological treatment systems.

The *alkalinity* of wastewater is important since it provides a buffer against acids produced by bacterial action in anaerobic or nitrifying systems. As sewage ages its pH tends to drop due to production of acids, but rises upon treatment.

Table 20-1 Typical domestic sewage characteristics, mg/l

Parameter	Weak	Medium	Strong
Total suspended solids	100	200	350
Volatile suspended solids	75	135	210
BOD	100	200	400
COD	175	300	600
TOC	100	200	400
Ammonia-N	5	10	20
Organic-N	8	20	40
PO_4-P	7	10	20

Table 20-1 presents typical variations in the strength of domestic sewage in the U.S. Techniques for the chemical analysis of wastewater are presented in Reference 1.

20-5 Microbiology of Sewage and Sewage Treatment

By its nature, domestic sewage contains enormous quantities of microorganisms. Depending upon its age and the quantity of dilution water, bacterial counts in raw sewage may be expected to range from 500,000/ml to 5,000,000/ml. Viruses, protozoans, worms, etc., are also present but their concentration is seldom important enough to require measurement.

Bacteria are single-celled plants which metabolize soluble food and reproduce by binary fission. Bacteria are capable of solubilizing food particles outside the cell by means of extracellular enzymes, and hence can remove soluble, colloidal, and solid organic matter from wastewater.

In the presence of adequate food and a suitable environment (temperature, pH, etc.) bacteria will reproduce as illustrated in Fig. 20-1 which shows numbers

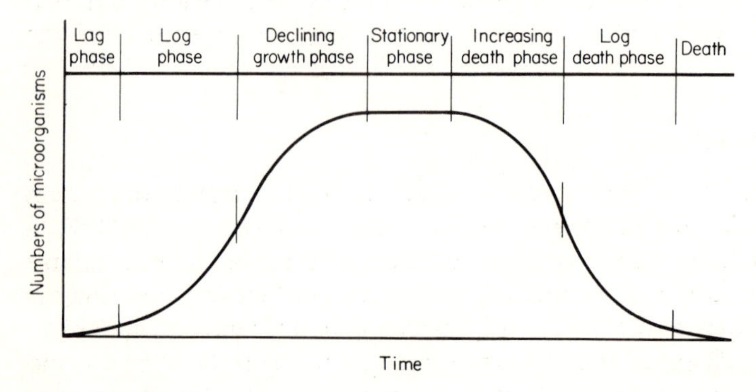

Figure 20-1 Growth pattern based on number of organisms. *(From Microbiology for Sanitary Engineers by Ross E. McKinney. Copyright 1962, McGraw-Hill Book Company, Inc. Used with permission of McGraw-Hill Book Company.)*

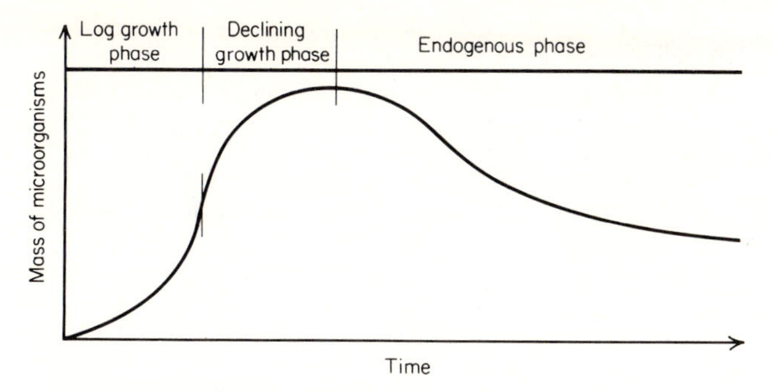

Figure 20-2 Growth pattern based on mass of organisms. *(From Microbiology for Sanitary Engineers by Ross E. McKinney. Copyright 1962, McGraw-Hill Book Company, Inc. Used with permission of McGraw-Hill Book Company.)*

of organisms vs. time. The end of log growth and beginning of declining growth indicates the point at which the available food supply has been largely depleted and food becomes a limiting element. Engineers seldom count bacteria, but infer their number from the concentration of suspended solids. Figure 20-2 presents a plot of mass rather than number of bacteria vs. time. It should be noted that this curve exhibits only three, rather than seven, phases and that the beginning of growth is immediate.

Although the log growth phase coincides with the maximum rate of substrate (food) removal, this is not the optimum zone of operation for waste treatment systems. In order to maintain log growth, the food must be in ample supply, but a low concentration of food is desired in the effluent. Further, maximum rate of utilization requires a maximum rate of supply of other nutrients, such as oxygen, and this may be difficult to supply. Finally, bacteria in the log growth phase of development have a great deal of energy available, have limited accumulation of waste products, and hence are likely to be motile, dispersed, and difficult to remove by sedimentation.

The declining growth phase is generally used for biological treatment systems since the problems listed above can be avoided. Some systems such as extended aeration (Chap. 24) and sludge digestion (Chap. 25) are operated in the endogenous phase.

20-6 Anaerobic Processes

Anaerobic bacteria oxidize organic matter utilizing electron acceptors other than oxygen. In carrying out their metabolic processes they produce CO_2, H_2O, H_2S, CH_4, NH_3, N_2, reduced organics, and more bacteria. A large part of the available energy appears in the form of end products, hence cell production is low and the by-products, such as methane, may be utilized as an energy source (Fig. 20-3).

The end products of an anaerobic fermentation are likely to be odorous, and intermediates such as the volatile acids may be toxic to bacteria, thus promoting upset of the process. The production of a stable effluent is unlikely since wastes do not usually contain sufficient electron acceptors to permit complete oxidation.

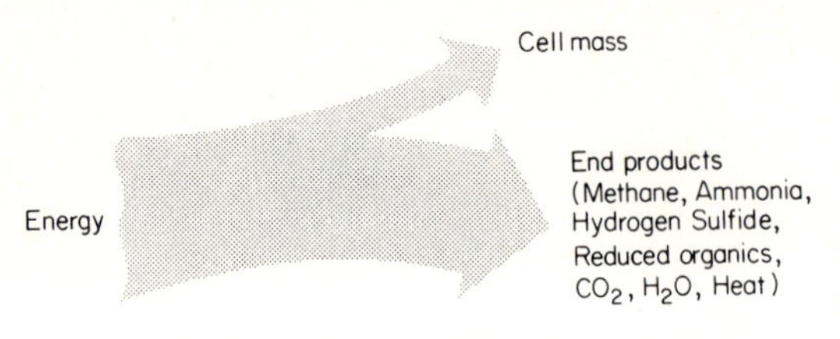

Figure 20-3 Energy conversion in anaerobic processes.

20-7 Aerobic Processes

Aerobic bacteria utilize free oxygen as an electron acceptor. The end products of aerobic activity are CO_2, H_2O, SO_4^{--}, NO_3^-, NH_3, and more bacteria. The bulk of the available energy finds its way into cell mass or heat, yielding a stable effluent which will not undergo further decomposition. The energy division is shown schematically in Fig. 20-4. The oxygen required may be furnished naturally from the atmosphere or mechanically by fine bubble aeration, thin film aeration, or droplet aeration.

The oxidation of ammonia to nitrate may or may not occur depending upon the retention time, oxygen available, temperature, bacterial predominance, and other factors (Chap. 26).

In addition to *strict* or *obligate* aerobes and anaerobes there is a third group of bacteria, called *facultative*, which can carry on their life processes under either aerobic or anaerobic conditions. The biological processes which these microbes utilize are identical to those of the obligate aerobes and anaerobes save that there are certain specific reactions which they cannot effect, such as reduction of CO_2 to CH_4 and oxidation of NH_3 to NO_3^-.

20-8 Other Microorganisms

Algae are photosynthetic microorganisms which can produce oxygen and protoplasm from inorganic chemicals. Algae are not significant in most waste treatment processes, but play a role in oxidation ponds where a symbiotic relationship exists between them and the saprophytic bacteria which oxidize the organic matter in the waste (Chap. 24).

Protozoa are single-celled animals which reproduce by binary fission. There are many species with differing size, shape, motility, and substrate. Protozoa may be aerobic, anaerobic, or facultative. Although many species can utilize soluble organic food, the concentration must be far higher than that in ordinary sewage,

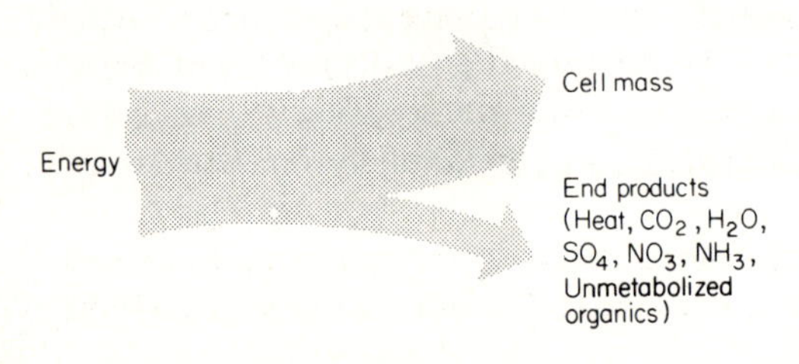

Figure 20-4 Energy conversion in aerobic processes.

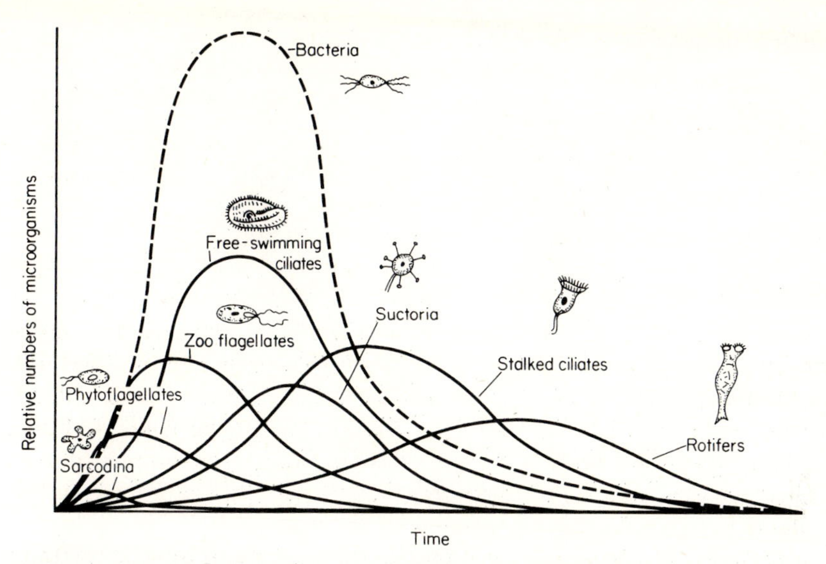

Figure 20-5 Relative growth of microorganisms in stabilization of wastes—individual curves are not to the same scale. *(From Microbiology for Sanitary Engineers by Ross E. McKinney. Copyright 1962, McGraw-Hill Book Company, Inc. Used with permission of McGraw-Hill Book Company.)*

hence the major food source for the protozoa is the bacteria. By reducing the number of bacteria the protozoa alter the food/mass ratio, thus stimulating further bacterial growth and further waste stabilization.

Fungi are multicellular nonphotosynthetic plants. Most fungi are aerobic, but anaerobic species are known. Due to their different cellular composition, fungi tend to predominate over bacteria in wastes which are deficient in nitrogen or low in pH. Because of their relatively large, filamentous shape, fungi tend to settle poorly, and are thus difficult to remove by sedimentation, and undesirable in biological treatment processes.

Rotifers are the simplest multicellular animal. Rotifers feed upon bacteria and small protozoa, thus further stabilizing the waste. They require a relatively high dissolved-oxygen concentration, hence their presence is a good indication of relative stability of a treated waste. The relative predominance of microorganisms in waste stabilization is illustrated in Fig. 20-5.

20-9 Biochemical Oxygen Demand (BOD)

Bacteria placed in contact with organic material will utilize it as a food source. In the utilization the organic material will eventually be oxidized to stable end products such as CO_2 and water. The amount of oxygen used in this process is called the biochemical oxygen demand (BOD) and is considered to be a measure of the organic content of the waste.

The BOD determination has been standardized[1] and measures the amount of oxygen utilized by microorganisms in the stabilization of wastewater for 5 days at

20°C. For domestic sewage the 5-day value or BOD_5 represents approximately $\frac{2}{3}$ of the demand which would be exerted if all the biologically oxidizable material were, in fact, oxidized.

In running the test on ordinary domestic sewage it can be assumed that a suitable bacterial inoculum will be present (Art. 20-5). If relatively clean industrial wastes are to be analyzed an inoculum may be required.[1,2]

The solubility of oxygen in water is distinctly limited (App. II), hence measurement of high BOD values requires that the waste be diluted. The dilution water is carefully manufactured and contains a mixture of salts providing all the trace nutrients necessary for biological activity plus a phosphate buffer to maintain a neutral pH. The water is aerated to saturate it with oxygen before mixing it with the sewage sample.

The exertion of BOD is considered to be a first order reaction defined by:

$$\frac{dy}{dt} = -K_1 y \tag{20-1}$$

in which y is the BOD remaining at time t and K_1 is a constant. Integrating Eq. (20-1) and letting L equal the BOD at $t = 0$,

$$y = Le^{-K_1 t} \tag{20-2}$$

The BOD exerted at time t is, of course, the difference between that initially present and that remaining, whence,

$$\text{BOD}_t = L - y = L(1 - e^{-K_1 t}) \tag{20-3}$$

The values of K_1 and L may be determined from a series of BOD measurements. Thomas[3] recognized the similarity between the series expansions for $1 - e^{-K_1 t}$ and $K_1 t[1 + (K_1 t/6)]^{-3}$ and developed the approximate formula:

$$\text{BOD} = LK_1 t\left(1 + \frac{K_1 t}{6}\right)^{-3} \tag{20-4}$$

which can be linearized as

$$\left(\frac{t}{\text{BOD}}\right)^{1/3} = (K_1 L)^{-1/3} - \frac{K_1^{2/3}}{6L^{1/3}}\, t \tag{20-5}$$

By plotting $(t/\text{BOD})^{1/3}$ vs. t a straight line may be obtained with intercept $(K_1 L)^{-1/3}$ at $t = 0$, and slope $K_1^{2/3}/6L^{1/3}$.

Example The following data are to be used to determine the value of K_1 and L for the waste tested:

Time, days	$\frac{1}{2}$	1	2	3	4	5	7	10	15
BOD, mg/l	5	20	90	160	200	220	260	285	320

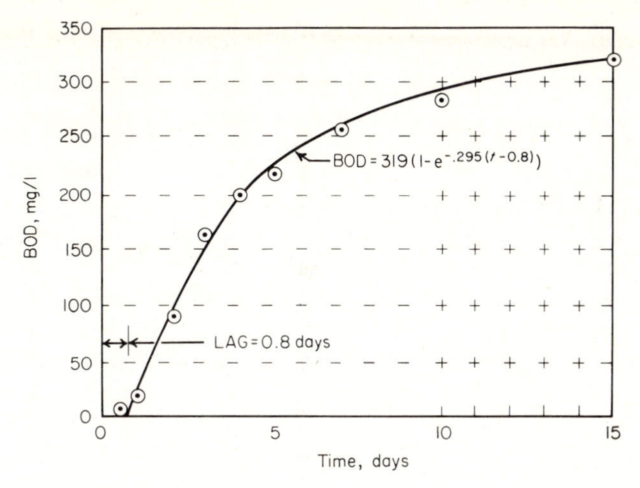

Figure 20-6 BOD exerted vs. time.

Solution The data is first plotted as shown in Fig. 20-6 to determine if the oxygen uptake began immediately or if a lag was evident. From the plot it appears that the time of beginning which best fits the data is $t_0 = 0.8$. The times at which the BOD data were taken are thus corrected by this amount:

Time, days	Corrected time, days	BOD, mg/l	$(t/\text{BOD})^{1/3}$
0.5	.....	5	
1.0	0.2	20	0.22
2.0	1.2	90	0.24
3.0	2.2	160	0.24
4.0	3.2	200	0.25
5.0	4.2	220	0.27
7.0	6.2	260	0.29
10.0	9.2	285	0.32
15.0	14.2	320	0.35

$(t/\text{BOD})^{1/3}$ is plotted vs. t as shown in Fig. 20-7. From the figure, $(K_1 L)^{-1/3} = 0.22$, and $K_1^{2/3}/6L^{1/3} = 0.0108$. Solving these identities simultaneously, $K_1 = 0.295$/day and $L = 319$ mg/l. The derived BOD equation is thus $\text{BOD} = 319[1 - e^{-0.295(t-0.8)}]$ which is shown superposed on Fig. 20-6. Not all experimental data will fit the theoretical curve as well as that of the example. The data which is best is generally that which falls between 2 and 10 days.

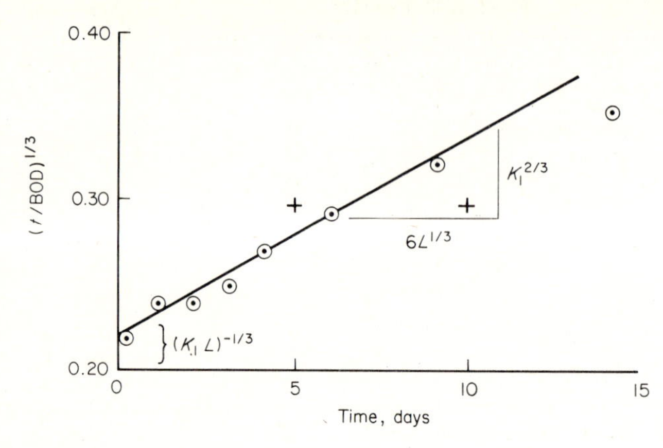

Figure 20-7 Graphical determination of K_1 and L.

The constant K_1 varies with temperature in accordance with:

$$K_{1(T)} = K_{1(20)}[1.047^{(T-20)}] \tag{20-6}$$

in which $K_{1(20)}$ is the value at 20°C determined in the BOD test and T is the actual temperature in degrees centigrade. Equation (20-6) can be used to determine the rate of oxygen demand exertion at temperatures other than 20°C, which is important in assessing the effects of waste discharges upon streams.

20-10 Chemical Oxygen Demand (COD)

The BOD test, while perhaps a good representation of what will occur in a stream, requires a minimum of 5 days, and is thus not useful in control of treatment processes. The COD test involves an acid oxidation with potassium dichromate. A measured amount of dichromate is added, the acidified sample is boiled for 2 hours, cooled, and the amount of dichromate remaining is measured by titration with ferrous ammonium sulfate. No clear correlation exists between BOD and COD in general, but at specific treatment plants a correlation is possible. COD results are generally higher than BOD values since the test will oxidize materials such as fats and lignins which are only slowly biodegradable.

20-11 Total Organic Carbon (TOC)

The total organic carbon test or TOC involves acidification of the sample to convert inorganic carbon to CO_2 which is then stripped. The sample is then injected into a furnace where it is oxidized in the presence of a catalyst. The CO_2 produced is measured by infra-red analysis, and converted instrumentally to original organic carbon content. The test is rapid, accurate, and correlates moderately well with BOD. The major obstacle to widespread use of this procedure is the cost of the equipment and the skill necessary in its operation.

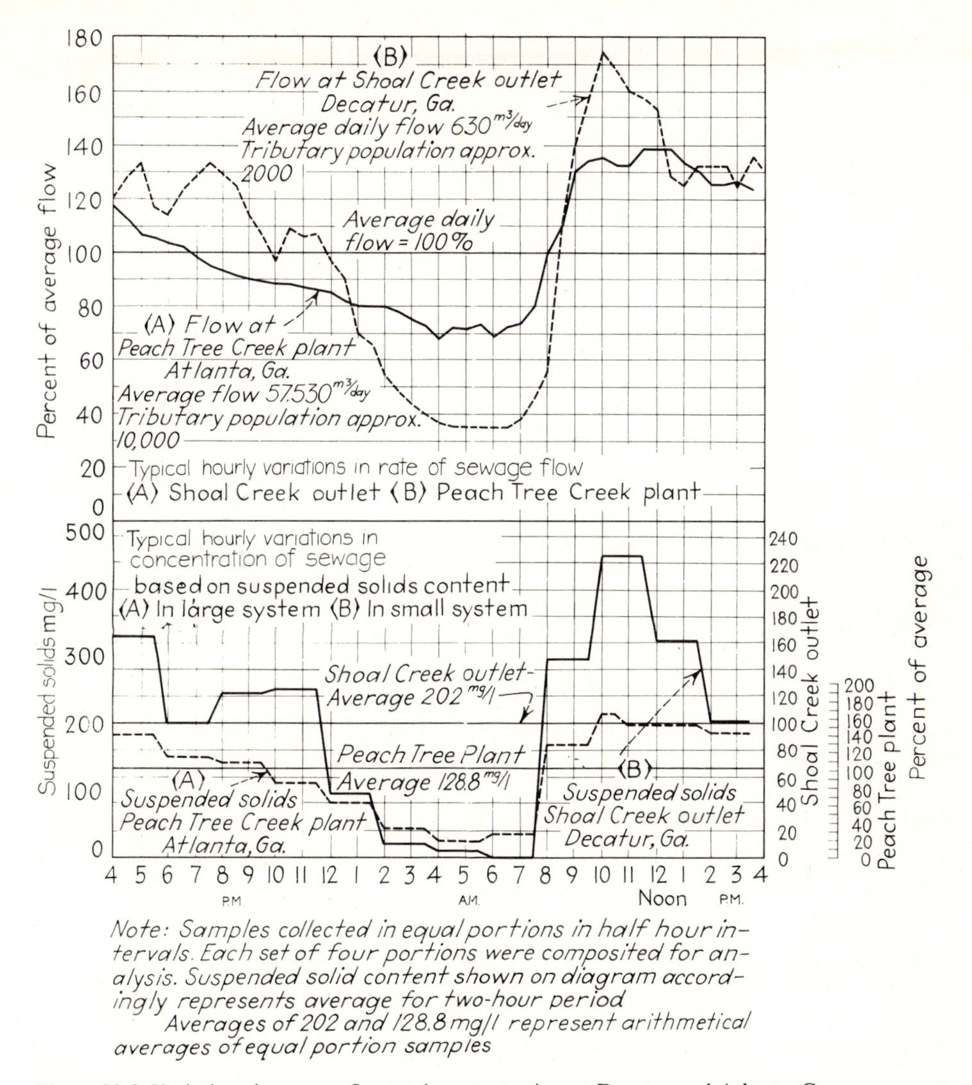

Figure 20-8 Variations in sewage flow and concentration at Decatur and Atlanta, Ga.

20-12 Sampling

Sewage is quite variable in content and it is important that representative samples be obtained. Grab samples are necessary to determine the maximum and minimum values which are likely to occur (Fig. 20-8). Composite samples can be obtained either manually, by taking a sample at regular intervals with the volume being proportional to flow, or automatically, using a proportional sampler. Samples should be iced to limit biological activity between time of collection and analysis.

A long period of record is desirable if the data is to be used in design of new facilities. If a special sampling program is necessary it should cover at least two weeks and should be continued as necessary to insure data from at least three wet and three dry days. The definition of a "wet" day for this purpose is arbitrary, but has been taken to mean any 24-hour period during which at least 25 mm (1 in) of rain falls.

20-13 Population Equivalents

Since the contribution of solids to sewage should be nearly constant on a per capita basis, the BOD contribution in grams/person per day should also be relatively uniform. This is, in fact, the case and the per capita contributions of suspended solids and BOD in the United States are approximately 90 and 80 g/day, respectively. The variations in concentration shown in Table 20-1 are due to variations in flow and the inclusion of industrial wastes.

The population equivalent of a waste may be determined by dividing the total mass per day of a pollutant by the appropriate per capita figure. For example a waste with a BOD of 300 mg/l and a flow of 10^6 l/day would contain 0.3×10^6 g of BOD, and would have a population equivalent of $300{,}000/80 = 3750$ persons.

Population equivalent has been used as a technique for assessing industrial waste treatment costs. Population equivalents may be determined for any constituent (Flow, BOD, COD, SS, P, N, etc.) and used to determine the proportional contribution and thus the proportional cost of treating industrial wastes in municipal plants. The calculation can be made equally well, however, by comparison of total municipal and industrial contributions. Assessment of industrial costs is required for new construction financed in part by federal funds,[4] and as a matter of equity is desirable in any event.

PROBLEMS

20-1 The 5-day BOD of a waste is 190 mg/l. Determine the ultimate oxygen demand (L). Assume $K_1 = 0.25$/day.

20-2 A BOD determination yields the following data. Determine the value of K_1 and L for this waste.

Time, days	1	2	4	5	6	7	9	11	13
BOD, mg/l	4	12	19	21	23	27	30	35	35

20-3 The average sewage flow of a city is 68,000 m^3/day. If the average BOD_5 is 285 mg/l, compute the total daily BOD_5 in kilograms and the population equivalent of the waste.

20-4 If the waste of Prob. 20-1 is discharged to a stream at an average temperature of 30°C what fraction of the BOD would be exerted in 5 days? How long would be required for the same degree of stabilization if the stream had a temperature of 10°C?

20-5 Sewage is applied to a trickling filter at a rate of 3000 m^3/day. The average BOD_5 of the influent is 140 mg/l, and it contains no dissolved oxygen. The effluent has a BOD_5 of 30 mg/l and contains 2.5 mg/l of dissolved oxygen. If K_1 is 0.25/day, how many kilograms of oxygen does the filter transfer per day?

REFERENCES

1. *Standard Methods for the Examination of Water and Wastewater* 14th ed., American Public Health Association, New York, 1976.
2. McKinney, Ross E.: *Microbiology for Sanitary Engineers*, McGraw-Hill, New York, 1962.
3. Thomas, H. A., Jr.: "Graphical Determination of BOD Curve Constants," *Water and Sewage Works* **97:**123, 1950.
4. "Federal Guidelines—Industrial Cost Recovery Systems," *MCD-45* U.S. Environmental Protection Agency, Washington, D.C., 1976.

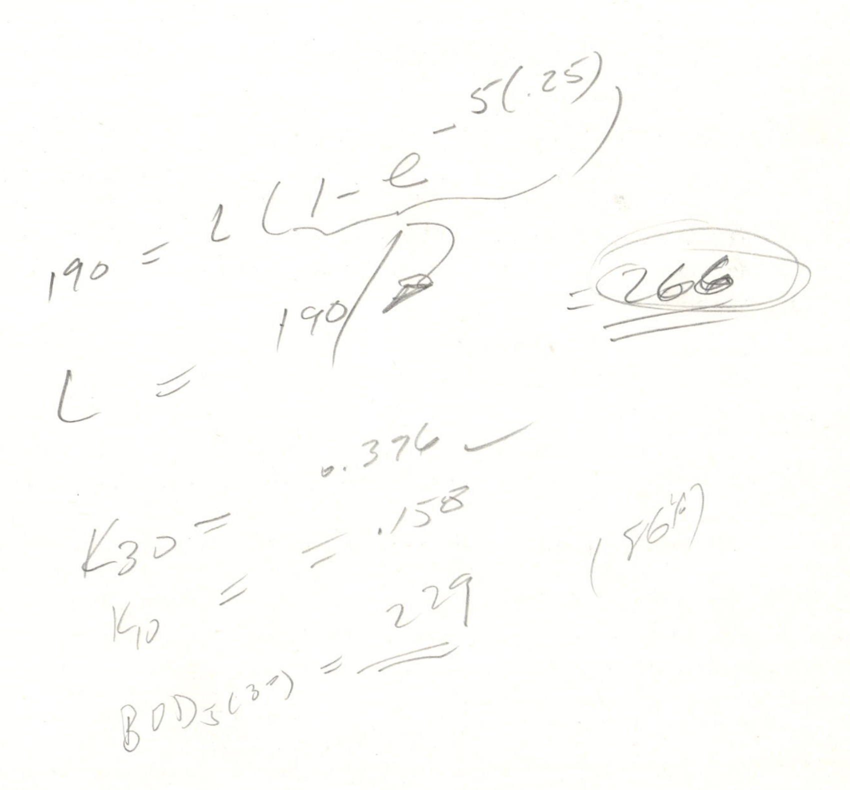

CHAPTER

TWENTY-ONE

SEWAGE DISPOSAL

21-1

Liquid wastes may be disposed of by discharge to surface waters, either directly or following passage over the soil surface; by discharge to groundwaters, either directly through deep wells or indirectly by percolation; or by evaporation to the atmosphere. Whatever the ultimate disposal technique, the waste must be treated first to at least the equivalent of secondary treatment as a matter of law, engineering necessity, or both.

21-2 Effects of Stream Discharge

In natural streams there is a balance between plant and animal life, with considerable interaction among the various life forms. Waters of good quality are characterized by multiplicity of species with no dominance. Organic matter which enters the stream is broken down by bacteria to ammonia, nitrates, sulfates, carbon dioxide, etc., which are utilized by plants and algae to produce carbohydrates and oxygen. The plant life is fed upon by microscopic animals (protozoa, rotifers, etc.) which, in turn, are fed upon by crustacea, insects, worms, and fish. Some of the animals feed upon the wastes of others, thus assisting bacterial degradation.

Introduction of excessive quantities of waste material can upset the cycle by causing rapid bacterial growth and resulting depletion of dissolved oxygen in the stream. Polluted waters are characterized by very large numbers of relatively few species. As the excess organic matter is stabilized the normal cycle will be reestablished in a process known as self-purification. Water quality standards are often based upon maintenance of some minimum dissolved oxygen concentration which will protect the natural cycle in the stream while taking advantage of its natural assimilative capacity.

21-3 Factors in Self-purification

Various physical factors affect the process of self-purification of streams.

Dilution with a large amount of water containing substantial dissolved oxygen will reduce the chance of significant pollutional effects. Conversely, wastes discharged to small streams may require very thorough treatment if water quality standards are to be met.

Currents assist in dispersion of the wastewater in the stream, preventing locally high concentrations of pollutants. Eddies and backwaters may permit sedimentation of suspended solids, formation of sludge banks, and production of odors. High velocity improves reaeration (Art. 21-7), reducing the time of recovery, but may increase the length of stream affected by the waste.

Sedimentation. Suspended solids, which may contribute to oxygen demand, are removed by settling if the stream velocity is less than the scour velocity of the particles. Such removal improves the water quality downstream of the sedimentation zone, but is detrimental at the point where solids accumulate.

Bottom deposits and runoff. Organic material which has settled or been adsorbed may be re-released or the products of its decomposition may be released. This phenomenon is often spread over a considerable length of stream, as is runoff of polluted storm waters from urban or agricultural land.

Sunlight acts as a disinfectant and stimulates the growth of algae. The algae produce oxygen during daylight, but utilize oxygen at night. Waters containing heavy algal growths may be supersaturated with dissolved oxygen during daylight hours and be anaerobic at night.

Temperature affects the solubility of oxygen in water, the rate of bacterial action, and the rate of reaeration. The critical condition is generally in warm weather when utilization rates are high and availability is low.

21-4 Self-purification of Lakes

In lakes self-purification is brought about by the same agents that operate in streams. Currents are less pronounced, however, and sedimentation will often cause heavy accumulations of sludge, dead algae, and other vegetation on the bottom. The inevitable decomposition, which may be slow because of the low temperatures of deep water, will use up all oxygen from the deeper layers. It is not uncommon to find in ponds and lakes that the upper layers of water have much dissolved oxygen and support clean-water plankton and fish, while the lower strata show the characteristics of pollution—no oxygen, anaerobic bacteria, and unpleasant odors. A shallow lake presents favorable conditions for quick self-purification—large water surface compared with the volume and much contact with algae both in floating mats and fixed upon aquatic plants and at margins.

The effect of waste discharges upon lakes and estuaries can be modeled by techniques similar to those presented below for streams. The models are considerably more complex than that presented here,[1,2] and their application is beyond the scope of this text.

21-5 The Oxygen Sag Curve

The general equation for transport by one-dimensional continuous flow is:

$$\frac{\partial C}{\partial t} = \frac{1}{A(x, t)} \frac{\partial}{\partial x}\left[E(x, t)A(x, t)\frac{\partial C}{\partial x}\right] - \frac{1}{A(x, t)} \frac{\partial}{\partial x}[Q(x, t)C] - S(C, x, t) \tag{21-1}$$

where C = the concentration of interest
A = the area of flow
E = a longitudinal dispersion coefficient
x = distance
t = time
Q = flow, and
S = any sources or sinks of the parameter being measured

If longitudinal dispersion is neglected ($E = 0$), this reduces to:

$$\frac{\partial C}{\partial t} = -\frac{1}{A(x, t)} \frac{\partial}{\partial x}[Q(x, t)C] - S(C, x, t) \tag{21-2}$$

The equation can be applied to the transport of both BOD and dissolved oxygen in a stream. Considering *dissolved oxygen*, the term $S(C, x, t)$ includes the utilization of oxygen to satisfy BOD, reaeration from the atmosphere, and any other sources and sinks (such as benthal utilization or algal production).

The rate of oxygen depletion by BOD exertion is equal to the rate of BOD removal, or,

$$\frac{dC}{dt} = \frac{dL}{dt} = -K_1 L(x, t) \tag{21-3}$$

For gases of low to moderate solubility the rate of change of concentration due to reaeration is given by:

$$\frac{dC}{dt} = K_L \frac{A}{V}(C_s - C) \tag{21-4}$$

where K_L = a mass transfer coefficient
A = the surface area
V = the liquid volume, and
C_s = the saturation concentration of the gas

The product $K_L A/V$ may be considered as a volumetric mass transfer coefficient, giving:

$$\frac{dC}{dt} = K_2(C_s - C) \tag{21-5}$$

For dissolved oxygen, in a reach of a stream in which flow and area are constant, Eq. (21-2) may thus be written as

$$\frac{\partial C}{\partial t} = -\frac{Q}{A}\frac{\partial C}{\partial x} + K_2(C_s - C) - K_1 L(x, t) - S_R(x, t) \tag{21-6}$$

If $(C_s - C)$ is replaced by D, the oxygen deficit,

$$\frac{\partial D}{\partial t} = -\frac{Q}{A}\frac{\partial D}{\partial x} - K_2 D + K_1 L(x, t) + S_R(x, t) \tag{21-7}$$

Under steady-state conditions $\partial D/\partial t = 0$ at any point in the stream, hence

$$U\frac{dD}{dx} = -K_2 D + K_1 L(x) + S_R(x) \tag{21-8}$$

where U = the velocity of flow (Q/A)
D = the oxygen deficit
$L(x)$ = the BOD, and
$S_R(x)$ = all other sources or sinks of oxygen

In similar fashion Eq. (21-2) may be applied to the transport of BOD. The term $S(C, x, t)$ in this case includes the removal of BOD by biological action,

$$\frac{dL}{dt} = -K_1 L(x, t) \tag{21-9}$$

the removal of BOD by sedimentation or adsorption,

$$\frac{dL}{dt} = -K_3 L(x, t) \tag{21-10}$$

and addition of BOD by resuspension of sediments or from surface runoff:

$$\frac{dL}{dt} = L_a(x, t) \tag{21-11}$$

For BOD transport then,

$$\frac{\partial L}{\partial t} = -\frac{Q}{A}\frac{\partial L}{\partial x} - K_1 L(x, t) - K_3 L(x, t) + L_a(x, t) \tag{21-12}$$

which for steady-state conditions $(\partial L/\partial t = 0)$ yields

$$U\frac{dL}{dx} = -(K_1 + K_3)L(x) + L_a(x) \tag{21-13}$$

For the condition in which K_1, K_2 and L_a are constant, Eq. (21-13) may be integrated to

$$L(x) = L_0 e^{-(K_1+K_3)x/U} + \frac{L_a}{K_1 + K_3}(1 - e^{-(K_1+K_3)x/U}) \tag{21-14}$$

in which L_0 is the BOD in the stream at $x = 0$, that is, at the point of discharge. Substituting Eq. (21-14) in Eq. (21-8) and integrating assuming that K_1, K_2, K_3, U, L_a, and S_R are constant,

$$D(x) = D_0 e^{-K_2 x/U} + \frac{K_1}{K_2 - (K_1 + K_3)}\left(L_0 - \frac{L_a}{K_1 + K_3}\right)\left(e^{-(K_1+K_3)x/U} - e^{-K_2 x/U}\right) + \left(\frac{S_R}{K_2} + \frac{K_1 L_a}{K_2(K_1 + K_3)}\right)\left(1 - e^{-K_2 x/U}\right) \qquad (21\text{-}15)$$

D_0 is the deficit at $x = 0$ and the other terms are as defined above. Equation (21-14) permits calculation of the BOD remaining at any point in the stream, while Eq. (21-15) yields the deficit at any point. The variation of these factors with distance is shown graphically in Fig. 21-1.

21-6 The Critical Deficit

The point of major interest in the oxygen sag analysis is the point of minimum dissolved oxygen, or point of maximum deficit. The maximum or critical deficit, labeled D_c in Fig. 21-1, occurs at the inflection point of the oxygen sag curve and may thus be found directly by taking a derivative of Eq. (21-15) with respect to x, setting it equal to zero, and solving for x_c. The value of x_c may then be substituted in Eq. (21-15) to find the critical deficit. In this manner,

$$x_c = \frac{U}{K_2 - (K_1 + K_3)} \ln \left\{\frac{K_2}{K_1 + K_3} + \frac{K_2 - (K_1 + K_3)}{(K_1 + K_3)L_0 - L_a}\left[\frac{L_a}{K_1 + K_3} - \frac{K_2 D_0 - S_R}{K_1}\right]\right\} \qquad (21\text{-}16)$$

21-7 Evaluation of Rate Constants

K_1, the deoxygenation constant, is usually measured in the laboratory as detailed in Art. 20-9. K_2, the reaeration constant, has been found to be a function of stream turbulence,[3] and may be calculated from:

$$K_2 = \left(\frac{D_m U}{H^3}\right)^{1/2} \qquad (21\text{-}17)$$

where U = the average stream velocity
H = the average depth, and
D_m = a molecular diffusion coefficient equal to 2.037×10^{-5} cm²/s at 20°C

K_2 varies with temperature in accord with

$$K_{2(T)} = K_{2(20°C)}(1.025)^{T-20} \qquad (21\text{-}18)$$

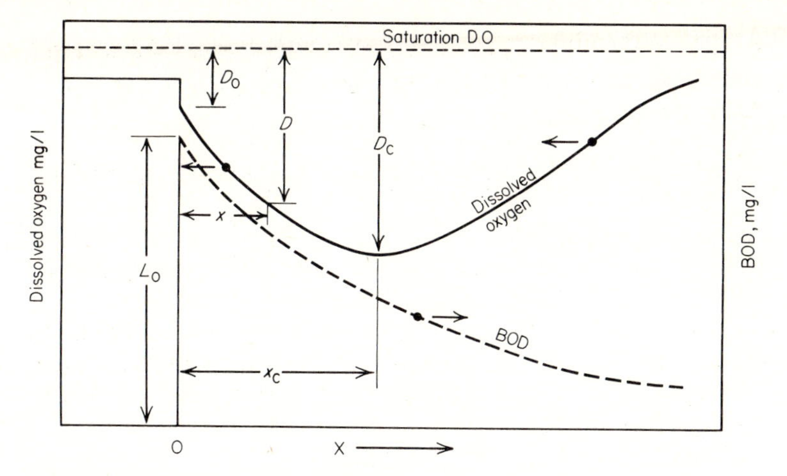

Figure 21-1 Oxygen sag and BOD removal in streams.

K_2 may also be measured directly by a radioactive tracer technique which involves the simultaneous release of an homogeneous mixture of three tracers: tritium in the form of tritiated water, dissolved Krypton-85, and a fluorescent dye. The fluorescent dye indicates when to sample for the other two tracers and provides a measure of the time of flow between sampling stations. The tritiated water provides a measure of the total dispersion of the tracer mixture. Since it is in the form of water molecules, the tritium is not adsorbed or otherwise lost in any significant amount, hence the decrease in its concentration between sampling stations is attributable to the total dispersion in the stream.

Because of the simultaneous release of the three tracers, the dissolved Krypton-85 is dispersed to the same extent as the tritiated water, but is, in addition, lost to the atmosphere as a result of turbulent mixing and surface replacement. The rate of Krypton-85 loss has been shown to be directly related to the rate of reaeration[4] and the reaeration constant K_2 may be calculated from the concentrations of tritium and Krypton-85 measured at two stations a and b and the time of flow between them:

$$K_2 = \frac{\ln \left[\dfrac{C_{\mathrm{Kr}_a}}{C_{\mathrm{tr}_a}} \Big/ \dfrac{C_{\mathrm{Kr}_b}}{C_{\mathrm{tr}_b}} \right]}{0.83t} \tag{21-19}$$

K_3 and L_a are measures of independent physical properties which are often considered to be negligible. Values of K_3 and L_a may be determined from field data as follows. If the BOD decreases along a reach and the BOD at some point, x, is less than predicted by $L(x) = L_0 e^{-K_1 x/U}$ then K_3 is positive and exceeds L_a. L_a is assumed to be zero, and K_3 is calculated from:

$$L(x) = L_0 e^{-(K_1 + K_3)x/U} \tag{21-20}$$

If the BOD increases with x or decreases less than predicted by $L(x) = L_0 e^{-K_1 x/U}$ then L_a is positive and exceeds K_3. K_3 is assumed to be zero and L_a is determined from:

$$L(x) = L_0 e^{-K_1 x/U} + \frac{L_a}{K_1}(1 - e^{-K_1 x/U}) \tag{21-21}$$

S_R is a measure of oxygen sources and sinks other than reaeration and BOD exertion. It is often neglected, but may be determined from field measurements of dissolved oxygen once L, K_1, K_2, K_3, and L_a are known.

21-8 Application of Formulas

The equations developed above may be used to predict the effect of a particular waste load upon a receiving stream or to determine the waste load which can be discharged without reducing the dissolved oxygen below some prescribed level.

Example A community discharges a waste flow of 1000 m^3/day to a small stream. The 7-day, 10-year low flow in the stream is 5.74 m^3/min. The maximum stream temperature is 30°C and this coincides with low flow. The BOD_5 in the stream is 5 mg/l above the sewage outfall. The values of K_1 and K_2 at 20°C are 0.23/day and 0.46/day respectively. K_3, L_a, and S_R are neglected. Determine the effluent BOD_5 from the sewage treatment plant which will not deplete the dissolved oxygen concentration to less than 4 mg/l at the 7-day, 10-year low flow.

SOLUTION In the absence of values for K_3, L_a, and S_R Eqs. (21-15) and (21-16) may be reduced to the classic Streeter-Phelps[5] formulation:

$$D(x) = D_0 e^{-K_2 x/U} + \frac{K_1}{K_2 - K_1} L_0(e^{-K_1 x/U} - e^{-K_2 x/U}) \tag{21-22}$$

and,

$$x_c = \frac{U}{K_2 - K_1} \ln \frac{K_2}{K_1}\left[1 - \frac{K_2 - K_1}{K_1 L_0}(D_0)\right] \tag{21-23}$$

$t = x/U$ is often substituted in these equations to yield:

$$D(t) = D_0 e^{-K_2 t} + \frac{K_1}{K_2 - K_1} L_0(e^{-K_1 t} - e^{-K_2 t}) \tag{21-24}$$

and,

$$t_c = \frac{1}{K_2 - K_1} \ln \left\{\frac{K_2}{K_1}\left[1 - \frac{K_2 - K_1}{K_1 L_0}(D_0)\right]\right\} \tag{21-25}$$

The solution requires assumption of a value of L_0 and calculation of t_c and D_c with repeated trials until a value of D_c equal to that specified is attained.

At the maximum stream temperature the saturation DO is 7.63 mg/l (App. II), but with a BOD_5 of 5 mg/l in the stream it is unlikely to be saturated. Actual dissolved oxygen measurements could be made if necessary. For this example it is assumed that the stream is 80 percent saturated, i.e., has DO equal to $0.8 \times 7.63 = 6.10$ mg/l. The waste, following a minimum of secondary treatment, is assumed to contain 2.0 mg/l DO. The dissolved oxygen in the combined flow (waste plus stream) is calculated from:

$$\text{DO}(Q_s + Q_w) = 6.10Q_s + 2.0Q_w$$

$$\text{DO} = \frac{6.10(8265.6) + 2.0(1000)}{9265.6}$$

$$= 5.66 \text{ mg/l}$$

The initial deficit, D_0, is then

$$D_0 = C_s - C = 7.63 - 5.66 = 1.97 \text{ mg/l}$$

At $T = 30°\text{C}$ K_1 and K_2 are equal to:

$$K_{1(30)} = 0.23(1.047)^{10} = 0.364/\text{day}$$

$$K_{2(30)} = 0.46(1.025)^{10} = 0.589/\text{day}$$

Whence,

$$D_c = 1.97e^{-0.589t_c} + 1.618L_0(e^{-0.364t_c} - e^{-0.589t_c})$$

$$t_c = 4.444 \ln 1.618\left[1 - \frac{1.218}{L_0}\right]$$

The maximum allowable value of D_c if a minimum DO concentration of 4 mg/l is to be maintained is $7.63 - 4 = 3.63$ mg/l. Values of L_0 are assumed and t_c and D_c are calculated.

L_0	t_c	D_c
15	1.76	4.88
10	1.56	3.50
11	1.62	3.77
10.5	1.59	3.64

The maximum allowable ultimate BOD in the stream is thus approximately 10.5 mg/l. In terms of 5-day BOD this equals

$$BOD_5 = L_0(1 - e^{-K_1 t})$$

$$= 10.5[1 - e^{-0.23(5)}]$$

$$= 7.18 \text{ mg/l}$$

Note that this calculation is made using $K_{1(20)}$ not $K_{1(30)}$. The BOD of the waste is then calculated from:

$$7.18(Q_s + Q_w) = 5Q_s + \text{BOD}_5(Q_w)$$

$$\text{BOD}_5 = \frac{7.18(9265.6) - 5(8265.6)}{1000}$$

$$= 25.20 \text{ mg/l}$$

In this case the water quality standard (minimum DO = 4 mg/l) governs rather than the effluent standard of BOD_5 = 30 mg/l. The treatment plant for this community must be designed and operated to yield a BOD_5 of approximately 25 mg/l. This can be routinely achieved in some secondary-type systems.

21-9 The Multiple Correlation Technique

Churchill and Buckingham[6] noted that the exertion of BOD and depletion of dissolved oxygen in any reach of a stream were functions of flow, temperature, and the BOD at the point of beginning. Multiple measurements of these factors and the BOD and DO at the downstream location of interest can usually be statistically correlated to yield equations for the downstream BOD and DO as a function of the upstream variables. The equations can then be used to predict the effect of other discharges. The usefulness of the technique is enhanced if the measurements used in obtaining the correlation equations are made at near maximum temperature and near minimum flow. The applications of this and other stream modeling techniques are treated in detail by Nemerow.[7]

21-10 Ocean Discharges

The saturation concentration of dissolved oxygen in water decreases with increasing salt content (App. II). In seawater the saturation concentration is approximately 80 percent of that in freshwater.

Saltwater is denser than freshwater, hence sewage may tend to spread, without mixing, over the surface. The lack of dilution when mixing is limited by density differences, coupled with the lower dissolved oxygen available in saltwater, may lead to nuisance conditions which would not occur in a freshwater impoundment.

The prediction of the effects of waste discharges upon tidal waters such as estuaries[2] is extremely complex and is beyond the scope of this text. The student should consult references 1 and 2 for general information and other references on this topic.

21-11 Submarine Outfalls

Cities located along coastlines may elect to discharge their wastewater to the sea. The degree of treatment required is dictated by the NPDES permit and will never be less than the equivalent of secondary treatment for new construction.

Submarine outfalls are expensive. First costs are high and pumping is nearly always required. The prevention of nuisance conditions on beaches may require outfalls several kilometers in length.

Outfall pipe lines are constructed of reinforced concrete, cast iron, ductile iron, or steel. Cast iron is sometimes given a cement mortar lining. Steel is more likely to be lined with mortar or bituminous material and is sometimes provided with cathodic protection. Joints in the pipe should have substantial mechanical strength and be resistant to chemical or biological corrosion. Ball-and-socket joints (Fig. 5-2) have been used in iron pipe, while steel pipe is usually welded. Several ingenious joints have been employed in concrete outfalls (Fig. 14-5). The pipe may be placed in trenches on bottoms of soft rock, sand, or gravel. On unstable bottoms piling is necessary. Outfalls may employ single outlets or a variety of diffusers. Typical diffusers employ a number of ports on the sides or top distributed over a long length of the pipe—perhaps as much as a third of its total length. The ports may be plain or may be fitted with Tees to discharge the sewage in two directions.

21-12 Land Disposal and Treatment

Wastewaters may be discharged to the land either for ultimate disposal or for additional treatment prior to discharge to surface waters. In modern practice land disposal is not an alternative to treatment, although untreated domestic and industrial wastewaters have been disposed of on the soil without production of nuisance conditions.[8,9]

It is generally accepted that wastes should be given the equivalent of secondary treatment prior to land disposal. The reasons for this include the reduced stress upon the soil system, reduced likelihood of production of nuisance conditions, and the probable need to store the wastewater for extended periods of time when local conditions are unfavorable for disposal. In northern states land disposal may be physically impossible or prohibited by regulatory agencies during much of the winter, while in portions of the south and northwest the rainy season may saturate the soil without any supplemental addition of wastewater. Required storage periods vary with climate and soil characteristics and range from as little as one week's to as much as four months' flow.

21-13 Disposal Site Selection

The major technical consideration in land disposal is a long-term ability to move liquid. Three application techniques are currently in use: Spray Irrigation (SI), Rapid Infiltration (RI), and Overland Runoff (OR). The first two depend upon moving the water downward through the soil and thus are limited by infiltration and percolation capacity. Percolation capacity is a function of soil characteristics while infiltration depends upon the degree of clogging at the soil surface. If the waste is sufficiently pretreated, clogging will be minimized and percolation will limit the rate at which liquid can be applied.

Although it is not truly suitable, the conventional field percolation test (Art. 27-9) is used to estimate percolation capacity. For percolation rates of 6 to 25 mm/min ($\frac{1}{4}$ to 1 in/min) rapid infiltration is practicable, for 2 to 6 mm/min ($\frac{1}{12}$ to $\frac{1}{4}$ in/min) spray irrigation is suitable, and below 2 mm/min ($\frac{1}{12}$ in/min) overland runoff should be considered.

Soils are classified in a variety of ways. The *Department of Agriculture classification* is based upon the total profile. Soils which are members of a given series are identical save for the surface horizon and will thus have similar percolation characteristics. The *unified soil classification*, which is commonly used by engineers, is based upon particle size distribution, and can be related to drainage characteristics and application techniques as shown in Table 21-1.

Table 21-1 Soil types and drainage characteristics[10]

Symbol	Type	Drainage characteristics	Potential land disposal mode
GW	Well graded gravels or gravel-sands, little or no fines	Excellent	RI
GP	Poorly graded gravels, gravel-sands, little or no fines	Excellent	RI
GM-d	Silty gravels, gravel-sand-silt mixtures	Fair to poor	SI
GM-u	Silty gravels, gravel-sand-silt mixtures	Poor to practically impervious	OR
GC	Clayey gravels, gravel-sand-clay mixtures	Poor to practically impervious	OR
SW	Well graded sands or gravelly sands, little or no fines	Excellent	RI
SP	Poorly graded sands or gravelly sands, little or no fines	Excellent	RI
SM-d	Silty sands, sand-silt mix	Fair to poor	SI
SM-u	Silty sands, sand-silt mix	Poor to practically impervious	OR
SC	Clayey sands, sand-clay mixtures	Poor to practically impervious	OR
ML	Inorganic silts, very fine sands, silty or clayey sands	Fair to poor	SI
CL	Inorganic clays, gravelly clays, sandy clays, silty clays, lean clays	Practically impervious	OR
OL	Organic silts, organic silt clays of low plasticity	Poor	SI-OR
MH	Inorganic silts, micaceous or diatomaceous fine sandy or silty soils	Fair to poor	SI
CH	Inorganic clays of high plasticity, fat clays	Practically impervious	OR
OH	Organic clays of medium to high plasticity, organic silts	Practically impervious	OR
PT	Peat and other highly organic soils	Fair to poor	SI

Available soil depth depends upon natural conditions and engineering control of groundwater depth (Art. 21-14). Maximum possible depth is desirable for storage of phosphorus and heavy metals. For fine textured soils depths to groundwater of 1.5 m (5 ft) may be adequate although 1.5 to 3 m (5 to 10 ft) is usually recommended. This depth must be maintained in an unsaturated condition by proper design and operation of the system.

21-14 Disposal Site Preparation

Site preparation is generally not necessary save for control of surface runoff to and from the disposal area. Natural runoff may result in hydraulic overloads while runoff from the site to other property can create both legal and aesthetic problems.

Recovery of product water is required in overland flow systems but is not necessary in spray irrigation or rapid infiltration. It might be justifiable in some circumstances if the water were necessary to supplement surface supplies.

Groundwater level control may be necessary to prevent its contamination. Recovery of product water will lower the water table as will French drains or well points. Site preparation of this sort is very expensive and may make land disposal uneconomical.

21-15 Spray Irrigation

Wastewater has been applied to both forest and agricultural land by spray systems.

Forested sites utilize pipe distribution systems with fixed nozzles installed on risers about 1 m (3 ft) above the ground surface. *Application rates* are 25 to 50 mm/week (1 to 2 in/week) on such species as white and Norway spruce, white and red oak, European and Japanese larch, and white and Austrian pine.[10] Red pine, fir, and mixed oak forests do not respond well to wastewater irrigation, probably due to saturation of the root zone.

Agricultural sites may employ immense center pivot systems capable of irrigating a full quarter section (0.65 km^2). Such a device is shown in Fig. 21-2. Buried pipe networks and conventional aluminum irrigation pipe left in place above ground are also used. Design of agricultural irrigation systems is detailed elsewhere.[11] Crops which respond well to irrigation include nearly all grains and grasses. Corn and hay grasses remove significant quantities of nutrients from the waste and are potentially salable. Corn requires annual plowing and planting while grasses need only be harvested. Additionally, the grasses have a fully established root system at the start of the season and can provide immediate nutrient uptake.

Droplet losses may result from spray irrigation since a fine spray is desirable to minimize erosion and compaction of the soil surface, and maximize coverage of the crop. Fine sprays may be carried long distances by wind currents—as much as 7 m per km/h (35 ft per mph) increase in wind velocity.[10] Buffer zones of 60 to 70 m (200 to 225 ft) may be necessary at agricultural sites. Forested sites provide better interception of droplets, and buffer zones are smaller.

Figure 21-2 Center pivot irrigation system. *(Courtesy Valmont Industries, Inc.)*

The application rate normally ranges from 5 to 150 mm/week with 50 mm/week (2 in) being typical. The application is made at a rate of about 6 mm ($\frac{1}{4}$ in) per hour for 8 hours, followed by 160 hours of resting or 6 mm/h for 4 hours twice a week. For preliminary purposes, such values are adequate. Actual design values depend upon climate, soil characteristics, and plant species. In a demonstration project in Michigan[12] local considerations led to a design rate of 100 mm (4 in) per week including natural precipitation.

21-16 Rapid Infiltration

Rapid infiltration may be used either for waste disposal, groundwater recharge, or both (Art. 4-13). Little treatment is provided by rapid infiltration systems. Their major advantage is their ability to handle large quantities of water.

Loading rates in Arizona[13] range from 300 to 1200 mm (1 to 4 ft) per day. A typical cycle averages 600 mm (2 ft) per day for 2 weeks followed by 10 or 20 days drying in summer and winter respectively. Total annual application in rapid infiltration reaches 120 m (400 ft) while that in spray or runoff systems is closer to 2.5 m (8 ft).

Wastewater is discharged into large basins underlain by sand and soils of high permeability. The bottom of the basin may be covered with a grass like bermuda or reed canary which can persist in wet or dry conditions. The grass assists in nitrogen removal and helps maintain the infiltration capacity of the surface.

Before design of a rapid infiltration system the soil profile must be investigated to considerable depth [30 m (100 ft) or more] and the groundwater condition and flow pattern must be established. Preliminary design rates of 150 mm (6 in) per day for 2 weeks followed by 2-weeks rest are appropriate. Actual design rates depend upon climate and soil conditions.

21-17 Overland Runoff

Overland runoff is not a true disposal system since the wastewater must be collected after passage over the soil. The technique is applied when soils have poor permeability and constitutes a tertiary treatment process in the sense that its purpose is a reduction in waste strength beyond that achievable in secondary systems. Plant or tree cover is essential to minimize erosion and assist in nutrient removal. Hay grasses are usually employed.

Site preparation in overland runoff is more extensive than in spray irrigation since surface grading is required to eliminate gullies and low spots and to construct terraces which will intercept the flow. Liquid is usually applied through spray nozzles which distribute it over the ground at the top of a sloped surface. The slope depends upon the application rate, since retention time is important for complete treatment and the velocity must be sufficiently low to prevent erosion. On the other hand, too gentle a slope can lead to ponding and development of anaerobic conditions. Research at Paris, Texas led to recommended slopes of 2 to 6 percent.[14]

Application rates have ranged up to 175 mm (7 in) per week,[10] but practical systems operate at closer to 50 mm (2 in) per week on an intermittent basis similar to that employed in spray irrigation. The length of the flow path should be at least 90 m (300 ft)[10] although shorter distances have been considered adequate in some cases.[14]

21-18 Soil Response to Wastewater Disposal

The major activity, both biological and chemical, occurs in the upper 300 mm (12 in) of the soil mantle. Adsorption of phosphates and heavy metals may occur at deeper levels as the capacity of the upper layer is exhausted.

Biodegradable compounds are oxidized in the upper few millimeters of the soil. Between 450,000 and 1 million kg of organic matter per square kilometer per year are required for general soil equilibrium. In disposing of secondary effluent, these rates are not approached. A waste with a BOD_5 of 30 mg/l, if applied at a rate of 50 mm per week year round, would add only 78,000 kg/km^2 per year.

Other organic compounds such as pesticides, cellulose, polysaccharides, and humic materials may be present in wastewater. Pesticides are adsorbed by the soil and may be slowly degraded. At the low concentrations present in domestic wastewater no difficulties have been encountered. The other organic compounds mentioned are natural plant degradation products and their addition is beneficial to the soil structure. They are, of course, subject to slow biological degradation.

Nitrogen may be added as ammonia, nitrate, or a combination of both depending upon the degree of pretreatment. *Nitrate* will enter the groundwater if it is not removed by plant uptake. *Ammonia* may be adsorbed on soil or fixed in clays. It is removed by plants and is oxidized to nitrate by soil bacteria. The oxidation is slow and the delay in passage through the soil increases the likelihood of plant utilization. A limited degree of nitrification-denitrification (Chap. 26) may also occur in the soil.

Phosphorus is utilized by plants as a nutrient and is fixed by adsorption and exchange reactions with aluminum and iron containing compounds in the soil. The adsorptive capacity of fine-textured soils is high (up to 2,250,000 kg/km^2), and offers a useful site life of as much as 100 years under favorable conditions. Overland runoff systems do not provide efficient phosphorus removal since little of the soil is involved.

Inorganic compounds which may be harmful include heavy metals and monovalent cations. *Heavy metals* are removed by adsorption on soil particles. While no clear limit to capacity is evident, the adsorptive capacity is thought to exceed the tolerance level of plants, hence the first effect would be upon the crop, not upon the groundwater. It should be noted, however, that the agricultural use of land may be destroyed by heavy metal poisoning. Under acid conditions heavy metals may be leached from the soil.

Monovalent ions tend to exchange for divalent ions in the soil matrix. In *Montmorillonite clays* such exchange leads to swelling and loss of permeability. Additionally, increased salinity of water decreases the ease of water utilization by plants. Water which has an ionic composition suitable for general agricultural use will not cause swelling of clay materials.

21-19 Evaporation (Total Retention) Systems

In some climates it is possible to dispose of wastewater by evaporation. Most so-called total retention systems, however, discharge a portion of the flow to the soil.

The most common evaporation system is an oxidation pond with no outlet. The design criteria and biological processes of oxidation ponds are discussed in Chap. 24; in this article it is sufficient to know that the liquid in the pond is more or less equivalent to that provided by secondary treatment.

Average annual precipitation and average annual pan evaporation for different areas of the United States are presented in Figs. 3-3 and 3-8. Broadly speaking, if the annual precipitation exceeds 70 percent of the pan evaporation (Art. 3-17) an evaporation system is not feasible. The area of the United States most suitable for such systems covers the Great Plains from the 100th meridian to the Rocky Mountains.

In some states, so-called total retention system design criteria permit percolation of 3 mm ($\frac{1}{8}$ in) of water per day through the bottom of the impoundment. This extends the practical range of the system to the upper Missouri River valley.

Total retention systems are designed on the basis of a mass balance. Since the total inflow must equal the water lost, in the long term,

$$A(Q_i) + Q_w = (Q_e + Q_p)A \tag{21-26}$$

where Q_i = the annual precipitation
Q_w = the annual waste flow
Q_e = the annual evaporation (0.7 × pan evaporation), and
Q_p = the allowable annual percolation

Example Determine the area required for a total retention system for a location in eastern Nebraska for which the annual precipitation is 600 mm/year and the average pan evaporation is 1200 mm/year. The waste flow is 190,000 l/day and the pond may be constructed to permit 3 mm of loss per day by percolation.

SOLUTION Converting all values to commensurate units, $Q_i = 0.6$ m/year, $Q_w = 69{,}350\ \text{m}^3/\text{year}$, $Q_e = 0.7(1.2) = 0.84$ m/year, $Q_p = 1.10$ m/year.

$$A(0.6) + 69350 = (0.84 + 1.10)A$$

$$A = 51{,}753\ \text{m}^2$$

An equivalent pond designed for treatment alone (see Chap. 24) would have an area of approximately 20,000 m^2.

In an actual case the designer would need to consider extreme as well as average values. Oxidation ponds are constructed so that their area varies significantly with depth, hence some variability in flow, rainfall, and evaporation can be accommodated.

21-20 Selection of a Disposal System

There is no single system which is best suited to the disposal of all wastewaters. The engineer must investigate each system which is physically practicable to determine the cheapest technique which is environmentally and socially acceptable.

Stream disposal is the commonest technique in the United States and is generally cheapest, provided water quality standards do not require advanced treatment of the waste.

Land disposal is often socially and politically desirable, and may be economical in water-poor areas where suitable land is available and stream standards are restrictive. Viewed simply as a disposal technique, it is generally quite expensive compared to discharge to surface waters.

Evaporation is practicable only in limited areas, and in those areas the water might be more profitably used to recharge groundwater or irrigate crops. The designer in such a case must weigh the cost of the more complex system (land disposal) against the benefit gained.

PROBLEMS

21-1 In the example of Art. 21-8 assume that the waste flow is 2000 m^3/day. What effluent BOD_5 must be provided by the treatment plant?

21-2 A stream has an average depth of 2 m, a velocity of 0.5 m/s, a flow of 30 m^3/s, and a temperature of 30°C, all at low flow. Determine the value of the reaeration coefficient.

21-3 Determine the land area required for a spray irrigation disposal system for a community with a population of 100,000 persons. The design criteria are as follows:

Average daily flow = 50,000 m^3/day
Application rate = 50 mm/week
Application period = March to November (9 months)

Assume that the irrigation area is square and is surrounded by a buffer zone 50 m wide. Storage basins are inside the site, have a useful depth of 3 m, and do not contribute to the irrigation area.

21-4 In the example of Art. 21-19, what pond area would be required if no percolation were permitted? Does this appear to be a practical system?

REFERENCES

1. Thomann, Robert V.: *Systems Analysis and Water Quality Management*, Environmental Research and Applications, Inc., New York, 1971.
2. Hinwood, J. B. and I. G. Wallis: "Classification of Models of Tidal Waters," *Journal Hydraulics Division, Proceedings American Society of Civil Engineers*, **101:H11:**1315, 1975.
3. O'Connor, D. and W. Dobbins: "The Mechanism of Reaeration in Natural Streams," *Journal Sanitary Engineering Division, Proceedings American Society of Civil Engineers*, **82:SA6:**655, 1956.
4. Tsivoglou, E. C.: "Symposium on Direct Tracer Measurement of the Reaeration Capacity of Streams and Estuaries," U.S. Environmental Protection Agency, Water Pollution Series, *Report No. 16050*, 1972.
5. Streeter, Harold W. and Earl Phelps: *U.S. Public Health Service Bulletin 146* 1925.
6. Churchill, M. A. and R. A. Buckingham: "Statistical Method for Analysis of Stream Purification Capacity," *Sewage and Industrial Wastes*, **28:**517, 1956.
7. Nemerow, Nelson Leonard: *Scientific Stream Pollution Analysis*, McGraw-Hill, New York, 1974.
8. Bendixen, T. W., et al.: "Ridge and Furrow Liquid Waste Disposal in a Northern Latitude," *Journal Sanitary Engineering Division, Proceedings American Society of Civil Engineers*, **94:SA-1,** 1968.
9. Schraufnagel, F. H.: "Ridge and Furrow Irrigation for Industrial Waste Disposal," *Journal Water Pollution Control Federation*, **34:**11, 1962.
10. Reed, S., et al.: "Wastewater Management by Disposal on the Land," *Special Report 171*, U.S. Army Corps of Engineers Cold Regions Research and Engineering Laboratory, Hanover, N.H., 1972.

11. *National Engineering Handbook*, U.S. Department of Agriculture, Soil Conservation Service.
12. "The Muskegon County Wastewater Management System," Bauer Engineering, Inc., Chicago, 1971.
13. Bouwer, J. C., et al.: "Renovating Sewage Effluent by Groundwater Recharge," Water Conservation Laboratory, U.S. Department of Agriculture, 1971.
14. Gilde, L. C., et al.: "A Spray Irrigation System for Treatment of Cannery Wastes," *Journal Water Pollution Control Federation*, **43**:2011, 1971.

CHAPTER

TWENTY-TWO

PRELIMINARY TREATMENT SYSTEMS

22-1

Wastewater contains varying quantities of floating and suspended solids, some of considerable size. Materials such as rags, pieces of wood, metal, plastic, or rubber, or fragments of masonry enter sewers and eventually may reach the treatment plant. The headworks of wastewater plants usually incorporate a flow measurement device such as a Parshall flume[1] and mechanical or physico-chemical systems designed for removal of large floating solids, grit, and perhaps grease. These constituents are removed when their presence would interfere with subsequent treatment processes or mechanical equipment.

22-2 The Parshall Flume

Measurement of flows containing large solids and flowing under open channel conditions is most readily done using a horizontally constricted vertical throat such as that employed in the Parshall flume (Fig. 22-1). A flume of this sort has a low headloss and a predictable discharge-head relationship:

$$Q = 2.23 \times 10^{-2} W \left(\frac{H_a}{304.8} \right)^{1.522(W/304.8)^{0.026}} \tag{22-1}$$

where Q = the flow in m^3/min and
W = the width of the throat and
H_a = the upstream depth in millimeters.

Equation (22-1) applies to flumes 300 mm to 3 m in width. Typical dimensions for Parshall flumes are given in Table 22-1. Smaller Parshall flumes are manu-

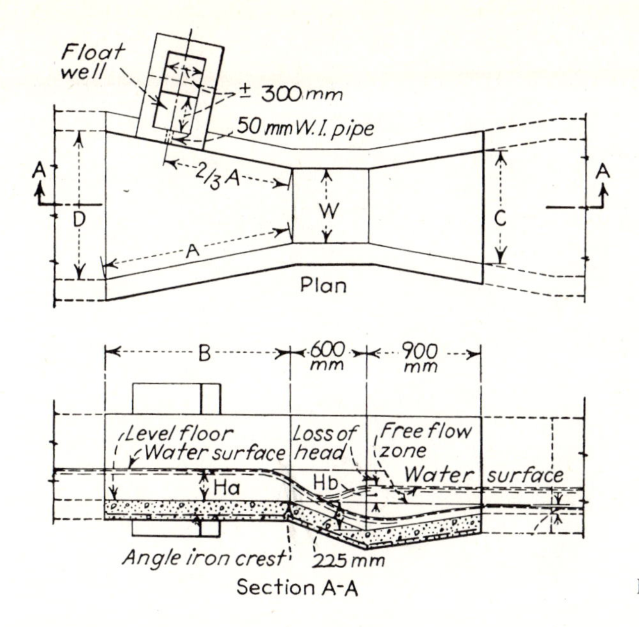

Figure 22-1 The Parshall flume.

factured, often of fiberglass (Fig. 22-2), and are supplied with rating curves. The depth of flow can be measured by a float in a stilling well or by an electronic pressure sensor.

If a flume is to measure the flow accurately there must be no disturbance of the normal flow pattern immediately up- or downstream. A straight uniform channel should be maintained for a reasonable distance in both directions. Approach zones equal to at least 10 times the throat width are desirable. Downstream changes in direction must be sufficiently distant that their backwater curves do not reach the flume.

Table 22-1 Standard dimensions and capacities of Parshall flumes. (See Fig. 22-1)

Dimensions, mm					Capacity, m^3/min	
W	A	B	C	D	Max	Min
300	1370	1345	600	845	27.4	0.6
600	1525	1495	900	1205	56.3	1.1
900	1675	1645	1200	1570	85.7	1.7
1200	1830	1795	1500	1935	115.4	2.1
1500	1980	1945	1800	2300	145.5	3.8
1800	2130	2090	2100	2665	176.0	4.5
2100	2290	2240	2400	3030	206.4	6.9
2400	2440	2390	2700	2400	237.2	7.9

(mm × 0.04 = in, m^3/min × 264.2 = gpm.)

Figure 22-2 Manufactured Parshall flume. *(Courtesy Free Flow, Inc.)*

22-3 Racks and Screens

The general principles of screening of water and wastewater are presented in Art. 9-3. *Mechanically cleaned racks* may be used in large plants (Fig. 9-1) and the rack may incorporate a grinding mechanism which reduces the solids to a size which will pass the barrier (Fig. 22-3).

Small plants often utilize *hand cleaned racks* like that illustrated in Fig. 22-4. These are installed as a supplement to a comminutor in a parallel channel, and serve to protect the downstream units when the comminutor is out of service for maintenance or repair. Large plants have multiple units and do not require standby channels.

Fine screens are seldom used in municipal wastewater treatment. Their major use is in treatment of industrial wastes which contain solids more uniform and predictable in character than those in domestic sewage.

The use of fine screens as a substitute for primary or secondary clarification, for removal of algae from oxidation ponds, and for tertiary treatment of wastewater are applications suggested by the manufacturers. The suitability of screens for these purposes is not established and the engineer should be cautious in using them. Pilot plant studies should be made before heavy capital investments in unproven equipment are considered.

The *quantity of material* removed on screens depends upon the size of the opening and the characteristics of the flow. Reported quantities removed on rela-

Figure 22-3 Barminutor. *(Courtesy FMC Corp.)*

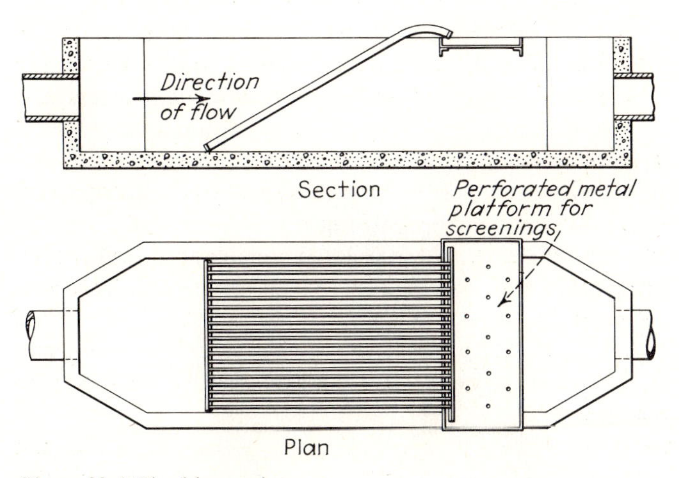

Figure 22-4 Fixed bar rack.

tively coarse screens range from 1.3×10^{-6} to 3.7×10^{-5} m^3/m^3 of flow, averaging about 1.5×10^{-5} m^3/m^3.[2]

Screenings are putrescible, are often impregnated with fecal matter, and must be disposed of promptly. The moisture content of screenings is about 80 percent and their bulk density approximately that of water. Disposal techniques include burial, incineration, and digestion. It is often simpler to pass the flow through a comminutor than to handle screenings separately.

22-4 Comminutors

Rather than removing large suspended solids, comminutors reduce them in size so that they will not harm subsequent equipment. The chopped or ground solids are then removed in sedimentation processes.

Comminutors consist of a fixed screen and a moving cutter. The rack of Fig. 22-3 is a comminuting device, but a more typical system employs a curved screen and a rotating or oscillating cutting blade (Fig. 22-5). Some rotating comminutor designs have a tendency to draw flexible materials through the screen rather than chop them. This can produce nuisance conditions in unskimmed clarifiers, trickling filters, and aeration basins.

Selection of comminution equipment is based upon flow rate. In small plants a single unit rated for peak flow may be used in parallel with a manually cleaned bar rack. In larger facilities multiple identical units are used, with the units sized so that the remaining machines will handle the peak flow with one or two out of service. Headloss across comminutors depends upon screen details and flow. Normal values are on the order of 50 to 100 mm (2 to 4 in). Values for particular devices are available from the manufacturer. Grit removal equipment should be installed ahead of comminutors when possible in order to protect them from unnecessary wear.

22-5 Grit Removal

A portion of the suspended solids load of municipal wastewater consists of inert inorganic material such as sand, metal fragments, eggshells, etc. This grit is not benefited by secondary treatment or sludge processing techniques and can promote excessive wear of mechanical equipment.

Grit removal devices rely upon the difference in specific gravity between organic and inorganic solids to effect their separation. In *standard gravity separators* the principles developed in Chap. 9 are applied. All particles are assumed to settle in accord with Newton's Law:

$$v_s = \left[\frac{4\mathbf{g}(\rho_s - \rho)\, d}{3C_D\rho}\right]^{1/2} \tag{9-1}$$

and to be scoured at a velocity:

$$V_h = \left[\frac{8\beta(s - 1)\mathbf{g}\, d}{f}\right]^{1/2} \tag{9-13}$$

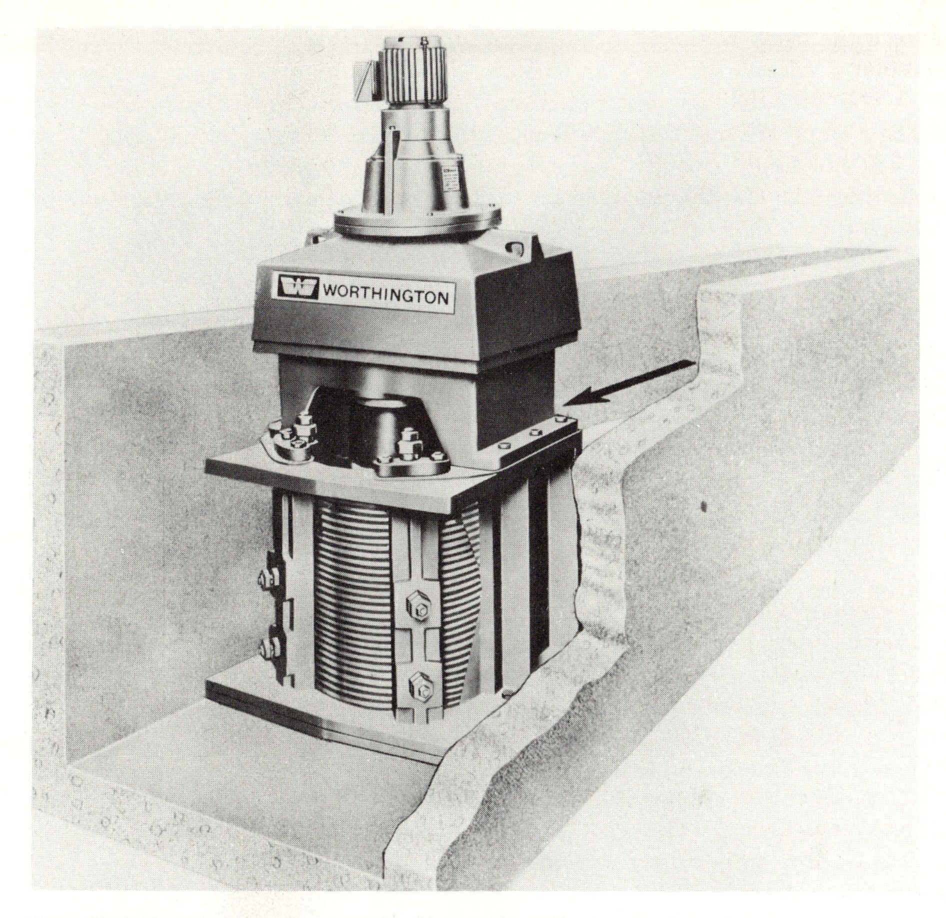

Figure 22-5 Comminutor. *(Courtesy Worthington Pump Corporation.)*

To assure removal of grit, while permitting such organic matter as might settle to be resuspended by scour, the necessary condition may be calculated as follows. For particles of grit with a diameter of 0.2 mm and specific gravity = 2.65,

$$v_s = \left[\frac{4(980)(2.65 - 1)(0.02)}{3 \times 10}\right]^{1/2} = 2.1 \text{ cm/s}$$

The scour velocity of these particles is given by:

$$V_h = \left[\frac{8(0.06)(2.65 - 1)(980)(0.02)}{0.03}\right]^{1/2} = 23 \text{ cm/s}$$

The scour velocity of organic particles with a specific gravity of 1.10 and the same diameter is:

$$V_h = \left[\frac{8(0.06)(1.10 - 1)(980)(0.02)}{0.03}\right]^{1/2} = 5.6 \text{ cm/s}$$

Thus if the basin is designed to have a surface overflow rate of 0.021 m/s and a horizontal velocity greater than 0.056 m/s and less than 0.23 m/s it will remove grit without removing organic material. To assure that the grit is reasonably clean, the horizontal velocity is generally close to the scour velocity of the grit.

The horizontal velocity in grit chambers is usually governed by either of two control sections. If the velocity in the basin is to be constant its cross section must be such that

$$V = C = \frac{Q}{\int_0^y x\, dy} \tag{22-2}$$

In the control section, in general,

$$Q = K'y^n \tag{22-3}$$

Equating (22-2) and (22-3),

$$K'y^n = C \int_0^y x\, dy \tag{22-4}$$

Which, when differentiated, yields

$$y^{n-1} = Kx \tag{22-5}$$

Thus the condition of constant velocity is maintained in the basin if its width varies so that $y^{n-1} = Kx$, where n is the discharge coefficient of the control section.

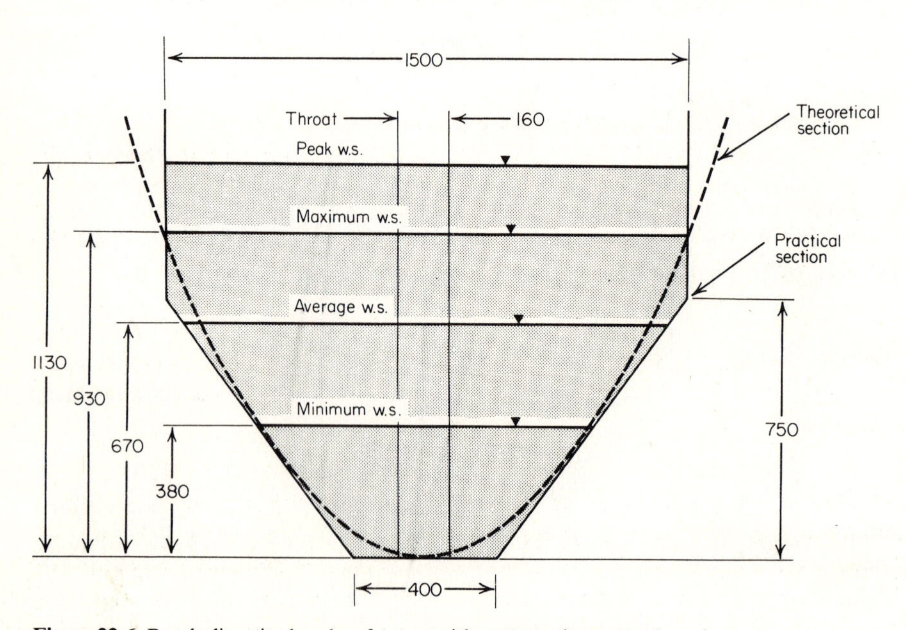

Figure 22-6 Parabolic grit chamber for use with rectangular control section.

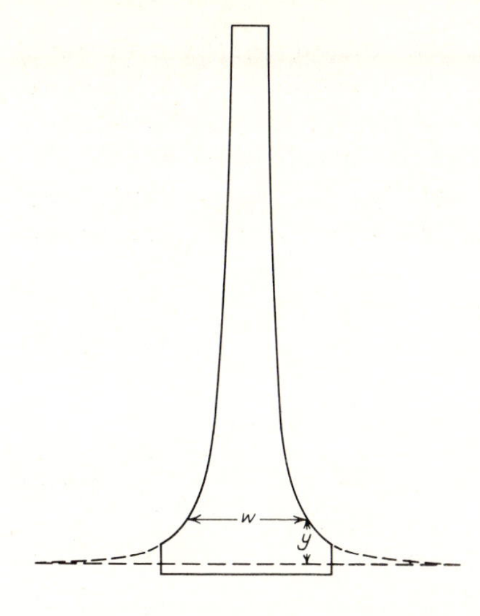

Figure 22-7 Proportional flow weir for use with rectangular grit chamber.

For a rectangular control section such as a Parshall flume, n is approximately 1.5, see Eq. (22-1), thus the channel cross section must be parabolic with

$$y = Cx^2 \tag{22-6}$$

as shown in Fig. 22-6.

If a rectangular channel is used, then $x = C = y^{n-1}/K$, $y^{n-1} = C'$, $n - 1 = 0$, and $n = 1$. The control section shown in Fig. 22-7 is variously called a proportional flow, Rettger, or Sutro Weir and has the discharge equation

$$Q = 8.18 \times 10^{-6} Ky \tag{22-7}$$

where Q = the flow in m^3/min
y = the head on the weir in millimeters, and
K = a weir constant equal at any point to $w\sqrt{y}$ where w is the weir opening in millimeters.

Proportional flow weirs require a free discharge with a fully ventilated nape and thus have a relatively high headloss.

Example Design a grit chamber for a horizontal velocity of 250 mm/s and a flow which ranges from a minimum of 20,000 m^3/day to a maximum of 80,000 m^3/day. Average flow is 50,000 m^3/day.

SOLUTION The grit removal system will consist of four mechanically cleaned channels, each designed to carry a peak flow of 26,666 m^3/day, a normal maximum of 20,000 m^3/day, an average of 12,500 m^3/day, and a minimum of 5000 m^3/day. A vertically controlled flume will be used to maintain the velocity. The flow in the control section is assumed to be at critical depth.

1. For a parabolic channel, $A = \frac{2}{3}\, WD$.

$$\text{at } Q_{peak},\ A = Q/V = 1.23\ \text{m}^2$$

$$Q_{max},\ A = 0.93\ \text{m}^2$$

$$Q_{avg},\ A = 0.58\ \text{m}^2$$

$$Q_{min},\ A = 0.23\ \text{m}^2$$

limiting the maximum width of the channel to 1.5 m at Q_{max},

$$D_{max} = 3/2(0.93/1.5) = 0.93\ \text{m}$$

The total head upstream of the control section is

$$h_1 = V^2/2\mathbf{g} + D_{max} = 0.93\ \text{m}$$

In the control section, at critical depth,

$$d_c = V_c^2/\mathbf{g}$$

Assuming the headloss in the control section is 10 percent of the velocity head, Bernoulli's equation may be written as

$$D = V_c^2/\mathbf{g} + V_c^2/2\mathbf{g} + 0.1V_c^2/2\mathbf{g}$$

$$D = 3.1V_c^2/2\mathbf{g}$$

At maximum flow, the velocity in the control section is thus

$$V_c = \sqrt{\frac{0.93(2)(9.8)}{3.1}}$$

$$= 2.42\ \text{m/s}$$

and

$$d_c = V_c^2/\mathbf{g}$$

$$= 0.60\ \text{m}$$

The area of the control section is $wd_c = Q/V_c$ and its width is,

$$w = 0.16\ \text{m}$$

2. For the other flow conditions,

$$d_c = \sqrt[3]{\frac{Q^2}{w^2\mathbf{g}}}$$

and from Bernoulli's equation,

$$D = (3.1/2)d_c$$

For the parabolic section,

$$W = (3/2)(A/D)$$

The resulting values are tabulated below and shown in Fig. 22-6.

Q	d_c	w	D	W
5,000	0.24	0.16	0.38	0.91
12,500	0.43	0.16	0.67	1.30
20,000	0.60	0.16	0.93	1.50
26,666	0.73	0.16	1.13	1.63[a]

[a] Theoretical value. The actual constructed width is as shown in Fig. 22-6.

The *length of grit chambers* depends upon the trajectory of the slowest settling particle and the depth of flow. For a settling rate of 21 mm/s and a horizontal velocity of 250 mm/s

$$V_h/v_s = L/D \qquad \text{[see Eq. (9-5)]}$$

$$L = D(V_h/v_s) = (250/21)D \approx 12D$$

For the maximum depth of 1.13 m in the example above, the length of the channel is about 13.5 m. This will provide an excess length of 2.3 m at maximum flow with all channels in use which will permit some turbulence at the entrance.

Aerated grit chambers control the separation of inorganic and organic solids by producing a rolling flow pattern similar to that in spiral flow aeration basins (Chap. 24). The velocity of the bulk solution is kept above the settling velocity of the organic particles, but below that of the inorganics. In the rolling flow pattern, all solids are carried to the bottom of the basin with the flow resuspending the organics while the inorganics remain behind. Air flow is regulated to give the separation desired. Detention times are usually less than 3 min with air flows on the order of 0.3 to 0.5 m^3/min per meter of tank length (3 to 5 ft^3/min per ft). The depth is 70 to 100 percent of the width and ranges from 3 to 5 m (10 to 16 ft). Headloss through aerated grit chambers is negligible.

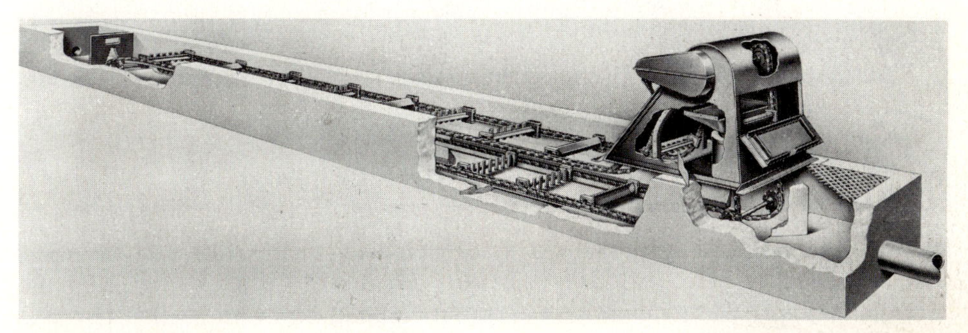

Figure 22-8 Rectangular gravity grit chamber with continuous chain removal. *(Courtesy Envirex, a Rexnord Company.)*

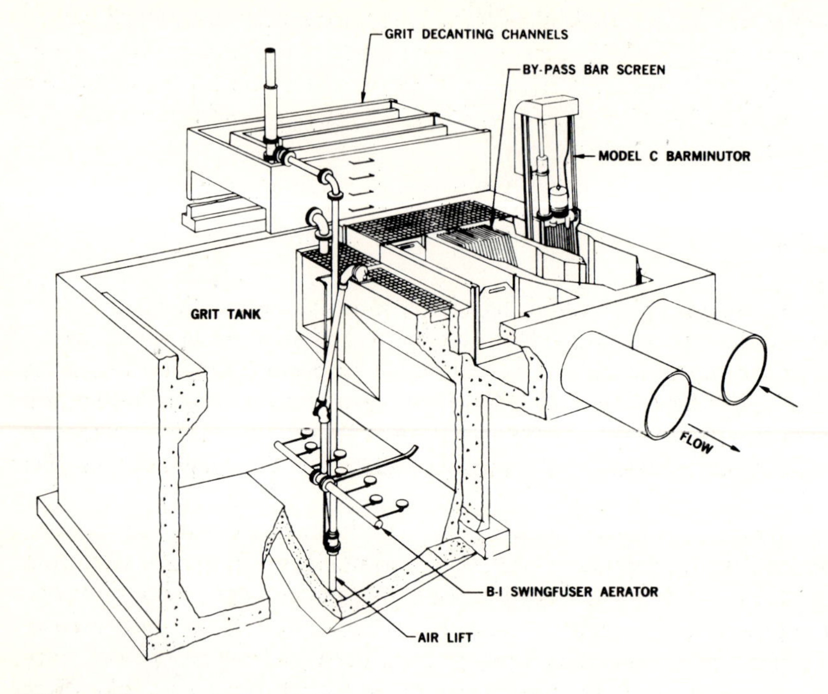

Figure 22-9 Aerated grit chamber. *(Courtesy FMC Corporation.)*

Grit from properly designed and operated gravity or aerated chambers is low in organic content and relatively innocuous. If equipment malfunctions occur it may contain up to 50 percent organic matter and be putrescible. Clean grit may be used for fill material. If contaminated, it is disposed of in sanitary landfills or by burial at the plant site.

The *quantity of grit* varies with the condition of the sewer system, the quantity of stormwater, and the proportion of industrial waste. Recorded quantities[2] range from as little as 2.5×10^{-6} to 1.8×10^{-4} m^3/m^3 of wastewater. An average value is about 6×10^{-5} m^3/m^3 but considerable variation should be anticipated in designing grit-handling facilities.[3]

Automatic equipment is usually employed to remove grit from the basins. Typical installations are illustrated in Figs. 22-8 and 22-9.

22-6 Grease Removal

Excessive quantities of grease may tend to plug trickling filters or coat biological floc in activated sludge processes. Grease is removed to a degree by skimming devices in primary sedimentation tanks, but communities with particularly high concentrations of grease or treatment plants which omit primary clarification may need other processes.

Skimming tanks employ baffled subsurface entrance structures which permit floating matter to be retained. Retention times are 15 min or less and continuous mechanical skimming is usually employed. The horizontal velocity should be kept in the range of 50 to 250 mm/s (0.2 to 0.8 ft/s) in order to prevent deposition of organic particles.

22-7 Flotation

Gases which are present in a liquid in concentrations in excess of their saturation value will come out of solution, and will come out of solution at a surface. In wastewaters this results in bubble formation about suspended particles with a net reduction in average density and an increase in particle size. Considering Stokes' Law, Eq. (9-4), it is evident that if the average density becomes less than that of water, the composite particle will rise and will rise more rapidly as its size increases.

Flotation is useful in grease removal after emulsions are chemically broken, and may be used in thickening waste biological sludges (Chap. 25). Flotation processes include *aeration alone*, which is not very effective; *injection* of air to the pressurized liquid which is then depressurized, releasing the air; and *vacuum flotation* in which the atmospheric pressure is reduced to draw gases from solution. Design procedures for flotation systems are presented elsewhere.[3] Pilot plant studies are usually necessary before detailed design is possible.

22-8 Preaeration

Aeration of sewage before primary treatment may offer a number of advantages. Grease removal may be modestly improved, odors from septic sewage will be reduced, a degree of flocculation may occur, and solids will be kept in uniform suspension. As noted in Art. 22-5, aeration may also be utilized to remove grit.

Preaeration generally consists of expanding aerated grit chambers to provide a retention time of about 30 min. Aeration rates range from 0.01 to 0.05 m^3 of air per m^3 of waste. Basin cross-sectional dimensions are similar to those given in Art. 22-5 for aerated grit chambers.

PROBLEMS

22-1 Design a single grit chamber utilizing a proportional flow weir for the following flows: $Q_{max} = 4000\ m^3/day$, $Q_{avg} = 1500\ m^3/day$, $Q_{min} = 500\ m^3/day$. The horizontal velocity is to be 250 mm/s and the maximum depth of flow 0.5 m. Determine the depth at average and minimum flow and dimension the weir plate. How long should the channel be?

22-2 Rework the example of Art. 22-5 for three channels with a maximum width of 2 m.

REFERENCES

1. Parshall, R. L.: "The Parshall Measuring Flume," *Bulletin 423*, Colorado Experiment Station, Fort Collins, Colorado, 1936.
2. "Sewage Treatment Plant Design," *Manual of Practice* **36**, American Society of Civil Engineers, New York, 1959.
3. Metcalf and Eddy, Inc., *Wastewater Engineering: Collection, Treatment, Disposal*. McGraw-Hill, New York, 1972.

CHAPTER

TWENTY-THREE

PRIMARY TREATMENT SYSTEMS

23-1

Primary treatment has traditionally implied a sedimentation process intended to remove suspended organic solids. Chemicals are sometimes added in primary clarifiers to assist in removal of finely divided and colloidal solids, or to precipitate phosphorus.

23-2 Plain Sedimentation

The theory of sedimentation of discrete, flocculent, and hindered suspensions is developed in Chap. 9. Sedimentation processes in wastewater treatment include all three types. Grit removal (Art. 22-5) and simple sedimentation approximate discrete settling processes, secondary settling following activated sludge processes is usually flocculent, and waste sludge thickeners exhibit hindered settling—as may the lower levels of the other clarification processes.

Primary clarifiers are usually designed to remove particles with settling rates of 0.3 to 0.7 mm/s. Within that range, suspended solids removals range from 30 to 60 percent depending in part upon the original concentration. Figure 23-1 presents reported solids removals in existing plants as a function of their average surface overflow rate. BOD removal is associated with the removal of solids and is often considered to vary with overflow rate as shown in Fig. 23-2. Actual removals vary considerably with the waste source and age.

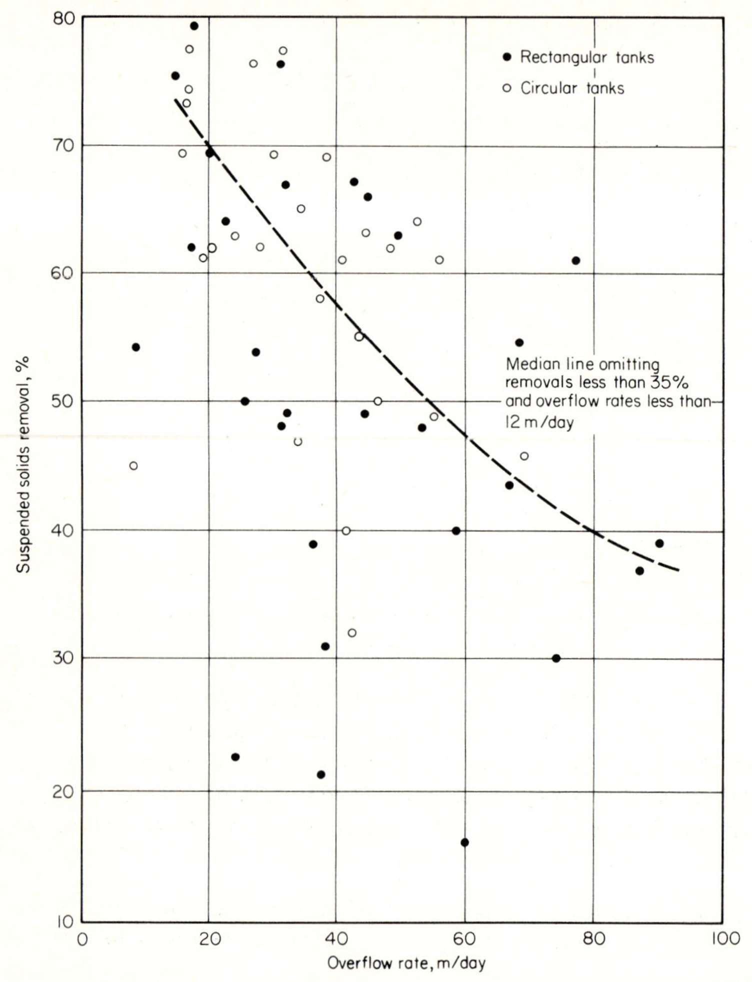

Figure 23-1 Suspended solids removal in primary clarifiers (*from "Sewage Treatment Plant Design"*[1] *copyright 1959 by the American Society of Civil Engineers and the Water Pollution Control Federation. Used with permission of the Water Pollution Control Federation*). (m/day × 24.5 = gal/ft^2 per day.)

Retention times in primary clarifiers are generally short, from 1 to 2 hours at peak flow. Combining this criterion with a surface overflow rate of 1 to 2.5 m^3/m^2 per hour (0.3 to 0.7 mm/s) yields a depth of 1 to 5 m (3 to 16 ft). Practical basins are seldom less than 2 m (6 ft) or more than 5 m (16 ft) in depth.

Multiple basins should be provided in all but the smallest plants. Primary clarifiers may be bypassed to secondary processes, but this is likely to overload the secondary, and is certain to produce a poorer overall effluent quality. Hydraulic overload of primary clarifiers designed in the usual range produces only modest decreases in their efficiency provided the channels and launders are capable of transporting the flow.

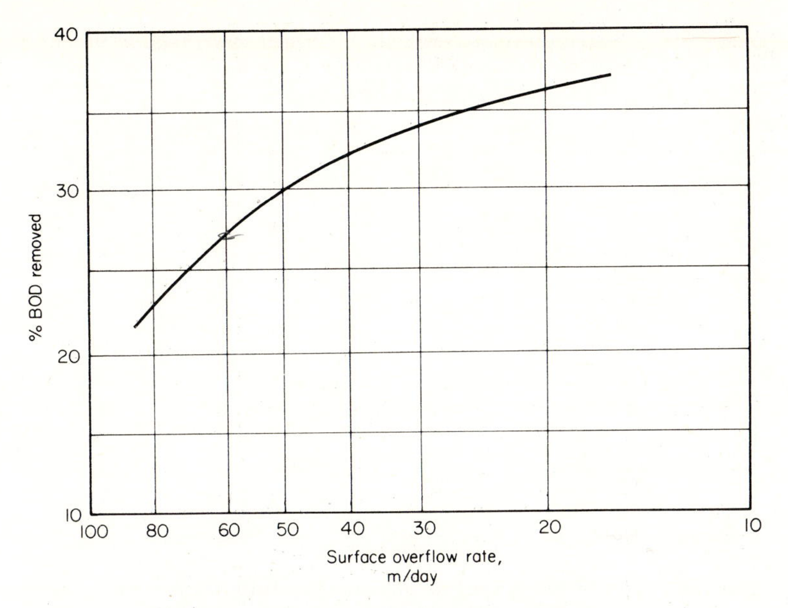

Figure 23-2 BOD removal in primary clarifiers *(modified from a figure in "Recommended Standards for Sewage Works," 1971 Edition[2])*. (m/day × 24.5 = gal/ft^2 per day.)

Example Design a primary clarification system for an average wastewater flow of 7500 m^3/day with a maximum of 18,000 m^3/day and a minimum of 4000 m^3/day. Use at least two basins and design for an estimated BOD removal of 35 percent at peak flow with all basins in operation and 30 percent at peak flow with one basin out of service. What is the efficiency at the other flow conditions?

SOLUTION From Fig. 23-2, for 35 percent removal the surface overflow rate is 26.5 m/day, and for 30 percent, 51 m/day. The total area required for all basins is thus:

$$A_n = \frac{18{,}000}{26.5} = 679 \text{ m}^2$$

and for all but one basin,

$$A_{n-1} = \frac{18{,}000}{51} = 353 \text{ m}^2$$

The limiting condition is $A_n = 679$ provided $n > 2$. Thus acceptable solutions would be:

3 basins, each 226 m^2 or 2 basins, each 353 m^2

The first alternative offers maximum flexibility, but actual selection would depend upon a cost analysis. It is certain in this case that it would be less expensive to build the two larger basins.

The two 353-m^2 basins would have overflow rates and estimated efficiencies (from Fig. 23-2) as shown below.

Flow	No. of basins	Area	SOR	Efficiency
18,000	2	706	25.5	35.1
18,000	1	353	51.0	30.0
7,500	2	706	10.6	> 37
7,500	1	353	21.3	36.0
4,000	2	706	5.7	> 37
4,000	1	353	11.3	> 37

The hydraulic retention time in a single basin at peak flow will establish the basin depth. Thus for a minimum retention time of one hour,

$$\text{Depth} = \frac{\text{Flow} \times \text{retention time}}{\text{Area}}$$

$$= 2.12 \text{ m}$$

The actual sidewall depth might be set at 2.5 m including freeboard.

23-3 Rectangular Tanks

Rectangular basins are thought by some engineers to offer a larger percentage of effective settling volume, but data such as that of Fig. 23-1 does not support the theory. The major advantages of rectangular designs are decreased likelihood of short-circuiting and more effective use of land area.

Inlets are designed to minimize the effect of density and velocity currents. Typical inlets consist of small pipe with upturned ells spaced along the basin end, perforated baffles, multiple ports discharging against baffles, single pipe turned back to discharge against the head wall, simple weirs, submerged weirs sloping upward to a horizontal baffle, etc. The inlet structure should not trap scum or settleable solids.

Baffles are installed 600 to 900 mm (2 to 3 ft) in front of inlets and are submerged 450 to 600 mm (18 to 24 in) with 50 mm (2 in) of water depth above the top of the baffle to permit floating solids to pass. *Scum baffles* are installed ahead of outlet weirs and extend 150 to 300 mm (6 to 12 in) below the water surface.[2] Rotating slotted scum pipes are sometimes used to incorporate a scum baffle and a scum removal channel. These devices are satisfactory only when coupled with positive mechanical skimming.

Outlets in rectangular tanks consist of weirs located toward the discharge end of the basin. *Weir loading rates* range from 120 to 370 m^3/day per meter of weir length (10,000 to 30,000 gal/ft per day). The upper value is used at peak flow conditions. For the flow and clarifier system of the example in Art. 23-2 the weir length in each clarifier would be 18,000/370 or 49 m. Multiple rectangular launders or serpentine weirs (Fig. 23-3) are used as necessary to develop the necessary length in narrow basins.

Figure 23-3 Serpentine weir pan. *(Courtesy Leopold Co., Division of Sybron Corporation.)*

Optimum tank proportions have not been established, but average length/width ratios are about 4 : 1 with a minimum depth of 2 m (6 ft). Average depth of rectangular tanks is approximately 3.5 m (12 ft). Figure 9-10 illustrates a typical installation with mechanical skimming and sludge removal. The floor is usually sloped gently toward the sludge hopper to facilitate draining the basin. For the example of Art. 23-2, appropriate dimensions might be 10 m wide by 36 m long by 2.5 m deep. The slight excess length provides for modest entrance and outlet zones even at peak flow. Development of weir length would require three launders and would cover 4 to 6 m of tank length at the outlet end.

Sludge and scum removal mechanisms may be chain driven (Fig. 9-10) or supported on bridges (Fig. 9-11) or floats. Velocity is usually less than 1 m/min (3 ft/min). Sludge is drawn to a sump from which it may flow by gravity or be pumped. Scum is skimmed to a trough and flows to a sump from which it is pumped. Vacuum sludge removal systems are not necessary for, nor are they well suited to, primary sludge handling.

23-4 Circular and Square Tanks

Circular and square basins employ similar scraping and skimming equipment save that square tanks require an additional corner sweep mechanism (Fig. 9-15).

Inlets in circular or square tanks are usually at the center and the flow is directed upward through a baffle which channels it toward the periphery

(Fig. 9-13). Inlet baffles have diameters 10 to 20 percent that of the basin and extend 1 to 2 m (3 to 6 ft) below the water surface.

Outlet weirs extend around the periphery of the tank with baffles extending 200 to 300 mm (8 to 12 in) below the water surface to retain floating solids. Weir length may be increased by supporting the launder on brackets, thus effectively doubling the length by permitting flow over two edges. Serpentine weir pans (Fig. 23-3) are also useful and will increase the effective length to 2.5 times the original. The weir pans shown extend 1.2 m from the effluent trough to which they are connected. Radial weirs are not common in sewage clarifiers since they complicate the skimming problem.

Tank proportions are dictated by flow. Side depths range from 2 to 4 m (6 to 13 ft). Floors slope toward the center at slopes of 1/12 to 1/24 with lower slopes in larger tanks. For the clarifier of Art. 23-2, the diameter would be 22 m to provide for some exit and entrance turbulence, and the side water depth 2 m with a wall depth of 2.2 m. The depth in the center might be 3.2 m. Slopes may have to be selected to match the scraping mechanism which is installed.

Sludge and scum are collected by rotating radial collectors which turn at peripheral speeds of 1.5 to 2.5 m/min (5 to 8 ft/min). Scum is swept outward to a hopper, while sludge is windrowed to a central pit. Removal techniques are similar to those in rectangular units.

23-5 Solids Quantity

The amount of *sludge produced* in a clarifier is a direct function of its efficiency. Figure 23-1 permits estimation of suspended solids removal as a function of surface overflow rate. In general, removals of 50 to 60 percent can be expected. Thus a wastewater containing 250 mg/l SS with a flow of 18,000 m^3/day might be expected to produce 2250 kg of solids per day. The moisture content of primary solids ranges from 93 to 98 percent, with an average value of 95 percent. The total mass of wet solids would therefore be 45,000 kg/day or about 2.5 kg/m^3 of flow (20,000 lb/million gallons).

If excess activated sludge is passed through the primary clarifier for wasting (Chap. 24) the mass of solids will be increased by the amount wasted and the moisture content will ordinarily be increased. Mixed waste activated sludge and primary solids seldom thicken to less than 96 percent moisture content in gravity separation systems.

Scum removed averages about 7.5×10^{-6} m^3/m^3 of flow. It is desirable that the scum handling facilities be capable of handling peak loads of up to 4.5×10^{-5} m^3/m^3 for brief periods.[3]

23-6 Chemical Addition

Addition of the common metallic coagulants or polymers (Chap. 9) will increase the removal of suspended solids in clarifiers. If the dosage of metallic coagulants is sufficiently large, significant quantities of phosphate may also be precipitated (Chap. 26).

The justification for chemical addition is based upon special conditions. Historically, it has been used when seasonal loads required special treatment to avoid plant expansion, when room for expansion was lacking, or when a degree of treatment intermediate between primary and secondary was desired. The latter condition is no longer likely in the United States.

The principles of chemical coagulation are developed in Chap. 9. When applied to wastewater in modern practice the major goal is usually phosphate precipitation. Suspended solids removals in such processes range from 70 to 90 percent.[3] Similar solids removals were obtained in the past with somewhat lower chemical dosages (100 mg/l or less).[1] BOD removals can be as high as 70 percent on fresh sewage, but decrease with age and degree of staleness.

Chemical treatment processes employ the same basic criteria in both water and wastewater plants. Chemicals must be flash mixed, flocculation is required, and surface overflow rates are similar to those in plain sedimentation.

Chemical sludges naturally contain more solids due to the precipitation of metallic hydroxides and increased efficiency of the process. The sludge volume, however, may not be increased, and may even be less due to the inclusion of less water in the mass. Moisture content can be as low as 90 percent.

23-7 Other Primary Processes

A number of manufacturers provide proprietary equipment which incorporates a flocculation or gentle mixing zone prior to sedimentation. The mixing may be effected mechanically or by aeration, and settled solids may be returned to the mixing zone to increase contact opportunity. Suspended solids and BOD removals are reportedly increased by about 10 percent over that obtainable in plain sedimentation. The systems are not widely used.

Imhoff tanks incorporate solids separation and digestion (Chap. 25) in a single unit. Modern versions of the process employ mixing and heating in the digestion zone, but the advantage of these units over separate processes is questionable. Imhoff tanks have been widely used in the past, particularly in small plants designed to provide only primary treatment. Old Imhoff tanks have been converted to other uses in recent years. Design criteria are presented in reference 1, but the use of this process is not recommended.

23-8 Secondary Clarifiers

Secondary clarifier design is dependent upon the preceding process. General criteria are listed in Art. 9-17, but the designer should use the specific values recommended under the various biological processes in Chap. 24.

PROBLEMS

23-1 A raw wastewater containing 250 mg/l SS and 200 mg/l BOD passes through a clarifier with a surface area of 500 m^2. The flow ranges from 10,000 m^3/day to 30,000 m^3/day. Estimate the effluent BOD and SS at maximum and minimum flow, and the maximum and minimum rate of sludge production. Assume the sludge is 95 percent water.

23-2 Design a primary clarifier system for a community with a population of 50,000 persons. Size the clarifiers so that the SOR will be not more than 40 m/day at peak flow with one unit out of service and not more than 25 m/day at average flow. The minimum retention time is to be 1 hour and the maximum weir overflow rate 370 m^3/day per meter.

23-3 For the system of Prob. 23-2, estimate the quantity of sludge which must be removed. If the sludge pumping cycle is 15 min/hour at maximum, what flow must the pump be sized to handle?

REFERENCES

1. "Sewage Treatment Plant Design," *Manual of Practice* **36,** American Society of Civil Engineers, New York, 1959.
2. *Recommended Standards for Sewage Works*, Great Lakes–Upper Mississippi Board of State Sanitary Engineers, Health Education Service, Albany, New York, 1971.
3. *Process Design Manual for Suspended Solids Removal*, U.S. Environmental Protection Agency, 1975.

CHAPTER

TWENTY-FOUR

SECONDARY TREATMENT SYSTEMS

24-1

Secondary treatment systems are intended to remove the soluble and colloidal organic matter which remains after primary treatment. While removal of this material can be effected by physico-chemical means, secondary treatment is usually understood to imply a *biological treatment process.*

It has been noted in Chaps. 20 and 21 that wastewater, in addition to containing organic matter, also carries a large number of microorganisms which are able to stabilize the waste in a natural purification process. Biological treatment consists of application of a controlled natural process in which microorganisms remove soluble and colloidal organic matter from the waste and are, in turn, removed themselves.

In order to carry out this natural process in a reasonable time it is necessary that a very large number of microorganisms be available in a relatively small container. Biological treatment systems are designed to maintain a large active mass of bacteria within the system confines. While the basic principles remain the same in all secondary processes, the techniques used in their application may vary widely, but may be broadly classified as either *attached (film) growth* or *suspended growth processes.*

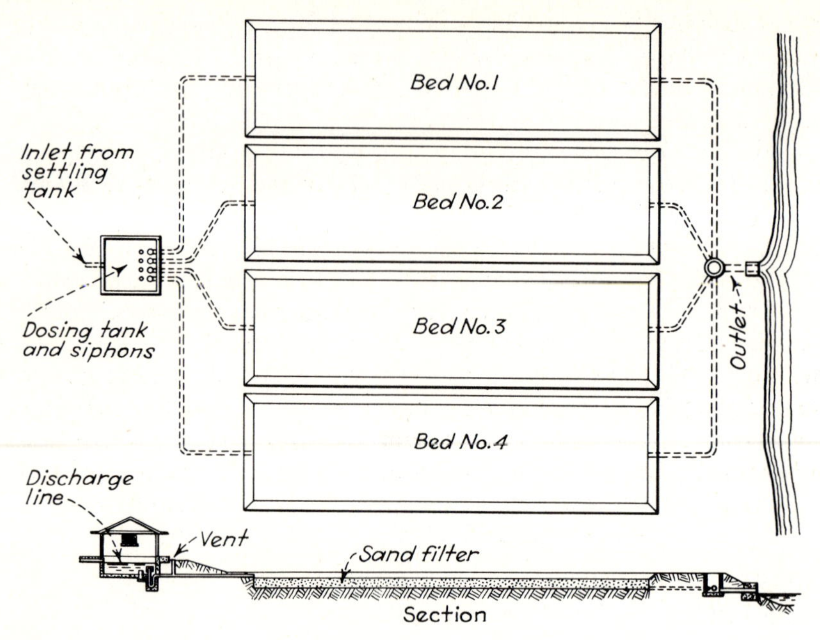

Figure 24-1 Layout of intermittent sand filter installation.

ATTACHED GROWTH PROCESSES

24-2

Attached growth processes utilize a solid medium upon which bacterial solids are accumulated in order to maintain a high population. The area available for such growth is an important design parameter, and a number of processes have been developed which attempt to maximize area as well as other rate limiting factors. Surface growth processes include intermittent sand filters, trickling filters, rotating biological contacters, and a variety of similar proprietary devices.

24-3 Intermittent Sand Filters

Intermittent sand filters are no longer used by major cities because of the large area required. They may, however, still have application in rural areas, particularly in upgrading oxidation pond effluent[1] (Art. 24-21). The advantages of intermittent sand filtration include low headloss, simple operation, a satisfactory effluent, and limited sludge production.

Operation consists of intermittent application of primary effluent or other wastewater to the sand surface. Solids are trapped in the sand while bacterial growth developed upon the surface of the grains adsorbs the soluble and colloidal organic matter. Between dosing cycles air penetrates the bed to permit biological oxidation of the bulk of the accumulated organics.

The sand used normally has an effective size of 0.2 to 0.5 mm and a uniformity coefficient of 2 to 5 although unsieved sand is sometimes used.[1] The depth of the bed ranges from 460 to 760 mm (18 to 30 in) with deeper beds yielding a somewhat better effluent. The sand is underlain by approximately 300 mm (12 in) of graded gravel ranging from 6 to 50 mm ($\frac{1}{4}$ to 2 in) in which perforated pipe or unjointed drain tile are placed in order to collect the treated waste. Typical filter bed details are shown in Figs. 24-1 and 24-2.

Hydraulic loading rates for settled sewage range from 70 to 235 mm/day (2 to 6 gal/ft^2 per day) with the flow being applied once or twice per day. BOD reduction may reach 95 percent in treating ordinary domestic sewage in ripened filters, but is lower when operation is first begun. Eventually the filter will become clogged with accumulated solids to the point that the hydraulic loading can no longer be maintained. When this occurs the upper 50 to 75 mm (2 to 3 in) of the sand is removed and replaced with clean material. The dirty sand may be disposed of by land spreading or filling. It cannot be cleaned economically. The duration of filter runs is typically several months, but depends upon temperature, mode of operation, sand gradation, and influent BOD and suspended solids.

24-4 Trickling Filters

Trickling filters utilize a relatively porous bacterial growth medium such as rock or formed plastic shapes. Bacterial growth occurs upon the surface while oxygen is provided by air diffusion through the void spaces. The wastewater is applied to the surface, usually in an intermittent fashion, and percolates through the filter, flowing over the biological growth in a thin film.

The process can be represented as shown in Fig. 24-3. Nutrients and oxygen are transferred to the fixed water layer, and waste products are transferred to the moving layer, primarily by diffusion. As the bacteria on the filter surface metabolize the waste they will reproduce, gradually producing an increase in the depth of

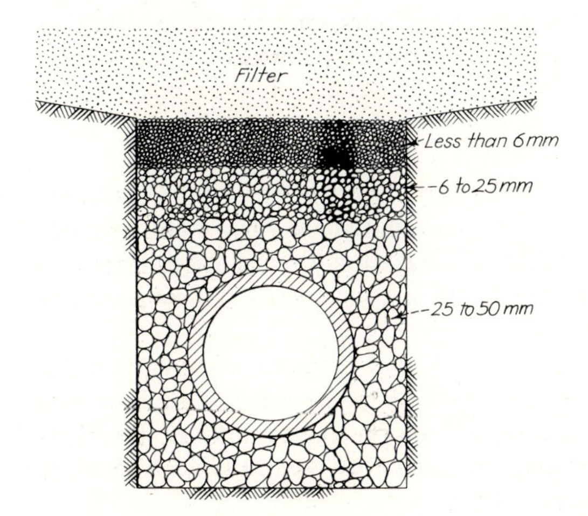

Figure 24-2 Underdrain of an intermittent sand filter. (mm × 0.04 = in.)

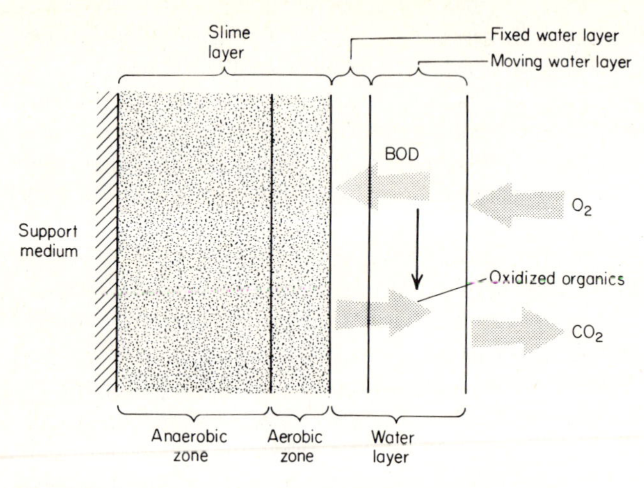

Figure 24-3 Schematic diagram of attached growth process.

the slime layer. With thickening of the biological layer, the bacteria in the interior layers find themselves in a nutrient limited situation since the organic matter and oxygen are utilized near the surface. Eventually these cells die and lyse, breaking the contact between the slime layer and the support medium. When sufficient cells have lysed the slime layer will slough off and be carried from the filter by the waste flow. The solids in the filter effluent are removed from the flow in a secondary clarifier.

Although trickling filters have been used for a great many years their operation is still not readily mathematically described. The process rate is affected by mass transfer of oxygen to the liquid from the air, and from the liquid to the bacterial slime; by transfer of biodegradable matter from the liquid to the slime; and by the rate of utilization of the degradable material by the bacteria.[2]

Mathematical models have been suggested which are empirical,[3,4] based upon a postulated reaction,[5,6] a simplified analogy,[7] or more complicated two-phase systems.[2,8] The theoretical models are generally impractical for design purposes although they may aid considerably in understanding the interplay of process variables.[9,10] In particular, it may be concluded that the Velz and NRC equations (see below), which are based upon an implicit assumption that oxygen is the rate limiting factor,[11] are not invalid upon that account provided the waste is not excessively dilute.[10,11] It has been concluded that oxygen is, in fact, the limiting variable for wastes with a soluble BOD greater than 40 mg/l,[10] which includes most wastes of interest.

24-5 Filter Classification

Trickling filters may be described by their hydraulic or organic loading rate or by the medium provided for bacterial growth. *Low-rate* or *conventional* filters are seldom used in modern practice. The organic loading rate on these units ranges from 0.3 to 1.5 kg BOD/m^3 of bulk filter volume per day, and the hydraulic rate from 1870 to 3740 mm/day applied to the filter plan area (45 to 90 gal/ft^2 per day). Recirculation of waste flow is not common in low-rate filters and either the hydraulic or organic loading rate may govern depending upon the waste strength.

High-rate filters employ recirculation of wastewater around the filter and are operated at higher organic loading rates as well. The organic load ranges from 1.5 to 18.7 kg BOD/m^3 of bulk volume per day and the hydraulic rate, including recirculation, from 9360 to 28,000 mm/day (230 to 690 gal/ft^2 per day), or more on the plan area. Plastic media, in particular, are often loaded hydraulically at over 90 m/day.

Roughing filters may be employed as a pretreatment device to reduce the strength of particularly strong wastewaters. Loadings exceed those on high-rate filters and have no arbitrary upper limit. The effluent of such a process requires further treatment.

Standard *rock* filters utilize crushed stone, slag, and gravel. The rock ranges from 60 to 90 mm ($2\frac{1}{4}$ to $3\frac{1}{2}$ in) and should be composed of traprock, granite, quartzite, or slag and be free of sand and clay.[12]

Plastic media may be either interlocking sheets which are carefully placed within the basin to assure uniform flow (Fig. 24-4) or molded or extruded shapes which may be dumped at random to fill the basin (Fig. 24-5). Plastic media are

Figure 24-4 Sheet plastic trickling filter medium. *(Courtesy Imperial Chemical Industries, Ltd.)*

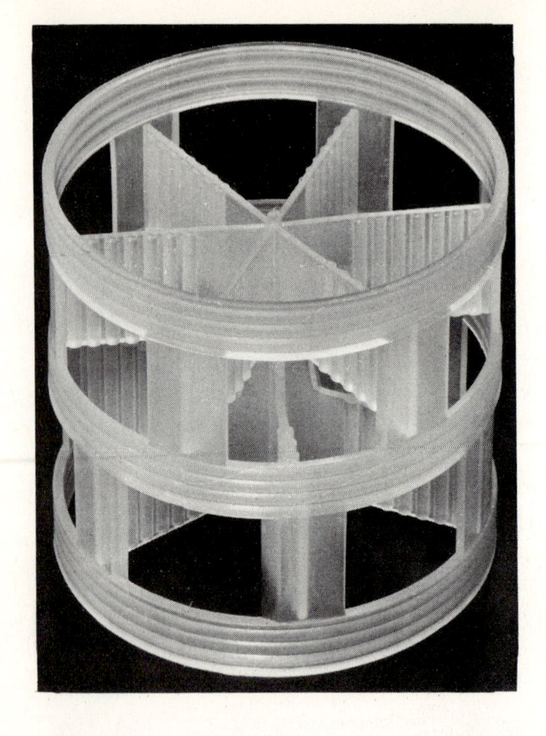

Figure 24-5 Random dumped plastic trickling filter medium. *(Courtesy Koch Engineering Co. Inc.)*

often cheaper than rock in areas where suitable rock does not occur naturally, and are significantly lighter.

Wood media have been used as a substitute for rock or plastic. Redwood slats are fabricated into flats which are stacked upon one another to fill the basin. The physical properties of various media are presented in Table 24-1.

24-6 Recirculation

Recirculation of wastewater is generally practiced in modern trickling filter plants. Techniques of recirculation vary widely, with at least fourteen different configurations being in use (Fig. 24-6). The procedure used reportedly has no

Table 24-1 Physical properties of trickling filter media (after Ref. 13)

Medium	Size, mm	Unit weight, kg/m^3	Specific surface, m^2/m^3	Void space, %
Plastic sheet	600 × 600 × 1200	32–96	82–115	94–97
Redwood	1200 × 1200 × 500	165	46	76
Granite	25–75	1440	62	46
Granite	100	1440	47	60
Slag	50–75	1090	67	49

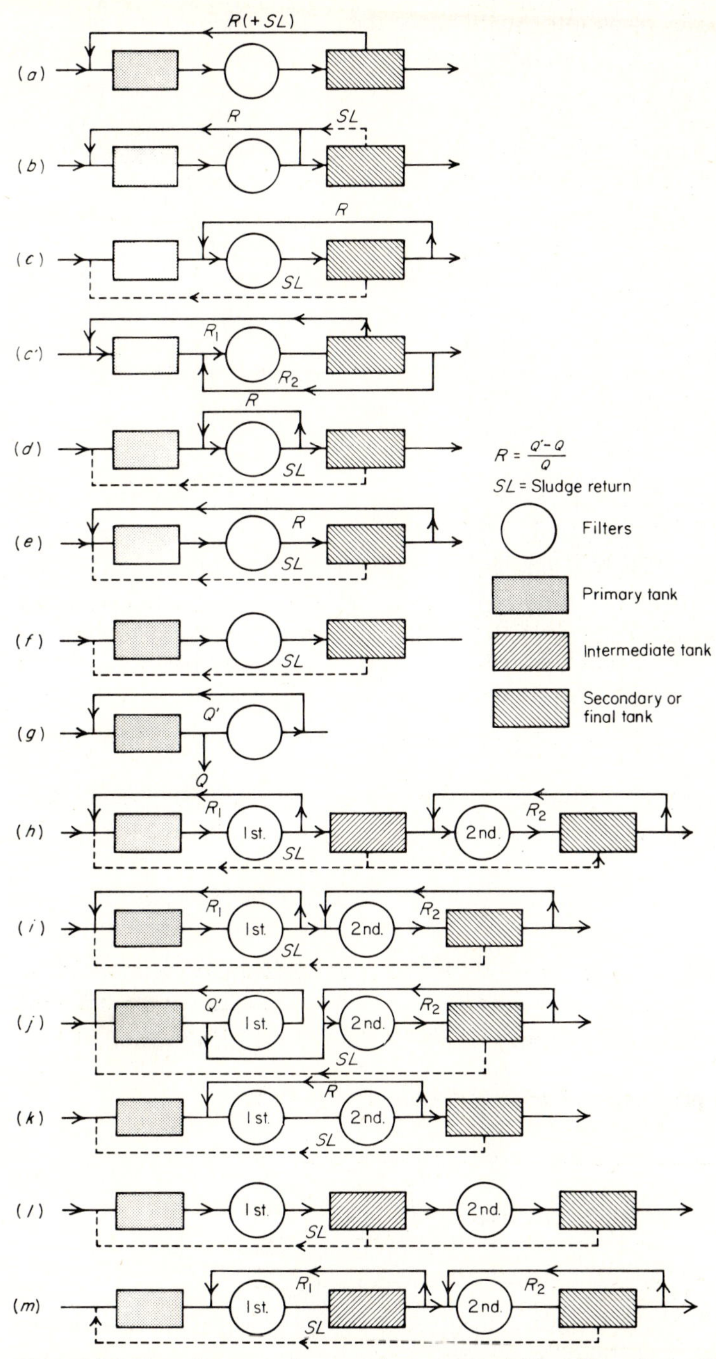

Figure 24-6 Flow diagrams of one- and two-stage trickling filter plants. *(From "Sewage Treatment Plant Design"[14] Copyright 1959 by the American Society of Civil Engineers and the Water Pollution Control Federation. Used with permission of the Water Pollution Control Federation.)*

effect upon the efficiency of the process,[14] hence the choice of configuration must be based upon other considerations. Flow patterns such as those in *(c)*, *(d)*, *(k)*, and *(m)* have the advantage of not introducing additional flow into the primary clarifier.

The advantages of recirculation include an increase in biological solids in the system and continuous seeding with recirculated sloughed solids; maintenance of a more uniform hydraulic and organic load; dilution of influent with better quality water; and thinning of the biological slime layer. It should be noted that recirculation may not increase efficiency in all circumstances,[11,15] particularly with relatively dilute wastes. This is usually not reflected in empirical or simplified formulas for trickling filter performance.

Recirculation rates range from 50 to 1000 percent of the raw wastewater flow with usual rates being 50 to 300 percent.

24-7 Efficiency of Trickling Filters

As noted in Art. 24-4, the design of trickling filters is based upon empirical or semi-empirical formulas. These formulas are used with satisfactory results as long as the designer does not try to apply them in conditions different from those for which they were derived.

The NRC formula[3] is based upon data collected at military bases within the United States during World War II. It may be applied to single- or two-stage systems and is:

$$\frac{C_i - C_e}{C_i} = \frac{1}{1 + 0.532\sqrt{\dfrac{QC_i}{VF}}} \tag{24-1}$$

for a single-stage filter or the first stage of a two-stage system. In the formula C_i is the influent BOD, C_e is the effluent BOD from the first stage, Q is the flow of wastewater in m^3/min, V is the filter volume in m^3 and F is a recirculation factor given by:

$$F = \frac{1 + r}{(1 + 0.1r)^2} \tag{24-2}$$

in which r is the ratio of the recirculated flow to the raw waste flow.

For the second stage the formula becomes:

$$\frac{C_e - C_e'}{C_e} = \frac{1}{1 + \dfrac{0.532}{1 - [(C_i - C_e)/C_i]}\sqrt{\dfrac{QC_e}{V'F'}}} \tag{24-3}$$

in which C_e' is the effluent BOD of the second stage, V' and F' are the volume and recirculation factor for the second stage, and the other terms are as defined above. The formulas include the effect of secondary clarifiers which must be properly designed (Art. 24-9).

Example Calculate the effluent BOD_5 of a two-stage trickling filter with the following flows, BOD_5, and dimensions:

$Q = 3.15\ m^3/min$
$BOD_5 = 290\ mg/l$
Volume of filter no. 1 = $830\ m^3$
Volume of filter no. 2 = $830\ m^3$
Filter depth = 2 m
Recirculation (filter no. 1) = 125 percent Q
Recirculation (filter no. 2) = 100 percent Q

SOLUTION For the first stage:

$$\frac{C_i - C_e}{C_i} = \frac{1}{1 + 0.532\sqrt{\dfrac{QC_i}{VF}}}$$

$$F = \frac{1 + r_1}{(1 + 0.1r_1)^2} = \frac{1 + 1.25}{(1.125)^2} = 1.78$$

$$\frac{290 - C_e}{200} = \frac{1}{1 + 0.532\sqrt{\dfrac{3.15(290)}{830(1.78)}}}$$

$$C_e = 85.5\ mg/l$$

For the second stage:

$$\frac{C_e - C'_e}{C_e} = \frac{1}{1 + \dfrac{0.532}{1 - [(C_i - C_e)/C_i]}\sqrt{\dfrac{QC_e}{V'F'}}}$$

$$F' = \frac{1 + r_2}{(1 + 0.1r_2)^2} = \frac{2}{1.21} = 1.65$$

$$\frac{85.5 - C_e}{85.5} = \frac{1}{1 + \dfrac{0.532}{1 - [(290 - 85.5)/290]}\sqrt{\dfrac{3.15(85.5)}{830 \times 1.65}}}$$

$$C'_e = 38\ mg/l$$

This would not be a satisfactory level of treatment, hence an increase in filter volume or in recirculation would be necessary. The maximum useful rate of recirculation predicted by the NRC formula is about 800 percent.

The *Velz formula*[5] takes the form:

$$C_e = \left(\frac{C_i + rC_e}{1 + r}\right)e^{-KD} \tag{24-4}$$

in which K is an experimental coefficient ranging from 0.49 for high-rate filters to 0.57 for low-rate systems and D is the filter depth in meters. For the example above, in the first stage,

$$C_e = \left(\frac{290 + 1.25C_e}{1 + 1.25}\right) e^{-0.49(2)}$$

$$C_e = 61.1 \text{ mg/l}$$

and in the second stage:

$$C'_e = \left(\frac{61.1 + 1.00C_e}{1 + 1.00}\right) e^{-0.49(2)}$$

$$C'_e = 14.1 \text{ mg/l}$$

The Velz equation is valid for BOD removals of 90 percent or less.

Eckenfelder[6] modified the Velz formula to include the effect of design variables such as specific surface (Table 24-1) and contact time, i.e., flow rate. Eckenfelder's formulation is

$$C_e = \left(\frac{C_i + rC_e}{1 + r}\right) e^{-K(A/(1+r)Q)^n D a_v{}^{1+m}} \tag{24-5}$$

in which K, n, and m are experimental coefficients which vary with media, quantity of slime growth, and, to a degree, flow. Evaluation of these constants is necessary for each material and hydraulic loading rate.

Rankin[5] has developed a number of empirical formulas, based upon the Ten States Standards[16] which are presented in detail in reference 14, and are summarized in Table 24-2. The loading criteria referred to in the table are:

I. High-rate processes with dosing rates between 9360 and 28,000 mm/day and organic loads, including recirculation, less than 1.75 kg/m^3 per day.
II. High-rate processes with loadings, recirculation included, of between 1.75 and 2.75 kg/m^3 per day. For loadings in excess of 2.75 kg/m^3 per day the Ten States Standards assume removal of 1.75 kg/m^3 per day.
III. First stage units without intermediate clarification prior to the second stage.

Table 24-2 Formulas for effluent BOD of trickling filter processes (after Ref. 14)

Flow diagram (Fig. 24-6)	Loading criterion I	II	III
a, b, c, e, and *f*	$C_e = (C_i - C_e) \times \left(\frac{1+r}{1+1.5r}\right)$	$C_e = (C_i - C_e) \times \left[\frac{1.78(1+r)}{2.78+1.78r}\right]$	$C_e = (C_i - C_e) \times \left(\frac{1+r}{2+r}\right)$

For second stage filters the effluent BOD_5 is given by the equation under III in Table 24-2 save that if the final BOD_5 is less than 30 mg/l the efficiency of the second stage cannot exceed 50 percent.

Applying the formulas of Table 24-2 to the example problem above, if the effluent BOD_5 from the first stage is assumed to be about 100 mg/l, the organic loading rate, including recirculation is:

$$290(3.15) + 100(3.15)(1.25) = 1307 \text{ g/min}$$
$$= 1882 \text{ kg/day}$$

For a filter with a volume of 830 m^3, the unit organic loading rate is 1882/830 = 2.27 kg/m^3 per day. Therefore the loading criterion is II and

$$C_e = (290 - C_e)\frac{1.78(1 + 1.25)}{2.78 + 1.78(1.25)} = 58 \text{ mg/l}$$

Checking the loading rate,

$$290(3.15) + 58(3.15)(1.25) = 1142 \text{ g/min}$$
$$= 1644 \text{ kg/day}$$

From which the loading rate is 1.98 kg/m^3 per day and the assumed loading condition is correct. The effluent BOD_5 of the second stage is given by

$$C'_e = (C_e - C'_e)\left(\frac{1 + r}{2 + r}\right) = (58 - C'_e)\left(\frac{2}{3}\right)$$
$$= 23 \text{ mg/l}$$

24-8 Design of Trickling Filter Systems

The sizing of trickling filter processes is based upon the application of suitable design formulas (Art. 24-7) or arbitrary loading standards established by state agencies. Filter depths are ordinarily about 2 m (6 ft) in standard rock or dumped plastic media designs. Packed towers containing sheet plastic or redwood media may be as much as 5 m (16 ft) in height, occasionally more. High hydraulic loads on the lesser plan area of deep filters, coupled with the smooth surface of plastic media may lead to a mode of operation that lies somewhere between attached and suspended growth. In a sense, a deep filter with a high recirculation rate might be considered a mechanically aerated activated sludge process.

Example Design a single-stage trickling filter to yield an effluent BOD_5 of 30 mg/l. The influent BOD_5 following primary clarification is 160 mg/l and the flow is 10^4 m^3/day. Maintain an hydraulic loading rate of 20 m/day and a filter depth of 2 m.

SOLUTION Applying the NRC formula, Eq. (24-1),

$$\frac{160-30}{160}=\frac{1}{1+0.532\sqrt{\dfrac{6.94(160)}{VF}}}$$

$$VF = 5901$$

For a depth of 2 m, $V = 2A$, and $A = 10^4(1 + r)/20$ for the specified hydraulic loading rate. Equating these,

$$5901\left(\frac{(1+0.1r)^2}{2(1+r)}\right)=\frac{10{,}000(1+r)}{20}$$

From which $r = 1.88$. If r is made equal to 2.0, $F = 2.08$, $V = 2837\ \text{m}^3$, $A = 1500\ \text{m}^2$, and the depth might be held constant at 2 m to yield

$$\frac{160-C_e}{160}=\frac{1}{1+0.532\sqrt{\dfrac{6.94(160)}{6000}}}$$

$$C_e = 29.8\ \text{mg/l}$$

The filter system would probably consist of two units, each 750 m^2 or 31 m in diameter, and suitable clarifiers. Checking this analysis with the Velz equation, Eq. (24-4), yields

$$C_e = 26.7\ \text{mg/l}$$

Manufacturers of artificial media may be able to furnish experimental data such as that in Fig. 24-7 which can be used in sizing biological systems employing their media.

Distribution systems may employ either rotary or fixed nozzles. The rotary systems consist of a hollow arm, pivoted at the center and fitted with nozzles so sized and placed that they provide an essentially uniform distribution of liquid over the filter surface. The arm may be driven by the liquid discharge if sufficient head is provided, or may be turned by an electric motor. Hydraulic drive is more common and has the advantage of automatically adjusting the rotor speed to the

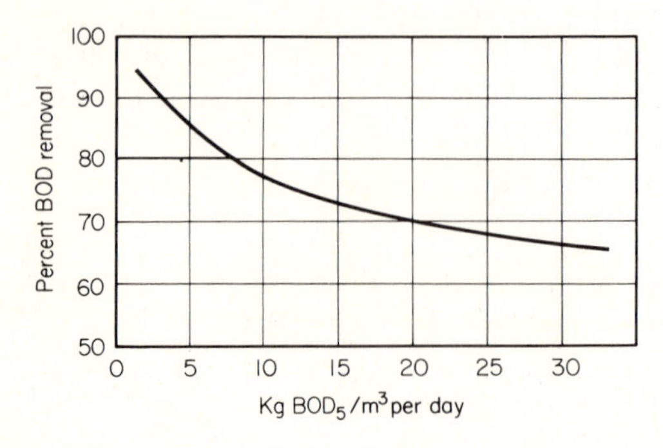

Figure 24-7 Efficiency of a plastic media filter.

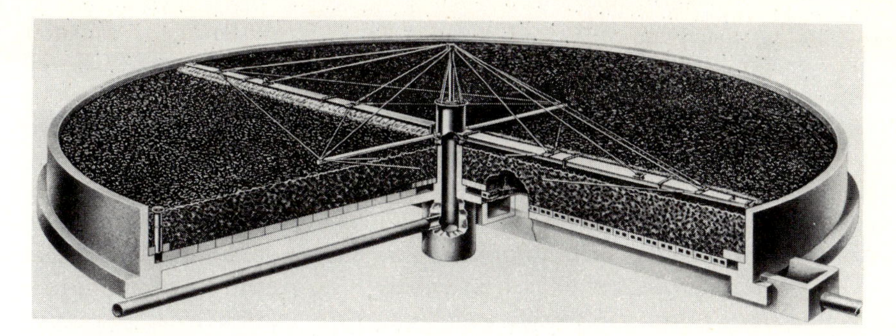

Figure 24-8 Rotary distributor trickling filter. *(Courtesy Dorr-Oliver Inc.)*

flow. The speed of rotation should be at least 6 revolutions/hour so that each area will be dosed at 5-min or shorter intervals. A standard rotary distributor is shown in Fig. 24-8.

Dosing tanks (Fig. 24-9) may be employed on low-rate filters which lack recirculation. The dosing mechanism assures that the flow will be sufficient to turn the rotor and prevents dribbling of low flows upon a small area. The tanks are

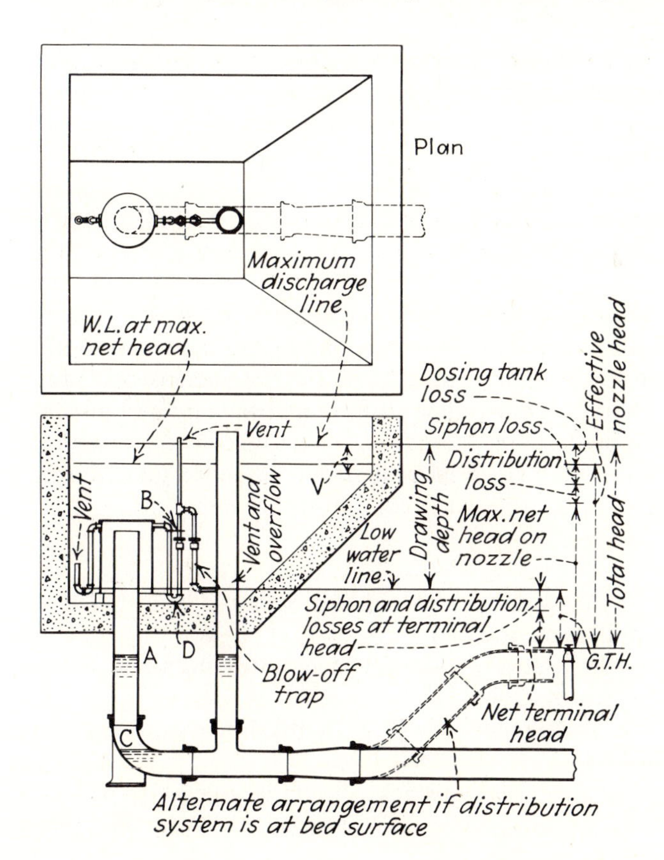

Figure 24-9 Dosing siphon and tank.

usually sized for 5-min retention at average flow. Recirculation is a better technique for maintaining rotor motion and provides other advantages as well (Art. 24-6).

Fixed nozzle trickling filters have generally employed dosing tanks in order to provide the intermittent application natural in rotor systems. The fixed nozzles are located so as to insure reasonably uniform distribution of the wastewater.

Modern trickling filters, particularly those employing artificial media, may be designed with fixed nozzles and continuous flow. Many laboratory models of trickling filters employ continuous flow and there is no reason why intermittent dosing is necessary on coarse media with large void spaces. In fact, in some circumstances continuous dosing has been demonstrated to produce a superior effluent.[17] A full-scale continuous flow plastic media trickling filter with fixed nozzles is shown in Fig. 24-4.

The *headloss* in distribution systems may be calculated using the principles of hydraulics. The total head required includes the entrance loss, the loss in the siphon and drop in head during dosing if a dosing tank is used, the friction loss in the pipe, the loss in the distributor itself [which is usually about 300 mm (12 in)] and the elevation of the rotor above the media. The influent flows to the distributor through a feed pipe which is either under the bed or supported on columns within it. The feed pipe terminates in a vertical riser which supports the distributor. Mercury seals, which were formerly used in rotary distributors are now prohibited due to the hazard associated with possible leakage.

The *underdrain system* serves to carry away the liquid effluent and sloughed biological solids and to distribute air through the bed. The underdrains are laid on a slope of about 1 percent toward a common collection point or channel and are sized so as to flow no more than half full at a velocity of 0.6 to 0.9 m/s (2 to 3 ft/s) at design flow (including recirculation). A variety of commercially available underdrains are shown in Fig. 24-10.

Air will be transported through the filter and underdrains by gravity except under unusual (and transient) conditions. When the air is approximately 3.4°C warmer than the sewage, air flow will be relatively low, but will increase with any change in either temperature. In warm weather air flow is downward and in cold weather upward through the bed.

Forced ventilation is not usually required or desirable provided underdrains are sized as described above and the following conditions are met:[18]

1. Ventilation stacks or manholes with open grated covers should be installed at either end of the central collection channel.
2. Branch collecting channels on large filters should have ventilating stacks at the filter periphery.
3. The open area in the top of the filter blocks (Fig. 24-10) should be at least 15 percent of the filter plan area.
4. The area of the open grating in ventilating stacks should be at least 0.4 percent of the filter plan area.

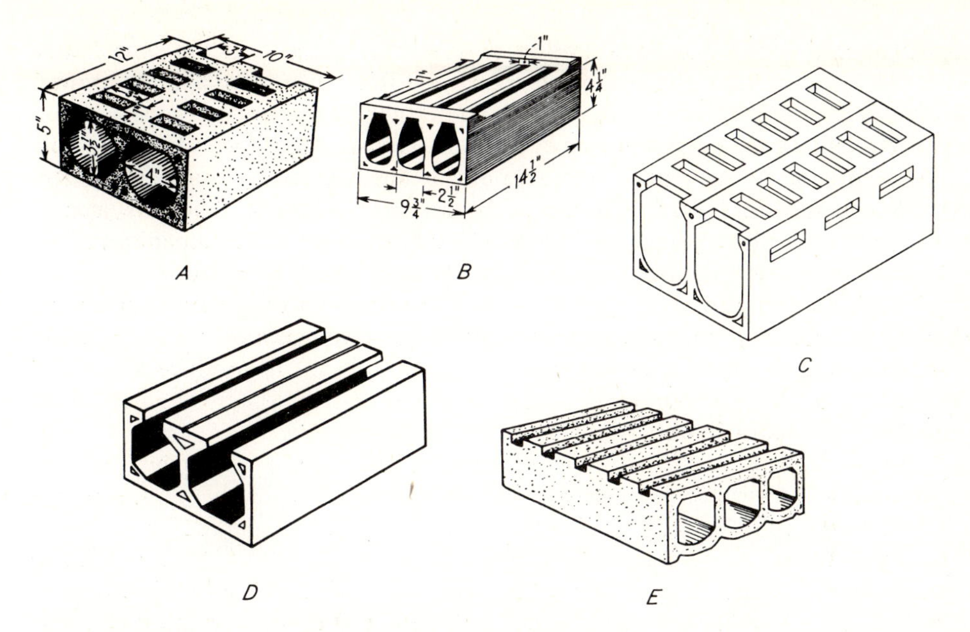

Figure 24-10 Blocks used in underdrain system of trickling filters. *A*, Armcre block; *B*, Natco tile; *C*, Armcre block with lateral ventilating slots; *D*, Metro monounit block; *E*, Cannelton block.

24-9 Trickling Filter Final Clarifier Design

Clarifiers following trickling filters are designed to remove relatively large particles of sloughed bacterial slime or humus. No thickening or hindered settling occurs, hence design criteria are based upon particle size and density.

Surface overflow rates are 25 to 33 m per day (600 to 800 gal/ft^2 per day) at average flow and should not exceed 50 m/day (1200 gal/ft^2 per day) at peak flow. Weir loading rates and retention times are similar to those in primary clarifiers (Chap. 23). Recirculated flow is included when sizing the clarifier if it actually passes through the basins—as it does in flow patterns (*a*), (*c*), (*c*′), (*e*), (*h*), (*i*), (*j*), and (*m*) in Fig. 24-6. It should also be noted that recirculated flow which passes through the primary clarifier will affect the design of that system. Sludge return is generally neglected in such calculations since it is relatively small in trickling filter systems.

In high-rate trickling filters, particularly those employing plastic media and very high recirculation rates, the designer may wish to consider the design standards for clarifiers following suspended growth systems to be applicable.

The *clarifier mechanism* employed in secondary clarifiers following trickling filters is similar to that used in primary systems (Chap. 23). Skimming is sometimes omitted but this is a questionable economy since the additional cost is small and process malfunctions may cause sludge to float.

The *quantity of waste sludge* produced in trickling filters is somewhat less than in suspended growth processes operated at similar loadings since cell mass is lost by anaerobic endogenous metabolism in the inner portion of the slime layer (Fig. 24-3) and the sloughed solids contain less water. Depending upon hydraulic and organic loading rates, solids production in trickling filters may be expected to range from 0.2 to 0.5 kg VSS per kg BOD_5 removed, with the lower production on low rate processes. The moisture content of trickling filter solids ranges from 90 to 98 percent with lower moisture content on low-rate processes. Sludge is usually returned to the plant influent wet well and reseparated in the primary clarifier. The return flow is negligible with respect to the raw waste flow and does not affect the design of the primary basin. The mixed primary and secondary solids are pumped from the primary clarifier for further processing (Chap. 25).

24-10 Operational Problems of Trickling Filters

The major operational problems of trickling filters are associated with cold weather operation. Efficiency in high-rate filters is reduced with decreased temperature by approximately 30 percent per 10°C and freezing may cause partial plugging of the filter medium and resulting overload in the open area.

In northern climates fiberglass covers or windbreaks have been employed to prevent ice formation. Covers have an additional advantage in that they help to contain odors which may be produced in the filter.

Psychoda alternata or filter flies breed in low-rate trickling filters but are less troublesome in high-rate systems in which the higher hydraulic rate carries the larvae from the filter before they can mature. It is important that the filter be uniformly dosed since areas of low flow may serve as breeding zones. Flooding the filter for 24 hours and application of insecticides have been effective techniques of filter fly control. If it is anticipated that flooding may be necessary, the filter must be designed to permit it.

24-11 Rotating Biological Contacters

Rotating biological contacters consist of a basin containing mechanically driven rotors which provide a large surface area for biological growth. The process is continuous flow and the rotating mechanism is covered with a slime film similar to that in trickling filters. The film is turned through the wastewater, which covers a little less than half the available surface, and through the air above the liquid. To the biological film the process is substantially the same as that in a trickling filter with a rotating distributor—save perhaps for limited oxygen transfer during the wet cycle.

Rotating contacters are usually disks (Fig. 24-11) typically about 3 m (10 ft) in diameter, 10 mm (0.4 in) thick, and placed 30 to 40 mm (1 to $1\frac{1}{2}$ in) on centers along a shaft of variable length. The shaft rotates at 1 to 2 r/min. A full-scale study of this process[19] indicated it was relatively stable and capable of producing a satisfactory effluent. The sludge produced settled to a moisture content of 95 to 98 percent alone and 93 to 95 percent when mixed with primary solids. The quantity

Figure 24-11 Rotating disk assemblies. *(Courtesy Biosystems Division, Autotrol Corporation.)*

of sludge produced amounts to approximately 0.4 kg per kg BOD_5 applied. The theoretical model of the process is similar to that for trickling filters,[15] but actual design is still empirical.

Hydraulic loading rates are 0.04 to 0.06 m/day and *organic loading rates* 0.05 to 0.06 kg BOD_5/m^2 per day based upon the disk surface area. Sloughing of biological solids is more or less continuous and the effluent contains a relatively constant concentration. The solids settle well and clarifier surface overflow rates of 33 m/day (800 gal/ft^2 per day) are reported to be satisfactory.[20]

Some research has indicated that oxygen might be limited in the bulk solution and recent modifications of RBC processes have included introduction of air or oxygen toward the bottom of the basin.

Another rotating contacter employs a wire drum filled with plastic spheres (Fig. 24-12). Liquid is carried out of the bulk solution, and air is carried into it, by open-tube sections on the periphery of the drum. The manufacturer claims that the splash aeration and bubble aeration afforded by this technique are significant, and improve the operation of the process. Recirculation of sludge from the secondary clarifier is a part of this process, which makes it more similar to an activated sludge (suspended growth) process.

Rotating biological contacters represent a relatively new technology, with most design information being empirical and developed by the process manufacturers. Capital costs are presently considered high, but operating costs may be somewhat lower than other biological processes. Rotating-disk systems must ordinarily be protected by a roof since heavy rains may strip off the slime growth, and hail may damage the plastic disks. In northern climates an enclosed heated building may be necessary to prevent freezing during the winter.[19]

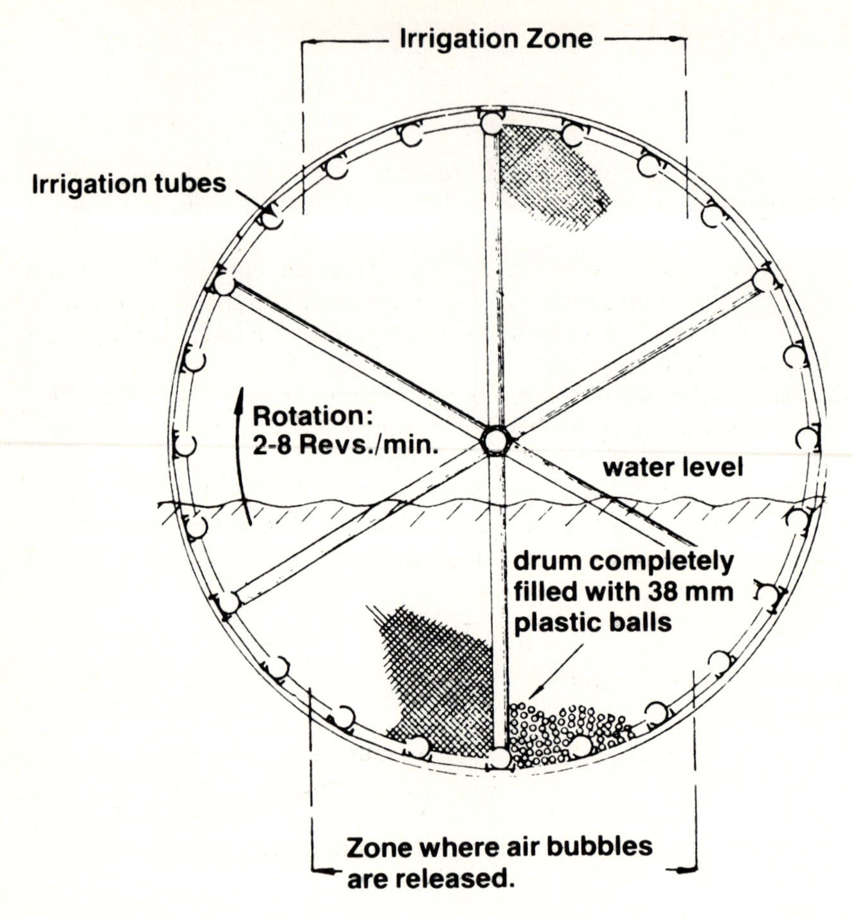

Figure 24-12 Rotating drum contacter. *(Courtesy Ralph B. Carter Co.)*

SUSPENDED GROWTH PROCESSES

24-12

Suspended growth processes maintain an adequate biological mass in suspension within the reactor by employing either natural or mechanical mixing. In most processes the required volume is reduced by returning bacteria from the secondary clarifier in order to maintain a high solids concentration. Suspended growth processes include activated sludge and its various modifications, oxidation ponds, and sludge digestion systems.

24-13 Principles of Suspended Growth Systems

The theory of suspended growth processes is well developed and generally accepted.[21,22] The basic factor in design, control, and operation of suspended growth systems is the mean cell residence time or sludge age (θ_c), defined by

$$\theta_c = \frac{X}{(\Delta X/\Delta t)} \tag{24-6}$$

in which X is the total microbial mass in a reactor and $\Delta X/\Delta t$ is the total quantity of solids withdrawn daily, including solids deliberately wasted and those in the effluent.

Two particular values of θ_c are of significance in design of biological processes. θ_c^m is defined as the lowest value of θ_c at which operation is possible. At retention times less than θ_c^m organisms are removed more quickly than they are synthesized, hence failure will occur.

θ_c^d is the design value of θ_c and must be significantly greater than θ_c^m. The ratio, θ_c^d/θ_c^m, gives the safety factor of the system. Required safety factors depend upon the anticipated variability of the waste and should be at least 4. In standard processes, safety factors of 20 or more are not unusual (Table 24-3).

For an equilibrium system the quantity of solids produced must equal that lost, and the quantity produced per day is given by

$$\mu = \frac{\hat{\mu}S}{K_s + S} - k_d \tag{24-7}$$

where μ = the net specific growth rate or growth per unit mass per unit time
$\hat{\mu}$ = the maximum rate of growth
S = the concentration of substrate surrounding the microorganisms
K_s = the half velocity constant, equal to the substrate concentration at which the rate of substrate removal is one-half the maximum rate, and
k_d = a microorganism decay coefficient (mass/unit mass/unit time) reflecting the endogenous burn-up of cell mass.

Application of mass balance equations and Eqs. (24-6) and (24-7) to specific process configurations permits general solutions for effluent quality in terms of experimental constants and sludge age.[22]

24-14 Completely Mixed Process with Solids Recycle

The completely mixed model presented below is applicable to activated sludge systems of any configuration. The model was specifically developed for completely mixed systems, but is conservative in its prediction of effluent quality for plug-flow systems which are not otherwise limited. The explicit assumptions of the model (Fig. 24-13) are that all waste utilization occurs in the biological reactor and that the total biological mass in the system is equal to the biological mass in the reactor. These imply that the clarifier volume is small and that recycle is continuous. The mean cell residence time, or sludge age, for this system by definition, (Eq. 24-6), is

$$\theta_c = \frac{X}{\left(\dfrac{\Delta X}{\Delta t}\right)} = \frac{xV}{Q_w x_r + (Q - Q_w)x_e} \tag{24-8}$$

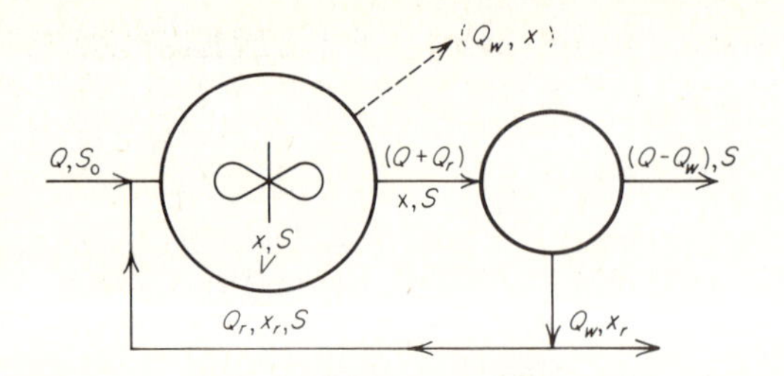

Figure 24-13 Completely mixed biological reactor with solids recycle.

where Q = the waste flow rate
x = the mixed liquor suspended solids concentration (MLSS)
x_r = the clarifier underflow suspended solids concentration
x_e = the effluent suspended solids concentration
Q_w = the waste sludge flow rate, and
Q_r = the return sludge flow rate

If solids are wasted from the reactor as shown in the dashed line, rather than from the clarifier, Eq. (24-8) becomes

$$\theta_c = \frac{xV}{Q_w x + (Q - Q_w)x_e} \tag{24-9}$$

If the system is operating properly, that is, if the clarifier is functioning, the bulk of the solids will be removed in the waste sludge rather than in the effluent. θ_c can thus be controlled by varying Q_w and is not completely dependent upon the reactor volume.

A mass balance equation written across the reactor yields

$$\frac{1}{\theta_c} = \frac{Q}{V}\left(1 + r - r\frac{x_r}{x}\right) \tag{24-10}$$

in which r is the ratio of the solids return flow to the raw waste flow, Q_r/Q. Thus θ_c is a function of both the recycle ratio and x_r/x, the ratio of return solids concentration in the clarifier underflow to the mixed liquor solids concentration. The value of x_r is dependent upon the efficiency of the clarifier and the settling characteristics of the biological mass (Art. 9-6).

Lawrence and McCarty[22] have shown that the total microbial mass in the reactor is given by

$$xV = \frac{YQ(S_0 - S)\theta_c}{1 + k_d\theta_c} \tag{24-11}$$

from which,

$$x = \frac{\theta_c}{\theta} \cdot \frac{Y(S_0 - S)}{1 + k_d\theta_c} \tag{24-12}$$

In Eqs. (24-11) and (24-12) Y is the growth yield coefficient relating cell yield to the material metabolized. S_0 is the influent BOD_5, S is the effluent soluble BOD_5 and θ is the liquid retention time in the reactor, V/Q.

When the required efficiency is known and the values of k, K_s, Y and k_d are established it is possible to determine the required reactor volume and microbial concentration for various values of r and x_r. From the possible systems the one with the lowest capitalized cost can then be selected.

Example Design an activated sludge process to yield an effluent BOD_5 of 20 mg/l and suspended solids of 25 mg/l. The influent BOD_5 following primary clarification is 160 mg/l. Assume $Y = 0.65$, $k_d = 0.05$, and $\theta_c = 10$ days. The waste flow is 10 m^3/min.

SOLUTION The BOD_5 of the effluent solids can be estimated to be 0.63(SS).[18] The soluble effluent BOD_5 must thus be reduced to $20 - 0.63(25) = 4$ mg/l.

The total biological mass is calculated from Eq. (24-11):

$$xV = \frac{YQ\theta_c(S_0 - S)}{1 + k_d\theta_c}$$

$$= \frac{0.65(10^4 \times 1440)(10)(160 - 4)}{1 + 0.05(10)}$$

$$= 9.73 \times 10^9 \text{ mg}$$

Assuming a mixed liquor volatile suspended solids concentration of 2500 mg/l,

$$V = \frac{9.73 \times 10^9}{2500} = 3.894 \times 10^6 \text{ l}$$

$$= 3894 \text{ m}^3$$

The rate of sludge (biological solids) production is obtained from Eq. (24-6) and is

$$\frac{dX}{dt} = \frac{xV}{\theta_c} = \frac{9.73 \times 10^9}{10} \text{ mg/day}$$

$$= 973.5 \text{ kg/day}$$

Assuming the solids are 80 percent volatile, total production would be

$$973.5/0.8 = 1217 \text{ kg/day}$$

The underflow solids concentration is unlikely to exceed 15,000 mg/l and could well be less. Assuming $x_r = 15{,}000$ mg/l,

$$Q_w = \frac{1217 \times 10^6 \text{ mg/day}}{15 \times 10^3 \text{ mg/l}}$$

$$= 81.1 \text{ m}^3\text{/day}$$

The recirculation flow can be calculated from Eq. (24-10) or directly from

$$Q_r \cdot x_r = (Q + Q_r)x$$

$$Q_r = \frac{Qx}{x_r - x} = \frac{10(2500)}{12{,}500}$$

$$= 2 \text{ m}^3/\text{min}$$

$$r = Q_r/Q = 0.2$$

The hydraulic retention time in the reactor is

$$t = V/Q = \frac{3.894 \times 10^3}{14{,}400} = 0.27 \text{ days}$$

$$= 6.5 \text{ hours}$$

The required amount of oxygen which must be provided in suspended growth processes is equal to the difference between the ultimate BOD of the waste which is removed and the ultimate BOD of the solids which are wasted. For ordinary domestic sewage this may be taken as equal to:[18]

$$O_2 \text{ demand} = 1.47(S_0 - S)Q - 1.14x_r(Q_w) \tag{24-13}$$

Applying this relation to the example above,

$$O_2 \text{ demand} = 1.47(160 - 4)(14.4 \times 10^6) - 1.14(15{,}000)(81{,}100)$$

$$= 1915 \times 10^9 \text{ mg/day}$$

$$= 1915 \text{ kg/day}$$

The volume of air required at STP in m^3 may be calculated from

$$Q_{air} = \frac{O_2 \text{ demand}}{0.232(1.20)}$$

$$= \frac{O_2 \text{ demand}}{0.278} \tag{24-14}$$

In the example above, if the actual efficiency of oxygen transfer is 7 percent, the air required will be:

$$Q_{air} = \frac{1915}{0.278} = 6856 \text{ m}^3/\text{day}$$

$$\text{Actual air required} = 6856/0.07$$

$$= 97{,}943 \text{ m}^3/\text{day}$$

$$= 68 \text{ m}^3/\text{min}$$

The air volume required per unit of BOD_5 removed is thus equal to

$$97{,}943/(160 - 4)(14.4) = 43.6 \text{ m}^3/\text{kg}$$

In the example above one may observe that maintenance of a higher value of x would permit a smaller reactor volume, that x is dependent upon x_e and Q_r, that the quantity of waste sludge depends upon θ_c, that the oxygen demand depends upon the quantity of sludge wasted (and thus upon θ_c), and, in short, that a great many solutions to problems of this sort are possible—subject to certain physical limitations on oxygen transfer and solids thickening. The general range of values encountered in design of activated sludge processes is presented in Table 24-3. The processes listed in the table are described below.

24-15 Activated Sludge Process Details

The activated sludge process originally was contained in long narrow tanks with air for oxygen supply and mixing being supplied through diffusers at the basin bottom. Many modifications in both basin configuration and aeration technique have been made in recent years. The more important processes are described below.

The *conventional process* consists of a rectangular basin, a clarifier, and a solids-return line from the clarifier bottom. Excess solids are usually wasted from the clarifier underflow, although separate sludge wasting from the reactor is possible and may be preferable.[18] The returned solids are mixed with the incoming waste and pass through the reactor in a plug flow fashion. Air is provided uniformly along the basin through porous diffusers, usually of the sort in Fig. 24-14. The high concentration of BOD and microbial solids at the head of the basin leads to rapid exertion of BOD and a high oxygen demand which may be difficult to meet. At the end of the basin the air supplied may be in excess of the demand (Fig. 24-15).

Tapered aeration processes attempt to match the oxygen supply to demand by introducing more air at the head end. This can be achieved by varying the diffuser

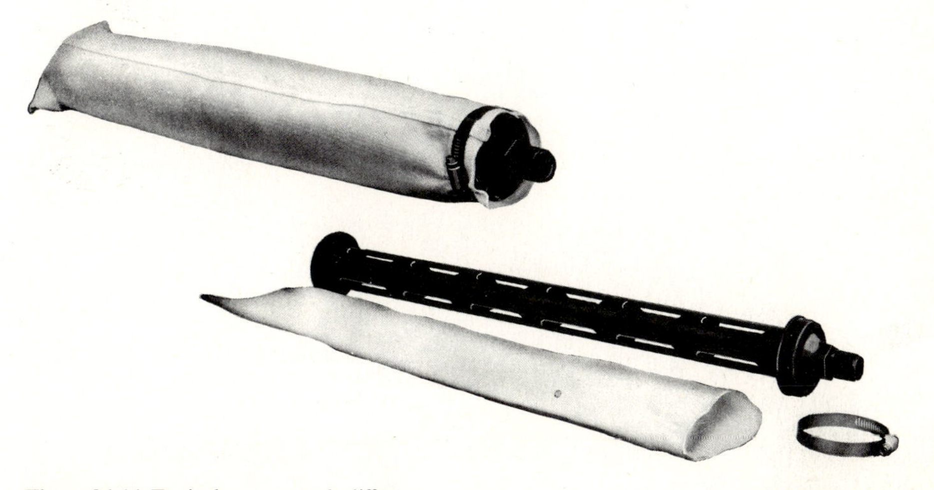

Figure 24-14 Typical porous sock diffuser.

Table 24-3 Typical process factors for activated sludge systems

Process	Normal loading per day (kg BOD_5/m^3)	Normal loading per day (kg BOD_5/kg MLVSS)	θ, days	θ_c^d, days	S.F.	r	Air supplied, m^3/kg BOD_5
Extended aeration	0.32	0.05–0.20	0.80–1.25	14–∞	$\geq$ 70	0.50–1.00	90–125
Conventional activated sludge	0.56	0.20–0.50	0.25–0.30	4–14	20–70	0.15–0.30	45–90
Tapered aeration	0.56	0.20–0.50	0.25–0.30	4–14	20–70	0.15–0.30	45–90
Step aeration	0.80	0.20–0.50	0.20–0.30	4–14	20–70	0.20–0.50	45–90
Contact stabilization	1.12	0.20–0.50	0.01–0.04	4–14	20–75	0.50–1.00	45–90
Short-term aeration (High-rate A.S.)	1.6–6.4	0.50–3.50	0.10–0.15	0.8–4	4–20	1.00–5.00	25–45

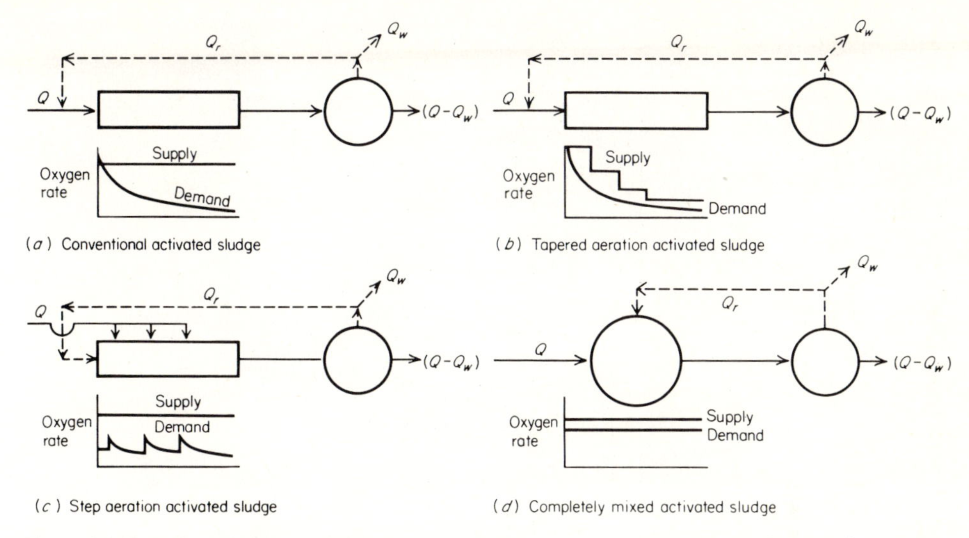

Figure 24-15 Activated sludge configurations and effect upon oxygen supply and demand.

spacing. The process is otherwise the same as described above and, like the conventional process, is sensitive to shock loadings and toxic materials.

Step aeration processes distribute the incoming flow to a number of points along the basin, thus avoiding the locally high oxygen demand encountered in conventional and tapered aeration. The distribution of flow lessens the effect of peak hydraulic and organic loadings and may provide sufficient dilution to protect against introduction of toxic materials.

Completely mixed processes disperse the incoming waste and return sludge uniformly throughout the basin. The shape of the reactor is not important, provided it is conducive to complete mixing. In such a process the oxygen demand is uniform throughout the basin (Fig. 24-15). In actual basins of any size true complete mixing is difficult to obtain, but for practical purposes it can generally be approximated with careful selection of mixing and/or aeration equipment. The effect of peak organic and hydraulic loads is minimized in completely mixed systems and toxic materials are normally diluted below their threshold concentration. In theory the effluent of a completely mixed system is somewhat inferior to that of a plug-flow process, but in practice the difference is not discernible.

Extended aeration is a completely mixed process operated at a long hydraulic retention time (θ) and a high sludge age (θ_c). The process is limited in application to small plants where its inefficiency is outweighed by its stability and simplicity of operation. Many extended aeration plants are prefabricated units ("package plants") which require little more than foundations and hydraulic and electrical connections. A typical extended aeration plant is shown in Fig. 24-16.

In selecting package plants the engineer should give careful consideration to the quality and capacity of pumps, motors, and compressors as well as the stated capacity of the system.

Figure 24-16 Extended aeration package plants. *(Courtesy Clow Corporation.)*

Short-term aeration or high-rate activated sludge is a pretreatment process similar in application to a roughing filter (Art. 24-5). Retention times and sludge age are low, which leads to a poor effluent and relatively high solids production. A possible application of this process is as the first stage of a two-stage nitrification process (Chap. 26).

Contact stabilization takes advantage of the observed adsorptive properties of activated sludge. Returned sludge which has been aerated for stabilization of previously adsorbed organics is mixed with incoming wastewater for a brief period—perhaps 30 min. The mixed flow is then separated in a clarifier, the treated liquid is released, and the sludge with its burden of adsorbed organics is aerated for 3 to 6 hours. During the aeration process the adsorbed organics are hydrolyzed and re-released to the liquid prior to final stabilization. If the retention time in the contact stage is too long, re-release may occur there with subsequent deterioration of effluent quality.

A significant reduction in required aeration capacity is possible in contact stabilization. If one assumes that the sludge will thicken to 1 percent solids

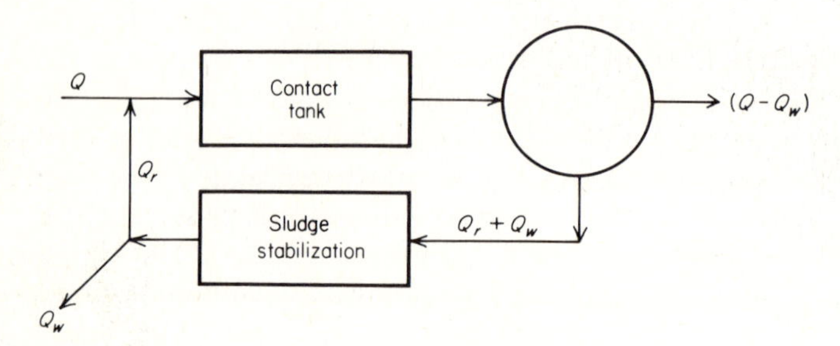

Figure 24-17 Contact stabilization process.

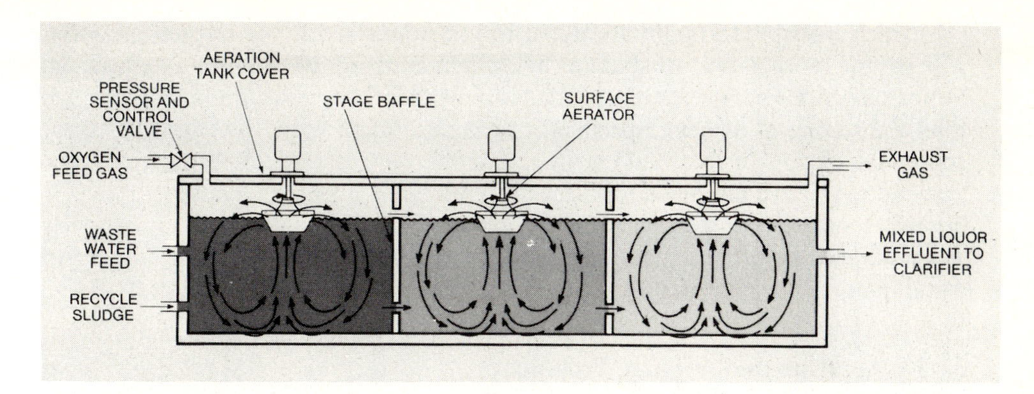

Figure 24-18 Closed reactor high purity oxygen system. *(Courtesy Union Carbide Corporation.)*

(10,000 mg/l) and that the concentration in the contact basin is 2500 mg/l, a simple calculation indicates that the total volume (contact plus aeration) is only one-third that of a standard process if both employ an aeration period of 6 hours. A schematic diagram of a contact stabilization process is shown in Fig. 24-17.

High purity oxygen activated-sludge systems have been developed in an attempt to permit easier matching of oxygen supply to oxygen demand and, perhaps, higher-rate processes through maintenance of higher concentrations of biological solids. Two process configurations are presently available: closed reactors with a high purity oxygen atmosphere (Fig. 24-18), and open reactors with fine bubble diffusion at the tank bottom (Fig. 24-19).

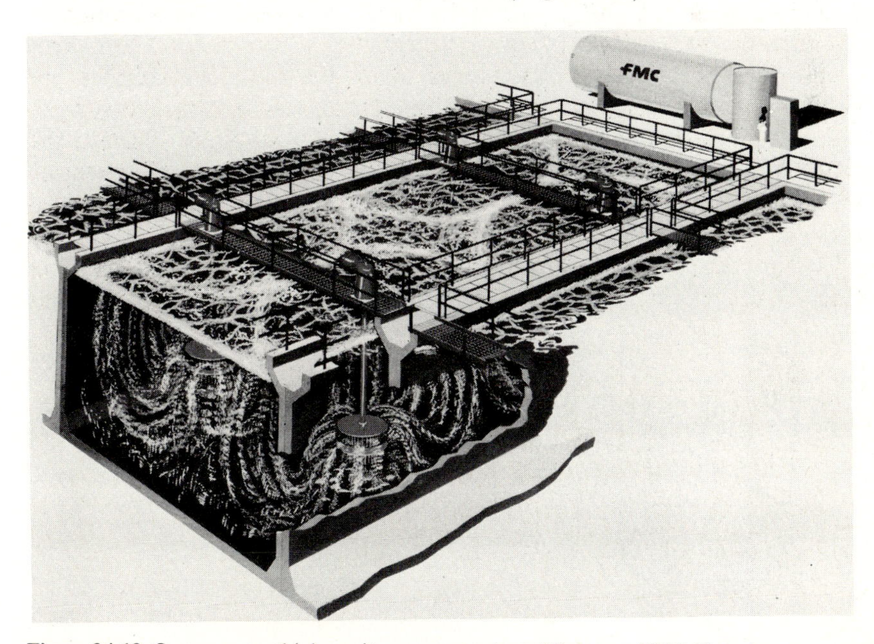

Figure 24-19 Open reactor high purity oxygen system. *(Courtesy FMC Corp.)*

If it is possible to maintain higher mixed liquor solids the oxygen processes will permit more rapid satisfaction of oxygen demand. Maintaining high mixed liquor suspended solids concentrations is dependent, however, upon factors other than the source of oxygen. Specifically, it is dependent upon the character of the solids, the design of the final clarifier, and the recycle rate. High purity oxygen systems may have certain advantages in coping with unanticipated increases in organic loading[23] and in treating certain industrial wastes, however, properly designed and operated air activated sludge systems can produce an effluent of equal quality at a comparable cost when treating domestic wastewater.[24]

24-16 Aeration and Mixing Techniques

A large number of proprietary combinations of mixing/aeration equipment and basin configurations have been developed in addition to the standard diffused air rectangular tank design.

Surface aerators consist of electrically driven propellers mounted in either a floating or fixed support. The propellers throw the bulk liquid through the air and oxygen transfer occurs both at the surface of the droplets and at the surface of the bulk solution which is mixed by the current generated by the aerators. Surface aerators are available in *high-speed* designs (Fig. 24-20) which are driven at 900 to

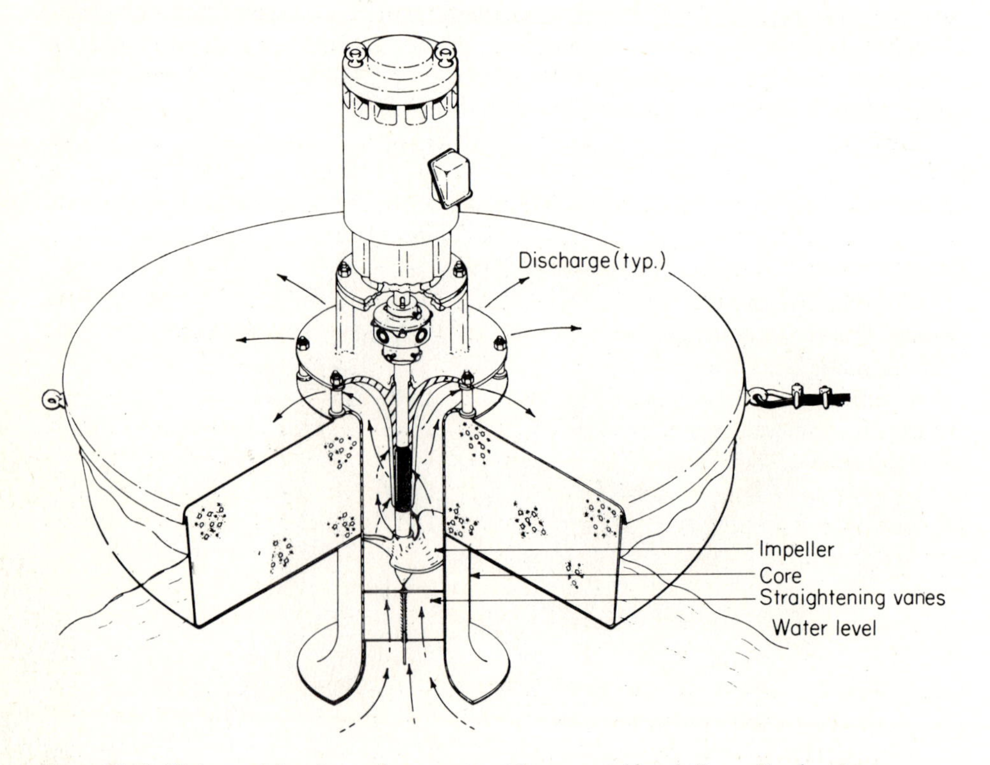

Figure 24-20 High-speed floating surface aerator. *(Courtesy Ashbrook-Simon-Hartley, Inc.)*

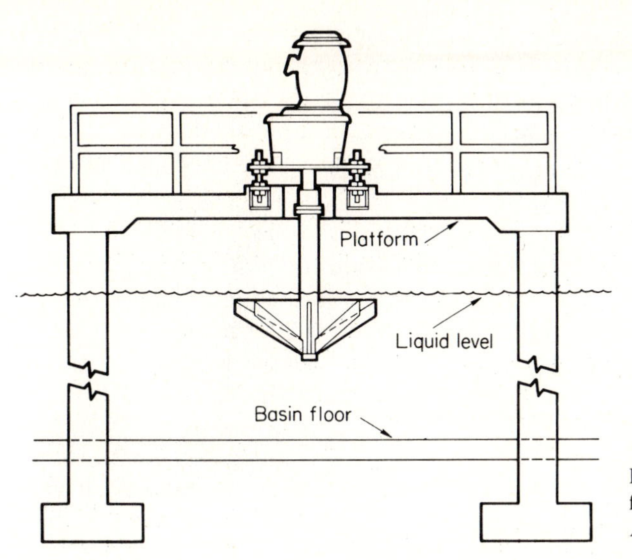

Figure 24-21 Low-speed fixed platform surface aerator. *(Courtesy Ashbrook-Simon-Hartley Inc.)*

1800 r/min, and *low-speed* designs employing a transmission (Fig. 24-21) which turn at 40 to 50 r/min. The latter are considerably more expensive than high-speed units but have fewer mechanical problems and are more desirable from a process standpoint since they produce less shearing of the biological floc. Finely divided solids tend to be difficult to remove and to thicken.

High-speed aerators may require addition of draft tubes to insure proper mixing in deep basins, or anti-erosion assemblies to prevent scour in shallow containers. The unmodified operating depth depends upon aerator design and power, ranging from 1.2 to 2.4 m (4 to 8 ft) for small units to 3 to 5 m (10 to 15 ft) for the largest available (about 90 kW).

The effective mixing zone is dependent upon depth of liquid, unit design, and power. Recommendations for proper spacing must be obtained from the manufacturer.

Selection of mechanical aerators is based upon consideration of both oxygen transfer and mixing requirements. The power required for *mixing* depends upon the basin configuration and normally ranges from 13 to 26 kW/1000 m^3 of basin volume. *Aeration* requirements depend upon the efficiency of the unit, the nature of the waste, the altitude, the temperature, and the desired dissolved oxygen concentration. These factors are related in Eq. (24-15).

$$M = M_0\left[\frac{\beta C_w - C_L}{9.17}(1.024)^{T-20}\alpha\right] \tag{24-15}$$

where M = the rate of oxygen transfer in kg/MJ under field conditions
β = a salinity correction, usually taken as 1
C_w = the oxygen saturation concentration at the given temperature and altitude

C_L = the operating oxygen concentration (at least 2.0 mg/l in activated sludge systems)
α = an oxygen transfer correction for the waste, usually 0.8 to 0.85
T = the temperature in °C; and
M_0 = the manufacturer's rating of the equipment at 20°C, 0 mg/l DO in pure water.

Example Determine the power required to provide 1000 kg of O_2 per day to a waste treatment process. The treatment plant is located at an elevation of 500 m, the desired dissolved oxygen concentration is 2.5 mg/l, $\alpha = 0.80$, and the maximum and minimum waste temperature are 30 and 10°C respectively. The manufacturer reports a transfer rate under standard conditions of 0.5 kg O_2/MJ.

SOLUTION The solubility of oxygen at 10 and 30°C may be obtained from App. II and is 11.2 and 7.6 mg/l respectively. The effect of altitude may be calculated from:

$$C_w = C_T\left(1 - \frac{\text{Altitude (m)}}{9450}\right) \tag{24-16}$$

Hence, at 500 m:

$$C_w = C_T(1 - 500/9450)$$
$$= 0.95\ C_T$$

At the lower temperature, then

$$M = 0.50\left[\frac{(1)(0.95 \times 11.2) - 2.5}{9.17}(1.025)^{-10}(0.80)\right]$$
$$= 0.28\ \text{kg/MJ}$$

At the higher temperature:

$$M = 0.50\left[\frac{(1)(0.95 \times 7.6) - 2.5}{9.17}(1.025)^{10}(0.80)\right]$$
$$= 0.26\ \text{kg/MJ}$$

The power required per day is thus

$$P = 1000/0.26 = 3846\ \text{MJ}$$

and the total connected load is 44 kW.

Mixing and aeration is also effected by *horizontally rotating* devices which act, in a sense, as paddlewheels. Machines of this sort are usually employed in relatively shallow basins and the mechanical drive serves primarily to keep the liquid in motion at a velocity sufficient to prevent deposition of solids. A large fraction of the oxygen is transferred through the free surface of the liquid.

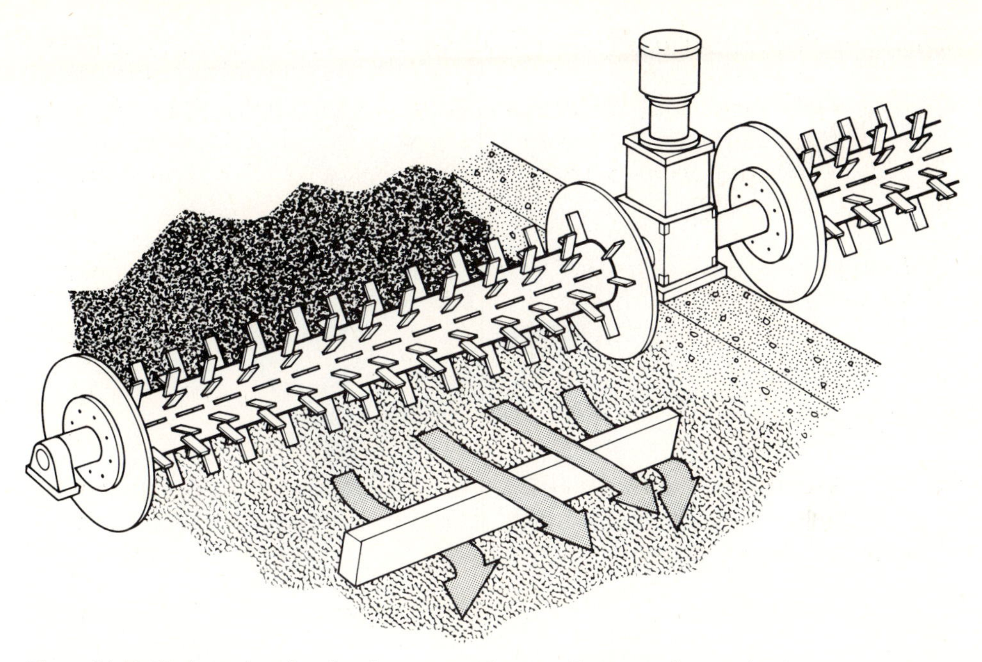

Figure 24-22 Horizontal surface brush aerator. *(Courtesy Passavant Corporation.)*

Systems employing shallow basins and horizontal rotors such as those in Figs. 24-22 and 24-23 are called *oxidation ditches* and are usually operated as extended aeration devices. Recommendations for unit size and power requirements are available from equipment manufacturers.

Any surface aerator may be adversely affected by ice formation during the winter months. Some designs are claimed to be less affected than others and some

Figure 24-23 Horizontal surface aerator. *(Courtesy Envirex, a Rexnord Company.)*

Figure 24-24 Jet aeration system. *(Courtesy Pentech Division, Houdaille Industries Inc.)*

incorporate electrical heating of critical areas. The manufacturers recommendations for cold weather operation should be adhered to.

Diffused air systems may rely entirely upon the mixing provided by the air or may employ air entrainment (Fig. 24-24) or a mechanically driven sparger like that in Fig. 24-25. Mechanical spargers are used in some high-purity oxygen systems.

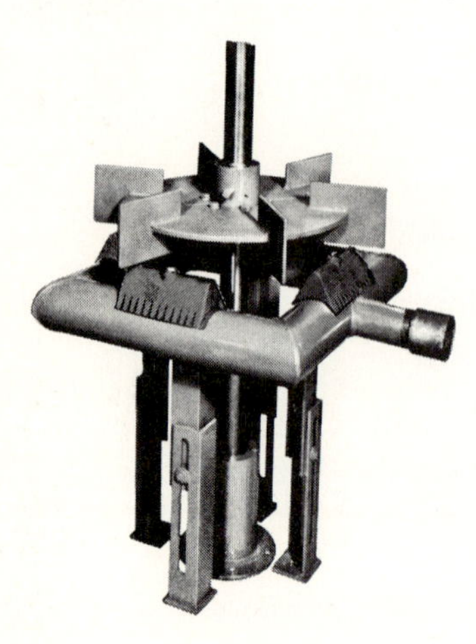

Figure 24-25 Mechanical sparger.

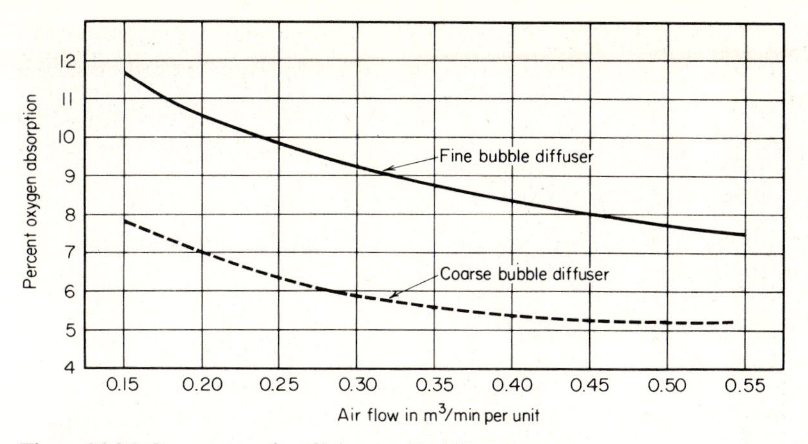

Figure 24-26 Oxygen transfer efficiency of bubble aerators.

Air requirements may be calculated as shown in the example of Art. 24-14. The efficiency of transfer of diffusion and sparging devices seldom exceeds 8 percent in actual practice. Figure 24-26 presents manufacturer's data for fine bubble and coarse bubble diffusers which are representative of available equipment.

The power required to transfer the air to the bottom of the aeration basin depends upon the pressure at the compressor, which in turn is a function of the depth of submergence of the diffuser and the headloss through the diffuser, air piping, and valves. Techniques for determining the headloss in air pipe and fittings are presented in references 14 and 18. Velocities range from 6 m/s (20 ft/s) in small (100 mm) pipe to 33 m/s (100 ft/s) in large (1520 mm) lines. Headlosses in pipe headers are generally on the order of 100 to 200 mm (4 to 8 in) of water while headlosses through diffusers range to 400 or 500 mm (16 to 20 in) of water. Overall pressure loss usually approximates 25 percent of diffuser submergence in an economical design.

As an approximation, the friction factor of steel pipe carrying air may be taken to be:

$$f = 0.029 \frac{D^{0.027}}{Q^{0.148}} \tag{24-17}$$

where D = the pipe diameter in meters and
Q = the air flow in m^3/min

The headloss in straight pipe can then be calculated from

$$H = f \frac{L}{D} h_v \tag{24-18}$$

where H = the head loss in millimeters
L = the pipe length in meters
D = the pipe diameter in meters, and
h_v = the velocity head in millimeters of water.

Alternately, the headloss may be approximated from:

$$H = 9.82 \times 10^{-8} \frac{fLTQ^2}{PD^5} \tag{24-19}$$

where T = the temperature in °K
P = the pressure in atmospheres

and the other terms are as given above. The absolute temperature may be estimated from the pressure rise using

$$T_2 = T_1(P_2/P_1)^{0.283} \tag{24-20}$$

Example Determine the headloss in 305 m of 380-mm steel pipe which carries 96.3 m^3/min of air at a pressure of 0.54 atm (gauge). Ambient temperature is 30°C.

SOLUTION The temperature in the pipe is:

$$T_2 = (273 + 30)(1.54/1)^{0.283}$$

$$= 340 \text{ °K}$$

The friction factor is approximately:

$$f = 0.029 \frac{(0.380)^{0.027}}{(96.3)^{0.148}} = 0.0144$$

and the head loss is

$$H = 9.82 \times 10^{-8} \frac{(0.0144)(305)(340)(96.3)^2}{(1.54)(0.380)^5}$$

$$= 110 \text{ mm of water}$$

Fittings in air distribution systems may be converted into equivalent length of straight pipe using the formula:

$$L = 55.4CD^{1.2} \tag{24-21}$$

where L = the equivalent length in meters
D = the pipe diameter in meters and
C = a factor given in Table 24-4.

Compressors for use in diffused air systems may be either positive displacement rotary lobe or centrifugal designs (Figs. 24-27 and 24-28), depending upon the specific application. For pressures up to 50 to 70 kPa (7 to 10 psi) and flows above 15 m^3/min (5000 ft^3/min) centrifugal blowers are preferable, particularly since the flow may be adjusted by throttling the inlet. Throttling the outlet of centrifugal blowers is not advisable since these machines will surge if throttled to close to their shutoff head. Throttling the inlet lowers the discharge by changing the characteristic curve rather than by increasing the head.

Table 24-4 Resistance factors for air fittings[14]

Fitting	C
Gate valves	0.25
Long radius ell or run of standard tee	0.33
Medium radius ell or run of tee reduced 25 percent	0.42
Standard ell or run of tee reduced 50 percent	0.67
Angle valve	0.90
Tee through side outlet	1.33
Globe valve	2.00

Positive displacement rotary lobe compressors are used primarily in small plants where air flow is low. The discharge of these machines can be altered only by varying their speed—which is a disadvantage. They are useful when pressures are over 40 to 50 kPa (6 to 7 psi) and flows are less than 15 m^3/min (5000 ft^3/min). Positive displacement blowers should not be throttled. Throttling will cause overheating and overpressure and result in damage to the machine.

In many wastewater applications single-stage centrifugal blowers can provide the head required [usually about 6 m (20 ft) of water]. If higher pressures are

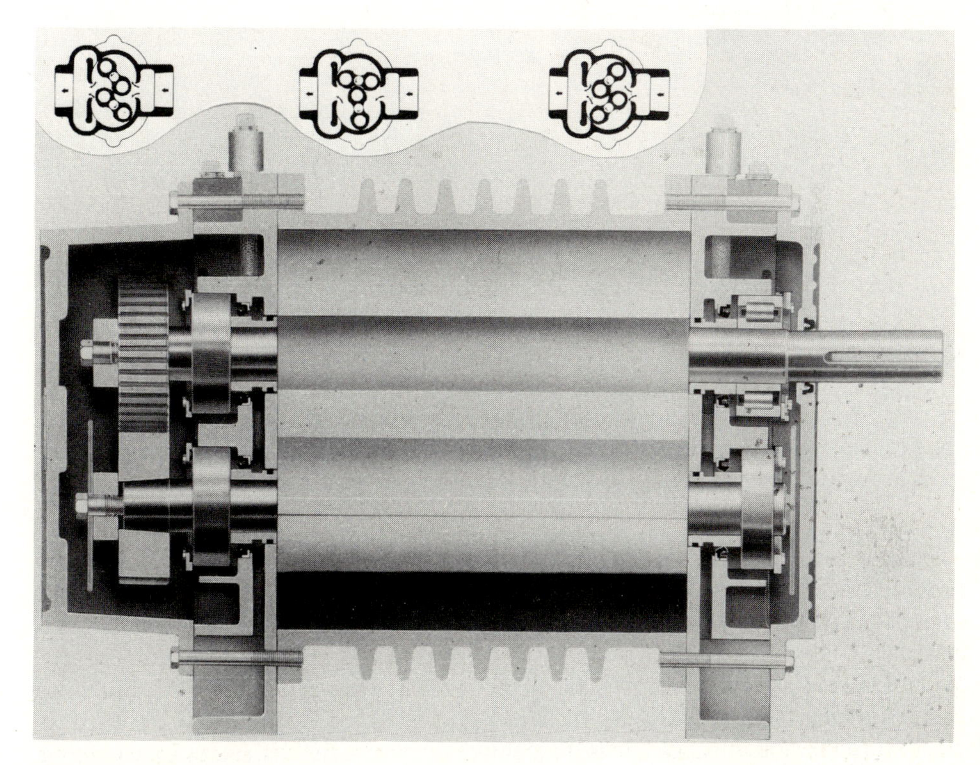

Figure 24-27 Rotary lobe compressor. *(Courtesy Dresser Industries, Inc.—Roots Blower Division.)*

Figure 24-28 Centrifugal blower. *(Courtesy Dresser Industries, Inc.—Roots Blower Division.)*

required, multistage blowers capable of pressures to 28 m (90 ft) of water are available.

Compressor power requirements may be estimated from the air flow, discharge and inlet pressures, and air temperature by using Eq. (24-22), which is based upon an assumption of adiabatic conditions:

$$P = \frac{wRT_1}{8.41e}\left[\left(\frac{P_2}{P_1}\right)^{0.283} - 1\right] \tag{24-22}$$

where P = the power required in kW
w = the air mass flow in kg/s
R = the gas constant (8.314)
T = the inlet temperature in °K
P_1 = the absolute inlet pressure in atm
P_2 = the absolute outlet pressure in atm, and
e = the efficiency of the machine (usually 70 to 80 percent)

Example Determine the power required to provide 1000 kg of O_2 per day through a diffused air system with a transfer efficiency of 7 percent. Assume the inlet temperature is 30°C, the discharge pressure is 5 m of water, and the efficiency of the compressor is 75 percent.

SOLUTION The oxygen required is 1000/0.07 = 14,286 kg/day, which represents an air flow of 14,286/0.232 = 61,578 kg/day. The temperature is 303°K, $P_1 = 1$ atm, $P_2 = 1.58$ atm, thus

$$P = \frac{(0.71)(8.314)(303)}{(8.41)(0.75)}\left[\left(\frac{1.58}{1}\right)^{0.283} - 1\right]$$

$$= 39.2 \text{ kW}$$

This value may be compared to that obtained above for a surface aerator under similar conditions.

The procedures outlined above will permit tentative selection of aeration equipment based upon the manufacturers' claimed efficiency at standard conditions. The specification of equipment should be made upon a performance basis in the actual system. The specification should address oxygen requirements in the system at a range of MLSS concentrations, dissolved oxygen concentrations, and temperatures; maximum power use allowable; mixing; noise, etc.[25]

24-17 Suspended Growth Process Clarifier Design

The design of the secondary clarifier is critically important to the operation of a suspended growth process. Not only do the solids which pass the clarifier contribute to the BOD of the effluent, but their loss may interfere with maintenance of the required sludge age in the biological reactor. The clarifier has a thickening function as well as one of clarification since a reasonable underflow density is necessary for solids recycle.

The designer must consider the peak flow rate which is likely to enter the clarifier and its effect upon surface overflow rate, weir loading rate, and solids loading rate. The sludge handling equipment should be sized to recirculate up to 100 percent Q in order to provide for short-term overloads.

Clarifiers may be rectangular or circular. The length of rectangular basins and the diameter of circular basins ordinarily does not exceed 10 times the depth, but dimensions are not critical. The important consideration is that the sludge collector be of high capacity to insure rapid removal in ordinary operation, high flow rates during unusual conditions, and capability to remove dense sludges which may accumulate during clarifier shutdown. Suction sludge removal systems (Fig. 24-29) are particularly suitable in secondary clarifiers and are available in configurations to fit both rectangular and circular basins.

Inlet baffles must be carefully designed (Chap. 23) since density currents are more pronounced in secondary than in primary clarifiers. Horizontal velocities

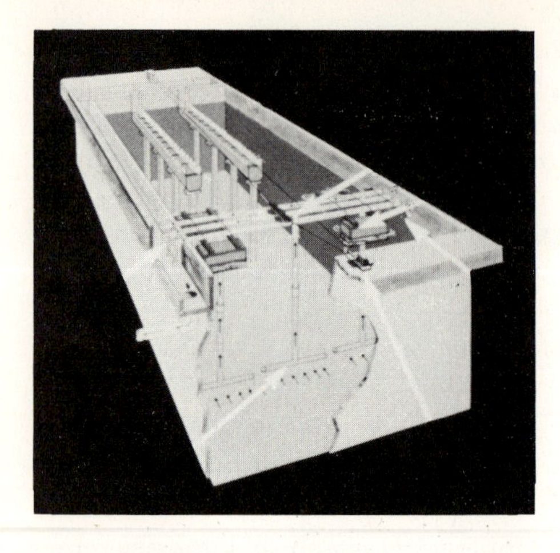

Figure 24-29 Vacuum sludge removal system. *(Courtesy Leopold Co., Division of Sybron Corp.)*

are limited to about 0.5 m/min (1.5 ft/min) in rectangular basins and the annular-inlet baffle in center-fed circular tanks should have a diameter of 15 to 20 percent that of the tank itself.

Solids loading is important in secondary clarifiers since hindered settling is likely to occur and the settling velocity of the discrete particles may not govern the basin design (Art. 9-6). The solids loading rate is expressed in kg SS/m^2 per hour and is related to the surface overflow rate by

$$\text{kg/m}^2 \text{ per hour} = \frac{\text{MLSS(mg/l)} \times \text{SOR(m/day)}}{24 \times 10^3} \tag{24-23}$$

Typical solids loading rates vary from 2.5 to 6.2 kg/m^2 per hour at average and peak loading conditions respectively. Surface overflow rates at peak flow, based upon wastewater flow not including recirculation, as recommended by Metcalf and Eddy[18] are presented in Table 24-5.

Table 24-5 Recommended surface overflow rates at various mixed liquor suspended solids concentrations and recycle rates (m/day)[18]

	Recycle percent	
MLSS, mg/l	25	50
500	57	57
1000	57	57
1500	49	49
2000	49	49
2500	47	39
3000	39	33
3500	33	28

These values are based upon a maximum solids loading of 6.2 kg/m^2 per hour

Weir loading rates should be kept below 370 m^3/m per day (30,000 gal/ft^2 per day) for weirs located away from the tank periphery in water at least 3.5 m (12 ft) deep, and below 250 m^3/m per day (20,000 gal/ft^2 per day) if located either against the wall where density currents might turn upward or in shallower basins. These loading rates are based upon peak flow rate. Loadings at average flow in small basins should be considerably less. The upward velocity in the vicinity of the weir should not exceed 3 to 5 m/hour (10 to 15 ft/hour).

Example Design a secondary clarifier for an activated sludge process with a recycle rate of 30 percent, a MLSS concentration of 3000 mg/l, and an anticipated peak flow rate of 10,000 m^3/day.

SOLUTION The solids loading per hour is

$$3000 \text{ mg/l} \times 10 \times 10^6 \text{ l/24 hour} \times 1.30 = 1625 \text{ kg/hour}$$

For a peak loading of 6.2 kg/m^2 per hour the basin area is:

$$A = 1625/6.2 = 262 \text{ m}^2$$

and the surface overflow rate based on the waste flow is

$$\text{SOR} = 10{,}000/262 = 38 \text{ m/day}$$

With a clarifier depth of 3.5 m the volume is 917 m^3 and the retention time 2.2 hours at peak flow. With a weir overflow rate of 250 m^3/m per day the weir length is 40 m. A circular clarifier 18.3 m in diameter will provide an area of 263 m^2 and a peripheral weir length of 57.5 m.

24-18 Operational Problems of Activated Sludge Systems

Operational problems of activated sludge systems generally hinge upon inability to maintain the desired sludge age. Sludge may be lost from a clarifier as a result of bulking or floating, even when the clarifier is properly designed from a standpoint of solids loading.

Floating sludge results from denitrification (Chap. 26) and is associated with high sludge age and long solids retention time in the clarifier. In activated sludge processes with a sludge age in excess of 10 days and dissolved oxygen in excess of 2 mg/l, a degree of conversion of ammonium to nitrate will occur. If the sludge remains in the clarifier too long, its endogenous metabolism will exhaust the oxygen available and the bacteria will then reduce any nitrate present to nitrogen gas. The resulting bubbles may then buoy the sludge to the surface.

Floating sludge can be controlled by increasing the recirculation rate, thereby shortening the solids detention time in the clarifier, or by increasing the rate of solids wasting, thereby reducing the sludge age.

Bulking sludge results from the presence of filamentous microorganisms or from entrained water in individual bacterial cells. Filamentous growth occurs in wastes deficient in nitrogen or other nutrients or at a low pH. The problem can

Table 24-6 Approximate nutritional requirements in activated sludge (mg/mg of BOD)

Element	Concentration
Nitrogen	0.050
Phosphorus	0.016
Sulfur	0.004
Sodium	0.004
Potassium	0.003
Calcium	0.004
Magnesium	0.003
Iron	0.001
Molybdenum	Trace
Cobalt	Trace
Zinc	Trace
Copper	Trace

usually be corrected by providing the lacking nutrients and a neutral pH. The nutritional requirements of activated sludge are approximately as presented in Table 24-6. The values listed are not absolute minima, but amounts which are known to be adequate.

Poorly settling bacterial suspensions may be associated with deficient dissolved oxygen or a sludge age which is either too short, or, more usually, too long. As a general rule the sludge age should be maintained in the range of 5 to 15 days.

Clarifiers which have pronounced density currents may permit short circuiting with loss of fresh sludge and accumulation of a mixed liquor with an average cell retention time which is different from that indicated by the sludge age.

Bulking sludge can be corrected in the short term by chlorination at a rate of about 0.5 to 1 percent of the sludge concentration. The organisms which are killed may appear in the plant effluent, producing a brief increase in BOD.

24-19 Completely Mixed Processes without Solids Recycle

The preceding articles have dealt with processes which control sludge age (θ_c) by returning solids from a secondary clarifier. Other, somewhat simpler, dispersed growth secondary processes do not employ solids return and may be represented schematically as shown in Fig. 24-30.

The soluble effluent BOD_5 from such a process is given by:

$$S = \frac{K_s(1 + k_d\theta_c)}{Yk\theta_c - k_d\theta_c - 1} \tag{24-24}$$

in which the terms are as defined in Arts. 24-12 and 24-13. The concentration of

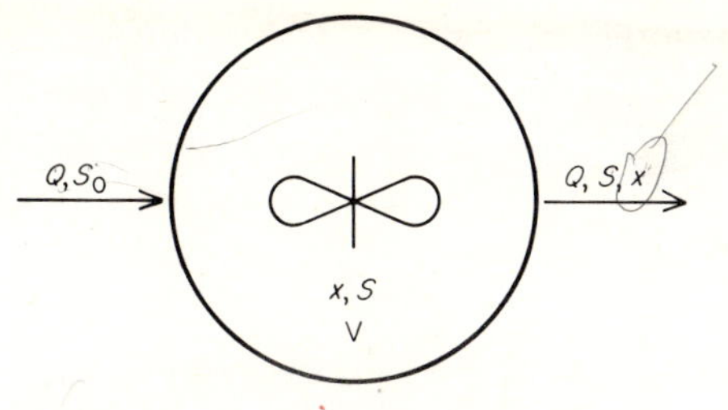

Figure 24-30 Completely mixed biological reactor without solids return.

solids in the reactor and effluent is given by:

$$x = \frac{Y(S_0 - S)}{1 + k_d\theta_c} \tag{24-25}$$

Equations (24-24) and (24-25) permit the following observations to be drawn. S, the effluent soluble BOD_5, is not dependent upon S_0, the influent BOD_5, but only upon various rate constants and the solids retention time (θ_c). Since no solids return is practiced, θ_c is equal to θ, the hydraulic retention time (V/Q). The effluent suspended solids concentration is, however, dependent upon both S_0 and θ_c. Hence the total effluent BOD_5 (to which the biological solids contribute) is affected by waste strength.

Processes which may be represented by the mathematical model above include oxidation ponds whether aerobic, anaerobic, or facultative, and biological sludge digestion systems (Chap. 25).

Example Determine the effluent BOD_5 to be expected from an oxidation pond treating raw domestic sewage with a BOD_5 of 250 mg/l. Assume $Y = 0.65$, $K_s = 20$, $k = 5$, $k_d = 0.05$ and that the hydraulic retention time is 30 days.

SOLUTION From Eq. (24-24), the effluent soluble BOD_5 is:

$$S = \frac{20(1 + 0.05[30])}{0.65(5)(30) - 0.05(30) - 1}$$

$$= 0.50 \text{ mg/l}$$

The suspended solids concentration, from Eq. (24-25) is:

$$x = \frac{0.65(250 - 0.5)}{1 + 0.05(30)}$$

$$= 65 \text{ mg/l}$$

The BOD_5 of the solids is estimated to be $0.63(x)$[18], thus the total BOD_5 of the effluent is:

$$BOD_5 = 0.50 + 0.63(65)$$

$$= 41.5 \text{ mg/l}$$

The actual measured BOD_5 at any time might be higher or lower, depending upon the instantaneous suspended solids concentration at the outlet. The latter could be affected by quiescent or windy periods, algal blooms, etc.

24-20 Oxidation Ponds

Oxidation or stabilization ponds are a relatively low-cost treatment system which has been widely used, particularly in rural areas. The ponds may be considered to be completely mixed biological reactors without solids return. The mixing is usually provided by natural processes (wind, heat, fermentation) but may be augmented by mechanical or diffused aeration.

Aerobic ponds are generally constructed to operate at a depth between 1 and 1.5 m (3 to 5 ft). Shallower levels will encourage growth of rooted aquatic plants, while greater depth may interfere with mixing and oxygen transfer from the surface.

Both aerobic and facultative ponds are biologically complex. The general reactions which occur are illustrated schematically in Fig. 24-31. Incoming organic matter is oxidized by bacteria to yield NH_3, NO_3^-, CO_2, SO_4^{--}, H_2O and the other end products of aerobic metabolism. These materials are then used by algae to produce more algal cells and, during daylight, oxygen which supplements that provided by wind action and is used by the bacteria to decompose the original waste.

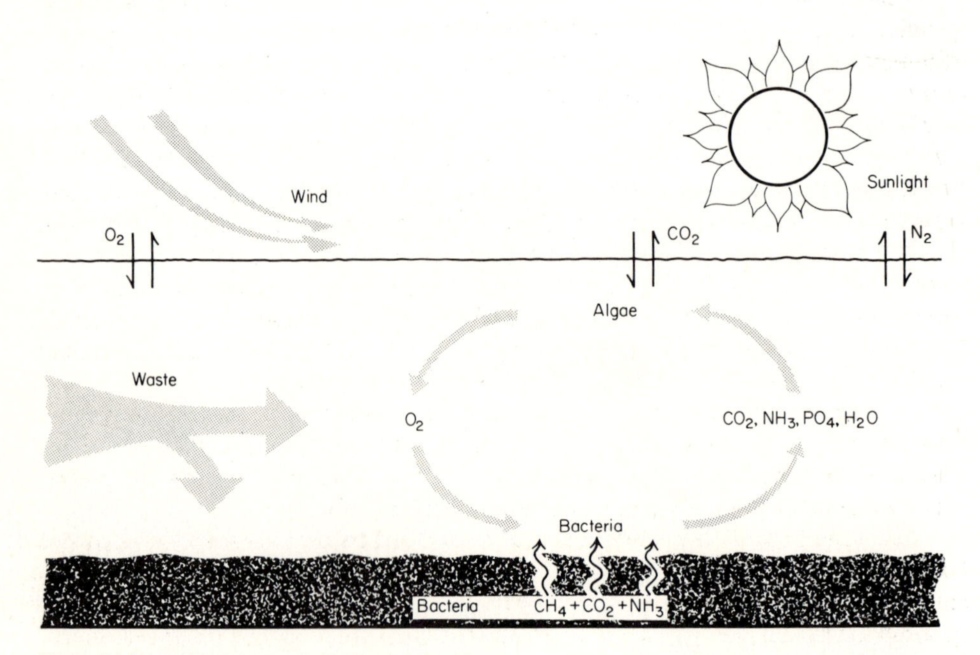

Figure 24-31 Oxidation pond schematic diagram.

The symbiotic relation between bacteria and algae leads to stabilization of the incoming waste as indicated in the example of Art. 24-19, but may not yield the effluent BOD_5 calculated from the model due to the presence of algal cells.

Design standards are usually specified by state regulatory agencies. Typical standards include embankment slopes (1:3 to 1:4), organic loading rate (2.2 to 5.5 g BOD_5/m^2 per day (20 to 50 lb/acre per day), depending on climate), hydraulic retention time (60 to 120 days), and permissible seepage through the basin bottom (0 to 6 mm/day). In some climates it may be possible to operate ponds with no discharge to surface waters (Art. 21-19).

As illustrated in the example of Art. 24-19, the soluble effluent BOD_5 from an aerobic pond will be quite low. The total BOD_5, however, will seldom meet the standards for secondary treatment specified by EPA, particularly if algal growth is heavy.

It has been argued that algal cell BOD is not comparable to raw waste BOD—in the same fashion that a dead fish is not comparable to a live one. The argument is simplistic since algal cells do exert a BOD (about 1.2 mg/mg cell mass),[26] may be toxic to other aquatic biota,[27] and can produce a variety of other problems in receiving waters. The desire to continue the use of oxidation ponds in rural areas is understandable, but such use must be justified by economic considerations and evaluation of actual environmental effects, not by claiming there are no effects. EPA has established somewhat looser effluent standards than 30 mg/l BOD and SS for oxidation ponds with flows less than 7500 m^3/day (2 mgd).

Aerated ponds are aerobic systems in which the natural oxygenation afforded by wind and algal action is supplemented by mechanical or diffused aeration. High-speed floating aerators (Fig. 24-20) with anti-erosion devices are used in this application in shallow ponds which have reached an overloaded condition. In ponds initially designed to be aerated both high- and low-speed aerators and static-tube aerators are employed. The latter consist of a cylinder containing a helical, or in some cases, a double helical core (Fig. 24-32). Air is released from pipe distributors through a coarse diffuser at the bottom of the tubes. The upward flow is constrained to follow the helical path, which extends its contact time and improves oxygen transfer. The airflow carries water with it and thus improves mixing in the basin. Simple perforated-tube diffusers have also been developed, but exhibit a tendency to plug and require considerable maintenance for that reason.

Air requirements in aerated ponds may be calculated by the technique illustrated in the example of Art. 24-13. Power requirements are determined as illustrated in Art. 24-15. In aerated ponds *mixing* generally governs the power required, since organic loading rates are moderately low.

Facultative ponds are sufficiently deep (> 2 m) to provide separation of the basin into three horizontal strata. The pond depth inhibits mixing, hence organic solids which settle will remain on the bottom and will be subjected to anaerobic decomposition. In the liquid above the sludge layer there will be a zone in which facultative bacteria oxidize the incoming organics and the products of anaerobic

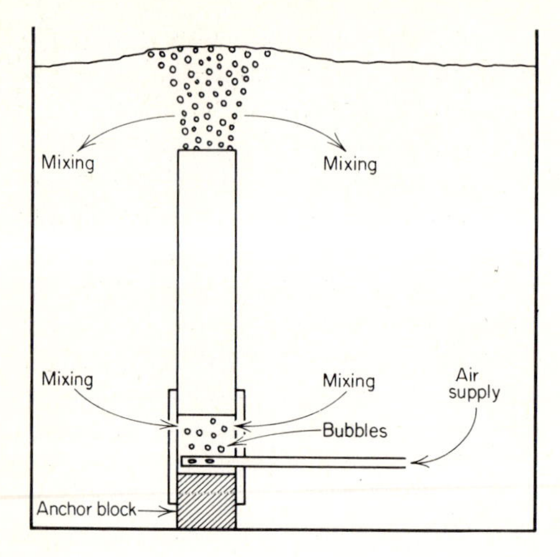

Figure 24-32 Static tube aerator.

decomposition, while at the surface the algae and bacteria will interact as illustrated above.

Surface aerators are sometimes used to assure that an aerobic zone will exist at the surface. The mixing provided must be sufficient to transfer the required oxygen, yet not so great as to mix the pond, or the benefit of anaerobic decomposition will be lost. The maximum applied power is limited to 1 to 2 kW/1000 m^3. BOD loadings are then based upon the installed aeration capacity.

Anaerobic ponds are normally deep and are subjected to relatively high organic loads. The principles of their operation are similar to those of anaerobic digesters (Chap. 25). The effluent is not suitable for discharge to a stream and must be subjected to further treatment. Anaerobic ponds are most commonly used as a pretreatment for strong industrial wastes, particularly those which are warm.

Metcalf and Eddy[18] present detailed theoretical calculations for the design of stabilization ponds. The use of such an approach is preferable to the application of simple design criteria specified by regulatory agencies, although the latter must be adhered to in most instances.

24-21 Upgrading Stabilization Pond Effluent

As noted above, the effluent of oxidation ponds will not ordinarily meet the standards for secondary treatment established by EPA. In recent years considerable research has been conducted and techniques for polishing lagoon effluent have been developed. The best demonstrated procedures include intermittent sand filtration,[1] dissolved air flotation,[28] coagulation and sedimentation, and filtration.[29] None of these processes is inexpensive and the combined cost of lagoon plus polishing may equal or exceed that of other secondary treatment systems. The major advantages of lagoon systems are the minor skill required in their operation and the minimal power required. Adding complex physical-chemical processes following the lagoon removes these advantages and will often reverse the relative cost of lagoons and secondary systems.

A possible moderate cost technique involves a two-stage lagoon with the second stage serving as an occasional coagulation and sedimentation basin.[26] Chemicals are distributed by boat, and the clarified effluent is discharged before the solids content can increase again. This procedure has not been proven by long-term operation. Multistage lagoon systems provide better, but somewhat erratic, bacterial reduction than single-cell lagoons, however they do not provide significant reduction in suspended solids or BOD.

Microscreens or microstrainers have been suggested as a possible means of upgrading lagoon effluent. Since the bulk of the algal cells are smaller than the smallest screen opening (23 μm) efficient removal by this technique is not likely and has not been demonstrated.[30]

PROBLEMS

24-1 Determine the filter volume required in each of two identical filters if the waste in the NRC example of Art. 24-7 is to be treated to yield an effluent BOD of 30 mg/l.

24-2 Rework the example of Art. 24-7 which presents the application of the Velz equation to determine the depth required to yield an effluent BOD of 30 mg/l.

24-3 Find the diameter required for a single-stage trickling filter which is to yield an effluent BOD_5 of 20 mg/l when treating settled domestic sewage with a BOD_5 of 120 mg/l. The wastewater flow is 2200 m^3/day and the recirculation is constant at 4000 m^3/day. The filter depth is 1.5 m.

24-4 Determine the effluent quality of the filter of Prob. 24-3 when it is subjected to a peak flow of 3800 m^3/day with a settled BOD_5 of 150 mg/l. The recirculation flow is unchanged.

24-5 An activated sludge plant treats a wastewater flow of 5000 m^3/day with a BOD_5 of 180 mg/l. The organic loading rate is 0.3 kg BOD_5 per kg MLSS and the hydraulic retention time is 5 hours. Determine the concentration of MLSS in mg/l.

24-6 If the reactor in Prob. 24-5 is operated at a sludge age of 10 days determine the sludge production rate. If the sludge is concentrated in the clarifier underflow to 1.5 percent solids content, what will be the waste sludge flow?

24-7 Design an activated sludge process to treat a waste flow of 15,000 m^3/day with a BOD_5 of 180 mg/l following primary treatment. The effluent BOD_5 and SS are to be less than 20 mg/l. Assume $x_r = 15{,}000$ mg/l, $x = 2500$ mg/l, $\theta_c = 10$ days, $Y = 0.60$, $k_d = 0.05$. Determine the reactor volume, the sludge production rate, the recirculation rate, the hydraulic retention time, and the oxygen required.

24-8 A wastewater has an average BOD_5 of 250 mg/l. The average flow is 8000 m^3/day, the minimum 3000 m^3/day, and the maximum 13,500 m^3/day. Design a complete activated sludge system, including primary and secondary clarifiers, to yield an effluent BOD_5 and SS of 30 mg/l at peak flow. What do you estimate the effluent BOD_5 will be at average flow? Assume $x_r = 1$ percent (min), MLSS = 3000 mg/l, $\theta_c = 7.5$ days (min), $Y = 0.65$, $k_d = 0.04$.

24-9 Determine the power required to aerate a wastewater which has a calculated oxygen demand of 1800 kg O_2 per day. The plant is at 1000-m elevation, the required DO is 2.2 mg/l, $\alpha = 0.85$, and the temperature ranges from 5 to 35°C. The equipment manufacturer claims a rate of 0.6 kg/MJ at standard test conditions.

24-10 Rework Prob. 24-9 for sea level..

24-11 Determine the pressure drop in 275 m of 305-mm steel pipe which contains 6 long radius ells, two standard ells, and two gate valves. The air flow is 50 m^3/min and the pressure in the pipe is 1.6 atm. The temperature range is 5 to 35°C.

24-12 How long a retention time is necessary in an oxidation pond to provide an effluent soluble BOD_5 of 1.0 mg/l? 10 mg/l? Why are oxidation ponds not designed to operate at such retention times?

REFERENCES

1. Harris, Steven E., et al.: "Intermittent Sand Filtration for Upgrading Waste Stabilization Pond Effluents," *Journal Water Pollution Control Federation*, **49:**1:83, 1977.
2. Mistry, Khurshed J. and David M. Himmelblau: "Stochastic Analysis of Trickling Filter," *Journal Environmental Engineering Division Proceedings American Society of Civil Engineers*, **101:EE3:**333, 1975.
3. "Sewage Treatment at Military Installations," *Sewage Works Journal*, **18:**5:787, 1946.
4. Rankin, R. S.: "Evaluation of the Performance of Biofiltration Plants," *Transactions, American Society of Civil Engineers*, **120:**823, 1955.
5. Velz, C. J.: "A Basic Law for the Performance of Biological Beds," *Sewage Works Journal*, **20:**4, 1948.
6. Eckenfelder, W. W., Jr.: "Trickling Filter Design and Performance," *Transactions, American Society of Civil Engineers*, **128,** 1963.
7. Jank, Bruce E. and W. Ronald Drynan: "Substrate Removal Mechanism of Trickling Filters," *Journal Environmental Engineering Division, Proceedings American Society of Civil Engineers*, **99:EE3:**187, 1973.
8. Ames, W. F., V. C. Behn, and W. Z. Collings: "Transient Operation of the Trickling Filter," *Journal Sanitary Engineering Division, Proceedings American Society of Civil Engineers*, **88:SA3:**21, 1962.
9. Atkinson, Bernard and John A. Howell: "Slime Holdup, Influent BOD, and Mass Transfer in Trickling Filters," *Journal Environmental Engineering Division, Proceedings American Society of Civil Engineers*, **101:EE4:**585, 1975.
10. Williamson, Kenneth and Perry L. McCarty: "A Model of Substrate Utilization by Bacterial Films," *Journal Water Pollution Control Federation*, **48:**1:9, 1976.
11. Mueller, James A.: "Oxygen Theory in Biological Treatment Plant Design—Discussion," *Journal Environmental Engineering Division, Proceedings American Society of Civil Engineers*, **99:EE3:**381, 1973.
12. "Filtering Materials for Sewage Treatment Plants," *Manual of Practice No. 13*, American Society of Civil Engineers, New York.
13. "Process Design Manual for Upgrading Existing Wastewater Treatment Plants," U.S. Environmental Protection Agency, Office of Technology Transfer, Washington, D.C., 1974.
14. "Sewage Treatment Plant Design," *Manual of Practice No. 36*, American Society of Civil Engineers, New York.
15. Schroeder, Edward D.: *Water and Wastewater Treatment*, McGraw-Hill, New York, 1977.
16. "Recommended Standards for Sewage Works," Great Lakes–Upper Mississippi Board of State Sanitary Engineers, Health Education Service, Albany, N.Y., 1971.
17. Cook, Echol E. and Leonard Crane: "Effects of Dosing Rates on Trickling Filter Performance," *Journal Water Pollution Control Federation*, **48:**12:2723, 1976.
18. Metcalf and Eddy, Inc.: *Wastewater Engineering: Collection, Treatment, Disposal*, McGraw-Hill, New York, 1972.
19. Antonie, Ronald L., David L. Kluge, and John H. Mielke: "Evaluation of a Rotating Disk Wastewater Treatment Plant," *Journal Water Pollution Control Federation*, **46:**3:498, 1974.
20. Antonie, Ronald L.: "Evaluation of a Rotating Disk Wastewater Treatment Plant—Discussion," *Journal Water Pollution Control Federation*, **46:**12:2792, 1974.
21. Jenkins, D. and W. E. Garrison: "Control of Activated Sludge by Mean Cell Residence Time," *Journal Water Pollution Control Federation*, **40:**11:1905, 1968.
22. Lawrence, Alonzo W. and Perry L. McCarty: "Unified Basis for Biological Treatment Design and Operation," *Journal Sanitary Engineering Division, Proceedings American Society of Civil Engineers*, **96:SA3:**757, 1970.
23. Parker, Denny S. and M. Steve Merrill: "Oxygen and Air Activated Sludge: Another View," *Journal Water Pollution Control Federation*, **48:**11:2511, 1976.
24. Kalinske, A. A.: "Comparison of Air and Oxygen Activated Sludge Systems," *Journal Water Pollution Control Federation*, **48:**11:2472, 1976.

25. Stukenberg, John R., Valery N. Wahbeh, and Ross E. McKinney: "Experiences in Evaluating and Specifying Aeration Equipment," *Journal Water Pollution Control Federation*, **49:** 1:66, 1977.
26. Friedman, A. A., David A. Peaks, and R. L. Nichols: "Algae Separation from Oxidation Pond Effluent," *Journal Water Pollution Control Federation*, **49:**1:111, 1977.
27. Slawson, Guenton C., Jr. and Lorne G. Everett: "Segmented Population Model of Primary Productivity," *Journal Environmental Engineering Division, Proceedings American Society of Civil Engineers*, **101:EE1:**127, 1976.
28. Stone, R. W., D. S. Parker, and J. A. Cotteral: "Upgrading Lagoon Effluent for Best Practical Treatment," *Journal Water Pollution Control Federation*, **47:**8:2019, 1975.
29. Middlebrooks, E. Joe, et al.: "Techniques for Algae Removal from Wastewater Stabilization Ponds," *Journal Water Pollution Control Federation*, **46:**12:2676, 1974.
30. Parker, Denny S. and Warren R. Uhte: "Discussion—Technique for Algae Removal from Oxidation Ponds," *Journal Water Pollution Control Federation*, **47:**9:2330, 1975.

CHAPTER

TWENTY-FIVE

SLUDGE TREATMENT AND DISPOSAL

25-1 Importance

The bulk of the suspended solids which enter a waste-treatment plant and the waste solids generated in biological treatment must be handled as sludge at some point in the treatment process. The character and amount of solids depends to some extent upon the primary and secondary processes employed, and thus the choice of these systems may hinge upon anticipated problems with sludge handling. Vesilind[1] has noted that solids handling accounts for 30 to 40 percent of capital costs, 50 percent of operating costs, and 90 percent of the operational problems at sewage treatment plants.

25-2 Amount and Characteristics of Sludge

Sewage sludge consists of the organic and inorganic solids present in the raw waste and removed in the primary clarifier, plus organic solids generated in secondary treatment and removed in the secondary clarifier. The inorganic fraction may be assumed to have a specific gravity of about 2.5 while the organic matter has a specific gravity of 1.01 to 1.06 depending upon its source.

The quantity of suspended solids in raw domestic wastewater is typically 90 g/day per capita, and the concentration about 200 to 250 mg/l depending upon the flow. Of this, approximately 60 percent will be removed in primary clarification. The remainder is either oxidized in secondary treatment or incorporated in the biological mass. The solids generated in secondary processes average 0.4 to 0.5 kg/kg BOD_5 applied in fixed growth processes, and depend upon sludge age in suspended growth processes—ranging from 0.2 to 1.0 kg/kg BOD_5 applied.

Example Estimate the solids production from a trickling filter plant treating 1000 m^3/day with a BOD_5 of 210 mg/l and SS of 260 mg/l. Assume that primary clarification removes 30 percent of the BOD and 60 percent of the influent solids.

SOLUTION The removal in the primary is

$$0.60(260) = 156 \text{ mg/l}$$

Production in the secondary is:

$$0.70(210)(0.5) = 74 \text{ mg/l}$$

and the total solids production will be approximately

$$\frac{(156 + 74) \times 10^6}{10^6} = 230 \text{ kg/day}$$

The sewage solids, as noted in Chap. 24, will contain a greater or lesser amount of water—depending in part upon the processes involved. Trickling filter plus primary sludges may range from 5 to 10 percent solids, activated sludge plus primary from 2 to 5 percent solids, while activated sludge by itself may contain less than 1 percent solids. Expected solids production and moisture content for different processes are presented in Chap. 24.

The moisture associated with waste sewage solids is in part free, and separable by sedimentation; in part trapped in the interstices of floc particles, and separable by mechanical dewatering; in part held by capillary action, and separable by compaction; and in part chemically bound within or without the bacterial cell, and separable only by destruction of the cell. The relative proportions of water and solids in waste activated sludge are presented in Table 25-1.

The effect of moisture content upon sludge volume is tremendous, and sludge-handling techniques are directed toward reducing the moisture content, and thereby the volume of the sludge. The effect of various stages in dewatering, digestion, and incineration may be illustrated as follows.

Example A wastewater plant produces 1000 kg of dry solids per day at a moisture content of 95 percent. The solids are 70 percent volatile with a

Table 25-1 Constituents of waste activated sludge

Constituent	Percentage by weight
Free water	70–75
Floc water	20–25
Capillary water	1–2
Bound water	1–2
Solids	0.5–1.5

specific gravity of 1.05, and 30 percent nonvolatile with a specific gravity of 2.5. Determine the sludge volume

(*a*) as produced

(*b*) after digestion reduces the volatile solids content by 50 percent and decreases the moisture content to 90 percent.

(*c*) after dewatering to 75 percent moisture.

(*d*) after drying to 10 percent moisture.

(*e*) after incineration (only nonvolatile solids remain).

SOLUTION

(*a*) The mass of sludge is 1000/0.05 = 20,000 kg. 1000 kg is solid, of which 70 percent is volatile. 700 kg at s.g. = 1.05 occupy a volume of 667 l. 300 kg at s.g. = 2.50 occupy a volume of 120 l. The total volume is thus

$$19{,}000 + 667 + 120 = 19{,}787 \text{ l.}$$

(*b*) After digestion the volatile solids are reduced to 350 kg. The total solids content is therefore 650 kg and the total mass 650/0.10 = 6500 kg. Proceeding as above, the total volume is

$$5850 + 333 + 120 = 6303 \text{ l.}$$

(*c*) After dewatering the total mass is 650/0.25 = 2600 kg, and the volume is

$$1950 + 333 + 120 = 2403 \text{ l.}$$

(*d*) After drying the total mass is 650/0.90 = 722 kg, and the volume is

$$72 + 333 + 120 = 525 \text{ l.}$$

(*e*) After incineration, with only the nonvolatile solids remaining, the total mass is 300 kg and the volume is 120 l.

The volume as a percent of original is:

(*a*) 100 percent, (*b*) 32 percent, (*c*) 12 percent, (*d*) 3 percent, (*e*) 1 percent.

25-3 Sludge Conditioning

Sludge conditioning includes a variety of processes, some biological, some chemical, and some physical, which may be applied to favorably alter the chemical characteristics of sludge to improve its dewaterability. The biological processes also provide some decrease (about 35 percent) in total solids content.

Conditioning processes include sludge digestion (both aerobic and anaerobic), chemical coagulation, and heat treatment. The selection of a conditioning or digestion process should be made solely upon the basis of present costs, not upon consideration of past practice. Modern conditioning techniques are more stable and may be less expensive than digestion processes, which in any event may produce a sludge requiring further chemical treatment.

25-4 Anaerobic Digestion

Digestion processes may be modeled as shown in Art. 24-19. Solids return in these systems would be ineffectual due to the high proportion of inert material. Solids retention time in anaerobic digestion is strongly affected by temperature: θ_c^m ranges from about 2 days at 35°C to 10 days at 20°C. θ_c^d is usually larger by a factor of 2 to 4.[2]

Modern digesters are both heated and mixed. The temperature is maintained at 35°C and mixing is sufficient to completely intermix the contents once daily. Most digesters are mixed more thoroughly, but complete mixing in the sense in which it exists in activated sludge is not necessary. Mixing is provided by recirculated gas from the head space above the mixed liquor or by mechanically driven propellers with draft tubes. A modest amount of mixing is provided by withdrawal and return of sludge for heating, but this is insufficient by itself.

The addition of waste sludge should be relatively uniform in order to avoid upsetting the process.[3] Addition should be made at least twice daily, and more frequent loading is desirable.

Design of anaerobic digesters is still largely empirical. Criteria for low rate (unmixed) and high rate (mixed) systems are presented in Table 25-2. Modern digestion systems are usually two-stage high-rate processes in which the first stage is heated and mixed and the second is quiescent and serves primarily as a thickener for the digested sludge (Fig. 25-1).

Table 25-2 Anaerobic digestion design criteria

	Process	
Design criterion	High rate	Low rate
Solids retention time, days (θ_c)	10–20	30–60
Volatile solids loading (kg/m^3 per day)	2.4–6.4	0.6–1.6
Volume (m^3 per capita)		
Primary only	0.04–0.06	0.06–0.09
Primary and trickling filter	0.08–0.09	0.12–0.14
Primary and activated sludge	0.08–0.12	0.12–0.17
Digested solids concentration, %	4–6	4–6
Volatile solids reduction, %	50	60
Gas production (m^3/kg VSS added)	0.53	0.65
Methane content, %	65	65

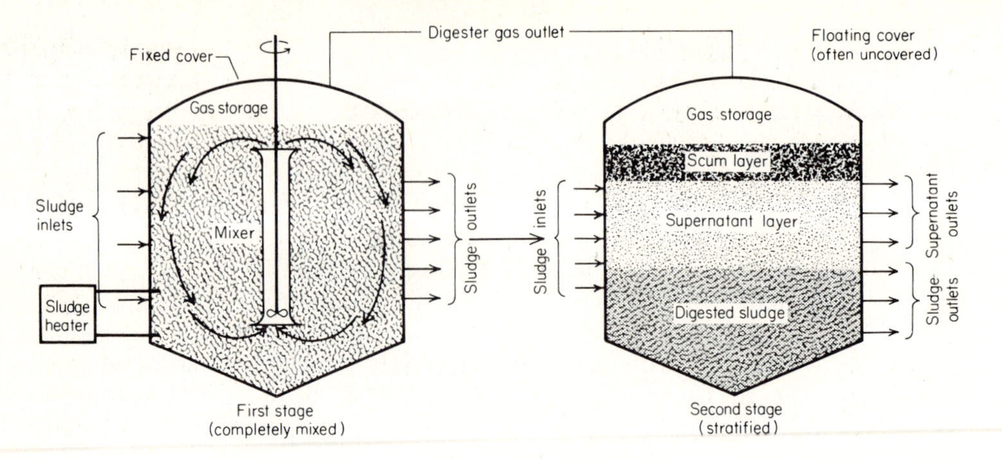

Figure 25-1 High-rate anaerobic digester. *(From "Wastewater Engineering" by Metcalf & Eddy, Inc. Copyright 1972, McGraw-Hill Book Company, Inc. Used with permission of McGraw-Hill Book Company.)*

Anaerobic digestion is a two-stage process in which the complex organics in the waste solids are first broken down to simpler compounds—notably the short-chain volatile fatty acids. These intermediates are then further reduced, by a separate group of bacteria, to methane and carbon dioxide. The overall anaerobic process may be depicted schematically as shown in Fig. 25-2.

The intermediate formation of organic acids can lead to toxicity problems due either to pH or the cations associated with the acid. Additionally, heavy metals may exhibit toxic effects in anaerobic systems since they may be solubilized under the chemically reduced conditions therein. The metals may be precipitated as sulfides by chemical addition if they are known to be present.[5] The obligate anaerobes which convert organic acids to methane and CO_2 are more sensitive to

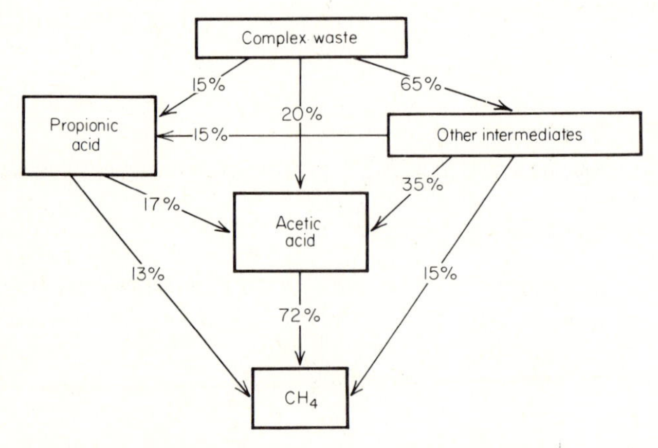

Figure 25-2 Metabolic pathways in anaerobic digestion.[4]

the surrounding environment than are the facultative organisms which effect the first step, hence toxic constituents will affect them first and cessation or slowing of their activity can lead to pickling of the digester.

The methane produced in anaerobic digestion is nearly universally used to heat the digesters and, in some instances, to provide mechanical power for other plant processes. Heat requirements depend upon climate, construction of the digester walls, and retention time. Procedures for calculating heat losses in anaerobic digesters are presented in reference 6.

The digested solids from anaerobic processes may be dewatered, without further treatment, upon open or covered drying beds. Chemical addition is likely to be required for satisfactory mechanical dewatering. Anaerobic processes are avoided by some designers because of their high first cost, susceptibility to biological upset, and mechanical complexity. The choice in a given circumstance should be made on the basis of estimated total costs and anticipated operational problems. Ordinary domestic sludges which do not contain significant quantities of industrial waste are readily digested in properly designed and operated anaerobic digesters.

25-5 Aerobic Digestion

Aerobic processes, including digestion, are much less susceptible to biological upsets than anaerobic systems. Aerobic systems contain many different organisms which can oxidize complex organics using different biochemical pathways, and hence are not so readily affected by toxic substances.

Aerobic digestion produces volatile solids reductions comparable to those in anaerobic digestion, has lower BOD in the supernatant, fewer operational problems, and lower capital cost. Operating cost for energy is higher and no useful by-product such as methane is produced. The digested sludge may be more difficult to dewater than anaerobically digested solids.

Design criteria are similar to those for high-rate anaerobic systems and are presented in Table 25-3.

Table 25-3 Aerobic digestion design criteria

Parameter	Value
Retention time (θ_c^d)	
Activated sludge only	15–20 days
Activated sludge plus primary	20–25 days
Air required (diffused air)	
Activated sludge only	20–35 l/min per m^3
Activated sludge plus primary	55–65 l/min per m^3
Power required (surface air)	0.02–0.03 kW/m^3
Solids loading	1.6–3.2 kg VSS/m^3 per day

Example Design an aerobic digester for a treatment system consisting of a primary clarifier producing 750 kg of solids per day and an activated sludge system producing 450 kg solids per day. The solids concentration in the mixed sludge is 4 percent.

SOLUTION The mass of the waste sludge is

$$(750 + 450)/0.04 = 30{,}000 \text{ kg/day}$$

and its volume, approximately, 30,000 l/day. For a retention time of 20 days the basin volume is $30 \times 20 = 600 \text{ m}^3$, and the air required about 36 m^3/min. The solids loading, assuming that the solids are 70 percent volatile will be

$$\frac{1200 \times 0.7}{600} = 1.4 \text{ kg/m}^3 \text{ per day}$$

A higher loading would be possible at this retention time only if the solids were first thickened.

The solids retention time for satisfactory dewatering must be kept in excess of 10 days. If the sludge is digested in a high-shear regime such as that produced by a high-speed aerator or if it is permitted to become anaerobic after treatment, it will be difficult to dewater.[7] Properly handled aerobically digested sludge will dewater on drying beds, vacuum filters, or the other processes discussed below. Chemical treatment is likely to be required for satisfactory handling in mechanical dewatering systems. *Pure oxygen* aerobic digestion has been evaluated for stabilization of thickened sludges. A temperature increase is likely due to the high oxidation rate and this may enhance the dewatering characteristics of the sludge. Retention times are similar to those in air systems, but solids loadings may be higher.

25-6 Chemical Conditioning

Chemical conditioning may be applied to sludges which have been digested or to raw sludges. The chemicals used include the metallic and polymeric coagulants discussed in Chap. 9. The mechanism of chemical conditioning consists of neutralization of charge and formation of a polymeric bridge which incorporates the individual particles into a lattice structure with sufficient rigidity and porosity to permit the water to escape.

The chemical dosage required is a function of pH, alkalinity, phosphate concentration, solids concentration, sludge-storage time, and sludge make-up, i.e., whether primary, secondary, or a mixture of both.[8] Typical chemical dosages for ferric chloride and lime (which are used together) are presented in Table 25-4. For a sludge containing 5 percent solids a dosage on the order of 1000 mg/l $FeCl_3$ plus 3500 mg/l lime is required. Polymer dosages are on the order of 100 mg/l—which may reduce chemical costs and will certainly reduce the mass of the final sludge and its proportion of inerts. In modern practice polymers are used nearly exclusively, although provision for addition of $FeCl_3$ is generally made.

Table 25-4 Dosages of $FeCl_3$ and CaO for sludge conditioning (% of sludge conc.)

	Pretreatment					
	None		Digestion		Digestion plus elutriation	
Sludge	$FeCl_3$	CaO	$FeCl_3$	CaO	$FeCl_3$	CaO
Primary	1–2	6–8	1.5–3.5	6–10	2–4	
Primary and trickling filter	2–3	6–8	1.5–3.5	6–10	2–4	
Primary and activated sludge	1.5–2.5	7–9	1.5–4	6–12	2–4	
Activated sludge only	4–6					

Polymeric coagulants, as noted in Chap. 9, may be anionic, cationic, or nonionic. For any sludge suitable polymers and appropriate dosages must be determined by jar tests. It is likely that the polymer used will fail to coagulate some fraction of the sludge solids and that this fraction will tend to build up in the plant. Occasional short-term changes in coagulant will usually remove these accumulations and thereby improve the overall operation of the plant.

25-7 Elutriation

Elutriation is, literally, a "washing" of the sludge. In Art. 25-6 it is noted that chemical conditioning is dependent in part upon alkalinity. In fact, the dosages of quicklime listed in Table 25-4 are required primarily to react with the alkalinity of the sludge in order to reduce the requirement for ferric chloride.

Sludges, particularly anaerobically digested sludges, have high alkalinity (several thousand mg/l) which can be reduced by washing or elutriating the sludge in water of low alkalinity—usually plant effluent.

The relative water flow required depends upon the process configuration and the relative alkalinities of sludge, washwater, and wastewater. Typical flow rates are 4 to 5 times the sludge flow.[6] Design criteria for elutriation systems usually specify a mixing time of about 1 min followed by settling in a gravity thickener (Art. 25-10) for 3 to 4 hours. Surface loading rates are on the order of 25 m/day (600 gal/ft^2 per day)[9] and solids loadings 1.5 to 2 kg/m^2 per hour (7 to 10 lb/ft^2 per day).

Elutriation can separate fine solids from the sludge and recycle them through the plant. This may eventually result in their loss in the effluent and will create operational problems throughout the treatment process. The cost of treating the washwater and rethickening the sludge must be weighed against the cost of the chemicals saved. The widespread use of polyelectrolytes for sludge conditioning has reduced the desirability of elutriation.

25-8 Heat Treatment

As noted in Art. 25-2, the water in sludge is in part free, in part trapped in the floc, and in part associated with the solids. Heat treatment actually destroys the cell structure releasing even the bound water. This permits mechanical dewatering to a degree not attainable with other conditioning techniques. Solids concentrations of 40 to 50 percent are routinely attained from heat-treated sludges dewatered on vacuum filters or in centrifuges.

A schematic of the Zimpro thermal sludge conditioning process is presented in Fig. 25-3. The sludge solids are mixed with air and heated to 180 to 200°C. The retention time in the process is 20 to 30 min at pressures on the order of 1500 kPa (200 psi).

The liquid effluent of heat conditioning processes is strong. The COD is on the order of 0.6 mg/l per mg/l solids, and, in effect returns to the biological process the bulk of the BOD removed by solids wasting (Art. 24-14). The design capacity of aeration basins or other biological processes must be increased by as much as 15 percent to compensate for this return.[1]

25-9 Other Conditioning Processes

Experimental work has been carried out on conditioning sludge with such materials as newspaper pulp and flyash.[1] The result of such treatment is a relatively dry sludge after mechanical dewatering (40 to 50 percent solids), but the sludge volume may be only slightly decreased due to the large quantities of additives employed.

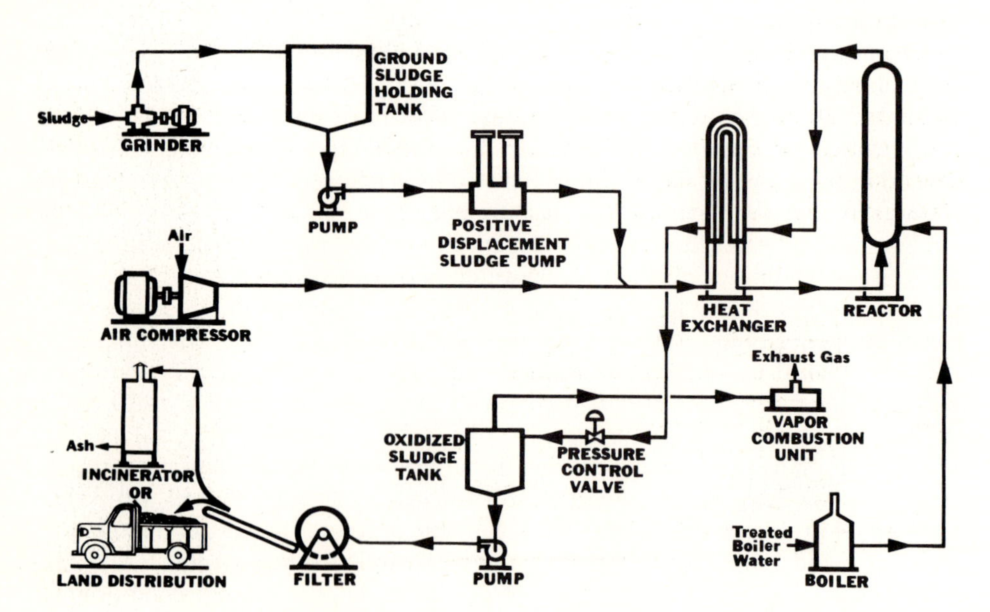

Figure 25-3 Thermal sludge conditioning process schematic. *(Courtesy Zimpro, Inc.)*

Freezing can be useful in dewatering sludges (Art. 11-26) but mechanical freezing is generally too rapid for biological sludges and results in compaction of the floc particles and poor drainage following thawing.[1]

25-10 Thickening

Thickening is a concentration technique in which relatively thin sludges such as waste activated sludge are increased in solids content in order to reduce the total volume and thus the size of subsequent treatment units. As may be noted from the example of Art. 25-2, a tremendous reduction in volume can be achieved through a modest increase in solids content.

Gravity thickeners are similar to sedimentation basins but may incorporate vertical pickets on the scraping mechanism (Fig. 25-4). These pickets serve to gently agitate the sludge and may aid in its concentration by promoting release of trapped water and gases.

The underflow concentration is controlled by the height of the sludge blanket in the basin. In general, the higher the blanket is, the thicker the solids will be. However, excessive solids retention times may lead to gasification and buoying of solids. This may change the chemical characteristics of aerobically digested sludge or waste activated sludge and interfere with its further dewatering.[7]

Design of gravity sludge thickeners should ideally be based upon settling tests such as those described in Art. 9-6 and reference 1. In the absence of an existing plant, design is based upon solids flux rates as shown in Table 25-5. The surface overflow rate is typically 15 to 35 m/day (350 to 850 gal/ft^2 per day) and is provided by recirculated secondary effluent which helps maintain aerobic conditions in the thickener.[10] Detention times are 3 to 4 hours.

Flotation thickeners[11] offer significant advantages in thickening light sludges such as waste activated sludge. Such sludges have a density very close to that of water and are thus readily buoyed to the surface. Waste activated sludge can be concentrated to approximately 4 percent solids by this technique. Surface overflow rates in flotation thickeners range from 10 to 45 m/day (250 to 1000 gal/ft^2 per day) at retention times of 30 min to 1 hour. Typical solids flux

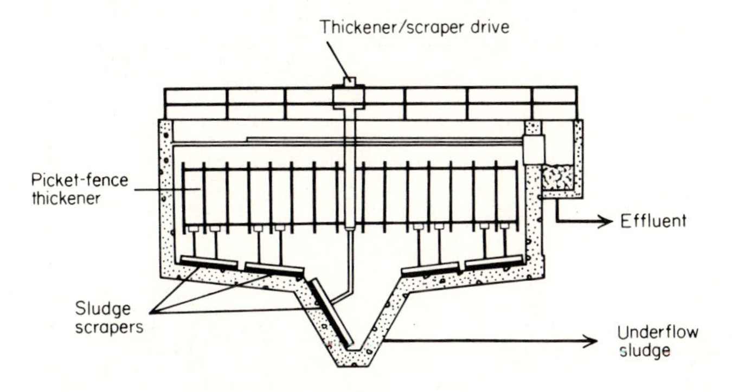

Figure 25-4 Gravity sludge thickener.

Table 25-5 Loadings and concentrations for gravity thickeners

Type of sludge	Loading (kg/m² per hour)	Thickened conc., %
Primary	4–5	8–10
Primary plus A.S.	1.6–3.8	4–9
Primary plus high purity O_2 A.S.	2.0	5–9
Primary plus T.F.	2.5–4.0	7–9
Trickling filter only	1.8	7–9
High purity oxygen activated sludge only	1.4–6.9	4–8
Activated sludge only	0.8–1.0	2.5–3.5

rates and anticipated concentrations are presented in Table 25-6. It should be noted that reference 12 specifies considerably higher loading rates (9.8 kg/m² per hour minimum) and the values in the table may be rather conservative.

Flotation systems should be designed, when possible, upon the basis of laboratory tests.[1] The design considerations are the air/solids ratio, the basin surface area, the detention time, and the recycle rate. Typical air/solids ratios range from 0.005 to 0.06 with 0.02 being most common for activated sludge thickeners.[12] The air/solids ratio may be calculated from

$$\frac{A}{S} = \frac{S_a(fP - 1)R}{C_0} \tag{25-1}$$

where A/S = the air/solids ratio

S_a = the saturation concentration of air in the waste flow at 1 atm (Table 25-7)

f = the fraction of saturation achieved (normally 0.5 to 0.8)

P = the pressure in atm (usually 3.7 to 4.8)

R = the recycle rate (the ratio of pressurized flow to waste flow (normally greater than 1), and

C_0 = the concentration of the solids in the influent sludge

Table 25-6 Loadings and float concentrations for flotation thickeners (after Ref. 9)

Type of sludge	Loading (kg/m² per hour)	Thickened conc., %
Primary	8–25	8–10
Primary plus activated sludge	4–8	6–8
Activated sludge only	2–4	4

Table 25-7 Solubility of air in water

Temp, °C	S_a, mg/l
0	37.2
10	29.3
20	24.3
30	20.9

Example Design a flotation thickener to concentrate waste activated sludge from 0.7 to about 4 percent. Assume that $A/S = 0.02$, Temp = 20°C, $P = 4.5$ atm, $f = 0.6$, SOR = 20 m/day, and sludge flow rate = 400 m^3/day.

SOLUTION

$$A/S = 0.02 = \frac{24.3(0.6(4.5) - 1)R}{7000}$$

$$R = 3.39$$

$$Q_r = 3.39(400) = 1355 \text{ m}^3/\text{day}$$

$$\text{Area} = (Q + Q_r)/\text{SOR} = 1755/20 = 87.8 \text{ m}^2$$

$$\text{Solids flux} = \frac{7000(400)}{1000(87.8)24} = 1.33 \text{ kg/m}^2/\text{hour}$$

The short hydraulic retention time and introduction of air in flotation thickeners are both favorable from the standpoint of further handling of the thickened sludge. Gravity thickeners may require return of plant effluent to maintain aerobic conditions and the long detention time will still produce a loss in dewaterability. A typical flotation unit is shown in Fig. 25-5.

25-11 Dewatering

Dewatering systems range from outdoor sludge drying beds to extremely sophisticated mechanical systems. A number of moderately priced machines have been developed in recent years which fill the gap between drying beds and vacuum or pressure filters.[13]

Drying beds consist of 150 to 300 mm (6 to 12 in) of coarse sand underlain by layers of graded gravel ranging from 3 to 6 mm ($\frac{1}{8}$ to $\frac{1}{4}$ in) at the top to 20 to 40 mm ($\frac{3}{4}$ to $1\frac{1}{2}$ in) at the bottom. The total gravel thickness is about 300 mm (12 in). The bottom is usually natural earth graded to 100 or 150 mm (4 to 6 in) draintile placed on 6 to 9 m (20 to 30 ft) centers. Side walls and partitions between bed sections are concrete and extend 300 to 400 mm (12 to 16 in) above the sand surface (Fig. 25-6).

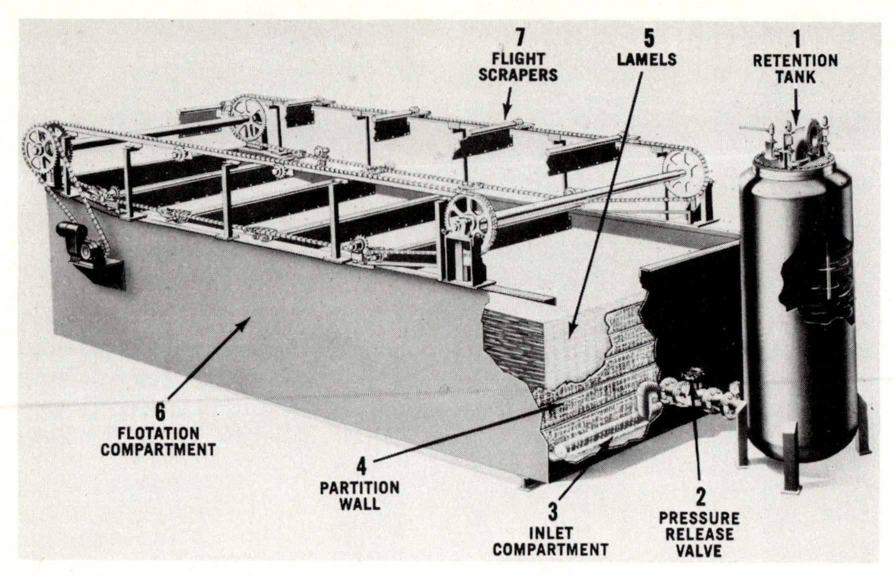

Figure 25-5 Flotation thickener. *(Courtesy Permutit Company, Inc.)*

Beds are 6 to 10 m (20 to 33 ft) wide and up to 40 m (130 ft) long. At least two beds should be provided in even the smallest plants. Dewatering occurs as a result of drainage and evaporation and is heavily dependent upon climate. Covering drying beds with glass or translucent plastic is helpful in wet climates. Actual testing on small drying beds is desirable, but in the event this is impossible the design criteria of Table 25-8 may be used. Some reduction in these areas, perhaps as much as 50 percent, is possible in the southern United States.

The beds are operated by filling them with digested sludge to a depth of 200 to 300 mm (8 to 12 in). After dewatering the volume will be reduced by some 60 percent and the cake may be removed with front-end loaders or similar light machines. A small amount of sand may be lost with each drying cycle. The time required for dewatering depends upon season and climate and ranges from several months to a few weeks. Solids content of the dewatered sludge ranges from 25 to 35 percent with 30 percent ordinarily being achieved.

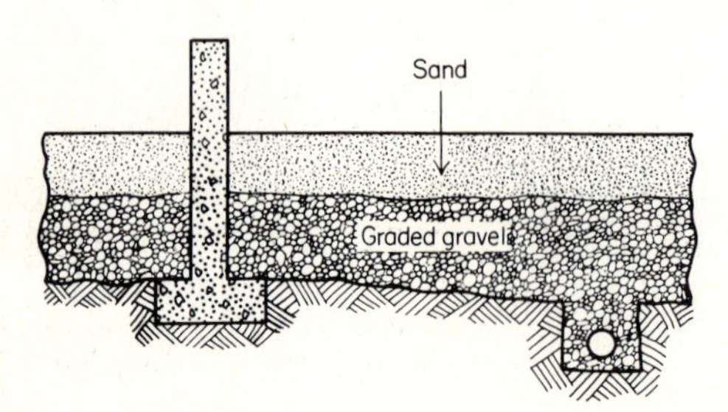

Figure 25-6 Section of sludge drying bed.

Table 25-8 Area required for sludge drying beds[6] (m^2/capita)

Type of sludge[a]	Area, m^2 per capita	
	Open beds	Covered beds
Primary	0.09–0.14	0.07–0.09
Primary plus trickling filter	0.12–0.16	0.09–0.12
Primary plus activated sludge	0.16–0.23	0.12–0.14
Primary plus chemically precipitated solids	0.19–0.23	0.12–0.14

[a] All sludges must be digested or heat treated prior to dewatering on sand beds. Chemically conditioned sludges are not suitable for sand dewatering.

Vacuum filters may be used to dewater either digested or conditioned sludge. Some chemical addition is normally required for digested solids as well as for raw sludge.

The filter consists of a rotating drum covered with a filter medium (usually cloth) and partially immersed in a basin in which the sludge is gently mixed. As the drum rotates, sludge is drawn against the screen and, as it leaves the liquid, the water is drawn through the medium leaving the solids behind. Further rotation carries the solids to a removal mechanism which may involve scraping the medium with a doctor blade, passing it about a small radius roller, or both. The medium is then rinsed before being re-immersed in the sludge.

Variables which can be used to control the filter yield include the vacuum applied, the depth of submergence, the speed of rotation, the degree of agitation of the sludge, the selection of the filter medium, and the pretreatment of the sludge.

Increasing the influent-solids concentration increases the filter yield and may reduce the amount of chemical required. Thickening, however, is not always justified since holding sludge for any period of time always decreases its filterability.[1]

High vacuum may compress the cake and reduce its filterability, but up to a point increased vacuum will increase yield. Yield is also increased by increasing the depth of submergence or the speed of rotation, however these expedients also tend to produce a wetter cake.

The filter medium on vacuum filters may be cotton, wool, felt, nylon, polyethylene, wire screens or wire coils. Selection of a filter medium and an appropriate rate should be made based upon laboratory tests of the sludge. Appropriate test procedures are presented elsewhere.[1] The values given in Table 25-9 may be expected in ordinary operation. The solids content of vacuum de-watered sludge ranges from as little as 5 percent for unthickened waste activated sludge to 40 percent for heat treated sludges. A typical vacuum filter is illustrated in Fig. 25-7.

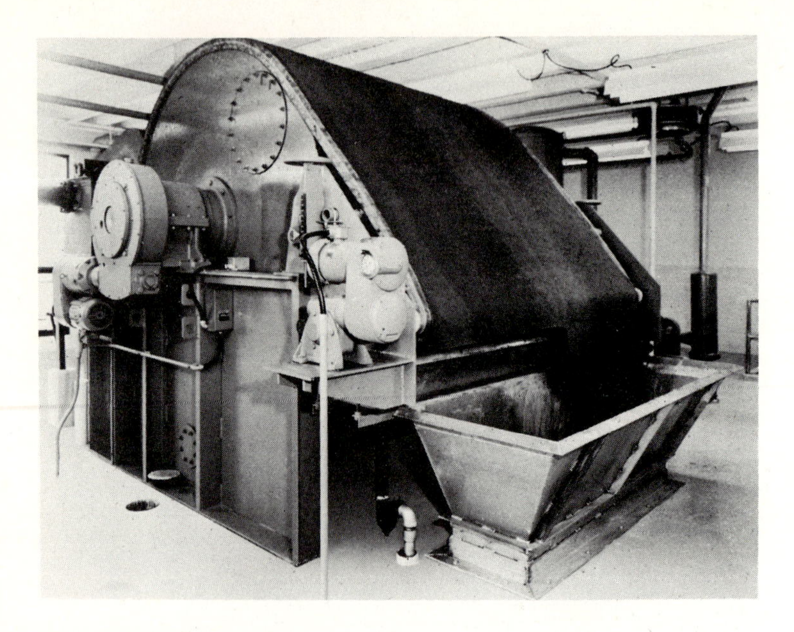

Figure 25-7 Vacuum filter. *(Courtesy Envirex, a Rexnord Company.)*

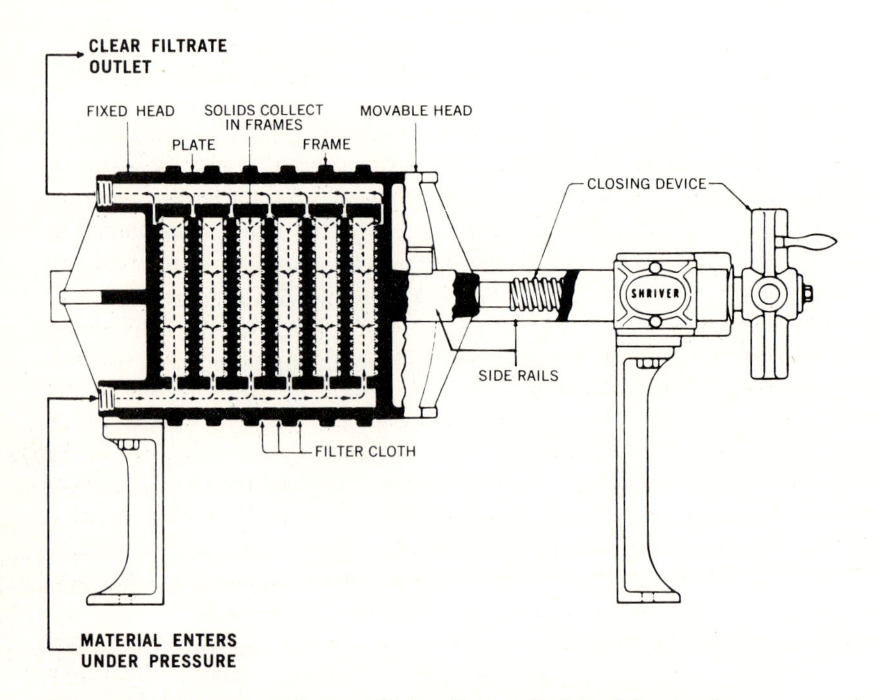

Figure 25-8 Filter press. *(Courtesy Shriver Division Envirotech Corporation.)*

Table 25-9 Typical yields of vacuum filters (after Refs. 6 and 12)

	Chemical conditioner			
	$FeCl_3$ plus CaO		Polyelectrolyte	
Type of sludge	Yield kg/m^2 per hour	% solids	Yield kg/m^2 per hour	% solids
Primary	20–60	25–38	40–50	25–38
Digested primary	20–40	25–32	35–40	25–32
Elutriated and digested primary	20–40	25–32		
Primary plus trickling filter	20–30	20–30	20–30	20–30
Digested primary plus trickling filter	20–25	20–30		
Elutriated and digested primary plus trickling filter	20–25	20–30		
Primary plus activated sludge	20–25	16–25	20–25	16–25
Digested primary plus activated sludge	20–25	14–22	17–30	14–22
Primary plus high purity oxygen activated sludge	25–30	20–28	20–30	20–28
Activated sludge alone	12–17	10–15	10–15	10–15

Vacuum filters generally are a source of odors since the vacuum extracts volatile fractions from the sludge and the agitation in the filter basin promotes degasification. Blinding of the filter cloth is a problem in some installations but can generally be remedied with better conditioning or a change in medium. In some industrial installations a precoat is used, but this is not a usual practice.

Pressure filters consist, in most instances, of a filter press similar to that shown in Fig. 25-8. Filter support plates hold a medium similar to that used on vacuum filters and the sludge is pumped against the medium and its support. Pumping is continued until flow virtually ceases. The pressure within the device ranges from 4 to 12 mPa (600 to 1750 psi) and the cycle time is 1 to 3 hours. At the end of the pressure cycle the frame is opened, the plates separated, and the cake removed. Solids content of the cake may be as high as 50 percent, but 35 to 40 percent is more likely on waste activated sludge. Addition of fly ash is necessary in amounts up to 250 percent of the sludge mass if high solids content is required. Polymers do not appear to be suitable conditioning agents for pressure filtration.[12]

The cyclic operation of filter presses is undesirable because it requires that solids be held while the filter is cycling. This may be acceptable if the storage is in a functioning digester rather than in a separate container. Chemical conditioning of

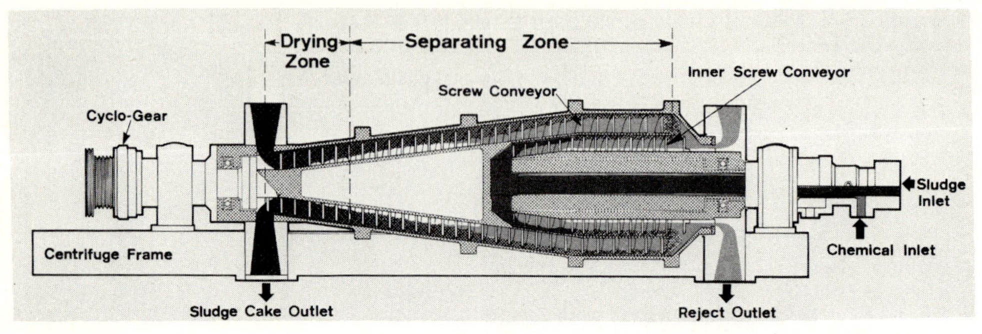

Figure 25-9 Horizontal solid bowl centrifuge. *(Courtesy Environmental Division, Ingersoll-Rand Co.)*

the sludge is always required and precoating the filter medium appears to be desirable.

Centrifuges may be used as thickening devices for activated sludge or as dewatering devices for digested or conditioned sludges. A typical solid bowl horizontal centrifuge is illustrated in Fig. 25-9. The screw conveyor or scroll rotates at a speed slightly more than that of the bowl and thus carries the solids through the device and up the ramp to the sludge-cake outlet.

The variables subject to control include the bowl diameter, length, and speed; the ramp slope and length; the pool depth; the scroll speed and pitch; the feed point of sludge and chemicals; the retention time; and the sludge conditioning provided. Vesilind[1] presents a complete discussion of the interrelation of these variables and their effect upon solids recovery and moisture content. Recovery of fine particles and light sludges can be improved by precoating the ramp with gypsum, thus producing zero clearance between the scroll and ramp. In general, increased solids recovery in centrifuges is associated with increased moisture content in the dewatered sludge. The effect of any process change may be seen in a

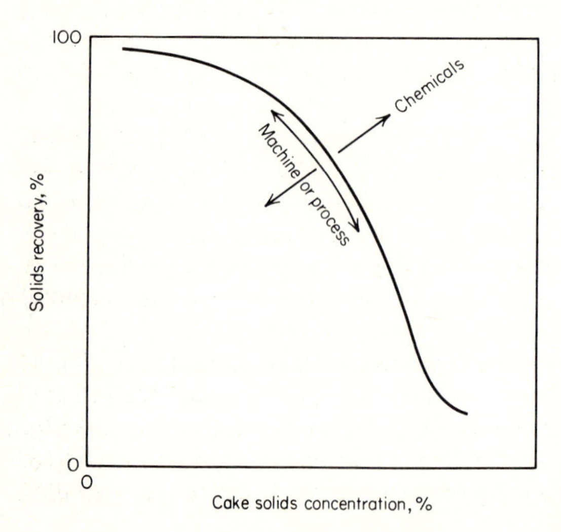

Figure 25-10 General solids recovery vs. cake solids curve for a centrifuge. (*From "Treatment and Disposal of Wastewater Sludge" by A. Vesilind, Ann Arbor Science Publishers, Inc. 1974. Used with permission of Ann Arbor Science Publishers.*)

Figure 25-11 Basket centrifuge. *(Courtesy Ametek, Inc.)*

plot like that of Fig. 25-10. The machine must operate along the curve at whatever combination of concentration and recovery is acceptable. The only way in which the operator can move off the machine curve is by changing the characteristics of the sludge through physical, chemical, or biological treatment.

Typical solids recovery in centrifuges ranges from 70 to 90 percent with cake moisture content of 20 percent being typical. Heat-treated sludges may dewater to 40-percent solids while waste activated sludge by itself may only reach 10- to 15-percent solids.

Disk and *basket* centrifuges have also been applied to dewatering waste sludges. Disk centrifuges tend to clog and require laborious disassembly. Basket centrifuges operate in a semicontinuous fashion with solids being removed when the effluent begins to deteriorate. A basket centrifuge is illustrated in Fig. 25-11.

Belt filters include a large variety of proprietary devices, some relying solely upon gravity; some pressure and gravity; some pressure, gravity, and vacuum; and some including absorption as well.

The *dual cell gravity filter* (Fig. 25-12) consists of a continuous belt shaped into two cylindrical cells. Waste sludge is dewatered by gravity in the first cell and, when it is sufficiently dry, is carried by the belt to the second cell where it is rolled into a ball expressing further moisture. Chemical conditioning is critical in the operation of this device. Inadequate treatment will permit solids to pass the filter,

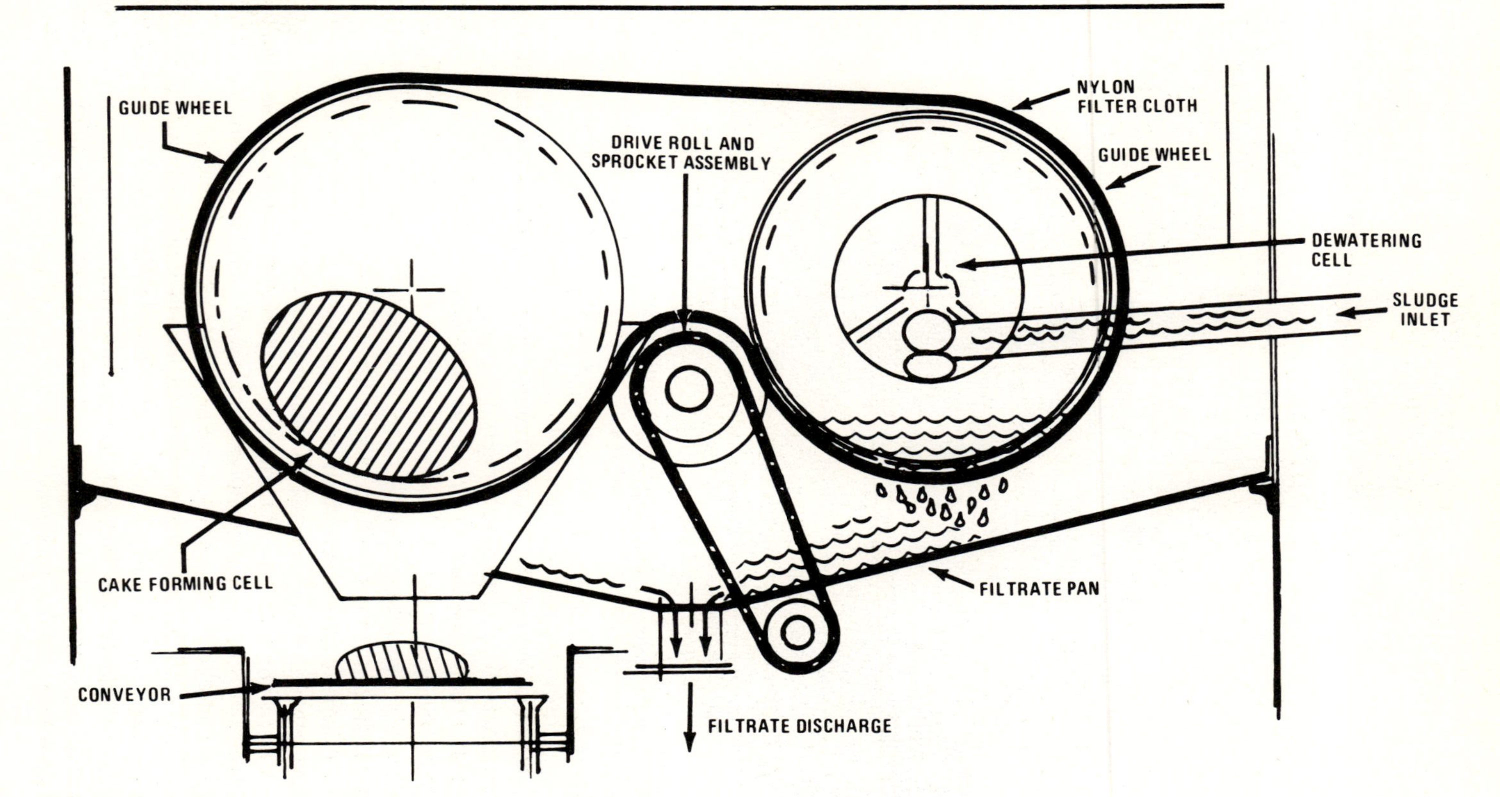

Figure 25-12 Dual cell gravity filter. *(Courtesy Permutit Company, Inc.)*

Figure 25-13 Multi-roll sludge press. *(Courtesy Permutit Company, Inc.)*

while overconditioning may blind the belt. Expected solids concentrations from this device are 10 to 15 percent.

The sludge exiting from the dual cell gravity filter may be further dewatered by passage through a *multi-roll sludge press* such as that of Fig. 25-13. The influent sludge must be at about 10-percent solids and is concentrated to 15 to 20 percent by being pressed between two endless polyester belts.

A combination gravity-pressure system is illustrated in Fig. 25-14. The upper belt offers gravity separation while the lower carries the sludge under a series of rollers. Expected solids concentrations from this process, like all others, depend upon pretreatment and conditioning. Aerobically digested activated sludge will thicken to about 9-percent solids with adequate polymer dosage.[12]

A superior device of the same general sort is illustrated in Fig. 25-15. This system employs gravity and vacuum, low-pressure filtration, high-pressure filtration, and provides continuous belt backwashing like that in vacuum filters. Solids concentrations comparable to those obtained on vacuum filters (25 to 30 percent) are attainable.

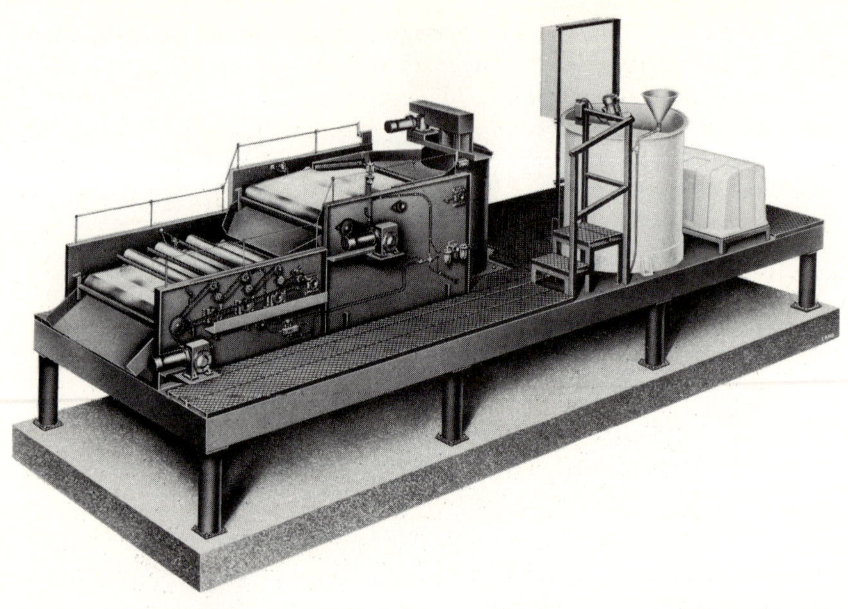

Figure 25-14 Sludge concentrator. *(Courtesy Ecodyne Corporation, Smith and Loveless Division.)*

The *squeegee* dewatering system employs a pair of endless belts, one carrying the sludge, and the other consisting of a sponge-like material which absorbs moisture from the sludge and is removed from contact and squeezed between opposed rollers at intervals along the flow path. The upper belt passes between compression rollers thus providing pressure dewatering before the cake is discharged. Effluent solids concentrations of 15 to 18 percent have been obtained at loading rates of 10 to 22 kg/m^2 per hour.[12]

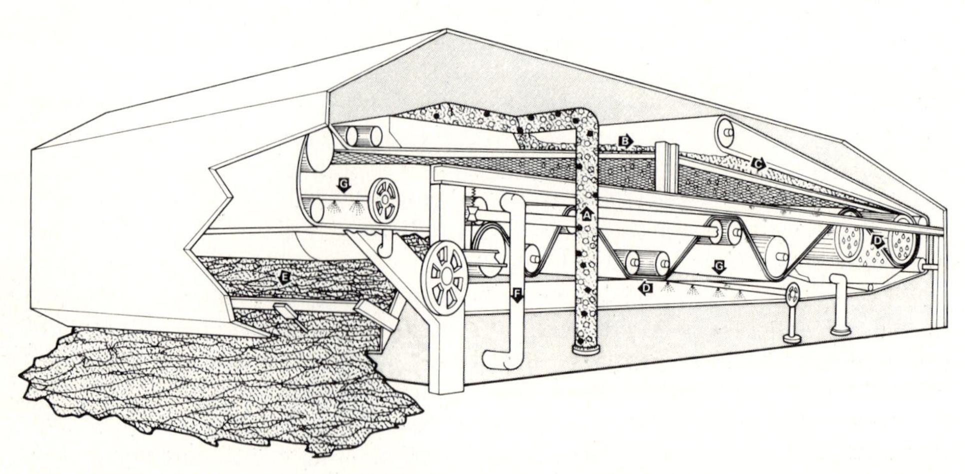

Figure 25-15 Combination gravity, vacuum, and pressure sludge dewatering system. *(Courtesy Passavant Corp.)*

The large variety of moderate-cost dewatering devices which have been developed in recent years offers the engineer considerably more flexibility than did the earlier choice between sand beds and vacuum filters. Systems which will dewater to 10-percent solids may be adequate in some instances, depending upon the ultimate disposal technique, and dewatering to that level can be done at relatively low cost. It should be remembered, however, that a 10-percent sludge is a liquid and must be handled as such.

Sludges which can be handled as solids can be produced on drying beds, vacuum filters, centrifuges, and several of the devices described above. The selection of an appropriate dewatering system is dependent in large part upon preceding and subsequent handling and the location and mode of ultimate disposal. In some instances dewatering may be omitted entirely.

25-12 Drying and Combustion

The process of sludge volume reduction begun in dewatering can be continued through processes which will reduce the moisture content to as little as 5 to 10 percent. Incineration can even reduce the sludge to a moisture-free ash. The cost of such processing is seldom justifiable unless disposal sites are very limited and haul distances are long.

Drying consists of dispersing the sludge in a stream of hot dry air and then separating the dried solids in a cyclone. The solids, depending upon their make-up, may be utilized to provide part or all of the heat required for drying. In this case the process becomes a partial or complete incineration technique. A schematic diagram of such a process is shown in Fig. 25-16.

Wet oxidation is a process which schematically is similar to that of Fig. 25-3. Typical temperatures range from 180 to 270°C and pressures from 7 to 20 mPa (1000 to 3000 psi). Under these conditions the sludge is oxidized with production of a clear liquid stream which is far lower in BOD than that from the sludge-conditioning process, and an inert ash which dewaters readily to 40- to 50-percent solids.

Pyrolysis is a destructive distillation technique which produces a carbon residue and a variety of gas and liquid products which may be economically recoverable. Processes of this sort have been evaluated on mixed-solid wastes,[14] but not specifically for sewage sludges.

Cyclonic incinerators introduce preheated air tangentially, at high velocity, into a combustion chamber with very hot walls. The sludge is sprayed radially toward the walls and is caught up in the tangential flow and rapidly burned to ash. Temperatures are over 800°C so no odors are produced in ordinary operation. Systems available in the United States have capacities up to 60 kg (130 lb) solids per hour, but foreign processes are available with ratings up to 5500 kg (6 tons) per hour.[12]

Electric (infrared) furnaces employ infrared lamps to initiate and maintain combustion of dewatered sludge. The process is thought to be applicable primarily in areas with limited fossil fuel supplies although it has some advantages in first and maintenance costs with respect to other systems.[12]

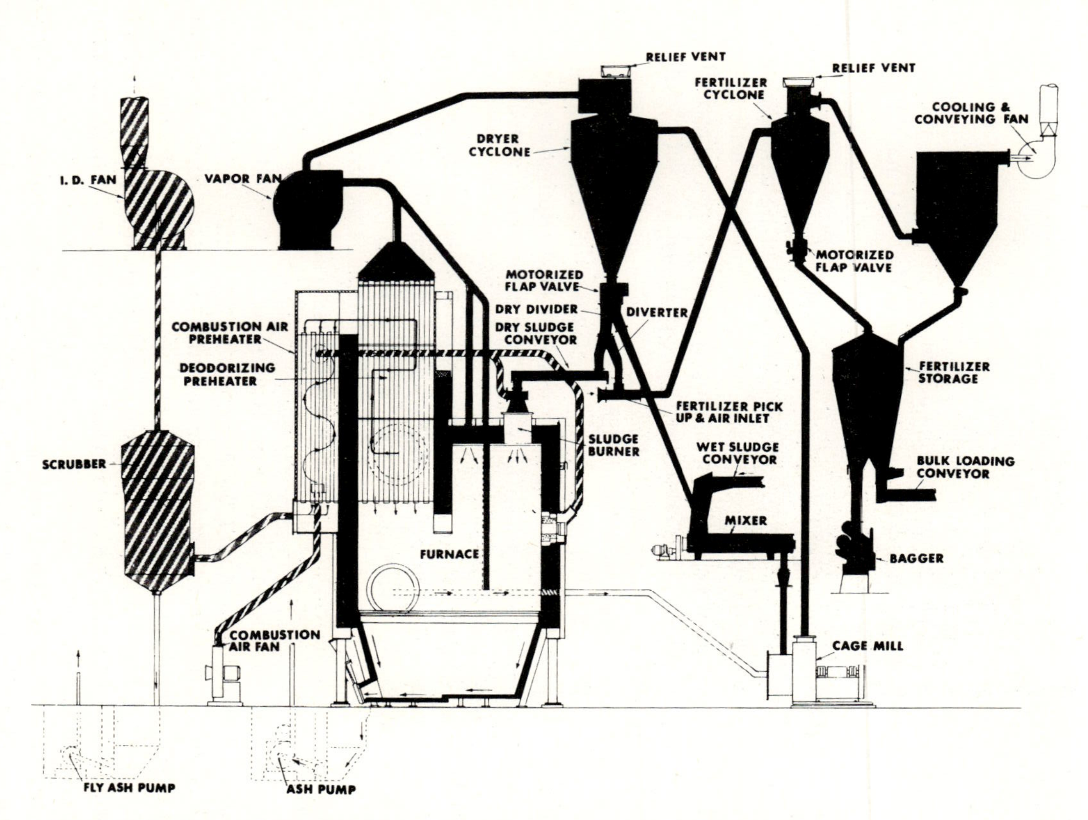

Figure 25-16 Flash drying incineration system. *(Courtesy Combustion Engineering, Inc.)*

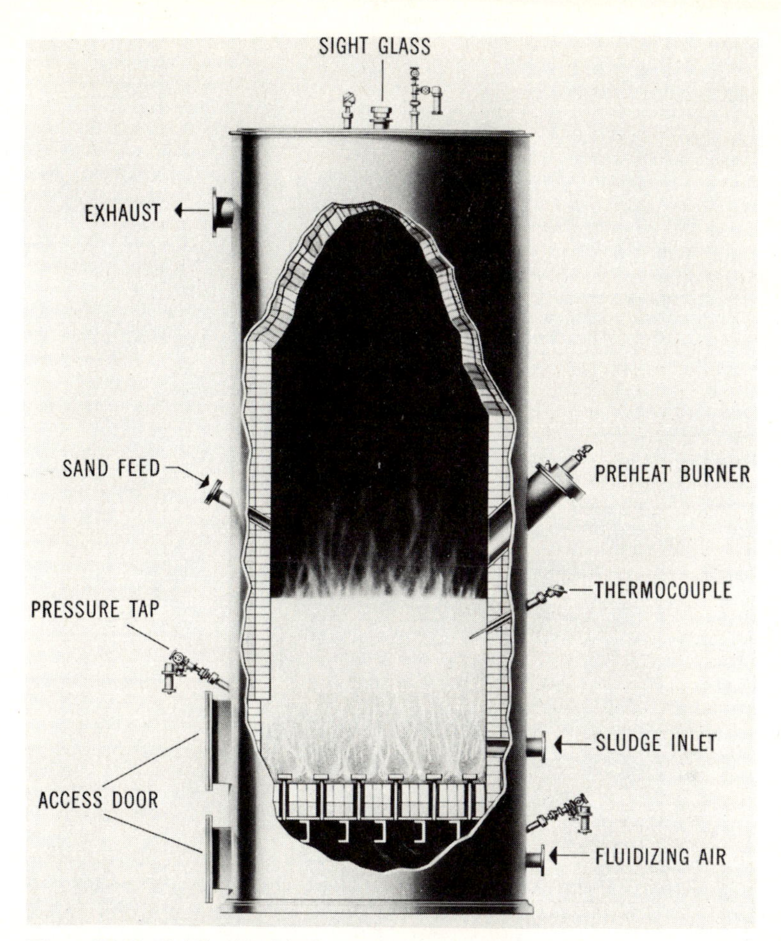

Figure 25-17 Fluidized bed incinerator. *(Courtesy Dorr-Oliver, Inc.)*

Fluidized bed incinerators[15] contain a bed of sand which is fluidized by the upward flow of injected sludge and air. Ash is carried out by the air flow and is removed by air-pollution-control systems. The sand is preheated to about 800°C with auxiliary fuel prior to injection of the sludge. Depending upon the fuel value of the sludge and its moisture content, auxiliary fuel may be required during routine operation. A fluidized bed incinerator is illustrated in Fig. 25-17.

Multiple hearth incinerators (Fig. 25-18) are the most widely used sludge incineration device and are available in capacities of 90 to 3600 kg (200 to 8000 lb) solids per hour. Sludge enters at the top and is windrowed across each hearth, falling then to the next. As it passes through the furnace it is dried and its temperature rises to the ignition point. The temperature within the furnace ranges from 550°C at the top (drying) zone to 1000°C in the middle (combustion) zone and 350°C at the bottom where the ash is removed. As in fluidized bed incinerators, auxiliary fuel is required for start-up, and may be required in routine operation.

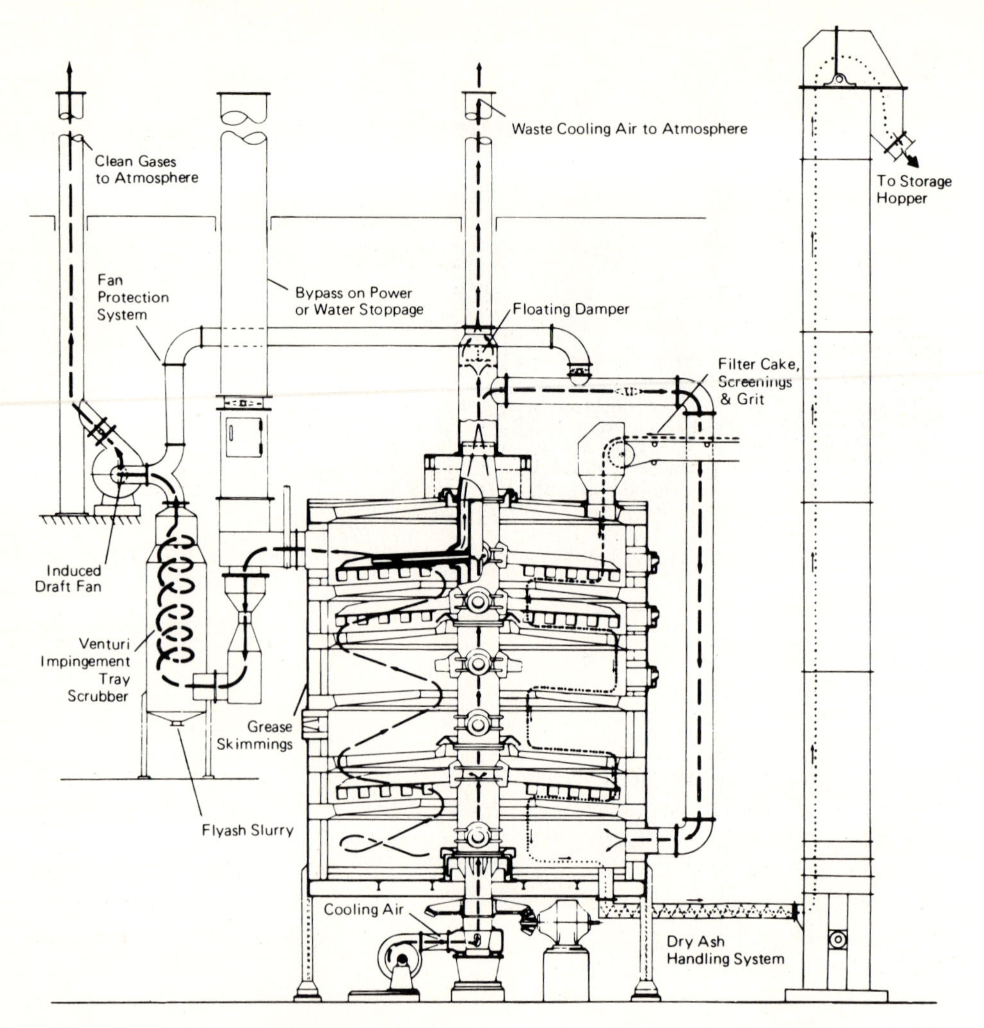

Figure 25-18 Multiple hearth incinerator. *(Courtesy Nichols Engineering and Research Corporation, a subsidiary of Neptune International Corporation.)*

25-13 Ultimate Disposal

Whatever processes are used in handling the solids from sewage treatment a residue will still remain which must be disposed of. This material may range from raw solids at a moisture content of over 95 percent to incinerator ash, and its handling will depend, in part, upon its nature.

Sanitary landfills are the preferred disposal technique for solid wastes in general, and are often used for sewage-solids disposal. Dewatered, dried, or incinerated sludges may be landfilled directly, while wet sludges, with careful handling,

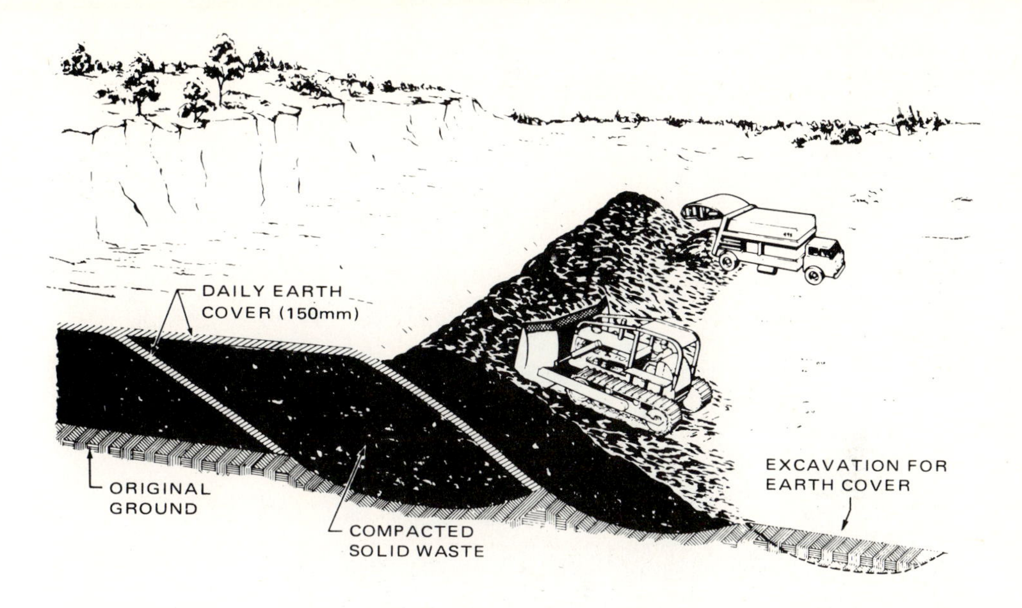

Figure 25-19 Sanitary landfill. *(From "Sanitary Landfill Design and Operation," Environmental Protection Agency, 1972.)*

can be spread on the working face and mixed with other refuse. A typical landfill operation is illustrated in Fig. 25-19. Design considerations and proper operation of solid-waste disposal sites are presented elsewhere.[16,17]

Land disposal of sludges, like land disposal of wastewater (Art. 21-12) is subject to biological, chemical, and physical constraints. Disease transmission, odors, heavy metal accumulation, and social and aesthetic problems must be considered before land application is selected.

Sewage sludge, whether digested or not, is not a particularly good fertilizer, hence no major benefit to the soil can be anticipated. Loading rates are governed by nitrogen—which is limited to the amount removable by crop growth; by heavy metals—which in sufficient accumulation are toxic to plant life; and by organics—which are limited by odor and the assimilative capacity of the soil.

State standards for sludge disposal vary considerably. Illinois[18] has recommended an accumulative application of 2.2 kg of dry solids/m^2 or 11 g of cadmium/m^2, whichever occurs first. Annual applications are limited by estimated nitrogen uptake and are in the range of 0.45 to 1.10 kg dry solids/m^2. The solids are applied as a slurry by spraying or injection from a tractor or truck (Fig. 25-20). Dewatered solids may be spread using ordinary farm equipment.

Ocean disposal has been commonly practiced by coastal cities with, in many cases, no discernable ill effects. Sludge may be pumped to sea or carried out in barges and dumped. It appears possible that this type of disposal will be severely restricted in the future, particularly on the east coast.[19]

Figure 25-20 Soil injection sludge disposal. *(Courtesy Big Wheels, Inc.)*

PROBLEMS

25-1 Estimate the quantity of solids produced in an activated sludge plant with a flow of 5500 m^3/day and BOD_5 and SS equal to 250 mg/l. Assume the primary clarifier removes 30 percent of the BOD and 50 percent of the solids, and that the sludge production in the secondary is equal to 0.7 × BOD applied.

25-2 Determine the reduction in volume if the solids of Prob. 25-1 are taken from an initial moisture content of 96- to 82-percent moisture on a vacuum filter.

55-3 Determine the digester volume required to aerobically digest the solids of Prob. 25-1 at 96-percent moisture and at 92-percent moisture content obtained by thickening in a centrifuge.

25-4 Determine the dimensions for a gravity thickener intended to concentrate mixed primary and trickling filter solids to 8 percent prior to digestion. Use a surface overflow rate of 20 m/day and a solids flux of 3.3 kg/m^2 per hour. How much plant effluent should be recirculated through the thickener to maintain the surface overflow rate and aerobic conditions?

25-5 Determine the dimensions of a flotation thickener to take waste activated sludge from 1- to 4-percent solids. Use an air/solids ratio of 0.03 and a system pressure of 4 atm. Assume the temperature is 20°C and the water is 70 percent saturated. The sludge flow is 1500 m^3/day.

25-6 What area would be required for a vacuum filter intended to dewater 3000 m^3 of sludge daily from 5- to 20-percent solids. The sludge is mixed primary and activated and the total flow is to be dewatered during a 12-hour period.

REFERENCES

1. Vesilind, P. Aarne: *Treatment and Disposal of Wastewater Sludge*, Ann Arbor Science, Ann Arbor, Mich., 1974.
2. Lynam, B., G. McDonnell, and M. Krup: "Start-up and Operation of Two New High-Rate Digestion Systems," *Journal Water Pollution Control Federation*, **39:**4:518, 1967.
3. McGhee, T. J.: "Volatile Acid Variation in Batch Fed Anaerobic Digesters," *Water and Sewage Works*, **117,** 5:130, 1971.
4. Jeris, J. S. and P. L. McCarty: "Biochemistry of the Methane Fermentation," *Proceedings* 17th Industrial Waste Conference, Purdue University, Lafayette, Ind., 1963.
5. Regan, Terry M. and Mercer M. Peters: "Heavy Metals in Digesters: Failure and Cure," *Journal Water Pollution Control Federation*, **42:**10:1832, 1970.
6. "Sewage Treatment Plant Design," *Manual of Practice No. 36*, American Society of Civil Engineers, New York, 1959.
7. Randall, C. W., J. K. Turpin, and P. H. King, "Activated Sludge Dewatering: Factors Affecting Drainability," *Journal Water Pollution Control Federation*, **43:**1:102, 1971.
8. Tenney, M. W., W. F. Echelberger, J. T. Coffey, and T. J. McAloon, "Chemical Conditioning of Biological Sludges for Vacuum Filtration," *Journal Water Pollution Control Federation*, **42:**2:R1, 1970.
9. Metcalf and Eddy, Inc., *Wastewater Engineering: Collection, Treatment, Disposal*, McGraw-Hill, New York, 1972.
10. McCarty, Perry L.: "Sludge Concentration—Needs, Accomplishments, and Future Goals," *Journal Water Pollution Control Federation*, **38:**4:493, 1966.
11. Katz, W. J. and A. Geinopolos: "Sludge Thickening by Dissolved Air Flotation," *Journal Water Pollution Control Federation*, **39:**6, 1967.
12. *Process Design Manual for Sludge Treatment and Disposal*, U.S. Environmental Protection Agency, Washington, D.C., 1974.
13. "Selecting Sludge Thickening and Dewatering Equipment," *Pollution Equipment News*, **17:**2:41, 1976.
14. Sussman, D. B.: "Baltimore Demonstrates Gas Pyrolysis: Resource Recovery from Solid Waste," U.S. Environmental Protection Agency, Washington, D.C., 1975.
15. Alford, J. M.: "Sludge Disposal Experience at North Little Rock, Arkansas," *Journal Water Pollution Control Federation*, **41:**2:175, 1969.
16. Brunner, D. R. and Daniel J. Keller: "Sanitary Landfill Design and Operation," U.S. Environmental Protection Agency, Washington, D.C., 1972.
17. Tchobanoglous, G., H. M. Theisen, and R. Eliassen: *Solid Wastes: Engineering Principles and Management*, McGraw-Hill, New York, 1977.
18. "Report of Illinois Advisory Committee on Sludge and Wastewater Utilization on Agricultural Land," Illinois Environmental Protection Agency, Springfield, Ill., 1975.
19. "Ocean Dumping Phase-Out in EPA Region II," *Journal Water Pollution Control Federation*, **48:**10:2246, 1976.

CHAPTER

TWENTY-SIX

ADVANCED WASTE TREATMENT

26-1 Scope

Advanced waste treatment encompasses those techniques which are applied in order to reduce the strength of wastewater below that usually achieved in secondary treatment. The soluble BOD_5 of a secondary effluent is only a few mg/l, and with proper design (Chap. 24) can easily be kept at that level. The total BOD_5, however, is also affected by the suspended solids, and reduction of suspended solids to levels less than 20 mg/l is difficult using sedimentation alone. *Suspended solids removal* is one application of advanced waste treatment, and is perhaps the most common. Other advanced waste-treatment techniques are directed toward reduction in *ammonia* and *organic nitrogen, total nitrogen, phosphorus, refractory organics,* and *dissolved solids.* This chapter discusses the principles of advanced waste treatment, but does not present detailed design procedures.

26-2 Suspended Solids Removal

The suspended solids in treated wastewater are in part colloidal and in part discrete, ranging from 10^{-3} to 100 microns (μm). None of this material settles readily or it would be removed in secondary clarification. The techniques which have been suggested for reduction of suspended solids concentration include microscreens, diatomaceous-earth filters, granular media filters, and chemical coagulation followed by sedimentation or dissolved-air flotation.

Chemical coagulation has been described in Chap. 9 and the principles developed there are applicable in wastewater treatment. The metallic coagulants also have a role in phosphorus removal (Art. 26-8) and may thus serve two purposes in advanced wastewater treatment. Coagulants used include lime, alum, iron, and polymers. Subsequent solids removal may be in solids contact units or ordinary sedimentation basins. In order to obtain very low suspended solids levels, filtration is required. Coagulation and clarification yields suspended solids concentrations between 2.5 and 25 mg/l.[1]

Microscreens consist of motor-driven drums which rotate about a horizontal axis in a rectangular basin. A fine screen with openings ranging from 23 to 60 μm covers the drum surface and removes particles larger than the screen opening. Wastewater is brought to the inside of the drum and passes outward, leaving the solids inside. As the drum rotates, the screened solids are carried to the top where they are cleaned from the screen by a high-velocity backwash spray. The backwash is collected in a hopper and recirculated through the plant. Design parameters include mesh size, submergence, allowable headloss, and peripheral speed.

Effluent suspended solids from microstrainers following activated sludge and trickling filter plants can be expected to range from 2 to 21 mg/l,[2,3] representing a 43- to 85-percent reduction from that in the process influent. Typical hydraulic loadings are 0.06 to 0.44 m/min (1.5 to 10 gal/ft^2 per min) upon the submerged area. Backwash flow is ordinarily constant and ranges up to about 5 percent of the product water flow. About 50 percent of this actually passes through the screen, the remainder running down the outside. A typical microscreen installation is shown in Fig. 26-1.

Diatomaceous-earth filters are described in Art. 10-19. In application to wastewater filtration, precoats amount to about 0.5 kg/m^2 and body coats to 5 to 6 mg/l per unit of influent turbidity. Acceptable operation is reported to be possible only with influent suspended solids less than 13 mg/l, and process costs are high.[1] The effluent is comparable in turbidity to potable waters.

Ultrafiltration is a system similar in operation to reverse osmosis (Art. 11-9) save that the membrane is far coarser and the pressure lower. The process removes finely divided suspended solids rather than dissolved solids, but has openings sufficiently small to remove colloidal particles as well. Ultrafiltration will remove all of the suspended solids and virtually 100 percent of the BOD, COD, and TOC.[1] The liquid flux rate in long-term operation is about 0.33 m/day (8 gal/ft^2 per day) at a pressure of 170 kPa (25 psi). Design variables include the membrane area, configuration, and material and the pressure applied. Membranes have a useful life of about six months.

Granular filters include the several systems discussed in Chap. 10. Filter configurations which constitute a cross-connection in waterworks practice can be used advantageously in wastewater filtration. Upflow filters (Art. 10-20) and biflow filters (Art. 10-21) may be particularly useful in filtering waters containing significant suspended solids concentrations. Design considerations include the filter configuration, media, and filtration rate. Filtration rates range from 0.06 to 0.50 m/min (1.5 to 12 gal/ft^2 per min) in ordinary practice with effluent suspended solids from 1 to 10 mg/l.[1,5] This represents a reduction of 20 to 95 percent from the concentration in the filter influent. Experimental rates up to 1.3 m/min

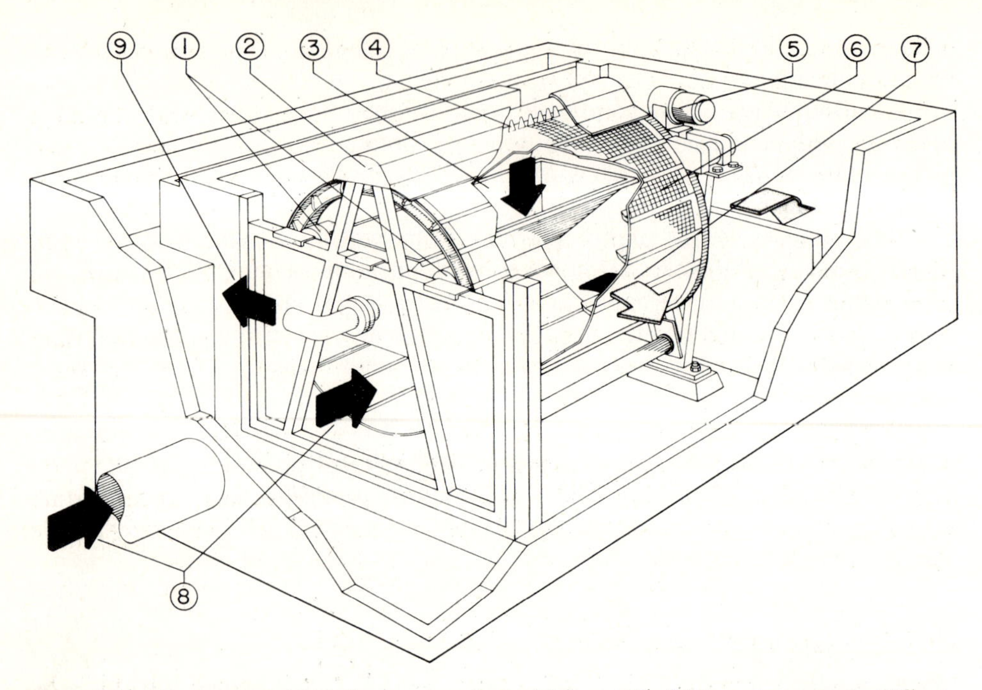

Figure 26-1 Microstrainer. (1) Drum support wheels; (2) Drum lift; (3) Screenings trough; (4) Backwash spray; (5) Variable speed drive; (6) Straining grid; (7) Effluent; (8) Influent; (9) Screenings return. *(Courtesy Zurn Industries, Inc., Envirosystems Division.)*

(32 gal/ft^2 per min) have been applied with suspended solids reductions of 50 percent.[6] Reference 1 presents a complete discussion of design and economic considerations which must be evaluated in the design of wastewater filtration systems.

The *filterability* of secondary effluents depends to a considerable extent upon the sludge age in the system. Culp and Culp[7] suggest that effluent suspended solids from a granular filter might range from 10 to 20 mg/l for a high-rate trickling filter to 1.5 mg/l for an extended aeration plant. The use of coagulants on secondary effluents prior to filtration will yield final turbidities comparable to those in drinking water.

26-3 Nitrogen Removal

Nitrogen in wastewater is found as organic nitrogen, ammonia, and nitrate. Minor amounts of nitrite may also occur. When present as organic-N or ammonia, nitrogen exerts an *oxygen demand* in accordance with:

$$NH_4^+ + 2O_2 \rightarrow NO_3^- + 2H^+ + H_2O \tag{26-1}$$

Each milligram of ammonia (as N) exerts an oxygen demand of about 4.6 mg, if nitrogen uptake by bacterial synthesis is neglected. It is apparent from this that the nitrogenous oxygen demand of a waste can be a large fraction of the total demand since total Kjeldahl nitrogen or TKN (ammonium plus organic-N) in wastewater amounts to 20 to 30 mg/l on average.

Nitrogen in the form of ammonia is also toxic to some fish, with the effect being dependent in part upon ammonia, carbon dioxide, and dissolved oxygen concentrations, pH, and temperature.[8] Toxic levels as low as 0.01 mg/l have been reported.[9]

Most nitrogen which enters the environment will eventually be oxidized to nitrate in which state it may be used as a *nutrient* by plants. Excess quantities of nitrogen in any form can thus contribute to eutrophication of surface waters. Nitrate itself, as noted in Art. 8-3, is the causative agent of *methemoglobinemia* and thus nitrogen discharges may constitute a public health hazard in some circumstances.

Depending upon the point of discharge and the applicable water quality standard, NPDES permits may specify no limitation on nitrogen, a limit on ammonia or TKN or both, or a limit on total nitrogen. Techniques for converting or removing nitrogen in wastewater are both biological and chemical and are discussed below.

26-4 Biological Nitrogen Removal

Nitrogen may be removed by incorporation in biological cell mass or by biological nitrification-denitrification. The first step in the latter process, nitrification, is adequate to meet some water quality limitations since the nitrogenous oxygen demand is satisfied.

Biological assimilation may be carried out by either algal or bacterial cultures. *Algal cultures* incorporate nitrogen, carbon dioxide, and trace nutrients to form algal cell mass.[10] The algae may also provide sufficient oxygen for stabilization of the organic matter present in the raw waste, however this process has not yet been demonstrated in a full-scale application. *Bacterial assimilation* can be encouraged by providing carbonaceous matter in a quantity commensurate with the nitrogen content (BOD_5 : N approximately 100 : 5). The disadvantages of bacterial assimilation include the cost of the carbon source—usually methanol or glucose—and the cost of handling the large quantity of sludge produced. Assimilation processes are generally not practical, but it should be noted that some assimilation will occur in any biological treatment process (Table 26-1). As noted in Art. 26-7, processes intended for assimilation of nitrogen will also provide some removal of phosphorus.

Nitrification is a biological process in which the species *Nitrosomonas* and *Nitrobacter* respectively, convert ammonia to nitrite and nitrite to nitrate. The nitrification process can be modeled mathematically in a fashion similar to that shown in Chap. 24. Complete design procedures are presented in reference 8. The process is dependent upon sludge age, as are other biological systems, but is also strongly affected by temperature, dissolved-oxygen concentration, final ammonia concentration, and pH. The reaction is enhanced at higher temperature, higher dissolved oxygen, and higher pH. The minimum DO for satisfactory nitrification

Table 26-1 Effect of various treatment processes on nitrogen compounds[8]

Treatment process	Effect on constituent			Removal of total nitrogen entering process, percent[a]
	Organic N	NH_3/NH_4^+	NO_3^-	
Conventional treatment processes				
Primary	10–20% removed	no effect	no effect	5–10
Secondary	15–25% removed[b] urea → NH_3/NH_4^+	< 10% removed	nil	10–20
Advanced wastewater treatment processes				
Filtration[c]	30–95% removed	nil	nil	20–40
Carbon sorption	30–50% removed	nil	nil	10–20
Electrodialysis	100% of suspended organic N removed	40% removed	40% removed	35–45
Reverse osmosis	100% of suspended organic N removed	85% removed	85% removed	80–90
Chemical coagulation[c]	50–70% removed	nil	nil	20–30
Land application				
Irrigation	→ NH_3/NH_4^+	→ NO_3^- → plant N	→ plant N	40–90
Infiltration/percolation	→ NH_3/NH_4^+	→ NO_3^-	→ N_2	0–50

Major nitrogen removal processes				
Nitrification	limited effect	$\rightarrow NO_3^-$	no effect	5–10
Denitrification	no effect	no effect	80–98% removed	70–95
Breakpoint chlorination	uncertain	90–100% removed	no effect	80–95
Selective ion exchange for ammonium	some removal, uncertain	90–97% removed	no effect	80–95
Ammonia stripping	no effect	60–95% removed	no effect	50–90
Other nitrogen removal processes				
Selective ion exchange for nitrate	nil	nil	75–90% removed	70–90
Oxidation ponds	partial transformation to NH/NH_4^+	partial removal by stripping	partial removal by nitrification-denitrification	20–90
Algae stripping	partial transformation to NH_3/NH_4^+	$\rightarrow$ cells	$\rightarrow$ cells	50–80
Bacterial assimilation	no effect	40–70% removed	limited effect	30–70

[a] Will depend on the fraction of influent nitrogen for which the process is effective, which may depend on other processes in the treatment plant.

[b] Soluble organic nitrogen, in the form of urea and amino acids, is substantially reduced by secondary treatment.

[c] May be used to remove particulate organic carbon in plants where ammonia or nitrate are removed by other processes.

is at least 2.0 mg/l at peak flow and load, although in some circumstances nitrification will occur at lower levels.

The *optimum pH range* for nitrification has been identified as being from 7.2 to 8.0.[8] The effect of pH is important since, as may be observed from Eq. (26-1), free hydrogen ion is produced in biological nitrification. This has a tendency to depress the pH and slow the oxidation process. As a rough guideline, one may assume that the alkalinity destroyed by nitrification will be about 7.2 mg as $CaCO_3$ per mg NH_3 as N. This is particularly significant in closed reactors such as those used in some high-purity oxygen systems, since in these there is also an increase in CO_2 concentration which further depresses the pH.

The effect of *temperature* has been studied extensively and it has been concluded that nitrification is more strongly inhibited at low temperature than is carbonaceous BOD removal.[11] Nitrification rates at 10°C are only about one quarter of those at 30°C even in attached growth systems,[8] and the latter are less affected by temperature than suspended growth systems.

Safety factors (Art. 24-13) in nitrification processes are particularly dependent upon desired effluent ammonia and fluctuations in waste strength and flow.[12] Flow equalization will permit reduction of the safety factor as shown in reference 8. Safety factors in the absence of equalization should be at least 2.5.

Nitrification can be obtained in *separate processes* following secondary treatment, or in *combined processes* in which both carbonaceous and nitrogenous demand are satisfied. In combined processes the ratio of BOD to TKN is greater than 5, while in separate processes the ratio in the second stage is less than 3. Any aerobic biological system can be used to achieve nitrification. In addition to activated sludge, rock-media trickling filters,[13] plastic-media trickling filters,[14] and rotating biological contacters[15,16] have been used.

Denitrification is a biological process which may be applied to nitrified wastewater in order to convert nitrate to nitrogen. The process is anaerobic, with the nitrate serving as the electron acceptor for the oxidation of organic material. Design considerations are presented in reference 8.

Methanol has been used as a carbon source in denitrification processes. The overall reaction requires methanol for the reduction of nitrite, nitrate, and dissolved oxygen in the nitrified wastewater. St. Amant and McCarty[17] have suggested the following equation for methanol demand:

$$C_m = 2.47(NO_3) + 1.53(NO_2) + 0.87(DO) \tag{26-2}$$

in which C_m is the methanol required in mg/l, and NO_3, NO_2, and DO are the concentrations, respectively, of nitrate, nitrite, and dissolved oxygen in mg/l as N and O_2.

A number of processes in which wastewater serves as the carbon source for denitrification have been developed[18,19] and industrial wastes have also been employed.[20] The denitrification process can be effected in packed bed or fluidized reactors with roughly comparable efficiency.[21,22] The packed beds can also serve as a filtration system if the medium is sufficiently fine.[23] A wide variety of process configurations are presented in reference 8, the advantages and disadvantages of which are summarized in Table 26-2.

Table 26-2 Comparison of denitrification alternatives[8]

System type	Advantages	Disadvantages
Suspended growth using methanol following a nitrification stage	Denitrification rapid, small structures required Demonstrated stability of operation Few limitations in treatment sequence options Excess methanol oxidation step can be easily incorporated Each process in the system can be separately optimized High degree of nitrogen removal possible	Methanol required Stability of operation linked to clarifier for biomass return Greater number of unit processes required for nitrification-denitrification than in combined systems
Attached growth (column) using methanol following a nitrification stage	Denitrification rapid, small structures required Demonstrated stability of operation Stability not linked to clarifier as organisms on media Few limitations in treatment sequence options High degree of nitrogen removal possible Each process in the system can be separately optimized	Methanol required Excess methanol oxidation process not easily incorporated Greater number of unit processes required for nitrification-denitrification than in combined system
Combined carbon oxidation-nitrification-denitrification in suspended growth reactor using endogenous carbon source	No methanol required Lesser number of unit processes required	Denitrification rates very low; very large structures required Lower nitrogen removal than in methanol based system Stability of operation linked to clarifier for biomass return Treatment sequence options limited when both N and P removal required No protection provided for nitrifiers against toxicants Difficult to optimize nitrification and denitrification separately
Combined carbon oxidation-nitrification-denitrification in suspended growth reactor using wastewater carbon source	No methanol required Lesser number of unit processes required	Denitrification rates low; large structures required Lower nitrogen removal than in methanol based system Stability of operation linked to clarifier for biomass return Tendency for development of sludge bulking Treatment sequence options limited when both N and P removal required No protection provided for nitrifiers against toxicants Difficult to optimize nitrification and denitrification separately

26-5 Chemical Nitrogen Removal

Nitrogen can be removed from wastewater or converted to other forms by a variety of chemical techniques. Those of major interest are breakpoint chlorination, ion exchange, and air stripping. The latter process may be applied to the wastewater itself or to the waste stream from an ammonium ion-exchange system.

Breakpoint chlorination converts 95 to 99 percent of the ammonia in wastewater to nitrogen gas with the remaining ammonia being converted to nitrate and nitrogen trichloride.[24] The overall reaction may be represented as:[8]

$$NH_4^+ + 1.5HOCl \rightarrow 0.5N_2 + 1.5H_2O + 2.5H^+ + 1.5Cl^- \qquad (26\text{-}3)$$

which indicates a weight ratio of chlorine to nitrogen of 7.6 to 1. In practice the actual dosage ranges from 8 to 10 to 1.[8] The chlorine dosage is affected by the presence of reduced chemicals, both organic and inorganic, and by pH.

The residual chlorine remaining after breakpoint treatment may exceed that permissible in the effluent,[25] and in such an event *dechlorination* is required. Demonstrated dechlorination techniques include application of sulfur dioxide:

$$SO_2 + HOCl + H_2O \rightarrow Cl^- + SO_4^{--} + 3H^+ \qquad (26\text{-}4)$$

or activated carbon:

$$C + 2HOCl \rightarrow CO_2 + 2H^+ + 2Cl^- \qquad (26\text{-}5)$$

Practical dosages of SO_2 range from 0.9 to 1.0 mg/mg Cl_2. Activated carbon has a capacity of 370 to 2400 kg/m^3 depending upon the hydraulic application rate.[26]

Ion-exchange processes for ammonium removal employ *clinoptilolite*, which is relatively specific for ammonium. The exchange bed is typically 1.5 m (5 ft) deep and is loaded at rates of 12 to 30 m/hour (5 to 12 gal/ft^2 per min) or $7\frac{1}{2}$ to 20 bed volumes/hour.[8] The total volume treated between regeneration cycles depends upon the ammonium concentration in the waste and the desired concentration in the effluent. At influent ammonium concentrations of 15 to 17 mg/l, 100 to 120 bed volumes can be treated before reaching an effluent concentration of 1 mg/l. This represents a 13- to 16-hour cycle at a loading rate of 7.5 bed volumes per hour. Design procedures for sizing clinoptilolite beds and selecting the process configuration are presented in reference 8.

Regeneration of the ion-exchange medium may be effected with alkaline or neutral sodium or calcium salts. Lime or sodium hydroxide will yield a waste stream containing NH_4OH, which can be readily removed by stripping (see below). Alkaline regeneration, however, will precipitate calcium and magnesium within the bed and this can coat the exchange medium and plug the bed.

Sodium chloride is also effective as a regenerant, although larger volumes (25 to 30 bed volumes) are required. The regenerant is then dosed with sodium hydroxide, elevating the pH and permitting stripping. Magnesium hydroxide is precipitated and is removed in a clarifier prior to stripping. The stripped ammonia may be released to the atmosphere or bubbled through an acid solution to permit its recovery. The stripped regenerant may be reused in the process.[8]

Stripping of waste regenerant can be effected with air at an air/liquid ratio of 1100 to 2200 m^3/m^3, or with steam[27] at rates of about 1.8 kg/m^3. Electrolytic destruction of the ammonium in the regenerant has also been evaluated,[8] but has considerably higher energy consumption than does stripping.

Stripping of the waste flow can be achieved in much the same fashion as in stripping ion-exchange regenerant. The theoretical basis for design of stripping towers is presented elsewhere;[8,28] in general air/liquid ratios range from 2240 to 6000 m^3/m^3 depending upon temperature and the pH of the waste.

The desirability of stripping is dependent largely upon whether or not the requisite pH (10.8 to 11.5) can be achieved at moderate cost. The dosage of alkali required is dependent in part upon the wastewater alkalinity, and stripping the bulk flow is generally practical only in association with lime precipitation of phosphorus (Art. 26-8). In this circumstance the pH required is produced in conjunction with phosphate removal.

The stripped gas, as in the stripping process in ion-exchange regenerant recovery, may be absorbed in acid solution for other use.

26-6 Phosphorus Removal

Phosphorus, like nitrogen, contributes to the eutrophication of surface waters and may be undesirable on that account. It is important to recognize however, that the concentration of phosphorus necessary to support an algal bloom is only 0.005 to 0.05 mg/l as P,[29] and that this level may be exceeded from natural sources in many surface waters.

Phosphorus may be removed biologically and chemically. In some instances chemicals may be added to biological reactors rather than in separate processes, while in others biologically concentrated phosphorus may be chemically precipitated. The various techniques are discussed below.

26-7 Biological Phosphorus Removal

Bacteria require phosphorus for their metabolic activities in a ratio of about 1 part phosphorus to 100 parts carbon, hence some phosphorus removal can be anticipated in any biological process. The removal is seldom adequate, however, and ordinarily does not exceed 20 to 40 percent.

The *activated algae* process discussed in Art. 26-4 and bacterial-assimilation processes have the potential of removing additional phosphorus, but if stoichiometric ratios are maintained it would be necessary to add both carbon and nitrogen to the latter, and the resultant sludge would be perhaps five times that produced in an ordinary plant.

Some treatment plants remove far more phosphorus than the theoretical bacterial requirement, and this phenomenon is termed *luxury uptake*. The mechanism of luxury uptake has been related to cellular uptake rather than to bacterially mediated precipitation, and the cellular uptake has been reliably induced in laboratory experiments by manipulation of the relative concentrations

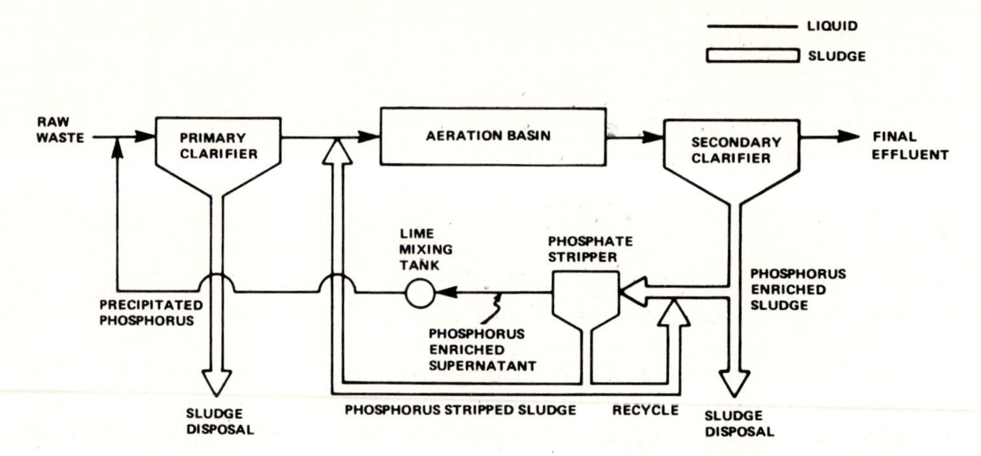

Figure 26-2 Biological-chemical phosphate-removal process. *(From "Operation of Full-Scale Biological Phosphorus Removal Plant" by G. V. Levin, G. J. Topol, and A. G. Tarnay in Journal Water Pollution Control Federation* **47** *(1975). Copyright 1975, Water Pollution Control Federation. Used with permission of the Water Pollution Control Federation.)*

of potassium and sodium.[30] No full-scale applications of this process have been reported, although some treatment plants provide natural luxury uptake.[31]

A novel biological-chemical process utilizes an activated sludge system and an anaerobic cell in which phosphorus taken up in the aerobic basin is released to the liquid phase, producing a concentrated phosphate stream and a phosphate-deficient sludge which is returned to the aeration basin where the uptake of phosphorus is repeated. The concentrated phosphate stream is dosed with lime (Art. 26-8) to precipitate the phosphorus—which is then removed from the liquid stream in the primary clarifier. The advantage of this process lies in the concentration provided since, as noted below, lime dosage for phosphate removal is relatively insensitive to phosphate concentration. A schematic of this treatment process is shown in Fig. 26-2.

26-8 Chemical Phosphorus Removal[33]

Phosphorus can be precipitated in a variety of forms, the least expensive ordinarily being as calcium, iron, or aluminum salts. Phosphorus in wastewater may exist as *organic phosphate*, *polyphosphate*, or *orthophosphate*, the latter consisting of four different ionic forms. For simplicity, the phosphorus is considered to be present as phosphate ion (PO_4^{3-}) and chemical reactions are presented upon that basis. The actual reactions may be more complex and chemical calculations based upon stoichiometric considerations will give only rough approximations of required dosages.

Aluminum phosphate can be precipitated in accord with:

$$Al^{3+} + PO_4^{3-} \rightarrow AlPO_4 \tag{26-6}$$

Table 26-3 Aluminum/phosphate ratios for chemical precipitation

Percent P removal	Al : P	
	Mole ratio	Weight ratio
75	1.38 : 1	1.2 : 1
85	1.72 : 1	1.5 : 1
95	2.30 : 1	2.0 : 1

The aluminum may be provided from alum ($Al_2(SO_4)_3 \cdot \times H_2O$) or sodium aluminate ($Na_2O \cdot Al_2O_3 \cdot \times H_2O$). Alum, as noted in Art. 9-10, reacts with water to produce free hydrogen ion, and thus will lower the pH of the waste. Sodium aluminate, on the other hand, is alkaline and will raise the pH. Since the solubility of aluminum phosphate is a minimum at pH 6,[34] judicious use of the appropriate chemical or combination of chemicals may reduce the required dosage. Estimated ratios of aluminum to phosphate required for various degrees of removal are presented in Table 26-3.[33] Note that Eq. (26-6) indicates a required theoretical Al : P ratio of 1 : 1.

Ferric and *ferrous phosphate* can be precipitated according to

$$Fe^{3+} + PO_4^{3-} \rightarrow FePO_4 \tag{26-7}$$

and

$$3Fe^{++} + 2PO_4^{3-} \rightarrow Fe_3(PO_4)_2 \tag{26-8}$$

The latter reaction is possible, but does not appear to be accurate since iron/phosphate ratios are independent of the valance of the iron. Iron salts also react with alkalinity to reduce the pH. The optimum pH for phosphorus removal with iron is 4 to 5 for ferric and 7 to 8 for ferrous ion. The iron/phosphorus ratio for ferric chloride is approximately[33] as listed in Table 26-4.

Hydroxyapatite may be precipitated by the reaction between calcium, hydroxyl ion, and phosphate:

$$5CaO + 5H_2O + 3PO_4^{3-} \rightarrow Ca_5(OH)(PO_4)_3 + 9OH^- \tag{26-9}$$

Table 26-4 Iron/phosphate ratios for chemical precipitation

Percent P removal	Fe : P	
	Mole ratio	Weight ratio
70	0.67 : 1	1.2 : 1
80	0.89 : 1	1.6 : 1
90	1.28 : 1	2.3 : 1
95	1.67 : 1	3.0 : 1

The precipitation is pH dependent and the lime will, of course, react with other chemical species as well. The lime dosage is primarily dictated by side reactions with alkalinity, hardness, and heavy metals rather than by the phosphate concentration. The pH required for substantial phosphate removal is about 9 and removal increases with increased pH.

Chemical addition for phosphate removal can be effected in the primary clarifier, in secondary biological processes, or in the final clarifier. The choice of point of addition and chemical is dependent upon the other treatment processes in the plant.

Addition in the primary is generally advantageous since removals of 80 percent will leave minor amounts which can be scavanged in biological treatment. The efficiency of biological systems is far greater than that of chemical systems in removal of low levels of pollutants.

If stripping of ammonia is to be practiced, addition of lime in the final clarifier is more logical since this will permit raising the pH for stripping while precipitating PO_4^{3-}. Nitrification processes, on the other hand, might be enhanced by alum addition in the primary. This would serve to reduce both the PO_4^{3-} concentration and the BOD_5 load on a combined nitrification system.

Chemical addition in biological processes can yield removal of phosphate[35] and, perhaps, assist in coagulation and separation of biological solids.[36] Phosphorus has been removed in this fashion in both trickling filters and activated sludge processes,[37] however good removal in trickling filter plants is not assured by this technique.[33]

26-9 Refractory Organics

Refractory organics are those which are not removed in biological systems operating under ordinary conditions of sludge age, liquid retention time, etc. The concentration of refractory organics may be roughly approximated by the difference between BOD and COD or TOC. The materials which contribute to this difference include organic molecules which are limited in solubility and those which contain resonant-ring structures. Examples of these substances include lignin, cellulose, and other polysaccharides, fat, phenolic material, detergents, etc.

Removal of refractory organics can be obtained by *adsorption* on activated carbon. The theory of adsorption is presented in Art. 11-12, but the application to wastewater treatment differs from that in water treatment plants.[38,39]

Wastewater adsorption systems ordinarily employ closed containers holding a bed of granular carbon through which the wastewater is passed. The rate of adsorption is affected by pH, temperature, waste concentration, molecular size, molecular weight, and molecular structure.

Carbon beds may be operated in either upflow or downflow configurations.[40] Although a downflow system appears to offer a simple scheme of obtaining both adsorption and filtration, this is not the case. When backwashed the spent carbon, being denser, will settle to the bottom (effluent) end where some desorption may occur. Carbon filters must be backwashed with high-quality water, while ordinary granular filters may be backwashed with primary or secondary effluent.

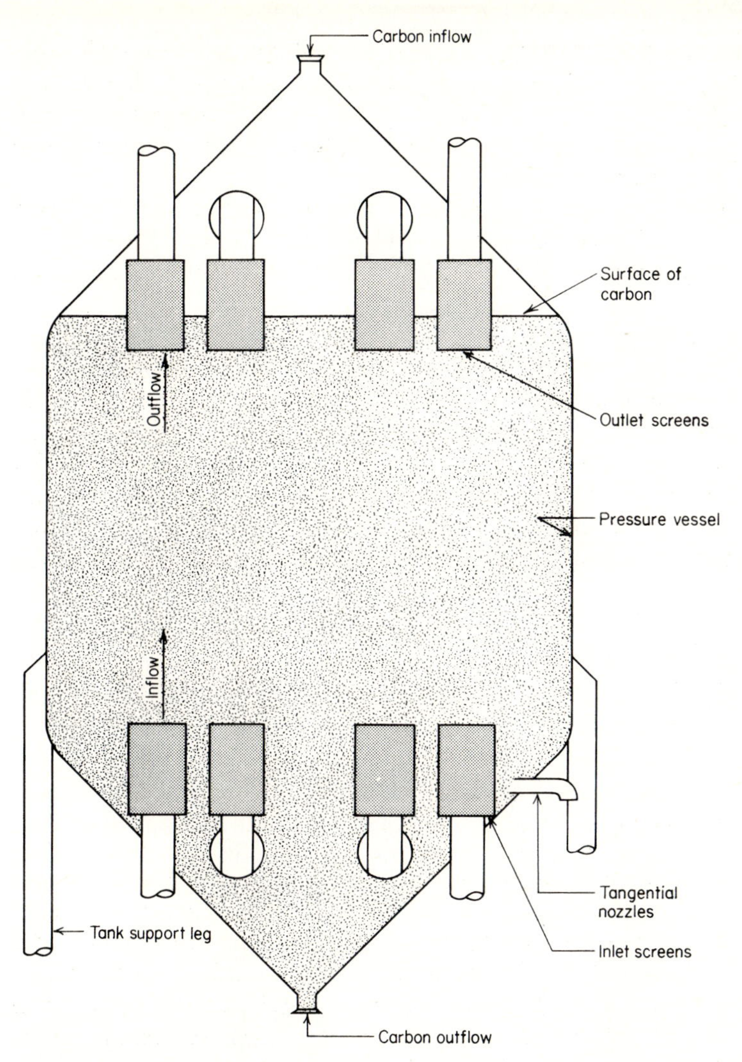

Figure 26-3 Upflow carbon column. *(From "Process Design Manual for Carbon Adsorption," Environmental Protection Agency, 1971.)*

Downflow units must be taken out of service for emptying and refilling, while upflow units can be operated on a more or less continuous basis. Upflow rates on the order of 0.2 to 0.4 m/min (5 to 10 gal/ft^2 per min), depending upon the carbon size, will expand the bed by about 10 percent. In this expanded condition spent carbon will move progressively toward the bottom, maintaining a countercurrent operation in which the adsorptive capacity of the carbon is most fully utilized and in which spent carbon can be readily removed from the process for regeneration. An upflow carbon column is illustrated in Fig. 26-3.[40]

Regeneration of spent carbon can be effected by washing with organic solvents, washing with mineral acid, washing with caustic, steam, or dry heat. The

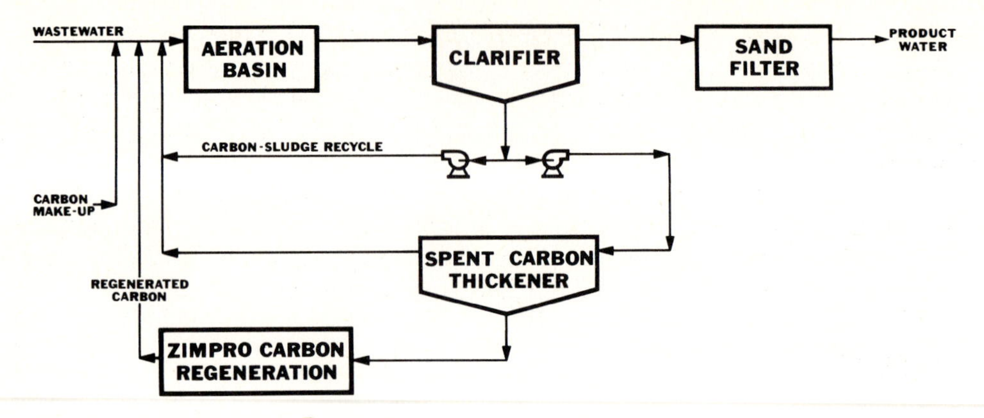

Figure 26-4 Biophysical wastewater treatment. *(Courtesy Zimpro, Inc.)*

first four techniques produce a liquid waste which is usually fairly concentrated while the last produces only gases and a quantity of ash and fines.

Thermal regeneration consists of three steps: drying at 100°C, pyrolysis of adsorbates at less than 800°C, and oxidation of decomposed adsorbates at temperatures above 800°C. Carbon losses are due to ignition in the furnace and to mechanical attrition and loss of fines. In addition, some loss of adsorption capacity will result from incomplete reactivation. The net loss is normally 5 to 10 percent per cycle from all causes.

Materials of construction and handling techniques are discussed in reference 40. Unit dimensions are generally based upon a ratio of depth to diameter of 2 : 1 to 10 : 1. Hydraulic retention time is 15 to 40 min based upon the volume of the empty container, but actual design criteria, including carbon selection, should always be based upon pilot studies.[7]

An interesting application of activated carbon to wastewater treatment is illustrated in Fig. 26-4. The process employs powdered activated carbon which serves both as an adsorbent and a surface for biological growth. Mixed-liquor suspended-solids concentrations approaching 1 to 1.5 percent can be maintained and treatment is commensurately thorough. Waste solids (carbon and biological) are passed through a low-pressure wet oxidation unit (Fig. 25-3) in which the carbon is regenerated and the biological solids are destroyed. The solids stream from this process is small and is inert and readily dewaterable.

26-10 Dissolved Solids

Dissolved mineral species such as nitrate or nitrite may cause health problems, while others such as chloride or hardness may be aesthetically or economically undesirable. Materials which are in true solution may be removed, in general, by the processes discussed in Arts. 11-18 and 11-19, but specific techniques have been developed for wastewater treatment.

Metals may be removed by ion exchange, precipitation, reverse osmosis, etc.[41] Polyelectrolytes with a soluble fraction precipitated by metal have been used to remove heavy metals from solution and coagulate them in a single-step process,[42] while reverse osmosis has been applied to removal of dissolved organics as well as inorganics.[43]

In general, the effluent of a well-designed and well-operated secondary-treatment plant is comparable to, and in some ways, superior to many surface waters. The water-treatment techniques discussed in Chaps. 9, 10, and 11 can ordinarily be applied to such wastewaters with satisfactory results.

REFERENCES

1. *Process Design Manual for Suspended Solids Removal*, U.S. Environmental Protection Agency, Washington, D.C., 1975.
2. Mixon, Forrest O.: "Filterability Index and Microscreener Design," *Journal Water Pollution Control Federation*, **42:**11:1944, 1970.
3. Lynam, Bart, Gregory Ettelt, and Timothy McAloon: "Tertiary Treatment at Metro Chicago by Means of Rapid Sand Filtration and Micro-Strainers," *Journal Water Pollution Control Federation*, **41:**2:247, 1969.
4. Biskner, Charles D. and James C. Young: "Two-stage Filtration of Secondary Effluent," *Journal Water Pollution Control Federation*, **49:**2:319, 1977.
5. Ripley, P. G. and G. L. Lamb, "Filtration of Effluent from a Biological-Chemical System," *Water and Sewage Works*, **12:**2:67, 1973.
6. Nebolsine, R., I. Poushine, and C. Y. Fan, *Ultra-High Rate Filtration of Activated Sludge Plant Effluent*, U.S. Environmental Protection Agency, Washington, D.C., 1973.
7. Culp, Russell L. and Gordon L. Culp: *Advanced Wastewater Treatment*, Van Nostrand Reinhold, New York, 1971.
8. *Process Design Manual for Nitrogen Control*, U.S. Environmental Protection Agency, Washington, D.C., 1975.
9. *Nitrogenous Compounds in the Environment*, U.S. Environmental Protection Agency, Washington, D.C., 1973.
10. Regan, Raymond W.: "An Update on the Activated Algae Process," *Transactions*, 25th Annual Conference on Sanitary Engineering, University of Kansas, Lawrence, Kansas, 1975.
11. Sutton, P. M., et al.: "Efficacy of Biological Nitrification," *Journal Water Pollution Control Federation*, **47:** 11: 2665, 1975.
12. Poduska, Richard A. and John F. Andrews: "Dynamics of Nitrification in the Activated Sludge Process," *Proceedings* 29th Annual Industrial Waste Conference **II:**1005, Purdue University, Lafayette, Ind., 1974.
13. *Process Design Manual for Upgrading Existing Wastewater Treatment Plants*, U.S. Environmental Protection Agency, Washington, D.C., 1974.
14. Stenquist, R. J., D. S. Parker, and T. J. Dosh: "Carbon Oxidation-Nitrification in Synthetic Media Trickling Filters," *Journal Water Pollution Control Federation*, **46:**10:2327, 1974.
15. Antoinie, R. L.: "Nitrification of Activated Sludge by Bio-Surf Process," *Water and Sewage Works*, **121:**11:44, 1974.
16. Murphy, Keith L., et al.: "Nitrogen Control: Design Considerations for Supported Growth Systems," *Journal Water Pollution Control Federation*, **49:**4:549, 1977.
17. St. Amant, P. and P. L. McCarty: "Treatment of High Nitrate Waters," *Journal American Water Works Association*, **61:**10:659, 1969.
18. Barnard, J. L.: "Cut P and N Without Chemicals," *Water and Wastes Engineering*, **14:**7:33, 1974.
19. Matsche, N. F. and G. Spatzierer: "Austrian Plant Knocks Out Nitrogen," *Water and Sewage Works*, **122:**1:18, 1975.

20. Wilson, Thomas E. and Donald Newton: "Brewery Wastes as a Carbon Source for Denitrification at Tampa, Florida," *Proceedings* 28th Annual Industrial Waste Conference **1:**138, Purdue University, Lafayette, Ind., 1973.
21. Jeris, J. S. and R. W. Owens: "Pilot-Scale High-Rate Biological Denitrification," *Journal Water Pollution Control Federation*, **47:**8:2043, 1975.
22. Michael, R. P. and W. J. Jewell: "Optimization of the Dentrification Process," *Journal Environmental Engineering Division, Proceedings American Society of Civil Engineers*, **101:EE4:**643, 1975.
23. English, John N. et al.: "Denitrification in Granular Carbon and Sand Columns" *Journal Water Pollution Control Federation*, **46:**1:28, 1974.
24. Pressley, T. A. et al.: "Ammonia Removal by Breakpoint Chlorination," *Environmental Science and Technology*, **6:**7:662, 1972.
25. Collins, H. F. and D. G. Desner: "Sewage Chlorination versus Toxicity—A Dilemma?" *Journal Environmental Engineering Division, Proceedings American Society of Civil Engineers*, **99:EE5:**761, 1973.
26. Stasiuk, W. N., L. J. Hetling, and W. W. Shuster: "Removal of Ammonia Nitrogen by Breakpoint Chlorination Using an Activated Carbon Catalyst," *Technical Paper No. 26*, New York State Department of Environmental Conservation, Albany, New York, 1973.
27. "Physical-Chemical Plant Treats Sewage Near the Twin Cities," *Water and Sewage Works*, **120:**9:86, 1973.
28. Metcalf and Eddy, Inc.: *Wastewater Engineering: Collection, Treatment, Disposal*, McGraw-Hill, New York, 1972.
29. Mulligan, H. F.: "Effects of Nutrient Enrichment on Aquatic Weeds and Algae," in *Relationship of Agriculture to Soil and Water Pollution*, Cornell University, Ithaca, New York, 1970.
30. Carberry, Judith B. and Mark W. Tenney: "Luxury Uptake of Phosphate by Activated Sludge," *Journal Water Pollution Control Federation*, **45:**12:2444, 1973.
31. Milbury, William F., Donald McCauley, and Charles H. Hawthorne: "Operation of Conventional Activated Sludge for Maximum Phosphorus Removal," *Journal Water Pollution Control Federation*, **43:**9:1890, 1971.
32. Levin, Gilbert V., George J. Topol, and Alexandra G. Tarnay: "Operation of Full-Scale Biological Phosphorus Removal Plant," *Journal Water Pollution Control Federation*, **47:**3:577, 1975.
33. *Process Design Manual for Phosphorus Removal*, U.S. Environmental Protection Agency, Washington, D.C., 1976.
34. Stumm, W. and J. J. Morgan: *Aquatic Chemistry*, John Wiley and Sons, Inc., New York, 1970.
35. Long, D. A. and J. B. Nesbitt: "Removal of Soluble Phosphorus in an Activated Sludge Plant," *Journal Water Pollution Control Federation*, **47:**1:170, 1975.
36. Unz, Richard F. and Judith A. Davis: "Microbiology of Combined Chemical-Biological Treatment," *Journal Water Pollution Control Federation*, **47:**1:185, 1975.
37. Lin, S. S. and D. A. Carlson: "Phosphorus Removal by the Addition of Aluminum (III) to the Activated Sludge Process," *Journal Water Pollution Control Federation*, **47:**7:1979, 1975.
38. Suhr, L. G. and G. L. Culp: "State of the Art—Activated Carbon Treatment of Wastewater," *Water and Sewage Works*, **121:**R104, 1974.
39. Rickert, D. A. and J. V. Hunter: "Adsorption of MBAS from Wastewaters and Secondary Effluents," *Journal Water Pollution Control Federation*, **46:**5:911, 1974.
40. *Process Design Manual for Carbon Adsorption*, U.S. Environmental Protection Agency, Washington, D.C., 1971.
41. Colman, T. W. and R. W. Dellinger: "Techniques for Removing Metals from Process Wastewater," *Chemical Engineering*, **81:**8:79, 1974.
42. Wing, Robert E., et al.: "Heavy Metal Removal with Starch Xanthate—Cationic Polymer Complex," *Journal Water Pollution Control Federation*, **46:**8:2043, 1974.
43. Cruver, J. E. and I. Nusbaum: "Application of Reverse Osmosis to Wastewater Treatment," *Journal Water Pollution Control Federation*, **46:**2:301, 1974.

CHAPTER

TWENTY-SEVEN

MISCELLANEOUS SEWAGE TREATMENT PROBLEMS

27-1 Disinfection

The use of chlorine for disinfection of treated sewage has caused some concern in light of the discovery of chlorinated organics in public water supplies (Chap. 11). It has been demonstrated that halogenated hydrocarbons are produced by chlorination of sewage[1,2] and regulatory agencies have therefore relaxed their wastewater disinfection requirements in some instances.

Combined *ultrasonation* and *ozonation* have proven effective in disinfection of wastewater[3] as has ozonation alone.[4] A proprietary process utilizes ozone as a disinfectant and then applies the resulting oxygen in a high-purity oxygen-activated-sludge system. Ozone is not competitive in cost even on this basis, but other considerations may require its use.

Chlorination of wastewater generally requires dosages of 2 to 15 mg/l, depending upon the preceding treatment processes. The actual dosage is set to maintain a residual of 0.2 to 0.5 mg/l in the plant effluent. *Contact chambers* are sized for 15-min retention at peak hourly flow and are designed as baffled plug-flow systems. Horizontal velocities are maintained at 1.5 m/min (5 ft/min) at minimum flow. In some circumstances a long outfall line may be used to reduce the size of, or entirely eliminate, a chlorine-contact basin. Chlorination equipment used in waste treatment is identical to that described in Chap. 11. Design considerations are presented in references 5 and 6.

27-2 Odor Control

Odor at waste-treatment plants is normally associated with the products of anaerobic decomposition: reduced organics and hydrogen sulfide. These materials can be made less offensive with any strong oxidizing agent. Chlorine, hydrogen peroxide,[7] and ozone[8] have all been successfully applied. Adsorption on activated carbon has also been employed.[9,10]

Odors arising from overloaded biological processes can only be corrected permanently by modification or expansion of the system. For short-term problems, the addition of nitrate will supply an additional source of hydrogen acceptors, but may increase the nitrogen content of the effluent.

27-3 Garbage Disposal With Sewage

Water carriage of ground garbage is increasingly common. Most modern residences have in-sink garbage grinders and many food-handling industries employ large grinders for their wastes. The effect of garbage grinding upon wastewater treatment is ordinarily not significant. Wastewater flow may be expected to increase by about 2 percent, suspended solids by 50 percent, BOD by 30 percent, and grit by 40 percent when grinders are in widespread use.[11] Sewer systems which are initially adequate will not be affected, while treatment facilities will be primarily affected in their solids-handling capability. The increased solids loading will not affect the hydraulic design of preliminary or primary systems.

A recently proposed enlarged domestic refuse grinder might produce more significant effects. In order to decrease the frequency of refuse pick-up, the grinder has the capability of handling most packaging materials (such as cardboard and plastic containers) as well as garbage. The anticipated increases in waste strength resulting from this system are: BOD 70 percent, VSS 1040 percent, TSS 825 percent, and grit 2800 percent.[12] If such a system were widely implemented the solids-handling capacity (digestion, dewatering, disposal) of the wastewater-treatment plant would require major expansion, although the physical size of primary and secondary systems might well be unchanged.

27-4 Industrial Wastes

Industrial process wastes, washwaters, and cooling waters are subject to discharge restrictions similar to those applied to municipal wastes. Individual industries have specific NPDES requirements which must be met and which depend upon particular processes and/or products and the size of the manufacturing plant.

Industries often have the option of treating their wastes to meet the NPDES requirements or discharging them as is, or following pretreatment, to the municipal treatment plant. If the wastes are treated by the municipality the industry ordinarily must expect to pay its fair share of the cost of construction and operation of the plant. If the construction is financed in part with federal funds recovery of the industrial share is mandatory.[13]

The cost of treatment is generally assigned to parameters involved in plant design such as flow, BOD, and SS. If the plant includes nitrogen or phosphorus removal, or if a particular waste is unusually high in some substance such as grease or floating solids, these other parameters might be included. The cost of each treatment process is then assigned, as far as possible, to these parameters. The cost of pumping would be charged to flow, while the cost of the secondary process might be charged partially to flow and partially to BOD, and the cost of solids handling partially to BOD and partially to SS. Industrial users would then pay in proportion to their contribution of the chargeable parameters.

Despite the requirement for cost recovery, it is often advantageous for the industry to allow the municipality to treat its wastes. First, the provisions of reference 13 do not require assessment of interest on the federal share of construction costs—in effect providing the industry with an interest-free loan and thereby reducing its total cost. Other advantages include[14] the placing of responsibility with one owner and one operator—who is generally better paid and better qualified than would otherwise be the case; reduction in unit cost of construction and operation due to economy of scale; improved treatability due to mixture of different waste streams which individually might be toxic or nutrient deficient; and finally, transfer of problem and responsibility to the municipality.

Pretreatment or *exclusion* may be required, depending upon federal and local regulations, for wastes which might cause problems in either the collection or treatment system. Examples of such materials include inflammables or explosives, excessive suspended solids, temperature, grease, acidity, alkalinity, or toxic materials such as phenol, cyanide, heavy metals, or radioactive substances.

Industrial waste treatment is an extensive topic which encompasses procedures such as *plant surveys* to determine individual waste sources and strengths, *process changes* to reduce waste flow and strength, *housekeeping* improvements, *pretreatment* to meet either sewer or other standards, and finally, *treatment* either separately or jointly with the municipality.[14]

Treatment processes employed are dependent upon the nature of the waste and may be physical, chemical, biological, or any combination of the three. Food-processing wastes are usually high in BOD while other industrial wastes may be low in BOD but contain heavy metals, cyanide, or arsenic or be at very low or high pH. Characteristics and possible treatment techniques are summarized in Table 27-1.

Industrial wastes are often deficient in trace materials, notably nitrogen and phosphorus, which are necessary for biological treatment. Since municipal sewage ordinarily contains a considerable excess of these materials, joint treatment is advantageous. Where joint treatment is not practical, the necessary trace chemicals must be added. Industrial wastes do not have the same characteristics as domestic sewage and the design criteria and rate constants presented in Chap. 24 are not necessarily applicable. Pilot studies should be conducted before heavy investments are made in process equipment. The *Journal of the Water Pollution Control Federation* and the *Proceedings of the Annual Industrial Waste Conferences* held at Purdue University are excellent sources of information on industrial waste treatment.

Table 27-1 Industrial wastes and treatment processes

Source	Characteristics	Treatment
Food industries		
breweries	high nitrogen and carbohydrates	biological treatment, recovery, animal feed
canneries	high solids and BOD	screens, lagoons, irrigation
dairies	high fat, protein, and carbohydrate	biological treatment
fish	high BOD, odor	evaporation, burial, animal feed
meat processing	high protein and fat (may be warm)	screens, sedimentation, flotation, biological treatment
pickles	high BOD, high or low pH, high solids	screening, flow equalization, biological treatment
soft drinks	high BOD and solids	screening, biological treatment
sugar	high carbohydrate	lagoons, biological treatment
Chemical industries		
acids	low pH, low organics	neutralization
detergents	high BOD and phosphate	flotation, precipitation
explosives	high organics and nitrogen, trinitrotoluene	chlorination or TNT precipitation
insecticides	high organics, toxic to biological systems	dilution, adsorption, chlorination at high pH
Materials industries		
foundries	high solids (sand, clay, and coal)	screens, drying
oil	dissolved solids, high BOD, odor, phenols	injection, recovery of oils
paper	variable pH, high solids	sedimentation, biological treatment
plating	acid, heavy metals, cyanide toxic to biological systems	oxidation, reduction, precipitation, neutralization
rubber	high BOD and solids, odor	biological treatment
steel	low pH, phenols, high suspended solids	neutralization, coagulation
textiles	high pH, and BOD, high solids	neutralization, precipitation, biological treatment

EXCRETA DISPOSAL IN UNSEWERED SECTIONS

27-5

Rural areas and the outskirts of urban areas may have insufficient population to support sewer systems and central treatment. In such areas it is likely that unsanitary disposal techniques will be used unless local regulations are both well thought out and scrupulously enforced.

Satisfactory techniques will insure that water supplies, particularly shallow wells, are not polluted; that flies and vermin have no access to excreta; and that nuisances such as odors are minimized. Acceptable systems, depending upon circumstances, include septic tanks and subsurface percolation; extended aeration, alone or following a septic tank; intermittent or subsurface sand filters following septic tanks or extended aeration; or intermittent sand filters alone. In areas without running water (primarily, but not exclusively, camp grounds) the pit privy is still used.

27-6 The Pit Privy

Improved pit privies still exist in large numbers.[15] They should be located at least 30 m (100 ft) away from, and preferably downslope from wells. Privies should not be located where cracks in the subsurface material may permit travel from the pit toward wells or springs. As illustrated in Fig. 27-1, a pit privy consists of a pit, 1 m (3 ft) square and 1.25 m (4 ft) deep, lined with rough boards on the sides and covered with a reinforced-concrete slab. A concrete riser supports the seat, and a galvanized ventilator pipe conveys odors through the roof. The slab rests on a concrete curb placed on the ground surface and the house is bolted to the slab. Earth is banked around the curb and slab to prevent surface water from entering. A privy of this type will serve an average family for 10 years. Cleaning is not practical and a new pit must be dug when the old one is full. The house and slab may be moved to the new pit.

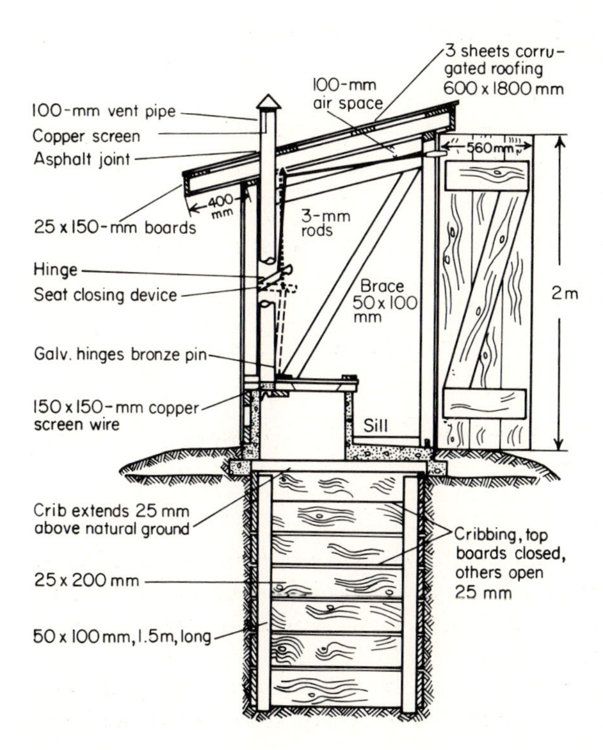

Figure 27-1 Concrete slab pit privy.

Table 27-2 Waste flow. (after Ref. 16)

Type of establishment	Liters per person per day
Small dwellings and cottages with seasonal occupancy	190
Single-family dwellings	280
Multiple-family dwellings (apartments)	225
Rooming houses	150
Boarding houses	190
Additional kitchen wastes for nonresident boarders	40
Hotels without private baths	190
Hotels with private baths (2 persons per room)	225
Restaurants (toilet and kitchen wastes per patron)	25–40
Restaurants (kitchen wastes per meal served)	10–12
Additional for bars and cocktail lounges	8
Tourist camps or trailer parks with central bathhouse	130
Tourist courts or mobile home parks with individual bath units	190
Resort camps (night and day) with limited plumbing	190
Luxury camps	380–570
Work or construction camps (semipermanent)	190
Day camps (no meals served)	55
Day schools without cafeterias, gymnasiums, or showers	55
Day schools with cafeterias, but no gymnasiums or showers	75
Day schools with cafeterias, gyms, and showers	95
Boarding schools	280–380
Day workers at schools and offices (per shift)	55
Hospitals	570–950 +
Institutions other than hospitals	280–470
Factories (flow per person per shift, exclusive of industrial wastes)	60–130
Picnic parks with bathhouses, showers, and flush toilets	40
Picnic parks (toilet wastes only), (flow per picnicker)	20
Swimming pools and bathhouses	40
Luxury residences and estates	380–570
Country clubs (per resident member)	380
Country clubs (per nonresident member present)	95
Motels (per bed space)	150
Motels with bath, toilet, and kitchen wastes	190
Drive-in theaters (per car space)	20
Movie theaters (per auditorium seat)	20
Airports (per passenger)	10–20
Self-service laundries (flow per wash, i.e., per customer)	190
Stores (per toilet room)	1500
Service stations (per vehicle served)	40

(Liters × 0.264 = gal)

Table 27-3 Characteristics of rural household wastewater[17]

Average flow per capita	160 l/day
Peak flow per capita	272 l/day
BOD per capita	0.050 kg/day
SS per capita	0.035 kg/day

Pit privies at camp sites which receive heavy use are often lined with concrete and have an access door to the pit at the rear of the unit. This permits the contents to be removed and hauled to a municipal treatment plant or a suitable disposal site.

27-7 Septic Tanks

Septic tanks are primarily sedimentation basins although a minor degree of solids destruction may occur due to anaerobic digestion. Units are ordinarily sized to provide a 24-hour retention time at average daily flow. Anticipated flows from various types of residential and public buildings are presented in Table 27-2.[16] The flows in Table 27-2 are generous estimates since recent research[17] indicates that per capita contributions from rural households may be expected to be somewhat less (Table 27-3).

A typical residential septic tank is illustrated in Fig. 27-2. Tanks are usually

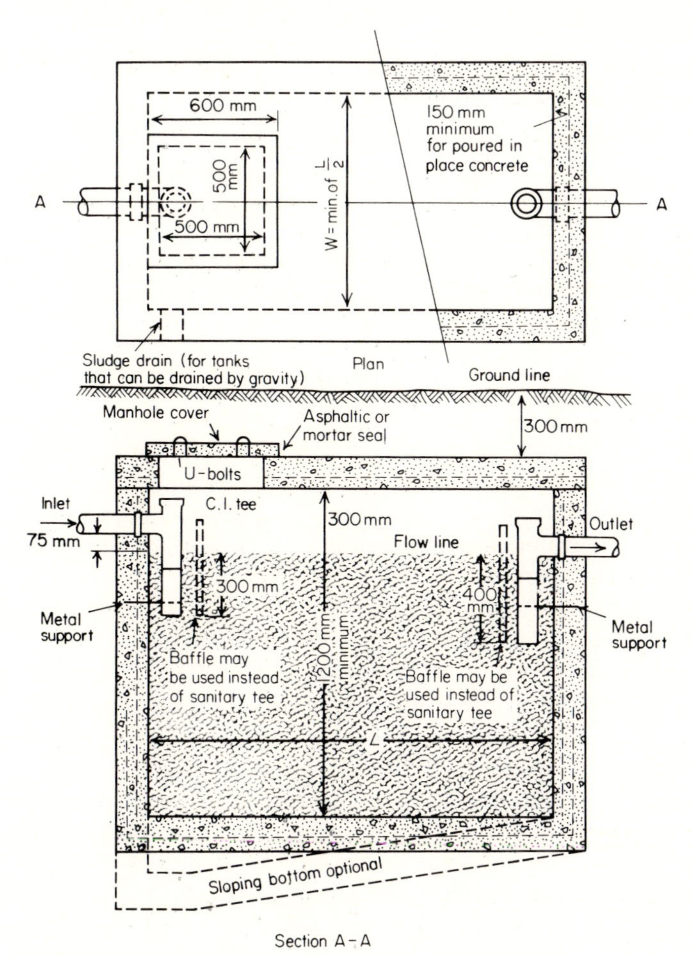

Figure 27-2 Residential septic tank.

made of concrete but steel and fiberglass have also been used. The effluent of a septic tank is offensive and potentially dangerous. The mean BOD_5 concentration observed in a number of septic tank installations ranged from 120 to 270 mg/l, and the mean SS from 44 to 69 mg/l.[18] Further treatment of septic tank effluent is required, either in an additional process or by soil disposal.

27-8 Aerobic Units

Aerobic biological treatment ordinarily provides a better effluent than anaerobic treatment, hence attempts have been made to apply the principle of extended aeration to single-family residence waste treatment. Available systems may employ a septic tank followed by an aerobic unit or an aerobic unit alone. Aeration is effected with diffused air supplied by a small compressor and solids recycle is provided either by gravity or air lift. These systems are installed below ground and their effluent may be discharged to a disposal field, a sand filter, or to surface drainage. The latter alternative is not desirable since the effluent quality may not be satisfactory under the best of circumstances, and there is no guarantee that the unit will be properly maintained. Mean effluent BOD_5 in a number of systems ranged from 38 to 57 mg/l and SS from 45 to 64 mg/l.[18]

Westfield and Smith[19] have suggested that an aerobic unit employing filter bags rather than a clarifier and followed by hypochlorination might provide an effluent suitable for direct discharge to storm sewers and surface drainage. The difficulty of assuring proper maintenance is a strong argument against such a system. Discharge to a percolation field or a subsurface filter, if the soil is unsuitable for percolation, appears to be a more dependable technique.

27-9 Subsurface Disposal Fields

Wastewater which has undergone some reduction in suspended solids and grease content in either an aerobic unit or a septic tank can be satisfactorily disposed of on many natural soils. The capability of the soil system is primarily limited by its long-term capacity to transmit water, hence the design of subsurface disposal systems is based upon a standard percolation test.[15,16]

The percolation test is conducted by boring a hole 100 mm (4 in) or more in diameter to the depth of the proposed disposal field (at least 500 mm). The sides of the hole are scratched and all loose earth is removed, after which 50 mm (2 in) of fine gravel or coarse sand is placed in the bottom. The hole is then filled with water to a depth of 300 mm (12 in) and that depth is maintained for at least 4 hours and preferably overnight by adding water. If the hole holds water overnight, the depth is adjusted to about 150 mm (6 in) above the gravel and the drop in the water surface in 30 min is recorded. If the hole is empty, it is refilled to about 150 mm (6 in) above the gravel and the drop in the water elevation is recorded at 30-min intervals for 4 hours. Water is added as necessary. The drop recorded during the last 30 min is used to determine the percolation rate. Table 27-4 may then be used to determine the required trench area. Alternately, Eq. (27-1) will

Table 27-4 Application rates for subsurface disposal (percolation)

Percolation rate (time required for water to fall 25 mm (1 in), min)	Maximum rate of waste application, l/m² per day
1 or less	204
2	143
3	118
4	102
5	90
10	65
15	53
30	37
45	33
60	24
> 60	Not suitable for percolation. Consider underground filter.

(l/m² per day × 0.0245 = gal/ft² per day.)

yield the flow which can be applied per unit area per day as a function of percolation rate:

$$Q = 204/\sqrt{t} \tag{27-1}$$

where Q = the flow in liters/m² per day and
t = the time required for the water surface to fall 25 mm, in minutes

If the subsidence rate is over 0.5 mm ($\frac{1}{4}$ in) per min a septic tank and tile disposal field will ordinarily prove to be an effective treatment method.[20] The subsurface disposal field is constructed of 100-mm (4-in) diameter pipe, either short lengths of solid pipe laid with open joints, or perforated plastic or fiber pipe. The slope of the laterals is 0.17 to 0.33 percent and individual lines are ordinarily 30 m (100 ft) or less in length. The pipe is placed in a ditch at least 500 mm (18 in) deep which has been excavated to a permeable stratum. The ditch is 300 to 900 mm (12 to 36 in) wide and is backfilled to a depth of 300 to 400 mm (12 to 16 in) with gravel before the pipe is placed. An additional 50 mm (2 in) of gravel is placed over the pipe before the remainder of the trench is filled with topsoil. The total length of pipe depends upon trench width since the product of these must equal the area obtained from the percolation rate. Laterals are laid about 2 m (6 ft) on centers.

Example Determine the size of a septic tank and percolation field for a mobile home park which has 210 residents. Percolation tests indicate the average subsidence rate is 6 mm/min.

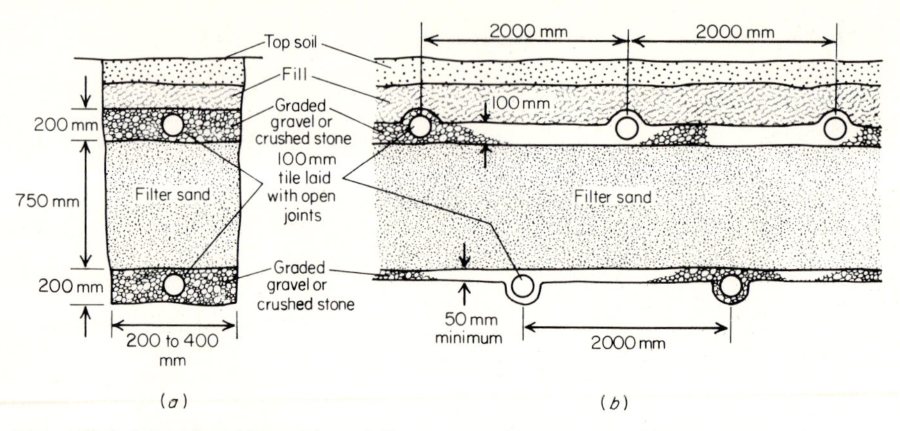

Figure 27-3 Subsurface filters: (*a*) sand-filter trench; (*b*) sand filter; (mm × 0.04 = in).

SOLUTION From Table 27-2, the anticipated flow is 190 l/day per capita or 39,900 l/day. The septic tank volume should thus be approximately 40 m^3. Dimensions might be 2.5 × 2.5 × 6.4 m.

The percolation rate (time for water to fall 25 mm) is 25/6 = 4.17 min, which yields an hydraulic-loading rate of 100 l/m^2 per day. The total trench area is therefore 400 m^2, and the length 444 m if the trenches are 900 mm wide. 15 laterals, each 30 m long, placed 2 m on centers might be used. The area dedicated to the field would be approximately 30 × 30 or 900 m^2.

27-10 Sand Filtration

Filtration of sewage upon either intermittent sand filters (Art. 24-3) or subsurface sand filters may be required when soils are relatively impermeable. Loading rates on intermittent sand filters treating septic tank or aerobic unit effluent have ranged from 0.20 to 0.16 m/day (5 to 4 gal/ft^2 per day), respectively.[21] The surface must be cleaned after 3 to 9 months, depending upon the degree of pretreatment achieved.

Subsurface sand filters are installed in place of impermeable material in a suitable excavation. A typical cross section is shown in Fig. 27-3. The sand should be relatively coarse (~ 1 mm) and uniform to permit thorough ventilation. Loading rates are about 0.04 m/day (1 gal/ft^2 per day). The effluent from either filtration process must be drained to surface water.

PROBLEMS

27-1 A treatment plant is to treat 15,000 m^3/day at design capacity and the peak-flow rate is 150 percent of this. It is expected that the effluent will have a 15-min chlorine demand of 15 mg/l and it is desired to maintain a 0.5 mg/l residual after this time. What is the required contact basin volume and what will be the average and peak rates of chlorine use?

27-2 The novel garbage grinder described in Art. 27-3 is expected to produce an average waste flow with total suspended solids of 1500 mg/l, 66 percent of which are volatile. What volume of primary sludge would you estimate would be produced by this flow? Assume the solids thicken to 5 percent and that the flow is 10,000 m^3/day.

REFERENCES

1. Jolley, Robert L.: "Chlorine-Containing Organic Constituents in Chlorinated Effluents," *Journal Water Pollution Control Federation*, **47:**3:601, 1975.
2. Glaze, W. H. and J. E. Henderson IV: "Formation of Organochlorine Compounds from the Chlorination of a Municipal Secondary Effluent," *Journal Water Pollution Control Federation*, **47:**10:2511, 1975.
3. "New Tertiary Sewage Process Needs No Chlorine to Disinfect," *Engineering News Record*, **192:**93, 1974.
4. Rosen, H. M.: "Use of Ozone and Oxygen in Advanced Wastewater Treatment," *Journal Water Pollution Control Federation*, **46:**12:2788, 1974.
5. Trussell, R. Rhodes and Junn-Ling Chao: "Rational Design of Chlorine Contact Facilities," *Journal Water Pollution Control Federation*, **49:**4:659, 1977.
6. Metcalf and Eddy, Inc.: *Wastewater Engineering: Collection, Treatment, Disposal*, McGraw-Hill, New York, 1972.
7. Cole, Charles A., Paul E. Paul, and Harold P. Brewer: "Odor Control with Hydrogen Peroxide," *Journal Water Pollution Control Federation*, **48:**2:297, 1976.
8. Allen, H. S. and D. R. Flett: "New Plant Solves Seasonal Problem," *Water and Wastes Engineering*, **12:**10:23, 1975.
9. Herr, E. and R. L. Poltorak: "Program Goal—No Odors," *Water and Sewage Works* **121:**10:56 1974.
10. Lovett, W. D. and R. L. Poltorak: "Activated Carbon Used to Control Odors," *Water and Sewage Works*, **121:**8:74, 1974.
11. American Public Works Association: *Municipal Refuse Disposal*, 3d ed., Public Administration Service, Chicago, Ill., 1970.
12. Kühner, Jochen and Peter M. Meier: "Hydraulic Collection and Disposal of Refuse," *Journal Environmental Engineering Division Proceedings American Society of Civil Engineers*, **102:EE4:**769, 1976.
13. "Federal Guidelines—Industrial Cost Recovery Systems," U.S. Environmental Protection Agency, Washington, D.C., 1976.
14. Nemerow, Nelson L.: *Liquid Waste of Industry—Theories, Practices, and Treatment*, Addison-Wesley, Reading, Mass., 1971.
15. Ehlers, M. V. and E. W. Steel: *Municipal and Rural Sanitation* 5th ed., McGraw-Hill, New York, 1958.
16. "Manual of Septic Tank Practice," *Publication No. 526*, U.S. Public Health Service, Washington, D.C., 1957.
17. Siegrist, Robert, Michael Witt and William C. Boyle: "Characteristics of Rural Household Wastewater," *Journal Environmental Engineering Division Proceedings American Society of Civil Engineers*, **102:EE3:**533, 1976.
18. Otis, Richard J. and William C. Boyle: "Performance of Single Household Treatment Units," *Journal Environmental Engineering Division Proceedings American Society of Civil Engineers*, **102:EE1:**175, 1976.
19. Westfield, James D. and Stephen C. Smith: "Discussion—Performance of Single Household Treatment Units," *Journal Environmental Engineering Division Proceedings American Society of Civil Engineers*, **102:EE6:**1297, 1976.
20. Ingham, A.: "Discussion—Efficiency of a Septic Tile System," *Journal Water Pollution Control Federation*, **49:**2:335, 1977.
21. Sauer, David K., William C. Boyle, and Richard J. Otis: "Intermittent Sand Filtration of Household Wastewater," *Journal Environmental Engineering Division Proceedings American Society of Civil Engineers*, **102:EE4:**789, 1976.

CHAPTER

TWENTY-EIGHT

FINANCIAL CONSIDERATIONS

28-1

The development of engineering plans for water and sewage works ordinarily involves economic evaluation of alternate schemes, selection of the least expensive satisfactory system, and development of a financing plan. The least expensive alternate must usually be evaluated from several points of view: total cost, total municipal cost, and total user cost. In many cases the federal or state government may pay a large part of capital costs, and other capital costs may be chargeable to property taxes. Operating costs, on the other hand, ordinarily must be paid by user fees which impinge more directly upon the public.

28-2 Cost Estimates

The cost of engineering works of any sort is generally estimated upon the basis of past costs for similar work. Unit costs for earthwork, pipe installation, reinforced concrete, piles, etc., are available from bid tabulations on other jobs and may be projected to new construction.

In making such a projection it is important that the engineer be certain that the work is actually comparable in difficulty and that the costs are adjusted to account for inflation and other fluctuations in market prices. Adjustments of this sort are made by using *cost indices.*

Engineering News Record publishes the *Building cost index* and *Construction cost index* for the U.S. weekly, and city by city *Building cost*, *Construction cost*, *Materials*, *Skilled labor*, and *Common labor indices* monthly in the second issue of the month. The building cost and construction cost indices are based upon a value of 100 in 1913 and in 1977 were in the vicinity of 1500 and 2500 respectively. The other indices are based upon a value of 100 in 1967 and ranged from 200 to 300 in 1977.

Example The cost of reinforced concrete placed in a 5-m deep clarifier was $\$325/m^3$ in April, 1974 in New Orleans, La. Estimate the cost of similar construction in May, 1977.

SOLUTION The construction cost index for New Orleans in March, 1974 was 1969. The index in April, 1977 was 2066. The estimated cost of concrete in place, in similar construction in May, 1977 is thus:

$$\text{Unit cost} = 325 \times \frac{2066}{1969} = \$341/\text{m}^3$$

Current costs of major equipment can be estimated in a similar fashion, but it is more usual to obtain estimates from manufacturers for such items as clarifier mechanisms, pumps, compressors, vacuum filters, centrifuges, chemical feeders, valves, controls, and so forth. A compilation of such costs for site preparation, construction, materials, equipment, and contingencies permits preparation of reasonably accurate cost estimates for projects represented by a complete set of plans. The Environmental Protection Agency also publishes a *Sewage treatment plant index* which may be used specifically in projections of such costs.

The making of *rough estimates* of treatment plant costs for the purpose of screening alternatives has been facilitated by the recent publication of EPA's *Guide to the Selection of Cost Effective Wastewater Treatment Systems.*[1] Similar publications have been made available which deal with specific topics such as land disposal systems.[2]

The process of estimating alternate costs may be illustrated by application, using Figs. 28-1 and 28-2 and Figs. AA through R in App. III which are taken from reference 1.

Figures 28-1 and 28-2 are popularly called "spiderwebs" and are applied as follows:

Given a requisite effluent quality established by the responsible regulatory agency, one follows the circular line coincident with that quality in Fig. 28-1 to find the final processes in the sequences which would be satisfactory. For example, an effluent quality corresponding to BOD = 15–20, SS = 15–20, P = 8, TKN = 1, Total N = 3 can be attained with four process sequences ending with process H or J. The process chains are obtained by moving inward along the more or less radial lines and are: (1) A_1–C_6–G_1–H; (2) A_1–C_1–J; (3) A_1–B_1–G_2–H; and (4) A_1–B_1–J which are respectively: (1) Primary settling–high-rate activated sludge–nitrification–denitrification; (2) Primary settling–conventional activated sludge–breakpoint chlorination; (3) Primary settling–trickling filter–nitrification–denitrification; and (4) Primary settling–trickling filter–breakpoint chlorination.

WASTEWATER TREATMENT UNIT PROCESSES

AA. Preliminary Treatment
Influent: Raw wastewater

AB. Raw Wastewater Pumping
Influent: Effluent from AA

A. Primary Sedimentation
Influent: Effluent from AA or AB
- A-1 Conventional
- A-2 Two-Stage Lime Addition
- A-3 Single Stage Lime Addition
- A-4 Alum Addition
- A-5 $FeCl_3$ Addition

B. Trickling Filter
- B-1 Influent: Effluent from A-1
- B-2 Influent: Effluent from A-3
- B-3 Influent: Effluent from A-4 or A-5

C. Activated Sludge
- C-1 Conventional
  Influent: Effluent from A-1
- C-2 Conventional
  Influent: Effluent from A-3
- C-3 Conventional
  Influent: Effluent from A-4 or A-5
- C-4 Alum Addition
  Influent: Effluent from A-1
- C-5 $FeCl_3$ Addition
  Influent: Effluent from A-1
- C-6 High Rate
  Influent: Effluent from A-1
- C-7 High Rate & Alum Addition
  Influent: Effluent from A-1
- C-8 High Rate & $FeCl_3$ Addition
  Influent: Effluent from A-1

D. Filtration
Influent: Effluent from A-2, B-2, B-3, C-2,
C-3, C-4, C-5, F-1 or F-2
G-1, G-2, G-3, G-4, H, J, K

E. Activated Carbon
Influent: Effluent from D

F. Two-Stage Tertiary Lime Treatment
- F-1 Influent: Effluent from B-1
- F-2 Influent: Effluent from C-1

G. Biological Nitrification
- G-1 Influent: Effluent from C-6
- G-2 Influent: Effluent from B-1
- G-3 Influent: Effluent from A-3, A-4 or A-5
- G-4 Influent: Effluent from A-2,C-7 or C-8

H. Biological Denitrification
Influent: Effluent from G-1, G-2, G-3 or G-4

I. Ion Exchanges
Associated with A-2, B-2, B-3, C-2, C-3, C-4,
C-5, F-1, or F-2

J. Breakpoint Chlorination
Influent: Effluent from A-2, B-1, B-2, B-3, C-1, C-2,
C-3, C-4, C-5, F-1 or F-2

K. Ammonia Stripping
Influent: Effluent from F-1 or F-2

R. Disinfection
Influent: Effluent from any treatment process

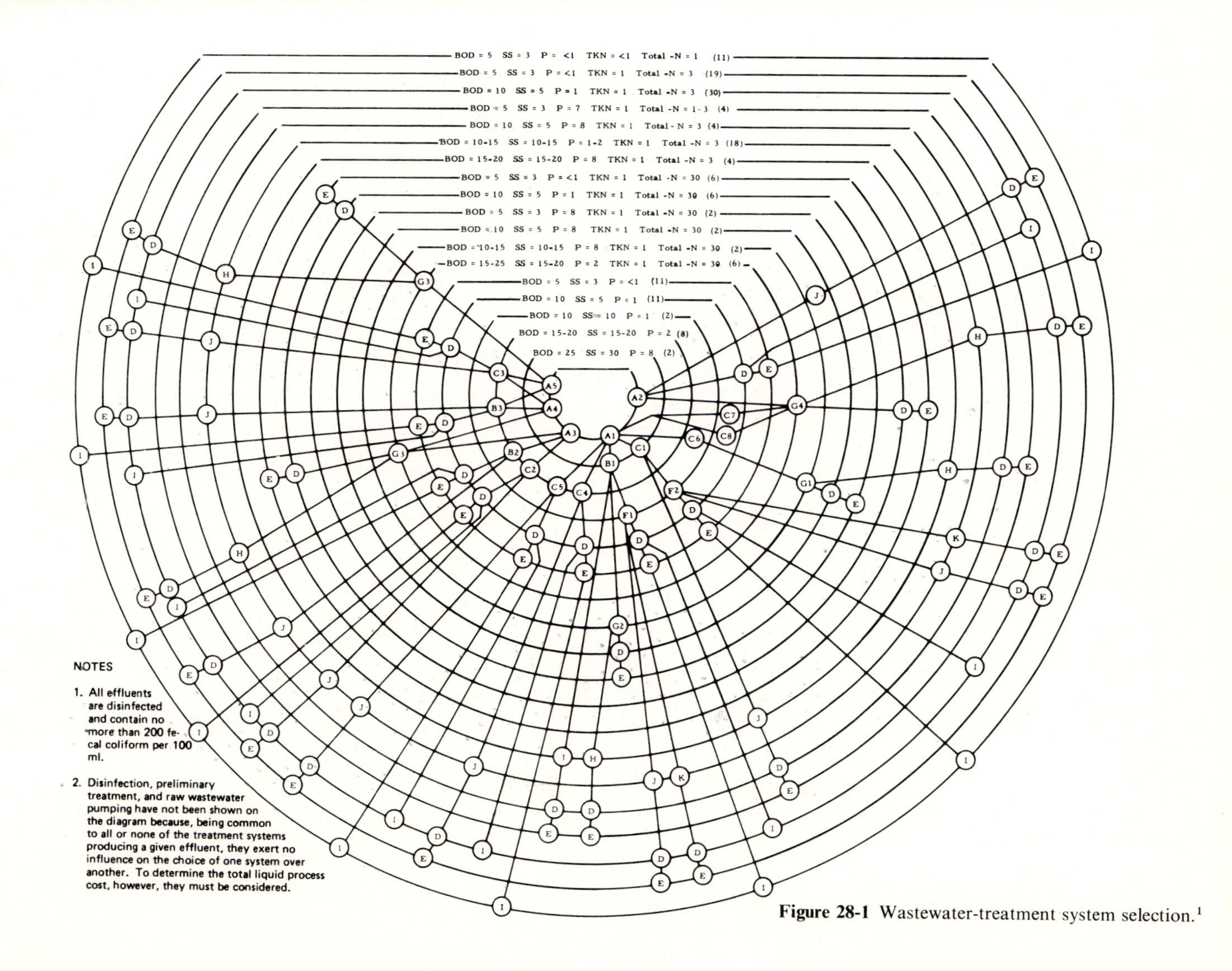

Figure 28-1 Wastewater-treatment system selection.[1]

SLUDGE HANDLING UNIT PROCESSES DESCRIPTION

L. Anaerobic Digestion
- L-1 Sludge Influent: Generated from A-1+B-1, C-1 or C-6
- L-2 Sludge Influent: Generated from A-1+C-4, or C-5, or C-7, or C-8, A-4+B-3 or C-3, A-5+B-3 or C-3

M. Heat Treatment
- M-1 Sludge Influent: Generated from A-1+B-1, C-1 or C-6
- M-2 Sludge Influent: Generated from A-1+C-4 or C-5, or C-7, or C-8, A-4+B-3 or C-3, A-5+B-3 or C-3

N. Air Drying
- N-1 Sludge Influent: Effluent Sludge from L-1
- N-2 Sludge Influent: Effluent Sludge from L-2

O. Dewatering
- O-1 Sludge Influent: Generated from A-1+B-1, C-1 or C-6
- O-2 Sludge Influent: Generated from A-1+C-4 or C-5, or C-7, or C-8, A-4+B-3 or C-3, A-5+B-3 or C-3
- O-3 Sludge Influent: Generated from A-2
- O-4 Sludge Influent: Generated from A-3+B-2 or C-2
- O-5 Sludge Influent: Effluent Sludge from L-1
- O-6 Sludge Influent: Effluent Sludge from L-2
- O-7 Sludge Influent: Generated from F-1 or F-2
- O-8 Sludge Influent: Effluent Sludge from M-1
- O-9 Sludge Influent: Effluent Sludge from M-2

P. Incineration
- P-1 Influent Sludge: Effluent Sludge from O-1
- P-2 Influent Sludge: Effluent Sludge from O-2
- P-3 Influent Sludge: Effluent Sludge from O-3
- P-4 Influent Sludge: Effluent Sludge from O-4
- P-5* Influent Sludge: Effluent Sludge from O-7+O-1
- P-6 Influent Sludge: Effluent Sludge from O-8
- P-7 Influent Sludge: Effluent Sludge from O-9

Q. Recalcination (includes chemical storage & feeding)
- Q-1 Sludge Influent: Effluent Sludge from O-3
- Q-2 Sludge Influent: Effluent Sludge from O-4
- Q-3 Sludge Influent: Effluent Sludge from O-7

*Note - Use pathway from 0-1 to P-5 only when F-1 or F-2 is included in the complete system.

**Sludge leaving this process may be recalcined in part. If this is the case, the remainder may be either incinerated or hauled to disposal.

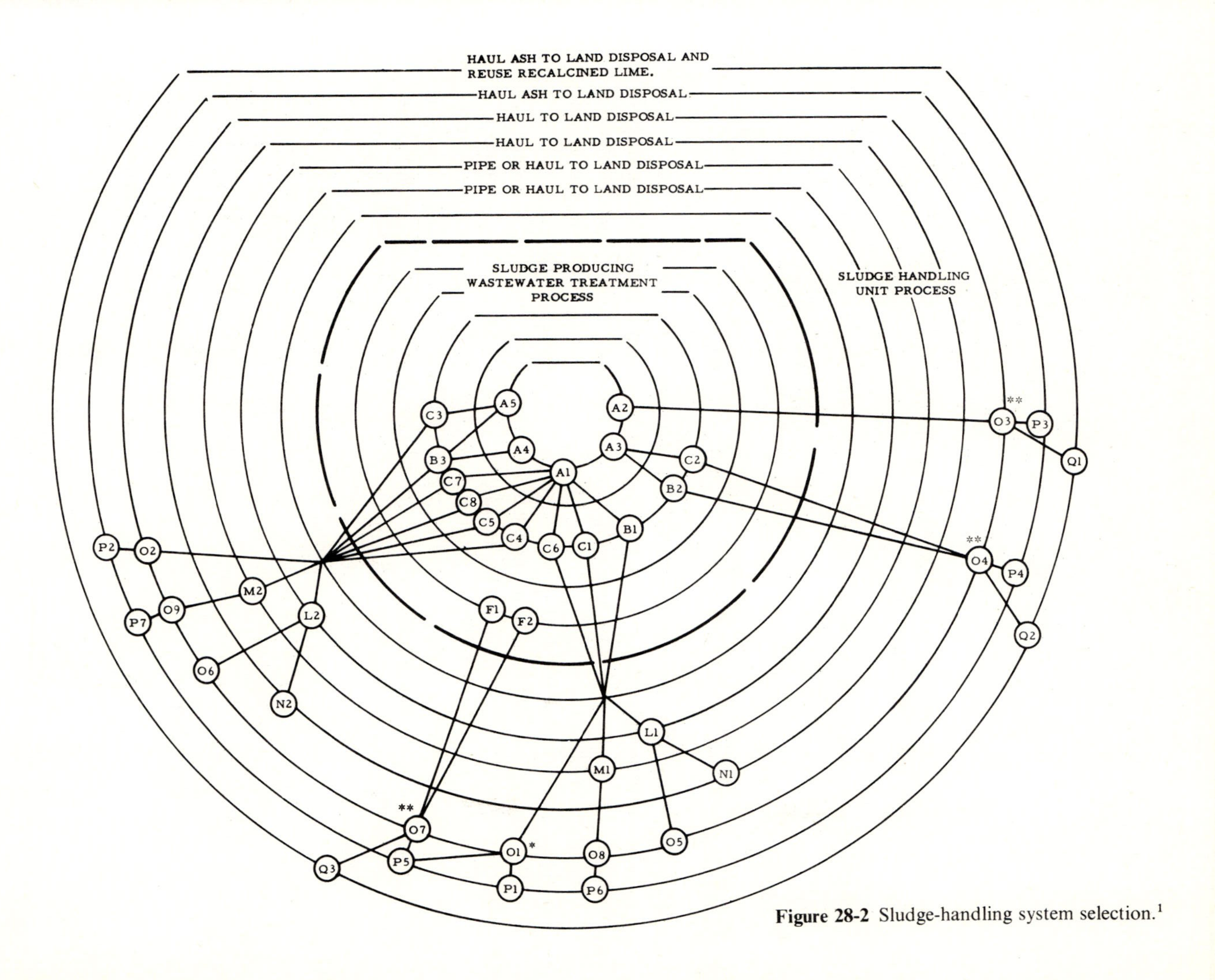

Figure 28-2 Sludge-handling system selection.[1]

Sludge-handling systems depend upon the preceding processes and may be selected from Fig. 28-2. The processes involved in alternate (1) above (A_1–C_6) lead to five possible solids handling trains: Moving outward from the center of Fig. 28-2 these trains are L_1–N_1; L_1–O_5; M_1–O_8–P_6; O_1–P_1; and O_1–P_5, all terminating in final land disposal. Alternates (2), (3), and (4) lead to the same solids-handling alternates: Anaerobic digestion–air drying; Anaerobic digestion–dewatering; heat treatment–dewatering–incineration; and dewatering–incineration.

The cost of these alternates can then be estimated for the given plant using the graphs of Appendix III. For a plant with a capacity of 37,850 m³/day the unit costs of the liquid treatment trains are:

(1) A_1–C_6–G_1–H = 1.7 + 5.8 + 4.9 + 4.3 = 16.70/3785 l

(2) A_1–C_1–J = 1.7 + 7.0 + 13.0 = 21.70/3785 l

(3) A_1–B_1–G_2–H = 1.7 + 4.9 + 4.6 + 4.3 = 15.50/3785 l

(4) A_1–B_1–J = 1.7 + 4.9 + 13.0 = 19.60/3785 l

yielding a minimum cost for alternate 3 of 4.10 cents/m³. This is not the actual cost of treatment since it does not include processes common to all trains such as pumping, grit removal, screening, etc. In similar fashion the least expensive sludge-handling train is found to be L_1–O_5 or anaerobic digestion plus mechanical dewatering, with a net cost of 0.79 cents/m³ of wastewater treated.

In this analysis only the total cost was taken from the graph. Since operation and maintenance do not necessarily fluctuate in the same fashion as capital cost, it is more appropriate to tabulate these separately and correct them to the current year using appropriate indices for skilled and unskilled labor, construction, and materials. Detailed information on the application of this technique is presented in reference 1. The curves are based upon a wholesale price index of 120, a sewage treatment plant index of 177.5 and an interest rate of $5\frac{5}{8}$ percent.

The spiderweb analysis yields a relatively rapid comparison between different alternatives but should not be used as a basis for estimating actual construction and operating costs for the selected alternate.

28-3 Cost Comparisons

In selecting treatment systems for projects involving federal funds it is generally necessary to demonstrate that the process train selected is "cost effective," that is, that it achieves the desired goal at minimum total cost.

The total cost involves both capital expenditure and recurring expenses for operation and maintenance, and it is necessary to add these costs in order to determine which process has the lowest present worth, annual cost, or unit cost. The principles of engineering economics are developed elsewhere[3] and only their direct application is discussed here.

Annual costs such as operating expenses can be converted to present worth by multiplying by the *Present Worth Factor* (P/A) given by:

$$(P/A, i, n) = \frac{(1+i)^n - 1}{i(1+i)^n} \tag{28-1}$$

in which i is the interest rate or cost of money and n is the estimated life of the plant. Capital costs can be converted to annual costs by multiplying by the Capital Recovery Factor:

$$(A/P, i, n) = \frac{i(1+i)^n}{(1+i)^n - 1} \tag{28-2}$$

Values of the interest factors are tabulated in many handbooks and texts and may be readily determined with pocket calculators. Unit costs of treatment are determined by dividing the total annual cost by the total annual flow.

Example A water-softening plant may be constructed as either a lime-soda process or an ion-exchange process. The plant is to have a capacity of 8000 m^3/day and the estimated costs and lives are as follows:

	Lime-soda	Ion-exchange
First cost (construction)	\$435,000	\$330,000
Operation and maintenance per year	\$61,320	\$52,560
Life, years	50	30
Interest rate	6%	6%

The comparison is most easily made upon the basis of annual cost since the lives of the alternatives differ. The annual costs are:

$$\text{Lime-soda} = 61{,}320 + 435{,}000 \times \frac{0.06(1+0.06)^{50}}{(1+0.06)^{50} - 1}$$

$$= \$88{,}918 \text{ per year.}$$

$$\text{Ion-exchange} = 52{,}560 + 330{,}000 \times \frac{0.06(1+0.06)^{30}}{(1+0.06)^{30} - 1}$$

$$= \$76{,}534 \text{ per year.}$$

The ion-exchange process is least expensive and has a unit cost of 2.62 cents/m^3.

28-4 Optimization of Process Selection

The techniques of Art. 28-2 may indicate that a particular sequence of treatment processes will be less expensive than other sequences, but do not establish the optimum proportion of the waste-treatment load which should be borne by each stage, nor the preferred selection among process variables for an individual system. For example, a primary clarifier and an activated sludge system will meet the ordinary secondary-treatment standards for domestic wastewater, irrespective of the primary clarifier size. As the surface overflow rate of the clarifier decreases its cost increases, but up to a point, its efficiency does as well, which will reduce the load upon the activated sludge process and reduce its cost. The activated sludge system can produce the required effluent quality by manipulation of hydraulic retention time and solids return to achieve the necessary sludge age. Reduction of hydraulic retention time decreases the basin volume but increases the cost of recycling solids and may affect the stability of the process.

The determination of the minimum cost combination of such systems requires complete analysis of all liquid- and solids-handling trains and the effects of changes in one process upon the others. Middleton and Lawrence[4] have presented such an analysis for a process train including primary settling, activated sludge, final settling, gravity thickening, anaerobic digestion, vacuum filtration, incineration, and ultimate disposal of sludge. The technique which they propose is general and may be modified to include other processes.

Tarrer, et al.[5] have prepared a computer program which allows optimization of the liquid-treatment train, but not solids handling, for activated sludge systems. The model, like that of Middleton and Lawrence,[4] incorporates practical constraints and also allows for uncertainty in waste parameters.

The design of the secondary clarifier has been found to exert a significant effect upon optimum liquid treatment design,[4,5] hence approaches to optimization of secondary clarifier design are also necessary. Lee, Fan, and Takamatsu[6] have presented such an analysis which indicates that under a given set of conditions there exists an optimal volume allocation in a multistage clarifier system and that a multistage clarifier is more efficient and stable than a single unit.

It is expected that further development of system and unit process optimization techniques will occur which will simplify the engineer's task of selecting among alternates.

28-5 Financing of Waterworks

Waterworks may be privately or publicly owned. Privately owned systems are usually incorporated, but may be individually owned. Corporations are financed by sale of stocks and bonds and may be regulated by public utility commissions. Before a private water company can operate it must obtain a franchise from the city which binds it to provide water to the citizens under specified conditions in return for use of the public right-of-way for its mains.

Publicly owned waterworks are in the majority. Water supply districts which include one or more cities and rural areas are occasionally organized, if the state permits, but most waterworks are owned by cities. In either case the procedure is much the same.

After the engineer's estimates have been received, a bond issue is authorized, generally by affirmative vote of the people of the city or district. The bonds are then sold, and the money is available to pay the contractors as construction work proceeds.

Bonds are of two types, sinking fund, also known as term bonds, which require the accumulation of a fund over their life to pay them when they are due; and serial bonds, of which a certain number are due and paid at intervals throughout the life of the issue. For example: An issue of 20-year term bonds would all become due and payable at the end of 20 years, and each year interest would be paid on the whole issue. If they were serial bonds, a proportion, usually one-twentieth, would be paid each year, and interest would be paid on those not yet matured. Serial bonds are the preferable type, since there is no need to accumulate and manage the sinking fund. Money for retiring bonds and paying interest is usually derived from the water revenues, although, with poor financial management, taxes may have to help out.

The municipally issued bonds, sometimes called general obligation bonds, mentioned above become a part of the total debt of a city. Most states impose debt limits upon their cities and these may prevent financing of adequate waterworks facilities. Some states permit the issuing of revenue bonds. These are issued with the water works as security, and the water revenues are used to pay interest and amortization. They require a higher rate of interest than tax-secured bonds but are not considered in relation to the state-imposed debt limit.

Where state laws permit, cities may pay part of the cost of waterworks by special assessments against the property served. This is entirely equitable, since the installation of water mains usually increases the value of the property served far more than their cost. The practice varies as to the size of the assessment. In some cities it is a flat charge made per front meter of abutting property. In others it is the actual cost of the main. The remainder of the money required is raised by a bond issue.

The financing of extensions presents a problem to the waterworks authorities, particularly during the present era of high construction costs, rapidly expanding cities, and the great numbers of suburban developments which are appearing outside but near to city limits. The privately owned water utilities are, in many states, subject to control by state utility commissions, and must be guided by their directions as to extensions. In any case they must maintain their capital structure and charge rates that will give a reasonable return on the capital invested, in addition to caring for operating expenses and depreciation. They do not, however, amortize indebtedness for capital improvements from the rates. When extensions are made the company will make no charge to the customer for, say, 15 m (50 ft) of 150-mm (6 in) main where fire protection is needed or for costs up to four times the expected annual revenue from the customer. If more is required the customers must pay the difference. As new customers connect, the difference paid is reimbursed to those who paid the excess cost. This reimbursement is made only for new connections made within a specified number of years, usually 10.

Municipally owned water systems amortize their bonded indebtedness, which means that the " old " customers have paid in their water rates for more or less of the first cost of the existing system. New customers within the city limits, who require main extensions in order to receive service should, therefore, pay an assessment for the extension. This may be the actual cost of a 150-mm (6-in) main of length equal to the abutting property. This, of course, would be justified from the standpoint of increased property value alone, and it would seem equitable to charge the new customer even more, since he also receives service from pumping station, trunk mains, treatment plant, and other important structures of the water works. Large real estate developments within the city may be required to install all the water mains, hydrants, valves, etc., required to care for the subdivision before connection is made to the system. Later the developers may be reimbursed by the water works for the difference between the usual front assessments and their actual expenditures. This results in the real estate developer adding the front assessment to the cost of the building lots.

Policies vary as to extension of water mains beyond the city limits by municipally owned waterworks. Many cities are reluctant to do so. Others foresee later annexation of neighboring areas and consider it desirable to exercise control of systems serving the real estate developments. This they can do by requiring compliance by the developer with city requirements as to pipe sizes, etc., and financing is by means described in the preceding paragraph.

28-6 Charges for Service

These charges, or water rates, made for water service must cover the following: interest on outstanding bonds; sinking fund contributions or other payments to retire indebtedness; operating expenses, which also include repairs; and a depreciation reserve, to replace worn-out equipment. Private water companies will pay these from the rates, except retirement of bonded indebtedness, in addition to dividends to stockholders and taxes. Publicly owned waterworks may also operate for a profit, which is placed in the general fund to reduce taxes. This, however, is not considered good practice. City councils may deplete reserves and allow the plant to deteriorate in their zeal to make all the money possible from a waterworks. Revenues may also be required to cover "free" water, which is that given by some cities to the public schools and other public buildings. This is also considered poor practice, as it encourages waste and places a burden upon the water rate payers. All city departments should be billed for the water that they use. Water that cannot be easily metered includes that used for fire fighting, street flushing, fountains, etc., and leakage.

Special charges are made to the city for fire service by a few publicly owned waterworks and usually by private companies. Such charges may take the form of a flat sum, a tax per capita, or an annual charge per fire hydrant. Such charges generally do not cover more than one-half the actual cost of fire protection, and the remainder must be made up from consumers' rates.

Water rates should be carefully constructed, a matter that many cities fail to recognize. Rate structures are too often copied from other cities; and if the gross revenues yielded are more than the expenses, nothing more is expected. Rates should distribute the burden of water costs as equitably as possible among the various classes of users, residential, commercial, and industrial, according to the demands that they make upon the system. If there is an adequate supply of water, the rates should encourage greater use so that more revenue will be obtained. Adequate depreciation and other reserves must be maintained. Rate systems should be adopted only after careful investigation by competent engineers of the needs and local conditions.

Several forms of rates are in use. The *flat rate* is a monthly or quarterly charge for each service. The charge may be varied according to the building use or number of water outlets. This type of rate is falling into disuse. It encourages waste and, since wasted water must be paid for, revenues must be obtained by high rates. Metering of all services is recognized as the economical and equitable procedure for arriving at the charges to the customers. The *step rate* of arriving at charges per month places the charge per unit used at some rate until some amount, say 40 m^3 (10,000 gal) is used. Then a lower rate is applied to all of the water used in that month. The objection to this is that near the change to the lower rate it may be of advantage to the customer to waste water so that his consumption reaches the next step. The *block rate* subdivides consumption into blocks, with the highest rate applying to the first 8 or 12 m^3 (2000 to 3000 gal) used. Each succeeding block carries a lower charge per unit volume until a minimum rate, at a very high consumption, is reached that approaches operating cost. The total charge is the sum of the charges for each block. The block rate is usually combined with the minimum bill. The minimum bill frequently includes a block of 8 m^3 (2000 gal), and it is paid whether the customer uses all, none, or any part of the block. The charge for the first block covers, at least in part, such administrative charges as meter reading, billing, and collection plus the cost of the water actually used. The other administrative costs and overhead charges are recovered in the charges for the succeeding blocks, which progressively decrease in amount per unit volume.

28-7 Accounting

Small cities sometimes neglect waterworks accounting and in consequence know little of production costs, frequently lose revenue, and receive many complaints. Accounting methods have been developed which will be of assistance to waterworks superintendents. The accounting system, briefly outlined, will include the following:

1. *Customers' ledger*. This shows the customer's name, address, account number, meter description, meter readings by month, amount used, charges, arrears, penalties and discounts, amount due, date, and amount of payment. Accounts must also be kept for other city departments that use water.
2. *Deposit ledger*. This has all essential information regarding the deposits usually required of all customers to guarantee payment of bills. Account number, name, and address of the customer will be shown, together with the amount

and date of the deposit on the debit side and the date and amount of refund on the credit side.

3. *Distribution ledger.* All cash receipts and disbursements are classified according to certain accounts. The receipts will be for operating or nonoperating revenues and will be further subdivided as to sources. For example, operating revenue would include nonmetered private service, metered private service, etc. Nonoperating revenue would include rents, interest, etc. The disbursements would cover a large variety of accounts whose main headings would be operating expenses and capital outlays. Operating expenses would include those for water collection, treatment, pumping, distribution, commercial and new business, general, and miscellaneous. All these will be further subdivided.

   Balance sheet accounts show assets and liabilities. The assets include cash, sinking funds, etc., but particularly the fixed capital. Disbursements for new equipment, extensions, etc., are capital outlays and increase fixed capital, while worn-out or discarded equipment must be written off. To show the true value of the assets, the depreciation reserves are shown on the liability side. Other liabilities are the funded debt, notes, customers' deposits, etc.
4. *Production data.* Proper operation of pumping and treatment plants will give the superintendent information as to the amount of water produced. From the customers' ledger the amount sold and otherwise used, or unaccounted for, can be found. From the disbursement accounts costs can be found, and the costs per unit volume can be computed.

28-8 Financing of Sewage Works

Treatment of wastewater is not a profitable venture, hence wastewater collection and treatment is a public responsibility. In recent years a large portion (up to 75 percent) of the capital cost of both collection and treatment systems has been financed with federal grants, and some states will contribute up to 50 percent of the remainder. Local capital costs may thus be as little as 12.5 percent of the total.

The capital cost provided by the municipality is raised in the same fashion as for water works: by bond issues supported either by general tax receipts or user charges, or by property assessments. General obligation bonds are used when the debt limit of the community permits this.

In communities or states in which revenue bonds are not permitted, frontage assessments may be required. This is a reasonable technique for recovering collection system costs, but may be inequitable if collection and treatment are lumped together.

28-9 Sewer Charges

The cost of maintaining and operating the collection and treatment systems should be, and in many cases, must be, paid by user charges. As noted in Art. 28-8 capital costs may have to be financed by revenue bonds, and this will increase the user fees.

Charges for wastewater treatment for individual residences may be based upon a flat fee, upon the volume of water used, or upon the volume of water used during the winter months. Industrial charges, as noted in Chap. 27, are based upon actual contribution of flow, BOD, SS, etc.

REFERENCES

1. *A Guide to the Selection of Cost Effective Wastewater Treatment Systems*, U.S. Environmental Protection Agency, Washington, D.C., 1975.
2. *Costs of Wastewater Treatment by Land Application*, U.S. Environmental Protection Agency, Washington, D.C., 1975.
3. Grant, Eugene L., W. Grant Ireson, and Richard S. Leavenworth: *Principles of Engineering Economy* 6th ed., Ronald Press, New York, 1976.
4. Middleton, Andrew C. and Alonzo W. Lawrence: "Least Cost Design of Activated Sludge Systems," *Journal Water Pollution Control Federation*, **48:**5:889, 1976.
5. Tarrer, Arthur R., et al.: "Optimal Activated Sludge Design under Uncertainty," *Journal Environmental Engineering Division, Proceedings American Society of Civil Engineers*, **102:EE3:**657, 1976.
6. Lee, Chin R., L. T. Fan, and T. Takamatsu: "Optimization of Multistage Secondary Clarifier Design," *Journal Water Pollution Control Federation*, **48:**11:2578, 1976.

APPENDIX

ONE

NATIONAL DRINKING WATER REGULATIONS

Title 40—Protection of Environment

CHAPTER I—ENVIRONMENTAL PROTECTION AGENCY

SUBCHAPTER D—WATER PROGRAMS

[FRL 464-7]

PART 141—NATIONAL INTERIM PRIMARY DRINKING WATER REGULATIONS

Subpart A—General

Sec.
141.1 Applicability.
141.2 Definitions.
141.3 Coverage.
141.4 Variances and exemptions.
141.5 Siting requirements
141.6 Effective date.

Subpart B—Maximum Contaminant Levels

141.11 Maximum contaminant levels for inorganic chemicals.
141.12 Maximum contaminant levels for organic chemicals.
141.13 Maximum contaminant levels for turbidity.
141.14 Maximum microbiological contaminant levels.

Subpart C—Monitoring and Analytical Requirements

141.21 Microbiological contaminant sampling and analytical requirements.
141.22 Turbidity sampling and analytical requirements.
141.23 Inorganic chemical sampling and analytical requirements.
141.24 Organic chemical sampling and analytical requirements.
141.27 Alternative analytical techniques.
141.28 Approved laboratories.
141.29 Monitoring of consecutive public water systems.

Subpart D—Reporting, Public Notification, and Record-keeping

141.31 Reporting requirements.
141.32 Public notification of variances, exemptions, and non-compliance with regulations.
141.33 Record maintenance.

AUTHORITY: Secs. 1412, 1414, 1445, and 1450 of the Public Health Service Act, 88 Stat. 1660 (42 U.S.C. 300g–1, 300g–3, 300j–4, and 300j–9).

SUBPART A—GENERAL

§ 141.1 Applicability

This part establishes primary drinking water regulations pursuant to section 1412 of the Public Health Service Act, as amended by the Safe Drinking Water Act (Pub. L. 93–523); and related regulations applicable to public water systems.

§ 141.2 Definitions

As used in this part, the term:

(a) "Act" means the Public Health Service Act, as amended by the Safe Drinking Water Act, Pub. L. 93-523.

(b) "Contaminant" means any physical, chemical, biological, or radiological substance or matter in water.

(c) "Maximum contaminant level" means the maximum permissible level of a contaminant in water which is delivered to the free flowing outlet of the ultimate user of a public water system, except in the case of turbidity where the maximum permissible level is measured at the point of entry to the distribution system. Contaminants added to the water under circumstances controlled by the user, except those resulting from corrosion of piping and plumbing caused by water quality, are excluded from this definition.

(d) "Person" means an individual, corporation, company, association, partnership, State, municipality, or Federal agency.

(e) "Public water system" means a system for the provision to the public of piped water for human consumption, if such system has at least fifteen service connections or regularly serves an average of at least twenty-five individuals daily at least 60 days out of the year. Such term includes (1) any collection, treatment, storage, and distribution facilities under control of the operator of such system and used primarily in connection with such system, and (2) any collection or pretreatment storage facilities not under such control which are used primarily in connection with such system. A public water system is either a "community water system" or a "non-community water system."

(i) "Community water system" means a public water system which serves at least 15 service connections used by year-round residents or regularly serves at least 25 year-round residents.

(ii) "Non-community water system" means a public water system that is not a community water system.

(f) "Sanitary survey" means an on-site review of the water source, facilities, equipment, operation and maintenance of a public water system for the purpose of evaluating the adequacy of such source, facilities, equipment, operation and maintenance for producing and distributing safe drinking water.

(g) "Standard sample" means the aliquot of finished drinking water that is examined for the presence of coliform bacteria.

(h) "State" means the agency of the State government which has jurisdiction over public water systems. During any period when a State does not have primary enforcement responsibility pursuant to Section 1413 of the Act, the term "State" means the Regional Administrator, U.S. Environmental Protection Agency.

(i) "Supplier of water" means any person who owns or operates a public water system.

§ 141.3 Coverage

This part shall apply to each public water system, unless the public water system meets all of the following conditions:

(a) Consists only of distribution and storage facilities (and does not have any collection and treatment facilities);

(b) Obtains all of its water from, but is not owned or operated by, a public water system to which such regulations apply:

(c) Does not sell water to any person; and

(d) Is not a carrier which conveys passengers in interstate commerce.

§ 141.4 Variances and Exemptions

Variances or exemptions from certain provisions of these regulations may be granted pursuant to Sections 1415 and 1416 of the Act by the entity with primary enforcement responsibility. Provisions under Part 142, *National Interim Primary Drinking Water Regulations Implementation*—subpart E (Variances) and subpart F (Exemptions)—apply where EPA has primary enforcement responsibility.

§ 141.5 Siting Requirements

Before a person may enter into a financial commitment for or initiate construction of a new public water system or increase the capacity of an existing public water system, he shall notify the State and, to the extent practicable, avoid locating part or all of the new or expanded facility at a site which:

(a) Is subject to a significant risk from earthquakes, floods, fires or other disasters which could cause a breakdown of the public water system or a portion thereof; or

(b) Except for intake structures, is within the floodplain of a 100-year flood or is lower than any recorded high tide where appropriate records exist.

The U.S. Environmental Protection Agency will not seek to override land use decisions affecting public water systems siting which are made at the State or local government levels.

§ 141.6 Effective Date

The regulations set forth in this part shall take effect 18 months after the date of promulgation.

SUBPART B—MAXIMUM CONTAMINANT LEVELS

§ 141.11 Maximum Contaminant Levels for Inorganic Chemicals

(a) The maximum contaminant level for nitrate is applicable to both community water systems and non-community water systems. The levels for the other inorganic chemicals apply only to community water systems. Compliance with maximum contaminant levels for inorganic chemicals is calculated pursuant to § 141.23.

(b) The following are the maximum contaminant levels for inorganic chemicals other than fluoride:

Contaminant	Level, milligrams per liter
Arsenic	0.05
Barium	1.
Cadmium	0.010
Chromium	0.05
Lead	0.05
Mercury	0.002
Nitrate (as N)	10.
Selenium	0.01
Silver	0.05

(c) When the annual average of the maximum daily air temperatures for the location in which the community water system is situated is the following, the maximum contaminant levels for fluoride are:

Temperature degrees Fahrenheit	Degrees Celsius	Level, milligrams per liter
53.7 and below	12.0 and below	2.4
53.8 to 58.3	12.1 to 14.6	2.2
58.4 to 63.8	14.7 to 17.6	2.0
63.9 to 70.6	17.7 to 21.4	1.8
70.7 to 79.2	21.5 to 26.2	1.6
79.3 to 90.5	26.3 to 32.5	1.4

§ 141.12 Maximum Contaminant Levels for Organic Chemicals

The following are the maximum contaminant levels for organic chemicals. They apply only to community water systems. Compliance with maximum contaminant levels for organic chemicals is calculated pursuant to § 141.24.

	Level, milligrams per liter
(a) Chlorinated hydrocarbons:	
Endrin (1,2,3,4,10, 10-hexachloro-6,7-epoxy-1,4, 4a,5,6,7,8,8a-octa-hydro-1,4-endo, endo-5,8-dimethano naphthalene).	0.0002
Lindane (1,2,3,4,5,6-hexachlorocyclohexane, gamma isomer).	0.004
Methoxychlor (1,1,1-Trichloro-2,2-bis [p-methoxyphenyl] ethane).	0.1
Toxaphene ($C_{10}H_{10}Cl_8$-Technical chlorinated camphene, 67–69 percent chlorine).	0.005
(b) Chlorophenoxys:	
2,4-D, (2,4-Dichlorophenoxyacetic acid).	0.1
2,4,5-TP Silvex (2,4,5-Trichlorophenoxypropionic acid).	0.01

§ 141.13 Maximum Contaminant Levels for Turbidity

The maximum contaminant levels for turbidity are applicable to both community water systems and non-community water systems using surface water sources in whole or in part. The maximum contaminant levels for turbidity in drinking water, measured at a representative entry point(s) to the distribution system, are:

(a) One turbidity unit (TU), as determined by a monthly average pursuant to § 141.22, except that five or fewer turbidity units may be allowed if the supplier of water can demonstrate to the State that the higher turbidity does not do any of the following:

(1) Interfere with disinfection;

(2) Prevent maintenance of an effective disinfectant agent throughout the distribution system; or

(3) Interfere with microbiological determinants.

(b) Five turbidity units based on an average for two consecutive days pursuant to § 141.22.

§ 141.14 Maximum Microbiological Contaminant Levels

The maximum contaminant levels for coliform bacteria, applicable to community water systems and non-community water systems, are as follows:

(a) When the membrane filter technique pursuant to § 141.21(a) is used, the number of coliform bacteria shall not exceed any of the following:

(1) One per 100 milliliters as the arithmetic mean of all samples examined per month pursuant to § 141.21(b) or (c);

(2) Four per 100 milliliters in more than one sample when less than 20 are examined per month; or

(3) Four per 100 milliliters in more than five percent of the samples when 20 or more are examined per month.

(b)(1) When the fermentation tube method and 10 milliliter standard portions pursuant to § 141.21(a) are used, coliform bacteria shall not be present in any of the following:

(i) more than 10 percent of the portions in any month pursuant to 141.21(b) or (c);

(ii) three or more portions in more than one sample when less than 20 samples are examined per month; or

(iii) three or more portions in more than five percent of the samples when 20 or more samples are examined per month.

(2) When the fermentation tube method and 100 milliliter standard portions pursuant to § 141.21(a) are used, coliform bacteria shall not be present in any of the following:

(i) more than 60 percent of the portions in any month pursuant to § 141.21(b) or (c);

(ii) five portions in more than one sample when less than five samples are examined per month; or

(iii) five portions in more than 20 percent of the samples when five or more samples are examined per month.

(c) For community or non-community systems that are required to sample at a rate of less than 4 per month, compliance with paragraphs (a), (b)(1), or (b)(2) of this section shall be based upon sampling during a 3-month period, except that, at the discretion of the State, compliance may be based upon sampling during a one-month period.

SUBPART C—MONITORING AND ANALYTICAL REQUIREMENTS

§ 141.21 Microbiological Contaminant Sampling and Analytical Requirements

(a) Suppliers of water for community water systems and non-community water systems shall analyze for coliform bacteria for the purpose of determining compliance with § 141.14. Analyses shall be conducted in accordance with the analytical recommendations set forth in "Standard Methods for the Examination of Water and Wastewater," American Public Health Association, 13th Edition, pp. 662–688, except that a standard sample size shall be employed. The standard sample used in the membrane filter procedure shall be 100 milliliters. The standard sample used in the 5 tube most probable number (MPN) procedure (fermentation tube method) shall be 5 times the standard portion. The standard portion is either 10 milliliters or 100 milliliters as described in § 141.14(b) and (c). The samples shall be taken at points which are representative of the conditions within the distribution system.

(b) The supplier of water for a community water system shall take coliform density samples at regular time intervals, and in number proportionate to the population served by the system. In no event shall the frequency be less than as set forth below:

Population served:	Minimum number of samples per month	Population served:	Minimum number of samples per month
25 to 1,000	1	18,101 to 18,900	21
1,001 to 2,500	2	18,901 to 19,800	22
2,501 to 3,300	3	19,801 to 20,700	23
3,301 to 4,100	4	20,701 to 21,500	24
4,101 to 4,900	5	21,501 to 22,300	25
4,901 to 5,800	6	22,301 to 23,200	26
5,801 to 6,700	7	23,201 to 24,000	27
6,701 to 7,600	8	24,001 to 24,900	28
7,601 to 8,500	9	24,901 to 25,000	29
8,501 to 9,400	10	25,001 to 28,000	30
9,401 to 10,300	11	28,001 to 33,000	35
10,301 to 11,100	12	33,001 to 37,000	40
11,101 to 12,000	13	37,001 to 41,000	45
12,001 to 12,900	14	41,001 to 46,000	50
12,901 to 13,700	15	46,001 to 50,000	55
13,701 to 14,600	16	50,001 to 54,000	60
14,601 to 15,500	17	54,001 to 59,000	65
15,501 to 16,300	18	59,001 to 64,000	70
16,301 to 17,200	19	64,001 to 70,000	75
17,201 to 18,100	20	70,001 to 76,000	80

(*continued*)

Population served:	Minimum number of samples per month	Population served:	Minimum number of samples per month
76,001 to 83,000	85	910,001 to 970,000	290
83,001 to 90,000	90	970,001 to 1,050,000	300
90,001 to 96,000	95	1,050,001 to 1,140,000	310
96,001 to 111,000	100	1,140,001 to 1,230,000	320
111,001 to 130,000	110	1,230,001 to 1,320,000	330
130,001 to 160,000	120	1,320,001 to 1,420,000	340
160,001 to 190,000	130	1,420,001 to 1,520,000	350
190,001 to 220,000	140	1,520,001 to 1,630,000	360
220,001 to 250,000	150	1,630,001 to 1,730,000	370
250,001 to 290,000	160	1,730,001 to 1,850,000	380
290,001 to 320,000	170	1,850,001 to 1,970,000	390
320,001 to 360,000	180	1,970,001 to 2,060,000	400
360,001 to 410,000	190	2,060,001 to 2,270,000	410
410,001 to 450,000	200	2,270,001 to 2,510,000	420
450,001 to 500,000	210	2,510,001 to 2,750,000	430
500,001 to 550,000	220	2,750,001 to 3,020,000	440
550,001 to 600,000	230	3,020,001 to 3,320,000	450
600,001 to 660,000	240	3,320,001 to 3,620,000	460
660,001 to 720,000	250	3,620,001 to 3,960,000	470
720,001 to 780,000	260	3,960,001 to 4,310,000	480
780,001 to 840,000	270	4,310,001 to 4,690,000	490
840,001 to 910,000	280	4,690,001 or more	500

Based on a history of no coliform bacterial contamination and on a sanitary survey by the State showing the water system to be supplied solely by a protected ground water source and free of sanitary defects, a community water system serving 25 to 1,000 persons, with written permission from the State, may reduce this sampling frequency except that in no case shall it be reduced to less than one per quarter.

(c) The supplier of water for a non-community water system shall sample for coliform bacteria in each calendar quarter during which the system provides water to the public. Such sampling shall begin within two years after the effective date of this part. If the State, on the basis of a sanitary survey, determines that some other frequency is more appropriate, that frequency shall be the frequency required under these regulations. Such frequency shall be confirmed or changed on the basis of subsequent surveys.

(d)(1) When the coliform bacteria in a single sample exceed four per 100 milliliters (§ 141.14(a)), at least two consecutive daily check samples shall be collected and examined from the same sampling point. Additional check samples shall be collected daily, or at a frequency established by the State, until the results obtained from at least two consecutive check samples show less than one coliform bacterium per 100 milliliters.

(2) When coliform bacteria occur in three or more 10-ml portions of a single sample (§ 141.14(b)(1)), at least two consecutive daily check samples shall be col-

lected and examined from the same sampling point. Additional check samples shall be collected daily, or at a frequency established by the State, until the results obtained from at least two consecutive check samples show no positive tubes.

(3) When coliform bacteria occur in all five of the 100-ml portions of a single sample (§ 141.14(b)(2)), at least two daily check samples shall be collected and examined from the same sampling point. Additional check samples shall be collected daily, or at a frequency established by the State, until the results obtained from at least two consecutive check samples show no positive tubes.

(4) The location at which the check samples were taken pursuant to paragraphs (d)(1), (2), or (3) of this section shall not be eliminated from future sampling without approval of the State. The results from all coliform bacterial analyses performed pursuant to this subpart, except those obtained from check samples and special purpose samples, shall be used to determine compliance with the maximum contaminant level for coliform bacteria as established in § 141.14. Check samples shall not be included in calculating the total number of samples taken each month to determine compliance with § 141.21(b) or (c).

(e) When the presence of coliform bacteria in water taken from a particular sampling point has been confirmed by any check samples examined as directed in paragraphs (d)(1), (2), or (3) of this section, the supplier of water shall report to the State within 48 hours.

(f) When a maximum contaminant level set forth in paragraphs (a), (b) or (c) of § 141.14 is exceeded, the supplier of water shall report to the State and notify the public as prescribed in § 141.31 and § 141.32.

(g) Special purpose samples, such as those taken to determine whether disinfection practices following pipe placement, replacement, or repair have been sufficient, shall not be used to determine compliance with § 141.14 or § 141.21(b) or (c).

(h) A supplier of water of a community water system or a non-community water system may, with the approval of the State and based upon a sanitary survey, substitute the use of chlorine residual monitoring for not more than 75 percent of the samples required to be taken by paragraph (b) of this section, *Provided*, that the supplier of water takes chlorine residual samples at points which are representative of the conditions within the distribution system at the frequency of at least four for each substituted microbiological sample. There shall be at least daily determinations of chlorine residual. When the supplier of water exercises the option provided in this paragraph (h) of this section, he shall maintain no less than 0.2 mg/l free chlorine throughout the public water distribution system. When a particular sampling point has been shown to have a free chlorine residual less than 0.2 mg/l, the water at that location shall be retested as soon as practicable and in any event within one hour. If the original analysis is confirmed, this fact shall be reported to the State within 48 hours. Also, if the analysis is confirmed, a sample for coliform bacterial analysis must be collected from that sampling point as soon as practicable and preferably within one hour, and the results of such analysis reported to the State within 48 hours after the results are known to the supplier of water. Analyses for residual chlorine shall be made in accordance with "Standard Methods for the Examination of Water and Wastewater," 13th Ed., pp. 129–132. Compliance with the maximum contaminant levels for coliform bacteria shall be determined on the monthly mean or quarterly mean basis specified in § 141.14, including those samples taken as a result of failure to maintain the required chlorine residual level. The State may withdraw its approval of the use of chlorine residual substitution at any time.

§ 141.22 Turbidity Sampling and Analytical Requirements

(a) Samples shall be taken by suppliers of water for both community water systems and non-community water systems at a representative entry point(s) to the water distribution system at least once per day, for the purpose of making turbidity measurements to determine compliance with § 141.13. The measurement shall be made by the Nephelometric Method in accordance with the recommendations set forth in "Standard Methods for the Examination of Water and Wastewater," American Public Health Association, 13th Edition, pp. 350–353, or "Methods for Chemical Analysis of Water and Wastes," pp. 295–298, Environmental Protection Agency, Office of Technology Transfer, Washington, D.C. 20460, 1974.

(b) If the result of a turbidity analysis indicates that the maximum allowable limit has been exceeded, the sampling and measurement shall be confirmed by resampling as soon as practicable and preferably within one hour. If the repeat sample confirms that the maximum allowable limit has been exceeded, the supplier of water shall report to the State within 48 hours. The repeat sample shall be the sample used for the purpose of calculating the monthly average. If the monthly average of the daily samples exceeds the maximum allowable limit, or if the average of two samples taken on consecutive days exceeds 5 TU, the supplier of water shall report to the State and notify the public as directed in § 141.31 and § 141.32.

(c) Sampling for non-community water systems shall begin within two years after the effective date of this part.

(d) The requirements of this § 141.22 shall apply only to public water systems which use water obtained in whole or in part from surface sources.

§ 141.23 Inorganic Chemical Sampling and Analytical Requirements

(a) Analyses for the purpose of determining compliance with § 141.11 are required as follows:

(1) Analyses for all community water systems utilizing surface water sources shall be completed within one year following the effective date of this part. These analyses shall be repeated at yearly intervals.

(2) Analyses for all community water systems utilizing only ground water sources shall be completed within two years following the effective date of this part. These analyses shall be repeated at three-year intervals.

(3) For non-community water systems, whether supplied by surface or ground water sources, analyses for nitrate shall be completed within two years following the effective date of this part. These analyses shall be repeated at intervals determined by the State.

(b) If the result of an analysis made pursuant to paragraph (a) indicates that the level of any contaminant listed in § 141.11 exceeds the maximum contaminant level, the supplier of water shall report to the State within 7 days and initiate three additional analyses at the same sampling point within one month.

(c) When the average of four analyses made pursuant to paragraph (b) of this section, rounded to the same number of significant figures as the maximum contaminant level for the substance in question, exceeds the maximum contaminant level, the supplier of water shall notify the State pursuant to § 141.31 and give notice to the public pursuant to § 141.32. Monitoring after public notification shall be at a frequency designated by the State and shall continue until the maximum contaminant level has not been exceeded in two successive samples or until a monitoring schedule as a condition to a variance, exemption or enforcement action shall become effective.

(d) The provisions of paragraphs (b) and (c) of this section notwithstanding, compliance with the maximum contaminant level for nitrate shall be determined on the basis of the mean of two analyses. When a level exceeding the maximum contaminant level for nitrate is found, a second analysis shall be initiated within 24 hours, and if the mean of the two analyses exceeds the maximum contaminant level, the supplier of water shall report his findings to the State pursuant to § 141.31 and shall notify the public pursuant to § 141.32.

(e) For the initial analyses required by paragraph (a)(1), (2) or (3) of this section, data for surface waters acquired within one year prior to the effective date and data for ground waters acquired within 3 years prior to the effective date of this part may be substituted at the discretion of the State.

(f) Analyses conducted to determine compliance with § 141.11 shall be made in accordance with the following methods:

(1) Arsenic—Atomic Absorption Method, "Methods for Chemical Analysis of Water and Wastes," pp. 95–96, Environmental Protection Agency, Office of Technology Transfer, Washington, D.C. 20460, 1974.

(2) Barium–Atomic Absorption Method, "Standard Methods for the Examination of Water and Wastewater," 13th Edition, pp. 210–215, or "Methods for Chemical Analysis of Water and Wastes," pp. 97–98, Environmental Protection Agency, Office of Technology Transfer, Washington, D.C. 20460, 1974.

(3) Cadmium—Atomic Absorption Method, "Standard Methods for the Examination of Water and Wastewater," 13th Edition, pp. 210–215, or "Methods for Chemical Analysis of Water and Wastes," pp. 101–103, Environmental Protection Agency, Office of Technology Transfer, Washington, D.C. 20460, 1974.

(4) Chromium—Atomic Absorption Method, "Standard Methods for the Examination of Water and Wastewater," 13th Edition, pp. 210–215, or "Methods for Chemical Analysis of Water and Wastes," pp. 105–106, Environmental Protection Agency, Office of Technology Transfer, Washington, D.C. 20460, 1974.

(5) Lead—Atomic Absorption Method, "Standard Methods for the Examination of Water and Wastewater," 13th Edition, pp. 210–215, or "Methods for Chemical Analysis of Water and Wastes," pp. 112–113, Environmental Protection Agency, Office of Technology Transfer, Washington, D.C. 20460, 1974.

(6) Mercury—Flameless Atomic Absorption Method, "Methods for Chemical Analysis of Water and Wastes," pp. 118–126, Environmental Protection Agency, Office of Technology Transfer, Washington, D.C. 20460, 1974.

(7) Nitrate—Brucine Colorimetric Method, "Standard Methods for the Examination of Water and Wastewater," 13th Edition, pp. 461–464, or Cadmium Reduction Method, "Methods for Chemical Analysis of Water and Wastes," pp. 201–206, Environmental Protection Agency, Office of Technology Transfer, Washington, D.C. 20460, 1974.

(8) Selenium—Atomic Absorption Method, "Methods for Chemical Analysis of Water and Wastes," p. 145, Environmental Protection Agency, Office of Technology Transfer, Washington, D.C. 20460, 1974.

(9) Silver—Atomic Absorption Method, "Standard Methods for the Examination of Water and Wastewater," 13th Edition, pp. 210–215, or "Methods for Chemical Analysis of Water and Wastes," p. 146, Environmental Protection Agency, Office of Technology Transfer, Washington, D.C. 20460, 1974.

(10) Fluoride—Electrode Method, "Standard Methods for the Examination of Water and Wastewater," 13th Edition, pp. 172–174, or "Methods for Chemical Analysis of Water and Wastes," pp. 65–67, Environmental Protection Agency, Office of Technology Transfer, Washington, D.C. 20460, 1974, or Colorimetric Method with Preliminary Distillation, "Standard Methods for the Examination of Water and Wastewater," 13th Edition, pp. 171–172 and 174–176, or "Methods for Chemical Analysis of Water and Wastes," pp. 59–60, Environmental Protection Agency, Office of Technology Transfer, Washington, D.C. 20460, 1974.

§ 141.24 Organic Chemical Sampling and Analytical Requirements

(a) An analysis of substances for the purpose of determining compliance with § 141.12 shall be made as follows:

(1) For all community water systems utilizing surface water sources, analyses shall be completed within one year following the effective date of this part. Samples analyzed shall be collected during the period of the year designated by the State as the period when contamination by pesticides is most likely to occur. These analyses shall be repeated at intervals specified by the State but in no event less frequently than at three year intervals.

(2) For community water systems utilizing only ground water sources, analyses shall be completed by those systems specified by the State.

(b) If the result of an analysis made pursuant to paragraph (a) of this section indicates that the level of any contaminant listed in § 141.12 exceeds the maximum contaminant level, the supplier of water shall report to the State within 7 days and initiate three additional analyses within one month.

(c) When the average of four analyses made pursuant to paragraph (b) of this section, rounded to the same number of significant figures as the maximum contaminant level for the substance in question, exceeds the maximum contaminant level, the supplier of water shall report to the State pursuant to § 141.31 and give notice to the public pursuant to § 141.32. Monitoring after public notification shall be at a frequency designated by the State and shall continue until the maximum contaminant level has not been exceeded in two successive samples or until a monitoring schedule as a condition to a variance, exemption or enforcement action shall become effective.

(d) For the initial analysis required by paragraph (a)(1) and (2) of this section, data for surface water acquired within one year prior to the effective date of this part and data for ground water acquired within three years prior to the effective date of this part may be substituted at the discretion of the State.

(e) Analyses made to determine compliance with § 141.12(a) shall be made in accordance with "Method for Organochlorine Pesticides in Industrial Effluents," MDQARL, Environmental Protection Agency, Cincinnati, Ohio, November 28, 1973.

(f) Analyses made to determine compliance with § 141.12(b) shall be conducted in accordance with "Methods for Chlorinated Phenoxy Acid Herbicides in Industrial Effluents," MDQARL, Environmental Protection Agency, Cincinnati, Ohio, November 28, 1973.

§ 141.27 Alternative Analytical Techniques

With the written permission of the State, concurred in by the Administrator of the U.S. Environmental Protection Agency, an alternative analytical technique may be employed. An alternative technique shall be acceptable only if it is substantially equivalent to the prescribed test in both precision and accuracy as it relates to the determination of compliance with any maximum contaminant level. The use of the alternative analytical technique shall not decrease the frequency of monitoring required by this part.

§ 141.28 Approved Laboratories

For the purpose of determining compliance with § 141.21 through § 141.27, samples may be considered only if they have been analyzed by a laboratory approved by the State except that measurements for turbidity and free chlorine residual may be performed by any person acceptable to the State.

§ 141.29 Monitoring of Consecutive Public Water Systems

When a public water system supplies water to one or more other public water systems, the State may modify the monitoring requirements imposed by this part to the extent that the interconnection of the systems justifies treating them as a single system for monitoring purposes. Any modified monitoring shall be conducted pursuant to a schedule specified by the State and concurred in by the Administrator of the U.S. Environmental Protection Agency.

SUBPART D—REPORTING, PUBLIC NOTIFICATION AND RECORD KEEPING

§ 141.31 Reporting Requirements

(a) Except where a shorter reporting period is specified in this part, the supplier of water shall report to the State within 40 days following a test, measurement or analysis required to be made by this part, the results of that test, measurement or analysis.

(b) The supplier of water shall report to the State within 48 hours the failure to comply with any primary drinking water regulation (including failure to comply with monitoring requirements) sct forth in this part.

(c) The supplier of water is not required to report analytical results to the State in cases where a State laboratory performs the analysis and reports the results to the State office which would normally receive such notification from the supplier.

§ 141.32 Public Notification

(a) If a community water system fails to comply with an applicable maximum contaminant level established in Subpart B, fails to comply with an applicable testing procedure established in Subpart C of this part, is granted a variance or an exemption from an applicable maximum contaminant level, fails to comply with the requirements of any schedule prescribed pursuant to a variance or exemption, or fails to perform any monitoring required pursuant to Section 1445(a) of the Act, the supplier of water shall notify persons served by the system of the failure or grant by inclusion of a notice in the first set of water bills of the system issued after the failure or grant and in any event by written notice within three months. Such notice shall be repeated at least once every three months so long as the system's failure continues or the variance or exemption remains in effect. If the system issues water bills less frequently than quarterly, or does not issue water bills, the notice shall be made by or supplemented by another form of direct mail.

(b) If a community water system has failed to comply with an applicable maximum contaminant level, the supplier of water shall notify the public of such failure, in addition to the notification required by paragraph (a) of this section, as follows:

(1) By publication on not less than three consecutive days in a newspaper or newspapers of general circulation in the area served by the system. Such notice shall be completed within fourteen days after the supplier of water learns of the failure.

(2) By furnishing a copy of the notice to the radio and television stations serving the area served by the system. Such notice shall be furnished within seven days after the supplier of water learns of the failure.

(c) If the area served by a community water system is not served by a daily newspaper of general circulation, notification by newspaper required by paragraph (b) of this section shall instead be given by publication on three consecutive weeks in a weekly newspaper of general circulation serving the area. If no weekly or daily newspaper of general circulation serves the area, notice shall be given by posting the notice in post offices within the area served by the system.

(d) If a non-community water system fails to comply with an applicable maximum contaminant level established in Subpart B of this part, fails to comply with an applicable testing procedure established in Subpart C of this part, is granted a variance or an exemption from an applicable maximum contaminant level, fails to comply with the requirement of any schedule prescribed pursuant to a variance or exemption or fails to perform any monitoring required pursuant to Section 1445(a) of the Act, the supplier of water shall given notice of such failure or grant to the persons served by the system. The form and manner of such notice shall be prescribed by the State, and shall insure that the public using the system is adequately informed of the failure or grant.

(e) Notices given pursuant to this section shall be written in a manner reasonably designed to inform fully the users of the system. The notice shall be conspicuous and shall not use unduly technical language, unduly small print or other methods which would frustrate the purpose of the notice. The notice shall disclose all material facts regarding the subject including the nature of the problem and, when appropriate, a clear statement that a primary drinking water regulation has been violated and any preventive measures that should be taken by the public. Where appropriate, or where designated by the State, bilingual notice shall be given. Notices may include a balanced explanation of the significance or seriousness to the public health of the subject of the notice, a fair explanation of steps taken by the system to correct any problem and the results of any additional sampling.

(f) Notice to the public required by this section may be given by the State on behalf of the supplier of water.

(g) In any instance in which notification by mail is required by paragraph (a) of this section but notification by newspaper or to radio or television stations is not required by paragraph (b) of this section, the State may order the supplier of water to provide notification by newspaper and to radio and television stations when circumstances make more immediate or broader notice appropriate to protect the public health.

§ 141.33 Record Maintenance

Any owner or operator of a public water system subject to the provisions of this part shall retain on its premises or at a convenient location near its premises the following records:

(a) Records of bacteriological analyses made pursuant to this part shall be kept for not less than 5 years. Records of chemical analyses made pursuant to this part shall be kept for not less than 10 years. Actual laboratory reports may be kept, or data may be transferred to tabular summaries, provided that the following information is included:

(1) The date, place, and time of sampling, and the name of the person who collected the sample;

(2) Identification of the sample as to whether it was a routine distribution system sample, check sample, raw or process water sample or other special purpose sample;

(3) Date of analysis;

(4) Laboratory and person responsible for performing analysis;

(5) The analytical technique/method used; and

(6) The results of the analysis.

(b) Records of action taken by the system to correct violations of primary drinking water regulations shall be kept for a period not less than 3 years after the last action taken with respect to the particular violation involved.

(c) Copies of any written reports, summaries or communications relating to sanitary surveys of the system conducted by the system itself, by a private consultant, or by any local, State or Federal agency, shall be kept for a period not less than 10 years after completion of the sanitary survey involved.

(d) Records concerning a variance or exemption granted to the system shall be kept for a period ending not less than 5 years following the expiration of such variance or exemption.

EPA has issued proposed regulations to serve as guidelines to the states with regard to the so-called secondary drinking water standards. These are: chloride—250 mg/l, color—15 color units, copper—1 mg/l, corrosivity—non-corrosive, Foaming agents—0.5 mg/l, Hydrogen Sulfide—0.05 mg/l, Iron—0.3 mg/l, Manganese—0.05 mg/l, Odor—3 Threshold Odor Number, pH—6.5–8.5, Sulfate—250 mg/l, Total Dissolved Solids 500 mg/l, Zinc—5 mg/l.

APPENDIX

TWO

SATURATION VALUES OF DISSOLVED OXYGEN IN FRESHWATER AND SEAWATER

Saturation values of dissolved oxygen in fresh- and seawater exposed to an atmosphere containing 20.9 percent oxygen under a pressure of 760 mm of mercury[a]

(Calculated by G. C. Whipple and M. C. Whipple from measurements of C. J. J. Fox)[b]

Temperature, °C	Dissolved oxygen (mg/l) for stated concentrations of chloride, mg/l					Difference per 100 mg/l chloride
	0	5000	10,000	15,000	20,000	
0	14.62	13.79	12.97	12.14	11.32	0.0165
1	14.23	13.41	12.61	11.82	11.03	0.0160
2	13.84	13.05	12.28	11.52	10.76	0.0154
3	13.48	12.72	11.98	11.24	10.50	0.0149
4	13.13	12.41	11.69	10.97	10.25	0.0144
5	12.80	12.09	11.39	10.70	10.01	0.0140
6	12.48	11.79	11.12	10.45	9.78	0.0135
7	12.17	11.51	10.85	10.21	9.57	0.0130
8	11.87	11.24	10.61	9.98	9.36	0.0125
9	11.59	10.97	10.36	9.76	9.17	0.0121
10	11.33	10.73	10.13	9.55	8.98	0.0118
11	11.08	10.49	9.92	9.35	8.80	0.0114
12	10.83	10.28	9.72	9.17	8.62	0.0110
13	10.60	10.05	9.52	8.98	8.46	0.0107
14	10.37	9.85	9.32	8.80	8.30	0.0104
15	10.15	9.65	9.14	8.63	8.14	0.0100

Temperature, °C	Dissolved oxygen (mg/l) for stated concentrations of chloride, mg/l					Difference per 100 mg/l chloride
	0	5000	10,000	15,000	20,000	
16	9.95	9.46	8.96	8.47	7.99	0.0098
17	9.74	9.26	8.78	8.30	7.84	0.0095
18	9.54	9.07	8.62	8.15	7.70	0.0092
19	9.35	8.89	8.45	8.00	7.56	0.0089
20	9.17	8.73	8.30	7.86	7.42	0.0088
21	8.99	8.57	8.14	7.71	7.28	0.0086
22	8.83	8.42	7.99	7.57	7.14	0.0084
23	8.68	8.27	7.85	7.43	7.00	0.0083
24	8.53	8.12	7.71	7.30	6.87	0.0083
25	8.38	7.96	7.56	7.15	6.74	0.0082
26	8.22	7.81	7.42	7.02	6.61	0.0080
27	8.07	7.67	7.28	6.88	6.49	0.0079
28	7.92	7.53	7.14	6.75	6.37	0.0078
29	7.77	7.39	7.00	6.62	6.25	0.0076
30	7.63	7.25	6.86	6.49	6.13	0.0075

[a] For other barometric pressures the solubilities vary approximately in proportion to the ratios of these pressures to the standard pressures.

[b] G. C. Whipple and M. C. Whipple, "Solubility of Oxygen in Sea Water," *Journal American Chemical Society*, **33**:362, 1911.

APPENDIX

THREE

COST ESTIMATION OF SEWAGE TREATMENT FACILITIES

WASTEWATER TREATMENT PROCESSES

COST CURVES

Preliminary Treatment	AA
Pumping	AB
Primary Sedimentation	A-1 thru A-5
Trickling Filter	B-1 thru B-3
Activated Sludge	C-1 thru C-8
Filtration	D
Activated Carbon	E
Two Stage Tertiary Lime	F-1 thru F-2
Biological Nitrification	G-1 thru G-4
Biological Denitrification	H
Ion Exchange	I
Breakpoint Chlorination	J
Ammonia Stripping	K
Disinfection	R

SLUDGE HANDLING PROCESSES

COST CURVES

Anaerobic Digestion	L-1 & L-2
Heat Treatment	M-1 & M-2
Air Drying	N-1 & N-2
Dewatering	0-1 thru 0-9
Incineration	P-1 thru P-7
Recalcination	Q-1 thru Q-3

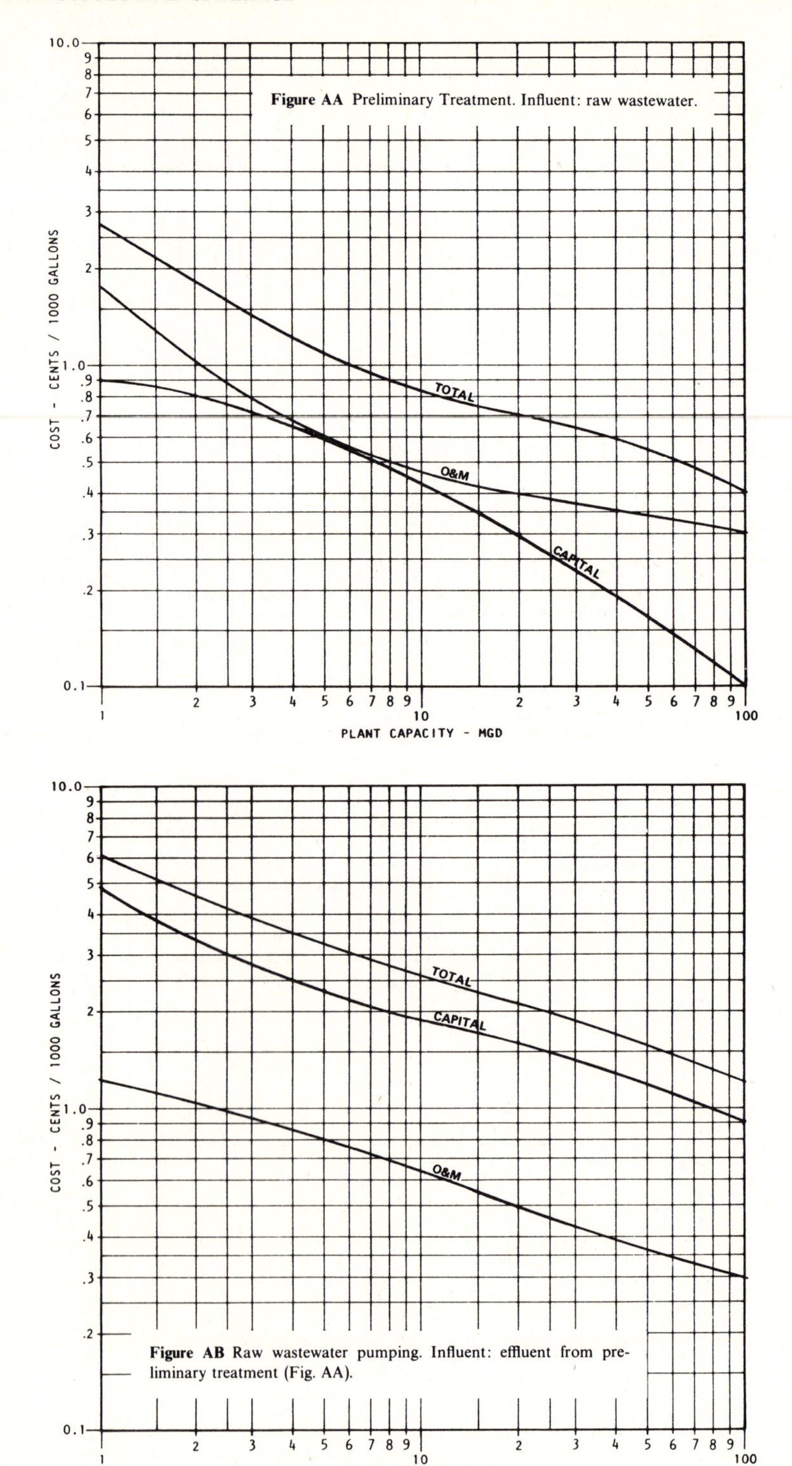

Figure AA Preliminary Treatment. Influent: raw wastewater.

Figure AB Raw wastewater pumping. Influent: effluent from preliminary treatment (Fig. AA).

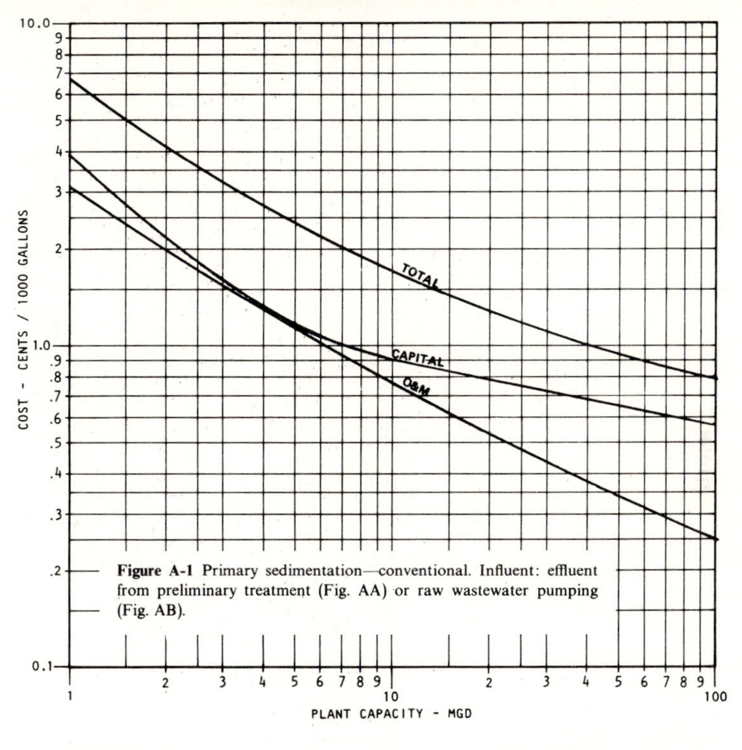

Figure A-1 Primary sedimentation—conventional. Influent: effluent from preliminary treatment (Fig. AA) or raw wastewater pumping (Fig. AB).

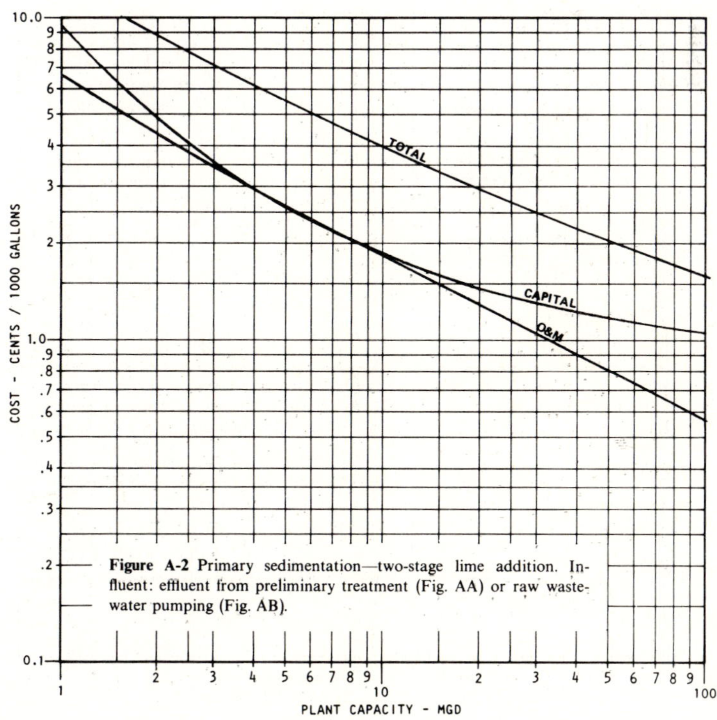

Figure A-2 Primary sedimentation—two-stage lime addition. Influent: effluent from preliminary treatment (Fig. AA) or raw wastewater pumping (Fig. AB).

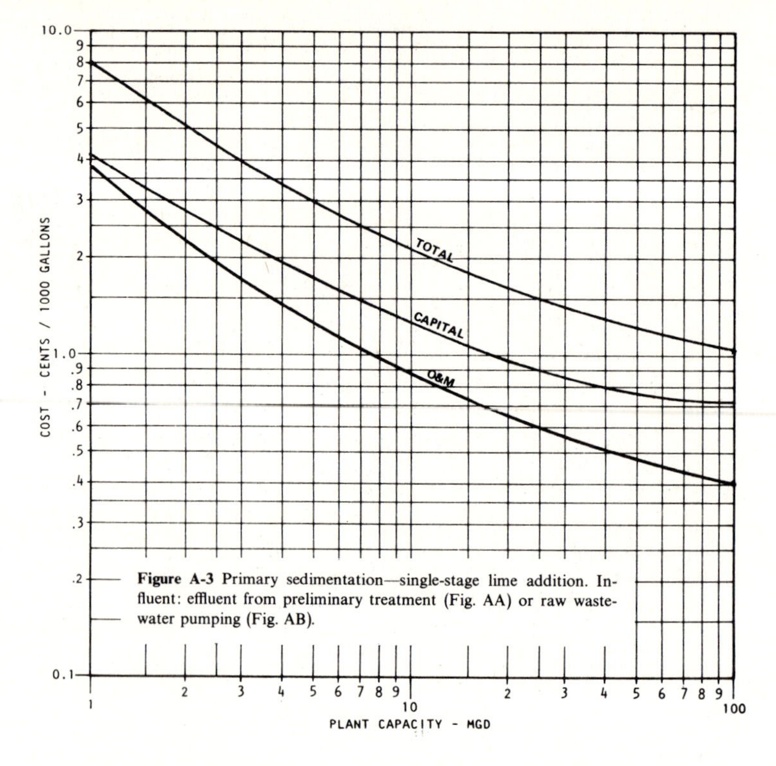

Figure A-3 Primary sedimentation—single-stage lime addition. Influent: effluent from preliminary treatment (Fig. AA) or raw wastewater pumping (Fig. AB).

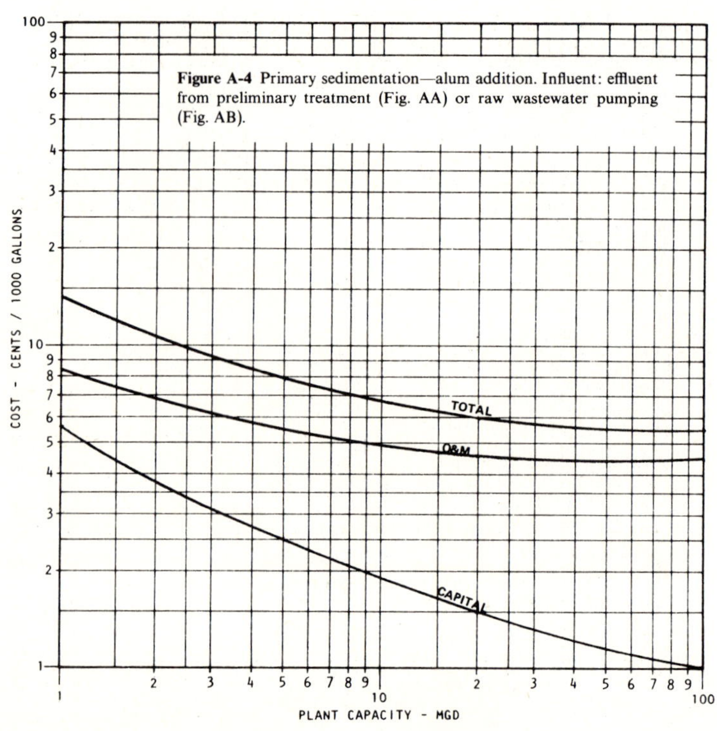

Figure A-4 Primary sedimentation—alum addition. Influent: effluent from preliminary treatment (Fig. AA) or raw wastewater pumping (Fig. AB).

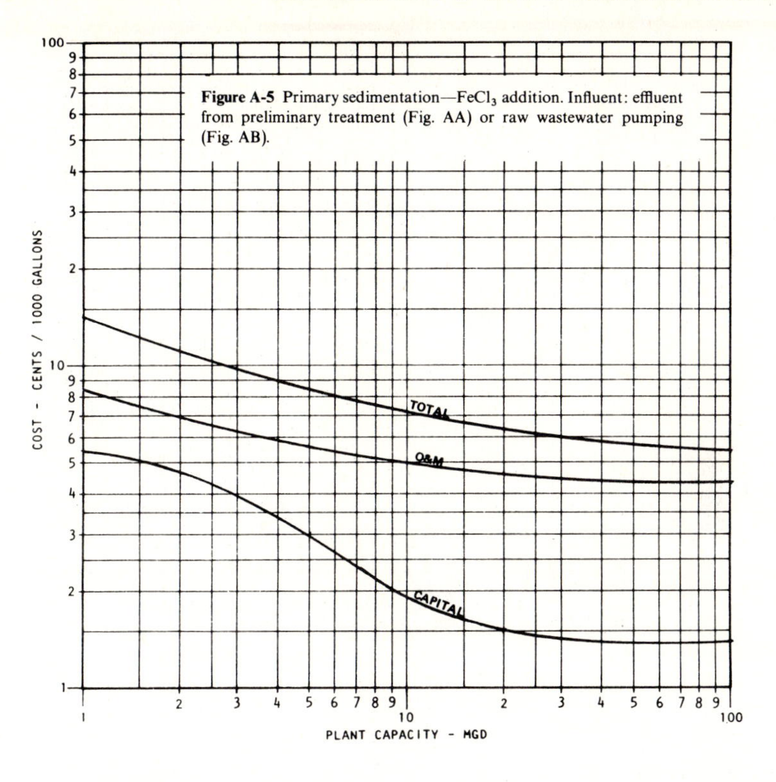

Figure A-5 Primary sedimentation—$FeCl_3$ addition. Influent: effluent from preliminary treatment (Fig. AA) or raw wastewater pumping (Fig. AB).

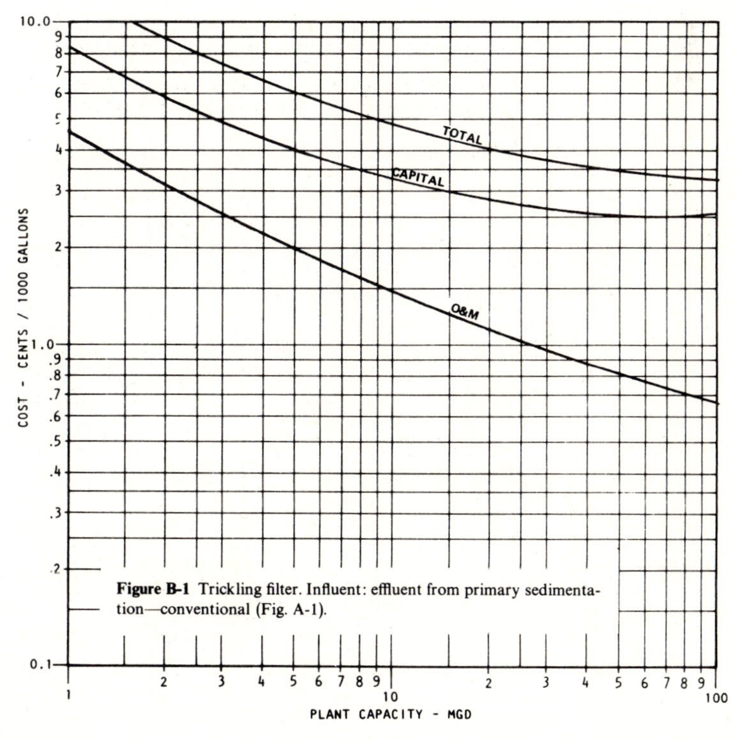

Figure B-1 Trickling filter. Influent: effluent from primary sedimentation—conventional (Fig. A-1).

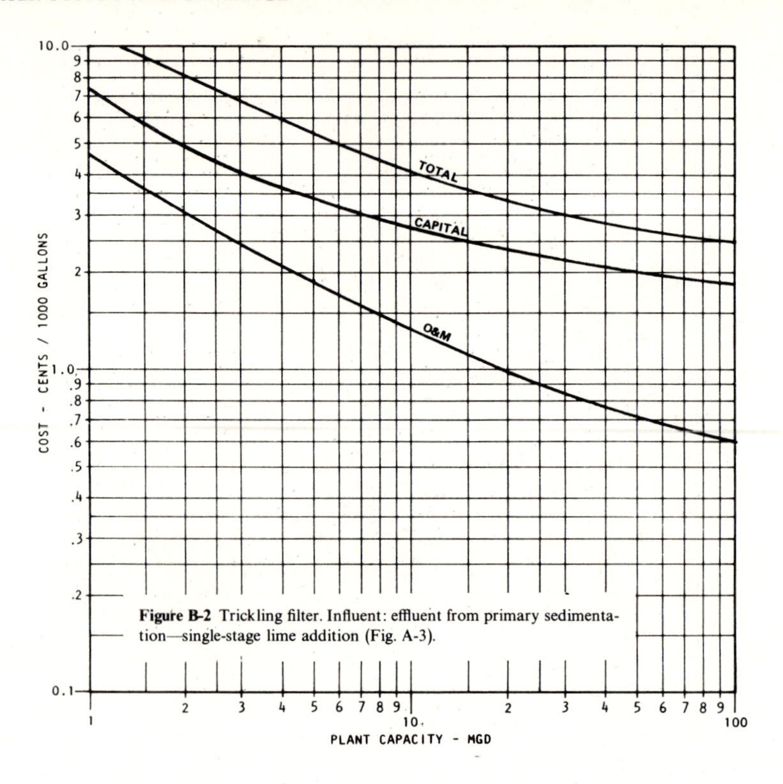

Figure B-2 Trickling filter. Influent: effluent from primary sedimentation—single-stage lime addition (Fig. A-3).

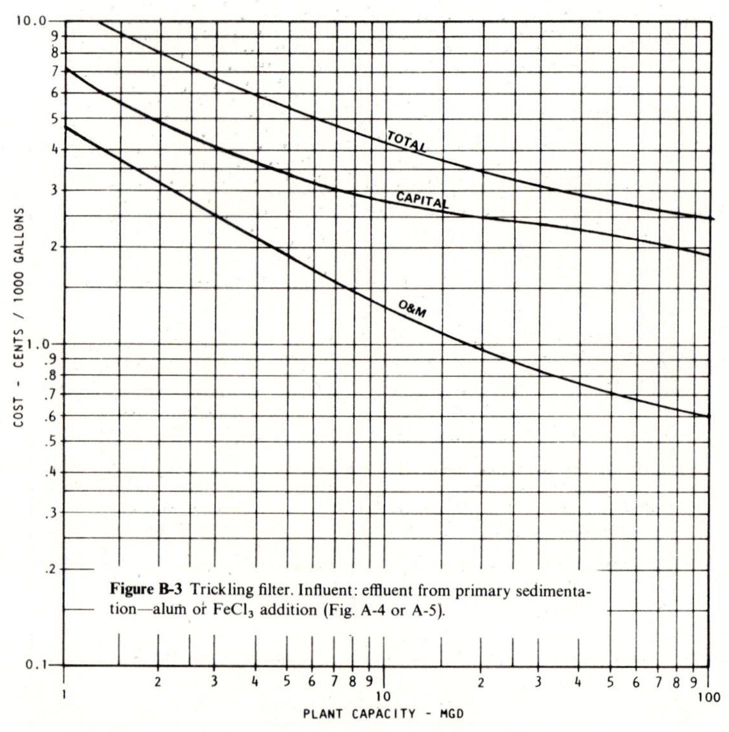

Figure B-3 Trickling filter. Influent: effluent from primary sedimentation—alum or $FeCl_3$ addition (Fig. A-4 or A-5).

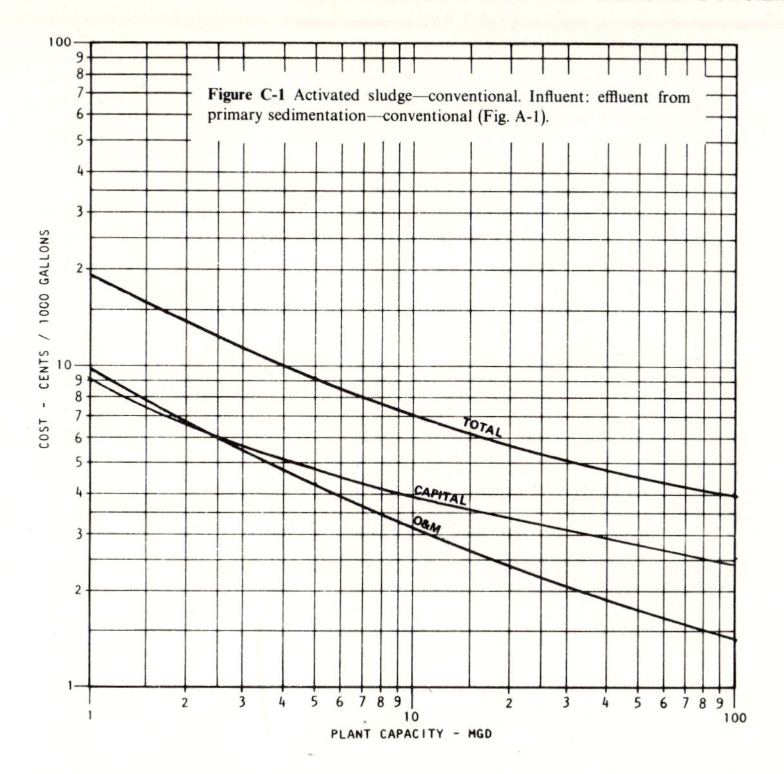

Figure C-1 Activated sludge—conventional. Influent: effluent from primary sedimentation—conventional (Fig. A-1).

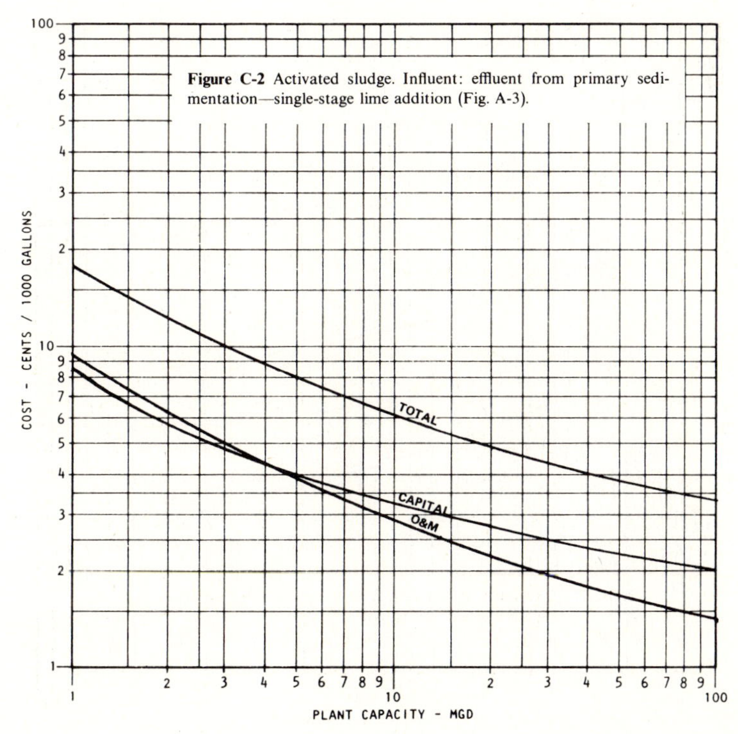

Figure C-2 Activated sludge. Influent: effluent from primary sedimentation—single-stage lime addition (Fig. A-3).

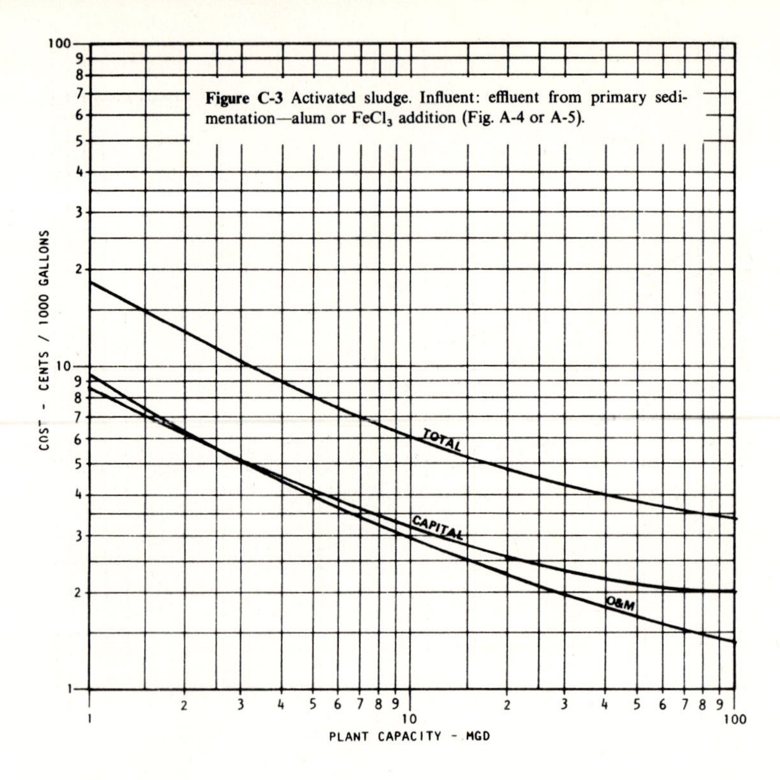

Figure C-3 Activated sludge. Influent: effluent from primary sedimentation—alum or $FeCl_3$ addition (Fig. A-4 or A-5).

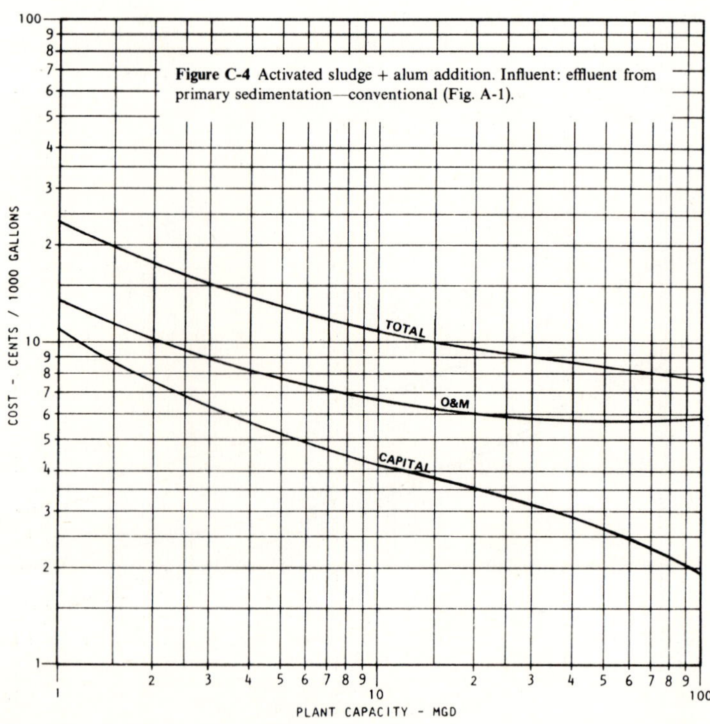

Figure C-4 Activated sludge + alum addition. Influent: effluent from primary sedimentation—conventional (Fig. A-1).

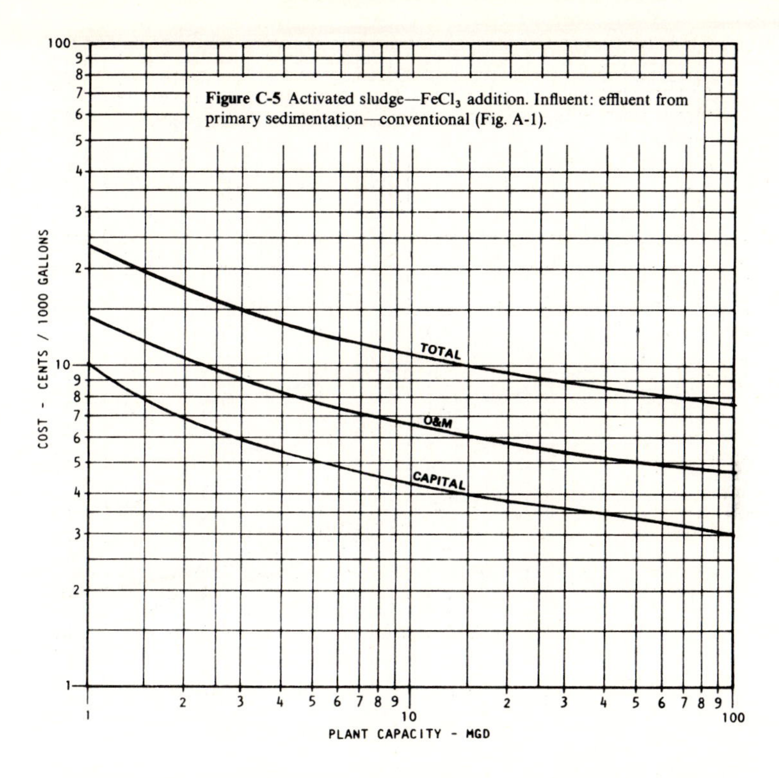

Figure C-5 Activated sludge—$FeCl_3$ addition. Influent: effluent from primary sedimentation—conventional (Fig. A-1).

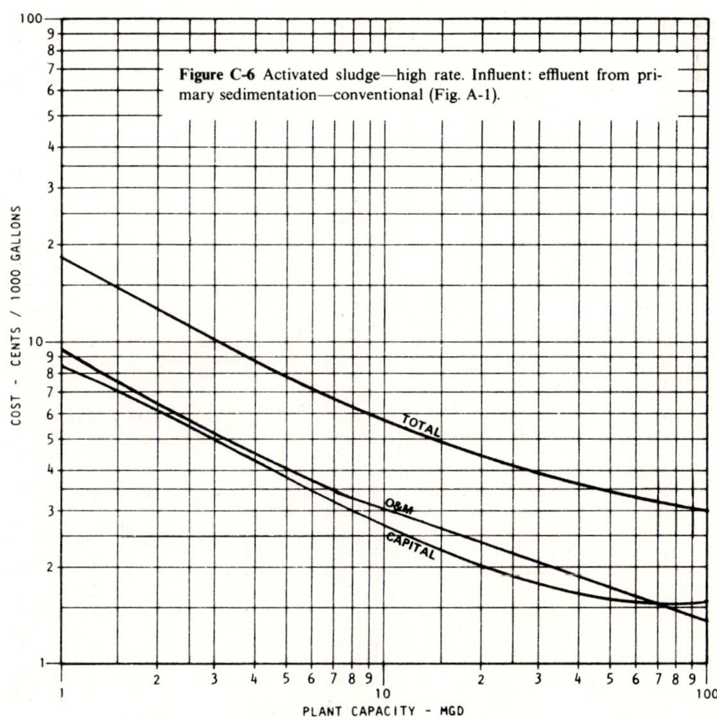

Figure C-6 Activated sludge—high rate. Influent: effluent from primary sedimentation—conventional (Fig. A-1).

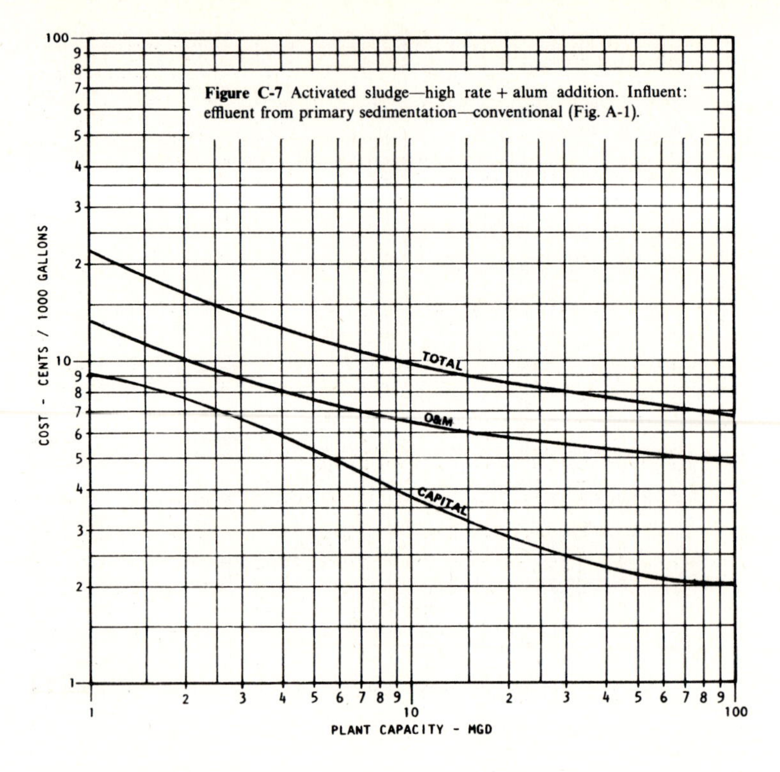

Figure C-7 Activated sludge—high rate + alum addition. Influent: effluent from primary sedimentation—conventional (Fig. A-1).

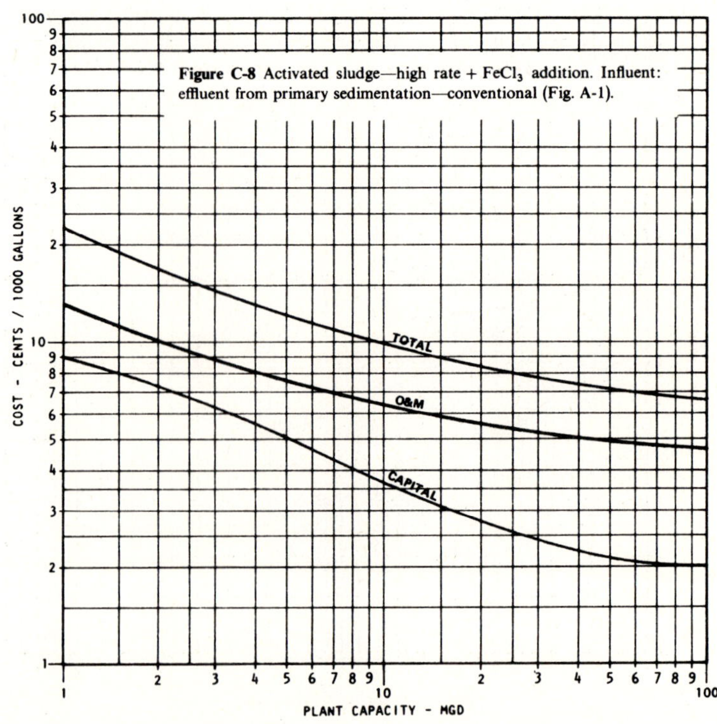

Figure C-8 Activated sludge—high rate + $FeCl_3$ addition. Influent: effluent from primary sedimentation—conventional (Fig. A-1).

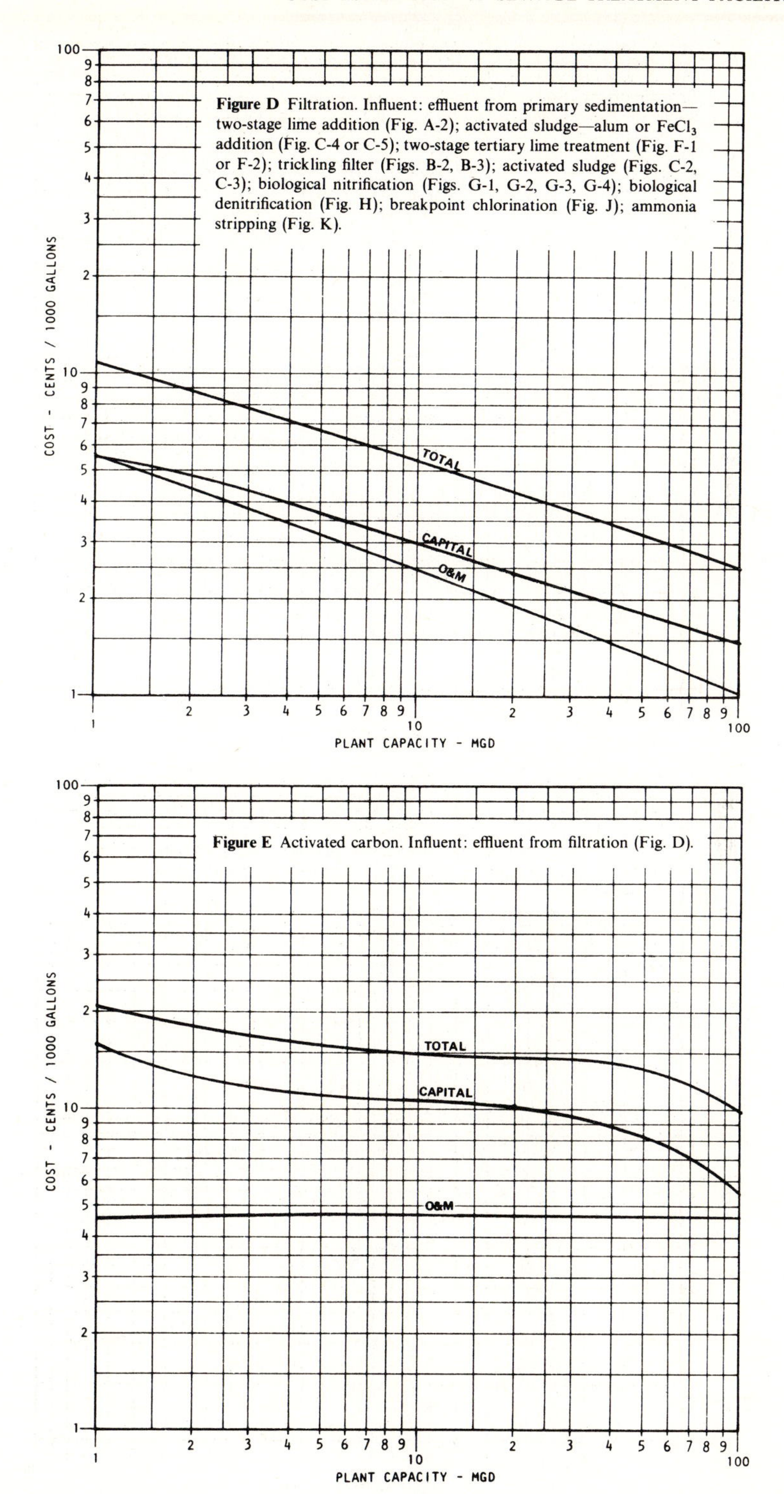

Figure D Filtration. Influent: effluent from primary sedimentation—two-stage lime addition (Fig. A-2); activated sludge—alum or $FeCl_3$ addition (Fig. C-4 or C-5); two-stage tertiary lime treatment (Fig. F-1 or F-2); trickling filter (Figs. B-2, B-3); activated sludge (Figs. C-2, C-3); biological nitrification (Figs. G-1, G-2, G-3, G-4); biological denitrification (Fig. H); breakpoint chlorination (Fig. J); ammonia stripping (Fig. K).

Figure E Activated carbon. Influent: effluent from filtration (Fig. D).

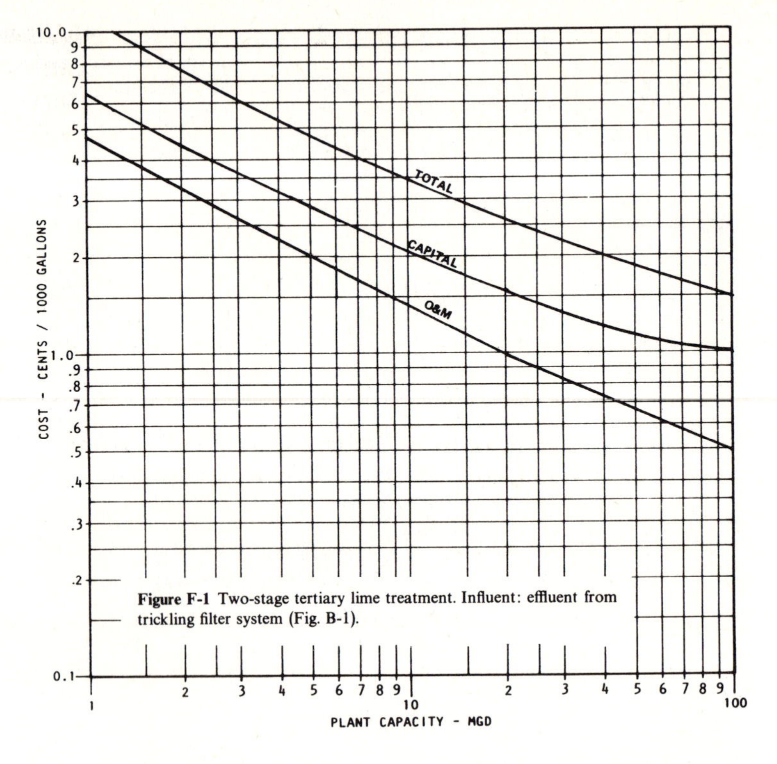

Figure F-1 Two-stage tertiary lime treatment. Influent: effluent from trickling filter system (Fig. B-1).

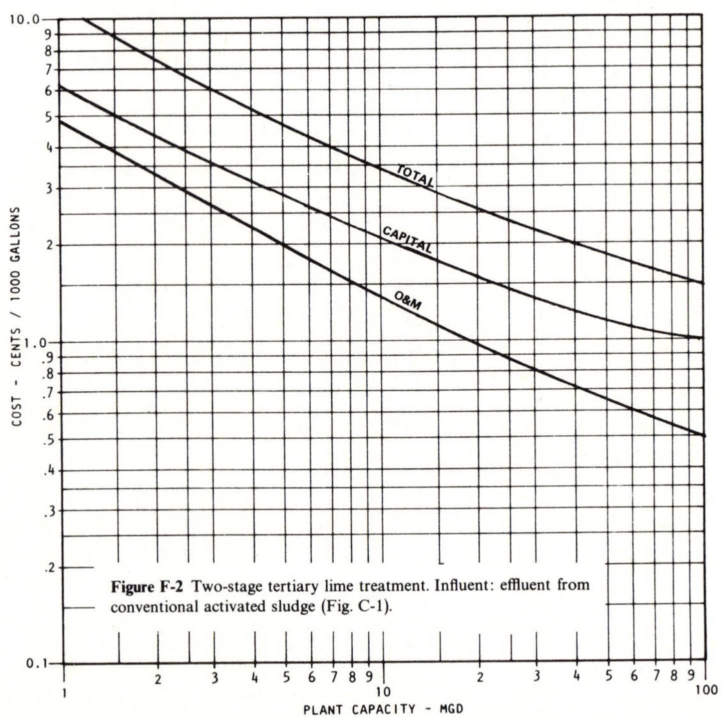

Figure F-2 Two-stage tertiary lime treatment. Influent: effluent from conventional activated sludge (Fig. C-1).

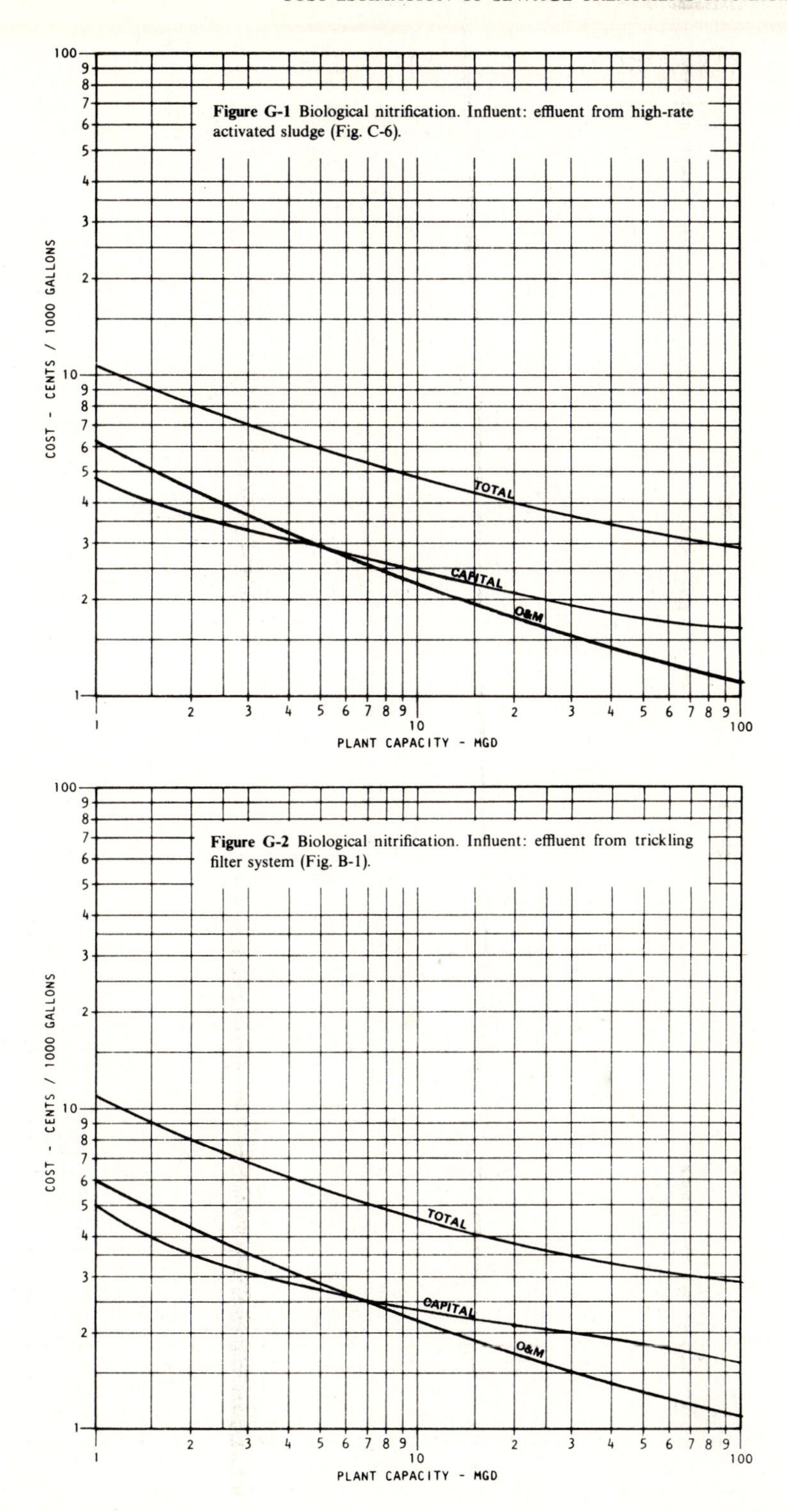

Figure G-1 Biological nitrification. Influent: effluent from high-rate activated sludge (Fig. C-6).

Figure G-2 Biological nitrification. Influent: effluent from trickling filter system (Fig. B-1).

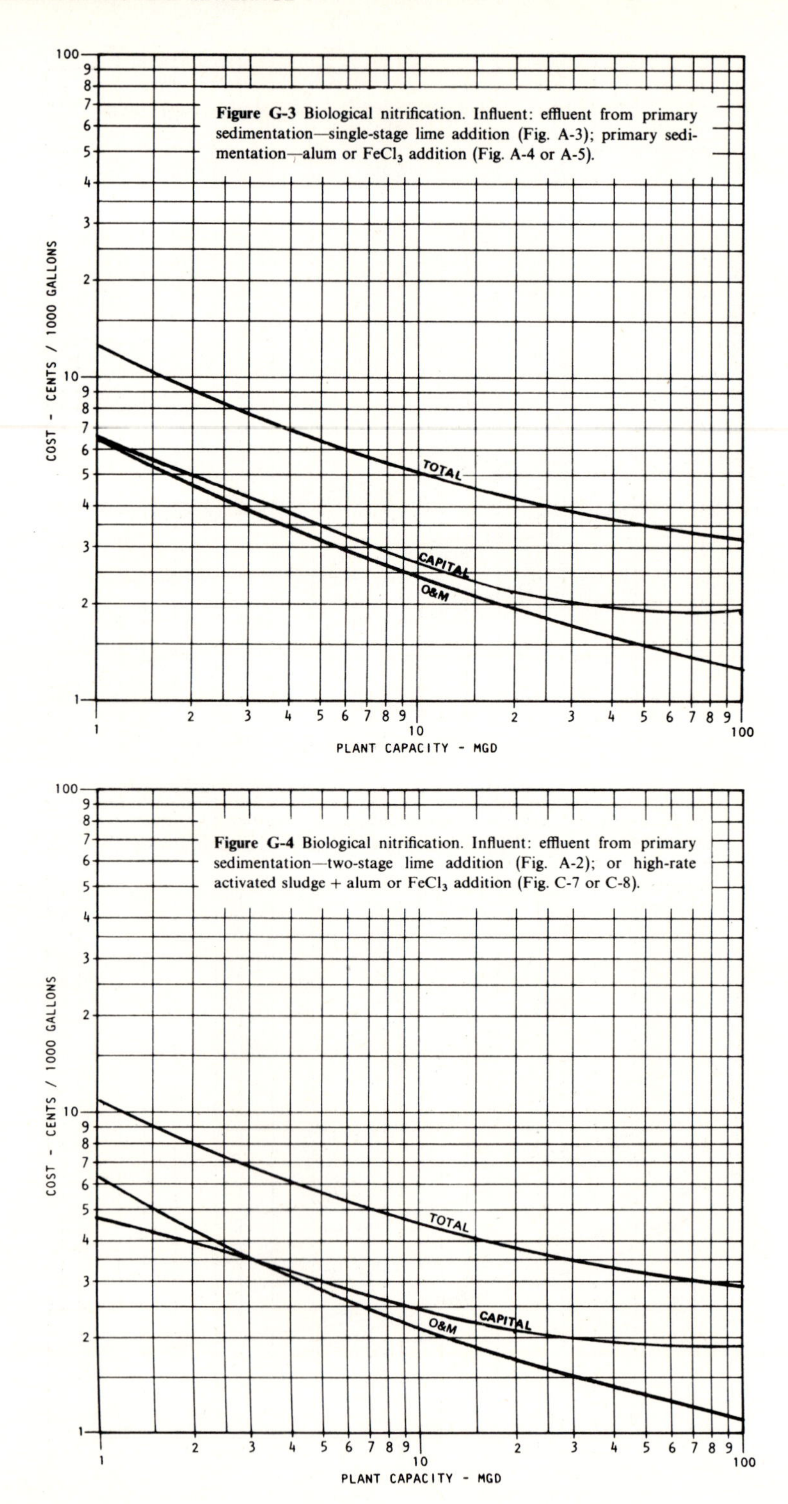

Figure G-3 Biological nitrification. Influent: effluent from primary sedimentation—single-stage lime addition (Fig. A-3); primary sedimentation—alum or $FeCl_3$ addition (Fig. A-4 or A-5).

Figure G-4 Biological nitrification. Influent: effluent from primary sedimentation—two-stage lime addition (Fig. A-2); or high-rate activated sludge + alum or $FeCl_3$ addition (Fig. C-7 or C-8).

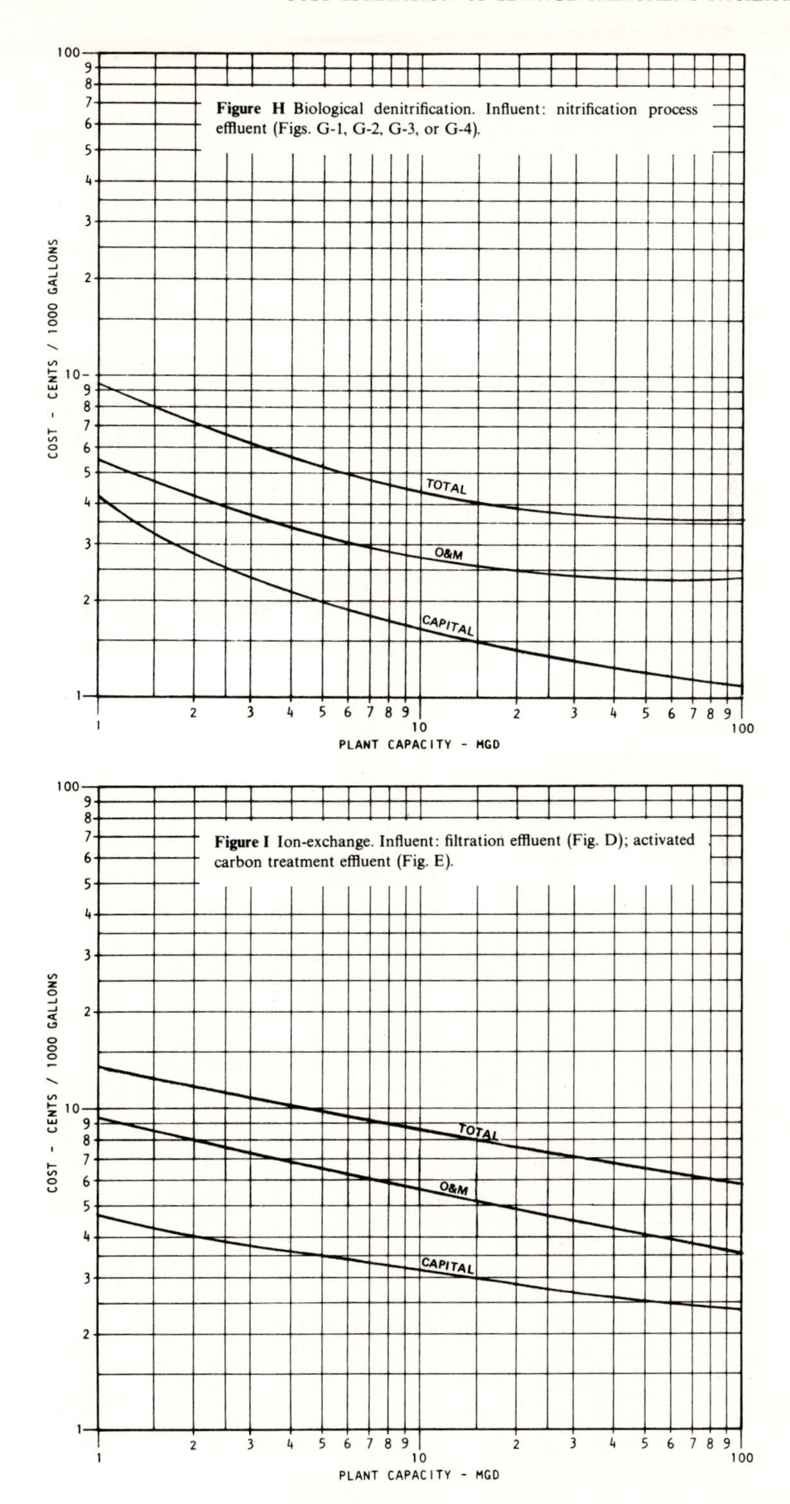

Figure H Biological denitrification. Influent: nitrification process effluent (Figs. G-1, G-2, G-3, or G-4).

Figure I Ion-exchange. Influent: filtration effluent (Fig. D); activated carbon treatment effluent (Fig. E).

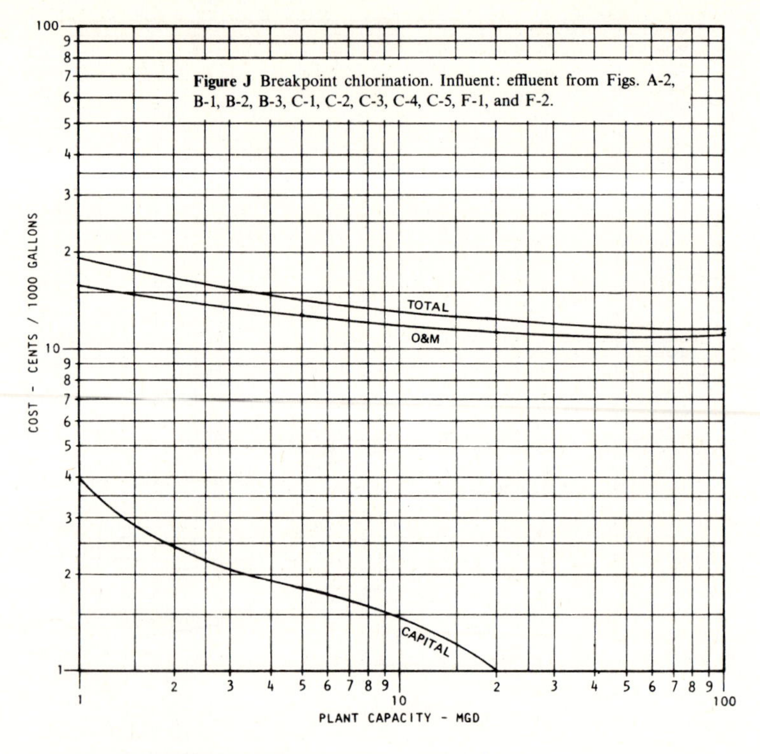

Figure J Breakpoint chlorination. Influent: effluent from Figs. A-2, B-1, B-2, B-3, C-1, C-2, C-3, C-4, C-5, F-1, and F-2.

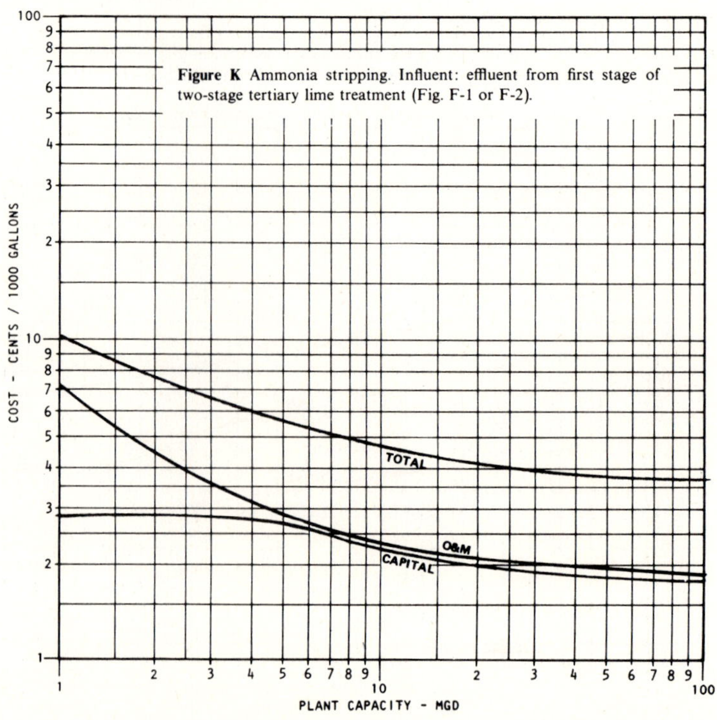

Figure K Ammonia stripping. Influent: effluent from first stage of two-stage tertiary lime treatment (Fig. F-1 or F-2).

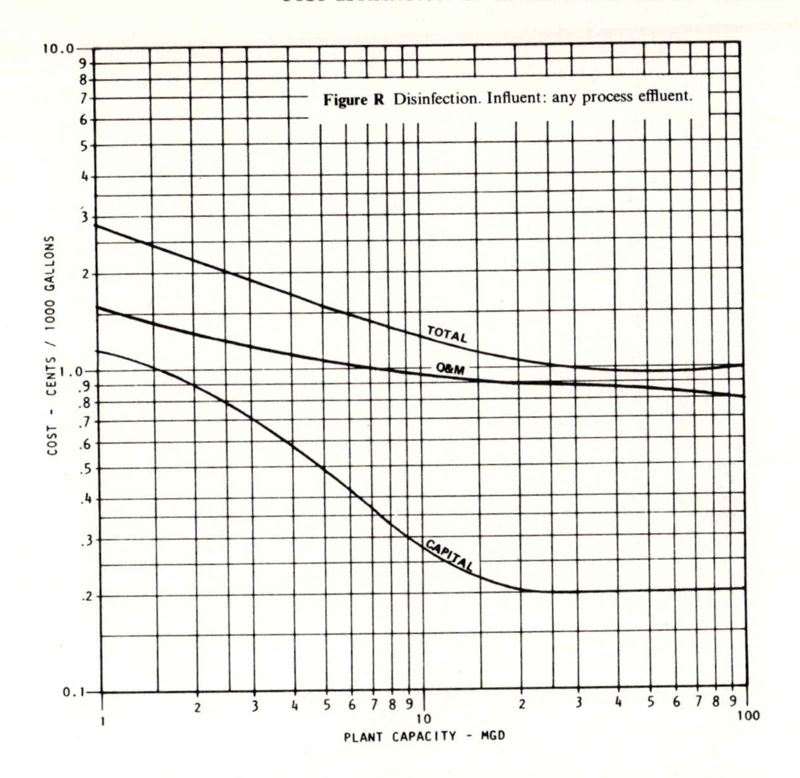

Figure R Disinfection. Influent: any process effluent.

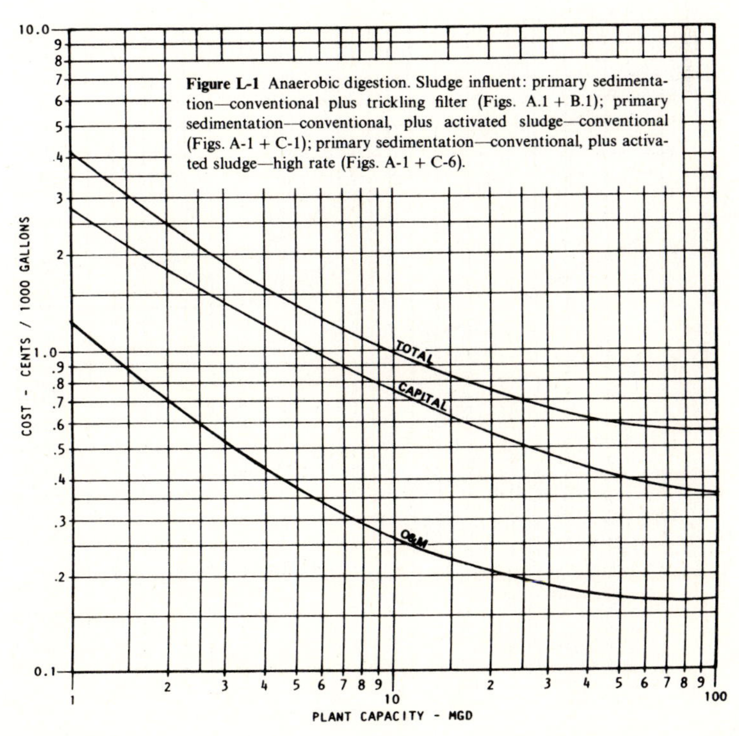

Figure L-1 Anaerobic digestion. Sludge influent: primary sedimentation—conventional plus trickling filter (Figs. A.1 + B.1); primary sedimentation—conventional, plus activated sludge—conventional (Figs. A-1 + C-1); primary sedimentation—conventional, plus activated sludge—high rate (Figs. A-1 + C-6).

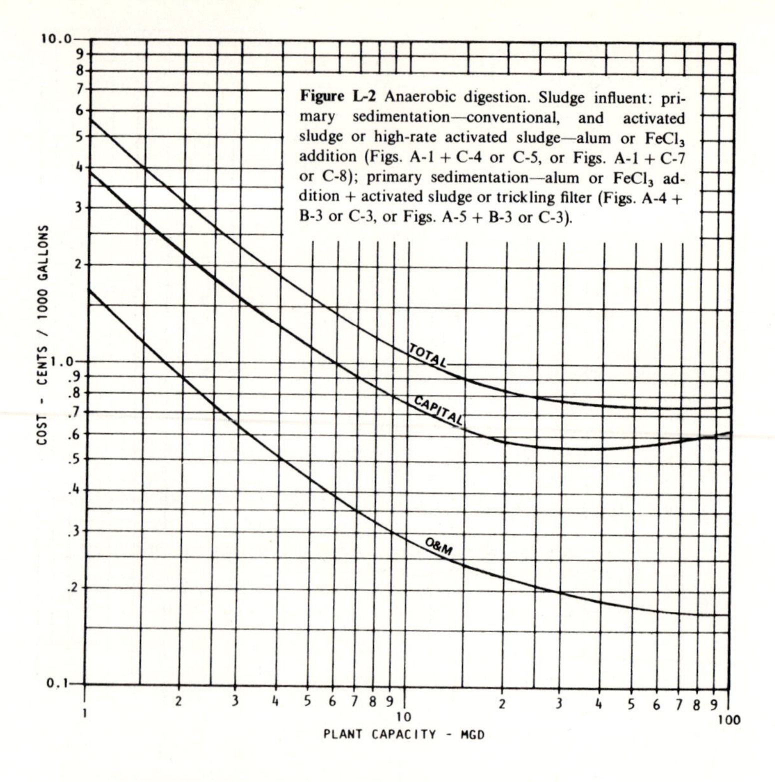

Figure L-2 Anaerobic digestion. Sludge influent: primary sedimentation—conventional, and activated sludge or high-rate activated sludge—alum or $FeCl_3$ addition (Figs. A-1 + C-4 or C-5, or Figs. A-1 + C-7 or C-8); primary sedimentation—alum or $FeCl_3$ addition + activated sludge or trickling filter (Figs. A-4 + B-3 or C-3, or Figs. A-5 + B-3 or C-3).

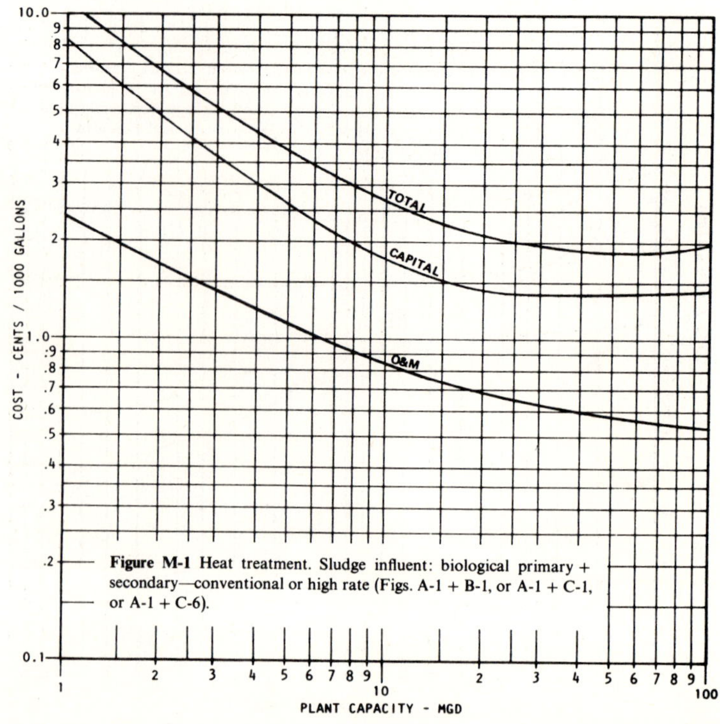

Figure M-1 Heat treatment. Sludge influent: biological primary + secondary—conventional or high rate (Figs. A-1 + B-1, or A-1 + C-1, or A-1 + C-6).

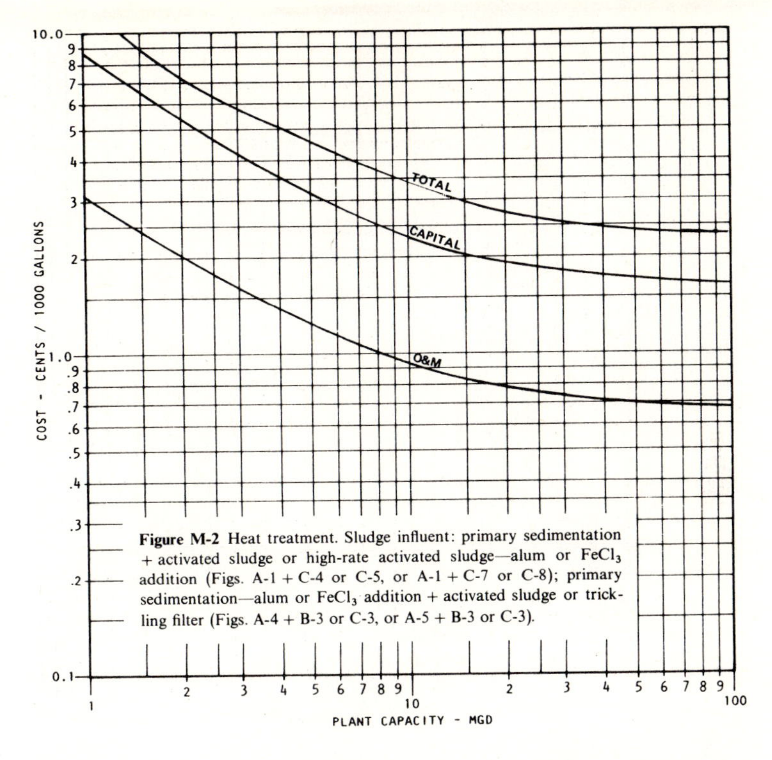

Figure M-2 Heat treatment. Sludge influent: primary sedimentation + activated sludge or high-rate activated sludge—alum or $FeCl_3$ addition (Figs. A-1 + C-4 or C-5, or A-1 + C-7 or C-8); primary sedimentation—alum or $FeCl_3$ addition + activated sludge or trickling filter (Figs. A-4 + B-3 or C-3, or A-5 + B-3 or C-3).

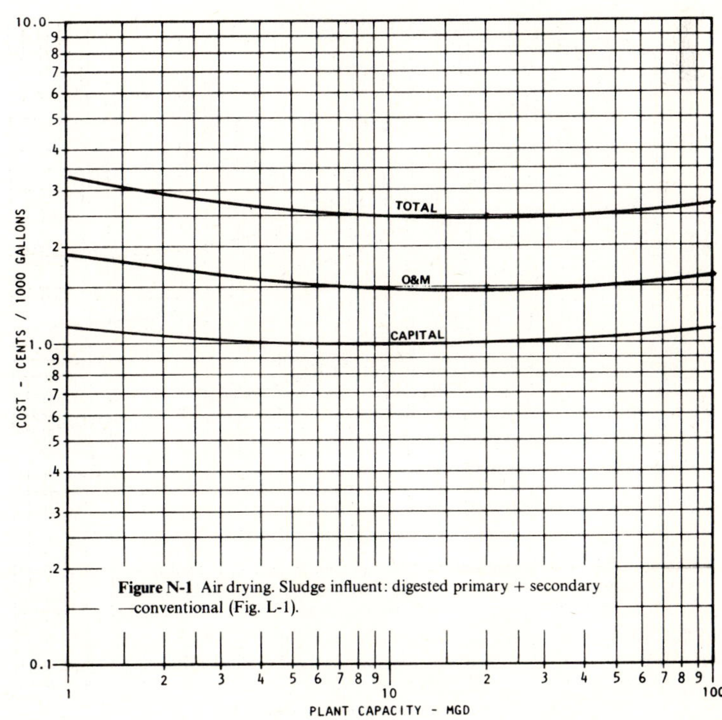

Figure N-1 Air drying. Sludge influent: digested primary + secondary —conventional (Fig. L-1).

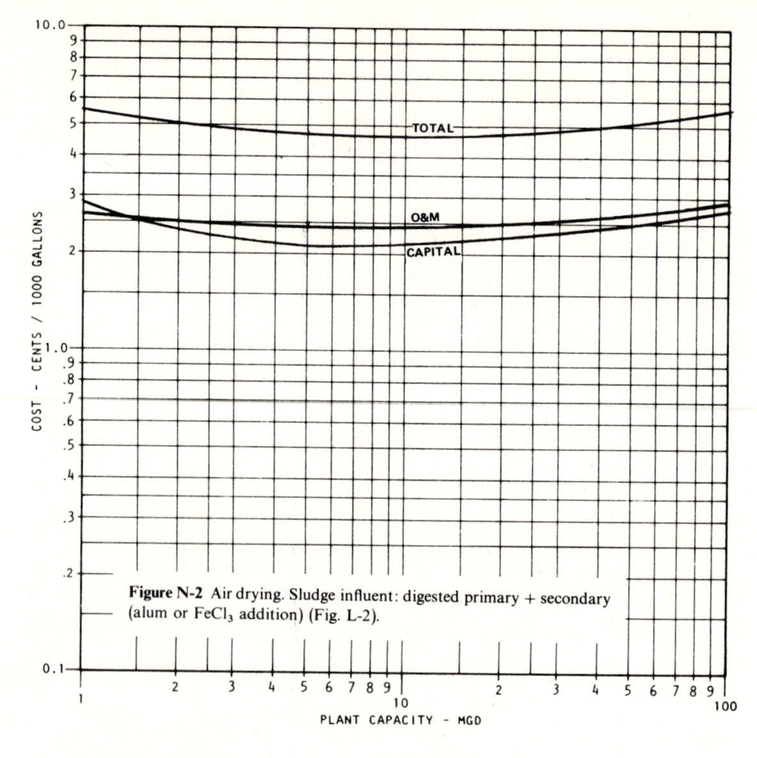

Figure N-2 Air drying. Sludge influent: digested primary + secondary (alum or $FeCl_3$ addition) (Fig. L-2).

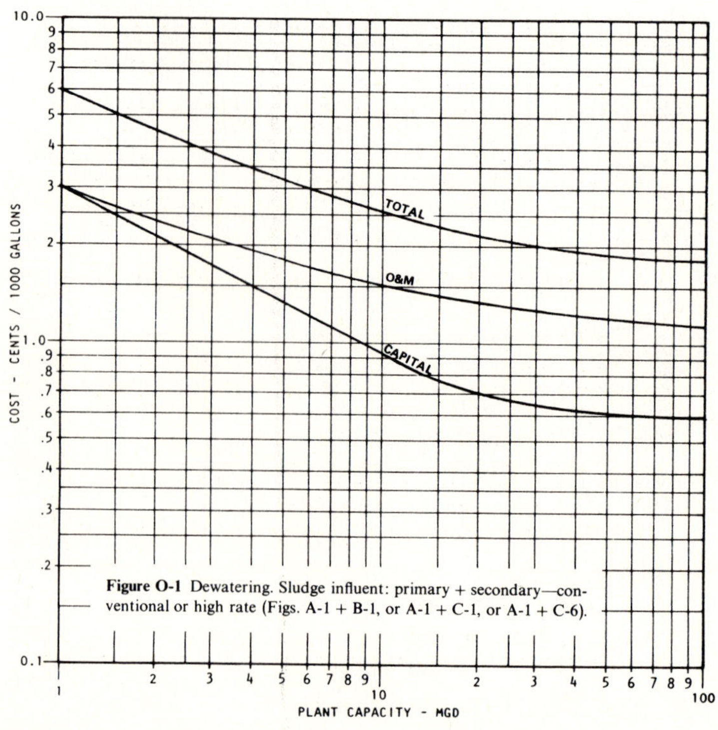

Figure O-1 Dewatering. Sludge influent: primary + secondary—conventional or high rate (Figs. A-1 + B-1, or A-1 + C-1, or A-1 + C-6).

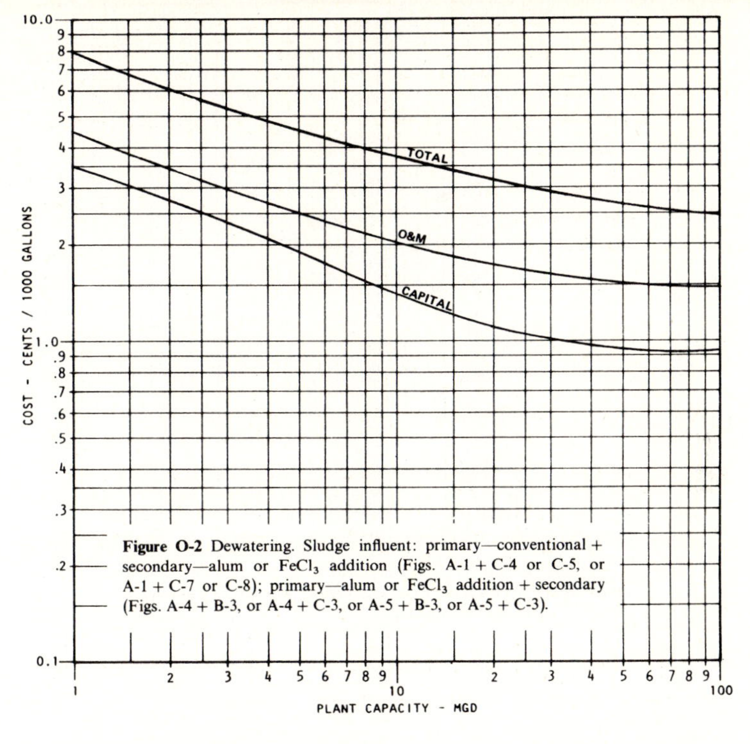

Figure O-2 Dewatering. Sludge influent: primary—conventional + secondary—alum or $FeCl_3$ addition (Figs. A-1 + C-4 or C-5, or A-1 + C-7 or C-8); primary—alum or $FeCl_3$ addition + secondary (Figs. A-4 + B-3, or A-4 + C-3, or A-5 + B-3, or A-5 + C-3).

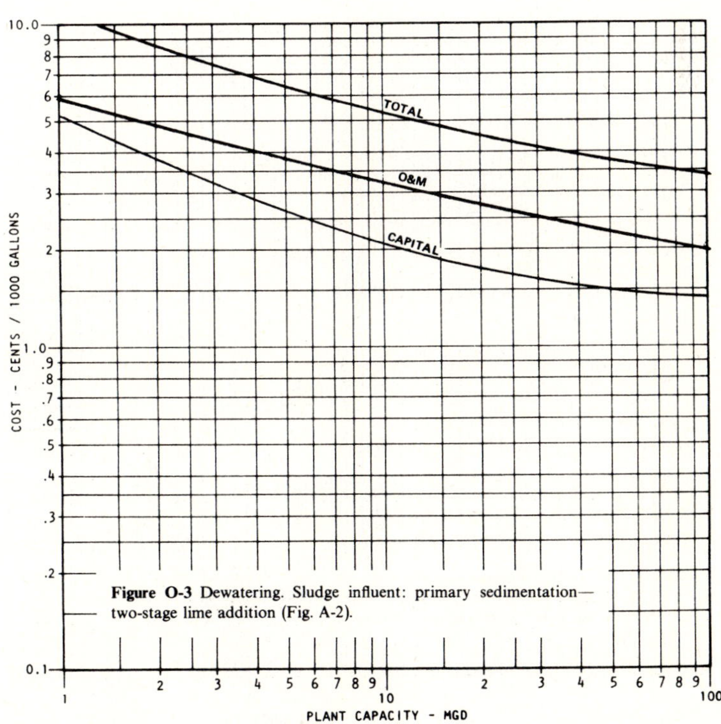

Figure O-3 Dewatering. Sludge influent: primary sedimentation—two-stage lime addition (Fig. A-2).

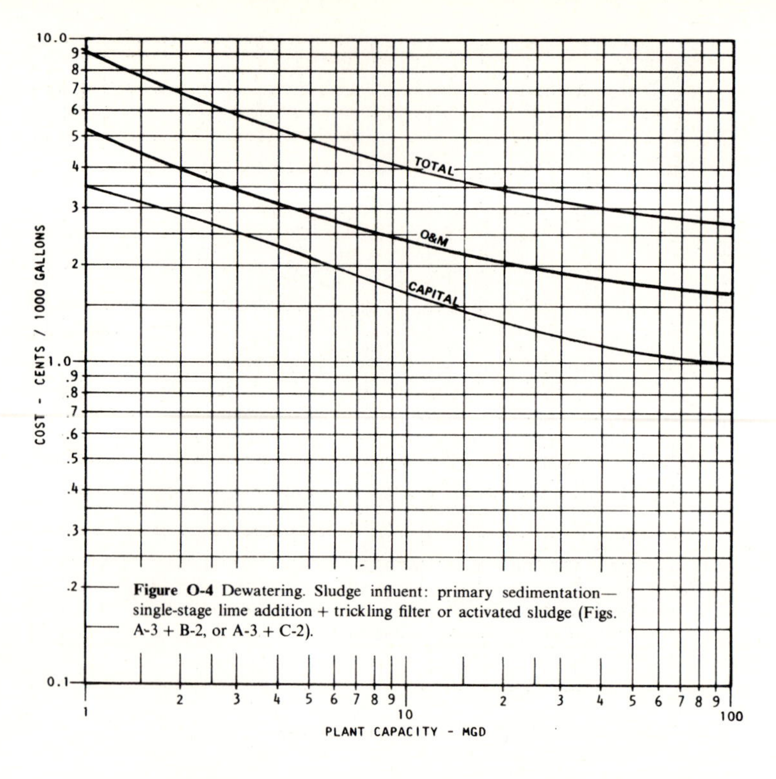

Figure O-4 Dewatering. Sludge influent: primary sedimentation—single-stage lime addition + trickling filter or activated sludge (Figs. A-3 + B-2, or A-3 + C-2).

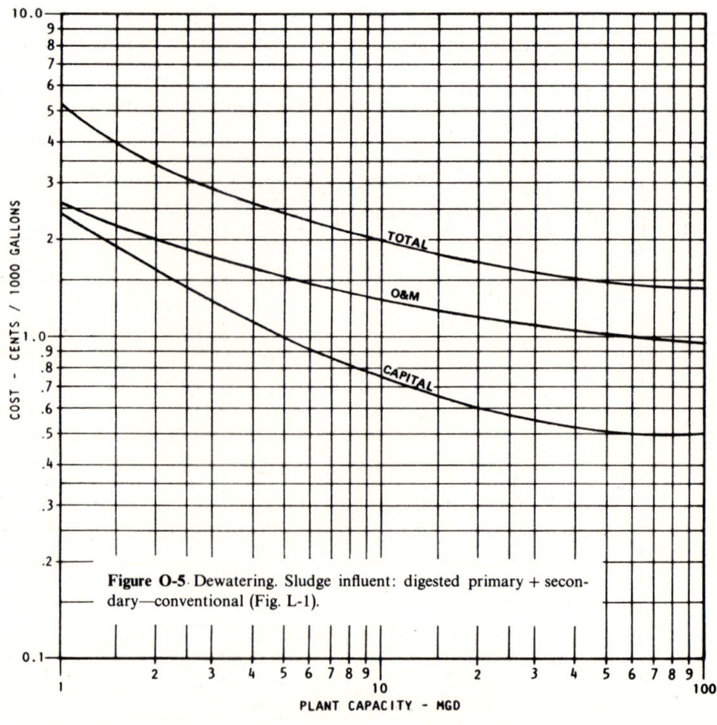

Figure O-5 Dewatering. Sludge influent: digested primary + secondary—conventional (Fig. L-1).

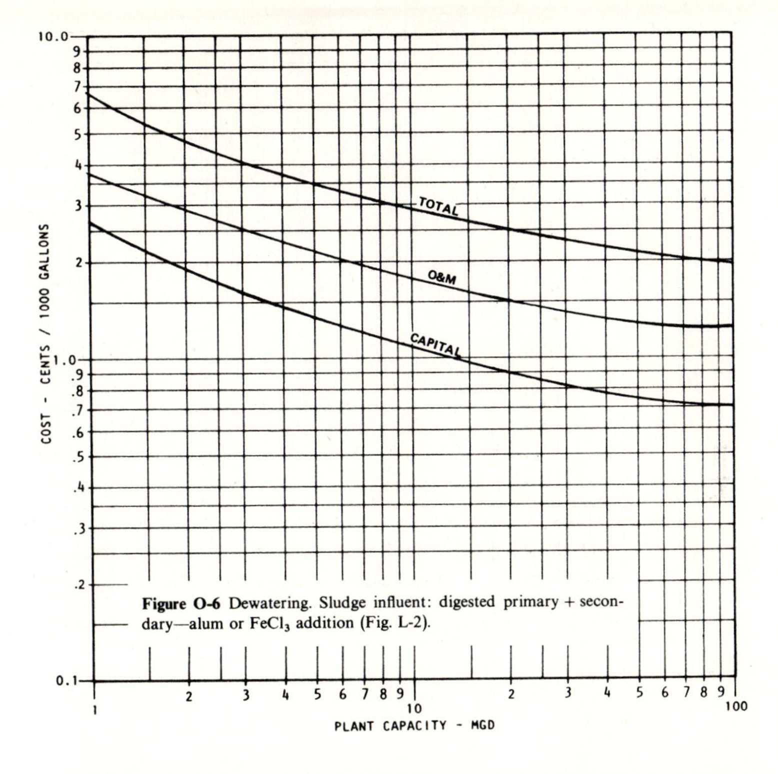

Figure O-6 Dewatering. Sludge influent: digested primary + secondary—alum or $FeCl_3$ addition (Fig. L-2).

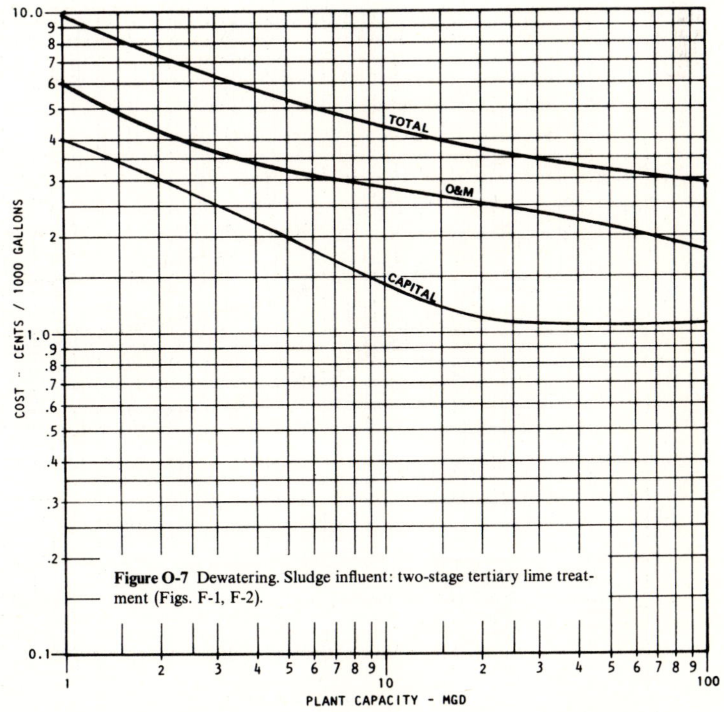

Figure O-7 Dewatering. Sludge influent: two-stage tertiary lime treatment (Figs. F-1, F-2).

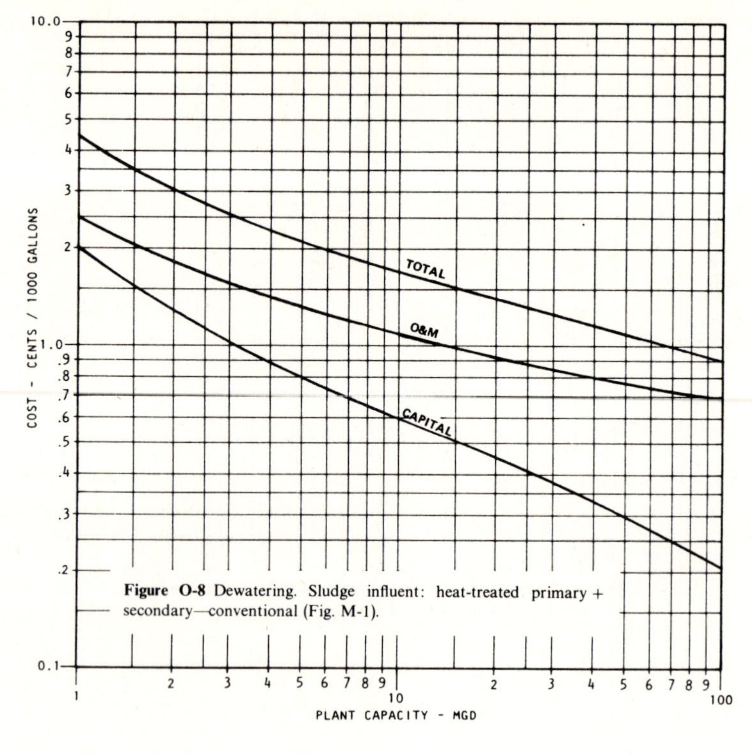

Figure O-8 Dewatering. Sludge influent: heat-treated primary + secondary—conventional (Fig. M-1).

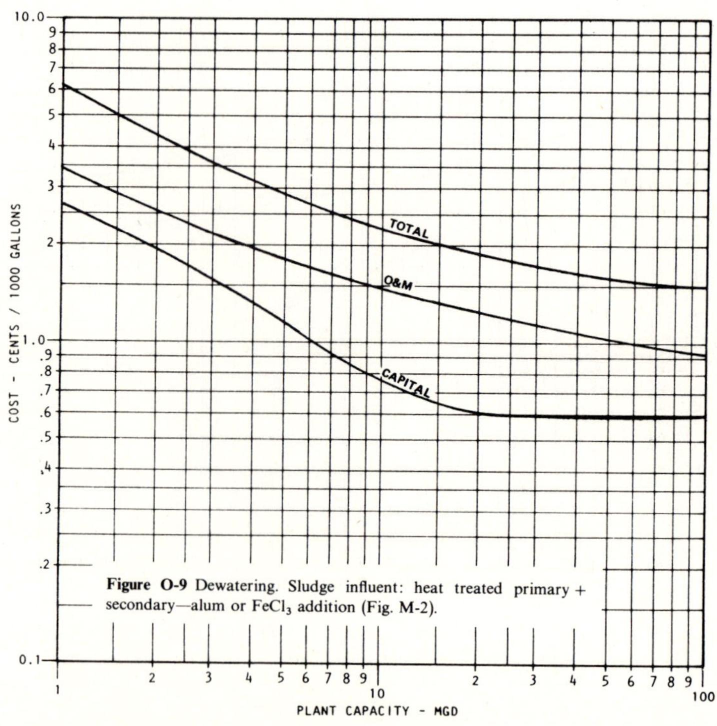

Figure O-9 Dewatering. Sludge influent: heat treated primary + secondary—alum or $FeCl_3$ addition (Fig. M-2).

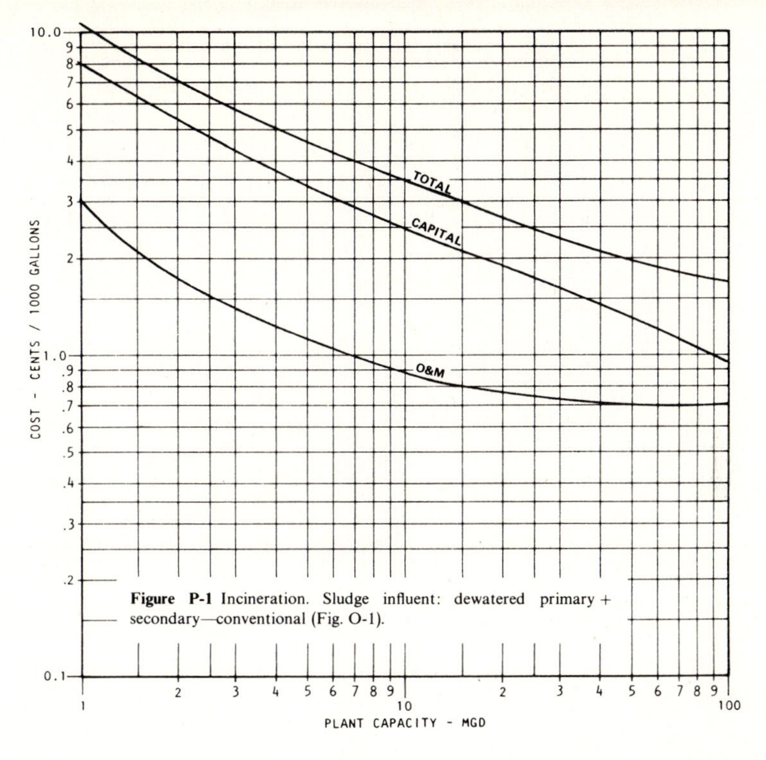

Figure P-1 Incineration. Sludge influent: dewatered primary + secondary—conventional (Fig. O-1).

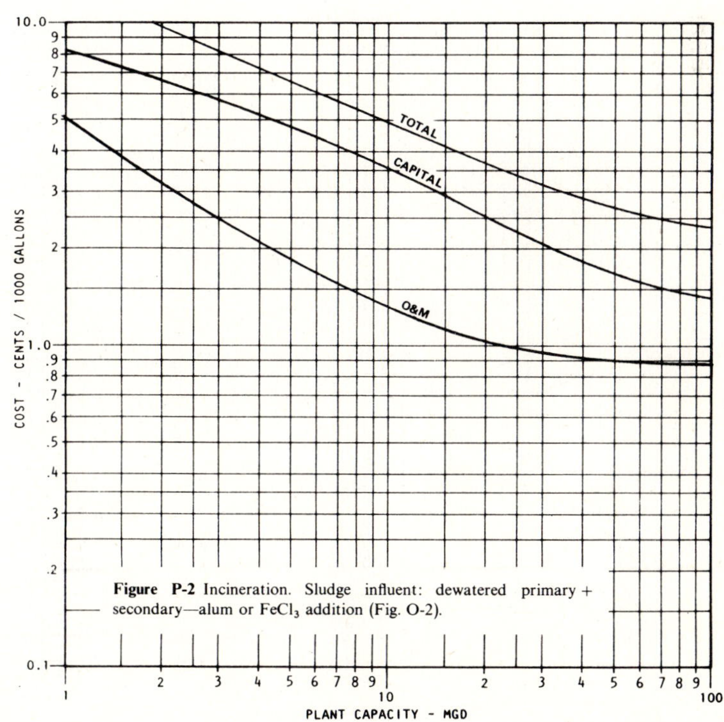

Figure P-2 Incineration. Sludge influent: dewatered primary + secondary—alum or $FeCl_3$ addition (Fig. O-2).

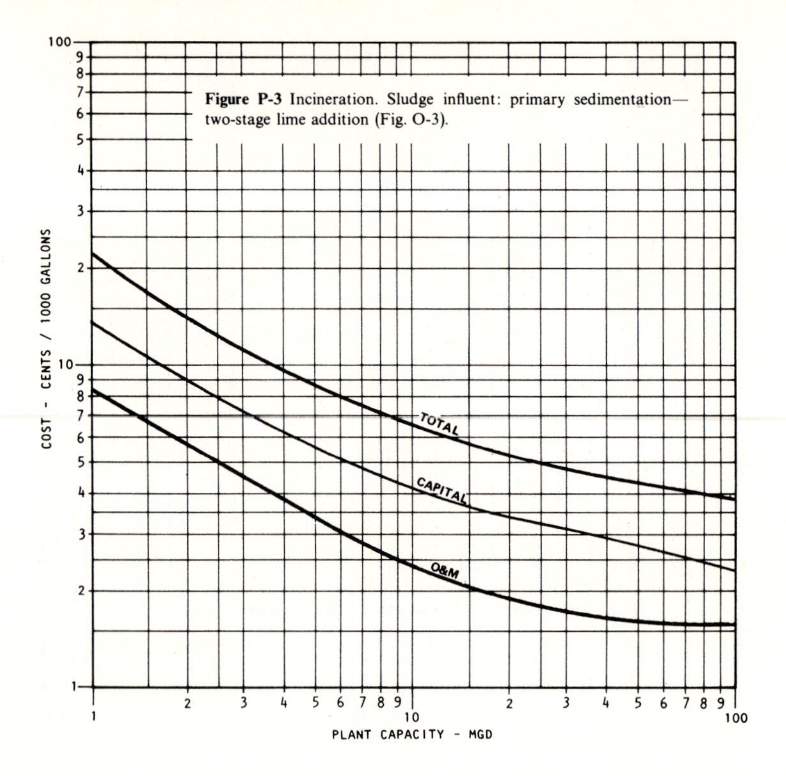

Figure P-3 Incineration. Sludge influent: primary sedimentation—two-stage lime addition (Fig. O-3).

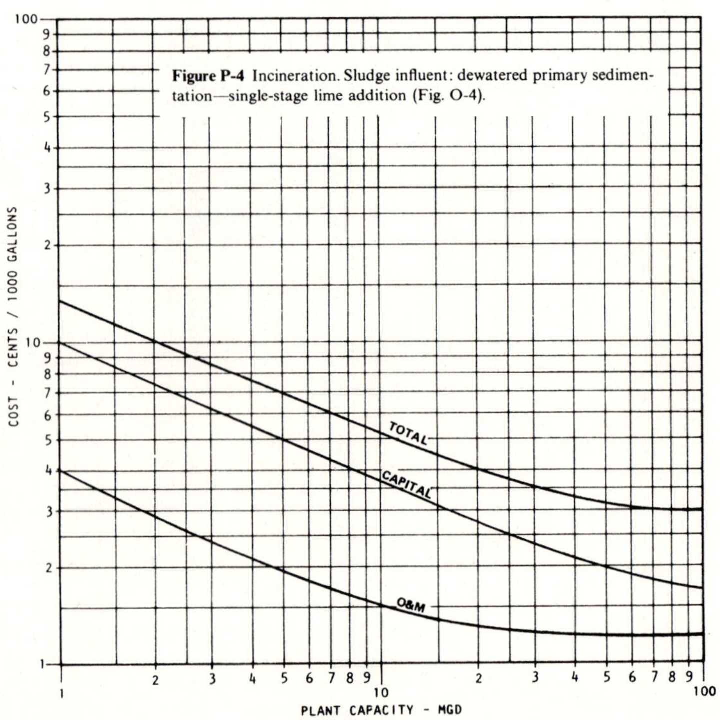

Figure P-4 Incineration. Sludge influent: dewatered primary sedimentation—single-stage lime addition (Fig. O-4).

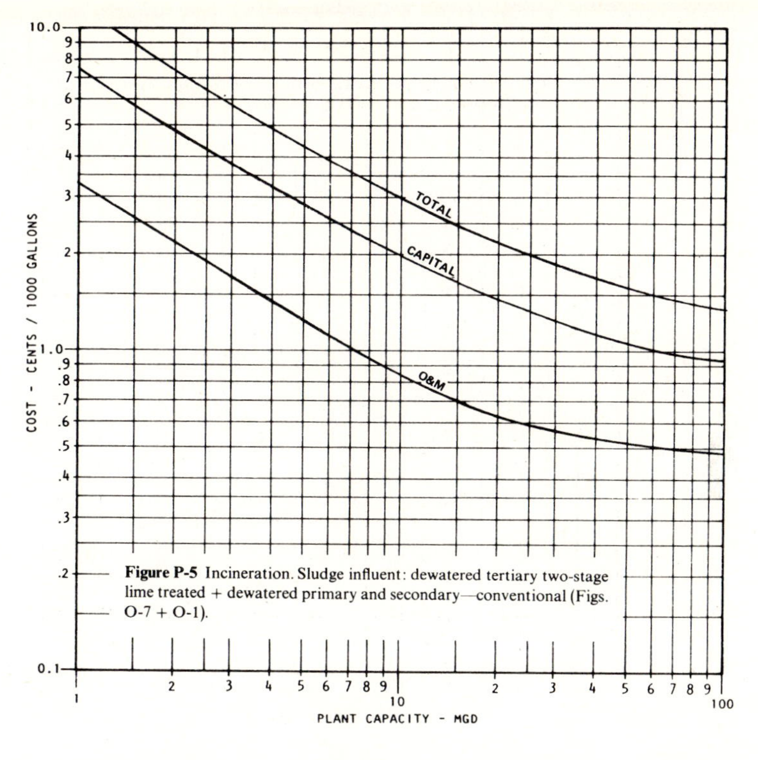

Figure P-5 Incineration. Sludge influent: dewatered tertiary two-stage lime treated + dewatered primary and secondary—conventional (Figs. O-7 + O-1).

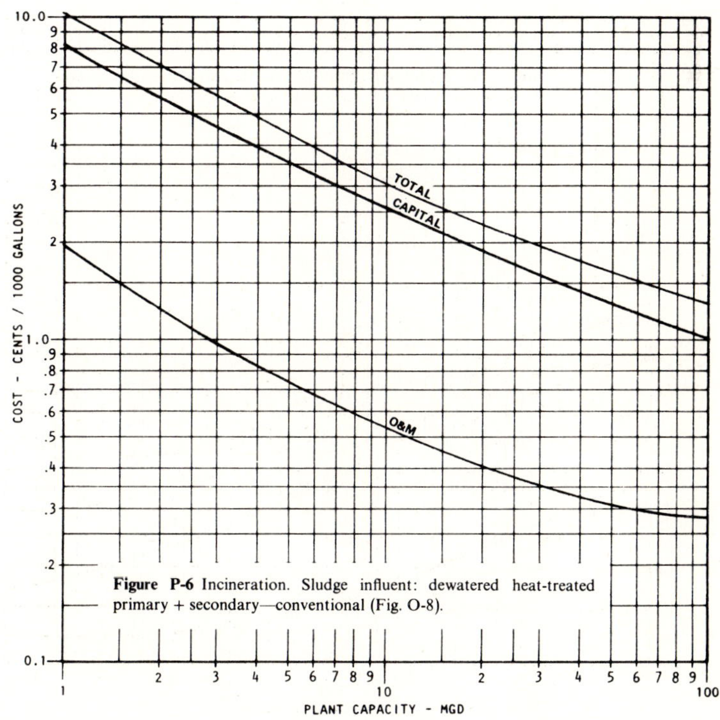

Figure P-6 Incineration. Sludge influent: dewatered heat-treated primary + secondary—conventional (Fig. O-8).

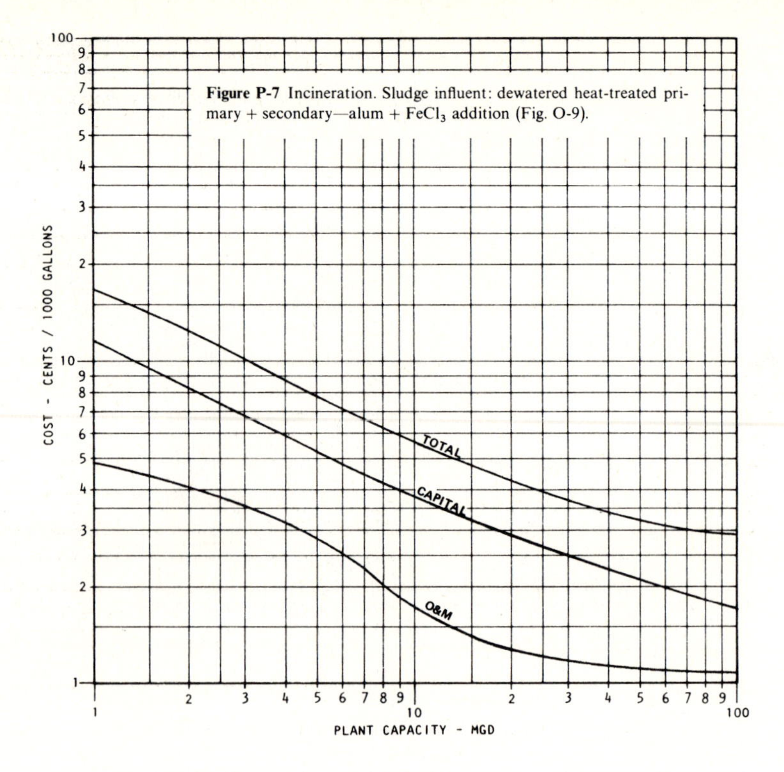

Figure P-7 Incineration. Sludge influent: dewatered heat-treated primary + secondary—alum + $FeCl_3$ addition (Fig. O-9).

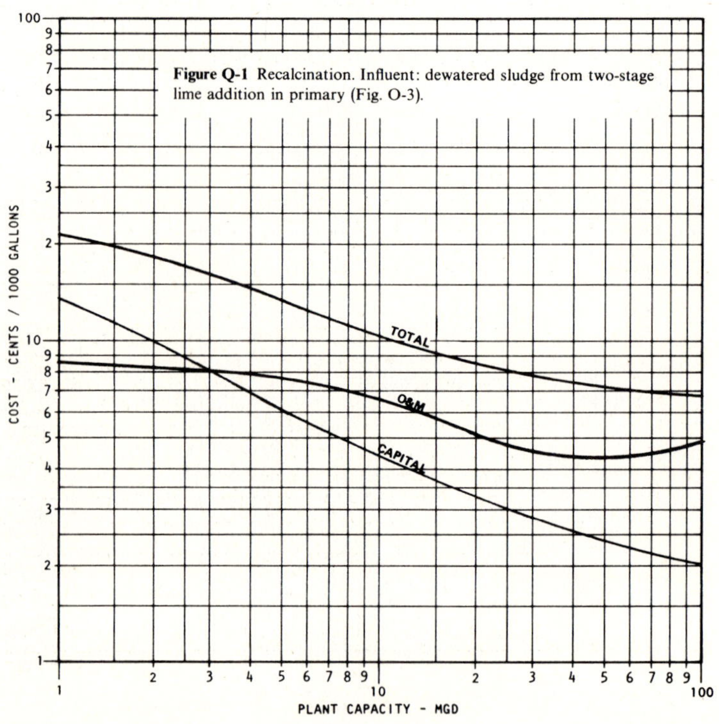

Figure Q-1 Recalcination. Influent: dewatered sludge from two-stage lime addition in primary (Fig. O-3).

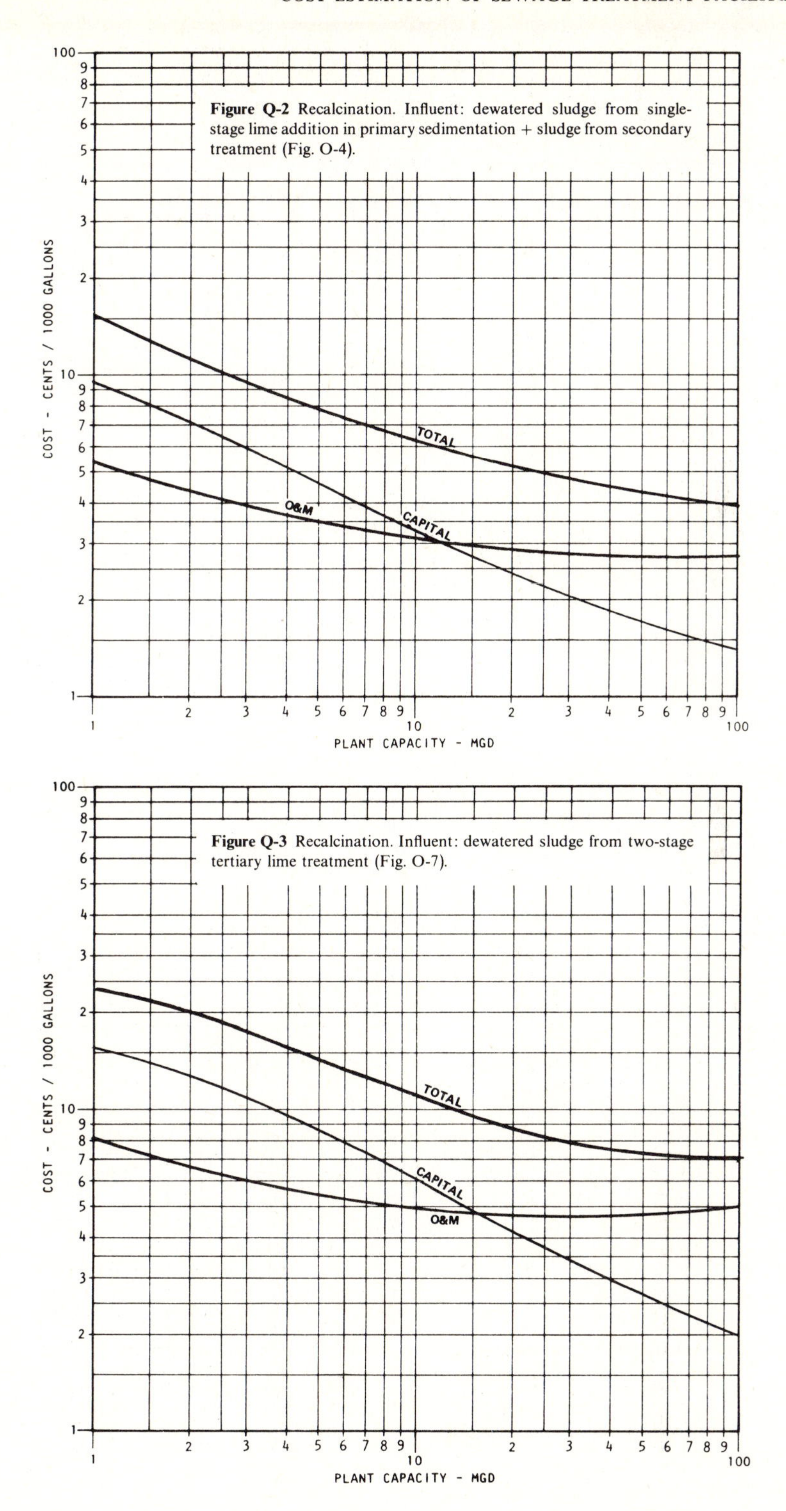

Figure Q-2 Recalcination. Influent: dewatered sludge from single-stage lime addition in primary sedimentation + sludge from secondary treatment (Fig. O-4).

Figure Q-3 Recalcination. Influent: dewatered sludge from two-stage tertiary lime treatment (Fig. O-7).

APPENDIX

FOUR

PHYSICAL PROPERTIES OF WATER

Physical properties of water at different temperatures

Temperature, °C	Density, g/cm^3	Absolute viscosity, cP	Kinematic viscosity, centistoke	Vapor pressure, kPa
0	0.99987	1.7921	1.7923	0.61
2	0.99997	1.6740	1.6741	0.71
4	1.00000	1.5676	1.5676	0.82
6	0.99997	1.4726	1.4726	0.94
8	0.99988	1.3872	1.3874	1.09
10	0.99973	1.3097	1.3101	1.23
12	0.99952	1.2390	1.2396	1.42
14	0.99927	1.1748	1.1756	1.61
16	0.99897	1.1156	1.1168	1.81
18	0.99862	1.0603	1.0618	2.02
20	0.99823	1.0087	1.0105	2.33
22	0.99780	0.9608	0.9629	2.66
24	0.99733	0.9161	0.9186	3.02
26	0.99681	0.8746	0.8774	3.38
28	0.99626	0.8363	0.8394	3.71
30	0.99568	0.8004	0.8039	4.24

$g/cm^3 \times 62.42 = lb/ft^3$
Centipoise $\times 2.088 \times 10^{-5} = lbf \cdot s/ft^2$
Centistoke $\times 1.075 \times 10^{-5} = ft^2/s$
kPa $\times 0.145 =$ psi

INDEX

Abandoned wells, closing of, 200
Accounting, 591–592
Acidity, units of, 190
Actinomycetes, 189
  odor production by, 189
Activated algae process, phosphorus removal by, 561
Activated carbon, 289–290, 564–566
  granular, 289, 565
    dosage of, 290
    regeneration of, 565–566
    size of, 290
  powdered, 289
    dosage of, 290
    size of, 290
  specific gravity of, 290, 565
Activated silica, 231–232
Activated sludge:
  aeration techniques for, 504–513
  air requirements in, 498
  completely mixed, 501
  contact stabilization, 502
  conventional, 499
  diffused air systems for, 508–513
  extended aeration, 501
  high purity oxygen, 503–504
  high rate, 502
  mixing techniques for, 504–513
  nutritional requirements, 516 (table)
Activated sludge:
  operational problems, 515–516
  oxygen requirements in, 498
  process details, 499–504
  process factors, 500 (table)
  short-term aeration, 502
  step aeration, 501
  tapered aeration, 499–500
  theory, 495–496
  waste sludge quantity, 499
Active mass (of bacteria), 477
Activity, chemical, 191
Adsorption, 289, 564–566
Advanced waste treatment, 552–567
Aeration of wastewater:
  bubble, 509
  efficiency of, 509 (illus.)
  diffused, 509–513
    power required for, 509–513
  static tube, 519, 520 (illus.)
  surface, 504–508
    high speed, 504–505
      anti-erosion assembly for, 505
      draft tube for, 505
    low speed, 505
    mixing zone in, 505
    power required for, 505–506
    rotating horizontal, 506–507
Aeration of water, 285–288

Aeration of water:
  cascade, 287
  diffused air, 288
  multiple tray, 287
  spray, 287
Aerobacter aerogenes, 184
Aerobacter cloaceae, 184
Aerobe:
  obligate, 430
  strict, 430
Aerobic processes, 430
  energy conversion in, 430 (illus.)
Aerobic units (individual household), 576
  effluent quality from, 576
Agglomeration, 232
Aggressive water, 309
Air agitation of mixed media filters, 256
Air binding of rapid filters, 243, 262, 264
  causes of, 264
  prevention of, 264
Air entrainment, 508
Air gap, 201
Air/liquid ratio in stripping processes, 561
Air piping, 509–511
  fittings in, 510
    equivalent length of, 511 (table)
  friction factor of, 509
  headloss in, 509
  velocity in, 509
Air scour, 253
Air, solubility in water, 535 (table)
Air transport systems, 509–513
Air wash of rapid filters, 247, 257
  M-Block, 257
  Camp, 257
  rate of air flow in, 257
  rate of water flow in, 257
  velocity of air in, 257
  velocity of water in, 257
Algae:
  blue-green, 188
  clogging of filters by, 188
  control of, 285
  effect on oxidation pond BOD, 519
  green, 188
  role in nitrogen cycle, 555
  role in oxidation pond, 519
  taste and odor from, 188
  turbidity from, 188
  in waste treatment, 430
Algal bloom, 561
  phosphorus concentration required for, 561
Algal production of oxygen, 440
Alkalinity, 190
Alkalinity:
  caustic, 190
  destruction of in nitrification, 558
  effect on coagulation, 229
  units of, 190
  of wastewater, 427
Alum sludge handling, 312
Ammonia:
  quaternary, 285
  removal by stripping, 561
Ammoniator, 278
Anaerobe:
  obligate, 430, 528
  strict, 430
Anaerobic bacteria, metabolic processes of, 429
Anaerobic processes, energy conversion in, 430
Annual cost, 586
Anodic protection, 101
Anthracite filter media, specification for, 244
Aqueduct, 85–86
  advantages of, 85
  disadvantages of, 86
  types of, 85
Aquiclude, 55
Aquifer, 55
  classes of, 57
  interference in, 66
  occurrence of, 56–57
  recharge of, 67–68
  velocity of flow in, 58
Army, United States, Corps of Engineers, 27, 52
Aromatic polyamides for reverse osmosis, 308
Ash, 427
Aspirator, 279
Assessment, 589
Assimilation of nutrients, 555
  algal, 555
  bacterial, 555
    disadvantages of, 555
Assimilative capacity, 438
Attached growth processes, 478–494
  importance of surface area in, 478
Auxiliary scour, 253
Auxiliary water supply, 201

Backfill material, unit weight of, 335 (table)
Backflow preventer:
  double check valve, 138
  inspection of, 201
  reduced pressure principle, 138
  testing of, 201
Backwash of rapid filter:
  length of, 255
  pumps for, 255

Backwash of rapid filter:
quantity of water for, 255
rate of:
for anthracite, 253
for sand, 253
Bacteria:
definition of, 428
enumeration of, 185–187
facultative, 430, 529
in groundwater, 184
saprophytic, 430
in surface water, 184
in water, 184–187
in water mains, 147
Bacterial assimilation, 555
Bacterial diseases, 180–181
Bacterial numbers in wastewater, 428
Bacterial pollution, 181
Bacteriological analysis, sampling for, 194
Barometric pressure, 162 (table)
Base flow, 50
Basin demand, 44
Basin, hypothetical, 214
Basin recharge, 44
Basin storage, 36
Batter board, 404–405
Belt filter (*see* Filter)
Benthal utilization of oxygen, 440
Bentonite clay, 231
Biflow filtration system, 271
Biochemical oxygen demand (BOD), 431–434
procedure for, 432
rate constant:
determination of, 432–434
variation with temperature, 434
vs. time, 433 (illus.)
Biodegradable compound removal in soil systems, 451
Biological treatment systems, 477–521
Biophysical wastewater treatment, 566
Bleach, laundry, 281
Blue baby, 181
Boiler compounds, 308
Boiler problems, 192
Boiler water, 308–309
Bond:
general obligation, 589
revenue, 589
Bottom deposits, 439
Bound water layer, 227
Braces, 408
Bracing:
stay, 408
of trenches, 407–411
Bracing:
removal of, 411
Brackish water, 306
Branch, blow-off, 88
Brine, disposal of, 67
Bromide, 283
Bromine, 283
Bromine chloride, use as disinfectant, 284
Bromodichloromethane, 282
Bromoform, 282, 283
Bryozoa, 189
Bubbler, for pump control, 385
Buffer zone in spray irrigation, 449

Calcining, 299
Calcium, removal of, 293
Calcium carbonate, solubility of, 293
Calgon, 299
Capacity:
specific, of wells, 61
spillway, 47
Capital recovery factor, 587
Carbon column:
design of, 566
downflow, 564
upflow, 564–565
Carbon dioxide, 190, 193
agency in corrosion, 194
half-bound, 193–194
removal of, 194, 286
Carbon tetrachloride, 282
Carbonic acid, 190
Cardiovascular disease, effect of dissolved solids in water upon, 180
Carrier of disease, 180, 196, (n.)
Cast-iron pipe, Manning's "*n*" for, 119
Catch basin, 373
cleaning of, 424
Cathodic corrosion of pipe, 102
Cathodic protection, 101
Cavitation, 164
avoidance of, 164
Center pivot irrigation, 449, 450 (illus.)
Centrifuge, 540–541
basket, 541
dewatering with, 540–541
disk, 541
solid bowl, 540
solids content obtained with, 541
solids recovery of, 540
thickening with, 540
Cercariae, 181, 269
Certification of public water supply, 196
Characteristic curve, combination of:

Characteristic curve, combination of:
in parallel, 167–168
in series, 168–169
Chemical coatings, 101
Chemical feeders:
gravimetric, 236
volumetric, 236
Chemical feeding methods, 235–237
Chemical feeding, pipe materials for, 237
Chemical oxygen demand (COD), 434
Chemical proportioning devices, 235
Chezy formula, 355
Chezy-Manning formula, 226, 356
Chezy-Manning "*n*," values of, 355 (table)
Chironomus, 189
Chloramines, 275, 277–278
Chloride indication of pollution, 193
Chlorinated hydrocarbon, 274
Chlorinated lime, 281
Chlorination:
breakpoint, 276
for nitrogen removal, 560
effect of pH upon, 275
marginal, 276
precautions in, 279
of wastewater, 569
dosages in, 569
residual in, 569
of water, 274–282
dosages in, 276
residual in, 276
Chlorinator, 278
Chlorinator room, 281
Chlorine, 275
Chlorine, combined available, 275
Chlorine contact chamber, 569
design criteria for, 569
Chlorine control of algae, 285
Chlorine demand, 275
Chlorine detector, 281
Chlorine dioxide, 284
manufacture of, 284
use in taste and odor control, 284
Chlorine, free available, 275
Chlorine gas, 278–281
application equipment, 278–280
Chlorine hydrate, 279
Chlorine ice, 279
Chlorine leaks, detection of, 281
Chlorine scale, 279
Chloroform, 282
Chlorophenol, 277
Chlorophyta, 188
Cholera, 180, 184
Churchill and Buckingham technique, 446
Circle method, 120–121
Clarification of wastewater:
primary, 467–474
BOD removal in, 467, 471 (illus.)
effect of hydraulic overload, 470
retention time in, 470
solids removal in, 469, 470 (illus.)
secondary, 475
density currents in, 516
following activated sludge, 513–515
following RBC process, 493
following suspended growth process, 513–515
following trickling filter, 491–492
importance to suspended growth process, 513
thickening function in, 513
Clarification of water, 206–239
Clarifier (*See also* sedimentation basin)
center feed, 221, 223 (illus.)
circular, 473–474
contact, 233–234, 235 (illus.)
design criteria, 234 (table)
peripheral feed, 221
primary, 469–474
rectangular, 472–473
square, 473–474
suction, 513–514
upflow, 233–234, 235 (illus.)
Cleaning of sewers, 422–423
Cleanout, 368–369
Clear well, 266
capacity, 111
Clinoptilolite, 560
Cloud seeding, 35
Coagulant aid, 231
Coagulant:
chemistry of, 227–230
optimum pH for, 230 (table)
purpose and action of, 226–232
Coagulation, 226–232
in advanced wastewater treatment, 552–553
chemicals used in, 229
control of, 229
effect of chlorination on, 276
effect of pH on, 228–229
for upgrading oxidation ponds, 520
Cock, corporation, 131
Coefficient of friction in pipe, determination of, 148
Coliform bacteria, 181–186
fecal, 186
non-fecal, 186

Coliform bacteria:
number in wastewater, 185
Coliform test, importance of, 185
Colloid, 226
area/volume ratio, 226–227
charge on, 227
stability of, 227
Colloidal stability, 227
Color, 190
platinum-cobalt standard for, 190
removal of, 308
Column settling test, 218–219
Combined sewer system, 317
Combustion of sludge, 545–548
Comminutor, 460
selection of, 460
Completed test, 187
Completely mixed biological process, 495
effect of peak load on, 501
effect of toxic material on, 501
Completely mixed biological reactor, 496 (illus.)
Compressor, 510–513
centrifugal, 511, 512 (illus.)
multistage, 512
surging of, 511
power required, 512–513
rotary lobe, 510, 511 (illus.)
Concentration, time of, 32–33, 48, 322–325
Condemnation of public water supply, 197
Confined water, 56
Confirmed test, 186
Connections to sewers, 423–424
Conservation of water, 67
Contact stabilization, 502–503
reduction in required aeration capacity in, 502
Contact time, 276
Contacter, rotating biological, 492–494
Contaminated water, 180
Copper sulfate, 285
dosage for algal control, 286 (table)
fish tolerance for, 287 (table)
Corps of Engineers (*see* Army)
Corrosion:
of metals, 99–101
protection against, 100–101
reactions, 99–100
of sewers, 348–352
control of, 351
of water mains, 192, 194
Cost comparison, 586–587
Cost estimate, 580–586
Cost estimation for wastewater treatment (App. III), 613–641
Cost index, 580–581
Cost projection, 580
Cost of sewerage, 6
Cost, unit, 580
Cost of water supply, 6
Cover for sewers, 394
Crenothrix, 80, 147, 184, 192, 278, 291
Cross connection, 196
definition of, 201
protection against, 276, 278
Crustacea, 188
Cunette, 364
Current:
density, in clarifier, 221
effect on water quality in streams, 439
Cyanophyta, 188
Cyclops, 188

Dam, diversion, 108
Daphnia, 188
Darcy's law, 58–60
Dechlorination:
with activated carbon, 277, 560
with sulfur dioxide, 560
Deficit, critical, 442
Defluoridation, 182, 310–311
by ion exchange, 310–311
with magnesium hydroxide, 310
Demineralization, 305–306
Denitrification, 558–559
comparison of alternatives, 559 (table)
use of methanol in, 558
Depression, cone of, 60, 62
Desalination, 306–308
cost of, 306
Design period, wastewater:
collection system, 24
pumps, 24
treatment, 24
Design period, water:
distribution system, 21
pipe lines, 20
pumping plant, 21
source, 20
storage, 21
treatment, 20
Design storm, return period of, 329
Detention time in sedimentation, 213
Detergents, removal of, 564
Dewatering:
of sludges, 535–545
of trenches, 411–412
Diatom, 188
Dibromochloromethane, 282
Dichloramine, 278

Dichloroethane, 282
Diesel engines for pumping, 172–173
Diffuser, air, 499
Digestion, aerobic, 529–530
  advantages of, 529
  design criteria, 529 (table)
  disadvantages of, 529
  pure oxygen, 530
  solids retention time in, 530
  toxicity in, 529
Digestion, anaerobic, 527–529
  design criteria, 527 (table)
  heating in, 527, 529
  high-rate, 527–528
  low-rate, 527
  mixing in, 527
  solids retention time in, 527
  solids return in, 527
Dilution of wastewater, 439
Discharge, ocean, 446
Disinfection:
  thermal, 284
  of wastewater, 569
  of water, 274, 283–284
Disposal:
  by evaporation, 454
  field, subsurface, 576–578
    application rates for, 577 (table)
    length of lines in, 577
    slope of laterals in, 577
  on land, 453–454
  ocean, 549
  of sludge, 548–558
  in streams, 453
  system selection, 453–454
  of wastewater, 438–454
Distribution, gravity, 109
Distribution, Gumbel, 51
Distribution, log normal, 51
Distribution, pumped, 110
Distribution system, 109–150
  analysis of present inadequacy, 116
  construction and maintenance of, 142–150
  design of, 115–116
  dual, 113
  pipe sizes in, 115
  prediction of future inadequacy, 116
  pressure in, 113–114
  velocity in, 115
Dosing tank, 489
Double mass curve, 32
Draft, 44
  maximum, 46
Draft tube, 262
Drag, coefficient of, 210
Drain, illegal connection to sanitary sewer, 318
Drawdown, 59 (n.), 62, 64–65
Drinking water regulations (App. I), 595–610
Drooping characteristic (of pump), 167–168
Droplet losses in spray irrigation, 449
Drying beds, sludge, 535–536
Drying of sludges, 545
Dysentery, 184, 269
  amoebic, 180, 188
  bacillary, 180
  epidemiology of, 195

Echo (virus), 187
Effluent, secondary, filterability of, 554
Ejector pump, 170
Ejector, sewage, 387, 388 (illus.)
Electric motor for pumping, 173–174
  slip-ring, 174
  squirrel-cage, 174
  starting equipment, 173
  synchronous, 174
Electrodialysis, 307
  cost of, 307
Electrolyte, hydrolyzing, 227
Electroneutrality, point of, 227
Elutriation, 531
Endamoeba histolytica, 188
Environmental Data Service, U.S., 27
Environmental Protection Agency:
  role of, 5
  standards for water supply, 194–195
Epidemiology, 181
Epsom salts, 192
Equivalent length of pipe fittings, 157 (table)
Equivalent pipe method, 119–120
Escherichia coli., 184
Evaporation:
  from land surfaces, 41
  pan, 41
  systems, 452–453
  techniques for reduction of, 41
  from water surfaces, 41
Evaporators, 306
  cost of, 306
Evapotranspiration, 43
  rates of, 44 (table)
Excavation:
  classification of, 406–407
  hand, 406–407
  machine, 406–407
  rock, 407
  of trenches, 406–407
Explosions in sewers, 424

Extended aeration, 501, 502 (illus.)

Fallout, 183
Fecal coliform/fecal strep ratio, 186
Feeding and mixing apparatus for lime-soda process, 298
Fiberglass, 102
Filamentous growth, role in sludge bulking, 515
Fill material:
- consolidation of, 415
- placement of, 415
- specification for, 415

Filter:
- American, 240
- appurtenances, headloss in, 255

Filter bags, 576
Filter:
- belt, 541–545
- biflow, 271, 553

Filter control system, 257–261
Filter:
- declining rate, 258–260
  - advantages of, 258
- depth, 245
- diatomaceous earth:
  - industrial, 270
  - for wastewater, 553
    - body coat, 553
    - precoat, 553
  - for water, 270
    - body coat, 270
    - precoat, 270
- dual cell gravity, 541–542

Filter flies, 492
Filter:
- granular in advanced treatment, 553–554
- Greenleaf, 258
- headloss through, 243
- intermittent sand, 478–479, 578
- mechanical, 240

Filter media:
- anthracite, 244
- effective size of, 244
- garnet, 245
- gradation, 245, 270
- ilmenite, 245
- mixed, 245–246
- sand, 244
- surface charge on, 243
- uniformity coefficient of, 244

Filter pipe gallery, 262
Filter press, 539–540, 538 (illus.)
Filter:
- pressure, 267–268
- rapid, 240–241
  - essential characteristics of, 242

Filter rate controller, 242
Filter, roughing, 481
Filter run, effect of prechlorination on length, 276
Filter:
- slow sand, 240, 271–272
- storage capacity of, 270
- subsurface, 578
- trickling, 479–492
  - air transport in, 490
  - classification of, 480–482
  - cold weather operation, 492
  - covers, 492
  - depth of, 487
  - design of, 487–492
  - design criteria for air transport, 490
  - distribution systems, 488–490
  - dosing siphons for, 489
  - efficiency of, 484–487
  - factors affecting rate, 480
  - final clarifiers for, 491–492
  - fixed distribution system, 488, 490, 481 (illus.)
  - flow patterns for, 483 (illus.)
  - forced ventilation in, 490
  - formulas for, 484–487
    - Eckenfelder, 486
    - NRC, 484
    - Rankin, 486
    - Ten States, 486
    - Velz, 485
  - headloss in, 490
  - high-rate, 481
  - low-rate, 480–481
  - mathematical models of, 480
  - media, 481–482
  - mode of operation, 487
  - operational problems of, 492
  - plastic media, 481–482
  - recirculation in, 482–484
    - advantages of, 484
    - effect upon efficiency, 484
    - rates of, 484
  - rock media, 481
  - rotary distributor, 488–489
    - speed of, 489
  - solids production in, 492
  - underdrains, 490–491

Filter underdrains (water treatment), 247–250
- Camp, 248
- design of, 247
- Infilco, 248

Filter underdrains (water treatment):
Leopold, 248
low headloss, 250
perforated pipe, 247
Filter units:
number of, 266
size of, 250, 266
Filter:
upflow, 270, 553
use in softening plants, 299
vacuum, 537–539
blinding of, 539
chemical addition in, 539
depth of submergence, 539
media for, 539
solids content produced by, 537
yield, 537
Filter washing, 252–257
expansion in, 252, 253
frequency of, 252
Filtration:
operational difficulties, 264–265
of oxidation pond effluent, 520
pretreatment for, 244
processes involved in, 240
rates, 246
theory of, 243
of wastewater, 578
of water, 240–272
Filtration gallery, 77
Financing, 580–593
of wastewater facilities, 592–593
assessments for, 592
bond issues for, 592
federal grants for, 592
of waterworks, 588–592
Finite element analysis of distribution systems, 130
Fire demand, 18–19
duration, 19 (table)
factors affecting, 18
residential, 18 (table)
Fire flow, 112–113, 121–123
maximum, 112
tests, 121–123
determination of headloss in, 121
effect of storage upon, 123
Fire hydrant, 112–113, 140–145
air removal at, 144
area served by, 113
capacity of, 141
classification of, 141
freezing of, 141, 145
location of, 113
Fire hydrant:
maintenance of, 145
placement, 144–145
sediment removal at, 144
spacing, 112
specifications for, 141
Fire pressure, 109, 110
Fire pumps, 110, 113
Fire service, charges for, 590
Fixed water layer, 479
Flatworms, 188
Flocculation, 227, 232–233
design factors for, 232
Flood flow, estimation of, 47–51
Flood frequency, 51–52
Flood, probable maximum, 52
Flooding (of granular materials) 236, 290
Flotation:
aeration, 467
dissolved air, 467, 520
use in grease removal, 467
vacuum, 467
Flow diagrams, 358, 359–361 (illus.)
Flow distribution, 213
Flow:
dry weather in streams, 37
flood, 37
Flow meter in filter, 261
Flow:
open channel, 354–356
in pipes, 116–119
correction for value of C, 119
nomogram for, 118
in sewers, 354–364
steady, 354
subsurface, 38
time of, 324
uniform, 354
Fluidization of filter media, 253
Fluistor, 129
Fluorescein, 204
Fluoride, 182–183, 310–311
dosage, 310
limit, 310
removal, 310–311
Fluoridation, 183, 310
Flush tank, automatic, 374
Flushing of sewers, 422
Flyash, use as sludge conditioner, 532, 539
Foaming in boilers, 309
Forms, all-metal collapsible for sewers, 415
Fountain, drinking, 203
Freezing:
for desalination, 307

Freezing:
  of sludges, 533
Fungi, 189, 431
  control of, 189
  taste and odor from, 189
  in waste treatment, 431

Galvanic protection, 101
Garbage disposal with wastewater, 570
Garbage grinders:
  effect upon sewer systems, 570
  effect upon wastewater treatment, 570
Garnet sand, 245
Gas masks, 281
Gases:
  in sewers, 424–425
  in water, 179, 193–194
Gasoline engines for pumping, 173
Gastroenteritis, 181, 184
Gauge:
  pressure, 177
  rain:
    float recording, 28
    tipping bucket, 28
    weather service standard, 27–28
  recording, 177
Giardiasis, 180
Glauber salts, 192
Gooseneck, 131
Government control of public water supply, 195
Gravel, filter:
  mounding of, 265
  purpose of, 246
  specification for, 246
Grease removal:
  need for, 466
  from sewers, 422
  from wastewater, 466–467
Grease trap, 375
Greensand, 301
Gridiron, pipe, 120
Grit chamber:
  aerated, 465–466
    air flow, 465
    detention time, 465
    headloss in, 465
  gravity, 460–465
    horizontal velocity, 462
    length of, 465
    scour in, 219
    surface overflow rate, 462
  parabolic, 463
  rectangular, 463
Grit, disposal of, 466
Grit quantity, 466
Grit removal:
  at sewage pumping stations, 382
  from sewers, 421–422
  from wastewater, 460–466
Grounding of electrical wiring, 101–102
Groundwater, 55–82
  contamination, 199–200
    causes of, 199
    prevention of, 200
  flow, 58
  level control in land disposal, 449
  projects, 78
  recharge, 36
  resources, 67
  table, 56
Growth pattern of microorganisms, 428, (illus.) 429
Growth yield coefficient, 496
Guniting of sewers, 423
Gutter storage, 330

Haloform, 283
Halogenated hydrocarbon production by wastewater chlorination, 569
Halogenated organics, 282
Halogens as disinfectants, 283
Hammer, water, 86, 88, 135
Hardness, 179, 191–192, 292
  carbonate, 191
  noncarbonate, 192
  permanent, 192
  temporary, 191–192, 194
Hardy Cross analysis, 123–129
  convergence of, 129, 130
  digital computer application, 129–130
Hazen-Williams coefficient, 116
  table of, 117
  variation in, 117
Hazen-Williams formula, 116
Head, 154
Head-discharge curve, 160–162
Head:
  friction, 155
  negative, 262–264
  specification of, 166
  suction, 154
  total dynamic, 154
  total static, 154
Headloss:
  at changes in sewer direction, 356–357
  in filter appurtenances, 254–255
  in filter backwash, 254–255
  in filter controller, 255

Headloss:
in filter gravel, 255
in filter sand, 254
in filter underdrains, 255
of pipe fittings, 156
Headloss gauge, 261
Heat treatment of sludges, 532
Hepatitis:
epidemic, 181
infectious, 180–181
Hexametaphosphate for water stabilization, 299, 310
High-consumption district, 110
High-value district, 113
Hollow fiber system (reverse osmosis), 308
Hose:
lines, length of, 120
stream, 112
Hydraulic grade line:
breaking of, 87
in open channel flow, 356
Hydraulic gradient, 56, 58
Hydraulics, groundwater, 58–60
Hydrocarbon, chlorinated, 195
Hydrogen sulfide:
control of in force mains, 386
removal of, 193, 285
role in sewer corrosion, 351
Hydrograph, 36
integrated, 45
overland flow, 48–49
summation, 45
unit, 49
Hydrologic cycle, 26
Hydrology, 26
significance of in water supply, 26
sources of information, 26–27
Hydrolysis:
of aluminum, 227–228
of iron, 228
Hydroxyapatite, 563
Hypochlorination, 281–282
Hypochlorite:
in algae control, 282
calcium, 281
emergency use of, 281–282
high-test, 281
sodium, 281
Hypochlorous acid, 275

Ice:
anchor, 106
frazil, 106
removal in intakes, 106
Ilmenite, 245
Images, method of, 66
Imhoff tank, 475
Impurities in water, 179–180
Incineration, 545–548
Incinerator, sludge, 545–548
cyclonic, 545
electric, 546
fluidized bed, 547
infrared, 546
multiple hearth, 547–548
Industrial waste, 316, 570–572
discharge restrictions, 570
disposal on land, 67
joint treatment, 570–571
advantages of, 571
pretreatment, 570
protective ordinances, 571
survey, 571
trace nutrients, 571
treatment, 571–572
costs, 436
criteria, 571
processes, 572 (table)
rate of, 571
studies, 571
Inert materials, 102
Infiltration gallery, 266–267
advantages of, 267
yield of, 267
Infiltration into sewers, 22, 316, 347–348
pumping, cost of, 348
rate of, 22
treating, cost of, 347
Infiltration of precipitation, 38–41
capacity of soils, 39
determination of rate, 39–41
factors affecting, 38–39
Inflow, 22, 316
Influence, circle of, 60
Influence, cone of, 68
Inhibition of corrosion, 101
Inlet:
capacity, 372 (table)
curb, 371
time, 323–324
Inspection:
of house sewer connections, 418
of sewers, 423
Intake, 104–109
conduit, 109
in impoundment, 104–105

Intake:
  in lake, 105–107
  location with respect to pollution sources, 104, 105
  navigation requirements, 104, 107, 108
  pipe, 107–108
  river, 107–108
  screen, 108
  shore, 108
  submerged crib, 107, 108
  velocity in, 105–106
  water, 84
Intensity-duration-frequency curve, 326
Intensity-duration-frequency data, use of, 329
Inorganic reactions in soil systems, 452
Inorganics in sewage, 427
Isohyet, 30
Isohyetal map, 30, 32 (illus.)
Iodine, 283
Iodine number, 289
Iodine, physiological effects of, 283
Ion exchange, 301–307, 560
  cost of, 307
  for desalination, 307
  media, capacity of, 304
  for nitrogen removal, 560
Iron, 192
  removal, 286, 291–292
    by aeration, 286, 291
    by adsorption, 292
    catalysis of, 291–292
    effect of pH on, 291–292
    with oxidizing agents, 292

Jacking and boring, 414
Jacob, well analysis technique of, 65–66
Jar test, 229, 231, 232
Jetting, 246, 247
Joints in sewer lines, 413–414
  bituminous, 413
  cement-mortar, 413
  plastic, 413–414
  rubber, 413–414
Junction, sewer, 376–377

Kjeldahl nitrogen, 555
Kutter formula, 355

Lag time, 49
Lagoon:
  aerated, 519
  facultative, 519–520
  upgrading, 520
Lagooning of water treatment wastes, 311
Lakes:
  bottom accumulations in, 439
  self-purification of, 439
Land disposal of sludge, 549–550
  disease transmission in, 550
  heavy metal accumulation in, 550
  loading rates for, 550
  State standards for, 550
  techniques for, 550
Land disposal of wastewater, 447–452
  site preparation, 449
  site selection, 447–449
Land treatment, 447–452
Langlier's index, 309
Laser alignment of sewers, 405 (illus.), 406
Launder, 224–226
  design of, 225–226
Lead:
  EPA limit in public water supplies, 182
  paint, 182
  poisoning, 182
Leak location, 148
Leak survey, 148
Ledger:
  customer, 591
  deposit, 591–592
  distribution, 592
Lignin, removal of, 564
Lime sludge handling, 312
Lime-soda method, 293–299
  advantages of, 299
  chemical dosages for, 293
  disadvantages of, 299
Log growth, 429

MPN (*see* most probable number)
McIlroy analyzer, 129
Magnesium hydroxide, solubility of, 296
Magnesium removal, 295
Magnetic detector, 148
Main, water:
  allowable leakage in, 147
  arterial, 114
  bacteriological tests of, 147
  breaks in, 147
  cleaning of, 148–149
  disinfection of, 146–147, 203
  fire requirements, 115
  flushing of, 146
  small, 115
  testing, 147
Maintenance of sewers, 418–425
  equipment for, 419–420

Manganese, 192
  removal, 291–292
    by aeration, 286
    by adsorption, 292
Manganese zeolite process, 292
Manhole, 365–369
  details, 365
  drop, 367
  frame, 368
  location, 365, 394
  spacing, 365
  for valve placement, 145
Manning formula, 356
  for gutters, 370
Marble test, 309–310
Mass diagram, 45
  residual, 47
Master plan, preparation of, 116
Mathematical modeling of rainfall-runoff events, 32
Maximum contaminant level (MCL), 194
Mean cell residence time, 495
Membrane filter:
  for bacterial enumeration, 185
  for solids determination, 427
Membrane, semipermeable, 307
Mercury, 195
Metal, heavy:
  adsorption by soil, 451
  precipitation as sulfide, 528
Metallic coatings, 101
Meter:
  accuracy, 133
  compound, 133
  damage to, 133
  magnetic, 177
  nutating disk, 132
  oscillating piston, 132
  ownership, 132
  reading, 133
    direct recording, 133
    direct transmission, 133
    visual, 133
  rotor, 132
  testing, 133
  venturi, 177
Metering:
  effect upon consumption, 16
  importance of, 15, 591
Methane production in anaerobic digestion, 529
Methemoglobinemia, 181–182
Microbiology of sewage, 428–434
Microbiology of sewage treatment, 428–434
Microorganisms, 187–189
Microorganisms:
  in waste stabilization, 431
Microscreen:
  in advanced waste treatment, 553
  backwash of, 554
  design parameters for, 554
  efficiency of, 554
  loading on, 554
  for oxidation pond upgrading, 521
Midge flies, 189, 287
Milk, infected, 195
Mineral impurities of water, 191–193
Minimum billing for water, 591
Minimum cost analysis of process trains, 588
Minimum route matrix analysis of water distribution systems, 130
Mixed media, 245–246
  backwash systems for, 246
  particle sizes for, 245
Mixers, 232
Mixing, 232
  turbulence in, 232
Molasses number, 289
Monochloramine, 278
Most probable number, 187
Motor pumper, 109, 113–114
  classes of, 114
Moving bed filter, 271
Moving water layer, 479
Mud accumulation in filters, 264–265
  correction of, 265
  prevention of, 265
Mudball, 255–257, 264–265
Multiple correlation technique, 446
Multiple tube fermentation technique, 186–187

NPDES, 4
NRC formula, 484
Natural cycle, 438
Negative test, 186
Negligence, 195
Net positive suction head, 162
  available, 163
  required, 162
Newton's law, 210
Nitrate, 181–182
Nitrification, 555–558
  combined, 558
  effect of dissolved oxygen on, 555
  effect of pH on, 555
  effect of temperature on, 555
  mathematical model for, 555
  separate, 558
  sludge age in, 555

Nitrogen:
  agency in methemoglobinemia, 555
  oxygen demand of, 554
  reactions in soil systems, 452
  removal, 554–561
    processes, 556–557 (table)
      biological, 555–559
      chemical, 560–561
  role in eutrophication, 555
  in sewage, 427
  toxicity, 555
  trichloride, 278
Nitrogenous compounds, oxygen demand of, 554

Ocean disposal of sludges, 549
Odor (*See also* taste and odor):
  causes of, 288
  control, 570
  in dead-end lines, 115
  destruction, 275
  prevention, 275, 278, 288–290
    with chloramines, 278
  relation to taste, 115
  removal, 288–290
  of sewage, 426
Oil trap, 375
Open channel, 84–85
  sections, 85, 86
Optimization of process selection, 587–588
Ordinances, protective, 418
Organics:
  chlorinated, 569
  interference with iron and manganese removal, 292
  refractory, removal of, 564–566
  in wastewater, 427
Orifice box, constant head, 235
Orthophosphate, 562
Osmosis, reverse, 307
Outfall:
  use as chlorine contact basin, 569
  submarine, 446–447
    cost of, 447
    construction of, 447
Outlet approach velocity (in clarifiers), 223
Outlet location, 223
Outlet, sewer, 377–378
Overland flow technique (Izzard), 324
Overland flow time, 325
Overland runoff, 451
Overturn, 198
Oxidation ditch, 507
Oxidation of iron and manganese, 291
Oxidation pond, 518–521
  aerated, 519
  aerobic, 518–519
  algae in, 518–519
  anaerobic, 520
  facultative, 519–520
  mathematical model for, 517
  surface aeration in, 520
  upgrading of, 520
Oxygen:
  use in activated sludge, 503–504
    advantages of, 504
Oxygen, dissolved:
  agency in corrosion, 193
  saturation values, 611–612 (table)
  in self-purification of streams, 193
Oxygen sag curve, 440–446
Oyster:
  contaminated, 195
  disease transmission through, 181
Ozonation:
  of wastewater, 569
  of water, 282–283
Ozone:
  destruction of organic matter by, 282
  dosages, 282
  effect of temperature on, 282
  manufacture of, 283
  oxidation of iron and manganese with, 282
  spontaneous breakdown of, 282

Package plant, 501–502
Packed biological reactor, 487
Paratyphoid, 180, 184
Parshall flume, 456–457
  design considerations, 457
  dimensions and capacity of, 457 (table)
Partial flow diagram, 358, 362, 363 (illus.)
Particle size distribution, 211
Pasteurization, 284
Percolation capacity, 447
Percolation field, 576–578
Percolation test, 576–577
Period of design (*see* Design)
Peripheral flow clarifier, 222
Permanganate:
  use as disinfectant, 285
  use in iron and manganese removal, 292
Permeability of aquifers, 59
Pesticide, 195
  adsorption by soil, 452
pH:
  disinfection by control of, 284
  measurement, 191

Phase change, 306
Phenol number, 289
Phenol removal, 564
Phosphate (*see* phosphorus)
Phosphorus:
  adsorption by soil, 451–452
  concentration for algal bloom, 561
  luxury uptake of, 561–562
    effect of cations on, 562
  organic, 562
  removal, 561–564
    with alum, 563
    with aluminate, 563
    with aluminum, 562, 563 (table)
    biological, 561–562
    biological and chemical process for, 562
    chemical, 562–564
      point of addition in, 564
    with iron, 563
    with lime, 563–564
    role of in eutrophication, 561
    scavenging in biological processes, 564
    utilization by plants, 452
    in wastewater, 427
Phreatic surface, 55
Piezometric level, 56
Piezometric surface, 61
Pig, 146
Pigging of sewers, 422
Pipe:
  asbestos cement, 97–99, 345–346
    corrosion of, 346
    fittings for, 346
    joints in, 97, 99, 345–346
    life of, 97
    loads on, 345
    specifications for, 345
  bitumenized fiber, 347
  buried, static load on, 333–338
    superficial load on, 337–338 (table)
  cast-iron, 88–93, 347
    amperage for thawing, 150
    flanged, 91
    joints in, 91–93
    life of, 88
    lined, 90
    manufacture of, 89
    for sewers, 347
  cement lining of, 149
  clay, 331–337
    bedding methods for, 336–337
    end configurations, 331
    fittings, 333–334
    specifications for, 331–333
Pipe:
  concrete:
    corrosion of, 345
    joints, 97–98
    life of, 96
    plain, 338–341
      bedding for, 339, 341
      joints in, 338–340
      limitations of, 338
      loads on, 339
      specifications for, 338
    reinforced:
      fittings for, 345
      joints in, 339, 342, 345
      loads on, 344–345
      specifications for, 339–341
  corrugated metal, 347
  couplings, 92–93
  ductile iron, 347
  fittings, cast-iron, 94 (illus.)
  joints:
    bell and spigot, 91
    cast-iron, 92 (illus.)
    flexible, 93, 143
    mechanical, 93
    precalked, 93
    victaulic, 93
  lines, 87–88
  loads on in unsheeted trenches, 408
  plastic, 99
    life of, 99
    solid wall, 347
    truss, 346–347
  service, amperage for thawing, 149 (table)
  sewer, 331–352
    handling of, 412
    inspection of, 412
    jointing of, 413–414
    laying, 412
      specifications for, 412
  steel:
    joints in, 95
    life of, 95
    maintenance of, 96
  systems, 114–115
  thawing, 149–150
    electrical, 149–150
    steam, 149
  water:
    backfill for, 142
    bedding of, 143–144
    corrosion of, 291
    cover for, 142
    crushing load, 143

Pipe, water:
excavation for, 142
handling, 143
laying, 143
placement near sewers, 203
submerged, 143–144
testing of, 144
trench width for, 142
Piping in filter plants, 261–262
Plate clarifier, 214, 215 (illus.)
Plug-flow system, 495
Plumbing defects, 201–203
Poling board, 408
Polio (virus), 187
Poliomyelitis, 180–181
Polluted water, 180
characterization of, 438
Pollution, tracing of, 204
Polyelectrolyte (*see* polymeric coagulant)
Polymeric coagulant, 227
use with activated carbon, 290
ampholytic, 231
anionic, 230
bridging effect of, 231–232
cationic, 230
dosages, 231
in filtration, 246
Polymerization of hydrolyzing electrolytes, 228
Polyphosphate, 562
dosage of, 292
as sequestering agent, 292
Population density, effect upon water supply and sewerage, 19
Population estimation, 7–11
arithmetic, 8
comparison, 9–10
curvilinear, 9–10
declining growth, 10
logistic, 10
ratio, 11
uniform percentage, 9
Population equivalent, 436
Population, saturation, 10
Porosity of soil and rock, 56–57
Positive presumptive test, 186
Potable water, 180
Power, of pumps:
input, 157
water, 155
Preaeration, 467
advantages of, 467
aeration rates in, 467
retention time in, 467
Precipitation, 27–35
Precipitation:
data, adjustment of, 30–34
deficiency, 34
mean, 30
measurement, 27–35
by radar, 28
point, 30
Preliminary treatment, 456–467
Presedimentation, 226
Present worth, 586
Present worth factor, 587
Pressure filter (*see* filter)
Pressure, osmotic, 307
Pressure sewerage system, 388
Primary clarification, chemical addition in, 474
Primary feeders, 114
looping of, 114
Primary treatment systems, 469–475
Prime, loss of, 164
Priming, 309
Privy, 573, 575
location of, 573
Product water recovery in overland flow, 449
Production data for waterworks, 592
Proportional flow weir, 463
Prospect hole, 78
Protozoa, 188, 430–431
control of, 188
Psychoda alternata, 492
Public water systems, 194
Pulp, use as sludge conditioner, 532
Pump:
air lift, 170
axial flow, 160
booster, 111, 153
capacity, 166–168
classification, 154
centrifugal, 158–168
action of, 158
characteristic curves, 160–162
effect of speed, 159–160
multistage diffuser, 168
curves, 165
drooping head, 161–162
flat, 162
rising head, 162
steep, 162
efficiency, 157–158
emergency, 176
fire, 176
fish, 381
grinder, 388–389
helical rotor, 170
jet, 170

Pump:
  nonclog, 381
  open line-shaft, 168
  parallel installation, 167–168
  positive displacement, diaphragm, 281
  propellor, 160
  radial flow, 160
  reciprocating, 171
  reliability, 153
  rotary, 170–171
  screw, 171
  selection, 164–168
  series, 160
  sewage, 381–382
    emergency power for, 382
    submersible, 382, 383 (illus.)
    switching of, 385
  sludge, 171
  suction lift of, 162–164
  sump, 176, 386–387
  throttling of, 167
  vertical turbine, 168
    characteristic curve of, 168–169
  water, 153–177
  water lubricated, 168
  work of, 154–158
Pumping of wastewater, 381–390
Pumping, power for, 172–174
Pumping station:
  wastewater, 383–386
    dry pit, 383
  water, 174–177
    architecture of, 174–175
    auxiliary, 176
    capacity of, 174
    location of, 174
    operation of, 174–175
Pyrolusite, 291, 292
Pyrolysis of sludge, 545

Quality:
  of groundwater, 179
  of surface water, 179
  of water supplies, 179–205
  of water, variation of, in reservoirs, 106
Quicksand, control of, 412

Rack, 458–460
  hand-cleaned, 458–459
  mechanically cleaned, 458
Radiation, limit in public water supplies, 183
Radioactive tracer technique for reaeration constant, 443
Radioactivity in water, 183–184
Radioisotope, 183
  concentration of, 184
  removal of, 311
Radium removal, 311
Rain:
  effective, 49
  gauge, relocation of, 32
Rainfall (*See also* precipitation), 27–35
  artificial control of, 34–35
  depth-area relationship, 32–33
  excess, 51
  intensity, 325–329
  intensity-duration-frequency, 32–33
  net, 51
  point, 34
  probable maximum, 52
  records, supplementation of, 32
Rainstorms, types of, 29
Ranger, 408
Ranney method of well construction, 76–77
Rapid filter (*See also* filter)
  depth of, 243
  kinetics of particle removal in, 243
Rapid infiltration, 450–451
  application rates in, 450–451
  waste disposal by, 450–451
Rate constants in stream modeling, evaluation of, 442–444
Rates:
  sewer, 592–593
  water, 590–591
Rational method, 40, 48, 320–329
Reaeration constant, determination of, 442–443
Recarbonation, 298
Recharge, artificial, 67
  with cooling water, 15
  with heated water, 67
  of water-bearing formations, 40
Recurrence interval, 33
Reclamation, Bureau of, 27
Records, importance of, 199
Recycle ratio, 496
Red water, 115
Reduced pressure principle (RPP), 201
Refrigerant, secondary, 307
Regeneration:
  of activated carbon, 565–566
  of ion-exchange media, 303
Regulator, sewer, 376
Reo (virus), 187
Repair of sewers, 423–424
Reservoir:
  cover, 200
  recreational use, 197

Reservoir:
  sanitation, 197
  silting of, 52
  storage by mass diagram, 45
Retention time, liquid, 496
Rettger weir, 463
Return solids, 496
Reverse osmosis, cost of, 308
Reynolds number, 210
Roots, removal from sewers, 421
Rotating biological contacter (RBC) (*see* contacter)
Rotifers, 188, 431
Roughness, 355
Runoff, 39, 439
  coefficient, 320
    for different areas, 323 (table)
    for different surfaces, 321 (table)
  control in land disposal, 449
  factors affecting, 36–44
  groundwater, 37
  infiltration approach to calculation, 40
  maximum rate of, 323
  measurement of, 35–36
  multiple correlation method of forcasting, 44
  surface, 36–37
Rural waste disposal, 572–578
Rust in mains, 148

Safe Drinking Water Act of 1974, 194
Safety factor:
  for nitrification processes, 558
  for suspended growth processes, 495
Saline water, 306
Salmonellosis, 180
Saltwater, intrusion of, 67
Sample:
  composite, 435
  grab, 435
  proportional, 435
Sampling:
  of wastewater, 435–436
  of water, 194
Sand:
  enlargement of, 265
  expansion gauge, 261
  filter media:
    depth of, 244
    specifications for, 244
  incrustation of, 265
  leakage detector, 261
  removal from sewers, 421–422
  torpedo, 246
  trap, 375

Sanitary landfill, 548–549
Sanitary survey, 196–197
Saturation, zone of, 55
Scale formation, 308
Schistosomiasis, 180–181, 269
Scour, 219
Screen:
  approach velocity, 207
  bar, 207–208
  fine, 207, 209 (illus.), 458
  headloss through, 207–208
  sewage, 458–460
  at sewage pumping stations, 382
Screenings:
  characteristics of, 460
  disposal techniques for, 460
  quantity of, 458, 460
Scum:
  quantity removed in primary clarification, 474
  removal mechanisms, 473
    speed of, 474
Secondary feeders, looping of, 114–115
Secondary treatment systems, 477–521
  purpose of, 477
Sedgwick-Rafter test, 187
Sediment in mains, 148
Sediment problems in intakes, 105, 108
Sedimentation, 206, 210–239
  of discrete particles, 210
  in streams, 439
  of wastewater, 469–475
Sedimentation basin:
  currents in, 213
  design of, 210, 234, 472–474
  details:
    circular, 221–223, 473–474
    rectangular, 219–221, 472–473
    square, 221, 224, 473–474
  detention time, 213, 470, 491
  equipment in, 219–221, 491, 513
  horizontal velocity in, 211
  ideal, 210–211
  inlets, 221–224, 473, 513
  outlets, 221–226, 474, 515
  short-circuiting in, 213, 516
Seepage pit, 68
Self-purification, 438–439
  factors in, 439
Semidiesel engine for pumping, 172–173
Separate system (wastewater), 317
Septic tank, 575–576
  sizing of, 575
  effluent from, 576
Sequestration of divalent cations, 299

Service charges, 590–591
Service line, 131–132
  plastic, 131
  size of, 132
Settling analysis, 216, 217 (illus.)
Settling column analysis, 211
Settling:
  discrete, 210
  flocculent, 216
  hindered, 217–219
Sewage, 316
  characteristics, 426–436, 428 (table)
    chemical, 427–428
    physical, 426–427
  collection system:
    gravity, 392–403
    pressure, 389
    vacuum, 387–388
  disposal on land, 67
  domestic, 21
  flow:
    relation to water consumption, 21–22
    variation in rate of, 23–24, 435 (illus.)
  industrial, 21
  liability for damages from, 318–319
  sampling of, 435–436
  sanitary, 21, 316
  sources of, 21
  storm, 316
  strength variation, 435
  variability of, 435
Sewer, 316
  appurtenances, 365–390
  built in place, 348
  charges, 592–593
  cleaning tools, 421
  combined, 21, 317
  common, 316
  concrete, 415–416
  construction, 404–417
    in place, 415–417
  crossing, 380
  flight, 367
  grade, accuracy of, 406
  house, 317
  intercepting, 317
  lateral, 317
  line, accuracy of, 406
  lines and grades, 404–406
  main, 317
  outfall, 317
  relief, 317
  sanitary, 21, 317
  shapes, 363–364
Sewer:
  storm, 21, 317
    velocity in, 357
  submain, 317
  survey, 393
  system:
    design, 392–403
    layout, 393
    mapping, 393
    preliminary investigations, 392
    profile, 394–395
    sanitary, 395–399
      design of, 395–399
    storm, 400
      design of, 400–403
  underground survey, 393
Sewerage, 316
  general considerations, 316–319
  historical development of, 3–4
Shear gate, 140
Shear surface, 227
Sheet piling, 410
Sheeting, 407–411
  box, 408
  open, 410
  removal of, 411
  skeleton, 410
  vertical, 409
Sheeting and bracing, need for, 407
Shigellosis, 180
Short-circuiting in sedimentation basins, 213
Silica:
  activated, 231
    dosage of, 232
    preparation of, 231–232
  powdered, 231
Silt load, 52
Silver, disinfection with, 284
Siphon:
  dosing, 489
  inverted, 378–380
    velocity in, 378
Siphonage, back, 203
Skimming, 466–467
  tanks, 467
    horizontal velocity in, 467
Slaking chamber, 236–237
Slime layer, 480
Slip-lining of sewers, 424
Slope of sewers, 357
Sludge age, 495, 513
  in nitrification processes, 555
Sludge blanket, 233, 299
Sludge bulking, 515–516

Sludge bulking:
causes of, 515–516
correction of, 516
Sludge, chemical, 475
Sludge, coagulation, handling of, 311–313
Sludge combustion, 545–548
Sludge conditioning, 526–533
selection of process, 526
thermal, 532
effect upon aeration capacity, 532
liquid effluent from, 532
Sludge conditioning, chemical, 530–531
chemicals used for, 530
dosages, 530, 531 (table)
jar tests for, 531
mechanism of, 530
use of polymers, 531
Sludge dewatering, 535–545
choice of system, 545
Sludge digestion, mathematical model for, 517
Sludge disposal, 524–550
land, 548–550
ocean, 549
sanitary landfill, 549–550
Sludge drying, 545, 546 (illus.)
Sludge drying bed, 535–537
covering of, 536
design criteria for, 537
details of, 535–536
sand in, 535
solids content obtained, 536
time for dewatering, 536
Sludge elutriation, 531
design criteria for, 531
disadvantages of, 531
Sludge, floating, control of, 515
Sludge handling system selection, 584–585 (illus.)
Sludge, heat-treated, 532
dewatering of, 532
Sludge press, multi-roll, 543
Sludge pumping, 171
Sludge quantity:
in activated sludge, 499
in primary clarification, 474
in RBC processes, 492–493
in suspended growth, 499
in trickling filters, 491–492
Sludge removal equipment, 473
speed of, 474
Sludge, softening, handling of, 311–313
Sludge thickener, 217
Sludge transport, 171
Sludge treatment, 524–548
Sludge, waste
moisture content of, 525
solids content of, 525
Sludge, waste activated, constituents of, 525 (table)
Sluice gate, 140
Snow, measurement of, 29
Soda-ash, use in softening, 294
Softening, 292–304
optimum pH for, 295
for iron and manganese removal, 291
Soil classification, 448
Soil Conservation Service, 27, 52
Soil depth, for wastewater disposal, 449
Soil mantle, 451
Soil response to wastewater disposal, 451–452
Soil types and characteristics, 448 (table)
Solar radiation, effect upon evaporation, 38
Solar still, 307
Solids:
concentration in wastewater, 524
contact clarifier, 300, 302
determinations, 427
dissolved:
removal of in advanced wastewater treatment, 566–567
standards for, 192
total, 192
floating, 456
generated in secondary treatment, 524
handling, cost of, 524
quantity in wastewater, 524
suspended, 427, 456
removal in advanced wastewater treatment, 552–554
total, 427
total volatile, 427
volatile, 427
volatile dissolved, 427
volatile suspended, 427
waste:
character, 524–526
moisture content, 492
quantity, 524–526
in activated sludge, 499
in primary clarification, 474
in RBC processes, 493
in suspended growth, 499
in trickling filter, 491–492
specific gravity of, 524
Source, water, 84
Sparger, 508
Specific speed, 160
Spiderweb analysis, 581–586

Spiderweb analysis:
  limitations of, 586
Split sleeve, 93
Split treatment, 296
Spray irrigation, 449–450
  of agricultural land, 449
  application rates, 449, 450
  of woodland, 449
Spring, 81–82
Squeegee dewatering system, 544
Stabilization, 298
Stabilization of water, 309–310
Stabilization ponds, 518–521
  upgrading of, 520
Stagnant water, 198
Standards, Public Health Service, 194–195
Standpipe, 88, 110
Static level, 56
Steam pumping, 172
Steam turbine, 172
Step aeration, 501
Sterilization, 274
Stern potential, 227
Stokes' law, 210
Stoppage, clearing of, 420–422
Storage:
  freezing of, 176
  location of, 110
  required, 45–46, 110–112
    to equalize pumping, 111
    for fire, 109, 111
  undesirable effects of, 206
  water treatment by, 206
Storage capacity, 111
  coefficient of, 62
  desirable effects of, 206
  determination from mass diagram, 111
  effect upon treatment capacity, 266
  elevated, 109–112
    effect upon pressure, 110–111
    electrode control of, 176
  to equalize pumping, 110
  to equalize supply and demand, 110
  to furnish fire flow, 110
STORET system, 27
Storm:
  convectional, 29
  cyclonic, 30
  frontal, 30
  orographic, 30
  thunder, 29
Storm flow, estimation of, 320–330
Storm runoff, miscellaneous techniques for, 330
Storm sewage:
  amount of, 320–330
  first-flush of, 318
  treatment of, 318
Stormwater:
  inlet, 370–373
    location of, 370
  pumping, 160
Stream:
  discharge, effects of, 438
  flow records, 27
  gauging records, 47
    necessary period of, 36
  gauging station, 27
  rating curve, 35
  stage, 35
Streeter-Phelps formula, 444
Streptococci, fecal, 186
Stress in pipes, 86–87
  at change in direction, 87
  internal pressure, 86
  thermal, 87
  due to vertical load, 87
  water hammer, 86
Stripping of ammonia, 561
Strontium removal, 311
Subsidence, rate of, 218
Sul-biSul process, 307
Sulfide precipitation of heavy metals, 528
Sunlight, effect on water quality, 439
Superchlorination, 277
Surcharging of sewers, 354
Surface wash, 253, 255–257
  Bayliss, 255–256
  Palmer, 256
  water required for, 256
Surface overflow rate, 211
Surge tank, 88, 311
Surgical sterilizers, 203
Surging:
  of centrifugal blower, 511
  of filter, 260
  of well, 75
Suspended growth process:
  aeration techniques for, 504–513
  air required, 498
  diffused air systems, 508–513
  mixing techniques, 504–513
  oxygen required, 498, 499
  principles of, 494–495
  waste sludge quantity, 499
Suspended solids:
  contact unit, 301
  particle size of, 215–216

Suspended solids:
  specific gravity of, 214–216
  types of, 214
Suspension, flocculent, 216
Sutro weir, 463
System analysis of hydrologic processes, 26
System curve, 165
System head curve, 155
System headloss curve, 155

Tapered aeration, 499–501
Taste and odor, 184, 199, 207
  reduction by prechlorination, 276
  removal by aeration, 285–286
Temperature, effect on water quality, 439
Tertiary treatment (*see* advanced waste treatment)
Theis method of well analysis, 61
Thermocline, 198
Thickener, flotation, 533–535
Thickener, gravity, 533–534
Thickening of waste sludges, 533–535
Thiem formula, 59
Thiessen method, 30
Thiobacillus, 351
Thomas' method for BOD constants, 432–434
Threshold odor, 290
Total Kjeldahl nitrogen (*see* Kjeldahl)
Total organic carbon (TOC), 434
Total retention systems, 452–453
Tracer studies, 213
Transmissibility, coefficient of, 62
Transpiration, 43
Treatment:
  acid, of wells, 80
  advanced, 552–567
  polyphosphate, of wells, 81
  process selection, 587–588
Trench, backfilling of, 415
Trench, drainage of, 412
Trickling filter (*see* filter)
Tube settler, 214
Tuberculation, 90
  of water mains, 148
Tunneling, 416–417
Turbidity, 189–190, 194, 195
  EPA standard, 190
  formazin standard, 190
  inorganic, 189
  organic, 190
  sensor, 261
  units, 190
Two-main system, 130–131
Type curve, 63–65
Typhoid fever, 180, 184
  epidemiology of, 195
  incidence of, 196

USGS, 27, 52
Ultrafiltration, 553
Ultrasonation:
  of wastewater, 569
  of water, 284
Ultraviolet irradiation, 284
Unit cost, 586
Unit hydrograph, application to storm runoff, 330
Unsafe water, liability for, 195
Upflow clarifier, 300, 302
Uranin, 204

Vacuum breaker, 201
  for filter surface wash, 257
Vacuum filter (*see* filter)
Vacuum sewerage system, 387–388, 389 (illus.)
Vacuum sludge withdrawal system, 223 (illus.)
Valve, 133–140
  actuation, 135
  air and vacuum, 88, 139
  air relief, 88, 114, 139
  altitude, 139, 176
  angle, 137
  blowoff, 114
  box, 134–135
  butterfly, 138, 139 (illus.)
  bypass, 135, 136
  check, 135–137, 176
    double, 201
    single, 201
  cone, 137
  connection, 135
  disk track, 135
  foot, 135, 176
  in filter pipe galleries, 262
  in filter plants, 135
  in filters, 242
  gate, 133–135
    location of, 133
  globe, 137
  inaccessibility, 146
  inspection, 145
  installation, 135
  leaks, 146
  location:
    in distribution systems, 115
    in pipe systems, 114
  maintenance, 145–146
  placement bases, 145

Valve:
 plug, 137
 poppet, 139
 pressure regulating, 114, 138
 pressure relief, 177
 in pumping stations, 385
 radio-controlled, 176
 remote controlled, 176
 standardization, 134, 146
 stems:
  left-hand, 134
  nonrising, 134
  right-hand, 134
  rising, 134
 swing-check, 136, 137 (illus.)
 vault, 145
Van der Waals attraction, 289
Vapor pressure of water, 163 (table)
Velocity gradient:
 in flocculation, 233
 in mixing, 232
Velocity in sewers, 357–358
Velocity in water lines, 88
Velz formula, 485
Viral disease, 180
Viruses, 187
Volatile acid, role in anaerobic digestion, 528

Wale, 408
Washer, bedpan, 203
Washer, glass, 203
Washwater:
 gullet, 250, 251 (illus.)
 tank, 242
 trough, 250–252
  construction, 250
  dimensions, 251–252
  elevation, 250
  freeboard, 250
  location, 250
Wastewater:
 flow:
  rural, 574–575
  from unsewered buildings, 574–575
 irrigation, response of crops to, 449
 treatment:
  for land disposal, 447
  system selection, 582–583 (illus.)
Water consumption:
 commercial, 11–12
 domestic, 11
 effect of climate upon, 15
 effect of metering upon, 16
 effect of zoning upon, 14–15
Water consumption:
 estimated future, 6, 12
 factors affecting, 13
 industrial, 11–12, 14
 loss and waste, 12
 maximum rate of, 17
 minimum rate of, 17
 per capita, 7
  in sewered homes, 14
  in unsewered homes, 14
 present, 5–6
 public, 12
 ratio of maximum to average, 13
 recorded, 13
 records of, 17
 variations in, 16–20
Water:
 bound, 525
 capillary, 525
 density, 642
 floc, 525
 free, 525
 ionization of, 191
 physical properties of, 642 (table)
 pipes, organic growths in, 276
 quality:
  model of lakes, 439
  in wells, 69–70
   deep, 70
   shallow, 69
 standards, primary, 194
 supply district, 588
 supply, historical development of, 2–3
 table, 55
  perched, 56
 treatment:
  degree of, 206
  reasons for, 206
  records required, 266
  testing required, 265
  wastes, 311–313
 vapor pressure of, 642 (table)
 viscosity of, 642 (table)
Waterborne disease, 180–182
 cause of outbreak, 196
 incidence of, 2
Waterborne epidemic, characteristics of, 195–196
Watershed sanitation, 197
Waterwaste survey, 147–148
Waterworks ownership, 588
Weather Service, U.S. National, 27
Weighting agent, 231
Weir:
 free-falling, 224

Weir:
  loading rate, 223
  serpentine, 472–474
  v-notch, 223
Wet oxidation:
  in regeneration of activated carbon, 290
  of sludge, 545
Wet pit, 383, 385 (illus.)
Well:
  analysis, 60–66
    Dupuit technique, 60
    equilibrium, 60
    modified non-equilibrium, 65–66
    non-equilibrium, 61–66
  artesian, 57, 58
  bored, 69
  casing, 80
  cementing of, 77–78
  construction, 69–72
    California method, 71
    core drilling method, 72
    jetting method, 72
    rotary method, 71
    standard method, 70
    stovepipe method, 71
  development, 75
  driven, 69
  dug, 69
Well:
  function, 63
  gravel-packed, 75
  log, 78
  lowering of water level in, 55
  pump, 168–169
    capacity of, 169
  screen, 72–75, 80
    velocity through, 72
  shallow, 69
  specifications, 78–80
  troubles, 80–81
  types, 57–58
  water table, 57
Wellhole, 367
Wellpoint, 412
Worms, 189

Yield, of impounding reservoir, 44–45
  safe, 44

Zeolite, 301
  carbonaceous, 305
  regeneration of, 303
Zeta potential, 227
Zoning, effect upon water supply and sewerage, 20

PRINCIPAL SI UNITS USED IN SANITARY ENGINEERING

Quantity	Unit	Symbol	Formula
Area	square meter	—	m^2
	square kilometer	—	km^2
Density	grams per cubic centimeter	—	g/cm^3
	kilograms per cubic meter	—	kg/m^3
Energy	joule	J	$N \cdot m$
Force	newton	N	$kg \cdot m/s^2$
Frequency	hertz	Hz	s^{-1}
Length	meter	m	*
Mass	kilogram	kg	*
Power	watt	W	J/s
Pressure	pascal	Pa	N/m^2
Time	second	s	*
Velocity	meters per second	—	m/s
Volume	cubic meter	—	m^3
	liter	l	$10^{-3}m^3$
Work	joule	J	$N \cdot m$

* Base unit

SI PREFIXES

Multiplication factor	Prefix†	Symbol
$1\,000\,000\,000\,000 = 10^{12}$	tera	T
$1\,000\,000\,000 = 10^{9}$	giga	G
$1\,000\,000 = 10^{6}$	mega	M
$1\,000 = 10^{3}$	kilo	k
$100 = 10^{2}$	hecto‡	h
$10 = 10^{1}$	deka‡	da
$0.1 = 10^{-1}$	deci‡	d
$0.01 = 10^{-2}$	centi‡	c
$0.001 = 10^{-3}$	milli	m
$0.000\,001 = 10^{-6}$	micro	μ
$0.000\,000\,001 = 10^{-9}$	nano	n
$0.000\,000\,000\,001 = 10^{-12}$	pico	p
$0.000\,000\,000\,000\,001 = 10^{-15}$	femto	f
$0.000\,000\,000\,000\,000\,001 = 10^{-18}$	atto	a

† The first syllable for every prefix is accented so that the prefix will retain its identity. Thus, the preferred pronunciation of kilometer places the accent on the first syllable, not the second.

‡ The use of these prefixes should be avoided, except for the measurement of areas and volumes and for the nontechnical use of centimeter, as for body and clothing measurements.